"Dr. Joseph Goebbels wrote that 'A lie told once remains a lie, but a lie told a thousand times becomes the truth.' Tragically for humanity, there are many, many untruths emanating from Fauci and his minions. RFK Jr. exposes the decades of lies."

—**Luc Montagnier**, Nobel laureate

"Bobby Kennedy is one of the bravest and most uncompromisingly honest people I've ever met. Someday he'll get credit for it. In the meantime, read this book."

—**Tucker Carlson**

"Throughout history, fear has been used to manipulate and control populations. In a democracy, we have the privilege and responsibility to question the things we are encouraged to fear. Robert F. Kennedy Jr. provides something denied to most Americans in recent years: the opportunity to hear more than one perspective. You can accept or reject the new information in this book—but at least listen."

—**Gavin de Becker**, bestselling author of *The Gift of Fear*

"As a trial lawyer, Robert F. Kennedy Jr. has taken on the world's most powerful corporations and held them accountable for harming people and the environment. Those companies denied any wrongdoing—but time and again, judges and juries were persuaded that Kennedy's position was the right one. [His] information should always be considered, and agree or disagree, we all learn from listening."

—**Tony Robbins**, bestselling author

"Bobby Kennedy and I famously disagree about many aspects of the current debates surrounding Covid and vaccines. We also disagree about Dr Fauci. But I always learn when I read or hear Bobby's take. So read this book and challenge its conclusions."

—**Alan Dershowitz**

"Kennedy's book proves beyond a shadow of doubt what many Americans have come to learn about Fauci: that he has stifled open debate to the point of utter stagnation of biomedical science."

—**James Lyons-Weiler**, biomedical research scientist

"If you've ever wondered why so many good scientists and doctors have been silenced for discoveries that don't fit the mainstream Big Pharma narrative, look no further than Robert F. Kennedy Jr.'s tour de force exposé of Anthony Fauci."

—**Christiane Northrup, MD**, former assistant clinical professor of Ob/Gyn, University of Vermont College of Medicine

"I thought I understood what was going on from an insider POV . . . But what this book clearly documents are the deeper forces and systemic, pervasive governmental corruption, that have led us to this point. . . ."

—**Robert W. Malone, MD**, virologist, immunologist, molecular biologist

"If you have any interest in doing a deep dive into the more than 100-year history of what led up to the COVID-19 pandemic, *The Real Anthony Fauci* is an absolute must-read. In addition to exposing Fauci, the book reveals the complex web of connections between Gates and Big Pharma and many of the most important players that were responsible for seeking to implement global tyranny and profit enormously from the propaganda behind the COVID injections, masks, and lockdowns."

—**Dr. Joseph Mercola**, founder of Mercola.com

"[T]his book unveils the astonishing, twisted truth about a man (Fauci) and a corrupt institution (NIH) that have betrayed humanity at every turn in order to achieve profits and power. If the American people knew the truth that's documented here, they would be marching by the millions, demanding criminal prosecutions of all those who are complicit."

—**Mike Adams**, founder of NaturalNews.com

"It is impossible to read Kennedy's book on Anthony Fauci without your jaw dropping to the floor. . . . It is a shocking tale of greed, corruption, and malpractice at the highest levels of government. Once Americans wake up from their hopefully brief infatuation with medical tyranny, this little Josef Stalin of medicine will go down in history as the country's most corrupt government servant."

—**Rob Schneider**

"RFK Jr.'s story of Fauci's failure as the government's AIDS coordinator is a highly disturbing prologue to his COVID mandate as head of NIAID. So, who is Dr. Fauci in the end? Has American medicine truly become a 'racket,' as corrupt as a mafia organization? . . . RFK Jr. has written a strong, strong book."

—**Oliver Stone**, award-winning director, producer, and screenwriter

"As I read Kennedy's book I thought a discreet and thorough criminal investigation into Fauci should occur. . . . It brought back memories of criminal bid-rigging schemes conducted by Organized Crime."

—**Mike Campi**, former FBI agent and retired coordinator of the Organized Crime branch of the New York Division

"RFK Jr. is a tireless champion of Americans' rights to be informed about their medical choices and has been way out front in warning us of the dangers of an out-of-control pharmaceutical industry."

—**Naomi Wolf**, bestselling author *The Beauty Myth*, *Give Me Liberty*, and *Outrages*

"The revelations in this book are disturbing and shocking, exposing how our political system and government agencies can be compromised, and how the mainstream media are being used to manipulate and control our everyday existence. . . ."

—**Randy Jackson**, award-winning musician and producer

The Real Anthony Fauci

The Real Anthony Fauci

Bill Gates, Big Pharma, and the Global War on Democracy and Public Health

Robert F. Kennedy Jr.

Skyhorse Publishing

Children's
Health Defense

Copyright © 2021 by Robert F. Kennedy Jr.

All rights reserved. No part of this book may be reproduced in any manner without the express written consent of the publisher, except in the case of brief excerpts in critical reviews or articles. All inquiries should be addressed to Skyhorse Publishing, 307 West 36th Street, 11th Floor, New York, NY 10018.

Skyhorse Publishing books may be purchased in bulk at special discounts for sales promotion, corporate gifts, fund-raising, or educational purposes. Special editions can also be created to specifications. For details, contact the Special Sales Department, Skyhorse Publishing, 307 West 36th Street, 11th Floor, New York, NY 10018 or info@skyhorsepublishing.com.

Skyhorse® and Skyhorse Publishing® are registered trademarks of Skyhorse Publishing, Inc.®, a Delaware corporation.

Visit our website at www.skyhorsepublishing.com.

14

Library of Congress Cataloging-in-Publication Data is available on file.

Print ISBN: 978-1-5107-6680-8
Ebook ISBN: 978-1-5107-6681-5

Cover design by Brian Peterson

Printed in the United States of America

CONTENTS

DEDICATION & ACKNOWLEDGMENTS

Dr. Anthony Fauci's opinions and proclamations have been omnipresent in American media, and some people might assume his ideas and actions are universally supported by scientists or that he somehow represents science and medicine.

To the contrary, many leading scientists and scholars around the world oppose lockdowns and other aspects of Dr. Fauci's pandemic management. They include Nobel laureates and other distinguished, accomplished, widely published, and internationally celebrated scientists.

I dedicate this book to that battle-hardened cadre of heroic scientists and physicians who have risked their careers, their livelihoods, and their reputations to champion evidence-based science and ethical medicine. By steadfastly prioritizing truth, the welfare of their patients, and the cause of public health above their own career ambitions, these brave men and women have succeeded, at great cost, in preserving their own integrity. They may one day restore from shame the shattered souls of the medical profession and the scientific establishment. Each of these individuals has emerged as a voice of sanity and a symbol of clarity and truth to those idealists across the globe who love democracy and resist the rising medical authoritarianism. Thanks to all of you for inspiring me with your character, your courage, your brilliant insights, and your passion for empiricism and critical thinking.

Dr. Harvey Risch, Yale University Professor of Epidemiology, Editor, Journal of the National Cancer Institute, Board of Editors, American Journal of Epidemiology (2014–2020), biostatistician

Dr. Robert Malone, original inventor mRNA and DNA vaccination technologies, NIH Accelerating COVID-19 Therapeutic Interventions and Vaccines (ACTIV) Clinical Working Group (2020), Editor-in-Chief, *Journal of Immune Based Therapies and Vaccines* (2007–2012), Salk Institute (1986–1989)

Dr. Geert Vanden Bossche, Senior Ebola Program Manager, Global Alliance for Vaccines & Immunization (2015), Head of Vaccine Development for Germany's Center for Infection Research (2017), vaccine developer at GSK (1995–2006), Novartis (2006), virologist

Dr. Michael Yeadon, Chief Scientist and vice-president of Pfizer's allergy and respiratory research division (1995–2011), respiratory pharmacologist

Dr. Luc Montagnier, Virologist, 2008 Nobel Prize in Physiology/Medicine

Dr. Wolfgang Wodarg, Chair of the Parliamentary Assembly, Council of Europe Heath Committee (1998–2010), member of German Parliament (1994–2009), pulmonologist

Dr. Peter McCullough, clinical cardiologist, vice chief of internal medicine at Baylor University Medical Center (2014–2021)

Dr. Peter Doshi, University of Maryland School of Pharmacy associate professor pharmaceutical health services, and senior editor at the *British Medical Journal*

Dr. Paul E. Marik, Founder, Front-Line Covid-19 Critical Care Alliance, Professor of Medicine, Chief of Pulmonary and Critical Care Medicine, Eastern Virginia Medical School

Dr. Pierre Kory, President and Chief Medical Officer of the Front-Line Covid-19 Critical Care Alliance, Former Associate Professor, Chief of Critical Care Service, Medical Director of Trauma and Life Support Center at the University of Wisconsin (2015–2020)

Dr. Byram Bridle, University of Guelph associate professor of Viral Immunology

Dr. Tess Lawrie, World Health Organization consultant, physician

Dr. Didier Raoult, Director, Infectious and Tropical Emergent Diseases Research Unit (France), physician and microbiologist

Dr. Peter Breggin, National Institute of Mental Health (1966–1968), Harvard Medical School (1963–1964), doctor of psychiatry, author of more than 40 books

Dr. Meryl Nass, physician, vaccine-induced illnesses, toxicology, expert delegate to the US Director of National Intelligence bio-threat study program (2008)

Dr. Vladimir Zelenko, Medical Director Monsey Family Medical Center, physician

Dr. Charles Hoffe, physician

Dr. James Todaro, physician

Dr. Scott Jensen, University of Minnesota Medical School Clinical Associate Professor, Minnesota State Senator (2016–2020), physician

Dr. Ryan Cole, pathologist

Dr. Jacob Puliyel, Director Department of Pediatrics and Neonatology, St. Stephen's Hospital (India), past member of India's National Technical Advisory Group on Immunization

Dr. Christiane Northrup, University of Vermont College of Medicine Assistant Clinical Professor of Obstetrics & Gynecology (1982–2001), physician, three-time *New York Times* bestselling author

Dr. Richard Urso, MD Anderson Cancer Center assistant professor (1993–2005), Chief of Orbital Oncology, scientist

Dr. Joseph Ladapo, Surgeon General of Florida, professor University of Florida College of Medicine, associate professor at UCLA's David Geffen School of Medicine, assistant professor of Population Health and Medicine at NYU School of Medicine

Dr. Martin Kulldorff, Harvard University professor of medicine, biostatistician, epidemiologist, expert in vaccine safety evaluations and monitoring infectious disease outbreaks

Dr. Michael Levitt, Stanford University biophysicist and professor of structural biology, 2013 Nobel Prize in Chemistry

Dr. Satoshi Ōmura, biochemist, 2015 Nobel Prize in Physiology/Medicine

Dr. Paul E. Alexander, US Department of Health & Human Services Senior Covid Pandemic Advisor (2020), WHO Pan American Health Organization (2020)

Dr. Clare Craig, UK National Health Service (2000–2015), pathologist, Oxford University and Cambridge University trained

Dr. Lee Merritt, US Navy physician and surgeon (1980–1989), past president Association of American Physicians and Surgeons

Dr. Sucharit Bhakdi, Microbiologist, head of the Institute of Medical Microbiology and Hygiene at University of Mainz (1991–2012)

Dr. Jay Bhattacharya, Stanford University Medical School professor, physician, epidemiologist, health economist, and public health policy expert focusing on infectious diseases

Dr. David Katz, Yale University, founder of Yale's Prevention Research Center, physician

John P.A. Ioannidis, Stanford University Professor of Medicine, Epidemiology and Population Health, physician-scientist

Dr. Sunetra Gupta, Oxford University epidemiologist, immunology expert, vaccine development, infectious disease mathematical modeling

Dr. Catherine L. Lawson, Rutgers University research professor, Institute for Quantitative Biomedicine

Dr. Salmaan Keshavjee, Harvard Medical School professor of Global Health and Social Medicine

Dr. Laura Lazzeroni, Stanford University professor of biomedical data science, professor of psychiatry and behavioral sciences

Dr. Cody Meissner, Tufts University professor of pediatrics, expert on vaccine development, efficacy, and safety

Dr. Lisa White, Oxford University professor of epidemiology and modeling

Dr. Ariel Munitz, Tel Aviv University professor of clinical microbiology and immunology

Dr. Motti Gerlic, Tel Aviv University, clinical microbiology and immunology

Dr. Angus Dalgleish, University of London professor of infectious disease,

Dr. Helen Colhoun, University of Edinburg professor of medical informatics and epidemiology, public health physician

Dr. Simon Thornley, University of Auckland epidemiologist and biostatistician

Dr. Stephanie Seneff, Massachusetts Institute of Technology, Senior research scientist

The Heroic Healers Honor Roll

There is a much larger retinue of thousands of lesser-known front-line medical professionals and PhD. researchers who have also chosen to wager everything on their commitment to their patients, to uncorrupted scientific inquiry, and to the opposition to authoritarian COVID policies. Many of them have battled unheralded in the trenches for decades. Each one of them has endured various intensities of vilification, excommunication, delicensing, and censorship by the Pharma cartel's captive regulators, its corrupt medical associations, its media and social media allies and medical journals, and its government-sponsored fact checkers. They have weathered derision, gaslighting, scapegoating, retractions, career and reputational injuries, and financial ruin, to protect their patients, and nurture truth. My apologies to all of you whom I have necessarily omitted due to space restrictions but who belong on this Honor Roll. I regret that the only consolation for your sacrifices may be a clear conscience.

Dr. Robin Armstrong
Dr. Cristian Arvinte
Dr. David Ayoub
Dr. Alan Bain
Dr. Richard Bartlett
Dr. Cammy Benton
Dr. Robert Berkowitz
Dr. Andrew Berry
Dr. Harvey Bialy
Dr. Thomas Borody
Dr. Joseph Brewer
Dr. Kelly Brogan
Dr. David Brownstein
Dr. Adam Brufsky
Dr. Zach Bush
Dr. Dolores Cahill
Dr. Teryn Clarke
Dr. Tom Cowan
Dr. Andrew Cutler
Dr. Roland Derwand
Dr. Joyce Drayton
Dr. Peter Duesberg
Dr. Alieta Eck
Dr. John Eck
Dr. Richard Eisner
Dr. Christopher Exley
Dr. George Fareed
Dr. Angelina Farella
Dr. Richard Fleming
Dr. Ted Fogarty
Dr. Silvia N. S. Fonseca
Dr. C. Stephen Frost
Dr. Michael Geier
Dr. Charles Geyer

Dr. Simone Gold
Dr. Gary Goldman
Dr. Russell Gonnering
Dr. Karladine Graves
Dr. Kenneth Gross
Dr. Sabine Hazan
Dr. Kristin Held
Dr. H. Thomas Hight
Dr. LeTrinh Hoang
Dr. Douglas Hulstedt
Dr. Suzanne Humphries
Dr. Stella Immanuel
Dr. Michael Jacobs
Dr. Tina Kimmel
Dr. Lionel H. Lee
Dr. Sin Han Lee
Dr. John Littell
Dr. Ivette Lozano
Dr. Douglas Mackenzie
Dr. Carrie Madej
Dr. Marty Makary
Dr. Harpal Mangat
Dr. Ben Marble
Dr. David E. Martin
Dr. John E. McKinnon
Dr. Shira Miller
Dr. Kari Mullis
Dr. Liz Mumper
Dr. Eileen Natuzzi
Dr. James Neuenschwander
Dr. Hooman Noorchashm
Dr. Jane Orient
Dr. Tetyana Obukhanych
Dr. Ramin Oskoui

Dr. Larry Palevsky
Dr. Vicky Pebsworth
Dr. Don Pompan
Dr. Brian Procter
Dr. Chad Prodromos
Dr. Jean-Jacques Rajter
Dr. Juliana Cepelowicz Rajter
Dr. C. Venkata Ram
Dr. David Rasnick
Dr. Salete Rios
Dr. Michael Robb
Dr. Rachel Ross
Dr. Moll Rutherford
Dr. Ed Sarraf
Dr. Martin Scholz
Dr. Bob Sears
Dr. Christopher Shaw
Dr. Marilyn Singleton
Dr. Kenneth Stoller
Dr. Kelly Sutton
Dr. Sherri Tenpenny
Dr. Paul Thomas
Dr. James Tumlin
Dr. Brian Tyson
Dr. Michelle Veneziano
Dr. Kelly Victory
Dr. Elizabeth Lee Vliet
Dr. Craig Wax
Dr. Rachel West
Dr. James Lyons-Weiler
Dr. Alexandre Wolkoff
Dr. Vicki Wooll
Dr. Bob Zajac
Dr. Andrew Zimmerman

And to truth-tellers everywhere who reject propaganda, defy censorship, and who live and die for freedom and human dignity.

In Memory of Katie Weisman and Dr. Valerie Kennedy Chamberlain

Our ferocious fact checker, Katie Weisman, died while I wrote this book. During the day, Katie was a devoted wife and mom to three vaccine-injured children. At night, even during the worst periods of her chemotherapy, Katie became the most fearsome (and intractable) fact checker. She helped lay the groundwork upon which we built our organization into a vessel for the reckless pursuit of existential truth. Her reckless integrity inspired our movement and ensured the accuracy of many of my projects. I am grateful also for her friendship.

Dr. Valerie Kennedy Chamberlain was a retired paleontologist and university professor who typed and fact-checked most of my handwritten drafts of *Thimerosal: Let the Science Speak*, and played a key role as a writer, editor, and typist at World Mercury Project from its inception, and later at Children's Health Defense. She was a wonderful friend.

Acknowledgments

Judith Conley and Nancy Hokkanen deciphered my chicken-scratch scrawl, typed my handwritten manuscript, and helped me fact-check, spell-check, and edit with reliable insight, humor, and patience. Their advice was always invaluable.

Dr. Robert Malone, Celia Farber, Dr. Meryl Nass, GDB, Ken McCarthy, Charles Ortleb, and John Lauritsen read sections of this manuscript and provided criticism, citations, and insightful commentary. I am grateful to all for your corrections and recommendations, and for broadening my understanding of innumerable scientific and historic issues.

Laura Bono, Rita Shreffler, Brian Hooker, Ph.D., Lyn Redwood, R.N., Liz Mumper, M.D., Jackie Hines, and Cari Shagena were part of the Children's Health Defense team of researchers and fact checkers who sourced, cited, and fact-checked this manuscript with energy, diligence, patience, idealism, and perseverance. I'm indebted to all of them for their devotion to children's health, to democratic principles, to precision, and to the noble search for existential truth and accuracy.

Mary Holland, Esq., Rolf Hazlehurst, Esq., Heidi Kidd, Lynne Arnold, Kari Bundy, Angela Braden, Katrina Boudreau, Polly Tommey, Joyce Ghen, Karen McDonough, Cornelia Tucker Mazzan, Aimee Villella McBride, Stephanie Locricchio, Katherine Paul, Aerowenn Hunter, Janet McLean, John Stone, Divyanshi Dwivedi, Esq., Megan Redshaw, Brian Burrowes, Dr. Aaron Lewis, Curtis Cost, Wayne Rohde, Alix Mayer, Kristina Kristen and Karen Kuntz are the Children's Health Defense warriors who give their lives selflessly to the battle for democracy, justice, and public health. All of them contributed directly and indirectly to this book.

Robyn Ross, Esq., Eileen Iorio, Dr. Kristine Severyn, Dr. James Lyons-Weiler, Joel Smalley, Dr. Jessica Rose and Sofia Karstens contributed to the work with their invaluable and resourceful support and talent in quick responses 24/7 on critical research questions.

To David Whiteside for his reliability, hard work and for never turning off his phone.

Thanks also go to Tony Lyons and his team at Skyhorse Publishing, including Hector Carosso, Mark Gompertz, Kirsten Dalley, and Louis Conte. Their vigilance and their congenial harassment ensured that this book came out on time.

Finally, I could not have written this book without the patience and wisdom of my wife, Cheryl Hines, whose reliable love fills all my empty spaces and makes me impervious to the furies of the world. I'm grateful for her generous humor, wisdom, support, and love, even when she disagrees with me. To my children, Bobby, Kick, Conor, Kyra, Finn, Aiden, and Cat, who bear the burdens I place upon them with laughter and stoicism, and whose love I always feel.

PUBLISHER'S NOTE

Complex scientific and moral problems are not resolved through censorship of dissenting opinions, deleting content from the Internet, or defaming scientists and authors who present information challenging to those in power. Censorship leads instead to greater distrust of both government institutions and large corporations.

There is no ideology or politics in pointing out the obvious: scientific errors and public policy errors do occur—and can have devastating consequences. Errors might result from flawed analysis, haste, arrogance, and sometimes, corruption. Whatever the cause, the solutions come from open-minded exploration, introspection, and constant review.

Ideally, science and public policy are never static. They are a process, a collaboration, a debate and a partnership. If powerful people challenged by this book claim it contains misinformation, our response is simply this: Tell us where you believe something is incorrect, make your best arguments, and offer the best available support for those arguments. We encourage and invite dialogue, criticism, engagement—and every suggestion will be heard and considered.

Since *The Real Anthony Fauci* is being published in the middle of rapidly unfolding events, and since censorship and suppression of information is underway, it's best to approach this book as a living document. When new information emerges that can add to or improve the thousands of references and citations in this book, updates, additional notes, and new references will be provided via the QR code below, and the QR codes that appear throughout the book.

We've published authors with views on all sides of many controversies. That's what we do, because at its best, publishing is a town square that lets everyone be heard—and lets everyone else listen, if they choose to. As Alan Dershowitz says, "I always learn when I read or hear Bobby's take." I can go several steps further, knowing from my inside view how principled and careful Bobby is as an author—and how painstakingly this book was researched.

We look forward to taking this important journey with you.

Tony Lyons
Skyhorse Publishing

ChildrensHealthDefense.org/fauci-book
childrenshd.org/fauci-book

For updates, new citations and references, and new information:

Editor's note:
For ease of reference and reading, throughout this
manuscript, both the virus SARS CoV-2 and the disease
Covid-19 are referred to as COVID-19.

INTRODUCTION

"The first step is to give up the illusion that the primary purpose of modern medical research is to improve Americans' health most effectively and efficiently. In our opinion, the primary purpose of commercially funded clinical research is to maximize financial return on investment, not health."

—John Abramson, M.D., Harvard Medical School

I wrote this book to help Americans—and citizens across the globe—understand the historical underpinnings of the bewildering cataclysm that began in 2020. In that single *annus horribilis*, liberal democracy effectively collapsed worldwide. The very governmental health regulators, social media eminences, and media companies that idealistic populations relied upon as champions of freedom, health, democracy, civil rights, and evidence-based public policy seemed to collectively pivot in a lockstep assault against free speech and personal freedoms.

Suddenly, those trusted institutions seemed to be acting in concert to generate fear, promote obedience, discourage critical thinking, and herd seven billion people to march to a single tune, culminating in mass public health experiments with a novel, shoddily tested and improperly licensed technology so risky that manufacturers refused to produce it unless every government on Earth shielded them from liability.

Across Western nations, shell-shocked citizens experienced all the well-worn tactics of rising totalitarianism—mass propaganda and censorship, the orchestrated promotion of terror, the manipulation of science, the suppression of debate, the vilification of dissent, and use of force to prevent protest. Conscientious objectors who resisted these unwanted, experimental, zero-liability medical interventions faced orchestrated gaslighting, marginalization, and scapegoating.

American lives and livelihoods were shattered by a bewildering array of draconian diktats imposed without legislative approval or judicial review, risk assessment, or scientific citation. So-called Emergency Orders closed our businesses, schools and churches, made unprecedented intrusions into privacy, and disrupted our most treasured social and family relationships. Citizens the world over were ordered to stay in their homes.

Standing in the center of all the mayhem, with his confident hand on the helm, was one dominating figure. As the trusted public face of the United States government response to COVID, Dr. Anthony Fauci set this perilous course and sold the American public on a new destination for our democracy.

This book is a product of my own struggle to understand how the idealistic institutions our country built to safeguard both public health and democracy suddenly turned against our citizens and our values with such violence. I am a lifelong Democrat, whose family has had eighty years of deep engagement with America's public health bureaucracy and long friendships with key federal regulators, including Anthony Fauci, Francis Collins, and Robert Gallo. Members of my family wrote many of the statutes under which these men govern, nurtured the growth of equitable and effective public health policies, and defended that regulatory bulwark against ferocious attacks funded by industry—and often executed by Republican-controlled congressional committees intent on defunding and defanging these agencies to make them more "industry friendly." I built alliances with these individuals and their agencies during my years of environmental and public health advocacy. I watched them, often with admiration. But I also watched how the industry, supposedly being regulated, used its indentured servants on Capitol Hill to systematically hollow out those agencies beginning in 1980, disabling their regulatory functions and transforming them, finally, into sock-puppets for the very industry Congress charged them with regulating.

My 40-year career as an environmental and public health advocate gave me a unique understanding of the corrupting mechanisms of "regulatory capture," the process by which the regulator becomes beholden to the industry it's meant to regulate. I spent four decades suing the US Environmental Protection Agency (EPA), and other environmental agencies to expose and remedy the corrupt sweetheart relationship that so often put regulators in bed with the polluting industries they regulated. Among the hundreds of lawsuits I filed, perhaps a quarter were against regulatory officials making illegal concessions to Big Oil, King Coal, and the chemical and agricultural polluters that had captured their loyalties. I thought I knew everything about regulatory capture and that I had armored myself with an appropriate shield of cynicism.

But I was wrong about that. From the moment of my reluctant entrance into the vaccine debate in 2005, I was astonished to realize that the pervasive web of deep financial entanglements between Pharma and the government health agencies had put regulatory capture on steroids. The CDC, for example, owns 57 vaccine patents[1] and spends $4.9 of its $12.0 billion-dollar annual budget (as of 2019) buying and distributing vaccines.[2,3] NIH owns hundreds of vaccine patents and often profits from the sale of products it supposedly regulates. High level officials, including Dr. Fauci, receive yearly emoluments of up to $150,000 in royalty payments on products that they help develop and then usher through the approval process.[4] The FDA receives 45 percent of its budget from the pharmaceutical industry, through what are euphemistically called "user fees."[5] When I learned that extraordinary fact, the disastrous health of the American people was no longer a mystery; I wondered what the environment would look like if the EPA received 45 percent of its budget from the coal industry!

Today many of my liberal chums are still crouched in a knee jerk posture defending "our" agencies against Republican slanders and budget cuts, never quite realizing how thoroughly the decades of attacks succeeded in transforming those agencies into subsidiaries of Big Pharma.

In this book, I track the rise of Anthony Fauci from his start as a young public health researcher and physician through his metamorphosis into the powerful technocrat who helped orchestrate and execute 2020's historic coup d'état against Western democracy. I explore the carefully planned militarization and monetization of medicine that has left American health ailing and its democracy shattered. I chronicle the troubling role of the dangerous concentrated mainstream media, Big Tech robber barons, the military and intelligence communities and their deep historical alliances with Big Pharma and public health agencies. The disturbing story that unfolds here has never been told, and many in power have worked hard to prevent the public from learning it. The main character is Anthony Fauci.

During the 2020 COVID-19 pandemic, Dr. Fauci, who turned 80 that year, occupied center stage in a global drama unprecedented in human history. At the contagion's beginnings, the US still enjoyed its reputation as the universal standard-bearer in public health. As the world's faith in American leadership dwindled during the Trump era, the singular US institutions that were seemingly immune from international disillusionment were our public health regulators; HHS—and its subsidiary agencies CDC, FDA, and NIH—persisted as role models for global health policies and gold standard scientific research. Other nations looked to Dr. Fauci, America's most powerful and enduring public health bureaucrat, to competently direct US health policies, and rapidly develop countermeasures that would serve as state-of-the-art templates for the rest of the world.

Dr. Anthony Fauci spent half a century as America's reigning health commissar, ever preparing for his final role as Commander of history's biggest war against a global pandemic. Beginning in 1968, he occupied various posts at the National Institute of Allergy and Infectious Diseases (NIAID), serving as that agency's Director since November 1984.[6] His $417,608 annual salary makes him the highest paid of all four million federal employees, including the President.[7] His experiences surviving 50 years as the panjandrum of a key federal bureaucracy, having advised six Presidents, the Pentagon, intelligence agencies, foreign governments, and the WHO, seasoned him exquisitely for a crisis that would allow him to wield power enjoyed by few rulers and no doctor in history.

During the epidemic's early months, Dr. Fauci's calm, authoritative, and avuncular manner was Prozac for Americans besieged by two existential crises: the Trump Presidency, and COVID-19. Democrats and idealistic liberals around the globe, traumatized by President Trump's chaotic governing style, took heart from Dr. Fauci's serene, solid presence on the White House stage. He seemed to offer a rational, straight-talking, science-based counterweight to President Trump's desultory, narcissistic bombast. Navigating the hazardous waters between an erratic President and a deadly contagion, Dr. Fauci initially cut a heroic figure, like Homer's Ulysses steering his ship between Scylla and Charybdis. Turning their backs to the foreboding horizon, trusting Americans manned the oars and blindly obeyed his commands—little realizing they were propelling our country toward the desolate destination where democracy goes to die.

Throughout the first year of the crisis, Dr. Fauci's personal charisma and authoritative voice inspired confidence in his prescriptions and won him substantial—though not universal—affection. Many Americans, dutifully locked in their homes in compliance with Dr. Fauci's quarantine, took consolation in their capacity to join a Tony Fauci fan club, chillax on an "I heart Fauci" throw pillow, sip from an "In Fauci We Trust" coffee mug, warm cold feet in Fauci socks and booties, gorge on Fauci donuts, post a "Honk for Dr. Fauci" yard sign, or genuflect before a Dr. Fauci prayer candle. Fauci aficionados could choose from a variety of Fauci browser games and a squadron of Fauci action figures and bobbleheads, and could read his hagiography to their offspring from a worshipful children's book. At the height of the lockdown, Brad Pitt performed a reverential homage to Dr. Fauci on *Saturday Night Live*,[8] and Barbara Streisand surprised him with a recorded message during a live Zoom birthday party in his honor.[9] *The New Yorker* dubbed him "America's Doctor."[10]

Dr. Fauci encouraged his own canonization and the disturbing inquisition against his blasphemous critics. In a June 9, 2021 *je suis l'état* interview, he pronounced that Americans who questioned his statements were, per se, anti-science. "Attacks on me," he explained, "quite frankly, are attacks on science."[11] The sentiment he expressed reminds us that blind faith in authority is a function of religion, not science. Science, like democracy, flourishes on skepticism toward official orthodoxies. Dr. Fauci's schoolboy scorn for citation and his acknowledgement to the *New York Times* that he had twice lied to Americans to promote his agendas—on masks and herd immunity—raised the prospect that some of his other "scientific" assertions were, likewise, *noble lies* to a credulous public he believes is unworthy of self-determination.[12,13]

In August 2021, Dr. Fauci's acolyte—CNN's television doctor, Peter Hotez— published an article in a scientific journal calling for legislation to "expand federal hate crime protections" to make criticism of Dr. Fauci a felony.[14] In declaring that he had no conflicts, Dr. Hotez, who says that vaccine skeptics should be snuffed out,[15] evidently forgot the millions of dollars in grants he has taken from Dr. Fauci's NIAID since 1993,[16] and more than $15 million from Dr. Fauci's partner, Bill Gates, for his Baylor University Tropical Medicine Institute.[17,18] As we shall see, Dr. Fauci's direct and indirect control—through NIH, Bill & Melinda Gates Foundation, and the Wellcome Trust of some 57 percent of global biomedical research funding[19]—guarantees him this sort of obsequious homage from leading medical researchers, allows him to craft and perpetuate the reigning global medical narratives, and can fortify the canon that he, himself, is science incarnate.

High-visibility henchmen like Hotez—and Pharma's financial control over the media through advertising dollars—have made Dr. Fauci's pronouncements impervious to debate and endowed the NIAID Director with personal virtues and medical gravitas supported by neither science nor his public health record. By the latter metric, his 50-year regime has been calamitous for public health and for democracy. His administration of the COVID pandemic was, likewise, a disaster.

As the world watched, Tony Fauci dictated a series of policies that resulted in by far the most deaths, and one of the highest percentage COVID-19 body counts of any nation on the planet. Only relentless propaganda and wall-to-wall censorship could conceal his disastrous mismanagement during COVID-19's first year. The US, with 4 percent of the world's population, suffered 14.5 percent of total COVID deaths. By September 30, 2021, mortality rates in the US had climbed to 2,107/1,000,000, compared to 139/1,000,000 in Japan.

Anthony Fauci's Report Card

Death Rates from COVID per million population, as of September 30, 2021[20]:

United States	2,107 deaths/1,000,000	Iran	1,449 deaths/1,000,000
Sweden	1,444 deaths/1,000,000	Germany	1,126 deaths/1,000,000

Cuba	650 deaths/1,000,000	Pakistan	128 deaths/1,000,000
Jamaica	630 deaths/1,000,000	Kenya	97 deaths/1,000,000
Denmark	455 deaths/1,000,000	South Korea	47 deaths/1,000,000
India	327 deaths/1,000,000	Congo (Brazzaville)	35 deaths/1,000,000
Finland	194 deaths/1,000,000	Hong Kong	28 deaths/1,000,000[21]
Vietnam	197 deaths/1,000,000	China	3 deaths/1,000,000
Norway	161 deaths/1,000,000	Tanzania	0.86 deaths/1,000,000
Japan	139 deaths/1,000,000		

After achieving these cataclysmicly awful results, "Teflon Tony's" media savvy and his skills for deft maneuvering beguiled incoming President Joe Biden into appointing him as the new administration's COVID Response Director.

Blinded by generously stoked fear of deadly disease against which Dr. Fauci seemed the only reliable bulwark, Americans failed to see the mounting evidence that Dr. Fauci's strategies were consistently failing to achieve promised results, as he doggedly elevated Pharma profits and bureaucratic powers over waning public health.

As we shall see from this 50-year saga, Dr. Fauci's remedies are often more lethal than the diseases they pretend to treat. His COVID prescriptions were no exception. With his narrow focus on the solution of mass vaccination, Dr. Fauci never mentioned any of the many other costs associated with his policy directives.

Anthony Fauci seems to have not considered that his unprecedented quarantine of the healthy would kill far more people than COVID, obliterate the global economy, plunge millions into poverty and bankruptcy, and grievously wound constitutional democracy globally. We have no way of knowing how many people died from isolation, unemployment, deferred medical care, depression, mental illness, obesity, stress, overdoses, suicide, addiction, alcoholism, and the accidents that so often accompany despair. We cannot dismiss the accusations that his lockdowns proved more deadly than the contagion. A June 24, 2021 *BMJ* study[22] showed that US life expectancy decreased by 1.9 years during the quarantine. Since COVID mortalities were mainly among the elderly, and the average age of death from COVID in the UK was 82.4, which was above the average lifespan,[23] the virus could not by itself cause the astonishing decline. As we shall see, Hispanic and Black Americans often shoulder the heaviest burden of Dr. Fauci's public health adventures. In this respect, his COVID-19 countermeasures proved no exception. Between 2018 and 2020, the average Hispanic American lost around 3.9 years in longevity, while the average lifespan of a Black American dropped by 3.25 years.[24]

This dramatic culling was unique to America. Between 2018 and 2020, the 1.9 year decrease in average life expectancy at birth in the US was roughly 8.5 times the average decrease in 16 comparable countries, all of which were measured in months, not years.[25]

"I naïvely thought the pandemic would not make a big difference in the gap because my thinking was that it's a global pandemic, so every country is going to take a hit," said Steven Woolf, Director Emeritus of the Center on Society and Health at Virginia Commonwealth University. "What I didn't anticipate was how badly the US would handle the pandemic. These are numbers we aren't at all used to seeing in this research; 0.1 years is something that normally gets attention in the field, so 3.9 years and 3.25 years and even 1.4 years is just horrible," Woolf continued. "We haven't had a decrease of that magnitude since World War II."[26]

Cost of Quarantines—Deaths

As Dr. Fauci's policies took hold globally, 300 million humans fell into dire poverty, food insecurity, and starvation. "Globally, the impact of lockdowns on health programs, food production, and supply chains plunged millions of people into severe hunger and malnutrition," said Alex Gutentag in *Tablet Magazine*.[27] According to the Associated Press (AP), during 2020, 10,000 children died each month due to virus-linked hunger from global lockdowns. In addition, 500,000 children per month experienced wasting and stunting from malnutrition—up 6.7 million from last year's total of 47 million—which can "permanently damage children physically and mentally, transforming individual tragedies into a generational catastrophe."[28] In 2020, disruptions to health and nutrition services killed 228,000 children in South Asia.[29] Deferred medical treatments for cancers, kidney failure, and diabetes killed hundreds of thousands of people and created epidemics of cardiovascular disease and undiagnosed cancer. Unemployment shock is expected to cause 890,000 additional deaths over the next 15 years.[30,31]

The lockdown disintegrated vital food chains, dramatically increased rates of child abuse, suicide, addiction, alcoholism, obesity, mental illness, as well as debilitating developmental delays, isolation, depression, and severe educational deficits in young children. One-third of teens and young adults reported worsening mental health during the pandemic. According to an Ohio State University study,[32] suicide rates among children rose 50 percent.[33] An August 11, 2021 study by Brown University found that infants born during the quarantine were short, on average, 22 IQ points as measured by Baylor scale tests.[34] Some 93,000 Americans died of overdoses in 2020—a 30 percent rise over 2019.[35] "Overdoses from synthetic opioids increased by 38.4 percent,[36] and 11 percent of US adults considered suicide in June 2020.[37] Three million children disappeared from public school systems, and ERs saw a 31 percent increase in adolescent mental health visits,"[38,39] according to Gutentag. Record numbers of young children failed to reach crucial developmental milestones.[40,41] Millions of hospital and nursing home patients died alone without comfort or a final goodbye from their families. Dr. Fauci admitted that he never assessed the costs of desolation, poverty, unhealthy isolation, and depression fostered by his countermeasures. "I don't give advice about economic things,"[42] Dr. Fauci explained. "I don't give advice about anything other than public health," he continued, even though he was so clearly among those responsible for the economic and social costs.

Economic Destruction and Shifting Wealth Upward

During the COVID pandemic, Dr. Fauci served as ringmaster in the engineered demolition of America's economy. His lockdown predictably shattered the nation's once-booming economic engine, putting 58 million Americans out of work,[43] and *permanently bankrupting small businesses, including 41 percent of Black-owned businesses, some of which took generations of investment to build.*[44] The business closures contributed to a run-up in the national deficit—the interest payments alone will cost almost $1 trillion annually.[45] That ruinous debt will likely permanently bankrupt the New Deal programs—the social safety net that, since 1945, fortified, nurtured, and sustained America's envied middle-class. Government officials have already begun liquidating the almost 100-year legacies of the New Deal, New Frontier, the Great Society, and Obamacare to pay the accumulated lockdown debts. Will we find ourselves saying goodbye to school lunches, healthcare, WIC, Medicaid, Medicare, university scholarships, and other long standing assistance programs?

Enriching the Wealthy

Dr. Fauci's business closures pulverized America's middle class and engineered the largest upward transfer of wealth in human history. In 2020, workers lost $3.7 trillion while billionaires gained $3.9 trillion.[46] Some 493 individuals became new billionaires,[47] and an additional 8 million Americans dropped below the poverty line.[48] The biggest winners were the robber barons—the very companies that were cheerleading Dr. Fauci's lockdown and censoring his critics: Big Technology, Big Data, Big Telecom, Big Finance, Big Media behemoths (Michael Bloomberg, Rupert Murdoch, Viacom, and Disney), and Silicon Valley Internet titans like Jeff Bezos, Bill Gates, Mark Zuckerberg, Eric Schmidt, Sergey Brin, Larry Page, Larry Ellison, and Jack Dorsey.

The very Internet companies that snookered us all with the promise of democratizing communications made it impermissible for Americans to criticize their government or question the safety of pharmaceutical products; these companies propped up all official pronouncements while scrubbing all dissent. The same Tech/Data and Telecom robber barons, gorging themselves on the corpses of our obliterated middle class, rapidly transformed America's once-proud democracy into a censorship and surveillance police state from which they profit at every turn.

CEO Satya Nadella boasted that Microsoft, by working with the CDC and the Gates-funded Johns Hopkins Center for Biosecurity, had used the COVID pandemic to achieve "two years of digital transformation in two months."[49] Microsoft Teams users ballooned to 200 million meeting participants in a single day, averaged more than 75 million active users, compared to 20 million users in November 2019,[50] and the company's stock value skyrocketed. Larry Ellison's company, Oracle, which partnered with the CIA to build new cloud services, won the contract to process all CDC vaccination data.[51] Ellison's wealth increased by $34 billion in 2020; Mark Zuckerberg's wealth grew by $35 billion; Google's Sergey Brin by $41 billion; Jeff Bezos by $86 billion; Bill Gates by $22 billion[52] and Michael Bloomberg by nearly $7 billion.[53]

Ellison, Gates, and the other members of this government/industry collaboration used the lockdown to accelerate construction of their 5G network[54] of satellites, antennae, biometric facial recognition, and "track and trace" infrastructure that they, and their government and intelligence agency partners, can use to mine and monetize our data, further suppress dissent, to compel obedience to arbitrary dictates, and to manage the rage that comes as Americans finally wake up to the fact that this outlaw gang has stolen our democracy, our civil rights, our country, and our way of life—while we huddled in orchestrated fear from a flu-like virus.

With fears of COVID generously stoked, the dramatic and steady erosion of constitutional rights and fomenting of a global coup d'état against democracy, the demolition of our economy, the obliteration of a million small businesses, the collapsing of the middle class, the evisceration of our Bill of Rights, the tidal wave of surveillance capitalism and the rising bio-security state, and the stunning shifts in wealth and power going to a burgeoning oligarchy of high-tech Silicon Valley robber barons seemed, to a dazed and uncritical America, like it might be a reasonable price to pay for safety. And anyway, we were told, it's just for 15 days, or maybe 15 months, or however long it takes for Dr. Fauci to "follow the data" to his answer.

Failing Upward

Dr. Fauci's catastrophic failure to achieve beneficial health outcomes during the COVID-19 crisis is consistent with the disastrous declines in public health during his half-century running NIAID. For anyone who might have assumed that federal and public health bureaucrats survive and flourish by achieving improvements in public health, Dr. Fauci's durability at NIAID is a disheartening wake-up call. By any measure, he has consistently failed upward.

The "J. Edgar Hoover of public health" has presided over cataclysmic declines in public health, including an exploding chronic disease epidemic that has made the "Fauci generation"—children born after his elevation to NIAID kingpin in 1984—the sickest generation in American history, and has made Americans among the least healthy citizens on the planet. His obsequious subservience to the Big Ag, Big Food, and pharmaceutical companies has left our children drowning in a toxic soup of pesticide residues, corn syrup, and processed foods, while also serving as pincushions for 69 recommended vaccine doses by age 18—none of them properly safety tested.[55]

When Dr. Fauci took office, America was still ranked among the world's healthiest populations. An August 2021 study by the Commonwealth Fund ranked America's health care system dead last among industrialized nations, with the highest infant mortality and the lowest life expectancy. "If health care were an Olympic sport, the US might not qualify in a competition with other high-income nations,"[56] laments the study's lead author, Eric Schneider, who serves as Senior Vice President for Policy and Research at the Commonwealth Fund.

Following WWII, life expectancy in the US climbed for five decades, making Americans among the longest-lived people in the developed world. IQ also grew steadily by three points each decade since 1900. But as Tony Fauci spent the 1990s expanding the pharmaceutical and chemical paradigm—instead of public health—the pace of both longevity and intelligence slowed. The life expectancy decrease widened the gap between the US and its peers to nearly five years,[57] and American children have lost seven IQ points since 2000.[58]

Under Dr. Fauci's leadership, the allergic, autoimmune, and chronic illnesses which Congress specifically charged NIAID to investigate and prevent, have mushroomed to afflict 54 percent of children, up from 12.8 percent when he took over

NIAID in 1984.[59] Dr. Fauci has offered no explanation as to why allergic diseases like asthma, eczema, food allergies, allergic rhinitis, and anaphylaxis suddenly exploded beginning in 1989, five years after he came to power. On its website, NIAID boasts that autoimmune disease is one of the agency's top priorities. Some 80 autoimmune diseases, including juvenile diabetes and rheumatoid arthritis, Graves' disease, and Crohn's disease, which were practically unknown prior to 1984, suddenly became epidemic under his watch.[60,61,62] Autism, which many scientists now consider an autoimmune disease,[63,64,65] exploded from between 2/10,000 and 4/10,000 Americans[66] when Tony Fauci joined NIAID, to one in thirty-four today. Neurological diseases like ADD/ADHD, speech and sleep disorders, narcolepsy, facial tics, and Tourette's syndrome have become commonplace in American children.[67] The human, health, and economic costs of chronic disease dwarf the costs of all infectious diseases in the United States. By this decade's end, obesity, diabetes, and pre-diabetes are on track to debilitate 85 percent of America's citizens.[68] America is among the ten most overweight countries on Earth. The health impacts of these epidemics—which fall mainly on the young—eclipse even the most exaggerated health impacts of COVID-19.

What is causing this cataclysm? Since genes don't cause epidemics, it must be environmental toxins. Many of these illnesses became epidemic in the late 1980s, after vaccine manufacturers were granted government protection from liability, and consequently accelerated their introduction of new vaccines.[69] The manufacturer's inserts of the 69 vaccine doses list each of the now-common illnesses—some 170 in total—as vaccine side effects.[70] So vaccines are a potential culprit, but not the only one. Other possible perpetrators—or accomplices—that fit the applicable criterion—a sudden epidemic across all demographics beginning in 1989—are corn syrup, PFOA flame retardants, processed foods, cell phones and EMF radiation, chlorpyrifos, ultrasound, and neonicotinoid pesticides.

The list is finite, and it would be a simple thing to design studies that give us these answers. Tracing the etiology of these diseases through epidemiological research, observational and bench studies, and animal research is exactly what Congress charged Dr. Fauci to perform. But Tony Fauci controls the public health bankbook and has shown little interest in funding basic science to answer those questions.

Is this because any serious investigation into the sources of the chronic disease epidemic would certainly implicate the powerful pharmaceutical companies and the chemical, agricultural, and processed food multinationals that Dr. Fauci and his twenty-year business partner, Bill Gates, have devoted their careers to promoting? As we shall see, his capacity to curry favor with these merchants of pills, powders, potions, poisons, pesticides, pollutants, and pricks has been the key to Dr. Fauci's longevity at HHS.

Is it fair to blame Dr. Fauci for a crisis that, of course, has many authors? Due to his vast budgetary discretion, his unique political access, his power over HHS and its various agencies, his moral authority, his moral flexibility, and his bully pulpit, Tony Fauci has more power than any other individual to direct public energies toward solutions. He has done the opposite. Instead of striving to identify the etiologies of the chronic disease pandemic, we shall see that Dr. Fauci has deliberately and systematically used his staggering power over Federal scientific research, medical schools, medical journals, and the careers of individual scientists, to derail inquiry and obstruct research that might provide the answers.

Dr. Phauci's Pharmanation

While some Republicans bridled warily at Dr. Fauci's accumulating power and seemingly arbitrary pronouncements, the alchemies of political tribalism and the relentlessly stoked terror of COVID-19 persuaded spellbound Democrats to close their eyes to the damning evidence that his COVID-19 policies were a catastrophic and dangerous failure.

As an advocate for public health, robust science, and independent regulatory agencies—free from corruption and financial entanglements with Pharma—I have battled Dr. Fauci for many years. I know him personally, and my impression of him is very different from my fellow Democrats, who first encountered him as the polished, humble, earnest, endearing, and long-suffering star of the televised White House COVID press conferences. Dr. Fauci played a historic role as the leading architect of "agency capture"—the corporate seizure of America's public health agencies by the pharmaceutical industry.

Lamentably, Dr. Fauci's failure to achieve public health goals during the COVID pandemic are not anomalous errors, but consistent with a recurrent pattern of sacrificing public health and safety on the altar of pharmaceutical profits and self-interest. He consistently priortized pharmaceutical industry profits over public health. Readers of these pages will learn how in exalting patented medicine Dr. Fauci has, throughout his long career, routinely falsified science, deceived the public and physicians, and lied about safety and efficacy. Dr. Fauci's malefactions detailed in this volume include his crimes against the hundreds of Black and Hispanic orphan and foster children whom he subjected to cruel and deadly medical experiments and his role, with Bill Gates, in transforming hundreds of thousands of Africans into lab rats for low-cost clinical trials of dangerous experimental drugs that, once approved, remain financially out of reach for most Africans. You will learn how Dr. Fauci and Mr. Gates have turned the African continent into a dumping ground for expired, dangerous, and ineffective drugs, many of them discontinued for safety reasons in the US and Europe.

You will read how Dr. Fauci's strange fascination with, and generous investments in, so-called "gain of function" experiments to engineer pandemic superbugs, give rise to the ironic possibility that Dr. Fauci may have played a role in triggering the global contagion that two US presidents entrusted him to manage. You will also read about his two-decade strategy of promoting false pandemics as a scheme for promoting novel vaccines, drugs and Pharma profits. You will learn of his actions to conceal widespread contamination in blood and vaccines, his destructive vendettas against scientists who challenge the Pharma paradigm, his deliberate sabotaging of patent-expired remedies against infectious diseases, from HIV to COVID-19, to grease the skids for less effective, but more profitable, remedies. You will learn of the grotesque body counts that have accumulated in the wake of his cold-blooded focus on industry profits over public health. All his strategies during COVID—falsifying science to bring dangerous and ineffective drugs to market, suppressing and sabotaging competitive products that have lower profit margins even if the cost is prolonging pandemics and losing thousands of lives—all of these share a common purpose: the myopic devotion to Pharma. This book will show you that Tony Fauci does not do public health; he is a businessman, who has used his office to enrich his pharmaceutical partners and expand the reach of influence that has made him the most powerful—and despotic—doctor in human history. For some readers, reaching that conclusion will require crossing some new bridges; many readers, however, intuitively know the real Anthony Fauci, and need only to see the facts illuminated and organized.

I wrote this book so that Americans—both Democrat and Republican—can understand Dr. Fauci's pernicious role in allowing pharmaceutical companies to dominate our government and subvert our democracy, and to chronicle the key role Dr. Fauci has played in the current coup d'état against democracy.

Endnotes

1   Google Patents, Assignee: Centers for Disease Control and Prevention, https://www.google.com/searc h?tbo=p&tbm=pts&hl=en&q=vaccine+inassignee:centers+inassignee:for+inassignee:disease+inassigne e:control&tbs=,ptss:g&num=100

2   Centers for Disease Control and Prevention, *President's Budget FY 2020, 2019 Enacted Column,* 2020, https://www.cdc.gov/budget/documents/fy2020/fy-2020-detail-table.pdf

3   Centers for Disease Control and Prevention, *Dept. of HHS FY 2020 Centers for Disease Control and Prevention-Justification of Estimates for Appropriation Committees- FY 2019 Enacted,* 2020, p. 42-43, https://www.cdc.gov/budget/documents/fy2020/fy-2020-cdc-congressional-justification.pdf

4   Cornell Law School, Legal Information Institute, *15 U.S. Code § 3710c—Distribution of royalties received by Federal agencies,* https://www.law.cornell.edu/uscode/text/15/3710c

5   FDA, *Fact Sheet: FDA at a Glance,* FDA (Nov. 18, 2020), https://www.fda.gov/about-fda/fda-basics/ fact-sheet-fda-glance

6   Anthony S. Fauci, MD, Biography, NIAID https://www.niaid.nih.gov/about/anthony-s-fauci-md-bio

7   Adam Andrezejewski, "Dr. Anthony Fauci: The Highest Paid Employee in the Entire U.S. Federal Government," *FORBES* (Jan. 25, 2021), https://www.forbes.com/sites/ adamandrzejewski/2021/01/25/dr-anthony-fauci-the-highest-paid-employee-in-the-entire-us-federal-government/?sh=5ed2512386f0

8   *Saturday Night Live,* "Dr. Anthony Fauci Cold Open—SNL, YOUTUBE" (Apr. 25, 2020), https:// www.youtube.com/watch?v=uW56CL0pk0g

9   Zack Budryk, "AIDS activists recruit Barbra Streisand for surprise Fauci birthday party on Zoom," *THE HILL* (Dec. 24, 2020, 5:36 PM), https://thehill.com/policy/healthcare/531636-aids-activists-recruit-barbra-streisand-for-surprise-zoom-birthday-party

10   Michael Specter, "How Anthony Fauci Became America's Doctor," *The New Yorker* (Apr. 10, 2020), https://www.newyorker.com/magazine/2020/04/20/how-anthony-fauci-became-americas-doctor

11   Peter Sullivan, "Fauci: Attacks on me are really also 'attacks on science,'" *The Hill* (Jun. 9, 2021), https://thehill.com/policy/healthcare/557602-fauci-attacks-on-me-are-really-also-attacks-on-science

12   Donald G. McNeil Jr., "How Much Herd Immunity Is Enough?" *New York Times* (Dec. 24, 2020, updated Apr. 2, 2021), https://www.nytimes.com/2020/12/24/health/herd-immunity-covid-coronavirus.html

13   Tiana Lowe, "Fauci lies about lying about the efficacy of masks," MSN (Jun. 21, 2021), https://www. msn.com/en-us/health/medical/fauci-lies-about-lying-about-the-efficacy-of-masks/ar-AALhCrp

14   Peter Hotez, "Mounting antiscience aggression in the United States," *PLOS BIOLOGY* (Jul. 28, 2021), https://journals.plos.org/plosbiology/article?id=10.1371/journal.pbio.3001369

15   Peter Hotez, "Will an American-Led Anti-Vaccine Movement Subvert Global Health?" *Scientific American* (Mar. 3, 2017), https://blogs.scientificamerican.com/guest-blog/will-an-american-led-anti-vaccine-movement-subvert-global-health/

16   National Institutes of Health, *National Institutes of Health Awards by Location and Organization,* (2021), https://childrenshealthdefense.org/citation/niaid-grants-to-baylor-by-year-since-1993/

17   *Philanthropy News Digest,* "Sabin Institute Receives $12 Million From Gates Foundation to Develop Hookworm Vaccine" (Jul 1, 2011), https://philanthropynewsdigest.org/news/sabin-institute-receives-12-million-from-gates-foundation-to-develop-hookworm-vaccine

18   Vipul Naik, "Bill and Melinda Gates Foundation donations made to Baylor College of Medicine," https://donations.vipulnaik.com/donorDonee.php?donor=Bill+and+Melinda+Gates+Foundation&donee=Baylor+College+of+Medicine

19   Rebecca G. Baker, "Bill Gates Asks NIH Scientists for Help in Saving Lives And Explains Why the Future Depends on Biomedical Innovation," *THE NIH CATALYST* (Jan-Feb, 2014), https://irp.nih.gov/catalyst/v22i1/bill-gates-asks-nih-scientists-for-help-in-saving-lives

20   Statista, *Coronavirus (COVID-19) deaths worldwide per one million population as of September 30, 2021, by country* (Oct. 6, 2021), https://www.statista.com/statistics/1104709/coronavirus-deaths-worldwide-per-million-inhabitants/

21   *Reported Cases and Deaths by Country or Territory*, WORLDOMETER (Oct. 4, 2021), https://www.worldometers.info/coronavirus/

22   S H Woolf, et al, "Effect of the covid-19 pandemic in 2020 on life expectancy across populations in the USA and other high income countries: simulations of provisional mortality data," *BMJ* 2021;373:n1343 (June 24, 2021), https://www.bmj.com/content/373/bmj.n1343

23   Jemima Kelly, "Covid kills, but do we overestimate the risk?" *Financial Times* (Nov. 20, 2020), https://www.ft.com/content/879f2a2b-e366-47ac-b67a-8d1326d40b5e

24   S H Woolf et al, "Effect of the covid-19 pandemic in 2020 on life expectancy across populations in the USA and other high income countries: simulations of provisional mortality data," *BMJ* 2021;373:n1343 (June 24, 2021) https://www.bmj.com/content/373/bmj.n1343

25   Kaitlin Sullivan, "U.S. Life Expectancy Decreased by an 'alarming' amount during pandemic," NBC NEWS (Jun. 23, 2021), https://www.nbcnews.com/health/health-news/u-s-life-expectancy-decreased-alarming-amount-during-pandemic-n1272206

26   Ibid.

27   Alex Gutentag, "The War on Reality," *TABLET MAGAZINE* (June 28, 2021), https://www.tabletmag.com/sections/news/articles/the-war-on-reality-gutentag

28   Lori Hinnant and Sam Mednick, "Virus-linked hunger tied to 10,000 child deaths each month," *AP* (Jul. 27, 2020), https://apnews.com/article/virus-outbreak-africa-ap-top-news-understanding-the-outbreak-hunger-5cbee9693c52728a3808f4e7b4965cbd

29   BBC News, "Covid-19 disruptions killed 228,000 children in South Asia, says UN report, *BBC* (Mar. 17, 2021), https://www.bbc.com/news/world-asia-56425115

30   Megan Henney, "COVID's economic fallout could elevate US mortality rate for years, study shows," FOX BUSINESS (Jan. 5, 2021), https://www.foxbusiness.com/economy/economic-fallout-from-coronavirus-pandemic-could-elevate-us-mortality-rate-for-years

31   Francesco Bianchi, Giada Bianchi, and Dongho Song, "The Long-term Impact Of The Covid-19 Unemployment Shock On Life Expectancy And Mortality Rates," *National Bureau of Economic Research* (Dec. 2020, rev. Sep. 2021), https://www.nber.org/system/files/working_papers/w28304/w28304.pdf

32   Ohio State University, "A third of teens, young adults reported worsening mental health during pandemic," OSU Press Release (Jul 12, 2021), https://www.eurekalert.org/news-releases/545757

33   CDC, *Emergency Department Visits for Suspected Suicide Attempts Among Persons Aged 12–25 Years Before and During the COVID-19 Pandemic—United States, January 2019–May 2021,* (Jun. 18, 2021), https://www.cdc.gov/mmwr/volumes/70/wr/mm7024e1.htm

34   Sean CL Deoni et al, *Impact of the COVID-19 Pandemic on Early Child Cognitive Development: Initial Findings in a Longitudinal Observational Study of Child Health,* medRxiv 2021.08.10.21261846; doi: https://doi.org/10.1101/2021.08.10.21261846

35   Bill Chappell, *Drug Overdoses Killed A Record Number Of Americans In 2020, Jumping By Nearly 30%,* NPR (Jul. 14, 2021), https://www.npr.org/2021/07/14/1016029270/drug-overdoses-killed-a-record-number-of-americans-in-2020-jumping-by-nearly-30

36   CDC Health Alert Network, *Increase in Fatal Drug Overdoses Across the United States Driven by Synthetic Opioids Before and During the COVID-19 Pandemic,* CDC (Dec. 20, 2020), https://emergency.cdc.gov/han/2020/han00438.asp

37   Andrea Petersen, *Amid Pandemic, More U.S. Adults Say They Considered Suicide,* (Aug. 13, 2020 7:42 pm), https://www.wsj.com/articles/amid-pandemic-more-u-s-adults-say-they-considered-suicide-11597362131

38   Rebecca T. Leeb et al, *Mental Health–Related Emergency Department Visits Among Children Aged <18 Years During the COVID-19 Pandemic — United States, January 1–October 17, 2020,* CDC (Nov. 13, 2020), https://www.cdc.gov/mmwr/volumes/69/wr/mm6945a3.htm

39   Alex Gutentag, *The War on Reality,* TABLET MAGAZINE (June 28, 2021), https://www.tabletmag.com/sections/news/articles/the-war-on-reality-gutentag

40   Id.

41   Amarica Rafanelli, Growing Up in a Pandemic: How Covid is Affecting Children's Development, DIRECT RELIEF (Jan. 19, 2021, 10:41 AM), https://www.directrelief.org/2021/01/growing-up-in-the-midst-of-a-pandemic-how-covid-is-affecting-childrens-development/

42   James Freeman, *The Limits of Anthony Fauci's Expertise,* WALL STREET JOURNAL (May 13, 2020 1:52 pm) https://www.wsj.com/articles/the-limits-of-anthony-faucis-expertise-11589392347

43   Nigel Chiwaya & Jiachuan Wu, *Unemployment claims by state: See how COVID-19 has destroyed*

the job market, NBC NEWS (Apr. 14, 2020, updated Aug.27, 2020), https://www.nbcnews.com/business/economy/unemployment-claims-state-see-how-covid-19-has-destroyed-job-n1183686

44   Anne Sraders & Lance Lambert, *Nearly 100,000 establishments that temporarily shut down due to the pandemic are now out of business*, FORTUNE (Sep. 28, 2020), https://fortune.com/2020/09/28/covid-buisnesses-shut-down-closed/

45   Deficit Tracker, BIPARTISAN POLICY (Sept. 20, 2021), https://bipartisanpolicy.org/report/deficit-tracker/

46   *Viral Inequity: Billionaires Gained $3.9tn, Workers Lost $3.7tn in 2020*, TRT WORLD (Jan. 28, 2021), https://www.trtworld.com/magazine/viral-inequality-billionaires-gained-3-9tn-workers-lost-3-7tn-in-2020-43674

47   Chase Peterson-Withorn, *Nearly 500 People Became Billionaires During The Pandemic Year*, FORBES (Apr. 6, 2021), https://www.forbes.com/sites/kerryadolan/2021/04/06/forbes-35th-annual-worlds-billionaires-list-facts-and-figures-2021/?sh=4c7b81775e58

48   Heather Long, *Nearly 8 million Americans have fallen into poverty since the summer*, WASHINGTON POST (Dec. 16, 2020), https://www.washingtonpost.com/business/2020/12/16/poverty-rising/

49   Jared Spataro, *2 Years of Digital Transformation in 2 Months*, MICROSOFT (Apr. 30, 2020), https://www.microsoft.com/en-us/microsoft-365/blog/2020/04/30/2-years-digital-transformation-2-months/

50   Id.

51   Oracle Cloud Manages COVID-19 Vaccination Program in the United States, ORACLE PRESS RELEASE (Dec. 15, 2020), https://www.oracle.com/news/announcement/oracle-cloud-manages-covid-19-vaccination-program-121520.html

52   Chase Petersen-Withorn, *How Much Money America's Billionaires Have Made During The Covid-19 Pandemic*, FORBES (Apr. 30, 2021), https://www.forbes.com/sites/chasewithorn/2021/04/30/american-billionaires-have-gotten-12-trillion-richer-during-the-pandemic/?sh=461b1067f557

53   Samuel Stebbins and Grant Suneson, Jeff Bezos, Elon Musk among US billionaires getting richer during coronavirus pandemic, USA TODAY, (Dec 1, 2020). https://www.usatoday.com/story/money/2020/12/01/american-billionaires-that-got-richer-during-covid/43205617/

54   Sue Halpern, *The Terrifying Potential of the 5G Network,* THE NEW YORKER (Apr. 26, 2019), https://www.newyorker.com/news/annals-of-communications/the-terrifying-potential-of-the-5g-network

55   *Recommended Child and Adolescent Immunization Schedule for ages 18 years or younger, United States, 2021*, CDC, https://www.cdc.gov/vaccines/schedules/hcp/imz/child-adolescent.html

56   Joseph Guzman, *Stunning new report ranks US dead last in health care among richest countries-despite spending the most*, THE HILL (Aug. 6, 2021), https://thehill.com/changing-america/well-being/longevity/566715-stunning-new-report-ranks-us-dead-last-in-healthcare

57   Kaitlin Sullivan, *U.S. Life Expectancy Decreased by an 'alarming' amount during pandemic*, NBC NEWS (Jun. 23, 2021), https://www.nbcnews.com/health/health-news/u-s-life-expectancy-decreased-alarming-amount-during-pandemic-n1272206

58   Dr. Robert Gorter, Dr. Joseph Mercola, et al., "Why are IQ scores declining over the previous 20 years?," *The Gorter Model*, (Jul. 1, 2018), http://www.gorter-model.org/iq-scores-declining-previous-20-years/

59   *Could Goldman Sachs Report Be Exposing Pharma's Real End Game of Drug Dependency vs. Curing Disease*, CHD (Apr. 18, 2018), https://childrenshealthdefense.org/news/could-goldman-sachs-report-be-exposing-pharmas-real-end-game-of-drug-dependency-vs-curing-disease/

60   Lana Andelane, *Autism may be an autoimmune disorder - study,* NEWSHUB,)Oct 20, 2019). https://www.newshub.co.nz/home/lifestyle/2019/10/autism-may-be-an-autoimmune-disorder-study.html

61   Children's Health Defense, Campaign to Restore Child Health, CHILDREN'S HEALTH DEFENSE, (2018). https://childrenshealthdefense.org/campaign-restore-child-health/

62   Gianna Melillo, Study Highlights Prevalence of Comorbid Autoimmune Diseases, T1D in Pediatric Populations, AJMC, (Sep 9, 2020). https://www.ajmc.com/view/study-highlights-prevalence-of-comorbid-autoimmune-diseases-t1d-in-pediatric-populations

63   J.B. HANDLEY, HOW TO END THE AUTISM EPIDEMIC, (Chelsea Green Publishing, 2018).

64   Elizabeth Edmiston, et al, Autoimmunity, Autoantibodies, and Autism Spectrum Disorder, BIOLOGICAL PSYCHIATRY, (Mar 1, 2017). https://www.biologicalpsychiatryjournal.com/article/S0006-3223(16)32739-1/fulltext

65   Heather K. Hughes et al, Immune Dysfunction and Autoimmunity as Pathological Mechanisms in Autism Spectrum Disorders, FRONTIERS IN CELLULAR NEUROSCIENCE, (Nov 13, 2018). https://www.frontiersin.org/articles/10.3389/fncel.2018.00405/full

66   THOMAS F. BOAT & JOE T. WU, ED., MENTAL DISORDERS AND DISABILITIES AMONG LOW-INCOME CHILDREN, 241 National Academies Press, (Oct. 28, 2015), https://www.ncbi.nlm.nih.gov/books/NBK332896/

67   Elizabeth Mumper, MD, *Increasing Rates of Childhood Neurological Illness*, THE INSTITUTE FOR FUNCTIONAL MEDICINE, (2017). https://www.ifm.org/news-insights/increasing-rates-childhood-neurological-illness/

68   Adela Hruby and Frank B. Hu, *The Epidemiology of Obesity: A Big Picture*,

PHARMACOECONOMICS, (Jul 1, 2016). https://www.ncbi.nlm.nih.gov/pmc/articles/PMC4859313/

69   Michael E. McDonald and John F. Paul, *Timing of Increased Autistic Disorder Cumulative Incidence*, ENVIRONMENTAL SCIENCE & TECHNOLOGY, (Feb 16, 2010). https://pubs.acs.org/doi/abs/10.1021/es902057k

70   Centers for Disease Control and Prevention, *Table 1. Recommended Child and Adolescent Immunization Schedule for ages 18 years or younger, United States, 2021*, (2021), https://www.cdc.gov/vaccines/schedules/downloads/child/0-18yrs-child-combined-schedule.pdf

ChildrensHealthDefense.org/fauci-book
childrenshd.org/fauci-book

For updates, new citations and references, and
new information about topics in this chapter:

CHAPTER 1
MISMANAGING A PANDEMIC

"My friend, have you ever been in a quarantined city? Then you cannot realize what you are asking me to do. To place such a curse on San Francisco would be worse than a hundred fires and earthquakes and I love this city too well to do her such a frightful hurt."
—Rupert Blue, Public Health Service Officer in charge of dealing with the 1907 plague outbreak. Blue subsequently served as fourth Surgeon General of the US and President of the American Medical Association.

I: ARBITRARY DECREES: SCIENCE-FREE MEDICINE

Dr. Fauci's strategy for managing the COVID-19 pandemic was to suppress viral spread by mandatory masking, social distancing, quarantining the healthy (also known as lockdowns), while instructing COVID patients to return home and do nothing—receive no treatment whatsoever—until difficulties breathing sent them back to the hospital to submit to intravenous remdesivir and ventilation. This approach to ending an infectious disease contagion had no public health precedent and anemic scientific support. Predictably, it was grossly ineffective; America racked up the world's highest body counts.

Medicines were available against COVID—inexpensive, safe medicines—that would have prevented hundreds of thousands of hospitalizations and saved as many lives if only we'd used them in this country. But Dr. Fauci and his Pharma collaborators deliberately suppressed those treatments in service to their single-minded objective—making America await salvation from their novel, multi-billion dollar vaccines. Americans' native idealism will make them reluctant to believe that their government's COVID policies were so grotesquely ill-conceived, so unfounded in science, so tethered to financial interests, that they caused hundreds of thousands of wholly unnecessary deaths. But, as you will see below, the evidence speaks for itself.

Peer-reviewed science offered anemic if any support for masking, quarantines, and social distancing, and Dr. Fauci offered no citations or justifications to support his diktats. Both common sense and the weight of scientific evidence suggest that all these strategies, and unquestionably shutting down the global economy, caused far more injuries and deaths than they averted.

Dr. Fauci was clearly aware that his mask decrees were contrary to overwhelming science. In July 2020, after switching course to recommend national mask mandates, Dr. Fauci told Norah O'Donnell with *InStyle* magazine that his earlier dismissal of mask efficacy was correct "in the context of the time in which I said it," and that he intended to prevent a consumer run on masks that might jeopardize their availability for front-line responders.[1] But Dr. Fauci's emails reveal that he was giving the same advice privately. Moreover, his detailed explanations to the public and to high-level health regulators indicate he genuinely believed that ordinary masks had little to no efficacy against viral infection. In a February 5, 2020 email, for example, he advised his putative former boss, President Obama's Health and Human Services Secretary,

Sylvia Burwell, on the futility of masking the healthy.[2] On February 17, he invoked the same rationale in an interview with *USA Today*:

> A mask is much more appropriate for someone who is infected and you're trying to prevent them from infecting other people than it is in protecting you against infection. If you look at the masks that you buy in a drug store, the leakage around that doesn't really do much to protect you. Now, in the United States, there is absolutely no reason whatsoever to wear a mask.[3]

During a January 28 speech to HHS regulators, he explained the fruitlessness of masking asymptomatic people.

> The one thing historically people need to realize, that even if there is some asymptomatic transmission, in all the history of respiratory borne viruses of any type, asymptomatic transmission has never been the driver of outbreaks. The driver of outbreaks is always a symptomatic person. Even if there's a rare asymptomatic person that might transmit, an epidemic is not driven by asymptomatic carriers.[4]

Consistent with Dr. Fauci's earlier statements, the peer-reviewed scientific literature has steadfastly refused to support masking the healthy as an effective barrier to viral spread, and Dr. Fauci offered a citation to justify his change of heart. A December 2020 comprehensive study of 10 million Wuhan residents confirmed Fauci's January 28, 2020 assertion that asymptomatic transmission of COVID-19 is infinitesimally rare.[5] Furthermore, some 52 studies—all available on NIH's website—find that ordinary masking (using less than an N95 respirator) doesn't reduce viral infection rates, even—surprisingly—in institutional settings like hospitals and surgical theaters.[6,7] Moreover, some 25 additional studies attribute to masking a grim retinue of harms, including respiratory and immune system illnesses, as well as dermatological, dental, gastrointestinal, and psychological injuries.[8] Fourteen of these studies are randomized, peer-reviewed placebo studies. There is no well-constructed study that persuasively suggests masks have convincing efficacy against COVID-19 that would justify accepting the harms associated with masks. Finally, retrospective studies on Dr. Fauci's mask mandates confirm that they were bootless. "Regional analysis in the United States does not show that [mask] mandates had any effect on case rates, despite 93 percent compliance. Moreover, according to CDC data, 85 percent of people who contracted COVID-19 reported wearing a mask,"[9] according to Gutentag.

Dr. Fauci observed in March 2020 that a mask's only real efficacy may be in "making people feel a little better."[10] Perhaps he recognized that what masking lacked in efficacy against contagion, it compensated for with powerful psychological effects. These symbolic powers demonstrated strategic benefits for the larger enterprise of encouraging public compliance with draconian medical mandates. Dr. Fauci's switch to endorsing masks after first recommending against them came at a time of increasing political polarization, and masks quickly became important tribal badges—signals of rectitude for those who embraced Dr. Fauci, and the stigmata of blind obedience to undeserving authority among those who balked. Moreover, masking, by amplifying everyone's fear, helped inoculate the public against critical thinking. By serving as persistent reminders that each of our fellow citizens was a potentially dangerous and germ-infected threat to us, masks increased social isolation and fostered divisions and fractionalization—thereby impeding organized political resistance. The impact of masking on the national psyche reminded me of the subtle contribution of the "duck and cover drills" of my youth, drills that sustained and cemented the militaristic ideology of the Cold War. Those futile exercises reinforced what my uncle John F. Kennedy's Defense Secretary, Robert McNamara, called "National Mass Psychosis." By suggesting to Americans that full-scale nuclear war was possible, but also survivable, ruinous investments in that project were justified. For the government and mandarins of the Military Industrial Complex, this absurd narrative yielded trillions in appropriations.

Social distancing mandates also rested on a dubious scientific footing. In September 2021, former FDA Commissioner Dr. Scott Gotleib admitted that the six-foot distancing rule that Dr. Fauci and his HHS colleagues imposed upon Americans was "arbitrary," and not, after all, science backed. The process for making that policy choice, Gotleib continued, "Is a perfect example of the lack of rigor around how CDC made recommendations."[11,12]

Finally, the lockdowns of the healthy were so unprecedented that WHO's official pandemic protocols recommended against them. Some WHO officials were passionate on the topic, among them Professor David Nabarro, Senior Envoy on COVID-19, a position reporting to the Director General. On October 8, 2020, he said:

> We in the World Health Organization do not advocate lockdowns as a primary means of controlling this virus. We may well have a doubling of world poverty by next year. We'll have at least a doubling of child malnutrition because children are not getting meals at school and their parents in poor families are not able to afford it. This is a terrible, ghastly, global catastrophe, actually, and so we really do appeal to all world leaders: Stop using lockdown as your primary control method . . . lockdowns just have one consequence that you must never ever belittle—and that is making poor people an awful lot poorer.[13]

As discussed above, Dr. Fauci and other officials made no inquiry or claims as to whether lockdowns would cause more harm and death than they averted. Subsequent studies have strongly suggested that lockdowns had no impact in reducing infection rates. There is no convincing difference in COVID infections and deaths between laissez-faire jurisdictions and those that enforced rigid lockdowns and masks.[14]

Noble Lies and Bad Data

Dr. Fauci's mask deceptions were among several "noble lies" that, his critics complained, revealed a manipulative and deceptive disposition undesirable in an even-handed public health official. Dr. Fauci explained to the *New York Times* that he had upgraded his estimate of the vaccine coverage needed to insure "herd immunity" from 70 percent in March to 80–90 percent in September not based on science, but rather in response to polling that indicated rising rates of vaccine acceptance.[15] He regularly expressed his belief that post-infection immunity was highly likely (with occasional waffling on this topic) although he took the public position that natural immunity did not contribute to protecting the population. He supported COVID jabs for previously infected Americans, defying overwhelming scientific evidence that post-COVID inoculations were both unnecessary and dangerous.[16,17] Under questioning on September 9, 2021, Dr. Fauci conceded he could cite no scientific justification for this policy.[18] In September 2021, in a statement justifying COVID vaccine mandates to school children, Dr. Fauci dreamily recounted his own grade school measles and mumps vaccines—an unlikely memory, since those vaccines weren't available until 1963 and 1967, and Dr. Fauci attended grade school in the 1940s.[19] Dr. Fauci's little perjuries about masks, measles, mumps, herd immunity, and natural immunity attest to his dismaying willingness to manipulate facts to serve a political agenda. If the COVID-19 pandemic has revealed anything, it is that public health officials have based their many calamitous directives for managing COVID-19 on vacillating and science-free beliefs about masks, lockdowns, infection and fatality rates, asymptomatic transmission, and vaccine safety and efficacy, which took every direction and sowed confusion, division, and polarization among the public and medical experts.

Dr. Fauci's libertine approach to facts may have contributed to what, for me, was the most troubling and infuriating feature of all the public health responses to COVID. The blatant and relentless manipulation of data to serve the vaccine agenda became the apogee of a year of stunning regulatory malpractice. High-quality and transparent data, clearly documented, timely rendered, and publicly available are the *sine qua non* of competent public health management. During a pandemic, reliable and comprehensive data are critical for determining the behavior of the pathogen, identifying vulnerable populations, rapidly measuring the effectiveness of interventions, mobilizing the medical community around cutting-edge disease management, and inspiring cooperation from the public. The shockingly low quality of virtually all relevant data pertinent to COVID-19, and the quackery, the obfuscation, the cherry-picking and blatant perversion would have scandalized, offended, and humiliated every prior generation of American public health officials. Too often, Dr. Fauci was at the center of these systemic deceptions. The "mistakes" were always in the same direction—inflating the risks of coronavirus and the safety and efficacy of vaccines in order to stoke public fear of COVID and provoke mass compliance. The excuses for his mistakes range from blaming the public (now blaming the unvaccinated), blaming politics, and explaining his gyrations by saying, "You've got to evolve with the science."[20]

At the outset of the pandemic, Dr. Fauci used wildly inaccurate modeling that overestimated US deaths by 525 percent.[21,22] Scammer and pandemic fabricator Neal Ferguson of Imperial College London was their author, with funding from the Bill & Melinda Gates Foundation (BMGF) of $148.8 million.[23] Dr. Fauci used this model as justification for his lockdowns.

Dr. Fauci acquiesced to CDC's selective protocol changes for completing death certificates in a way that inflated the claimed deaths from COVID, and thus inflated its infection mortality rate. CDC later admitted that only 6 percent of COVID deaths occurred in entirely healthy individuals. The remaining 94 percent suffered from an average of 3.8 potentially fatal comorbidities.[24]

Regulators misused PCR tests that CDC belatedly admitted in August 2021 were incapable of distinguishing COVID from other viral illnesses. Dr. Fauci tolerated their use at inappropriately high amplitudes of 37 and up to 45, even though Fauci had told Vince Racaniello that tests employing cycle thresholds of 35 and above were very unlikely to indicate the presence of live virus that could replicate.[25] In July 2020, Fauci remarked that at these levels, a positive result is "just dead nucleotides, period,"[26] yet did nothing to modify testing so it might be more accurate. As America's COVID czar, Dr. Fauci never complained about CDC's decision to skip autopsies from deaths attributed to vaccines. This practice allowed CDC to persistently claim that all deaths following vaccination were "unrelated to vaccination." CDC also refused to conduct follow-up medical inquiries among people claiming vaccine injuries. Inspired by rich incentives to classify every patient as a COVID-19 victim—Medicare paid hospitals $39,000 per ventilator[27] when treating COVID-19 and only $13,000 for garden variety respiratory infections[28]—hospitals contributed to the deception. Once more, Dr. Fauci winked at the fraud.

Dr. Fauci's refusal to fix the HHS's notoriously dysfunctional vaccine injury surveillance system (VAERS) constituted inexcusable negligence. HHS's own studies indicate that VAERS may be understating vaccine injuries by OVER 99 percent.[29]

The public never received facts about infection fatality rates or age-stratified risks for COVID with the kind of clarity that might have allowed them and their physicians to make evidence-based personal risk assessments. Instead, federal officials relied on vagueness and deception to recklessly overestimate the dangers from COVID in every age group. All of these deceptions riddled virtually every mainstream media report—particularly those by CNN and the *New York Times*—leaving the public with a vastly inflated and cataclysmically inaccurate impression of its lethality. Public surveys showed that, just as Fox News audiences were shockingly misinformed following the 9/11 bombings, CNN viewers and *New York Times* readers were catastrophically misinformed about the facts of COVID-19 during 2020. Successive Gallup polling showed that the average Democrat believed that 50 percent of COVID infections resulted in hospitalizations. The real number was less than one percent.[30]

Trust the Experts

Instead of demanding blue-ribbon safety science and encouraging honest, open, and responsible debate on the science, badly compromised government health officials charged with managing the COVID-19 pandemic collaborated with mainstream and social media to shut down discussion on key public health questions. They silenced doctors who offered any early treatments that might compete with vaccines or who refused to pledge unquestioning faith in zero liability, shoddily tested, experimental vaccines.

The chaotic and confusing data collection and interpretation allowed regulators to justify their arbitrary diktats under the cloak of "scientific consensus." Instead of citing scientific studies or clear data to justify mandates for masks, lockdowns, and vaccines, our medical rulers cited Dr. Fauci or WHO, CDC, FDA, and NIH—captive agencies—to legitimize the medical technocracy's assumption of dangerous new powers.

Dr. Fauci's aficionados, including President Biden and the cable and network news anchors, counseled Americans to "trust the experts." Such advice is both anti-democratic and anti-science. Science is dynamic. "Experts" frequently differ on scientific questions and their opinions can vary in accordance with and demands of politics, power, and financial self-interest. Nearly every lawsuit I have ever litigated pitted highly credentialed experts from opposite sides against each other, with all of them swearing under oath to diametrically antithetical positions based on the same set of facts. Telling people to "trust the experts" is either naive or manipulative—or both.

All of Dr. Fauci's intrusive mandates and his deceptive use of data tended to stoke fear and amplify public desperation for the anticipated arrival of vaccines that would transfer billions of dollars from taxpayers to pharmaceutical executives and share-holders. Some of America's most accomplished scientists, and the physicians leading the battle against COVID in the trenches, came to believe that Anthony Fauci's do-or-die obsession with novel mRNA vaccines—and Gilead's expensive patented antiviral, remdesivir—prompted him to ignore or even suppress effective early treatments, causing hundreds of thousands of unnecessary deaths while also prolonging the pandemic.

Fortifying Immune Systems

I was struck, during COVID-19's early months, that America's Doctor, apparently preoccupied with his single vaccine solution, did little in the way of telling Americans how to bolster their immune response. He never took time during his daily White House briefings from March to May 2020 to instruct Americans to avoid tobacco (smoking and e-cigarettes/vaping double death rates from COVID);[31] to get plenty of sunlight and to maintain adequate vitamin D levels ("Nearly 60 percent of patients with COVID-19 were vitamin D deficient upon hospitalization, with men in the advanced stages of COVID-19 pneumonia showing the greatest deficit");[32] or to diet, exercise, and lose weight (78 percent of Americans hospitalized for COVID-19 were overweight or obese).[33] Quite the contrary, Dr. Fauci's lockdowns caused Americans to gain an average of two pounds per month and to reduce their daily steps by 27 percent.[34] He didn't recommend avoiding sugar and soft drinks, processed foods, and chemical residues, all of which amplify inflammation, compromise immune response, and disrupt the gut biome which governs the immune system. During the centuries that science has fruitlessly sought remedies against coronavirus (aka the common cold), only zinc has repeatedly proven its efficacy in peer-reviewed studies. Zinc impedes viral replication, prophylaxing against colds and abbreviating their duration.[35] The groaning shelves that commercial pharmacies devote to zinc-based cold remedies attest to its extraordinary efficacy. Yet Anthony Fauci never advised Americans to increase zinc uptake following exposure to infection.

Dr. Fauci's neglect of natural immune response was consistent with the pervasive hostility towards any non-vaccine intervention that characterized the federal regulatory gestalt. On April 30, 2021, Canadian Ontario College of Physicians and Surgeons threatened to delicense any doctor who prescribed non-vaccine health strategies including Vitamin D.[36] "They are trying to erase any notion of natural immunity," says Canadian vaccine researcher Dr. Jessica Rose, Ph.D., MSc, BSc. "Pretty soon the incessant lies and propaganda will have successfully instilled in the masses that the only hope for staying alive is via injection, pill-popping, so in sum, no natural immunity." In a podcast interview on October 1, 2021, *Washington Post* reporter Ashley Fetters Maloy pretended to expose "misinformation" about COVID-19 by broadcasting misinformation:

> There's a pervasive idea that your body and your immune system can be healthy enough to ward off COVID-19, which, of course, we know it's a novel coronavirus. No one's body can. No one's body is healthy enough to recognize and just totally ward this off without a vaccine.[37]

Clearly, this is false information. Throughout 2020, before vaccines were available, some 99.9 percent of people's natural immune systems protected their owners from severe illness and death. The CDC and World Health Organization, indeed all global health authorities, have recognized that healthy people, with healthy immune systems, bear minimal risk from COVID. Indeed, many people, according to our health authorities, have an immune response sufficient that they don't even know they have COVID. Maloy's pronouncement that humans cannot fight off COVID-19 without a vaccine is misinformation in its purest form.

Instead of urging calm and telling us, as FDR did during the depths of the Depression, that "we have nothing to fear but fear itself," all of Dr. Fauci's prescriptions and communications seemed intended to maximize stress and trauma: enforced isolation, mandated masking, business closures, evictions and bankruptcies, lockdowns, and separating children from parents and parents from grandparents.[38,39] We now know that fear, stress, and trauma wreak havoc on our immune systems.

Early Treatment

His critics argue that Dr. Fauci's "slow the spread, flatten the curve, wait for the jab" strategy—all in support of a long-term bet on unproven vaccines—represented a profound and unprecedented departure from accepted public health practice. But most troubling were Dr. Fauci's policies of ignoring and outright suppressing the early treatment of infected patients who were often terrified. "The Best Practices for defeating an infectious disease epidemic," says Yale epidemiologist Harvey Risch, "dictate that you quarantine and treat the sick, protect the most vulnerable, and aggressively develop repurposed therapeutic drugs, and use early treatment protocols to avoid hospitalizations."

Risch is one of the leading global authorities in clinical treatment protocols. He is the editor of two high-gravitas journals and the author of over 350 peer-reviewed publications. Other researchers have cited those studies over 44,000 times.[40] Risch points out a hard truth that should have informed our COVID control strategy: "Unless you are an island nation prepared to shut out the world, you can't stop a global viral pandemic, but you can make it less deadly. Our objective should have been to devise treatments that would reduce hospitalization and death. We could have easily defanged COVID-19 so that it was less lethal than a seasonal flu. We could have done this very quickly. We could have saved hundreds of thousands of lives."

Dr. Peter McCullough concurs: "Once a highly transmissible virus like COVID has a beachhead in a population, it is inevitable that it will spread to every individual who lacks immunity. You can slow the spread, but you cannot prevent it—any more than you can prevent the tide from rising." McCullough was an internist and cardiologist on staff at the Baylor University Medical Center and the Baylor Heart and Vascular Hospital in Dallas, Texas. His 600 peer-reviewed articles in the National Library of Medicine make McCullough the most published physician in history in the field of kidney disease related to heart disease, a lethal sequela of COVID-19. Before COVID-19, he was editor of two major journals. His recent publications include over 40 on COVID-19, including two landmark studies on critical care of the disease. His two breakthrough papers on the early treatment of COVID-19 in *The American Journal of Medicine*[41] and *Reviews in Cardiovascular Medicine*[42] in 2020 are, by far, the most downloaded documents on the subject. "I've had COVID-19 myself with pulmonary involvement," he told me. "My wife has had it. On my wife's side of the family, we've had a fatality . . . I believe I have as much or more medical authority to give my opinion as anybody in the world."

McCullough observes that, "We could have dramatically reduced COVID fatalities and hospitalizations using early treatment protocols and repurposed drugs including ivermectin and hydroxychloroquine and many, many others." Dr. McCullough has treated some 2,000 COVID patients with these therapies. McCullough points out that hundreds of peer-reviewed studies now show that early treatment could have averted some 80 percent of deaths attributed to COVID. "The strategy from the outset should have been implementing protocols to stop hospitalizations through early treatment of Americans who tested positive for COVID but were still asymptomatic. If we had done that, we could have pushed case fatality rates below those we see with seasonal flu, and ended the bottlenecks in our hospitals. We should have rapidly deployed off-the-shelf medications with proven safety records and subjected them to rigorous risk/benefit decision-making," McCullough continues. "Using repurposed drugs, we could have ended this pandemic by May 2020 and saved 500,000 American lives, but for Dr. Fauci's hard-headed, tunnel vision on new vaccines and remdesivir."

Pulmonary and critical care specialist Dr. Pierre Kory agrees with McCullough's estimate. "The efficacy of some of these drugs as prophylaxis is almost miraculous, plus early intervention in the week after exposure stops viral replication and prevents development of cytokine storm and entrance into the pulmonary phase," says Dr. Kory. "We could have stopped the pandemic in its tracks in the spring of 2020."

Risch, McCullough, and Kory are among the large chorus of experts (including Nobel Laureate Luc Montagnier) who argue that, by treating infected patients at home during the early stages of the illness, we could have averted cataclysmic lockdowns and found medicine resources for protecting vulnerable populations while encouraging the spread of the disease in age groups with extremely low-risk, in order to achieve permanent herd immunity. They point out that natural immunity, in all known cases, is superior to vaccine-induced immunity, being both more durable (it often lasts a lifetime) and broader spectrum—meaning it provides a shield against subsequent variants. "Vaccinating citizens with natural immunity should never have been our public health policy," says Dr. Kory.

Dr. Fauci's strategy committed hundreds of billions of societal resources on a high-risk gambit to develop novel technology vaccines, and virtually nothing toward developing repurposed medications that are effective against COVID. "That strategy kept the medical treatment on hold globally for an entire year as a readily treatable respiratory virus ravaged populations," says Kory. "It is absolutely shocking that he recommended no outpatient care, not even Vitamin D despite the fact he takes it himself and much of the country is Vitamin D deficient."

Dr. Kory[43] is president of Front Line COVID-19 Critical Care Alliance, a former associate professor, and Medical Director of the Trauma and Life Support Center at the University of Wisconsin Medical School Hospital, and the Critical Care Service Chief at Aurora St. Luke's Medical Center in Milwaukee. His milestone work on critical care ultrasonography won him the British Medical Association's President's Choice Award in 2015.

Risch, McCullough, and Kory are also among the hundreds of scientists and physicians who express shock that Dr. Fauci made no effort to identify repurposed medicines. Says Kory, "I find it appalling that there was no consultation process with treating physicians. Medicine is about consultation. You had Birx, Fauci, and Redfield doing press conferences every day and handing down these arbitrary diktats and not one of them ever treated a COVID patient or worked in an emergency room or ICU. They knew nothing."

"As I watched the White House Task Force on T.V.," recalls Dr. McCullough, "no one even said that hospitalizations and deaths were the bad outcome of COVID-19, and that they were going to put together a team of doctors to identify protocols and therapeutics to stop these hospitalizations and deaths."

Dr. McCullough argues that, as COVID czar, Dr. Fauci should have created an international communications network linking the world's 11 million front-line doctors to gather real-time tips, innovative safety protocols, and to develop the best prophylactic and early treatment practices. "He should have created hotlines and dedicated websites for medical professionals to call in with treatment questions and to consult, collect, catalogue, and propagate the latest innovations for prophylaxing vulnerable and exposed individuals, and treating early infections, so as to avert hospitalizations." Dr. Kory agrees: "The outcome we should have been trying to prevent is hospitalizations. You don't just sit around and wait for an infected patient to become ill. Dr. Fauci's treatment strategies all began once all these under-medicated patients were hospitalized. By that time, it was too late for many of them. It was insane. It was perverse. It was unethical."

Dr. McCullough says that Dr. Fauci should have created treatment centers for ambulatory patients and field clinics specializing in treating asymptomatic or early-stage COVID. "He should have been encouraging doctors to use satellite clinics to conduct small outpatient clinical trials to quickly identify the most effective protocols, drugs, and therapeutics."

Professor Risch concurs: "We should have deployed teams of doctors all over the world doing short-term clinical trials and testing promising drugs and reporting successful protocols. The endpoints were obvious: preventing hospitalizations and deaths. In addition to rapidly developing and continuously updating protocols and remedies, McCullough and Kory say that the government failed to perform the essential duty of a public health regulator during modern pandemics—to publish the best early treatment protocols on NIH's website and then establish communication lines call centers to foster consultation and information sharing and webpages to share, broadcast and update the latest remedies and continually escalate public knowledge about the most successful strategies.

Dr. McCullough adds, "We should have created information and communication centers where treating physicians and hospitals could get round-the-clock, up-to-date bulletins with data. Instead, doctors who wanted to provide their infected patients with early treatment were out of luck."

Nursing Homes and Quarantine Facilities

Dr. Risch says that in addition to developing early treatment protocols, public health officials should have made sure that elderly patients remained in quarantine hospitals until no longer contagious. "It's obvious that we should have had quarantine facilities so we wouldn't be sending infected patients to crowded nursing homes. Instead, we should have housed them in safe facilities and protected them with cutting-edge care." Risch points out that taxpayers spent $660 million building field hospitals across the country.[44] Democratic Governor Andrew Cuomo and other Democratic governors kept these facilities empty to maintain bed inventories in anticipation of the flood of patients inaccurately predicted by the fear-mongering models, ginned up by two Gates-funded organizations, IMHE and Royal College of London, and then anointed as gospel by Dr. Fauci—seemingly as part of the crusade to generate public panic. With those quarantine centers standing empty, those governors sent infected elderly back to crowded nursing homes, where they spread the disease to the most vulnerable population with lethal effect. Risch points out that, "Half the deaths, in New York, and one-third nationally,[45] were among elder care facility residents."

Dr. Fauci made another inexplicable policy choice of not supplying the nursing homes with monoclonal antibodies where they might have saved thousands of lives. "With Operation Warp Speed, we had monoclonal antibodies that were high tech and fully FDA-approved by November 2020—long before the vaccines," says Dr. McCullough.

"Monoclonal antibodies work great, but they're not suitable for outpatients because they are administered IV It's therefore perfect for nursing homes. About one-third of COVID deaths occurred in the nursing homes and ALFs across the US during the pandemic.[46] Dr. Fauci should have equipped both nursing homes and quarantine hospitals with monoclonal antibodies," said Risch. Instead, he obstructed these institutions from administering that medicine. "It was a kind of staggering savage act of malpractice and negligence to deny this remedy to elder care facilities at a time when the elderly were dying at a rate of 10,000 per week."

"You need, in short, to do the opposite of everything they did. It's difficult to identify anything they did that was right," says McCullough.

Independent Doctors into the Breach

Early in the pandemic, Kory and his mentor, Dr. Paul Marik, Professor of Medicine and Chief of Division of Pulmonary and Critical Care Medicine at Eastern Virginia Medical School, began assembling the world's most highly published and accomplished critical care specialists to rapidly develop functional COVID treatments. Each of the core five founders of FLCCC is globally renowned for having made significant pre-COVID contributions to the science of critical care and pulmonary illnesses. Some 1,693 front-line physicians globally now belong to their alliance.[47] Early in the pandemic, these doctors stepped into the breach left by the government agencies and pandemic centers and began coordinating the development of early treatments with repurposed drugs. They quickly proved that they could drastically reduce COVID's lethality. Instead of winning applause as medical healers, their success at treating COVID made them enemies of the State.

Long before he heard of Pierre Kory or FLCCC, Dr. Peter McCullough reached the same conclusions about the futility and immorality of the federal effort, and felt the same indignation and determination to change things. "By April and May, I noticed a disturbing trend," recalls McCullough. "The trend was, no effort to treat patients who are infected with COVID-19 at home or in nursing homes. And it almost seemed as if patients were intentionally not being treated, allowed to sit at home and get to the point where they couldn't breathe and then be admitted to the hospital."

Dr. Fauci adopted this unprecedented protocol of telling doctors to let patients diagnosed with a positive COVID test go home, untreated—leaving them in terror, and spreading the disease—until breathing difficulties forced their return to hospitals. There they faced two deadly remedies: remdesivir and ventilators.

I experienced my own personal frustrations with this bewildering policy. When, in December 2020, I asked my 93-year-old mother's physician to describe her treatment plan if she got a positive PCR, he told me, "There is really nothing we can do unless she starts having trouble breathing. Then we will send her up to Mass General for ventilation." When I asked him about using ivermectin or hydroxychloroquine, he shrugged his shoulders. He had never heard of their use in COVID patients. "There is no early treatment for COVID," he assured me.

Dr. Fauci's choice to deny infected Americans early treatment was not just a bad public health strategy; it was, McCullough avows, "Cruelty at a population level." Says McCullough, "Never in history have doctors deliberately treated patients with this kind of barbarism."

"I told myself, 'I am not going to tolerate that—in my practice, or on a national level or worldwide,'" Dr. McCullough told me. Realizing that COVID had to be fought on multiple fronts, McCullough began contacting physicians in other nations who were reporting success against the disease, including doctors in Italy, Greece, Canada, across Europe, and in Bangladesh and South Africa.

McCullough continues, "If this had been any other form of pneumonia, a respiratory illness, or any other infectious illness in the human body, we know that if we start early, we can actually treat much more easily than wait until patients are very sick." McCullough says that the rule holds true for COVID-19: "We learned quickly that it takes about two weeks for someone infected with COVID to get sick enough at home to require hospitalization."

Front-line clinical doctors quickly recognized that the disease was operating through multiple pathways, each requiring their own treatment protocol. "There were three major parts of the illness," says McCullough: "1) the virus was replicating for as long as two weeks, 2) there was incredible inflammation in the body, and 3) that was followed by blood clotting." He adds, "By April 2020, most doctors understood a single drug was not going to be enough to treat this illness. We had to use drugs in combination."

"We quickly developed three principles," says McCullough; his three-step protocol was as follows:

- Use medications to slow down the virus;
- Use medications to attenuate or reduce inflammation;
- Address blood clotting.

McCullough and his global partners quickly identified a pharmacopoeia of off-the-shelf treatments demonstrating extraordinary efficacy against each stage of COVID when administered early in the course of the disease.

McCullough chronicles the rapid pace with which front-line doctors uncovered rich apothecaries of effective COVID remedies. HHS's early studies supported hydroxychloroquine's efficacy against coronavirus since 2005, and by March 2020, doctors from New York to Asia were using it against COVID with extraordinary effect. That month, McCullough and other physicians at his medical center organized, with the FDA, one of the first prophylactic protocols using hydroxychloroquine. "We had terrific data on ivermectin, from the medical teams in Bangladesh and elsewhere by early summer 2020. So now we had two cheap generics." McCullough and his growing team of 50+ front-line doctors discovered that while HCQ and IVM work well against COVID, adding other medications boosts outcomes drastically. These included azithromycin or doxycycline, zinc, vitamin D, Celebrex, bromhexine, NAC, IV vitamin C, and quercetin. McCullough's team realized that, like hydroxychloroquine and ivermectin, quercetin—that ubiquitous health store nutraceutical—is an ionophore—meaning that it facilitates zinc uptake in the cells, destroying the capacity of coronavirus to replicate. "The Canadians came on with Colchicine in a high-quality trial based on an initial Greek trial," McCullough continued. "We learned more from experts at UCLA and elsewhere with respect to blood clotting and the need for aspirin and blood thinners. We got early approval for monoclonal antibodies. It was later learned that both fluvoxamine and famotidine could play roles in multidrug treatment." LSU Medical School professor Paul Harch discovered peer-reviewed papers from China where researchers there had been using hyperbaric chambers (HBOT) with stunning success.[48] Between April and May, a group of NYU researchers reproduced that success by getting patients off ventilators and quickly recovering 18 of 20 ventilator cases using HBOT.[49] (Yale is currently conducting Phase 3 with stellar early results.)

There were many other promising treatments. Asian nations were using saline nasal lavages to great effect to reduce viral loads and transmission.[50] McCullough discovered he could prophylax patients and drop viral load and prevent transmission with a variety of other oral/nasal rinses and dilute virucidal agents, including povidone iodine, hydrogen peroxide, hypochlorite, and Listerine or mouthwash with cetylpyridinium chloride. Mass General's infectious disease maven Dr. Michael Callahan had seen hundreds of patients in Wuhan in January 2020, and assessed the impressive efficacy of Pepcid, an over-the-counter indigestion medicine. The Japanese were already using Prednisone, Budesonide, and Famotidine with extraordinary results.

By July 1, McCullough and his team had developed the first protocol based on signals of benefit and acceptable safety. They submitted the protocol to the *American Journal of Medicine*. That study, titled "The Pathophysiologic Basis and Clinical Rationale for Early Ambulatory Treatment of COVID-19,"[51] quickly became the world's most-downloaded paper to help doctors treat COVID-19.

"It is extraordinary that Dr. Fauci never published a single treatment protocol before that," says McCullough, "and that 'America's Doctor' has never, to date, published anything on how to treat a COVID patient. It shocks the conscience that there is still no official protocol. Anyone who tries to publish a new treatment protocol will find themselves airtight blocked by the journals that are all under Fauci's control."

The Chinese published their own early treatment protocol on March 3, 2020,[52,53] using many of the same categories of prophylactic and early treatment drugs uncovered by McCullough—chloroquine (a cousin of hydroxychloroquine), antibiotics, anti-inflammatories, antihistamines, a variety of steroids, and probiotics to stabilize and fortify the immune system and apothecaries of traditional Chinese medicines, vitamins, and minerals, including a variety of compounds containing quercetin, zinc, and glutathione precursors.[54] The Chinese made early treatment the central priority of their COVID strategy. They used intense—and intrusive—track-and-trace surveillance to identify and then immediately hospitalize and treat every COVID-infected Chinese. Early treatment helped the Chinese to end their pandemic by April 2020. "We could have done the same," says McCullough.

Though now he is often censored, the AMA still lists Dr. McCullough's study as the most frequently downloaded paper for 2020. The Association of American Physicians and Surgeons (AAPS) downloaded and turned McCullough's AMA article into its official treatment guide.[55] AAPS Director Dr. Jeremy Snavely told me in August 2021 that the Guide had 122,000 downloads: "We figure it has been seen by over a million people. It's the only trusted guide. Our phone never stops ringing. Mostly the calls are from physicians and patients desperate for the help they cannot get from any HHS website."

By autumn, front-line physicians had assembled a pharmacopeia of repurposed drugs, all of which were effective against COVID.

By that time, more than 200 studies supported treatment with hydroxychloroquine, and 60 studies supported ivermectin. "We combined these medicines with doxycycline, azithromycin to suppress infection," says McCullough. Another meta-analysis supported the use of prednisone and hydrocortisone and other widely available steroids to combat inflammation.[56] Three studies supported the use of inhaled budesonide against COVID; an Oxford University study published in February 2021 demonstrated that that treatment could reduce hospitalizations by 90 percent in low-risk patients,[57] and a publication in April 2021 showed that recovery was faster for high-risk patients, too.[58] Furthermore, a very large study supported colchicine as an anti-inflammatory.[59] Finally, McCullough's growing array of physicians had observational data from late-stage treatment of hospitalized patients with full-dose aspirin and antithrombotics, including Enoxaparin, Apixaban, Rivaroxaban, Dabigatran, Edoxaban, and full-dose anticoagulation with low molecular weight heparin for blood clots.[60]

"We were able to show that doctors can work with four to six drugs in combination, supplemented by vitamins and nutraceuticals including zinc, vitamins D and C, and Quercetin. And they can guide patients at home, even the highest-risk seniors, and avoid a dreaded outcome of hospitalization and death," said McCullough.

Working with a large practice in the Plano/Frisco area north of Dallas, McCullough and his team administered this protocol to some eight hundred patients and demonstrated an 85 percent reduction in hospitalization and death. Another practice led by the legendary Dr. Vladimir Zelenko in Monroe, New York showed similar astonishing results.[61]

Independent physicians unaffiliated with the government or the universities that are so dependent on Dr. Fauci's good favor were discovering new COVID treatments by the day. Researchers treated 738 randomly selected Brazilian COVID-19 patients with another adjuvant, fluvoxamine, identified early in the pandemic for its potential to reduce cytokine storms.[62] Another 733 received a placebo between Jan. 20 and Aug. 6 of 2021. The researchers tracked every patient receiving fluvoxamine during the trial for 28 days and found about a 30-percent reduction in events among those receiving fluvoxamine compared to those who did not. Like almost all the other remedies, it is cheap and proven safe by long use. Fluvoxamine costs about $4 per 10-day course. Fluvoxamine has been used since the 1990s, and its safety profile is well known.[63]

"Hydroxychloroquine and ivermectin are not necessary nor sufficient on their own—there are plenty of molecules that treat COVID," says McCullough. "Even if hydroxychloroquine and ivermectin had become so politicized that no one wanted to allow these to be used, we could use other drugs, anti-inflammatories, antihistamines, as well as anti-coagulants and actually stop the illness and again, treat it to reduce hospitalization and death."

When the pandemic started, most of the other medical practices in the Detroit area shut down, Dr. David Brownstein told me. "I had a meeting with my staff and my six partners. I told them, 'We are going to stay open and treat COVID.' They wanted to know how. I said, 'We've been treating viral diseases here for twenty-five years. COVID can't be any different.' In all that time, our office had never lost a single patient to flu or flu-like illness. We treated people in their cars with oral vitamins A, C, and D, and iodine. We administered IV solution outside all winter with IV hydrogen peroxide and vitamin C. We'd have them put their butts out the car window and shot them up with intramuscular ozone. We nebulized them with hydrogen peroxide and Lugol's iodine. We only rarely used ivermectin and hydroxychloroquine. We treated 715 patients and had ten hospitalizations and no deaths. Early treatment was the key. We weren't allowed to talk about it. The whole medical establishment was trying to shut down early treatment and silence all the doctors who talked about successes. A whole generation of doctors just stopped practicing medicine. When we talked about it, the whole cartel came for us. I've been in litigation with the Medical Board for a year. When we posted videos from some of our recovered patients, they went viral. One of the videos had a million views. FTC filed a motion against us, and we had to take everything down." In July 2020, Brownstein and his seven colleagues published a peer-reviewed article describing their stellar success with early treatment. FTC sent him a letter warning him to take it down. "No one wanted Americans to know that you didn't have to die from COVID. It's 100 percent treatable," says Dr. Brownstein. "We proved it. No one had to die."

"Meanwhile," adds Dr. Brownstein, "we've seen lots of really bad vaccine side effects in our patients. We've had seven strokes—some ending in severe paralysis. We had three cases of pulmonary embolism, two blood clots, two cases of Graves' disease, and one death."

Repurposed medicines, the record shows, could also have drastically reduced death among hospitalized patients. One of Dr. Kory's cofounders of FLCCC, Houston Memorial Medical Center's Chief Medical Officer, Dr. Joe Varon, worked 400 days in a row, seeing between 20–30 patients/day. Using ivermectin and a cocktail of anti-inflammatories, steroids, and anticoagulants since Spring 2020, Dr. Varon lowered hospital mortality among ICU COVID patients to about 4.1 percent, compared to well over 23 percent nationally. "Even in the ICUs where patients were coming in undertreated, we were able to dramatically reduce mortality," says Dr. Kory.

"Almost anything you do in the nursing homes—basically, any combination of the various components of these protocols—reduces mortalities by at least 60 percent," McCullough told me. A 2021 paper in *Medical Hypotheses* supports McCullough's claim.[64] That study by twelve physician co-authors shows that diverse combinations of many of these and similar medications dramatically lower death rates in a variety of nursing homes. The study concludes that even the most modest early medical therapy combinations were associated with 60 percent reductions in mortality. Says Dr. McCullough, "Therapeutic nihilism was the real killer of America's seniors."

McCullough's findings may be conservative. Early in the pandemic, two Spanish nursing homes simultaneously experimented with early treatment with cheap, available repurposed drugs and achieved 100 percent survival among infected residents and staff. Between March and April 2020, COVID-19 struck two elder care facilities in Yepes, Toledo, Spain. The mean age of residents in those locations was 85, and 48

15

percent were over 80 years old. Within three months, 100 percent of the residents at both locations had caught the virus. By the end of June, 100 percent of residents and half the workers were seropositive for COVID, meaning they had endured infection and recovered. None of them went to the hospital and none died. None had adverse drug effects. Local doctors rapidly discovered early treatment with the same sort of remedies that McCullough was championing: antihistamines, steroids, antibiotics, anti-inflammatories, aspirin, nasal washes, bronchodilators, and blood thinners. In pooled data, 28 percent of the residents in similar nursing homes in the same region over the same time period died. That study supports the experience of front-line physicians that cheap available, repurposed drugs can easily prevent hospitalizations and deaths.[65]

Dr. McCullough and 57 colleagues published a second study in December of 2020 in a dedicated issue of *Reviews In Cardiovascular Medicine*. The article, "Multifaceted highly targeted sequential multidrug treatment of early ambulatory high-risk SARS-CoV-2 infection (COVID-19)," described a marvelous breadth of effective drugs that these physicians had, by then, developed.[66]

By collecting data from the vast network of doctors across the globe, they added dozens of new compounds to the arsenal—all proven effective against COVID-19. Dr. Kory told me that he was deeply troubled that the extremely successful efforts by scores of front-line doctors to develop repurposed medicines to treat COVID received no support from any government in the entire world—only hostility—much of it orchestrated by Dr. Fauci and the US health agencies. The large universities that rely on hundreds of millions in annual funding from NIH were also antagonistic. "We didn't have a single academic institution come up with a single protocol," said Dr. McCullough. "They didn't even try. Harvard, Johns Hopkins, Duke, you name it. Not a single medical center set up even a tent to try to treat patients and prevent hospitalization and death. There wasn't an ounce of original research coming out of America available to fight COVID—other than vaccines." All of these universities are deeply dependent on billions of dollars that they receive from NIH. As we shall see, these institutions live in terror of offending Anthony Fauci, and that fear paralyzed them in the midst of the pandemic.

"Dr. Fauci refused to promote any of these interventions," says Kory. "It's not just that he made no effort to find effective off-the-shelf cures—he aggressively suppressed them."

Instead of supporting McCullough's work, NIH and the other federal regulators began actively censoring information on this range of effective remedies. Doctors who attempted merely to open discussion about the potential benefits of early treatments for COVID found themselves heavily and inexplicably censored. Dr. Fauci worked with Facebook's Mark Zuckerberg and other social media sites to muzzle discussion of any remedies. FDA sent a letter of warning that N-acetyle-L-cysteine (NAC) cannot be lawfully marketed as a dietary supplement, after decades of free access on health food shelves, and suppressed IV vitamin C, which the Chinese were using with extreme effectiveness.

In September, Dr. McCullough used his own money to create a YouTube video showing four slides from his peer-reviewed American Medical Association articles to teach doctors the miraculous benefits of early treatment with HCQ and other remedies. His video went viral, with hundreds of thousands of downloads; YouTube pulled it two days later.

Leading doctors and scientists, including some of the nation's most highly published and experienced physicians and front-line COVID specialists like McCullough, Kory, Ryan Cole, David Brownstein, and Risch believe that Dr. Fauci's suppression of early treatment and off-patent remedies was responsible for up to 80 percent of the deaths attributed to COVID. All five doctors independently told me the same thing. The relentless malpractice of deliberately withholding early effective COVID treatments, of forcing the use of toxic remdesivir, may have unnecessarily killed up to 500,000 Americans in hospitals.

Dr. Kory says so plainly: "Dr. Fauci's suppression of early treatments will go down in history as having caused the death of a half a million Americans in the ICU."

Ryan Cole is one of the doctors who adopted McCullough's protocols early in the pandemic. Dr. Cole is a Mayo Clinic and Columbia University-trained Board Certified Anatomic/Clinical Pathologist and the CEO/Medical Director of Cole Diagnostics, the largest independent lab in Idaho. He has diagnosed more than 350,000 patients in his career. Dr. Cole discovered McCullough's research during his own investigation of early treatment remedies when his overweight brother called Dr. Cole from a neighboring state on his way to the ER with a positive PCR test, labored breathing, blood oxygen at 86, and chest discomfort that he rated nine out of ten. "He has Type 1 diabetes," explains Dr. Cole. Dr. Cole redirected his sibling to a local pharmacy and called in an ivermectin prescription. "Within six hours, my brother's chest pain was down to two out of ten due to the interferon effect of ivermectin, and within 24 hours after taking ivermectin, his oxygen was 98, and he then fully recovered." Cole told me, "A light bulb went off."

Dr. Cole has overseen or helped perform over 125,000 COVID tests during the pandemic. Since rescuing his brother, he has encountered many patients in early stages of the disease. "Almost none of them could find doctors in the community to treat them," he told me. "I intervened to provide early treatment to over 300 positive patients, half of whom were comorbid and high risk." Of this cohort, none were hospitalized and none died. "Early treatment of COVID-19, plain and simple, saves lives. If the medical profession had been forward thinking and hands-on, and focused on this disease, with an early outpatient multi-drug approach, knowing that COVID-19 is an inflammatory clotting disease, hundreds of thousands of lives could have been saved in the US."

"Never in the history of medicine," says Dr. Cole, "has early treatment, of any patient with any disease, been so overtly neglected by the medical profession on such a massive scale."

Cole adds, "To not treat, especially in the midst of a highly transmissible, deadly disease, is to do harm."

Cole says that the only truly deadly pandemic is "the pandemic of under treatment." He says, "The sacred doctor–patient relationship needs to be wrenched away from Anthony Fauci and the government/medical/pharmaceutical industrial complex. Doctors need to return to their oaths. Patients need to demand from medicine their right to be treated. This year has revealed the countless flaws of a medical system that has lost its direction and soul."

Cole points out that, "If you are under 70 years of age and have no severe preexisting illness, you can hardly die [from SARS-CoV-2 infection]. So, there is no fatality rate that can be reduced. . . . And for people who are elderly and have preexisting illness," he adds, "as we know from Dr. Peter McCullough and his colleagues' work, there are miraculously effective medicines to treat this virus so that the fatality rates go down another 70 to 80 percent, which means there is no ground for emergency use whatsoever. That's a huge threat to the vaccine cartel and to remdesivir."

It was only the independent doctors like Ryan Cole, who were not reliant on Dr. Fauci's largesse and who threw themselves into hand-to-hand combat against COVID-19, who discovered readily available treatment modes: "We had hero doctors that really had to break with the academic ivory tower," says McCullough. Finally, a group of independent organizations, including the Association of American Physicians and Surgeons, the Front-Line Critical Care Consortium, and America's Frontline Doctors, galvanized to organize the country into four national telemedicine services, and three regional telemedicine services. Following Dr. Kory's explosive Senate testimony, thousands of doctors and frightened COVID patients began calling the hotlines for treatment. "We took over health care," says McCullough.

"In numerous countries and regions around the world, repeated, striking temporally associated reductions in both cases and deaths occurred very soon after either ivermectin was distributed or health ministry ivermectin recommendations were announced." said Dr. Kory. It could be argued that a similar association occurred in the US.

Dr. Fauci and the industry propagandists later attributed the January decline in COVID cases, hospitalizations, and deaths to their vaccines, which began their rollout in mid-December 2020.

However, even mainstream media doctors reluctantly acknowledged that the drop could not possibly be a vaccine effect. By February 1, only 25.2 million, or 7.6 percent of Americans, had received a single vaccine dose.[67] The CDC acknowledges that there is no effect until many weeks after the second COVID jab.

Tony Fauci's decision to deny early treatments undoubtedly prolonged and intensified the pandemic. McCullough points out that early treatment does not just prevent hospitalization; it quickly starves pandemics to death by stopping their spread. "Early treatment reduces the infectivity period from 14 days to about four days," he explains. "It also allows someone to stay in the home so they don't contaminate people outside the home. And then it has this remarkable effect in reducing the intensity and duration of symptoms so patients don't get so short of breath, they don't get into this panic where they feel they have to break containment and go to the hospital." McCullough says that those hospital trips are tinder for pandemics, especially since, at that point, the patient is at the height of infectivity, with teeming viral loads. "Every hospitalization in America—and there's been millions of them—has been a super-spreader event. Sick patients contaminate their loved ones, paramedics, Uber drivers, people in the clinic and offices. It becomes a total mess." McCullough says that by treating COVID-19 at home, doctors actually can extinguish the pandemic.

"So this has been a story of American heroes. It's been a story of worldwide success." McCullough's group is now part of a worldwide network of front-line physicians using repurposed drugs to save lives around the globe. These doctors have built networks and information banks outside of the government agency and university hegemony allowing doctors to actually practice the art of healing. Their network includes the BIRD medical coalition in the UK and Treatment Domiciliare COVID-19 group in Italy, which conducts rallies to celebrate zero hospitalizations from this multidrug approach. "We have PANDA in South Africa, the Covid Medical Network in Australia. And so on," says McCullough. "Despite the various government agencies and the ivory tower medical institutions literally not lifting a finger, COVID-19 independent doctors and hero organizations kicked in."

"And to this day, we're in the middle of the Delta outbreak. Guess who's treating the Delta patients? It's again not the academic medical centers or the government or even the large group practices. They're not touching these patients. Once again, it is independent physicians." It's independent doctors who are actually compassionately reaching out and using what we call the precautionary principle. They are using their best medical judgment and scientific data to apply therapy now and to practice the art of healing. For any of our academic colleagues that have said, 'Dr. McCullough, we need to wait for large, randomized trials,' what I've always said is, 'Listen, this is a mass casualty event.' People are dying now. They're being hospitalized now. We can't wait for large, randomized trials. We need to be doctors. We need to start healing people."

II: KILLING HYDROXYCHLOROQUINE

Most of my fellow Democrats understand that Dr. Fauci led an effort to deliberately derail America's access to lifesaving drugs and medicines that might have saved hundreds of thousands of lives and dramatically shortened the pandemic. There is no other aspect of the COVID crisis that more clearly reveals the malicious intentions of a powerful vaccine cartel—led by Dr. Fauci and Bill Gates—to prolong the pandemic and amplify its mortal effects in order to promote their mischievous inoculations.

From the outset, hydroxychloroquine (HCQ) and other therapeutics posed an existential threat to Dr. Fauci and Bill Gates' $48 billion COVID vaccine project, and particularly to their vanity drug remdesivir, in which Gates has a large stake.[1] Under federal law, new vaccines and medicines cannot quality for Emergency Use Authorization (EUA) if any existing FDA-approved drug proves effective against the same malady:

> For FDA to issue an EUA (emergency use authorization), there must be no adequate, approved, and available alternative to the candidate product for diagnosing, preventing, or treating the disease or condition. . . .[2]

Thus, if any FDA-approved drug like hydroxychloroquine (or ivermectin) proved effective against COVID, pharmaceutical companies would no longer be legally allowed to fast-track their billion-dollar vaccines to market under Emergency Use Authorization. Instead, vaccines would have to endure the years-long delays that have always accompanied methodical safety and efficacy testing, and that would mean less profits, more uncertainty, longer runways to market, and a disappointing end to the lucrative COVID-19 vaccine gold rush. Dr. Fauci has invested $6 billion in taxpayer lucre in the Moderna vaccine alone.[3] His agency is co-owner[4] of the patent and stands to collect a fortune in royalties. At least four of Fauci's hand-picked deputies are in line to collect royalties of $150,000/year based on Moderna's success, and that's on top of the salaries already paid by the American public.[5,6]

So there was good reason that very powerful potentates of the medical cartel were already targeting HCQ long before President Trump began his infamous romance with the malaria remedy. President Trump's endorsement of HCQ on March 19, 2020[7] hyper-politicized the debate and gave Dr. Fauci's defamation campaign against HCQ a soft landing among Democrats and the media. Trump's critics relegated any further claims of HCQ efficacy to the same anti-science waste bin as Trump's notorious recommendation for bleach to cure COVID and his denial of climate change. But HCQ had a long history of safe medical use that got lost in the politics and propaganda.

HCQ Before Dr. Fauci's Smear Campaign

Dr. Fauci's challenge—to prove that HCQ is dangerous—was daunting because hydroxychloroquine is a 65-year-old formula that regulators around the globe long ago approved as both safe and effective against a variety of illnesses. HCQ is an analog of the quinine found in the bark of the cinchona tree that George Washington used to protect his troops from malaria. For decades, WHO has listed HCQ as an "essential medicine," proven effective against a long list of ailments.[8] It is a generally benign prescription medicine, far safer—according to the manufacturer's package inserts[9]—than many popular over-the-counter drugs.

Generations have used HCQ billions of times throughout the world, practically without restriction. During my many childhood trips to Africa, I took HCQ daily as a preventive against malaria, a ritual that millions of other African visitors and residents embrace. Long use has thoroughly established HCQ's safety and efficacy such that most African countries authorize HCQ as an over-the-counter medication. Africans call the drug "Sunday-Sunday"[10] because millions of them take it religiously, once a week, as a malaria prophylaxis. It's probably not a coincidence that these nations enjoyed some of the world's lowest mortality rates from COVID. HCQ is the #1 most used medication in India, the second-most populous nation on the planet, with 1.3 billion people. Prior to the COVID pandemic, HCQ and its progenitor, chloroquine (CQ), were freely available over the counter in most of the world, including France, Canada, Iran, Mexico, Costa Rica, Panama, and many other countries.

In the United States, the FDA has approved HCQ without limitation for 65 years, meaning that physicians can prescribe it for any off-label use. CDC's information sheet deems hydroxychloroquine safe for pregnant women, breastfeeding women, children, infants, elderly and immune-compromised patients and healthy persons of all ages.[11] The CDC sets no limits on the lengthy and indefinite use of hydroxychloroquine for the prevention of malaria. Many people in Africa and India take it for a lifetime. Since its recommended protocol as a remedy for COVID requires only one week's use, Dr. Fauci's sudden revelation that the drug is dangerous was specious at best.

According to Dr. Peter McCullough, "To date, there has not been a single credible report that the medication increases the risk of death in COVID-19 patients when prescribed by competent physicians who understand its safety profile."[12]

Efficacy Against Coronavirus with Early Intervention HCQ Protocol

Some 200 peer-reviewed studies (C19Study.com) by government and independent researchers deem HCQ safe and effective against Coronavirus, especially when taken prophylactically or when taken in the initial stages of illness along with zinc and Zithromax.

The chart below lists 32 studies of early outpatient treatment of COVID using hydroxychloroquine. Thirty-one of the studies showed benefit, and only one study showed harm. The study showing harm resulted from a single patient in the treatment group requiring hospitalization. When all the studies are collected together, despite having different outcome measures, the average benefit is 64 percent. This means that subjects who received hydroxychloroquine were only 36 percent as likely to reach the negative outcomes as subjects in the control groups.

The scientific literature first suggested that HCQ or CQ might be effective treatments for Coronavirus in 2004.[13] In that era, following an outbreak, Chinese and Western governments were pouring millions of dollars into an effort to identify existing, a.k.a. "repurposed," medicines that were effective against coronaviruses. With HCQ, they had stumbled across the Holy Grail. In 2004, Belgian researchers found that chloroquine was effective at viral killing at doses equivalent to those used to treat malaria, i.e., doses that are safe.[14] A CDC study published in 2005 in the *Virology Journal*, "Chloroquine is a Potent Inhibitor of SARS Coronavirus Infection and Spread" demonstrated that CQ quickly eliminated coronavirus in primate cell culture during the SARS outbreak. That study concludes: "We report . . . that chloroquine has strong antiviral effects on SARS-Coronavirus infection of primate cells . . . [both] before or after exposure to the virus, suggesting both prophylactic and therapeutic advantage."[15]

This conclusion was particularly threatening to vaccine makers since it implies that chloroquine functions both as a preventive "vaccine" as well as a cure for SARS coronavirus. Common sense would presume it to be effective against other coronavirus strains. Worse still for Dr. Fauci and his vaccine-making friends, a NIAID study[16] and a Dutch paper,[17] both in 2014, confirmed chloroquine was effective against MERS—still another coronavirus.

In response to their studies, physicians worldwide discovered early in the pandemic that they could successfully treat high-risk COVID-19 patients as outpatients, within the first five to seven days of the onset of symptoms, with a chloroquine drug

All 32 hydroxychloroquine COVID-19 early treatment studies —

Study	Improvement	RR [CI]		Treatment	Control	Dose (4d)
Gautret	66%	0.34 [0.17-0.68]	viral+	6/20	14/16	2.4g
Huang (RCT)	92%	0.08 [0.01-1.32]	no recov.	0/10	6/12	4.0g (c)
Esper	64%	0.36 [0.15-0.87]	hosp.	8/412	12/224	2.0g
Ashraf	68%	0.32 [0.10-1.10]	death	10/77	2/5	1.6g
Huang (ES)	59%	0.41 [0.26-0.64]	viral time	32 (n)	37 (n)	2.0g (c)
Guérin	61%	0.39 [0.02-9.06]	death	0/20	1/34	2.4g
Chen (RCT)	72%	0.28 [0.11-0.74]	viral time	18 (n)	12 (n)	1.6g
Derwand	79%	0.21 [0.03-1.47]	death	1/141	13/377	1.6g
Mitjà (RCT)	16%	0.84 [0.35-2.03]	hosp.	8/136	11/157	2.0g
Skipper (RCT)	37%	0.63 [0.21-1.91]	hosp./death	5/231	8/234	3.2g
Hong	65%	0.35 [0.13-0.72]	viral+	42 (n)	48 (n)	n/a
Bernabeu-Wittel	59%	0.41 [0.36-0.95]	death	189 (n)	83 (n)	2.0g
Yu (ES)	85%	0.15 [0.02-1.05]	death	1/73	238/2,604	1.6g
Ly	56%	0.44 [0.26-0.75]	death	18/116	29/110	2.4g
Ip	55%	0.45 [0.11-1.85]	death	2/97	44/970	n/a
Heras	96%	0.04 [0.02-0.09]	death	8/70	16/30	n/a
Kirenga	26%	0.74 [0.47-1.17]	recov. time	29 (n)	27 (n)	n/a
Sulaiman	64%	0.36 [0.17-0.80]	death	7/1,817	54/3,724	2.0g
Guisado-Vasco (ES)	67%	0.33 [0.05-1.55]	death	2/65	139/542	n/a
Szente Fonseca	64%	0.36 [0.20-0.67]	hosp.	25/175	89/542	2.0g
Cadegiani	81%	0.19 [0.01-3.88]	death	0/159	2/137	1.6g
Simova	94%	0.06 [0.00-1.13]	hosp.	0/33	2/5	2.4g
Omrani (RCT)	12%	0.88 [0.26-2.94]	hosp.	7/304	4/152	2.4g
Agusti	68%	0.32 [0.06-1.67]	progression	2/87	4/55	2.0g
Su	85%	0.15 [0.04-0.57]	progression	261 (n)	355 (n)	1.6g
Amaravadi (RCT)	60%	0.40 [0.13-1.28]	no recov.	3/15	6/12	3.2g
Roy	2%	0.98 [0.45-2.20]	recov. time	14 (n)	15 (n)	n/a
Mokhtari	70%	0.30 [0.20-0.45]	death	27/7,295	287/21,464	2.0g
Million	83%	0.17 [0.06-0.48]	death	5/8,315	11/2,114	2.4g
Sobngwi (RCT)	52%	0.48 [0.09-2.58]	no recov.	2/95	4/92	1.6g
Rodrigues (RCT)	-200%	3.00 [0.13-71.6]	hosp.	1/42	0/42	3.2g
Sawanpanyalert	42%	0.58 [0.18-1.91]	progression	n/a	n/a	varies
Early treatment	**64%**	**0.36 [0.29-0.46]**		**148/20,390**	**996/34,231**	

64% improvement

0 0.25 0.5 0.75 1 1.25 1.5 1.75

Tau² = 0.20; I² = 52.9%; Z = 8.20   Effect extraction pre-specified, see appendix   Favors HCQ   Favors control

alone or with a "cocktail" consisting of hydroxychloroquine, zinc, and azithromycin (or doxycycline). Multiple scholarly contributions to the literature quickly confirmed the efficacy of hydroxychloroquine and hydroxychloroquine-based combination treatment when administered within days of COVID symptoms. Studies confirming this occurred in China,[18] France,[19] Saudi Arabia,[20] Iran,[21] Italy,[22] India,[23] New York City,[24] upstate New York,[25] Michigan,[26] and Brazil.[27]

HCQ's first prominent champion was Dr. Didier Raoult, the iconic French infectious disease professor, who has published more than 2,700 papers and is famous for having discovered 100 microorganisms, including the pathogen that causes Whipple's Disease. On March 17, 2020, Dr. Raoult provided a preliminary report on 36 patients treated successfully with hydroxychloroquine and sometimes azithromycin at his institution in Marseille.[28]

In April, Dr. Vladimir (Zev) Zelenko , M.D., an upstate New York physician and early HCQ adopter, reproduced Dr. Didier Raoult's "startling successes" by dramatically reducing expected mortalities among 800 patients Zelenko treated with the HCQ cocktail.[29]

By late April of 2020, US doctors were widely prescribing HCQ to patients and family members, reporting outstanding results, and taking it themselves prophylactically.

In May 2020, Dr. Harvey Risch, M.D., Ph.D. published the most comprehensive study, to date, on HCQ's efficacy against COVID. Risch is Yale University's super-eminent Professor of Epidemiology, an illustrious world authority on the analysis of aggregate clinical data. Dr. Risch concluded that evidence is unequivocal for early and safe use of the HCQ cocktail. Dr. Risch published his work—a meta-analysis reviewing five outpatient studies—in affiliation with the Johns Hopkins Bloomberg School of Public Health in the *American Journal of Epidemiology*, under the urgent title, "Early Outpatient Treatment of Symptomatic, High-Risk COVID-19 Patients that Should be Ramped-Up Immediately as Key to Pandemic Crisis."[30]

He further demonstrated, with specificity, how HCQ's critics—largely funded by Bill Gates and Dr. Tony Fauci[31]—had misinterpreted, misstated, and misreported negative results by employing faulty protocols, most of which showed HCQ efficacy administered without zinc and Zithromax which were known to be helpful. But their main trick for ensuring the protocols failed was to wait until late in the disease process before administering HCQ—when it is known to be ineffective. Dr. Risch noted that evidence against HCQ used late in the course of the disease is irrelevant. While acknowledging that Dr. Didier Raoult's powerful French studies favoring HCQ efficacy were not randomized, Risch argued that the results were, nevertheless, so stunning as to far outweigh that deficit: "The first study of HCQ + AZ [. . .] showed a 50-fold benefit of HCQ + AZ vs. standard of care . . . This is such an enormous difference that it cannot be ignored despite lack of randomization."[32] Risch has pointed out that the supposed need for randomized placebo-controlled trials is a shibboleth. In 2014 the Cochrane Collaboration proved in a landmark meta-analysis of 10,000 studies, that observational studies of the kind produced by Didier Raoult are equal in predictive ability to randomized placebo-controlled trials.[33] Furthermore, Risch observed that it is highly unethical to deny patients promising medications during a pandemic—particularly those which, like HCQ, have long-standing safety records.

So, against all that I've shared here, Dr. Fauci offered up one answer: hydroxychloroquine should not be used because we don't understand the mechanism it uses to defeat COVID—another shibboleth transparently invoked to defeat common sense. Regulators do not understand the mechanism of action of many drugs, but they nonetheless license those that are effective and safe. The fact is that we know more about how HCQ beats COVID than we know about the actions of many other medicines, including—notably—Dr. Fauci's darlings, mRNA vaccines and remdesivir.

Furthermore, an August 2020 paper from Baylor University by Dr. Peter McCullough et al. described mechanisms by which the components of the "HCQ cocktail" exert antiviral effects.[34] McCullough shows that the efficacy of the HCQ cocktail is based on the pharmacology of the hydroxychloroquine ionophore acting as the "gun" and zinc as the "bullet," while azithromycin potentiates the anti-viral effect.

An even more expansive September 30, 2020 meta-review summarizes more recent research, concluding that ALL the studies on early administration of HCQ within a week following infection demonstrate efficacy, while studies of HCQ administered later in the illness show mixed results.[35]

In March, 2020 *Nature* published a paper demonstrating the specific mechanisms in tissue culture by which chloroquine stops viral reproduction.[36]

In April, 2020, a team of Chinese scientists published a preprint of a 62-patient placebo-controlled trial of hydroxychloroquine, resulting in demonstrably improved time to recovery and less progression to severe disease in the treated group.[37]

In May, 2020, a Chinese expert consensus group recommended doctors use chloroquine routinely for mild, moderate, and severe cases of COVID-19 pneumonia.[38]

A national study in Finland in May 2021 showed a 5x efficacy.[39] And national studies in Canada and Saudi Arabia showed 3x efficacy.[40]

I'll stop gilding the lily here and ask the reader: Was hydroxychloroquine some crazy baseless idea, or ought regulators to have honestly investigated it as a potential remedy during a raging pandemic?

Pharma's War on HCQ

The prospect of an existing therapeutic drug (with an expired patent) that could outperform any vaccine in the war against COVID posed a momentous threat to the pharmaceutical cartel. Among the features pharma companies most detest is low cost, and HCQ is about $10 per course.[41] Compare that to more than $3000 per course for Dr. Fauci's beloved remdesivir.[42]

No surprise, pharmaceutical interests launched their multinational preemptive crusade to restrict and discredit HCQ starting way back in January 2020, months before the WHO declared a pandemic and even longer before President Trump's controversial March 19 endorsement. On January 13, when rumors of Wuhan flu COVID-19 began to circulate, the French government took the bizarre, inexplicable, unprecedented, and highly suspicious step of reassigning HCQ from an over-the-counter to a prescription medicine.[43] Without citing any studies, French health officials quietly changed the status of HCQ to "List II poisonous substance" and banned its over-the-counter sales.[44] This absolutely remarkable coincidence repeated itself a few weeks later when Canadian health officials did the exact same thing, quietly removing the drug from pharmacy shelves.[45]

A physician from Zambia reported to Dr. Harvey Risch that in some villages and cities, organized groups of buyers emptied drugstores of HCQ and then burned the medication in bonfires outside the towns. South Africa destroyed two tons of life-saving hydroxychloroquine in late 2020, supposedly due to violation of an import regulation.[46] The US government in 2021 ordered the destruction of more than a thousand pounds of HCQ, because it was improperly imported.[47] "The Feds are insisting that all of it be destroyed, and not be used to save a single life anywhere in the world," said a lawyer seeking to resist the senseless order.

By March, front-line doctors around the world were spontaneously reporting miraculous results following early treatment with HCQ, and this prompted growing anxiety for Pharma. On March 13, a Michigan doctor and trader, Dr. James Todaro, M.D., tweeted his review of HCQ as an effective COVID treatment, including a link to a public Google doc.[48,49] Google quietly scrubbed Dr. Todaro's memo. This was six days *before* the President endorsed HCQ. Google apparently didn't want users to think Todaro's message was *missing*; rather, the Big Tech platform wanted the public to believe that Todaro's memo never even *existed*. Google has a long history of suppressing information that challenges vaccine industry profits. Google's parent company Alphabet owns several vaccine companies, including Verily, as well as Vaccitech, a company banking on flu, prostate cancer, and COVID vaccines.[50,51] Google has lucrative partnerships with all the large vaccine manufacturers, including a $715 million partnership with GlaxoSmithKline.[52] Verily also owns a business that tests for COVID infection.[53] Google was not the only social media platform to ban content that contradicts the official HCQ narrative. Facebook, Pinterest, Instagram, YouTube, MailChimp, and virtually every other Big Tech platform began scrubbing information demonstrating HCQ's efficacy, replacing it with industry propaganda generated by one of the Dr. Fauci/Gates-controlled public health agencies: HHS, NIH and WHO. When President Trump later suggested that Dr. Fauci was not being truthful about hydroxychloroquine, social media responded by removing his posts.

It was a March 2020 news conference where Dr. Fauci launched his concerted attack on HCQ. Asked whether HCQ might be used as a prophylaxis for COVID, he shouted back: "The answer is No, and the evidence that you're talking about is anecdotal evidence."[54] His reliable allies at the *New York Times* then launched a campaign to defame Dr. Raoult.[55]

In the midst of a deadly pandemic, somebody very powerful wanted a medication that had been available over the counter for decades, and known to be effective against coronaviruses, to be suddenly but silently pulled from the shelves—from Canada to Zambia.

In March, at HHS's request, several large pharmaceutical companies—Novartis, Bayer, Sanofi, and others—donated their inventory, a total of 63 million doses of hydroxychloroquine and 2 million of chloroquine, to the Strategic National Stockpile, managed by BARDA, an agency under the DHHS Assistant Secretary for Preparedness and Response.[56] BARDA's Director, Dr. Rick Bright, later claimed the chloroquine drugs were deadly, and he needed to protect the American public from them.[57] Bright colluded with FDA to restrict use of the donated pills to hospitalized patients. FDA publicized the authorization using language that led most physicians to believe that prescribing the drug for any purpose was off-limits.

But at the beginning of June, based on clinical trials that intentionally gave unreasonably high doses to hospitalized patients and failed to start the drug until too late, FDA took the unprecedented step of revoking HCQ's emergency authorization,[58] rendering that enormous stockpile of valuable pills off limits to Americans while conveniently indemnifying the pharmaceutical companies for their inventory losses by allowing them a tax break for the donations.

After widespread use of the drug for 65 years, without warning, FDA somehow felt the need to send out an alert on June 15, 2020 that HCQ is dangerous, and that it required a level of monitoring only available at hospitals.[59] In a bit of twisted logic,

Federal officials continued to encourage doctors to use the suddenly-dangerous drug without restriction for lupus, rheumatoid arthritis, Lyme and malaria. Just not for COVID. With the encouragement of Dr. Fauci and other HHS officials, many states simultaneously imposed restrictions on HCQ's use.

The Fraudulent Industry Studies

Prior to COVID-19, not a single study had provided evidence *against* the use of HCQ based on safety concerns.

In response to the mounting tsunami that HCQ was safe and effective against COVID, Gates, Dr. Fauci and their Pharma allies deployed an army of industry-linked researchers to gin up contrived evidence of its dangers.

By 2020, we shall see, Bill Gates exercised firm control over WHO and deployed the agency in his effort to discredit HCQ.[60]

Dr. Fauci, Bill Gates, and WHO financed a cadre of research mercenaries to concoct a series of nearly twenty studies—all employing fraudulent protocols deliberately designed to discredit HCQ as unsafe. Instead of using the standard treatment dose of 400 mg/day, the 17 WHO studies administered a borderline lethal *daily* dose starting with 2,400 mg.[61] on Day 1, and using 800 mg/day thereafter. In a cynical, sinister, and literally homicidal crusade against HCQ, a team of BMGF operatives played a key role in devising and pushing through the exceptionally high dosing. They made sure that UK government "Recovery" trials on 1,000 elderly patients in over a dozen British, Welsh, Irish and Scottish hospitals, and the U.N. "Solidarity" study of 3,500 patients in 400 hospitals in 35 countries, as well as additional sites in 13 countries (the "REMAP-COVID" trial), all used those unprecedented and dangerous doses.[62] This was a brassy enterprise to "prove" chloroquine dangerous, and sure enough, it proved that elderly patients can die from deadly overdoses. "The purpose seemed, very clearly, to poison the patients and blame the deaths on HCQ," says Dr. Meryl Nass, a physician, medical historian, and biowarfare expert.

In each of these two trials, SOLIDARITY and RECOVERY, the hydroxychloroquine arm predictably had 10–20 percent more deaths than the control arm (the control arm being those patients lucky enough to receive standard supportive care).[63]

The UK government and Wellcome Trust and the Bill and Melinda Gates Foundation (BMGF) jointly financed the Recovery Trial.[64] The principal investigator (PI), Peter Horby, is a member of SAGE and is the chairman of NERVTAG, the New and Emerging Respiratory Virus Threats Advisory Group, both important committees that give the UK government advice on mitigating the pandemic.[65,66] Horby's willingness to risk death of patients given toxic doses of HCQ fueled his subsequent rise in the UK medical hierarchy. Horby received a parade of extraordinary promotions after he orchestrated the mass poisonings of senior citizens. Queen Elizabeth recently knighted him.[67]

Gates's fingerprints are all over this sanguinary project. Despite suspiciously missing pages, the published minutes of WHO's part-secret March and April meetings show these medical alchemists establishing the lethal dosing of chloroquines (CQ and HCQ) for WHO's Solidarity clinical trial. Only four participants attended the second WHO meeting to determine the dose of HCQ and CQ for the Solidarity trial. One was Scott Miller, the BMGF's Senior Program Officer. The report admits that the Solidarity trial was using the highest dose of any recent trial.[68]

The report acknowledges that, "The BMGF developed a model of chloroquine penetration into tissues for malaria."[69] BMGF's unique dosing model for the studies deliberately overestimated the amount of HCQ that necessary to achieve adequate lung tissue concentrations. The WHO report confesses that, "This model is however not validated." Gates's deadly deception allowed FDA to wrongly declare that HCQ would be ineffective at safe levels.

The minutes of that March 13, 2020 meeting suggest that BMGF knew the proper drug dosing and the need for early administration. Yet their same researchers then participated in deliberately providing a potentially lethal dose, failing to dose by weight, missing the early window during which treatment was known to be effective, and giving the drug to subjects who were already critically ill with comorbidities that made it more likely they would not tolerate the high dose. The Solidarity trial design also departed from standard protocols by collecting no safety data: only whether the patient died, or how many days they were hospitalized. Researchers collected no information on in-hospital complications. This strategy shielded the WHO from gathering information that could pin adverse reactions on the dose.

The report of WHO's HCQ trial notes that WHO researchers did not retain any consent forms from the elderly patients they were overdosing, as the law in most countries requires, and makes the bewildering claim that some patients signed consent forms "in retrospect"—a stunning procedure that is unethical on its face. The WHO's researchers noted in their interim report on the trial, *"Consent forms were signed and **retained by the patients**; [An extremely unorthodox and suspicious procedure that suggests that there may have been no formal consents] but noted for record that, consent was generally prospective, but could (where locally approved) be **retrospective**."* One wonders if researchers notified their families of the high dose they were giving to their elderly parents and grandparents in locked COVID wards to which they denied family members access.

The researchers evinced their guilty knowledge by concealing the research records of the doses they used in Solidarity when they filed their trial reports. They also omitted dosing numbers from the report of WHO's meeting to determine the dose, and omitted details of dosing from the WHO's Solidarity trial registration.

Another group of researchers using overdose concentrations of chloroquine published their study as a preprint in mid-April 2020 (and quickly brought to print) in the preeminent journal, *JAMA* (*The Journal of the American Medical Association*) In this murder-for-hire scheme, Brazilian researchers used a dose of 1,200 mg/day for up to ten days of CQ.[70] According to a 2020 review of CQ and HCQ toxicity, "As little as 2–3 g of chloroquine may be fatal in adult patients, though the most commonly reported lethal dose in adults is 3–4 g." Predictably, so many subjects died in the Brazilian high dose study (39 percent, 16 of 41 of the subjects who took this dose) that the researchers had to halt the study. The subjects' mean age was only 55.[71] Their medical records revealed EKG changes characteristic of CQ toxicity.

The WHO and UK trial coordinators must have known this information, but they made no efforts to stop their own overdose trials, nor to lower the doses.

Although Gates did not fund the *JAMA* study directly (it's very possible he funded it indirectly through a nebulous list of funders), the senior and last author, Marcus Vinícius Guimarães Lacerda, has been a Gates-funded researcher on numerous projects. Further, the BMGF has funded multiple projects at the same medical foundation where he and the first, or "lead" author, Borba, work in Manaus, Brazil.[72] (Traditionally, the first listed author is generally seen as the senior and accountable author.)

Gates and his cabal used an arsenal of other deceptive gimmickry to assure that HCQ would appear not just deadly, but ineffective. Each of the studies that Gates funded failed to incorporate Zithromax and zinc—important components of HCQ protocols. All of the Fauci, Gates, WHO, Solidarity, Recovery and Remap-COVID studies administered HCQ at late stages of COVID infection, in contravention of the prevailing recommendations that deem HCQ effective only when doctors administer it early.[73,74] Viewing this orchestrated sabotage with frustration, critics accused the Gates grantees of purposefully designing these studies, at best, to fail and, at worst, to murder.[75] Brazilian prosecutors have accused the authors of the study of committing homicide by purposefully poisoning the elderly subjects in their study with high doses of chloroquine.[76]

All through 2020, Bill Gates and Fauci lashed out against HCQ every chance they got. During the early stages of the pandemic in March, Bill Gates penned an op-ed in The *Washington Post*.[77] Besides calling for a complete lockdown in every state, along with accelerated testing and vaccine development, Gates warned that: "Leaders can help by not stoking rumors or panic buying. Long before the drug hydroxychloroquine was approved as an emergency treatment for COVID-19, people started hoarding it, making it hard for lupus patients who need it to survive."[78]

This, of course, was a lie. The only ones hoarding HCQ were Dr. Fauci and Rick Bright, who had padlocked 63 million doses in the Strategic National Stockpile[79]—more than enough to supply virtually every gerontology-ward patient in America. Despite such efforts to create a shortage, none existed. HCQ is cheap, quick, and easy to manufacture, and since its patent is expired, dozens of manufacturers around the world can quickly ramp up production to meet escalating demand.

In July, Gates endorsed censorship of HCQ recommendations after a video touting its efficacy against coronavirus accumulated tens of millions of views.[80] Gates called the video "outrageous," and praised Facebook and YouTube for hastily removing it. He nevertheless complained "You can't find it directly on those services, but everybody's sending the link around because it's still out there on the internet."[81] This, Gates told *Yahoo News*, revealed a persistent shortcoming of the platforms. "Their ability to stop things before they become widespread, they probably should have improved that," Gates scolded.

Asked by *Bloomberg News* in mid-August about how the Trump White House had promoted HCQ "despite its repeatedly being shown to be ineffective and, in fact, to cause heart problems in some patients," Gates happily responded: "This is an age of science, but sometimes it doesn't feel that way. In the test tube, hydroxychloroquine looked good. On the other hand, there are lots of good therapeutic drugs coming that are proven to work without the severe side effects."[82] Gates went on to promote Gilead's remdesivir as the best alternative, despite its lackluster track record compared to HCQ. He didn't mention having a large stake in Gilead,[83] which stood to make billions if Dr. Fauci was able to run remdesivir through the regulatory traps.

Obsequious reporters consistently encouraged Gates to portray himself as an objective expert, and Gates used that interview to discredit HCQ, and also me. His *Bloomberg* questioner opened the door with a typical softball: "For years, people have said if anti-vaxxers had lived through a pandemic, the way their grandparents did, they'd think differently." Gates replied: "The two times I've been to the White House [since 2016], I was told I had to go listen to anti-vaxxers like Robert Kennedy, Jr. So, yes, it's ironic that people are questioning vaccines and we're actually having to say, 'Oh, my God, how else can you get out of a tragic pandemic?'"[84]

If he had only asked me, I could have told him!

*Lancet*gate

It remains an enduring mystery just which powerful figure(s) caused the world's two most prestigious scientific journals, *The Lancet* and the *New England Journal of Medicine* (*NEJM*), to publish overtly fraudulent studies from a nonexistent database owned by a previously unknown company. Anthony Fauci and the vaccine cartel celebrated the *Lancet* and *NEJM* papers on May 22, 2020 as the final nail in hydroxy-chloroquine's coffin.[85,86]

Both studies in these respected publications relied on data from the Surgisphere Corporation, an obscure Illinois-based "medical education" company that claimed to somehow control an extraordinary global database boasting access to medical information from 96,000 patients in more than 600 hospitals.[87] Founded in 2008, this sketchy enterprise had eleven employees, including a middling science fiction writer and a porn star/events hostess. Surgisphere claimed to have analyzed data from six continents and hundreds of hospitals that had treated patients with HCQ or CQ in real time. Someone persuaded the *Lancet* and the *New England Journal of Medicine* to publish two Surgi-sphere studies in separate articles on May 1 and 22. Like the other Gates-supported studies, the *Lancet* article portrayed HCQ as ineffective and dangerous. The *Lancet* study said that the Surgisphere data proved that HCQ increased cardiac mortality in COVID-19 patients. Based on this study, the FDA withdrew its EUA recommendation on June 15, 2020,[88] the WHO and UK suspended their hydroxychloroquine clinical trials on May 25.[89] Each resumed briefly, then stopped for good in June declaring HCQ unhelpful.[90] Three European nations immediately banned use of HCQ, and others fol-lowed within weeks.[91]

That would normally have been the end of it, if not for the 200 independent scien-tists who quickly exposed the *Lancet* and *NEJM* studies as shockingly clumsy con jobs.[92] The Surgisphere datasets that formed the foundation of the studies were so ridiculously erroneous that they could only have been a rank invention. To cite only one of many dis-crepancies, the number of reported deaths among patients taking hydroxychloroquine in one Australian hospital exceeded the total number of deaths for the entire country. An international brouhaha quickly revealed that the Surgisphere database did not exist, and soon enough, Surgisphere itself vanished from the Internet. The University of Utah terminated the faculty appointment of one of the article's authors, Amit Patel. Surgisphere's founder, Sapan S. Desai, disappeared from his job at a Chicago hospital.

Even the *New York Times* reported that "More than 100 scientists and clinicians have questioned the authenticity" of the database, as well as the study's integrity.[93] Despite the barrage of astonished criticism, the *Lancet* held firm for two weeks before relenting to the remonstrances. Finally, three of the four *Lancet* coauthors requested the paper be retracted. Both *The Lancet* and *NEJM* finally withdrew their studies in shame. Somebody at the very pinnacle of the medical cartel had twisted arms, kicked groins, and stoved in kneecaps to force these periodicals to abandon their policies, shred their ethics, and spend down their centuries of hard-won credibility in a desperate bid to torpedo HCQ. To date, neither the authors nor the journals have explained who induced them to coauthor and publish the most momentous fraud in the history of scientific publishing.

The headline of a comprehensive exposé in *The Guardian* expressed the global shock among the scientific community at the rank corruption by scientific publishing's most formidable pillars: "*The Lancet* has made one of the biggest retractions in modern history. How could this happen?"[94] *The Guardian* writers openly accused *The Lancet* of promoting fraud: "The sheer number and magnitude of the things that went wrong or missing are too enormous to attribute to mere incompetence." *The Guardian* commented, "What's incredible is that the editors of these esteemed journals still have a job—that is how utterly incredible the supposed data underlying the studies was."

The capacity of their Pharma overlords to strong-arm the world's top two medical journals, the *NEJM* and *The Lancet*, into condoning deadly research[95,96] and to simultaneously publish blatantly fraudulent articles in the middle of a pandemic, attests to the cartel's breathtaking power and ruthlessness. It is no longer controversial to acknowledge that drug makers rigorously control medical publishing and that *The Lancet*, *NEJM*, and *JAMA* are utterly corrupted instruments of Pharma. *The Lancet* editor, Richard Horton, confirms, "Journals have devolved into information laundering operations for the pharmaceutical industry."[97] Dr. Marcia Angell, who served as an *NEJM* editor for 20 years, says journals are "primarily a marketing machine."[98] Pharma, she says, has co-opted "every institution that might stand in its way."[99,100]

Cracking Down on HCQ to Keep Case Fatalities High

Referring to the *Lancet* Surgisphere study during a May 27 CNN interview, Dr. Fauci stated on CNN about hydroxychloroquine, "The scientific data is really quite evident now about the lack of efficacy."[101] And even after the scandal lay exposed and the journals retracted their articles, Dr. Fauci let his lie stand. Instead of launching an investigation of this momentous and enormously consequential fraud by the world's two leading medical journals and publicly apologizing, Dr. Fauci and the medical establishment simply ignored the wrongful conduct and persevered in their plan to deny global populations access to lifesaving HCQ.

The historic journal retractions went practically unnoticed in the slavish, scientifically illiterate mainstream press, which persisted in fortifying the COVID propaganda. Headlines continued to blame HCQ for the deaths instead of the deliberately treacherous researchers who gave sick, elderly, and compromised patients toxic drug dosages. And most remarkable of all, the FDA made no effort to change the recommendation it made against HCQ. Other countries persisted in demonizing the life-saving drug.

Once the FDA approves a prescription medication, federal laws allow any US physician to prescribe the duly approved drug for any reason. Twenty-one percent of all prescriptions written by American doctors, exercising their medical judgment, are for off-label uses.[102]

Even after the FDA withdrew its Emergency Use Authorization and posted the fraudulent warning on its website,[103] many front-line doctors across the country continued to prescribe and report strong benefits with appropriate doses of HCQ. In response, Dr. Fauci took even more unprecedented steps to derail doctors from prescribing HCQ.

In March, while people were dying at the rate of 10,000 patients a week, Dr. Fauci declared that hydroxychloroquine should only be used as part of a clinical trial.[104] For the first time in American history, a government official was overruling the medical judgment of thousands of treating physicians, and ordering doctors to stop practicing medicine as they saw fit. Boldly and relentlessly, Dr. Fauci kept declaring that "The Overwhelming Evidence of Properly Conducted Randomized Clinical Trials Indicate No Therapeutic Efficacy of Hydroxychloroquine (HCQ)."[105] Dr. Fauci failed to disclose that NONE of the trials he had used as the basis for that pronouncement involved medication given in the first five to seven days after onset of symptoms. Instead, all of those randomized controlled trials targeted patients who were already sick enough to be hospitalized.

People wanting to be treated in that first critical week of illness and avoid being hospitalized were basically out of luck as Dr. Fauci moved to foreclose patients from receiving the lifesaving remedy during the treatment window when science and previous experience showed it to be effective.

On July 2, following the humiliating journal retractions, Detroit's Henry Ford Health System published a peer-reviewed study showing that hydroxychloroquine significantly cut death rates even in mid-to-late COVID cases, and without any heart-related side effects.[106] Fauci leapt to the barricades to rescue his vaccine enterprise. On July 30, he testified before Congress that the Michigan results were "flawed."[107]

The FDA revocation of the EUA and Dr. Fauci's withering response to the Michigan trial provided cover for 33 governors whose states moved to restrict prescribing or dispensing of HCQ.[108]

In New York, Governor Andrew Cuomo drove up record death counts by ordering that physicians prescribe HCQ only for hospitalized patients.[109] In Nevada, Governor Steven Sisolak prohibited both prescribing and dispensing chloroquine drugs for COVID-19.[110] State medical licensing boards threatened to bring "unprofessional conduct" charges against non-complying doctors (a threat to their license) and to "sanction" doctors if they prescribed the drug.[111] Most pharmacists were afraid to dispense HCQ, and on June 15, state pharmacy boards in Arizona, Arkansas, Michigan, Minnesota, New Hampshire, New York, Oregon, and Rhode Island began refusing orders from physicians and retailers.[112] Hospitals commanded doctors to cease treating their patients with HCQ beginning June 15, 2020.[113] The NIAID halted a clinical trial of the drug in outpatients, in June 2020, only a month after it started, having enrolled only 20 of the planned 2,000 enrollees.[114] The FDA blocked access to the millions of doses of HCQ and CQ that Sanofi and other drug makers had donated to the Strategic National Stockpile (with appropriate tax benefits).[115] Sanofi announced it would no longer supply the drug for use treating COVID. Dr. Fauci and his HHS cronies decreed that the medication rot in warehouses while Americans unnecessarily sickened and died from COVID-19.

On June 17, the WHO—for which Mr. Gates is the largest funder after the US, and over which Mr. Gates and Dr. Fauci exercise tight control—called for the halt of HCQ trials in hundreds of hospitals across the world.[116] WHO Chief Tedros Adhanom Ghebreyesus ordered nations to stop using HCQ and CQ. Portugal, France, Italy, and Belgium banned HCQ for COVID-19 treatment.[117]

Foreign Experiences

In compliance with the WHO recommendation, Switzerland banned the use of HCQ; however, about 2 weeks into the ban, Switzerland's death rates tripled, for about 15 days, until Switzerland reintroduced HCQ. COVID deaths then fell back to their baseline.[118] Switzerland's "natural experiment" had provided yet another potent argument for HCQ.

Similarly, Panamanian physician and government advisor Sanchez Cardenas notes that when Panama banned HCQ, deaths shot up, until the government relented, at which point deaths dropped back to baseline.[119]

Seven months into the pandemic, nations that widely used HCQ and made it readily available to their citizens demonstrated overwhelming evidence that HCQ was obliterating COVID-19.

A June 2, 2020 court filing supporting the use of HCQ for COVID included an Association of American Physicians & Surgeons (AAPS) comparison of national death rates among countries with varying policies governing access to HCQ. Many countries with underdeveloped health care systems were using HCQ early and achieving far lower mortalities than in the United States, where HHS and the FDA impede access to HCQ.[120] AAPS General Counsel Andrew Schlafly observed that "Citizens of the Philippines, Poland, Israel, and Turkey all have greater access to HCQ than American citizens do," and they have superior morbidity outcomes. He added, "In Venezuela, HCQ is available over the counter without a prescription, while in the United States, pharmacists are prevented from filling prescriptions for HCQ."[121]

Other foreign studies support strong claims for HCQ. A study by Nova demonstrated that nations using HCQ have death rates 80 percent lower than those that banned it.[122]

A meta-review of 58 peer-reviewed observational studies by physician researchers in Spain, Italy, France, and Saudi Arabia found that hydroxychloroquine dramatically reduced mortality from COVID, while additional articles by doctors in Turkey, Canada, and the US found that HCQ's cardiac toxicity is negligible. (See c19study. com for a compilation of 99 (58 peer-reviewed) studies of the chloroquine drugs in COVID-19.)[123]

Furthermore, mortality and morbidity data from over six dozen nations indicate a strong relationship between access to HCQ and COVID-19 death rates.[124,125] While such a relationship does not prove cause/effect, it would be lunacy to simply ignore the reality and assume no relationship.

Country by country, data consistently links broader access to HCQ to lower mortality. The very poorest countries—if they used HCQ—had far lower case fatality rates than wealthy countries that did not. Even impoverished African nations, where "experts" like Bill Gates predicted the highest death rates, had drastically lower mortalities than in nations that banned HCQ. Senegal and Nigeria, for example, both use hydroxychloroquine and had COVID fatality rates that were significantly lower than those experienced in the United States.[126]

Similarly, despite the fact that hygiene in those countries is often far inferior, in Ethiopia,[127] Mozambique,[128] Niger,[129] Congo,[130] and Ivory Coast,[131] there are far fewer per capita deaths than in the US. In those nations, death rates vary between 8 and 47.2 deaths per million inhabitants as of September 24, 2021. In contrast, western countries that denied access to HCQ experienced numbers of coronavirus deaths per million inhabitants between 220 per million in Holland,[132] 2,000 per million in the US, and 850 deaths per million in Belgium.[133] Dr. Meryl Nass observed, "If people in these malaria countries would boost their immune system with zinc, vitamin C and vitamin D, the coronavirus death toll would even further decrease."

Similarly, Bangladesh CFR, Senegal, Pakistan, Serbia, Nigeria, Turkey, and Ukraine all allow unrestricted use of HCQ and all have miniscule case fatality rates compared to the countries that ban HCQ.[134] Wealthier democracies or countries with especially restrictive HCQ protocols—Ireland, Canada, Spain, the Netherlands, UK, Belgium, and France—are comparatively deadly environments.

Andrew Schlafly observed that, "The mortality rate from COVID-19 in countries that allow access to HCQ is only one-tenth the mortality rate in countries where there is interference with this medication, such as the United States. . . . In some areas of Central America, officials are even going door to door to distribute HCQ. . . . These countries have been successful in limiting the mortality from COVID-19 to only a fraction of what it is in wealthier countries."[135]

As the industry/government cartel ramped up its campaign to keep HCQ from the masses, many doctors fought back. On July 23, Yale virologist Dr. Harvey Risch persisted, this time with a *Newsweek* article titled "The key to defeating COVID-19 already exists. We need to start using it."[136] Dr. Risch beseeched the authorities: HCQ saves lives and its use could quickly end the pandemic. By then, Dr. Risch had updated his rigorous analysis of the early treatment of COVID-19 with hydroxychloroquine, zinc, and azithromycin. He now cited *twelve* clinical studies suggesting that the early administration of HCQ could lower death rates by 50 percent. In that case,

COVID-19 would have a lower case fatality rate than the seasonal influenza. "We would still have had a pandemic," Harvey Risch told me, "but we wouldn't have had the carnage."

Noting more than fifty HCQ studies, Dr. Meryl Nass, in June 2020, supported Risch's calculation: "If people were treated prophylactically with this drug (using only 2 tablets weekly) as is done in some areas and some occupational groups in India, there would probably be at least 50 percent fewer cases after exposure."[137] Stopping the pandemic in its tracks seemed to be the last thing Tony Fauci wanted. Thanks to Dr. Fauci, most US states had by then banned treatment with HCQ, including Dr. Nass's home state of Maine, which banned it for prophylaxis, but did allow it for acute treatment. Dr. Nass suggested that the "acts to suppress the use of HCQ [were] carefully orchestrated" and that "these events [might] have been planned to keep the pandemic going to sell expensive drugs and vaccines to a captive population."[138]

In the same article by Dr. Meryl Nass, published on June 27, 2020,[139] Nass—who has extensively studied HCQ—pointed out that with prophylactic treatment with HCQ "at the onset of their illness, over 99 percent would quickly resolve the infection, avoiding progression to the late-stage disease characterized by cytokine storm, thrombophilia, and organ failure. Despite claims to the contrary, this treatment is very safe, yet outpatient treatment is banned in the United States."

Beginning June 27, 2020, Dr. Nass began a list of deceptive strategies that the Fauci/Pharma/Gates cartel used to control the narrative on hydroxychloroquine and deny Americans access to this effective remedy. The list has grown to 58 separate strategies.[140] "It is remarkable," she observed, how "a large series of events taking place over the past months produced a unified message about hydroxychloroquine (HCQ) and produced similar policies about the drug in the US, Canada, Australia, New Zealand and western Europe. The message is that generic, inexpensive hydroxychloroquine (costing only $1.00 to produce a full course) is dangerous."[141]

Dr. Fauci's Hypocritical HCQ Games

In his early AIDS days, Dr. Fauci had thrashed FDA as inhumane for demanding randomized double-blind placebo studies at the height of the pandemic. Now, here he was doing what he had condemned by blocking an effective treatment simply because it would compete with his expensive patent-protected pharmaceutical, remdesivir, and vaccines.

* * *

Dr. Fauci repeatedly insisted he would not allow HCQ for COVID-19 until its efficacy is proven in "randomized, double-blind placebo studies."[142] Dr. Risch calls this position a "transparent sham." Dr. Fauci knew that neither industry nor its PI's would ever sponsor trials for a product with expired patents. It's noteworthy that while Dr. Fauci was bemoaning the lack of evidence of HCQ efficacy, he was refusing to commission his own trials to study early use of the hydroxychloroquine, zinc, and Zithromax remedy. Dr. Fauci himself, while spending 48 billion dollars on zero-liability vaccines, at first refused to allocate anything for a randomized placebo study of HCQ. Even worse, he cancelled two NIAID-sponsored trials of outpatient HCQ before completion.[143]

Dr. Fauci's hypocrisy about HCQ is evident to anyone who looks at his vacillating pronouncements throughout his long career. He has persistently insisted on double-blind randomized placebo trials for medicines he dislikes (those that compete with his patented remedies) and airily fixed the NIAID study of remdesivir by changing the endpoints midstream to favor the drug. Dr. Fauci did not sponsor or encourage randomized trials for masks, lockdowns, or social distancing. And in the decades since he took over NIAID, he has never demanded randomized studies to confirm safety of the combined 69 vaccine doses currently on the childhood schedule. Every one of these vaccines is regarded as so "unavoidably unsafe"—in the words of the 1986 Vaccine Act (NCVIA) and the Supreme Court—that their manufacturers have demanded—and received—immunity from liability.

During a 2013 *USA Today* interview, Dr. Fauci discussed remedies for another deadly coronavirus, MERS, which was causing an outbreak in Qatar and Saudi Arabia with over 30 percent mortality.[144] Dr. Fauci then sang an entirely different tune than he is singing now about hydroxychloroquine. He suggested using a combination of the antiviral drugs ribavirin and interferon-alpha 2b to treat MERS, even though the treatment had never been tested for safety or effectiveness against MERS in humans. In that circumstance, Dr. Fauci's NIAID had found that the treatment could stop MERS virus from reproducing in lab-grown cells. And, oh yes, NIAID had patented it.[145]

"We don't have to start designing new drugs," Dr. Fauci told journalists.[146] "The next time someone comes into an emergency room in Qatar or Saudi Arabia, you would have drugs that are readily available. And at least you would have some data."[147] Even though the treatment hadn't gone through any trials, Dr. Fauci urged its compassionate use: "If I were a physician in a hospital and someone were dying, rather than do nothing, you can see if these work."[148]

He played by all-new rules when it came to COVID, forcing doctors to stand on the sidelines while patients died and prohibiting them from trying combinations of repurposed therapeutics to "see if these work."[149] Back in 2013, when Dr. Fauci endorsed Ribavirin/Interferon for use against MERS, the two-punch hepatitis C remedy was, according to NIH, horrendously dangerous, with harms occurring in literally every patient who took the concoction. It causes hemolytic anemia chronic fatigue syndrome, and a retinue of birth defects and/or death of unborn children. Ribavirin is genotoxic, mutagenic, and a potential carcinogen.[150]

Nevertheless, in 2013, Dr. Fauci advocated the therapy, despite the total lack of randomized, placebo-controlled clinical trials, in fact, the lack of any human data on using the combination against MERS.

The COVID vaccines that qualified for Emergency Use Authorization include novel platforms like mRNA and DNA with no known safety profile. Others use toxic adjuvants like squalene and aluminum or novel adjuvants, with proven risks and potentially high rates of serious injuries. The two-month randomized clinical trials that justified the EUAs for COVID vaccines were far too brief to detect injuries with longer incubation periods.[151,152,153] The vaccines are so risky that the insurance industry has refused to underwrite them,[154] and the manufacturers refuse to produce them without blanket immunity from liability.[155] Bill Gates, who is the principal investor in many of these new COVID vaccines, stipulated that their risk is so great that he would not provide them to people unless every government shielded him from lawsuits.[156]

Why then should HCQ be the only remedy required to cross this artificially high hurdle? After all, HCQ is less in need of randomized placebo studies than any of these vaccines or remdesivir; the safety of HCQ has been established over more than six decades. While vaccines are given to healthy people who face small risk of catching the disease, HCQ is administered to people who are actually sick, with virtually no risk to the patient. If a drug is safe and might work, if people are dying and there are no other good options, must we not try it?

Dr. Fauci's on-again-off-again interest in drug safety is situational and self-interested. He claimed on July 31 about HCQ that "If that randomized placebo-controlled trial shows efficacy, I would be the first to admit it and to promote it, but I have not… So I just have to go with the data. I don't have any horse in the game one way or the other; I just look at the data."[157]

In fact, Dr. Fauci always had a stable of horses in the game. One of them is remdesivir, even after the WHO's randomized placebo trial showed remdesivir ineffective against COVID.[158] Furthermore, remdesivir has a catastrophic safety profile.[159]

His second nag is the Moderna vaccine, in which he invested years and six billion taxpayer dollars. He was thrilled to sponsor a human trial of a Moderna COVID vaccine (partly owned by his agency), before there were any safety and efficacy data from animal studies, which goes against FDA regulations. He then pushed for hundreds of millions of people to get EUA vaccines before the randomized placebo-controlled trials were complete. So much for Dr. Fauci's requirement for having high-quality evidence before risking use of drugs and vaccines in humans.

Dr. Fauci's ethical flip-flopping about the need for rigid safety testing is particularly troubling since he is championing a competitive product from which his agency and his employees expect a lucrative financial outcome.

In the midst of a pandemic, with hundreds of thousands of deaths attributed to COVID, and the economy in free fall, Dr. Fauci's suggestion that we withhold promising treatments that have an established safety profile—from patients who have a potentially lethal disease—pending the completion of randomized controlled clinical trials, is highly manipulative and utterly unethical. It is not medically ethical to allow a COVID-19 patient to deteriorate in the early stages of the infection when there is an inexpensive, safe, and demonstrably effective HCQ treatment that CDC's and NIAID's own studies show blocks coronavirus replication. It would be equally unethical to enroll sick individuals in such studies—as Dr. Fauci proposes—in which half the infected patients would receive a placebo.

Dr. Fauci's hypocrisy is particularly acute since the 21st Century Cures Act, which Congress passed in 2016, directs the FDA to accept precisely the type of "real world" evidence reported by treating physicians like Drs. Zelenko, Raoult, Risch, Kory, McCullough, Gold, and Chinese doctors, in lieu of controlled clinical trials, for licensing new products.[160]

The Cures Act[161] recognizes that doctors and scientists can obtain very useful information when treating patients and observing the results outside of a formal trial setting.

For Big Pharma, no milestone was more important during the current pandemic than neutralizing HCQ to prevent its widespread beneficial use.

Dr. Fauci's shocking inconsistency and ethical breaches are congruent with his long history of promoting Big Pharma's more profitable patented products and using his power and influence to advance its agenda without regard to public health. Dr. Fauci's leadership role in this deadly scandal is consistent with his long history of discrediting therapies that compete with vaccines and other patented pharmaceutical products.

Thanks to Dr. Fauci's strategic campaign, most Americans are still unable to obtain HCQ for early treatment of COVID-19, even fewer Americans are able to access it as preventive medicine, and fewer still are aware of its benefits.

His bizarre and inexplicable actions give credence to the suspicions held by many Americans that Dr. Fauci is working to prolong the epidemic in order to impose expensive patented drugs and vaccines on a captive population, during a pandemic that has crashed the world economy, caused famines, and destroyed lives. While Dr. Fauci held us hostage waiting for what turned out to be imperfect vaccines, his own agency attributed over half a million deaths in America to COVID.

Professor Risch believes that Dr. Fauci knowingly lied about the drug hydroxychloroquine and used his influence to get the FDA to suppress it because he and other bureaucrats are "in bed with other forces that are causing them to make decisions that are not based on the science [and are] killing Americans."[162]

Moreover, Dr. Risch specifically claims that Fauci and the FDA have caused " the deaths of hundreds of thousands of Americans who could have been saved by" HCQ.[163]

III: IVERMECTIN

By the summer of 2020, front-line physicians had discovered another COVID remedy that equalled HCQ in its staggering, life-saving efficacy.

Five years earlier, two Merck scientists won the Nobel Prize for developing ivermectin (IVM), a drug with unprecedented firepower against a wide range of human parasites, including roundworm, hookworm, river blindness, and lymphatic filariasis.[1] That salute was the Nobel Committee's only award to an infectious disease medication in 60 years. FDA approved IVM as safe and effective for human use in 1996. WHO includes IVM (along with HCQ) on its inventory of "essential medicines"—its list of remedies so necessary, safe, efficacious, and affordable that WHO deems easy access to them as essential "to satisfy the priority health care needs of the population."[2] WHO has recommended administering ivermectin to entire populations to treat people who might have parasitic infections—meaning they consider it safe enough to give to people who haven't even been diagnosed.[3] Millions of people have consumed billions of IVM doses as an anti-parasitic, with minimal side effects. Ivermectin's package insert suggests that it is at least as safe as the most popular over-the-counter medications, including Tylenol and aspirin.

Researchers at Japan's Kitasato Institute published a 2011 paper describing IVM in terms almost never used for any other drug:

There are few drugs that can seriously lay claim to the title of "Wonder drug," penicillin and aspirin being two that have perhaps had greatest beneficial impact on the health and wellbeing of Mankind. But ivermectin can also be considered alongside those worthy contenders, based on its versatility, safety, and the beneficial impact that it has had, and continues to have, worldwide—especially on hundreds of millions of the world's poorest people.[4]

Three statues—at the Carter Center, at the headquarters of the World Bank, and at the headquarters of the World Health Organization—honor the development of ivermectin.

Because since 2012, multiple in-vitro studies have demonstrated that IVM inhibits the replication of a wide range of viruses. *Nature Magazine* published a 2020 study reviewing 50 years of research finding IVM "highly effective against microorganisms including some viruses," and reporting the results in animal studies demonstrating "antiviral effects of ivermectin in viruses such as Zika, dengue, yellow fever, West Nile . . ."[5]

An April 3, 2020 article entitled "Lab experiments show anti-parasitic drug, ivermectin, eliminates SARS-CoV-2 in cells in 48 hours,"[6] by Australian researchers at Monash and Melbourne Universities and the Royal Melbourne Hospital, first won IVM global attention as a potential treatment for COVID. The international press initially raved that this safe, inexpensive, well-known, and readily available drug had demolished SARS-CoV-2 in cell cultures. "We found that even a single dose could essentially remove all viral RNA by 48 hours and that even at 24 hours there was a really significant reduction in it," said lead researcher Dr. Kylie Wagstaff.[7] Based on this study, on May 8, 2020, Peru—then under siege by a crushing COVID endemic—adopted ivermectin in its national guidelines. "Peruvian doctors already knew the medicine, widely prescribed it for parasites, and health authorities knew it was safe and were comfortable with it," recalls Dr. Pierre Kory. COVID deaths dropped precipitously—by 14-fold—in the regions where the Peruvian government effectively distributed ivermectin. Reductions in deaths correlated with the extent of IVM distributions in all 25 states. In December 2020, Peru's new president, under pressure from WHO, severely restricted IVM availability and COVID cases rebounded with deaths increasing 13-fold.[8]

In prophylaxis studies, ivermectin repeatedly demonstrated far greater efficacy against COVID than vaccines at a fraction of the cost.

In Argentina, for example, in the summer of 2020, Dr. Hector Carvallo conducted a randomized placebo-controlled trial of ivermectin as a preventative, finding 100 percent efficacy against COVID. Carvallo's team found no infections among the 788 workers who took weekly ivermectin prophylaxis, whereas 58 percent of the 407 controls had become ill with COVID-19.[9]

A later observational study[10] from Bangladesh—also investigating ivermectin as a pre-exposure prophylaxis against COVID-19 among health care workers—found nearly as spectacular results: only four of the 58 volunteers who took a minimal dose of ivermectin (12 mg once per month for four months) developed mild COVID-19 symptoms, compared to 44 of the 60 health care workers who had declined the medication.

Furthermore, a 2021 study suggested that a key biological mechanism of IVM—competitive binding with SARS-CoV-2 spike protein—was not specific to any coronavirus variant and therefore, unlike vaccines, ivermectin would probably be effective against all future variants.[11]

As early as March 1, 2020, some front-line ICU and ER doctors began using ivermectin in combination with HCQ in early treatment protocols. Dr. Jean-Jacques Rajter,[12] a Belgian physician working in Miami, began using the drug March 15 and immediately saw an uptick in recoveries. He published an excellent paper on June 9. Meanwhile, two Western physicians using ivermectin in Bangladesh also reported a very high rate of recoveries, even among patients in later states of illness.[13]

Since March 2020, when doctors first used IVM against COVID-19, more than 20 randomized clinical trials (RCTs) have confirmed its miraculous efficacy against COVID for both inpatient and outpatient treatment. Six of seven meta-analyses of IVM treatment RCTs completed in 2021 found notable reductions in COVID-19 mortality. The relevant studies "all showed significant benefit for high-risk outpatients," says the eminent Yale epidemiologist Dr. Harvey Risch. The only studies where its performance was anything short of stellar were those that investigated its efficacy in patients in very late stages of COVID.

But even late-stage patients showed benefits in almost all studies, although somewhat less dramatic. According to a 2020 review by McCullough et al., "Numerous clinical studies—including peer-reviewed randomized controlled trials—showed large magnitude benefits of ivermectin in prophylaxis, early treatment, and also in late-stage disease management. Taken together . . . dozens of clinical trials that have now emerged from around the world are substantial enough to reliably assess clinical efficacy and infer a signal of benefit with acceptable safety."[14]

Early in January 2021, Dr. David Chesler, a geriatric specialist who had treated 191 infected patients since the previous spring at seven Virginia nursing homes, wrote to Dr. Fauci claiming that he had achieved a mortality rate of 8 percent using ivermectin—half (and 146,000 deaths less than) the US average in elder-care facilities. In his letter to Dr. Fauci, Chesler attached a peer-reviewed case study documenting reports of similar efficacy from other countries. Neither Dr. Fauci nor anyone else from NIAID replied to Dr. Chesler's letter.[15]

The *Annals of Dermatology and Venereology* reported that in a French nursing home, all 69 residents—average age 90—and 52 staff survived a COVID-19 outbreak.[16,17] As it turns out, they had all taken ivermectin for a scabies infestation. COVID decimated the surrounding community, but only seven elder home residents and four staff were affected, and all had mild illness. None required oxygen or hospitalization.

Research suggests that ivermectin may work through as many as 20 separate mechanisms. Among them, ivermectin functions as an "ionophore," facilitating transfer of zinc into the cells, which inhibits viral replication. Ivermectin stops replication of COVID-19, seasonal flu, and many other viruses through this and other mechanisms. For example, a March 2021 study[18] by Choudhury et al., found that "Ivermectin was found as a blocker of viral replicase, protease and human TMPRSS2, which could be the biophysical basis behind its antiviral efficiency." The drug also reduces inflammation via multiple pathways, thereby protecting against organ damage. Ivermectin furthermore impairs the spike protein's ability to attach to the ACE2 receptor on human cell membranes, preventing viral entry. Moreover, the drug prevents blood clots through binding to spike protein, and also deters the spike protein from binding to CD147 on red blood cells, which would otherwise trigger clumping. When patients take IVM before exposure, the drug prevents infection, which halts onward transmission, and helps protect the entire community.

In March, 2021, a published study by Peter McCullough and 57 other front-line physicians from multiple countries found that "Our early ambulatory treatment regimen was associated with *estimated 87.6 percent and 74.9 percent reductions in hospitalization and death.*"[19]

Many other studies echo Dr. McCullough's results. The average reduction in mortality, based on 18 trials, is 75 percent,[20] according to a January 2021 meta-analysis presentation to the NIH COVID-19 Treatment Guidelines Panel. A WHO-sponsored meta-review[21] of 11 studies likewise suggests ivermectin can reduce COVID-19 mortality by as much as 83 percent. Below is a compilation of seven meta-analyses looking at ivermectin's effect on mortality. Each one found a large benefit, ranging between 57 percent and 83 percent reduction in deaths:

Ivermectin meta analysis mortality results — ivmmeta.com Oct 12, 2021

	Improvement, RR [CI]	
Kory et al.	69%	0.31 [0.20-0.47]
Bryant et al.	62%	0.38 [0.19-0.73]
Lawrie et al.	83%	0.17 [0.08-0.35]
Nardelli et al.	79%	0.21 [0.11-0.36]
Hariyanto et al.	69%	0.31 [0.15-0.62]
WHO (OR)	81%	0.19 [0.09-0.36]
ivmmeta	57%	0.43 [0.32-0.58]

0   0.25   0.5   0.75   1   1.25   1.5   1.75   2+
Favors ivermectin   Favors control

Below is a compilation of the studies of ivermectin for COVID prevention. On average, used prophylactically, ivermectin prevented 86 percent of the adverse outcomes. Over all these studies, ivermectin protected 6 of every 7 people who used it to prevent COVID.

And of 29 studies of early treatment of COVID using ivermectin, listed on opposite page, the average benefit was 66 percent. The 3 tables presented here and their adjacent forest plots can be found on the ivmmeta.org website. They are part of a much larger website that has compiled all completed, validated studies for each of 27 different treatments for COVID-19, at c19study.com.

A January 2021 study in *The Lancet* found that ivermectin dramatically reduced the intensity and duration of symptoms and viral loading.[22]

In March 2020, Dr. Paul Marik, chief of intensive care medicine at Eastern Virginia Medical School, began posting treatment guidelines for the care of COVID patients. Dr. Marik, one of the best known and well-published professors of intensive care medicine, recruited a team of the most highly respected and most published leading ICU physicians from across the globe to systematically research all possible approaches to this new virus. Soon, his organization, Front Line COVID-19 Critical Care Alliance (FLCCC), created a website and posted their first treatment protocols in mid-April 2020. By November 2020, the FLCCC doctors felt there was enough evidence to add ivermectin to their protocols.[23] "The data show the ability of the drug ivermectin to prevent COVID-19, to keep those with early symptoms from progressing to the hyper-inflammatory phase of the disease, and even to help critically ill patients recover."[24] Peer-reviewers green-lighted the clinical and scientific rationale for FLCCC's hospital protocols, and the *Journal of Intensive Care Medicine* published them in mid-December 2020.[25] FLCCC also published on its website a one-page summary (regularly updated) of the clinical trial evidence for ivermectin.[26]

In December 2020, FLCCC President and Chief Medical Officer, Dr. Pierre Kory, a pulmonary and critical care specialist, testified to the benefits of ivermectin before a number of COVID-19 panels, including the Senate Committee on Homeland Security and Governmental Affairs.[27] In riveting testimony, Kory described:

> Six studies with a total of over 2,400 patients—all showing near-perfect prevention of transmission of this virus in people exposed to COVID-19 . . . Three RCT's randomized controlled studies and multiple large case series—involving over 3,000 patients showing stunning recovery among hospitalized patients and four large randomized controlled trials involving 3,000 patients all showing large and statistically significant reductions in mortality when treated with ivermectin.

Two weeks later, on January 6, 2021, Dr. Kory spoke to the National Institutes of Health COVID-19 Treatment Guidelines Panel.[28] Along with *Éminence grise* Dr. Paul Marik, and other members of the Front Line COVID-19 Critical Care Alliance also presented positive data on ivermectin, as did the WHO's meta-analysis author, Dr. Andrew Hill who they had invited to present with them.

The *Financial Times* followed with an article citing Hill's research for the WHO at the University of Liverpool. Hill's meta-analysis of six ivermectin studies showed a cumulative 75 percent reduction of risk of death in a subset of moderate to severe COVID-19 patients, in whom the drug reduced inflammation and sped up elimination of the virus.[29]

Kory testified that "IVM could reduce hospitalizations by almost 90 percent and deaths by almost 75 percent." Kory is one of a multitude of leading front-line physicians, including McCullough, Florida's Surgeon General Joe Ladopo, Professor Paul Marik, Dr. Joseph Varone, and mRNA vaccine inventor, Dr. Robert Malone, and many, many others, who believe that early treatment with ivermectin would have avoided 75 percent-80 percent of deaths and saved our country a trillion dollars in treasure.

"COVID resulted in ~6 million hospitalizations and 700,000+ deaths in America," says Dr. Kory. "If HCQ and IVM had been widely used instead of systematically suppressed, we could have prevented 75 percent, or at least 500,000 deaths, and 80 percent of hospitalizations, or 4.8 million. We could have spared the states hundreds of billions of dollars."

Ten days after the FLCCC presentation, on January 14, the NIH's COVID-19 Treatment Guidelines Panel changed its previously negative recommendation to doctors regarding ivermectin to "neither for nor against," cracking open the door just a little for physicians to use IVM as a therapeutic option. That is the same neutral

	Improvement, RR [CI]		Treatment	Control	Dose (1m)		
Shouman (RCT)	91% 0.09 [0.03-0.23]	symp. case	15/203	59/101	36mg		
Behera	54% 0.46 [0.29-0.71]	cases	41/117	145/255	42mg		
Bernigaud	99% 0.01 [0.00-0.10]	death	0/69	150/3,062	84mg		
Alam	91% 0.09 [0.04-0.25]	cases	4/58	44/60	12mg		
Chahla (RCT)	95% 0.05 [0.00-0.80]	m/s case	0/117	10/117	48mg		CT²
Behera	83% 0.17 [0.12-0.23]	cases	45/2,199	133/1,147	42mg		
Seet (CLUS. RCT)	50% 0.50 [0.33-0.76]	symp. case	32/617	64/619	12mg		OT¹
Morgenstern (PSM)	80% 0.20 [0.01-4.15]	hosp.	0/271	2/271	56mg		
Mondal	88% 0.12 [0.01-0.55]	symp. case	128 (n)	1,342 (n)	n/a		
Prophylaxis	**84% 0.16 [0.09-0.31]**		**137/3,779**	**607/6,974**			**84% improvement**
Tau² = 0.58; I² = 87.5%							

recommendation the NIH committee members gave for monoclonal antibody and convalescent plasma treatments. Although the hopes were that both of these latter treatments would be effective when used early, convalescent plasma, "a favorite of nearly all academic medical centers in the country, failed miserably to show efficacy in numerous clinical trials" said Dr. Kory, while monoclonal antibodies did prove effective in preventing hospitalization.

NIH's neutral January 14, 2021[30] "non-recommendation,"[31] issued in the face of strong evidence of ivermectin's safety and efficacy for COVID-19, was the first obvious signal of the agency's determination to suppress IVM. NIH claimed that there was "Insufficient evidence . . . to recommend either for or against the use of ivermectin for the treatment of COVID-19."

NIH shrouded its process for reaching that non-recommendation in secrecy, refusing to disclose the panel members who took part in the ivermectin deliberations, and redacting their names from the documents that various Freedom of Information Act requests compelled the agency to produce. For a time, only Dr. Fauci, Francis Collins, and the panelists themselves knew their identities. NIH took extreme measures to keep the names secret, fighting all the way into federal court to shield the proceedings from transparency.[32,33]

As Collins and Dr. Fauci maneuvered to shade the process from sunlight, the Centers for Disease Control and Prevention (CDC), in response to a separate FOIA request, disclosed the group's nine members.[34] Three members of the working group, *Adaora Adimora, Roger Bedimo,* and *David V. Glidden,* had disclosed financial relationships with Merck.

A fourth member of the NIH Guidelines Committee, Susanna Naggie, received a $155 million grant[35] to conduct further studies of ivermectin following the NIH non-recommendation. NIAID's windfall payoff to Naggie would have been unlikely to go forward if the committee voted to approve IVM.

Today, as Dr. Fauci moves the US to eliminate all use of ivermectin, other countries are using more of it.

In February 2021, the head of the Tokyo Metropolitan Medical Association held a press conference to call for adding ivermectin to its outpatient treatment protocol. Several Indian states had added ivermectin to their list of essential medicines to fight COVID-19.[36] Indonesia's government not only authorized the use of the drug but also created a website showing its real-time availability.[37] After giving out 3rd booster doses of Pfizer's COVID-19 vaccine, but still seeing high rates of COVID-19 hospitalizations and deaths, Israel started using ivermectin officially in September 2021, with the health insurance companies distributing ivermectin to high-risk citizens. El Salvador distributes IVM for free to all of its citizens.[38]

Nations whose residents have easy access to ivermectin invariably see immediate and dramatic declines in COVID deaths. Hospitals in Indonesia started using ivermectin on July 22, 2021. By the first week of August, cases and deaths were plummeting.[39]

A December 2020 study showed that African and Asian countries that widely used ivermectin to treat and prevent various parasitic diseases enjoy some of the world's lowest-reported COVID case and mortality rates.[40] After controlling for confounding factors, including the Human Development Index (HDI), the eleven African nations with membership in the African Programme for Onchocerciasis (aka "river blindness," for which ivermectin is standard of care) APOC show 28 percent lower mortality than non-APOC African countries, and an 8 percent lower rate of COVID-19 infection.

On April 20, 2021, India's medical societies added ivermectin to the national protocol. According to Indian and international news, an aggressive campaign by the government of the Indian state of New Delhi, where COVID was raging, showed stunning success. The *Desert Review* reported that in April 2021, New Delhi was experiencing a COVID epidemic crisis. The state government obliterated 97 percent of Delhi cases by distributing ivermectin.[41] "IVM Crushed COVID in New Delhi," wrote Dr. Justus R. Hope, M.D.[42] Following IVM's introduction, according to *TrialSite News*, cases dropped dramatically. "At the national level, the massive surge that overtook the country at the beginning of April slowed exponentially after the new COVID-19 protocol was introduced, which includes the use of ivermectin and budesonide." [43] India showed that early combination therapy—budesonide, ivermectin, doxycycline, and zinc, costing between two and five dollars—made COVID symptoms disappear within three to five days. By January 2021, a country of more than 1.3 billion people and a vaccine uptake of almost 7.6 percent nationally[44] had witnessed only 150,000 COVID deaths.[45] By comparison, the US, with a population of 331 million, had recorded 357,000 deaths.[46] Many Indian officials and doctors consider ivermectin a miracle drug for controlling the outbreak. A natural experiment involving two Indian states—Uttar Pradesh and Tamil Nadu—with opposite COVID strategies helped cement that impression.

With 241 million people, Uttar Pradesh has the equivalent of two-thirds of the United States population. According to the *Indian Express*:[47] "Uttar Pradesh was the first state in the country to introduce large-scale prophylactic and therapeutic use of ivermectin. In May-June 2020, a team at Agra [Uttar Pradesh's fourth largest city], led by Dr. Anshul Pareek, administered ivermectin to all RRT team members in the district on an experimental basis. None of them developed Covid-19 despite being in daily contact with patients who had tested positive for the virus.[48] Uttar Pradesh State Surveillance Officer Vikssendu Agrawal added that, based on the findings from Agra, the state government sanctioned the use of ivermectin as a prophylactic for all the contacts of COVID patients and began administering doses to infected persons.

By September, the Uttar Pradesh government announced that the state's 33 districts are virtually devoid of active cases, despite having a vaccination rate of only 5.8 percent.[49] The *Hindustan Times* reported, "Overall, the state has a total of 199 active cases, while the positivity rate came down to less than 0.01 per cent. The recovery rate, meanwhile, has improved to 98.7 per cent."[50] When America's vaccination rate was at 54 percent, cases were still rising and governments were still imposing draconian restrictions. As of August 10, 2021, the United States saw 161,990 new cases and 1,049 new deaths.[51] Uttar Pradesh, in contrast, saw only 19 new cases and *one* death—more than 1,000 times lower than the US.[52]

Dr. Agrawal attributes the timely introduction of ivermectin to ending the first COVID wave: "Despite being the state with the largest population base and a high population density, we have maintained a relatively low positivity rate and cases per million of population."[53]

According to *TrialSite News*, despite the Indian government's success in using ivermectin and budesonide, "the media hasn't shown interest in sharing this news. Instead, the comments continue to promote remdesivir as an effective drug, and the few media outlets that do refer to ivermectin call it an 'unproven medicine' or an 'outdated treatment.' It is as if there are two different treatment realities, one on the ground and one in the local health systems. Millions of patients are now receiving ivermectin, yet one would never know by the media topics."[54]

Meanwhile, the Indian state of Tamil Nadu continued using Anthony Fauci's protocol of administering remdesivir, outlawing ivermectin, and discouraging early treatment. According to the *Indian Times*, Tamil Nadu continues to experience cases and fatalities that perfectly match the US catastrophe.[55]

The massive and overwhelming evidence in favor of ivermectin includes scientist Dr. Tess Lawrie's highly regarded, peer-reviewed meta-analysis.

Dr. Lawrie assessed 15 trials, finding a cumulative benefit of IVM in reducing deaths of 62 percent. Although the data quality of the ivermectin for prevention studies was less strong, they showed that ivermectin prophylaxis reduced COVID infections by 86 percent.[56]

Dr. Lawrie, a world-renowned data researcher and scientific consultant, is an iconic eminence among global public health scientists and agencies. *The Desert Review* has deemed her "The Conscience of Medicine"[57] because of her reputation for competence, precision, and integrity. Lawrie's consulting group, the Evidence-Based Medicine Consultancy, Ltd. performs the scientific reviews that develop and support guidelines for global public health agencies, including the WHO and European governments, as well as international scientific and health consortia like the Cochrane Collaboration. Her clients have included a retinue of virtually all the larger government regulators now involved in the suppression of IVM and other repurposed drugs.

At the end of December 2000, Dr. Lawrie happened on a YouTube video of Pierre Kory's Senate testimony on ivermectin. Her interest piqued, Dr. Lawrie conducted a "pragmatic rapid review" between Christmas and New Year's to validate the 27 studies from the medical literature that Kory cited, assessing each of them for quality and power.

"After a week, I realized it was a go. IVM's safety was well-established as a widely

48 ivermectin COVID-19 studies after exclusions

	Improvement	RR [CI]		Treatment	Control	Dose (4d)	
Chowdhury (RCT)	81%	0.19 [0.01-3.96]	hosp.	0/60	2/56	14mg	OT¹·CT²
Espitia-Hernandez	70%	0.30 [0.16-0.55]	recov. time	28 (n)	7 (n)	12mg	CT²
Mahmud (DB RCT)	86%	0.14 [0.01-2.75]	death	0/183	3/183	12mg	CT²
Ahmed (DB RCT)	85%	0.15 [0.01-2.70]	symptoms	0/17	3/19	48mg	
Chaccour (DB RCT)	96%	0.04 [0.00-1.01]	symptoms	12 (n)	12 (n)	28mg	
Afsar	98%	0.02 [0.00-0.20]	symptoms	0/37	7/53	48mg	
Babalola (DB RCT)	64%	0.36 [0.10-1.27]	viral+	40 (n)	20 (n)	24mg	OT¹
Ravikirti (DB RCT)	89%	0.11 [0.01-2.05]	death	0/55	4/57	24mg	
Bukhari (RCT)	82%	0.18 [0.07-0.46]	viral+	4/41	25/45	12mg	
Mohan (DB RCT)	62%	0.38 [0.08-1.75]	no recov.	2/40	6/45	28mg	
Biber (DB RCT)	70%	0.30 [0.03-2.76]	hosp.	1/47	3/42	36mg	
Elalfy	87%	0.13 [0.06-0.27]	viral+	7/62	44/51	36mg	CT²
Chahla (CLUS. RCT)	87%	0.13 [0.03-0.54]	no disch.	2/110	20/144	24mg	
Mourya	89%	0.11 [0.05-0.25]	viral+	5/50	47/50	48mg	
Loue (QR)	70%	0.30 [0.04-2.20]	death	1/10	5/15	14mg	
Merino (QR)	74%	0.26 [0.11-0.57]	hosp.	population-based cohort		24mg	
Faisal (RCT)	68%	0.32 [0.14-0.72]	no recov.	6/50	19/50	48mg	
Aref (RCT)	63%	0.37 [0.22-0.61]	recov. time	57 (n)	57 (n)	n/a	
Krolewiecki (RCT)	-152%	2.52 [0.11-58.1]	ventilation	1/27	0/14	168mg	
Vallejos (DB RCT)	-33%	1.33 [0.30-5.72]	death	4/250	3/251	24mg	
Buonfrate (DB RCT)	-600%	7.00 [0.39-126]	hosp.	4/58	0/29	336mg	
Mayer	55%	0.45 [0.32-0.63]	death	3,266 (n)	17,966 (n)	151mg	
Early treatment	**73%**	**0.27 [0.20-0.37]**		37/4,500	191/19,166		**73% improvement**

Tau² = 0.18; I² = 42.7%

used dewormer," she told me. "I was startled by the magnitude of its benefits. Its efficacy against COVID was consistently clear in multiple studies. I thought that all these people were dying and this was a moral obligation—this drug should have been rolled out." Dr. Lawrie dispatched an urgent letter to British Health Minister Matt Hancock on January 4 with her Rapid Review attached. She never heard back from Hancock. But in a suspicious coincidence, someone leaked a meta-review by WHO researcher Andrew Hill to the *Daily Mail*.[58] Three days later, Hill posted a preprint of his study. In the one month since he testified enthusiastically beside Dr. Kory in favor of ivermectin before the January 13 NIH panel, Hill had made a neck-wrenching *volte face*. Cumulatively, the seven studies in Hill's original meta-review still showed a dramatic reduction in hospitalizations and deaths among patients receiving IVM. The leaked version of Hill's meta-review included all the same papers that formerly supported his gung-ho promotion of IVM as a miraculous cure for COVID. Hill had altered only his conclusions. Now he claimed that those studies comprised a low quality of evidence, and so although they yielded a highly positive result, Hill assigned the result a "low certainty." He could then declare that WHO should not recommend IVM without first performing long-term, randomized placebo-controlled studies that would require many months if not longer. "Someone got to him," suggests Kory. "Someone sent him the memo. Andrew Hill has been captured by some really dark forces."

On January 7, Dr. Lawrie summarized the overwhelming evidence from her Rapid Review in a video directed at British Prime Minister Boris Johnson, urging him to break the logjam and roll out IVM immediately. Her video, says Dr. Kory, was "absolutely convincing." She forwarded the video appeal to the British and South African Prime Ministers on January 7. She heard nothing from either.

On January 13, 2021, Dr. Lawrie used her convening power to assemble an invitation-only symposium of twenty of the world's leading experts, including researchers, physicians, patient advocates, and government consultancy advisers, to review her meta-analysis and make evidence-based recommendations on the use of ivermectin to prevent and treat COVID-19. She called the conference the British ivermectin Recommendation Development (BIRD) study.

"Tess Lawrie did exactly what WHO should have done," says Dr. Kory. "She made a thorough, open, and transparent review of all the scientific evidence."

During the daylong conference, the conferees reviewed each study in Dr. Lawrie's rapid meta-review, agreeing that the evidence supported an immediate rollout. Before adjourning, Dr. Lawrie and the scientific panel committed to conducting a full-scale Cochrane-style meta-review of all the scientific literature. Due to the mortal urgencies, they pledged to reconvene in a much larger group on January 14.

In the meantime, Dr. Lawrie managed to reach Andrew Hill by phone on January 6, two days after the *Daily Mail* leaked his meta-review. She informed him that some of the leading lights of science had agreed to collaborate on the Cochrane-style meta-review, and she proposed that Hill should join the effort as a collaborator. She offered to share her data with Hill and, after the call, she sent him her spreadsheets. Dr. Lawrie had coordinated many Cochrane Reviews for WHO and was indisputably among the world's ranking experts in systematically reviewing study data. Dr. Lawrie invited Hill to co-author the Cochrane Review and to attend the next BIRD meeting on January 13.[59] It was an exciting opportunity. Under normal circumstances, Hill should have pounced on this chance to serve as lead author with some of the world's most prestigious researchers in creating a professional, bulletproof Evidence-to-Decision framework for the WHO. He was nevertheless noncommittal. He did agree to review Dr. Lawrie's spreadsheet.

Dr. Lawrie and her colleagues launched a marathon effort to conduct a brand new review of all published studies in the medical literature from scratch, assessing each for power and bias. She presented her draft to the exclusive BIRD group in mid-January. All agreed that the common-sense approach was to release ivermectin. She submitted the protocol to Cochrane for external scientific review.

British and Scandinavian scientists founded the Cochrane Collaboration in 1993 to address pharmaceutical industry corruption that had become pervasive in clinical trials for new drugs. Today, the Cochrane Collaboration is a coalition of 30,000 independent scientists and 53 large research institutions who volunteer to routinely review industry data using evidence-based science to advise regulatory agencies.[60] Cochrane seeks to restore integrity and standardized scientific methodologies to the crooked realm of drug development trials. Cochrane uses standardized parameters and rigorous methodologies for evaluating evidence. Cochrane reviewers systematically assess the power of each individual study within the meta-review, interpreting data to identify and discount for bias, and to score each study as "high," "moderate," or "low" certainty evidence and to determine whether it's acceptable to pool the data in a single meta-review.

Dr. Lawrie knew that to make its ivermectin determination, WHO would rely on Hill's study and another study from McMaster University known as the "Together Trial." McMaster was hopelessly and irredeemably conflicted. NIH gave McMaster $1,081,541 in 2020 and 2021.[61] A separate group of McMaster University scientists was, at that time, engaged in developing their own COVID vaccine—an effort that would never pay dividends if WHO recommended ivermectin as Standard of Care. The Bill and Melinda Gates Foundation was funding the massive "Together Trial" testing ivermectin, HCQ, and other potential drugs against COVID, in Brazil and other locations. Critics accused Gates and the McMaster researchers of designing that study to make ivermectin fail. Among other factors, the study targeted a population that was already heavily utilizing ivermectin, creating a confounding variable (placebo recipients could obtain over the counter ivermectin) that would clearly hide efficacy. McMaster University researchers would certainly know that a positive recommendation for IVM would cost their university hundreds of millions. The Together Trial organizer was Gates' trial designer, Ed Mills, a scientist with heavy conflicts with Pharma and a reputation as a notorious industry biostitute.

Dr. Lawrie knew that the only way to salvage the WHO Guidelines and produce a high-quality scientific study was to persuade Mills to do a full-scale Cochrane Collaboration meta-analysis. The following week, she spoke to Hill again, this time by Zoom.

The Zoom call was recorded.

During her first conversation with Hill, Dr. Lawrie had concluded that the techniques that Hill employed throughout his meta-review were "deeply flawed," and that Hill lacked the experience to perform a systematic review or a meta-analysis: "I was surprised he had been given the job."

In fact, the transcript of her January 18 conversation suggests that Hill was completely unfamiliar with the requirements of a systematic review, which requires researchers to evaluate and score each study using uniform criteria to assess power and the risk of bias, and to conduct a "sensitivity analysis" to exclude studies with high risk of bias. This kind of review necessarily judges the reliability of the authors of each participating study. The Cochrane reviewers must be prepared to make harsh judgments about the work quality, integrity, and potential prejudices of each listed co-author of all the studies included in their review, based in part on their individual competence, and the financial conflicts of interest potentially affecting each researcher. But Hill, bizarrely, had included the names of all the authors of all of his seven accumulated studies on the list of the co-authors of his meta-review. "That's the equivalent of asking the catcher in a baseball team to also play the umpire," says Dr. Kory. "No one with any familiarity with the game would make that mistake. Hill was supposed to be judging these authors. Instead, he treated them as his collaborators."

Dr. Lawrie gently informed Hill that that was "irregular for a meta-analysis," adding, "When you do a systematic review, you usually don't include the authors of the studies because that inherently biases your conclusions. It's got to be independent."

Dr. Lawrie explained that Hill's paper, in addition to listing as co-authors the researchers whose work he was supposed to be evaluating, makes no pretense of systematically grading evidence according to standardized protocols. Those deficiencies make it utterly useless, she explained, for providing "clinical guidelines to the WHO." Furthermore, Hill's meta-review looked at only one outcome, the deaths of COVID patients, which was only a small subset of the criteria and endpoints in the studies he had analyzed. She told Hill: "You don't just do a meta-analysis . . . when there's all those other outcomes that you didn't even meta-analyze. You just meta-analyzed the death outcome [using only a fraction of the available evidence], and then [said], 'Oh, we need more studies.'"

Dr. Lawrie asked Hill to explain his U-turn on ivermectin, which his own analysis found overwhelmingly effective. "How can you do this?" she inquired politely. "You are causing irreparable harm."

Hill explained that he was in a "tricky situation," because his sponsors had put pressure on him. Hill is a University of Liverpool virologist who serves as an advisor to Bill Gates and the Clinton Foundation. "He told me his sponsor was Unitaid." Unitaid is a quasi-governmental advocacy organization funded by the BMGF and several European countries—France, the United Kingdom, Norway, Brazil, Spain, the Republic of Korea, and Chile—to lobby governments to finance the purchase of medicines from pharmaceutical multinationals for distribution to the African poor. Its primary purpose seems to be protecting the patent and intellectual property rights of pharmaceutical companies—which, as we shall see, is the priority passion for Bill Gates—and to insure their prompt and full payment. About 63 percent of its funding comes from a surtax on airline tickets. The Bill & Melinda Gates Foundation holds a board seat and chairs Unitaid's Executive Committee, and the BMGF has given Unitaid $150 million since 2005.[62] Various Gates-funded surrogate and front organizations, like Global Fund, Gavi, and UNICEF also contribute, as does the pharmaceutical industry. The BMGF and Gates personally own large stakes in many of the pharmaceutical companies that profit from this boondoggle. Gates also uses Unitaid to fund corrupt science by tame and compromised researchers like Hill that legitimizes his policy directives to the WHO. Unitaid gave $40 million to Andrew Hill's employer, the University of Liverpool, four days before the publication of Hill's study.

Hill, a PhD, confessed that the sponsors were pressuring him to influence his conclusion. When Dr. Lawrie asked who was trying to influence him, Hill said, "I mean, I, I think I'm in a very sensitive position here. . . ."

Dr. Tess Lawrie, MD, PhD: "Lots of people are in sensitive positions; they're in hospital, in ICUs dying, and they need this medicine."

Dr. Hill: "Well. . . ."

Dr. Tess Lawrie: "This is what I don't get, you know, because you're not a clinician. You're not seeing people dying every day. And this medicine prevents deaths by 80 percent. So 80 percent of those people who are dying today don't need to die because there's ivermectin."

Dr. Andrew Hill: "There are a lot, as I said, there are a lot of different opinions about this. As I say, some people simply. . . ."

Dr. Tess Lawrie: "We are looking at the data; it doesn't matter what other people say. We are the ones who are tasked with . . . look[ing] at the data and reassur[ing] everybody that this cheap and effective treatment will save lives. It's clear. You don't have to say, well, so-and-so says this, and so-and-so says that. It's absolutely crystal clear. We can save lives today. If we can get the government to buy ivermectin."

Dr. Andrew Hill: "Well, I don't think it's as simple as that, because you've got trials. . . ."

Dr. Tess Lawrie: "It is as simple as that. We don't have to wait for studies . . . we have enough evidence now that shows that ivermectin saves lives, it prevents hospitalization. It saves the clinical staff going to work every day, [and] being exposed. And frankly, I'm shocked at how you are not taking responsibility for that decision. And you still haven't told me who is [influencing you]? Who is giving you that opinion? Because you keep saying you're in a sensitive position. I appreciate you are in a sensitive position, if you're being paid for something and you're being told [to support] a certain narrative . . . that *is* a sensitive position. So, then you kind of have to decide, well, do I take this payment? Because in actual fact, [you] can see [your false] conclusions . . . are going to harm people. So maybe you need to say, I'm not going to be paid for this. I can see the evidence, and I will join the Cochrane team as a volunteer, like everybody on the Cochrane team is a volunteer. Nobody's being paid for this work."

Dr. Andrew Hill: "I think fundamentally, we're reaching the [same] conclusion about the survival benefit. We're both finding a significant effect on survival."

Dr. Tess Lawrie: "No, I'm grading my evidence. I'm saying I'm sure of this evidence. I'm saying I'm absolutely sure it prevents deaths. There is nothing as effective as this treatment. What is your reluctance? Whose conclusion is that?"

Hill then complains again that outsiders are influencing him.

Dr. Tess Lawrie: "You keep referring to other people. It's like you don't trust yourself. If you were to trust yourself, you would know that you have made an error and you need to correct it because you know, in your heart, that this treatment prevents death."

Dr. Andrew Hill: "Well, I know, I know for a fact that the data right now is not going to get the drug approved."

Dr. Tess Lawrie: "But, Andy—know this will come out . . . It will come out that there were all these barriers to the truth being told to the public and to the evidence being presented. So please, this is your opportunity just to acknowledge [the truth] in your review, change your conclusions, and come on board with this Cochrane Review, which will be definitive. It will be the review that shows the evidence and gives the proof. This was the consensus on Wednesday night's meeting with 20 experts."

Hill protests that NIH will not agree to recommend IVM.

Dr. Tess Lawrie: "Yeah, because the NIH is owned by the vaccine lobby."

Dr. Andrew Hill: "That's not something I know about."

Dr. Tess Lawrie: "Well, all I'm saying is this smacks of corruption and you are being played."

Dr. Hill: "I don't think so."

Dr. Tess Lawrie: "Well then, you have no excuse because your work in that review is flawed. It's rushed. It is not properly put together."

Dr. Lawrie points out that Hill's study ignores a host of clinical outcomes that affect patients.

She scolds Hill for ignoring the beneficial effects of IVM as prophylaxis, its effect on speed to PCR negativity, on the need for mechanical ventilation, on reduced admissions to ICUs, and other outcomes that are clinically meaningful.

She adds, "This is bad research . . . bad research. So, at this point, I don't know . . . you seem like a nice guy, but I am really, really worried about you."

Dr. Andrew Hill: "Okay. Yeah. I mean, it's, it's a difficult situation."

Dr. Tess Lawrie: "No, you might be in a difficult situation. I'm not, because I have no paymaster. I can tell the truth . . . How can you deliberately try and mess it up . . . you know?"

Dr. Andrew Hill: "It's not messing it up. It's saying that we need, we need a short time to look at some more studies."

Dr. Tess Lawrie: "So, how long are you going to let people carry on dying unnecessarily—up to you? What is, what is the timeline that you've allowed for this, then?"

Dr. Andrew Hill: "Well, I think . . . I think that it goes to WHO and the NIH and the FDA and the EMEA. And they've got to decide when they think enough's enough."

Dr. Tess Lawrie: "How do they decide? Because there's nobody giving them good evidence synthesis, because yours is certainly not good."

Dr. Andrew Hill: "Well, when yours comes out, which will be in the very near future . . . at the same time, there'll be other trials producing results, which will nail it with a bit of luck. And we'll be there."

Dr. Tess Lawrie: "It's already nailed."

Dr. Andrew Hill: "No, that's, that's not the view of the WHO and the FDA."

Dr. Tess Lawrie: "You'd rather... risk loads of people's lives. Do you know if you and I stood together on this, we could present a united front and we could get this thing. We could make it happen. We could save lives; we could prevent [British National Health Service doctors and nurses] people from getting infected. We could prevent the elderly from dying."

Dr. Tess Lawrie: "These are studies conducted around the world in several different countries. And they're all saying the same thing. Plus there's all sorts of other evidence to show that it works. Randomized controlled trials do not need to be the be-all and end-all. But [even] based on the randomized controlled trials, it is clear that ivermectin works... It prevents deaths and it prevents harms and it improves outcomes for people . . . I can see we're getting nowhere because you have an agenda, whether you like it or not, whether you admit to it or not, you have an agenda. And the agenda is to kick this down the road as far as you can. So . . . we are trying to save lives. That's what we do. I'm a doctor and I'm going to save as many lives as I can. And I'm going to do that through getting the message [out] on ivermectin. . . . Okay. Unfortunately, your work is going to impair that, and you seem to be able to bear the burden of many, many deaths, which I cannot do."

Then she asks again.

Dr. Tess Lawrie: "Would you tell me? I would like to know who pays you as a consultant through WHO."

Dr. Andrew Hill: "It's Unitaid."

Dr. Tess Lawrie: "All right. So who helped to . . . ? Whose conclusions are those on the review that you've done? Who is not listed as an author? Who's actually contributed?"

Dr. Andrew Hill: "Well, I mean, I don't really want to get into, I mean, it . . . Unitaid . . . ."

Dr. Tess Lawrie: "I think that . . . It needs to be clear. I would like to know who, who are these other voices that are in your paper that are not acknowledged. Does Unitaid have a say? Do they influence what you write?"

Dr. Andrew Hill: "Unitaid has a say in the conclusions of the paper. Yeah."

Dr. Tess Lawrie: "Okay. So, who is it in Unitaid, then? Who is giving you opinions on your evidence?"

Dr. Andrew Hill: "Well, it's just the people there. I don't . . . ."

Dr. Tess Lawrie: "So they have a say in your conclusions."

Dr. Andrew Hill: "Yeah."

Dr. Tess Lawrie: "Could you please give me a name of someone in Unitaid I could speak to, so that I can share my evidence and hope to try and persuade them to understand it?"

Dr. Andrew Hill: "Oh, I'll have a think about who to, to offer you with a name.... But I mean, this is very difficult because I'm, you know, I've, I've got this role where I'm supposed to produce this paper and we're in a very difficult, delicate balance...."

Dr. Lawrie interjects: "Who are these people? Who are these people saying this?"

Dr. Andrew Hill: "Yeah . . . it's a very strong lobby . . ."

Dr. Tess Lawrie: "Okay. Look I think I can see [we're] kind of [at] a dead end, because you seem to have a whole lot of excuses, but, um, you know, that to, to justify bad research practice. So I'm really, really sorry about this, Andy. I really, really wish, and you've explained quite clearly to me, in both what you've been saying and in your body language that you're not entirely comfortable with your conclusions, and that you're in a tricky position because of whatever influence people are having on you, and including the people who have paid you and who have basically written that conclusion for you."

Dr. Andrew Hill: "You've just got to understand I'm in a difficult position. I'm trying to steer a middle ground and it's extremely hard."

Dr. Tess Lawrie: "Yeah. Middle ground. The middle ground is not a middle ground… [Y]ou've taken a position right to the other extreme calling for further trials that are going to kill people. So this will come out, and you will be culpable. And I can't understand why you don't see that, because the evidence is there and you are not just denying it, but your work's actually actively obfuscating the truth. And this will come out. So I'm really sorry . . . As I say, you seem like a nice guy, but I think you've just kind of been misled somehow."

Hill promised he would do everything in his power to get ivermectin approved if she would give him six weeks.

Dr. Andrew Hill: "Well, what I hope is that this, this stalemate that we're in doesn't last very long. It lasts a matter of weeks. And I guarantee I will push for this to last for as short amount of time as possible."

Dr. Tess Lawrie: "So, how long do you think the stalemate will go on for? How long do you think you will be paid to [make] the stalemate… go on?"

Dr. Andrew Hill: "From my side. Okay . . . I think end of February, we will be there six weeks."

Dr. Tess Lawrie: "How many people die every day?"

Dr. Andrew Hill: "Oh, sure. I mean, you know, 15,000 people a day."

Dr. Tess Lawrie: "Fifteen thousand people a day times six weeks . . . Because at this rate, all other countries are getting ivermectin except the UK and the USA, because the UK and the USA and Europe are owned by the vaccine lobby."

Dr. Andrew Hill: "My goal is to get the drug approved and to do everything I can to get it approved so that it reaches the maximum. . . ."

Dr. Tess Lawrie: "You're not doing everything you can, because everything you can would involve saying to those people who are paying you, 'I can see this prevents deaths. So I'm not going to support this conclusion anymore, and I'm going to tell the truth.'"

Dr. Andrew Hill: "What, I've got to do my responsibilities to get as much support as I can to get this drug approved as quickly as possible."

Dr. Tess Lawrie: "Well, you're not going to get it approved the way you've written that conclusion. You've actually shot yourself in the foot, and you've shot us all in the foot. All of . . . everybody trying to do something good. You have actually completely destroyed it."

Dr. Andrew Hill: "Okay. Well, that's where we'll, I guess we'll have to agree to differ."

Dr. Tess Lawrie: "Yeah. Well, I don't know how you sleep at night, honestly."

At the conclusion of the January 14 BIRD conference, Dr. Lawrie delivered a monumental closing address that should be recorded among the most important speeches in the annals of medical history. Dr. Lawrie spoke out at considerable personal risk, since her livelihood and career largely rely on the very agencies she targeted for criticism.

Dr. Lawrie began by endorsing the miraculous efficacy of IVM.

> Had ivermectin been employed in 2020 when medical colleagues around the world first alerted the authorities to its efficacy, millions of lives could have been saved, and the pandemic with all its associated suffering and loss brought to a rapid and timely end.

Dr. Lawrie told the audience that the suppression of ivermectin was a signal that Pharma's pervasive corruption had turned a medical cartel against patients and against humanity.

> The story of ivermectin has highlighted that we are at a remarkable juncture in medical history. The tools that we use to heal and our connection with our patients are being systematically undermined by relentless disinformation stemming from corporate greed. The story of ivermectin shows that we as a public have misplaced our trust in the authorities and have underestimated the extent to which money and power corrupts.

Dr. Lawrie called for reform of the method used to analyze scientific evidence.

> They who design the trials and control the data also control the outcome. So, this system of industry-led trials needs to be put to an end. Data from ongoing and future trials of novel COVID treatments must be independently controlled and analyzed. Anything less than total transparency cannot be trusted.

Dr. Lawrie called out the corruption of modern medicine by Big Pharma and other interests and attributed the barbaric suppression of IVM to the single-minded obsession with more profitable vaccines.

> Since then, hundreds of millions of people have been involved in the largest medical experiment in human history. Mass vaccination was an unproven novel therapy. Hundreds of billions will be made by Big Pharma and paid for by the public. With politicians and other nonmedical individuals dictating to us what we are allowed to prescribe to the ill, we as doctors have been put in a position such that our ability to uphold the Hippocratic oath is under attack.

She hinted at Gates' role in the suppression.

> At this fateful juncture, we must therefore choose: will we continue to be held ransom by corrupt organizations, health authorities, Big Pharma, and billionaire sociopaths, or will we do our moral and professional duty to do no harm and always do the best for those in our care? The latter includes urgently reaching out to colleagues around the world to discuss which of our tried and tested safe older medicines can be used against COVID.
>
> Never before has our role as doctors been so important, because never before have we become complicit in causing so much harm.

Finally, Dr. Lawrie suggested that physicians form a new World Health Organization that represents the interests of the people, not corporations and billionaires, a people-centered organization.

* * *

On October 1, 2021, Hill resurfaced on Twitter touting his upcoming lecture, ironically titled, "Effects of Bias and Potential Medical Fraud in the Promotion of Ivermectin." Says Pierre Kory in disgust, "Andrew is apparently making a living now accusing the doctors and scientists who support ivermectin of medical fraud." Dr.

Kory adds, "Hill and his backers are some of the worst people in human history. They are responsible for the deaths of millions."

* * *

Andrew Hill's emergence is only one front in the war by NIH and the medical/media cartel to block doctors from using IVM. FDA issued its first warning about IVM on April 10, 2020, in reaction to ivermectin studies by Australia's Monash University and American physician Dr. Jean-Jacques Rajter, claiming on its website "Additional testing is needed to determine whether ivermectin might be safe or effective to prevent or treat coronavirus or COVID-19."

When Dr. Kory's explosive December 8, 2020 Senate testimony[63] describing the peer-reviewed science supporting ivermectin went viral, prescriptions for ivermectin from US doctors exploded. Americans were getting legitimate prescriptions filled at pharmacies, up to 88,000 scripts in a single week.

The truth of the drug's benefits was going viral, and the last thing Dr. Fauci et al. could tolerate was an effective treatment for COVID. Something needed to be done.

The government moved aggressively to block its use. On December 24, in what seemed like a trial balloon, the South African government quietly banned the importation of ivermectin. YouTube soon scrubbed Kory's video[64] and Facebook blocked him. Then in March 2021 the US FDA, the European Medicines Association (EMA), and the WHO issued statements advising against the use of ivermectin for COVID-19. The EMA said it should not be used at all. The WHO, echoing its strategy for tanking hydroxychloroquine, said ivermectin's use should be limited to clinical trials (the high costs of running a clinical trial and their reliance on NIH, NIAID, Gates, or pharma funding means that their results may be easily controlled). FDA issued a much firmer directive: "You should not use ivermectin to treat or prevent COVID-19."[65]

Here are the FDA guidelines:

> The FDA has not authorized or approved ivermectin for use in preventing or treating COVID-19 in humans or animals. Ivermectin is approved for human use to treat infections caused by some parasitic worms and head lice and skin conditions like rosacea.
>
> Currently available data do not show ivermectin is effective against COVID-19. Clinical trials[66] assessing ivermectin tablets for the prevention or treatment of COVID-19 in people are ongoing.
>
> Taking large doses of ivermectin is dangerous.
>
> If your health care provider writes you an ivermectin prescription, fill it through a legitimate source such as a pharmacy, and take it *exactly* as prescribed.

On July 28, 2021, a front-page *Wall Street Journal* headline asked, "Why is the FDA Attacking a Safe, Effective Drug?"[67]

On August 16, 2021, two weeks after the *Wall Street Journal* article, CDC ordered doctors to stop prescribing IVM. On August 17, 2021, the NIAID recommended against ivermectin's use to combat the novel coronavirus. On August 26, 2021, CDC sent out an emergency warning using its Health Alert Network.[68]

In early September 2021, following the FDA/CDC/NIAID's lead, the American Medical Association (AMA), the American Pharmacists Association (APhA), and the American Society of Health-System Pharmacists (ASHP) called on doctors to immediately stop prescribing ivermectin for COVID outside of clinical trials.[69] These influential organizations are largely dependent on pharmaceutical industry largesse.

On September 2, 2021 on *MSNBC Tonight*, Chris Hayes interviewed the president of the AMA, Dr. Gerald Harmon, who said that the AMA now advises doctors against prescribing ivermectin except in clinical trials. He explained that the AMA is taking this unprecedented step because ivermectin isn't "approved" by the FDA for treatment of COVID-19. He failed to mention that up to 30 percent of prescriptions written by America's doctors are for off-label uses not approved by the FDA. The AMA, meanwhile, ignored the cascading toll of injuries and deaths from Big Pharma's injections, while endorsing the revolutionary notion that FDA should be the arbiter of what doctors can and cannot use to heal their patients. Physicians traditionally have had unlimited authority to prescribe FDA-approved medications for any purpose as long as they explain the risks and benefits to their patients. Suddenly, the AMA and its industry patrons and captive regulators moved to limit the doctor's authority to treat patients. FDA has no authority to regulate the practice of medicine. As Stephen Hahn, FDA's last Commissioner (no one has been appointed to the role since he left) pointed out in October 2020, off-label prescribing is between a doctor and his/her patient.

The sad episode, still ongoing, raises questions one expects doctors to be asking:

- Is ivermectin a safe drug?
- Will it do harm?
- Are we in a situation in which authorities have not provided a proven therapeutic for COVID-19?
- Do treating physicians have the freedom to try medicines they have reason to believe might be helpful, particularly when there is no reason to believe the medicine will be hurtful?

Doctors who answered those questions for themselves and prescribed ivermectin after early September faced growing scrutiny and heavy-handed tactics including censorship, threats to their license and board certification, and other repressive policies from governments and medical boards. Pharmacists, including the large chains like CVS and Walmart, refused to fill prescriptions. "For the first time in history, pharmacies were telling doctors what they can and cannot prescribe," says Dr. McCullough. The directives shattered the traditional sacred relationship between doctors and patients that the profession had nurtured and protected since Hippocrates. The medical profession has long told doctors that their single obligation is to their patients. The AMA's declaration helped march doctors into their new role as agents of state policy. The state policy is to prescribe treatments, not based upon the health interests of the individual patient but based upon the perceived best interests of the state.

"The suppression of HCQ and IVM is one of the greatest tragedies and crimes of

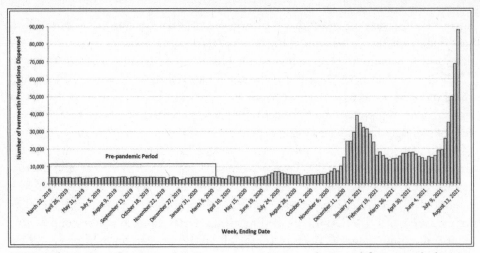

Estimated number of outpatient ivermectin prescriptions dispensed from retail pharmacies—United States, March 16, 2019–August 13, 2021*

the modern era," Dr. Peter Breggin told me. Dr. Breggin, who has been called "The Conscience of Psychiatry," by author Candace Pert, is the author of *Talking Back to Prozac* and *COVID-19 and the Global Predators*.[70]

In Florida and South Carolina, Blue Cross Blue Shield (BCBS) sent mass mailings to physicians notifying them that BCBS would no longer pay insurance claims for IVM, and threatened audits of any physician who wrote prescriptions for the drug.

In January 2021, Syracuse, New York attorney Ralph Lorigo filed for an injunction on behalf of a critically ill hospital patient—the mother of one of his clients—against a local hospital that was resisting family requests to treat her with ivermectin. A State Superior Court judge immediately granted Lorigo's request. Within 12 hours of taking ivermectin, the dying woman miraculously began to recover. Two weeks later, Lorigo obtained a second injunction for a similarly situated client, who also made a preternatural recovery. When local news organizations reported Lorigo's IVM victories, his law office telephone began ringing off the hook. Within a few weeks, he was working twenty-hour days struggling to keep up with a new cottage industry filing injunctions in New York and Ohio courts to help dying patients get access to ivermectin. To date, Lorigo has been in thirty courts. "The people who I've been able to get the ivermectin to on time have all lived; the others have died." He has obtained dozens of injunctions for patients, precipitating a host of sudden recoveries. "The hospitals are so arrogant. They are letting the people die. They get $37,000 to put them on the vent, and they just let them die."

Merck's Steps to Kill Its Baby

During the early industry offensive against HCQ, one of the drug's principal manufacturers, Sanofi, suddenly detected "safety concerns" with HCQ that it had never noticed during decades of profitable pre-pandemic production. In a remarkable coincidence, on February 4, 2021, Merck similarly discovered "a concerning lack of safety data in the majority of studies" regarding IVM.[71] Merck was ivermectin's original manufacturer and had formerly boasted of ivermectin as its "wonder drug."

During the 40 years that it marketed the drug worldwide, Merck had never spoken of these worrisome safety signals. Since 1987, Merck has given billions of doses to the developing world for scabies, river blindness,[72] lymphatic filariasis, elephantiasis, and assorted parasites without any safety alarms. In 2016, Merck distributed 900 million doses in Africa alone. "The drug is safe and has minor side effects," a Merck spokesman said at the time.

> Unlike previous treatments, which had serious—sometimes fatal—side effects, ivermectin is safe and can be used on a wide scale. It is also a very effective treatment, and has single-handedly transformed the lives of millions of people. . . .

What prompted Merck's sudden safety concerns?

Merck's exclusive ivermectin patent rights expired in 1996,[73] and dozens of generic drug companies now produce IVM, for about 40¢/dose, badly diminishing ivermectin's profit profile for Merck. Furthermore, only ten days before Merck discovered its concerns about IVM, Merck signed a manufacturing partnership for the Novavax and Emergent BioSolutions COVID vaccine as it moved into final trials.

Furthermore, in December 2020, Merck had announced a $356 million supply deal by which NIAID agreed to purchase 60,000 to 100,000 doses of an experimental COVID pill called MK 7110. Merck paid $425 million to buy the Oncoimmune company which developed the drug as part of the deal. Bill Gates's quasi-governmental organization, the International AIDS Vaccine Initiative (IAVI), agreed to distribute the product in developing nations.

But most importantly, ivermectin is also a low-profit competitor for another new Merck product for COVID-19—a high-cost antiviral drug, molnupiravir, for which Merck had the highest financial ambitions. Ironically, molnupiravir, a copycat formula, utilized an identical mechanism of action as ivermectin.[74] That drug will retail at around $700 per course[75] but only if Merck can kill its cheap rival.

It's worth a moment to consider molnupiravir's pedigree, because the drug emerged from a shadowy black market of spies, pharmaceutical mountebanks, biosecurity profiteers, and Pentagon contractors who played key roles in militarizing and monetizing the COVID pandemic, and whom you will meet later in this book. The CIA officer and bioweapons developer, Michael Callahan, one of molnupiravir's key patrons, has dubbed this group of shady bioweapons operators as his "Secret Handshake Club." Molnupiravir is a protease inhibitor that mimics the antiviral properties of ivermectin. Unlike ivermectin, molnupiravir showed safety signals so alarming that some of its codevelopers at Emory University protested its introduction into human Phase I trials. Among other problems, they cite the possibility that it will cause birth defects.

Callahan's boss, bioweapons enthusiast and former DHHS Assistant Secretary for Preparedness and Response Robert Kadlec, MD—an unabashed "gain-of-function" promoter with military and intelligence agency pedigrees, who built his career profiting from hyped pandemics—almost single-handedly created the $7 billion National Strategic Stockpile and runs it as a private fiefdom to enrich his friends and connections. Kadlec also runs the super-secretive P3CO Committee inside of NIH, which greenlights—and never denies—Tony Fauci's gain-of-function bioweapons research projects. Gain-of-function refers to experiments that intentionally modify a pathogen to create the ability to cause or worsen disease, enhance transmissibility, and/or create novel strains with potential to cause global spread in humans.[76] One of Kadlec's many dodgy business partners is John Clerici,[77] a Washington lawyer, lobbyist, and artful rogue who almost single-handedly created The Biomedical Advanced Research and Development Authority (BARDA), a new agency formed after 9/11 under the HHS Office of the Assistant Secretary for Preparedness and Response (ASPR) of which Kadlec was director during the Trump administration. BARDA is a taxpayer-infused investment fund that purchases and develops technology for Kadlec's Strategic Stockpile and postulated future threats. Clerici boasts that, "If someone wants to get BARDA money, they've gotta go through me." His LinkedIn profile crows that "John has assisted over three dozen companies in obtaining nearly $3 billion in funding and research for development and procurement of public health countermeasures for the federal government, including the majority of the awards made under Project BioShield, the US government's initiative for preparing the nation against biological attacks."

Clerici brandishes, also, his innovative authorship of the PREP Act, a corporate welfare boondoggle that bestows protection against liability upon manufacturers and providers of vaccines and all other pandemic countermeasures to shield them from lawsuits. Under the PREP Act, no matter how negligently or reprehensibly the company behaves and no matter how grievous the injuries to their victims, the companies cannot be held liable—unless the injured party can prove willful misconduct. Even then, a lawsuit can commence only with the approval of the Secretary of HHS.

The Defense Threat Reduction Agency (DTRA),[78] another Pentagon bioweapons agency and corporate welfare program for military contractors, provided $10 million in 2013 and 2015 to Emory University to develop molnupiravir as a veterinary drug for horses (against equine encephalitis). NIAID contributed $19 million[79] then transferred the toxic drug in a golden handoff to Merck and another drug company, Ridgeback Biotherapeutics, along with a guaranteed market and rich returns. As we shall see in later chapters, DTRA was a major funder of EcoHealth Alliance, Peter Daszak's "charity" that sought out lethal animal viruses around the world, retrieving the most deadly for the Pentagon.

In June 2021, as FDA and NIAID were cranking up the medical cartel's opposition against IVM, the HHS agreed to purchase 1.7 million 5-day treatment courses of molnupiravir from Merck for 1.2 billion dollars[80]—when the drug wins FDA approval, a contingency that can be virtually guaranteed while Anthony Fauci is Washington's drug kingmaker.

On June 9, 2021 President Biden dutifully reiterated the US government's commitment to procure approximately 1.7 million courses of the NIAID-funded drug from Merck.[81] BARDA collaborated with a confederacy of other shady Defense Department operatives, including the DoD Joint Program Executive Office for Chemical, Biological, Radiological and Nuclear Defense (JPEO-CBRND) and the

Army Contracting Command, on the $1.2 billion purchase. Not only was the drug developed with taxpayer money, but its $712 per dose price to the taxpayer is forty times more than Merck's $17.64 cost of production. Merck, which expects to make $7 billion per year on the new blockbuster, saw its stock price spike on news of the government contract and after President Biden's televised plug.

With so many powerful and important godfathers and the United States president fully committed, it would be unprecedented for FDA to deny authorization to molnupiravir, no matter how disastrous the clinical trial results may be. Merck is so certain of FDA's approval that by September 2020, it was already scaling up manufacturing, even though its clinical trials are still underway.

Merck announced in October 2021 that molnupiravir had shown "game-changing" results against COVID in clinical trials, reducing hospitalizations and deaths by 50 percent against a placebo. "The news of the efficacy of this particular antiviral is obviously very good news," trumpeted the White House's Chief Medical Advisor and Pharma spokesperson, Anthony Fauci. "The FDA will look at the data and in the usual very efficient, very effective way, will evaluate the data as quickly as they possibly can, and then it will be taken from there."

Horse Drugs

As Merck stood poised to release its new horse drug molnupiravir onto the market, the other US behemoth, Pfizer, was racing Merck neck and neck with its own antiviral pill, PF-07321332,[82] an ivermectin knockoff that is so similar to IVM (except, of course, in price point) that critics call it "Pfizermectin."[83] Like IVM, it is also a protease-inhibiting anti-parasitic. With these two new drugs teed up for a simultaneous FDA approval, the entire medical/media cartel launched a final coordinated coup de grâce against IVM—branding it a dangerous horse drug. Mainstream media outlets across the US and overseas obediently ran stories promoting the horse medicine propaganda scam.

In late August 2021, NIH, FDA, and CDC launched an innovative new campaign to slander ivermectin as a "horse dewormer" that only deluded foolhardy nincompoops would consume. Picking up on those themes, *The Independent* asked, "Ivermectin: Why Are US Anti-Vaxxers Touting a Horse Dewormer as a Cure for COVID?"

Business Insider warned that people were "poisoning themselves trying to treat or prevent COVID-19 with a horse de-worming drug."

Associated Press assures readers that, "No Evidence Ivermectin is a Miracle Drug Against COVID-19."

On August 15, the FDA instructed, on its website: "You are not a horse." In an August 21, 2021 Twitter post,[84] the FDA expanded the theme: "You are not a horse. You are not a cow. Seriously, y'all. Stop it." The White House and CNN also urged listeners that they should avoid veterinary products. CDC joined the chorus, warning Americans to not risk their health consuming a "horse de-wormer." Elsewhere on its website, the CDC urged black and brown human immigrants to load up on ivermectin. "All Middle Eastern, Asian, North African, Latin American, and Caribbean refugees should receive presumptive therapy with: ivermectin, two doses 200 mcg/Kg orally once a day for two days before departure to the United States.[85] Whether this was intended to deworm them or to prevent COVID transmission during travel to the US is unclear.

Only *Green Med Info*, a health news and information site, saw through the chicanery: "A Media Smear Campaign Timed to Clear Market for Pfizer's Ivermectin Clone Drug, Which Will Be Hailed as a 'Miracle.'"

Demonizing IVM as a "horse drug" was, of course, ironic, given that NIAID initially developed Merck's replacement therapy, molnupiravir, as a horse drug. Furthermore, calling ivermectin a horse drug is like calling antibiotics a horse drug. Many long-established basic drugs are, of course, effective in all mammals because they work on our shared biology. But facts be damned, media companies called all hands on deck to push these stories. Ivermectin's devastating effectiveness against infections from parasites and solid 40-year history of proven safety have made it, also, the world's most prescribed veterinary medicine—but the Nobel Prize was for those billions of times it helped humans, and the government's silly safety warnings were, of course, specious.

Compare ivermectin's safety record to Dr. Fauci's two chosen COVID remedies, remdesivir (which hospital nurses have dubbed "Run-death-is-near"), and the COVID vaccines. Over 30 years, ivermectin has been associated with only 379 reported deaths, an impressive death/dose reporting ratio of 1/10,584,408. In contrast, over the 18 months since remdesivir received an EUA, about 1.5 million patients have received remdesivir, with 1,499 deaths reported (a dire 1/1,000 D/D ratio). Meanwhile, among recipients of COVID jabs in the US during the ten months following their rollout, some 17,000 deaths have occurred following vaccination, a reported D/D ratio of 1/13,250. Ivermectin, therefore, is thousands of times safer than remdesivir and COVID vaccines. The science also indicates that it is far more effective than either.

Dr. Fauci himself took early charge of spreading the rumor that ivermectin was poisoning deluded Americans. "Don't do it," he told pharma propagandist Jake Tapper of CNN in an August 29, 2021 interview. "There's no evidence whatsoever that that works, and it could potentially have toxicity . . . with people who have gone to poison control centers because they've taken the drug at a ridiculous dose and wind up getting sick. There's no clinical evidence that indicates that this works."

Jake Tapper, who has sounded progressively more like a pharma rep than a journalist as the lockdown dragged on, slavishly parroted Dr. Fauci's new talking point:[86] "Poison control centers are reporting that their calls are spiking in places like Mississippi and Oklahoma, because some Americans are trying to use an anti-parasite horse drug called ivermectin to treat coronavirus, to prevent contracting coronavirus." It mattered not that both Mississippi and Oklahoma officials quickly denied that anyone in their state had been hospitalized for IVM poisoning.

An AP story claimed that 70 percent of calls to the Mississippi poison control center were for ivermectin overdoses; it turned out perhaps 2 percent of calls were. Barely anyone saw the grudging retractions.[87]

Additional news articles reported alleged rises in ivermectin-related overdoses in other states. These, too, were exaggerated. Kentucky poison control acknowledged a slight uptick in calls about veterinary ivermectin overdose—about six per year compared to an average of one per year. Despite claims of mass poisoning, the media could not find a single case of IVM leading to death or hospitalization. People were not dying from horse ivermectin overdoses. They were certainly not dying from appropriately dosed and prescribed oral ivermectin. But many were dying from untreated COVID-19.

Bill Gates's surrogate group GAVI asked in a press release: "How did a drug many used to treat parasites in cows come to be of interest to doctors treating humans with COVID-19?" The characterization was especially insincere. Gates' foundation and GAVI were, at that moment, distributing millions of doses of ivermectin annually to Indian children for filariasis, and to Africans for river blindness and filariasis.

It wasn't just the safe drug and caring physicians that were under attack. When, in September 2021, the popular comedian and podcast host Joe Rogan announced he'd kicked COVID in just a few days using a cocktail of drugs, including ivermectin, the global press, government, and pharmaceutical interests coalesced to denounce, vilify, and gaslight him. NPR, which has taken $3 million from the Bill & Melinda Gates Foundation, jumped on the dogpile and deceptively insinuated that Rogan took horse-level doses:[88]

> Joe Rogan has told his Instagram followers he has been taking ivermectin, a deworming veterinary drug formulated for use in cows and horses, to help fight the coronavirus. The Food and Drug Administration has warned against taking the medication, saying animal doses of the drug can cause nausea, vomiting and in some cases severe hepatitis.

But Rogan never took veterinary ivermectin paste. Rogan said he had talked with "multiple doctors" who advised him to take the drug. He followed their advice and he got well, remarkably quickly.

Rolling Stone, the onetime banner of the counterculture, had by 2021 devolved into a reliable mouthpiece for medical cartel orthodoxies.[89] In October 2021, *Rolling Stone* announced that it had removed from its website a 2005 article linking mercury in vaccines to brain injuries in children. *Rolling Stone* also reported that Oklahomans overdosing on ivermectin horse dewormer were causing emergency rooms to be "so backed up that gunshot victims were having hard times getting" access to health facilities. An accompanying photo purported to depict a long line of ambulatory gunshot casualties seeking hospital admission to an Oklahoma emergency room already filled to capacity with dingbats poisoned by horse wormer.[90]

The *Rolling Stone* story[91] spread like wildfire among the world's reigning media outlets, including the *Daily Mail*,[92] *Business Insider*, *Newsweek*,[93] *Yahoo News*, *The Guardian*,[94] and *The Independent*,[95] many of which rely on Gates Foundation largesse. MSNBC's news host, Rachel Maddow, told her audience that "Patients are overdosing on ivermectin backing up rural Oklahoma hospitals, ambulances."[96]

"Ivermectin is meant for a full-size horse," she explained. She repeated that the victims first gullibly swallowed the false claims of antivaxxers before guzzling down horse dewormer. "The ERs are so backed up that gunshot victims were having hard times getting to facilities where they can get definitive care and be treated."

The story, of course, was fraud. Days later, Oklahoma's Sequoyah Northeastern

Health System posted a categorical denial on its website, dismissing the entire story as mere fabrication. That *Rolling Stone* picture of the long lines was an Associated Press stock photo from the previous January, a photo of people waiting in line to get vaccines. As it turns out, not a single patient has been treated in Oklahoma for ivermectin overdose.

Instead of retracting the article, *Rolling Stone* simply posted an attention-dodging "update" at the top of the article reporting the hospital's denial.[97] *The Guardian* similarly published a nondescript update at the bottom of its article.[98]

The FDA doubled down with the claim that IVM may cause "serious harm," is "highly toxic" and may cause "seizures," "coma and even death."[99]

As we shall see, these kinds of warnings are far more applicable to COVID shots. The CDC issued an emergency memo on August 26, 2021 warning doctors and pharmacists not to prescribe ivermectin.[100]

As molnupiravir's debut approached, the war against IVM escalated. On September 23, the Colorado Department of Law issued a cease-and-desist order and fined a Loveland medical clinic $40,000 for "marketing and overstating the effectiveness of ivermectin." And pharmacists still willing to dispense ivermectin faced a new problem. The wholesalers began dribbling out a few pills at a time, but not enough for even one prescription per week. All their diabolical tricks seemed the work of winks and nods and a powerful hidden hand, with no corporation or federal agency taking clear responsibility for carrying out a deliberate policy to suppress a life-saving drug.

On September 28, the *New York Times* introduced a new tactic: reporting that the demand for ivermectin among the crackpots trying to treat COVID had created a shortage for veterinary purposes, warning that—any day now—animals might begin to suffer.[101]

Peter McCullough laughs at the propaganda: "Ivermectin is a molecule that is miraculously effective against parasites and viral infections along multiple pathways and mechanisms of action. It's a molecule. It doesn't care if it's used in a horse, or a cow, or a human. The rules of physics and chemistry are the same across species."

Pierre Kory concurs. "Ivermectin has multiple properties. It operates against COVID along a multitude of separate pathways. In addition to being antiparasitic, it also has potent antiviral properties and even "protects against SARS-CoV-2 spike protein damage."

The osteopath Dr. Joseph Mercola observed,[102] "This idea that ivermectin is a horse dewormer that poses a lethal risk to humans is pure horse manure, shoveled at us in an effort to dissuade people from using a safe and effective drug against COVID-19. . . . The intent is clear. What our so-called health agencies and the media are trying to do is confuse people into thinking of ivermectin as a 'veterinary drug,' which simply isn't true. Ultimately, what they're trying to do is back up the Big Pharma narrative that the only thing at your disposal is the COVID shot."

IV: REMDESIVIR

Anthony Fauci needed to use all his moxie and all his esoteric bureaucratic maneuvers—mastered during his half-century at NIH—to win FDA's approval for his vanity drug, remdesivir. Remdesivir has no clinical efficacy against COVID, according to every legitimate study. Worse, it is deadly poisonous, and expensive poison at $3,000 for treatment.[1] In fact, remdesivir's wholesale cost is roughly 1,000x more costly than hydroxychloroquine and ivermectin. The challenge required Dr. Fauci to first sabotage HCQ and IVM. Under federal rules discussed earlier, FDA's recognition of HCQ and IVM efficacy would automatically kill remdesivir's ambitions for EUA designation. And even if Dr. Fauci somehow finagled an FDA license for remdesivir, demand for the product, which doctors were administering late in the disease, as it had to be given through an IV in the hospital, would plummet if either HCQ or IVM stopped the COVID-19 infections early.

Why would Dr. Fauci care to undermine any medicine that might compete with remdesivir? Might it have something to with NIAID and CDC having just spent $79 million[2] developing remdesivir for Gilead, a company in which the Bill & Melinda Gates Foundation owns a $6.5 million stake?[3,4] The BMGF is engaged in other large drug development deals with the company, including a cofunded $55 million investment in a malaria treatment being developed by Lyndra Therapeutics. Gates has also funded the promotion of Gilead's Truvada in Kenya.[5,6] Another Gilead partner, the US Army Medical Research Institute of Infectious Diseases at Ft. Detrick, Maryland (USAMRIID), where the drug was studied in monkeys, also contributed millions to remdesivir's development.[7] At the outset of the coronavirus plague, remdesivir was just another pharma-owned molecule that FDA had never approved as safe and efficacious for any purpose. In 2016, remdesivir demonstrated middling antiviral properties against Zika, but the disease disappeared before the expensive non-remedy got traction.[8] After the Zika threat vanished, NIAID put some $6.9 million into identifying a new pandemic against which to deploy remdesivir. In 2018, Gilead entered remdesivir in a NIAID-funded clinical trial against Ebola in Africa.[9]

This is how we know that Anthony Fauci was well aware of remdesivir's toxicity when he orchestrated its approval for COVID patients. NIAID sponsored that project. Dr. Fauci had another NIAID-incubated drug, ZMapp, in the same clinical trial, testing efficacy against Ebola alongside two experimental monoclonal antibody drugs. Researchers planned to administer all four drugs to Ebola patients across Africa over a period of four to eight months.[10,11]

However, six months into the Ebola study, the trial's Safety Review Board suddenly pulled both remdesivir and ZMapp from the trial.[12] Remdesivir, it turned out, was hideously dangerous. Within 28 days, subjects taking remdesivir had lethal side effects including multiple organ failure, acute kidney failure, septic shock, and hypotension, and 54 percent of the remdesivir group died—the highest mortality rate among the four experimental drugs.[13] Anthony Fauci's drug, ZMapp, ran up the second-highest body count at 44 percent. NIAID was the primary funder of this study, and its researchers published the bad news about remdesivir in the *New England Journal of Medicine* in December 2019.[14] By then, COVID-19 was already circulating in Wuhan. But two months later, on February 25, 2020, Dr. Fauci announced, with great fanfare, that he was enrolling hospitalized COVID patients in a clinical trial to study remdesivir's efficacy.[15] For important context, this was a month before the WHO declared the new pandemic, a time that there were only fourteen confirmed COVID cases in the United States, most from the Diamond Princess cruise ship. These individuals were among the first wave of COVID-19 hospitalizations from whom NIAID recruited the 400 US volunteers for Dr. Fauci's remdesivir trial.[16] Dr. Fauci's press release said only that remdesivir "has shown promise in animal models for treating Middle East Respiratory Syndrome (MERS)."[17] It's unclear, then, if NIAID informed these frightened souls that, less than a year earlier, a safety review board had deemed remdesivir unacceptably toxic.

Its deadly effect on patients aside, remdesivir was a perfect strategic option for Dr. Fauci. Optics required that NIH devote some resources to antiviral therapeutic drugs; critics would complain if he spent billions on vaccines and nothing on therapeutics. However, any licensed, repurposed antiviral that was effective against COVID for prevention or early treatment (like IVM or HCQ) could kill his entire vaccine program because FDA wouldn't be able to grant his jabs Emergency Use Authorization. Remdesivir, however, was an IV remedy, appropriate only for use on hospitalized patients in the late stages of illness. It would therefore not compete with vaccines, allowing Dr. Fauci to support it without compromising his core business. Furthermore, while HCQ and IVM were off-patent and available generically, remdesivir was in the sweet spot of still being on patent. The potential profit upside was impressive. Remdesivir cost Gilead $10 per dose to manufacture.[18,19] But by granting Gilead an EUA, regulators could force private insurers, Medicare, and Medicaid to fork over around $3,120.00 per treatment—hundreds of times the cost of the drug.[20,21] Gilead predicted remdesivir would bring in $3.5 billion in 2020 alone.[22]

Dr. Fauci did not suddenly get the idea that remdesivir might work against coronavirus in January 2020. In one of his many extraordinary feats of uncanny foresight, beginning in 2017, Dr. Fauci paid $6 million to his gain-of-function guru, Ralph Baric—a University of North Carolina microbiologist—to accelerate remdesivir as a coronavirus remedy at China's biosecurity laboratory in Wuhan.[23,24] Baric used coronavirus cultures obtained from bat caves by Chinese virologists working with Peter Daszak's EcoHealth Alliance, another recipient of Dr. Fauci's funding.[25,26] Dr. Fauci demonstrated his personal interest in those experiments by dispatching his most trusted deputies, Hugh Auchincloss in 2018 and then Cliff Lane in 2020, to negotiate with the Chinese government and to supervise Baric's experiments at the Wuhan lab and elsewhere in China.[27] Baric claimed that his mouse studies showed remdesivir impeded SARS replication, suggesting that it might inhibit other coronaviruses. Chinese researchers at the Wuhan Lab and China's Military Medicine Institute of the People's Liberation Army Academy of Military Science submitted their own patent application for remdesivir.[28] China's military brass said the joint patent application was "aimed at protecting China's national interests."[29]

Early in March 2020, the Gates Foundation bankrolled $125 million of tax-deductible grants to support drug makers to develop coronavirus treatments.[30] Gates and/or his foundation had large equity stakes in many of the pharmaceutical companies that received these funds—including Gilead. On April 24, 2020, Gilead's volunteer spokesperson Bill Gates declared: "For the novel coronavirus, the leading drug candidate in this category is remdesivir from Gilead."[31]

For HCQ, Dr. Fauci demanded well-designed randomized double-blind placebo-controlled trials[32,33] and he warned against the use of IVM for treatment.[34] In contrast, Fauci green-lighted remdesivir following studies in which the control group did not receive a real placebo.[35] Instead, Fauci's researchers used no placebo in the more severely ailing patients and gave the remaining patients an "active comparator" containing the same treatment protocol agents as used in the remdesivir arm except for substituting sulfobutyl for remdesivir as the test agent.[36] Utilization of so-called "toxic" or "spiked" placebos—also known as "fauxcebos"—is a fraudulent gimmick that Dr. Fauci and his drug researchers have pioneered over forty years to conceal adverse side effects of toxic drugs for which they seek approval. Dr. Fauci eventually recruited 400 US hospitalized volunteers for NIAID's remdesivir trials, but despite this fauxcebo chicanery, Dr. Fauci's researchers just couldn't get remdesivir to show any improvement in COVID survival.[37]

Despite its disappointing performance, Dr. Fauci worked hand-in-hand with Gilead's remdesivir team to guide the trial to a satisfactory outcome. According to Vera Sharav, the President and founder of the Alliance for Human Research Protection (AHRP), "The National Institute of Allergy and Infectious Diseases (NIAID) had complete control over the trial and made all decisions regarding trial design and implementation. Gilead Sciences employees participated in discussions about protocol development and in weekly protocol team calls with NIAID."

Sharav's organization, Alliance for Human Research Protection (AHRP), monitors the quality and ethical performance of clinical trials. NIAID's remdesivir trial's original endpoint made sense: to win approval, the drug would need to demonstrate a "reduction in COVID mortality." However, the drug didn't show the hoped-for benefit. While fewer patients receiving remdesivir died, those receiving remdesivir were also a lot less sick than the placebo subjects when they entered the trial. So Dr. Fauci's team decided to move the goalposts. The researchers, in fact, had changed the trial "endpoints" twice in an effort to create a meager appearance of benefit. Dr. Fauci's new endpoints allowed the drug to demonstrate a benefit, not by improving the chances of surviving COVID, but by achieving shorter hospital stays.[38] Yet this too was a scam, because it turned out that almost twice as many remdesivir subjects as placebo subjects had to be readmitted to the hospital after discharge—suggesting that Fauci's improved time to recovery was due, at least in part, to discharging remdesivir patients prematurely. Altering protocols in the middle of an ongoing study is an interference commonly known as "scientific fraud" or "falsification." UCLA Epidemiology Professor Sander Greenland explains, "You're not supposed to change your endpoint mid-course. That's frowned upon." Vera Sharav agrees: "Changing primary outcomes after a study has commenced is considered dubious and suspicious."[39]

But Dr. Fauci had little reason to worry that insiders would complain about the corruption of the study, since his trusted deputy, Cliff Lane, chaired the NIH Treatment

Guidelines panel.[40] Lane was doubly conflicted, since he had personally overseen the remdesivir trials in China, and stood, potentially, to share in patent rewards and royalties for the drug.[41] In addition to Lane, seven of the panel members had financial relationships with Gilead—and eight additional panel members had had financial relationships with Gilead prior to the past eleven months, for which they were required to declare a relationship.[42] "Is it any wonder remdesivir is the only drug recommended for COVID?" asks Vera Sharav, a Holocaust survivor who has devoted her life to advocating for ethics in the notoriously corrupt clinical trial industry.[43,44,45,46,47]

Before his study was completed or peer-reviewed, much less published, Dr. Fauci learned that *The Lancet* had just published a placebo-controlled Chinese study that showed remdesivir utterly ineffective at keeping hospitalized patients alive OR reducing the duration of hospitalizations.[48] Even more importantly, remdesivir did not reduce the presence of the virus in the blood. Worst of all, the Chinese study confirmed remdesivir's deadly toxicity. The Chinese regulators and researchers shuttered that trial because of potentially lethal side effects. Remdesivir caused serious injuries in 12 percent of the patients, compared to 5 percent of patients in the placebo group.[49] Unlike Dr. Fauci's trial, the Chinese study was a randomized, double-blind, placebo-controlled, multi-center, peer-reviewed study, published in the world's premier scientific journal, *The Lancet*. All the underlying data was available to the incurious press and the uninformed public.

In contrast, Dr. Fauci's NIAID-Gilead study was at that point, still unpublished, not peer-reviewed, its details undisclosed. It employed a phony placebo and had suffered a sketchy mid-course protocol change. In April, the Chinese cancelled two ongoing clinical trials with NIAID in China because the Chinese had succeeded in ending the COVID epidemic in the country, and researchers could no longer identify enough COVID patients to enroll in the study.[50]

In any event, the Chinese study spelled certain doom for remdesivir. It was now D.O.A. at FDA—a poem title? But Dr. Fauci never accepted this. The inimitable maestro of regulatory combat responded to the crisis with savvy and bold action that would miraculously salvage his sinking product: He appeared at one of his regular White House press conferences, this one in the Oval Office. Seated on the couch next to Deborah Birx and opposite President Trump, Dr. Fauci made a surprise announcement.

From that lofty platform, Dr. Fauci, with great fanfare, declared victory. The data from NIAID's clinical trial for remdesivir shows "quite good news," he said, glossing over the drug's failure to demonstrate any mortality advantage.[51] He boasted that the median time for hospitalization was eleven days for patients taking remdesivir, compared to fifteen days in the placebo group. He told the credulous press: "The data shows that remdesivir has a clear-cut, significant, positive effect in diminishing the time to recovery." He claimed that his study had therefore proven remdesivir so remarkably beneficial to COVID patients that he had decided that it would be unethical to deny Americans benefits of this wonder drug. He was, he declared, unblinding and ending the study and giving remdesivir to the placebo group. Remdesivir would be America's new "standard of care"[52] for COVID. It was, of course, all a lie.

On May 1, the FDA granted the pandemic's first Emergency Use Authorization for a COVID drug, allowing remdesivir treatments for patients hospitalized with severe COVID-19.[53,54]

Based on Dr. Fauci's representation, President Trump purchased the world's entire stock of remdesivir for Americans.[55] The European Union signed a "joint procurement agreement" with Gilead to queue up in the pipeline for 500,000 treatment courses.[56] The day after Dr. Fauci's announcement at the White House, the University of North Carolina issued a press release headlined: "Remdesivir, developed through a UNC-Chapel Hill partnership, proves effective against COVID-19 in NIAID human clinical trials."[57] Dr. Fauci's gain-of-function wizard, Dr. Ralph Baric, called this "a game changer for the treatment of patients with COVID-19."[58]

Vera Sharav points out that in a rational universe, a poison like remdesivir would have no hope of winning regulatory approval—unless, of course, the company could somehow distract attention from the overwhelmingly catastrophic scientific evidence by getting the world's most powerful health official—the man who conducted the clinical trial—to pronounce the drug a "miracle cure" at a globally attended press conference while lounging on an Oval Office divan beside the president of the United States. Says Sharav, "What better free advertisement?"[59]

Sharav adds, "Dr. Fauci had a vested interest in remdesivir. He sponsored the clinical trial whose detailed results were not subject to the peer review he demanded for the drugs he regarded as rivals, like hydroxychloroquine and ivermectin. Instead of showing transparent data and convincing results, he did 'science' by fiat. He simply declared the disappointing results to be 'highly significant,' and pronounced remdesivir to be the new 'standard of care.' Fauci made the promotional pronouncement while sitting on a couch in the White House, without providing a detailed news release, without a briefing at a medical meeting, or peer review for publication in a scientific journal—as is the norm and practice, to allow scientists and researchers to review the data."

"Standard of Care"

FDA's recognition of remdesivir as the new "Standard of Care" for COVID meant that Medicaid and insurance companies could not legally deny it to patients and would have to fork over Gilead's exorbitant price tag on a product US taxpayers had, by then, spent at least $85 million to develop.[60] Improving Gilead's business even more, doctors and hospitals that failed to use remdesivir could now be sued for malpractice, leading some medical experts to believe that coercing the use of this worthless and dangerous drug on COVID patients almost certainly cost tens of thousands of Americans their lives.

As we shall see, Dr. Fauci copied the choreographed script for winning remdesivir's EUA from the worn rabbit-eared playbook that he developed during his early AIDS years, and then used repeatedly across his career to win approvals for deadly and ineffective drugs. Time and again, he has terminated clinical trials of his sweetheart drugs the moment they begin to reveal cataclysmic toxicity. He makes the absurd claim that his drug-du-jour had proven so miraculously effective that it would be unethical to deny it to the public, and then he strong-arms FDA to grant his approvals. This time only, the brazenness of the fraud earned Dr. Fauci some rare criticism even in mainstream science and press, and from academic institutions that customarily maintain silence about his shenanigans, given their addictions to whopping NIH and BMGF funding.

On October 24, 2020, Umair Irfan noted that "The FDA is once again promoting a Covid-19 therapy based on shaky evidence."[61]

The *British Medical Journal* pointed out, "None of the randomized controlled trials published so far, however, have shown that remdesivir saves significantly more lives than standard medical care."[62]

Eric Topol of Scripps Research Translational Institute scolded that, "This is a very, very bad look for the FDA, and the dealings between Gilead and EU make it another layer of badness."[63]

Angela Rasmussen, a virologist at Columbia University Mailman School of Public Health, told a reporter: "I was really surprised when I saw that news."[64]

Science Magazine said Dr. Fauci's move had, "baffled scientists who have closely watched the clinical trials of remdesivir unfold over the past 6 months—and who have many questions about remdesivir's worth."[65]

University of Oxford Professor of Clinical Therapeutics Duncan Richard scathingly observed that, "Research based on this kind of use should be treated with extreme caution because there is no control group or randomization, which are some of the hallmarks of good practice in clinical trials."[66]

Professor Stephen Evans in Pharmacoepidemiology, at the Gates-funded London School of Hygiene & Tropical Medicine, offered a particularly scathing assessment—"The data from this paper are almost uninterpretable. It is very surprising, perhaps even unethical, that the *New England Journal of Medicine* has published it. It would be more appropriate to publish the data on the website of the pharmaceutical company that has sponsored and written up the study. At least Gilead has been clear that this has not been done in the way that a high-quality scientific paper would be written."[67]

Even Bill Gates raised an eyebrow about the audacity of the caper. When *Wired* magazine in August 2020 asked Gates what therapeutic treatment he'd ask for if hospitalized with COVID-19, he did not hesitate. "Remdesivir," Gates replied, adding a comment that put daylight between him and the embarrassing clinical trial fiasco. "Sadly the trials in the US have been so chaotic that the actual proven effect is kind of small. Potentially the effect is much larger than that. It's insane how confused the trials here in the US have been."[68]

* * *

Then, on October 19, 2020, three days before remdesivir's FDA approval, the World Health Organization published a definitive study on remdesivir involving 11,266 COVID-19 patients in 405 hospitals and 30 countries.[69,70] The power of this study dwarfed the Fauci/Gilead project, which had recruited 1,062 patients. In the WHO's trial, remdesivir failed to reduce mortality, and failed to reduce the need for ventilators OR the length of hospital stays. WHO researchers found no detectable benefits from remdesivir and recommended against its use in COVID-19 patients.[71] WHO published its devastating indictment of remdesivir one month after FDA issued the remdesivir EUA for children less than 12 years of age. Dr. Fauci and the FDA knew about the WHO study before the FDA issued the EUA for remdesivir, and almost certainly read the preprints and understood the findings. It appears, in fact, that Dr. Fauci once again hurried the approval through FDA so as to beat the publication of a negative study.

On July 15, 2021, a large Johns Hopkins Study in *Original Investigation | Infectious Diseases* once again confirmed that "Remdesivir treatment was not associated with improved survival but was associated with <u>longer</u> hospital stays."[72] (Emphasis added.)

On October 2, 2020, the European Union released its own safety review of remdesivir. The study reported serious side effects.[73,74]

"Every independent randomized controlled trial of remdesivir has shown either a lack of benefit or a clear trend to harm," says Dr. Pierre Kory. "It's only those two Pharma studies (with Dr. Fauci) that show any benefits and even then, the benefits are minor."

"It makes no sense to give an antiviral in late stages of a viral infection," Dr. Kory adds. "The viral replication mainly takes place prior to day seven. If an antiviral works, that's when you administer antivirals. Remdesivir might work early on, but we don't know, because it's IV administered and you can't really do that to ambulatory patients."

A Remedy Worse than the Malady

From early in May 2020, doctors and hospitals began using remdesivir on hospitalized patients who tested positive for COVID in PCR tests. By November 9, 2021, the publication date of this book, CDC's website lists only two drugs approved for treating COVID-19, remdesivir and the corticosteroid dexamethasone.[75,76,77] Doctors often use the two drugs in conjunction. Assessing remdesivir's impact on hospitalized COVID-19 patients is difficult, in part, because—like COVID-19—remdesivir causes extreme toxicity to lungs and kidneys,[78] and mimics several of the other lethal symptoms of COVID, including multi-organ failure.[79] Many doctors believe our country's record COVID-19 fatalities are at least in part due to widespread use of remdesivir in 2020. "We had the most deaths worldwide," says Dr. Ryan Cole. "It's a haunting question: How many of these Americans were remdesivir casualties?"

For several months, we were the only country treating people with a drug proven to be lethal. That year, 2020, we had almost double the number of deaths per month compared to most other countries. Brazil, one of the first nations to widely use remdesivir, had the second highest death toll.[80,81]

In May of 2020, New York doctors repeatedly marveled at the tendency for COVID-19 to cause kidney failure, something that no other respiratory virus does. Doctors began seeing acute kidney failure on day three, four, and five after admission.[82] Hospitals short on ventilators also ran out of dialysis machines. Physician and laboratory CEO Dr. Ryan Cole is one of many doctors who believe that many of those cases were attributable to remdesivir. "COVID-19 can affect the kidneys," he says. "We know this because we can recover the spike protein from urine. But it's dubious that the sheer magnitude of acute renal failure we saw among hospitalized COVID patients can all be attributed solely to the coronavirus infection."

Dr. Cole told me that in the animal studies, one-fourth of the animals died from kidney failure. He explains that kidney collapse can lead to fluid accumulations in the lungs and everywhere and results in multi organ failure and sepsis—all of which are also sequelae of COVID. "Remdesivir shouldn't be on the market," he added.

Dr. Fauci's 2019 Ebola study proved that remdesivir, by day three, four, and five, caused acute kidney failure in upwards of 31 percent of patients. In less than five days of remdesivir treatment, 8 percent of all people died or experienced life-threatening multiple organ failure or kidney failure so severe they had to be taken off the drug. "So it may not be a coincidence that roughly the same number of hospitalized COVID patients—8–10x were dying in the first week," says Cole.

Dr. McCullough gives us a stark and clear summary: "Remdesivir has two problems. First, it doesn't work. Second, it is toxic and it kills people."

V: FINAL SOLUTION: VACCINES OR BUST

"The only means to fight a plague is honesty."

—Albert Camus, *The Plague* (1947)

During the spring of 2020, Dr. Fauci and Bill Gates carpet-bombed the airwaves, bullishly predicting that a "miraculous vaccine" would stop COVID transmission, prevent illness, end the pandemic, and release humanity from house arrest. Even vaccinology's most stalwart tub thumpers—true believers like Dr. Peter Hotez and Dr. Paul Offit—regarded those forecasts as far-fetched and foolhardy.[1,2] After all, for decades, two perilous and seemingly insurmountable impediments had thwarted every attempt to craft a coronavirus vaccine.

Leaky Vaccines

The first obstacle was the coronavirus's tendency to rapidly mutate, producing vaccine-resistant variants. Vaccine developers like Hotez and Offit doubted that, after decades of futile efforts, researchers could suddenly develop a COVID vaccine that would provide "sterilizing immunity," meaning that it would completely obliterate viral colonies in vaccinated individuals and prevent transmission and mutation.

As if to confirm such fears, in May of that year, Britain's top vaccinologist, Andrew Pollard, admitted that the Oxford University's government-funded and patriotically ballyhooed AstraZeneca vaccine had failed to achieve sterilizing immunity in monkeys; the inoculated macaques, even when asymptomatic, continued to support high viral loads in their nasal pharynxes.[3] Then in August, Dr. Fauci primped up the dismaying news of similar failures by all the competing candidates with a kind of celebratory bravado. Instead of declaring defeat and retreating to the drawing board, Dr. Fauci cheerfully announced that none of the first-generation COVID vaccines was likely to prevent transmission.[4] That news should have cratered the entire project. Leading virologists, including Nobel Laureate Luc Montagnier, pointed out that a non-sterilizing, or "leaky," vaccine could not arrest transmission and would therefore fail to stop the pandemic.[5] Even worse, vaccinated individuals, he warned, would become asymptomatic carriers and "mutant factories" blasting out vaccine-resistant versions of the disease that were likely to lengthen and intensify rather than abbreviate the pandemic.

But Tony Fauci and his partner, Bill Gates, seemed to have a strategy for neutralizing the variant threat. The two men had put billions of taxpayer and tax-deducted dollars into developing an mRNA platform for vaccines that, in theory, would allow them to quickly produce new "boosters" to combat each new "escape variant." This scheme was Big Pharma's holy grail. Vaccines are one of the rare commercial products that multiply profits by failing. Each new booster doubles the revenues from the initial jab. Since NIAID co-owned the mRNA patent,[6] the agency stood to make billions from its coronavirus gambit by producing successive boosters for every new variant; the more, the better! The good news for Pharma was that all of humanity would be permanently dependent on biannual or even triannual booster shots. Dr. Peter McCullough warned that mass vaccination with a leaky vaccine during a pandemic "would put the world on a never-ending booster treadmill."[7] That kind of talk had Pharma popping champagne corks. In October 2021, Pfizer announced that it was projecting an astonishing $26 billion in revenues from its COVID boosters.[8]

Pathogenic Priming

The even more daunting obstacle to coronavirus vaccines was their tendency to induce "pathogenic priming"—also known as "antibody-dependent enhancement" (ADE)—an overstimulation of immune system response that can cause severe injuries and death when vaccinated individuals subsequently encounter the wild viruses. In early experiments, coronavirus vaccines produced a robust immune response in both animals and children—temporarily heartening researchers—but then tragically killing the vaccine recipients upon re-exposure to the wild virus, or making them vulnerable to uniquely debilitating infections. Early in 2020, vaccinology's most brassbound commissars warned of this pitfall as Dr. Fauci unleashed the industry, with billions in federal lucre, to gin up COVID inoculations at record pace. In his March 5, 2020 testimony before the House Science, Space and Technology Committee on Coronavirus, Bill Gates's paid mouthpiece, Dr. Peter Hotez, cautioned:[9]

> One of the things we're not hearing a lot about is the unique potential safety problem of coronavirus vaccines. With certain types of respiratory virus vaccines you get immunized, and then when you get actually exposed to the virus, you get this kind of paradoxical immune enhancement phenomenon.[10]

Dr. Hotez confessed to the committee that his colleagues had killed a number of children from pathogenic priming during experiments with the respiratory syncytial virus (RSV) vaccines in 1966, and recounted that during his own earlier work on coronavirus vaccines, he saw the same effect on ferrets:

> We started developing coronavirus vaccines and our colleagues—we noticed in laboratory animals that they started to show some of the same immune pathology. So we said, "Oh my God, this is going to be problematic."

In an April 26, 2020 interview with Pharma troll Dr. Zubin "ZDogg" Damania, MD, Merck's top vaccine promoter, Dr. Paul Offit, amplified these concerns:[11]

> [B]inding antibodies can be dangerous and cause something called Antibody Dependent Enhancement. And we've seen that. I mean, we saw that with the [Gates-funded] dengue vaccine. But with the dengue vaccine, in children who had never been exposed to dengue before, it actually made them worse when they were then exposed to the natural virus. Much worse. Vaccinated children who were less than nine years of age, who had never been exposed to dengue before, **were more likely to die if they'd been vaccinated than if they hadn't been vaccinated**.[12]

And even Dr. Anthony Fauci, during his March 26, 2020, White House coronavirus briefing, acknowledged the perils of pathogenic priming:[13]

> The issue of safety is something I want to make sure the American public understands: does the vaccine make you worse? And there are diseases, in which you vaccinate someone, they get infected with what you're trying to protect them with [sic] and you actually enhance the infection. **That's the worst possible thing you could do—is vaccinate somebody to prevent infection and actually make them worse.** (emphasis added)

Dr. Fauci must have recognized that since vaccine makers had immunity from liability [which he had helped arrange] and were playing, as it were, with house money [which he diverted to them through NIH], these companies had little incentive to invest in the kind of long-term studies necessary to eliminate the pathogenic priming hazard. In retrospect, it seems that Dr. Fauci and his confederates had at least six strategies for dealing with this grim risk. All six tactics involved hiding the evidence of ADE if it did occur:

1) Dr. Fauci's first approach was to abort the three-year clinical trials at six months and then vaccinate the controls—a preemption that would prevent detection of long-term injuries, including pathogenic priming. Regulators initially intended the Pfizer vaccine trial to continue for three full years, until May 2, 2023.[14] Because the FDA allowed Pfizer to unblind and terminate its study after six months—and to offer the vaccine to individuals in the placebo group—we will never know whether vaccinated individuals in the trial suffered long-term injuries, including pathogenic priming, that cancelled out short-term benefits. Science and experience tell us that many vaccines can cause injuries like cancers, autoimmune diseases, allergies, fertility problems, and neurological illnesses with long-term diagnostic horizons or long incubation periods. A six-month study will hide these harms.

2) Second, as COVID czar, Dr. Fauci stubbornly refused to fix HHS's designed-to-fail vaccine injury surveillance system (VAERS), which systematically suppresses reporting of most vaccine injuries. The Vaccine Adverse Event Reporting System (VAERS) is a passive, voluntary system, jointly managed by the CDC and FDA, that accepts reports from anyone. A 2010 HHS study of the government's notoriously dysfunctional VAERS concluded that VAERS detects "fewer than 1 percent of vaccine injuries."[15] Put another way, VAERS misses OVER 99 percent of vaccine injuries, thereby lending the illusion of safety to even the most deadly inoculations. In 2010, the federal Agency for Health Care Research Quality (AHRQ) designed and field-tested a state-of-the-art machine-counting (AI) system as an efficient alternative to VAERS. By testing the system for several years on the Harvard Pilgrim HMO, AHRQ proved that it could capture most vaccine injuries. AHRQ initially planned to roll out the system to all remaining HMOs, but after seeing the AHRQ's frightening results—vaccines were causing serious injuries in 1 of every 40 recipients—CDC killed the project and stowed the new system on a dusty shelf. Dr. Fauci left that system safely cached, throughout the pandemic, allowing HHS's broken voluntary system to continue to conceal vaccine injuries, including any evidence of pathogenic priming.

3) Third, Dr. Fauci's trump card was his capacity to enlist mainstream and social media companies to make reporting of injuries and deaths disappear from the airwaves, newspapers, and the Internet, and therefore from the public consciousness. Facebook, Google, and the television networks purged doctors and scientists who reported pathogenic priming, and censored reports about the waves of other vaccine injuries. As a federal official sworn for four decades to uphold the Constitution, Dr. Fauci should have been the champion of free speech and vigorous debate during the pandemic. Instead, he worked hand in glove with Bill Gates, Mark Zuckerberg, and other Big Tech titans to censor criticism of his various mandates and suppress information about vaccine injuries, including discussions of pathogenic priming.[16,17] Email traffic shows that Dr. Fauci colluded directly with Mark Zuckerberg and the social media platforms to censor doctors who reported vaccine failures, harms, and deaths, to deplatform public health advocates like myself, and to evict and muzzle patients who reported their own injuries. The science journals, utterly dependent on Pharma advertising, obligingly refused to publish studies on the rash of deadly and debilitating jab reactions. The Bill Gates-funded fact-checking organization, Politifact,[18] worked with Pharma-funded fact-checkers like FactCheck, which receives, funding from the Robert Wood Johnson Foundation, and whose current CEO is Richard Besser, former acting head of the CDC, which owns $1.8 billion in Johnson & Johnson stock[19,20] to "debunk" stories and studies of vaccine injuries.

On October 7, 2021, Dr. Robert Malone, the inventor of the mRNA vaccine, complained in a tweet that America's people were almost utterly blind to the floods of adverse vaccine events that were killing and debilitating their countrymen: "The real problem here is the damn press and the internet giants. The press and these tech players act to manufacture and reinforce 'consensus' around selected and approved narratives. And then this is being weaponized to attack dissenters, including highly qualified physicians."[21]

4) Fourth, Dr. Fauci allowed CDC to discourage autopsies in deaths following vaccination. CDC refused to recommend autopsies on deaths reported to VAERS. That omission allowed the agency to repeatedly make the audacious, fraudulent declaration that all the 16,000 reported deaths following vaccination by October 2021 were "unrelated to the vaccines." The regulatory agencies thereby abolished vaccine deaths and injuries by fiat.

Instead of exposing this sort of rank deception by government authorities, media and social media enablers emboldened HHS to new nadirs in regulatory malpractice. In January of 2021, baseball superstar Hank Aaron, whom I knew, died seventeen days after receiving the COVID jab at a CDC-sponsored press conference in Atlanta. I observed, in a *Defender* article,[22] that Aaron's death was one of a wave of deaths among the elderly following COVID jabs. This was true, but the *New York Times* nevertheless vilified me for spreading "misinformation" and claimed that the Fulton County Coroner had determined that Aaron's death was "unrelated to vaccines." *USA Today, Newsweek, TIME, Daily Beast*, ABC, CNN, and CBS reported the *Times* claim.[23] But when I called to verify their claim, the Fulton County Coroner told me that the office has never seen Aaron's body and that no autopsy was ever performed. Aaron's family had buried the home-run hero without a postmortem. The *Times'* fabrication was part of the systematic campaign of deception, propaganda, and censorship by HHS regulators in partnership with mainstream media—almost unprecedented in the American experience—that helped conceal the tsunami of vaccine injuries and fatalities.

"Anthony Fauci is a great guy in the same way that Harvey Weinstein was a great guy," says Jeff Hanson, the chairman of a large publicly traded healthcare corporation. "It all changed when widespread private knowledge about him crossed the transom into public knowledge. Weinstein, too, had powerful mainstream media outfits watching his back.

Incidentally, autopsy reports from other nations are revealing exactly the sorts of information that CDC, understandably, wants to protect Americans from learning.

In September 2021, veteran German pathologists and professors Dr. Arne Burkhardt, who served as director of the Institute of Pathology in Reutlingen for 18 years, and Dr. Walter Lang, chief of a leading lung pathology institute for 35 years, performed autopsies on ten cadavers of individuals who died following vaccination, finding that five were very likely, and two more probably, related to the jab.[24]

In three cases, they found strong evidence of lethal multi-system inflammation and runaway autoimmunity, including rare autoimmune diseases, like Hashimoto's, an autoimmune-triggered hypothyroidism; leukoclastic vasculitis, an inflammatory reaction in the capillaries that leads to skin bleeding, and Sjögren's syndrome, an inflammation of the salivary and lacrimal glands. "Three autoimmune diseases in a total of ten is a strikingly high rate," said Professor Lang. The doctors also found large clusters of endothelial cells detached from the walls of blood vessels, and clumps of red blood cells that cause thrombosis, and giant cells that formed around trapped foreign bodies. Lang said he had not seen anything like these clusters of lymphocytes in hundreds of thousands of pathological studies: "The lymphocytes are running amok in all organs." Lang faulted government regulators for hindering autopsies on vaccine reactions: "We're missing out on 90 percent."

5) Fifth, Dr. Fauci populated the key FDA and CDC committees with NIAID, NIH, and Gates Foundation grantees and loyalists to insure rubber-stamp approvals for his mRNA vaccines, without any long-term injury studies. More than half of FDA's VRBPAC committee, which approved EUAs for Moderna, Johnson & Johnson, and Pfizer, and granted final licensure to the Pfizer vaccine, were grant recipients from NIH, NIAID, BMGF, and pharmaceutical companies.[25,26] More than half the CDC's ACIP committee participants were similarly compromised.

6) Sixth, by vaccinating the entire population, Dr. Fauci seems to be striving to eliminate the control group, to hide vaccine injuries. In a 2015 interview, Dr. Fauci said:

> I mean, if a parent really feels strongly against [vaccination], that parent can get an exemption. So there's never a situation where someone is going to tie you down and vaccinate you or say you can't go to any schools at all if you're not vaccinated. Nowhere should you force someone to do anything.[27]

In the run-up to the rollout, Dr. Fauci frequently repeated his ethical antipathy against mandating vaccination. But once the voluntary market reached saturation, those scruples melted away and, following his guidance, the federal policies began treating the vaccine-hesitant as dangerous public enemies. "Our patience is wearing thin," warned Joe Biden during a national address on September 9, 2021.[28]

Dr. Fauci presided over a progression of increasingly draconian forms of coercion to compel vaccination of the entire population. With his open encouragement, universities, schools, businesses, hospitals, public employers, and a litany of other societal power centers simultaneously launched numbing waves of strong-arm tactics to compel unwilling Americans to submit to vaccination, including threats of discrimination, job loss, exclusion from schools, parks, sports and entertainment venues, bars, restaurants, military service, public employment, travel, and health care. The unvaccinated experienced exclusion, marginalization, vilification, purges by social media platforms and mainstream media, as well as threats of violence, incarceration, legal reprisals, and deprivation of rights. New York City Mayor Bill de Blasio threatened to exclude the unvaccinated from subways, gyms, bars, and businesses. A Colorado hospital announced the removal of unvaccinated patients from its lists of those eligible for organ transplants. Observing that some 25 percent of African Americans were unvaccinated, civil rights leader Kevin Jenkins declared, "This is the new Jim Crow."

Whether intentional or not, the effect of this escalation was, increasingly, to eliminate the control group—which, coincidentally, would permanently hide the evidence of vaccine injuries. This motivation alone explains Dr. Fauci's reckless and ferocious drive to vaccinate every last American, even those who have natural immunity and nothing to gain from vaccination, Americans below fifty, even kindergarten-age children with zero risk from COVID, and pregnant women, despite a nearly complete lack of information about the jab's impact on the fetus. Dr. Fauci continued to insist that fully vaccinating the entire population was the only path to ending the pandemic. This assertion ignored the fact that COVID vaccines prevent neither transmission nor infection, nor reductions in viral loads. Overwhelming science has proven that vaccinated and unvaccinated individuals are equally likely to spread disease. A September 2021 Israeli study demonstrating that natural immunity provides 27x better protection against COVID than the Pfizer vaccine is just one of 29 recently published peer-reviewed studies that vouch for the superiority of natural immunity.[29,30] What, then, is motivating the fierce campaign to nevertheless coercively vaccinate the vaccine-resistant 25 percent, other than a strategy to eliminate the control group to hide the deaths and injuries?

* * *

By November 2021, that retinue of concerning devices largely succeeded in concealing from Americans the well-established facts that Dr. Fauci's vaccines neither prevented the disease nor its transmission, and that COVID vaccines were killing and injuring record numbers of Americans. The relentless broadcast of frightening and purposefully inflated COVID death reports stoked fears of the contagion that convinced many Americans to believe the government's mantra that COVID vaccines were "safe and effective" and that, to the extent they weren't, "the vaccines cause more good than harm."

Physicians and scientists complained that Dr. Fauci's vaccine promotions constituted a vast, unprecedented population-wide experiment, with shady recordkeeping and no control group. Meanwhile, the actual data suggested that the COVID vaccines were causing far more deaths than they were averting.

The Pfizer Vaccine: A Cold Look at the Shocking Data

At this book's November 2021 publication date, only Pfizer's COVID vaccine, known as Comirnaty, had won FDA approval. Although Comirnaty is not yet given in the United States, its counterpart—the Pfizer-BioNTech, the same vaccine under a different name—is, so I will focus on the Pfizer-BioNTech vaccine. As of October 6, US health officials had administered more than 230 million doses of Pfizer's COVID vaccine, compared to 152 million doses of Moderna, and 15 million doses of Johnson & Johnson.[31]

The final summary of the Pfizer's six-month clinical trial data—the document that Pfizer submitted to FDA to win approval—revealed one key data point that should have killed that intervention forever. Far more people died in the vaccine group than in the placebo group during Pfizer's clinical trials. The fact that FDA nevertheless granted Pfizer full approval, and that the medical community embraced and prescribed this intervention for their patients, is eloquent testimony to the resilience of even the most deadly and inefficacious products, and the breathtaking power of the pharmaceutical industry and its government allies to control the narrative through captive regulators, compliant physicians, and media manipulation, and to overwhelm the fundamental common sense of much of humanity.

The Pfizer vaccine trial offers a lesson on the perils of ignoring "all-cause mortality" as the governing endpoint for vaccine approval. But before we talk about "all-cause mortality," let's look at the evidence that convinced FDA to grant Pfizer its license.

Mathematical Chicanery: Relative Risk vs. Absolute Risk

On the next page is Pfizer's table S4 that summarizes death data from Pfizer's six-month clinical trial. This was Pfizer's final report to FDA; the study by then was unblinded and over.[32] As anyone can see, Pfizer won FDA's approval despite the rather pathetic showing that its vaccine might prevent one COVID death in every 22,000 vaccine recipients.

So, how did Pfizer transform its unimpressive record of eliminating a single COVID fatality among 22,000 vaccinated subjects into a $5 billion/year success story? By gulling the public with a deceptive measure called "relative risk," instead of the presumptive and far more useful measure of "absolute risk."

The table shows that during the six-month trial, two people in the placebo group numbering approximately 22,000 and only one in the similarly sized vaccine group died from COVID. Believe it or not, this data point is the source of Pfizer's claim that the vaccine is 100 percent efficacious against death. Since only one person died from COVID in the vaccine group and two died in the placebo group, Pfizer can technically represent that the vaccine is a 100 percent improvement over the placebo. After all, the number "2" is 100 percent greater than the number "1," right? The media winked at this canard, obligingly reporting Pfizer's extraordinary 100 percent efficacy claim. At least some reporters must have understood that most Americans hearing this statistic would naturally believe that the vaccine would prevent 100 percent of deaths. A more honest—and helpful—way of thinking about the Pfizer vaccine's efficacy is to consider that 22,000 vaccines must be given to save a single life from COVID. Equally concerning, every virologist and infectious disease expert knew that the true reduction in risk of 1/22,000—or about 0.01 percent, as the *BMJ* reported—was far too insignificant to make the vaccine even a minor barrier against the spread of COVID. It's axiomatic that any vaccine that does not prevent transmission and that spares only 1 in 22,000 from death from the target contagion has no ability to stop a pandemic.[33] "Because the clinical trial showed that vaccines reduce absolute risk less than 1 percent (See: Brown R. and colleagues from Waterloo in Canada), those vaccines can't possibly influence epidemic curves. It's mathematically impossible," explains Peter McCullough. Nevertheless, Dr. Fauci continued to promote the vaccine as the ultimate panacea.

The entire justification that Gates and Dr. Fauci had been trumpeting for a year—that their vaccines would end the pandemic—was now so much exploded shrapnel. Nevertheless, Dr. Fauci continued to claim that full vaccination of the entire population was the only way to end the pandemic. He thereby justified his insistence that Americans submit to mass vaccination.

But the story gets even worse. As table S4 shows, this entire meager advantage of preventing a single COVID death in every 22,000 vaccinated individuals (1/22,000)

Reported Cause of Death[a]	BNT162b2 (N=21,926) n	Placebo (N=21,921) n
Deaths	15	14
Acute respiratory failure	0	1
Aortic rupture	0	1
Arteriosclerosis	2	0
Biliary cancer metastatic	0	1
COVID-19	0	2
COVID-19 pneumonia	1	0
Cardiac arrest	4	1
Cardiac failure congestive	1	0
Cardiorespiratory arrest	1	1
Chronic obstructive pulmonary disease	1	0
Death	0	1
Dementia	0	1
Emphysematous cholecystitis	1	0
Hemorrhagic stroke	0	1
Hypertensive heart disease	1	0
Lung cancer metastatic	1	0
Metastases to liver	0	1
Missing	0	1
Multiple organ dysfunction syndrome	0	2
Myocardial infarction	0	2
Overdose	0	1
Pneumonia	0	2
Sepsis	1	0
Septic shock	1	0
Shigella sepsis	1	0
Unevaluable event	1	0

Table S4 | Causes of Death from Dose 1 to Unblinding (Safety Population, ≥16 Years Old). a. Multiple causes of death could be reported for each participant. There were no deaths among 12–15-year-old participants.

*Pfizer reported five additional deaths in the vaccinated group before unblinding the study that the company failed to tabulate in Table S4.

is entirely cancelled out by a fivefold increase in excess fatal cardiac arrests and congestive heart failures in vaccinated individuals (5/22,000). Pfizer and its regulatory magician, Dr. Fauci, used smoke and mirrors to divert public attention from this all-important question of all-cause mortality.

All-Cause Mortality

"All-cause mortality" should be the key metric in weighing the value of any medical intervention. That measure alone tells us whether vaccinated individuals enjoy better outcomes and longer lives than the unvaccinated. Drugs and vaccines that appear, at first glance, effective against the target disease may, over longer terms, trigger deaths from unexpected causes: accidents, cancers, heart attacks, seizures, even depression and suicide—or from pathogenic priming—which cancel out the short-term benefits of the intervention. As we shall see in the next two chapters, Dr. Fauci learned, at the outset of his career, to find excuses for abbreviating clinical trials of toxic medications to keep long-term mortalities invisible and to cloud overall cost/benefit assessments.

Pfizer's six-month clinical data for its COVID vaccine trials suggested that, while the vaccine would avert a single death from COVID-19, the vaccinated group suffered 4x the number of lethal heart attacks as the unvaccinated. In other words, there was no mortality benefit from the vaccines; for every life saved from COVID, there were four excess heart attack fatalities.[34] Twenty people died of "all-cause mortality" among the 22,000 recipients in Pfizer's vaccine group, versus only fourteen in the numerically comparable placebo group. (Pfizer was evidently so alarmed by the total number of deaths in its vaccine cohort that it omitted five of them from table S4, and only disclosed them in fine print buried in the body of its report.) That means there were 42.8 percent more deaths in the vaccine than in the placebo groups. Under FDA guidelines, researchers must attribute all injuries and deaths among the study group during clinical trials to the intervention (the vaccine) unless proven otherwise.[35] Under this rule, the FDA must assume people who take the vaccine have a 42.8 percent increased risk of dying.

This six-month safety report was so damning that it should have closed the case against this vaccine, but captured FDA officials nevertheless gave Pfizer their approval; the broken VAERS system and the mainstream and social media all conspired to conceal the evidence of the crime when vaccinated Americans began dying in droves, and CDC implemented its own retinue of enshrouding machinations to cloak the real-life carnage.

Did US Cases and Deaths Drop After the National Vaccination Campaign Began?

Dr. Fauci and the vaccine lobby began an opportunistic campaign of deception by claiming credit for their jabs when COVID-19 deaths dropped precipitously in mid-December, 2020, just after the vaccine rollout began. But the first Pfizer jab had reached only 27 million Americans (about 8 percent of the population) by February 1, and—according to the CDC—the jab takes at least sixty days to provide protection, so vaccines had little if anything to do with the drop. By mid-April, only 31 percent of

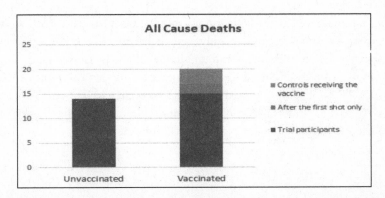

Americans were vaccinated and even by June 15, only 48 percent had been jabbed. The January drop-off was probably from natural herd immunity—thanks to the spread of natural infections over the previous year—and widespread use of ivermectin and hydroxychloroquine following Pierre Kory's December 5 Senate testimony,[36] and the proliferation of six nationwide telemedicine clinics and several large networks of independent physicians that began early treatment of about one-fourth to one-third of all new infections in January.

Americans wouldn't see the true impacts of vaccines on US mortalities until summer. But let's look, for a moment, at what happened in other countries with faster rollouts, less guileful regulators, and more scrupulous data collection and reporting.

International Databases: Infection Increases Following COVID Vaccines

Virtually all the countries that implemented rapid and aggressive COVID-19 vaccine campaigns experienced dramatic spikes in COVID infections. This documentation of increased susceptibility to COVID among highly vaccinated populations hints at the onset of the dreaded pathogenic priming in the months following mass vaccination.[37]

Gibraltar

The world's most vaccinated nation, Gibraltar, aggressively inoculated its 34,000 inhabitants, achieving 115 percent coverage (officials also vaccinated Spanish tourists) by July 2021. In December 2020, prior to the vaccine rollout, Gibraltar's health agency had experienced only 1,040 confirmed cases and five deaths from COVID-19. After the vaccination blitz, the number of new infections increased fivefold—to 5,314—and the number of deaths increased nineteen-fold.[38]

Malta

Malta, another of Europe's vaccine champions, administered 800,000 doses to its 500,000 inhabitants, achieving vaccine coverage of nearly 84 percent over six months. But beginning in July 2021, the epidemic and fatalities surged, forcing the authorities to impose new restrictions and to admit that vaccination cannot shield the population from COVID.[39]

Iceland

By July 2021, Iceland vaccinated 80 percent of its 360,000 inhabitants with one vaccine and 75 percent with two. But by mid-July, new daily infections had risen from about ten to about 120 before stabilizing at a rate higher than the pre-vaccination period. This sudden recurrence convinced Iceland's chief epidemiologist, Þórólfur Guðnason, of the impossibility of achieving herd immunity through vaccination.[40] "It's a myth," he publicly declared. "In Iceland, people no longer believe in herd immunity," according to oncologist and statistician Dr. Gérard Delépine.[41]

Belgium

By June 2021, Belgium had vaccinated nearly 75 percent of its 11.5 million population with one jab, and 65 percent with two. However, by the end of June 2021, new daily infections had risen from less than 500 to nearly 2,000. Belgian health officials acknowledged that the current vaccines cannot stop COVID, nor protect Belgium's citizens.[42]

Singapore

Singapore vaccinated nearly 80 percent of the population of 5,703,600 with at least one dose by the end of July 2021. But in late August, the country faced an exponential resumption of the epidemic. Daily cases increased from about ten in June to more than 150 at the end of July, and 1,246 cases on September 24.[43]

Britain

By July 2021, the United Kingdom had inoculated over 70 percent of its 67 million Brits with one shot, and 59 percent with both. Nevertheless, by mid-July Great Britain was suffering 60,000 new cases per day.[44] Faced with record viral surges, Britain's leading vaccinologist, Andrew Pollard, leader of the Oxford Vaccine Group, acknowledged before Parliament: collective immunity through vaccination is a myth.[45]

Even more worrying, British data compiled by Will Jones for the *Daily Sceptic* from August 2020 show a NEGATIVE VACCINE EFFECTIVENESS of -53 percent for the over-40 age group. Reported infections are highest in the double-vaccinated. This means that fully vaccinated individuals from this age group experienced a 53 percent HIGHER reported infection rate than the unvaccinated that month. ***Rather than preventing cases, the vaccine may be enhancing transmission***. This disproportionate number of vaccinated persons who seem to be sickening and dying strongly suggests that the world is beginning to see the predicted expression of pathogenic priming.[46]

Israel

Israel, champion of the Pfizer injection and pioneer of draconian mass vaccination mandates, inoculated 70 percent of its nine million people with at least one shot, and nearly 90 percent of those at risk with two, by June 2021. Israel, which formerly boasted itself the template for ruthless vaccine efficiency, is now the global model for vaccine failure.[47]

The epidemic rebounded in Israel stronger than ever in July, with a national record of 11,000 new cases recorded in a single day (September 14, 2021), surpassing by nearly 50 percent the previous peaks in January 2021 during the outbreak following the first Pfizer injections.[48]

On August 1, 2021, the director of Israel's Public Health Services, Dr. Sharon Alroy-Preis, announced half of all COVID-19 infections were among the fully vaccinated. Signs of more serious disease among fully vaccinated are also emerging, she said, particularly in those over the age of 60.[49]

68 Nations and 3,000 US Counties

An October 3, 2021 study by scientists at Harvard's T.H. Chan School of Public Health compared vaccination rates for 68 nations and 2,947 counties across America as of September 21, and compared them to COVID-19 cases per one million people. Their report concludes that nations and counties with higher vaccination rates do not experience lower per capita Sars-CoV-2 cases.[50,51]

Pathogenic Priming? COVID Vaccines Are Linked to Increased Deaths and Hospitalizations

By August 2021, Dr. Fauci, the CDC, and White House officials were reluctantly conceding that vaccination would neither stop illness nor transmission, but nevertheless, they told Americans that the jab would, in any case, protect them against severe forms of the disease or death. (It's worth mentioning that HCQ and ivermectin could have accomplished this same objective at a tiny fraction of its price.) Dr. Fauci and President Biden, presumably with Dr. Fauci's prompting, told Americans that 98 percent of serious cases, hospitalizations, and deaths were among the unvaccinated. This was a lie. Real-world data from nations with high COVID jab rates show the complete converse of this narrative; the resumption of infections in all those countries accompanied an explosion of hospitalizations, severe cases and deaths **among the vaccinated!** Mortalities across the globe, in fact, have tracked Pfizer's deadly clinical trial results, with the vaccinated dying in higher numbers than the non-vaccinated. These data cemented suspicions that the feared phenomenon of pathogenic priming has arrived, and is now wreaking havoc.

Gibraltar

Following its pioneering world-record vaccine rollout, Gibraltar saw an immediate spike in deaths, suffering 2,853 fatalities per million inhabitants, a European per capita mortality record. During the first days of the rollout—which began with senior citizens—some 84 people died shortly after vaccination.

England

Over a period of seven months preceding October 2021, some 60 percent of those 2,542 Brits who died from COVID were double vaccinated. Of people hospitalized in the UK for COVID in the last seven months, 157,000 were double-vaccinated.[52] There were more per capita deaths among the "fully" vaccinated than the unvaccinated.[53] The UK government's latest Office for National Statistics report on mortality rates by COVID vaccination status shows that for age-adjusted mortality rate, the death rate by October 2021 was higher among the vaccinated than the unvaccinated.[54]

Wales

According to October 2021 data from public health officials in Wales, UK, vaccinated individuals accounted for shocking 87 percent of all new COVID hospitalizations.[55]

Only 80 percent of Welsh were then fully vaccinated. In other words, only 13 percent of severe cases that required a trip to the hospital were unvaccinated, suggesting that those who have taken the experimental vaccine are more likely to experience adverse reactions and become hospitalized from COVID-19.

Scotland

In Scotland, official data on hospitalizations and deaths for October 2021 showed 87 percent of those who had died from COVID-19 in the third wave that began in early July were vaccinated. Only 70 percent of Scots were, at the time, fully vaccinated.[56]

Israel

In Israel, an increase in hospitalizations accompanied the epidemic's ferocious resumption. The vaccinated represented the majority of those hospitalized. By the end of July, some 71 percent of the 118 seriously and critically ill Israelis were fully vaccinated! This proportion of seriously ill people vaccinated is much higher than the proportion of fully vaccinated people: 61 percent. According to Israel's official report, August deaths were more frequent among fully vaccinated patients (679) than among non-vaccinated patients (390), belying official claims of a protective effect of the vaccine against dying. On August 5, 2021, Dr. Kobi Haviv, director of the Herzog Hospital in Jerusalem, reported on Channel 13 News that 95 percent of severely ill COVID-19 patients are fully vaccinated, and that vaccinated Israelis make up 85 percent to 90 percent of COVID-related hospitalizations overall.[57] As the doubly vaccinated overwhelmed Israeli hospitals, the government announced in August a new plan for coping with its "Pandemic of the Vaccinated." Israel said it will "update" its definition of "full vaccination" to require three, or even four, injections. "We are updating what it means to be vaccinated," said Israel's COVID czar, Salman Zarka.

Vermont

Vermont is America's most vaccinated state. On October 10, 2021, with 86 percent of its citizens fully vaccinated (according to COVID Dashboard), Vermont officials nevertheless reported the largest rate of infections ever—and revealed that more than three-quarters of Vermont's September COVID-19 deaths occurred in the "fully vaccinated." Unvaccinated accounted for only eight of the state's 33 virus deaths that month, and officials declined to reveal whether those eight were partially vaccinated. A department spokesman explained to *Lifesite News* that the breakthrough cases may reflect failing vaccine efficacy, as those who died were likely "among the very first to be vaccinated." As hospitalizations approached the pandemic peak, September turned into Vermont's second-deadliest month during the pandemic, according to the Associated Press.

Cape Cod

In my own hometown in Cape Cod, Massachusetts, a CDC investigation of an outbreak in Barnstable County, between July 6 through July 25, found 74 percent of

IRELAND

PORTUGAL

ISRAEL

THAILAND

MALAYSIA

UGANDA

UNITED KINGDOM

NEPAL

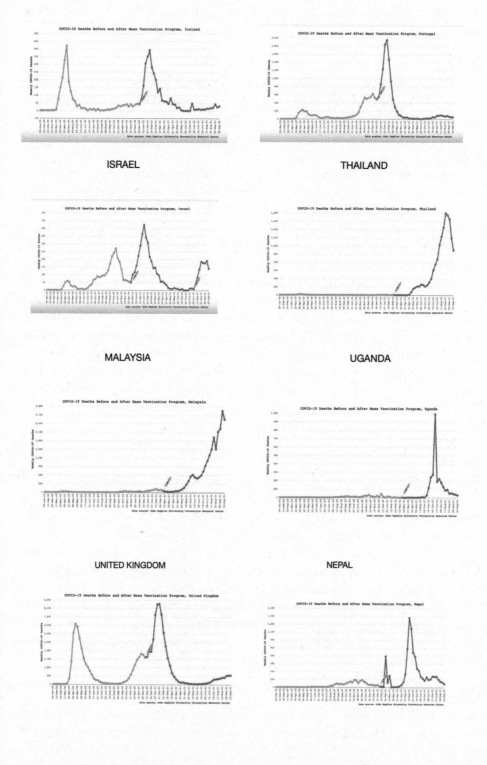

NAMIBIA

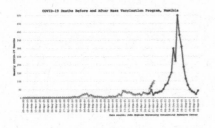

MONGOLIA

ZAMBIA

BAHRAIN

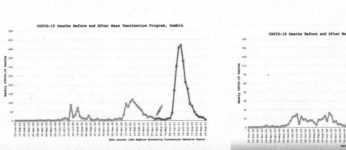

PARAGUAY

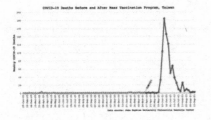

URUGUAY

TAIWAN

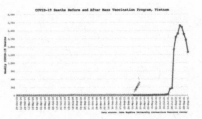

VIETNAM

BAHAMAS

SRI LANKA

AFGHANISTAN

TUNISIA

BANGLADESH

BURMA

FIJI

CAMBODIA

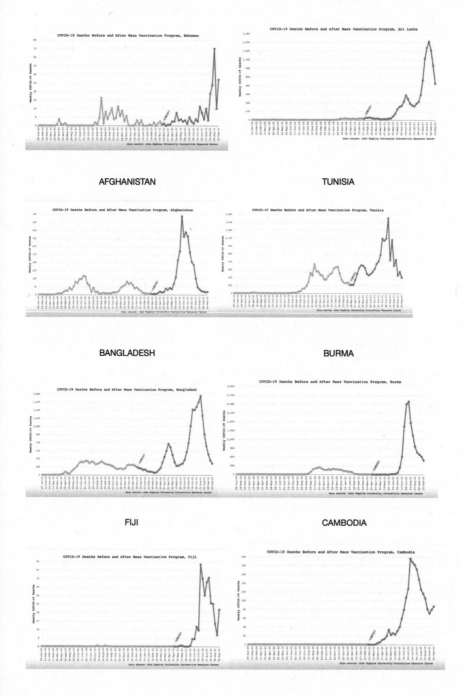

those who received a diagnosis of COVID-19, and 80 percent of hospitalizations, were among the fully vaccinated.[58] COVID resurgence and soaring breakthrough cases have plagued most of heavily-vaccinated New England, including Massachusetts, which has a vaccination rate nearly as perfect as Vermont's. COVID-19 cases were more than four times higher in the Bay State in September 2021 compared to the previous September. Half the deaths were among the fully jabbed and with an unknown number among partially vaxxed.

New England's COVID vaccine failure reflects an alarming national trend. A September report from the US Department of Defense revealed that 71 percent of recent cases of those hospitalized for COVID-19 in late August were fully vaccinated. DOD did not explain how many of the remainder were partially vaccinated.

Critics suggest that the shocking and predictable rise in COVID death following vaccination is evidence of long-feared pathogenic priming. Officials have offered no other compelling explanation as to why the vaccine consistently precipitates disproportionate injuries and deaths among the jabbed. It is not my intention to resolve this mystery here. Rather, I'm sharing the preceding graphs because the data trends they illustrate clash dramatically with official narratives. For that reason, you will not see reports about this alarming phenomenon on mainstream media. The Johns Hopkins University Coronavirus Resource Center collated the data for these graphs. Johns Hopkins is a central support column of mainstream medicine, and an aggressive promoter of COVID vaccines in particular. Johns Hopkins has received tens of millions of dollars from the Bill & Melinda Gates Foundation, and over a billion dollars from Tony Fauci's NIAID and NIH.[59,60] The Johns Hopkins data, nevertheless, clearly demonstrate that COVID deaths typically spike sharply in many country after country immediately after mass vaccination. The South African physicians group PANDA has assembled the Johns Hopkins data for every nation in an easy-to-view video.[61] PANDA's graphs illustrate this frightening "dead zone" that immediately followed vaccination drives in most of the world's nations.

In the US, COVID Vaccines Caused Record Deaths

Despite CDC's efforts to hide the carnage in the US, even the dysfunctional VAERS system has recorded unprecedented waves of documented deaths following COVID vaccines.

In 1976, US regulators pulled the swine flu vaccine after it was linked to 25 deaths.[62] In contrast, between December 14, 2020 and October 1, 2021, American doctors and bereaved families have reported more than 16,000 deaths and a total of 778,685 injuries to the Vaccine Adverse Event Reporting System (VAERS) following COVID vaccination.[63,64] The Europeans' surveillance sites tallied 40,000 deaths and 2.2 million adverse reactions. Due to chronic undercounting by VAERS and its European sister system, those numbers are almost certainly only a fraction of the true injuries. To illustrate how unprecedented this harm and death is, look at this "hockey stick" effect in CDC's own graph of the 30-year history of deaths reported to VAERS from all vaccines.

Health workers have administered many billions of vaccines during the past

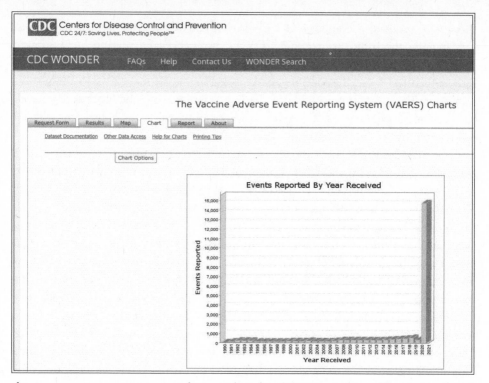

The Vaccine Adverse Event Reporting System (VAERS) Charts

Request Form | Results | Map | Chart | Report | About

Dataset Documentation   Other Data Access   Help for Charts   Printing Tips

Chart Options

Events Reported By Year Received

thirty-two years, yet in just eight months, the COVID vaccines have injured and killed far more Americans than all other vaccines combined over three decades. VAERS data show the huge spikes—69.84 percent[65]—of deaths occurring during the two weeks after vaccination, 39.48 percent within 24 hours of the injections.[66] According to CDC's fatality data, a COVID vaccine is 98 times more likely to kill than a flu vaccine.[67]

Other databases have, not surprisingly, yielded much higher projections of COVID vaccine deaths than VAERS.

A recent peer-reviewed study in the high-gravitas Elsevier journal *Toxicology Reports* found that COVID-19 vaccines kill more people in each age group than they save. According to that study the "best-case scenario" is five times the number of deaths attributable to each vaccination vs. those attributable to COVID-19 in the most vulnerable 65+ demographic.[68]

Similarly, a September 2021 analysis by a team of prominent scientists and mathematicians convened by Silicon Valley entrepreneur Steve Kirsch—of half a dozen population and surveillance system databases, including VAERS—using eight different independent methods, attributes 150,000 deaths to COVID vaccines in the United States since January 2020. Kirsch has offered a million-dollar reward for anyone who finds an error in this calculation.[69,70] Kirsch's study which found that the vaccines kill more people than they save in every age range was consistent with Pfizer's six-month clinical trial finding that people who took the vaccine were more likely to die than people who didn't take the vaccine (there were a total of twenty deaths in the people who took the vaccine vs. fourteen deaths in the people who didn't take the vaccine).[71]

In yet another effort to calculate excess deaths from vaccinations from a non-VAERS database, Ohio-based Attorney Thomas Renz used the Medicare database (Centers for Medicare & Medicaid Services) to calculate that there have been 48,465 deaths among Medicare/Medicaid beneficiaries within fourteen days of a first or second dose of a COVID-19 vaccine.[72,73] There are about 59.4 million Americans covered by Medicare, representing only 18.1 percent of the population, so these staggering numbers are roughly comparable to Steve Kirsch's population-wide estimate of 150,000.

How CDC Hid The Wave of Vaccine Deaths

According to Dr. Fauci, the Centers for Disease Control and Prevention, the White House, and most mainstream media, we now have a "pandemic of the unvaccinated,"[74] with 95 percent to 99 percent of COVID-related hospitalizations and deaths being attributed to the unvaccinated. As I mentioned above, these estimates are the product of systematic deception of the public—and presumably of the President—by America's top regulators. So how did CDC go about fooling President Biden?

One of CDC's bold deceptions is to hide vaccine mortalities in US data by counting all people as "unvaccinated" unless their deaths occur more than two weeks AFTER the second vaccine.[75] (Ironically, CDC doubles down on this fraud by counting many of these vaccine deaths as COVID deaths.) In this way, CDC captures that wave of deaths that occurs after vaccination and attributes them all to "unvaccinated." This is only one of many statistical chicaneries that the CDC employs to hide vaccine injuries and to stoke public fears of COVID.

The CDC utilized an even brassier canard to support President Joe Biden's claim that 98 percent of vaccine hospitalizations and deaths were among the unvaccinated. In an August 5 video statement, CDC director Dr. Rochelle Walensky inadvertently revealed the agency's principal gimmick for fabricating that statistic. Walensky sheepishly admitted that CDC included hospitalization and mortality data from January through June 2021 in its calculation.[76] The vast majority of the US population were, of course, unvaccinated during that time frame, so it makes sense that almost all hospitalizations would therefore be only among the unvaccinated. This is simply because there were almost no vaccinated Americans during that time period! By January 1, only 0.4 percent of the US population had received a COVID shot.[77] By mid-April, an estimated 37 percent had received one or more shots[78] and as of June 15, only 43.34 percent were fully "vaccinated."[79] Using these data was therefore pretty blatant fraud. Of course, CDC never let on that it was foisting eight-month-old data on Americans, allowing us instead to believe that these were current hospitalization rates as of August. To compound this flimflam, CDC perpetuated an even more audacious hustle. CDC omitted the current (as of August) data related to hospitalizations from the Delta variant, which disproportionately hospitalized vaccinated individuals in those other countries for which we have more reliable data.

CDC's promotion of this statistical bunko was obviously grossly misleading. Assuming President Biden wasn't deliberately lying to the American people, it's clear that CDC was lying to President Biden and using him to dupe the rest of us.

COVID Vaccines—Other Injuries

Despite the obstacles to reporting, VAERS recorded nearly 800,000 injuries by the 9½ months between December 14, 2020 and October 2021, with 112,000 classified as "serious." Pfizer either did not report several severe injuries—short of death—or deceptively deemphasized their severity, during clinical trials, including neurological harm, thrombocytopenia, blood clots, strokes, embolisms, aneurysms, myocarditis, Bell's palsy, Guillain-Barré syndrome, multi-organ failure, amputation, blindness, paralysis, tinnitus, and menstrual harms. More than 30,000 women in the UK[80] and 6,000 in the US have complained of the latter.[81]

On September 28, a scientific journal, *JAMA Neurology*, reported a new series of cases of cerebral venous sinus thrombosis (CVST) linked to COVID-19 vaccines,[82] confirming the severity of the reaction and the associated high mortality rate, and another journal confirmed the resumption of hepatitis C in a patient related to the jab.[83]

The numbers of and diversity of these serious injuries probably continue to be dramatically underreported. Steve Kirsch has investigated several broad deceptions Pfizer used to conceal injuries to the vaccine group during its clinical trial. We know, for example, due to the courage of Maddie and her parents, that Maddie de Garay, a 14-year-old who participated in the Pfizer trial, suffered severe neurological injuries including seizures and permanent paralysis. However, Pfizer reported only that Maddie suffered a stomach ache.

The Pfizer vaccine only gained emergency authorization for use in children because Pfizer manipulated trial data and committed serious offences, like hiding Maddie de Garay's injury.

Given that Maddie was only one of 2,300 teenagers in Pfizer's trial, her injury was potentially very significant. By extrapolating a one in 2,300 injury rate to the 86 million teens who Pfizer and Dr. Fauci have targeted for vaccination, some 36,000 of these potentially debilitating injuries could be expected to develop nationwide. While COVID may kill old people,[84] the vaccine, in Maddie's case, shows it also kills and harms the young.

Pfizer's clinical data predicted potentially fatal myocarditis in one in every 318 teens. Post-marketing data confirm astronomically high rates of myocarditis injuries. On October 1, 2021, a team of medical researchers and statisticians found that myocarditis rates reported in VAERS were significantly higher in teens than Pfizer had reported in its clinical data.

According to the Vaccine Adverse Event Reporting System, there have been 7,537 cases of myocarditis and pericarditis reported following COVID vaccines,[85] with 5,602 cases attributed to Pfizer.[86] Some 476 of these reports occurred in children from 12 to 17 years old.[87]

According to an article in *Current Trends in Cardiology*, "Within eight weeks of the public offering of COVID-19 products to the 12–15-year-old age group, we found 19 times the expected number of myocarditis cases in the vaccination volunteers over background myocarditis rates for this age group."[88] But even these alarming numbers may underreport myocarditis injuries. Israeli data and US data presented to CDC's advisory committee on June 23, 2021 similarly found the rate of reported cases of myocarditis in vaccinated teenage boys aged 12–17 is at least twenty-five times greater than expected, and is fifty times greater than the reported rate in vaccinated males over 65.

These astonishing numbers mean myocarditis is far from a "rare" side effect, as Dr. Fauci and Pfizer like to claim. Nor is it harmless. A recent study suggests that myocarditis is associated with a 50 percent mortality within five years.[89] A teen had effectively zero risk of dying from COVID and a substantial risk of death from vaccination.

In October 2021, Sweden, Denmark, and Finland announced that they will pause the use of Moderna's COVID vaccine for children under 18 years of age, after increased reports of inflammatory diseases like myocarditis and pericarditis.[90,91] That same week, Iceland banned Moderna's jab outright due to heart inflammation risk.

Furthermore, the VAERS data may also be dramatically underreporting myocarditis and other injuries.

Just before I published this book, in late October 2021, FDA made an extraordinary admission in a letter to Pfizer[92] to explain the chronic underreporting of serious but common vaccine-induced injuries and deaths. FDA, at last, admitted that VAERS is worthless for detecting vaccine injuries.

> We have determined that an **analysis** of spontaneous postmarketing adverse events [VAERS reports] reported under section 505(k)(1) of the FDCA [Federal Food, Drug and Cosmetic Act] **will not be sufficient to assess known serious risks** of myocarditis and pericarditis and identify an unexpected serious risk of subclinical myocarditis. Furthermore, **the pharmacovigilance system that FDA is required to maintain under section 505(k)(3) of the FDCA is not sufficient to assess these serious risks**.

At best, this letter is a shocking acknowledgement that regulators have no way to assess whether their vaccines are killing and injuring more humans than they are helping. In any rational regulatory environment, FDA's alarming admission would demand an instantaneous cessation of the vaccine rollout.

Only Dr. Anthony Fauci can answer the question, "Why—given FDA's stunning confession that America has no functional surveillance system—did HHS not immediately stop the COVID vaccine rollout?" The answer, of course, is that Dr. Fauci knows that America's bought, brain-dead, and scientifically illiterate media will never force him to answer this query.

Waning Vaccines

Compounding concerns over FDA's confession that Americans have no way to assess the risks from COVID vaccines is the uncontestable proof that COVID vaccine efficacy drops precipitously almost immediately after vaccination.

Pfizer and FDA may have opted to end the company's clinical trial after six months (the optional plan was a three-year trial ending in December 2023), after realizing that the vaccine was causing significant harms and that its fast-waning efficacy would make a cost/benefit analysis unsupportable if the study continued. In other words, the injury axis almost immediately crosses the benefits axis.

An October 3, 2021 study in the peer-reviewed journal *BioRxiv* by Stanford and Emory University scientists suggests that antibody levels generated by the Pfizer-BioNTech vaccine can suffer a ten-fold decrease seven months after the second vaccination.[93] The scientists warn that the precipitous drop in antibody levels will compromise the body's ability to defend itself against COVID-19 if the individual is exposed to COVID.

A second study published the same week confirms that the immune protection offered by two doses of Pfizer's COVID-19 vaccine drops off after only two months![94]

Another government-funded study in October confirms the decline in vaccine effectiveness in England[95] finding that the reduction in transmission "declined over time since second vaccination, for Delta reaching similar levels to unvaccinated individuals by 12 weeks for [the AstraZeneca vaccine] and attenuating substantially for [Pfizer]." In other words, within just three months, AstraZeneca did nothing to prevent transmission, and Pfizer was scarcely better.[96]

The study appearing in *The Lancet* confirms that vaccine effectiveness against infection disappears so fast that it is ephemeral. The heavily powered study involved 3,436,957 Kaiser Permanente Southern California customers and compared infections and COVID-19-related hospital admissions of fully vaccinated to unvaccinated people over the age of twelve for up to six months.[97]

The researchers found that vaccine effectiveness against infection plummeted from 88 percent during the first month after double vaccination to 47 percent after five months. The researchers found vaccine effectiveness against Delta infection was 93 percent during the first month after double vaccination but dropped to 53 percent after four months.[98]

This information should sicken every doctor who has ever given one of these jabs to a trusting patient. It means that these products confer no benefits to individuals or society and their long-term costs are foreboding and largely unknown. How could this have happened?

Vaccinating Children is Unethical

Our collective nausea can only amplify when we ask, "Why are we vaccinating children?" Kirsch's model estimates that 600 children have already died from COVID vaccines as of September 2021. A recent *Lancet* study shows that a healthy child has zero risk for COVID, suggesting that most of these kids are dying unnecessarily.[99] Some 86 percent of children suffered an adverse reaction to the Pfizer COVID vaccine in clinical trial. And one in nine children suffered a serious reaction grave enough to leave them unable to perform daily activities. How can we then justify forcing a healthy child to take a vaccine that is dead certain to injure many and kill some while bestowing no benefits? "How can anyone consider it ethical," asks Kirsch, "to put a child at risk, for the pretext that it might shield an adult? Show me any adult who thinks this is okay, and I'll show you a monster!"

COVID-19 vaccines have caused cardiac arrest, blindness, and paralysis in

Reporting rates of myopericarditis (per million doses administered), by manufacturer, sex, and dose number, 7-day risk period* (as of Aug 18, 2021)

Ages† (yrs)	Pfizer (All)		Moderna (All)		Janssen (All)	Pfizer (Males)		Moderna (Males)		Janssen (Males)	Pfizer (Females)		Moderna (Females)		Janssen (Females)
	Dose 1	Dose 2	Dose 1	Dose 2	Dose 1	Dose 1	Dose 2	Dose 1	Dose 2	Dose 1	Dose 1	Dose 2	Dose 1	Dose 2	Dose 1
12–15	2.6	20.9	0.0	not calc.	0.0	4.8	42.6	0.0	not calc.	0.0	0.5	4.3	0.0	0.0	0.0
16–17	2.5	34.0	0.0	14.6	0.0	5.2	71.5	0.0	31.2	0.0	0.0	8.1	0.0	0.0	0.0
18–24	1.1	18.5	2.7	20.2	2.7	2.4	37.1	5.1	37.7	3.0	0.0	2.6	0.7	5.3	1.6
25–29	1.0	7.2	1.7	10.3	1.9	1.8	11.1	3.2	14.9	2.0	0.3	1.3	0.4	6.3	0.0
30–39	0.8	3.4	1.0	4.2	0.4	1.1	6.8	1.6	8.0	0.0	0.6	1.0	0.4	0.7	1.0
40–49	0.4	2.8	0.5	3.2	1.2	0.7	4.4	0.6	4.6	2.2	0.1	1.8	0.4	2.1	0.0
50–64	0.2	0.5	0.6	0.8	0.2	0.2	0.5	0.4	1.0	0.0	0.3	0.8	0.8	0.7	0.5
65+	0.2	0.3	0.2	0.3	1.0	0.2	0.4	0.4	0.4	1.0	0.2	0.4	0.1	0.2	0.9

* Reports with time to symptom onset within 7 days of vaccination
† Reports among persons 12–29 years of age were verified by provider interview of medical record review

CDC — 13

1 in 317 boys (16-17) will get myocarditis from the vaccine

After the booster it could be 1 in 25

Note:
Two dose calc: 1000000/((5.2+71.5)*41)=317 (note 41 is the URF)
Third dose calc: 1000000/(71.5*13.75*41) = 25
Assumes each dose increases SAE rate by 13.75 (=71.5/5.2) in that age range

Reference: John Su, Safety update for COVID-19 vaccines: VAERS

15

American children. British Health Service reports emergency calls for cardiac arrest are at an all-time high since the government began offering teens the COVID-19 vaccine. COVID vaccines do not protect children from hospitalization or death associated with COVID-19 because healthy children are not being hospitalized or dying with COVID-19 [NHS statistics]. Children will not gain anything from having the jab because the vaccines do not prevent infection or transmission, as in three recent studies published by the CDC, UK government, and Oxford University. There is no evidence that the vaccines have prevented a single child's death.

Troubling statistics from Britain's Office for National Statistics (ONS) verify the expected: deaths among teenagers during the summer of 2021 increased significantly over the previous year, coinciding with the vaccine rollout. According to an analysis by *The Exposé*'s Will Jones,[100] between weeks 23 and 37 in 2021—simultaneous with

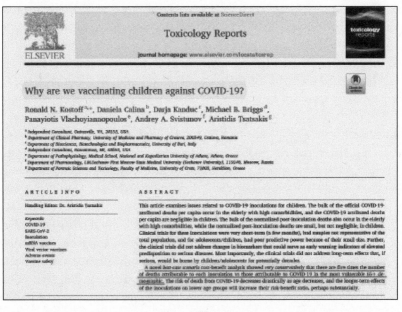

the vaccine rollout—there were 252 deaths among 15- to 19-year-olds in England and Wales, compared to 162 in the same period in 2020, an increase of ninety, or 56 percent—a very high number that deserves some kind of explanation.

Importantly, there is no similar rise among younger children aged one to fourteen, a cohort that was not vaccinated. Instead, 2020 was a low-mortality year for this age group. COVID cannot be blamed for the sudden rise in deaths among 15- to 19-year-olds in summer 2021, as the Office of National Statistics (ONS) data shows that over the period, there were only nine deaths with COVID in that age group. This real-world evidence suggests that over the summer, the vaccines killed nine times as many 15- to 19-year-olds as COVID did—eighty-one versus nine. "If not," asks Jones, "what are the other possible explanations, and how likely are they?"

Teen deaths among 15- to 19-year-olds have increased by 47 percent in the UK since they started getting the COVID-19 vaccine, according to official ONS data.[101]

Since the vaccine almost certainly causes more teen deaths and injuries than COVID-19, vaccinating this age group[102] is highly unethical, and any physician who inoculates a healthy child is committing serious medical malpractice.

Nevertheless, Anthony Fauci is urging that kids will be vaccinated in schools without parental consent, despite a mountain of evidence that the COVID-19 vaccines are killing American children and bestow on them no benefit.

Media Censors Reports of Vaccine Deaths

Most Americans are unaware of all this carnage because the mainstream and social media companies immediately scrub injuries reported by doctors, victims, and families. Media outlets like CNN and the *New York Times* ignore the tsunami of vaccine injuries and deaths while reflexively inflating those deaths they can blame on COVID. As part of a broad propaganda agenda, they report—with seeming glee—the occasional COVID death among the unvaccinated. Illustratively, on September 10, 2021, an ABC affiliate in Detroit solicited stories on its Facebook page about unvaccinated people who had died from COVID. Instead, the network got something they did not want: more than 230,000 messages containing heartbreaking stories of injuries and deaths from vaccines. None of these communications were reporting deaths among the unvaccinated. Readers shared the post over two hundred thousand times in ten days.[103]

Vaccinated Are Equally Likely to Spread COVID

Dr. Fauci's official theology makes "unvaccinated" America's national scapegoat, holding that they are more likely to spread disease and therefore should not be allowed to participate in civic life. The data across multiple sources and studies depict a very different reality.

In July 2021, the CDC found that fully vaccinated individuals who contract the infection have as high a viral load in their nasal passages as unvaccinated individuals who get infected. This means the vaccinated are just as infectious as the unvaccinated.

Another study from Indonesia supported this observation, noting that vaccinated individuals carry 251x the viral loads of Delta and other mutant variants than they did in the pre-vaccine era. Simply put, as Dr. Peter McCullough observed, "each vaccinated person is now a kind of Typhoid Mary for COVID, spreading concentrated viral loads of vaccine resistant mutants to vaccinated and unvaccinated alike."[104] CDC acknowledges that vaccinated individuals carry at least as many COVID germs in their noses as the unvaccinated.[105] CDC cited this revelation to justify its August 2021 mask mandate.[106]

An October 2021 investigation by Israel's medical authorities of a COVID-19 outbreak in a highly vaccinated population of health workers at the Meir Medical Center in Sheba recorded 23.3 percent of patients and 10.3 percent of staff infected, despite a 96.2 percent vaccination rate among exposed individuals.[107] Moreover, the researchers recorded multiple transmissions between two fully vaccinated individuals, both wearing surgical masks, and in one instance using full PPE, including N-95 mask, face shield, gown, and gloves.[108]

Endnotes - Arbitrary Decrees: Science-Free Medicine

1   Norah McDonnell, Dr. Fauci Says, "With All Due Modesty, I Think I'm Pretty Effective." *InStyle.* (July 15, 2020). https://www.instyle.com/news/dr-fauci-says-with-all-due-modesty-i-think-im-pretty-effective

2   Darragh Roche, "Fauci Said Masks 'Not Really Effective in Keeping Out Virus,' Email Reveals," Newsweek (Jun. 2, 2021), https://www.newsweek.com/fauci-said-masks-not-really-effective-keeping-out-virus-email-reveals-1596703

3   *USA Today*, 'Danger of getting coronavirus now is just miniscusely low', USA Today Video (Feb. 17, 2020), https://www.usatoday.com/story/opinion/2020/02/17/new-coronavirus-what-dont-we-know-dr-anthony-fauci-q-a-opinion/4790996002/

4   US Dept HHS, 00:44:12, "Update on the New Coronavirus Outbreak First Identified in Wuhan, China | January 28, 2020," *YouTube*, https://www.youtube.com/watch?v=w6koHkBCoNQ&t=2605s

5   Shaun Griffin, Covid-19: Asymptomatic cases may not be infectious, Wuhan study indicates, *BMJ*, (Dec. 1, 2020). https://www.bmj.com/content/371/bmj.m4695

6   Children's Health Defense, The Science of Masks, (2020-2021). https://childrenshealthdefense.org/the-science-of-masks/

7   Jeffrey D. Smith et al, Effectiveness of N95 respirators versus surgical masks in protecting health care workers from acute respiratory infection: a systematic review and meta-analysis, CMAJ. (May 17, 2016). https://childrenshealthdefense.org/wp-content/uploads/Effectiveness-of-N95-respirators-versus-surgical-masks-in-protecting-health-care-workers-from-acute-respiratory-infection-a-systematic-review-and-meta-analysis.pdf

8   Ibid.

9   Alex Gutentag, "The War on Reality," *Tablet* (Jun. 28, 2021), https://www.tabletmag.com/sections/news/articles/the-war-on-reality-gutentag

10   Edmund DeMarche, Fauci's controversial '60 Minutes' interview about mask-wearing was one year ago, *FOX News*, (Mar 8, 2021). https://www.foxnews.com/health/faucis-controversial-60-minutes-interview-about-mask-wearing-was-one-year-ago

11   Graison Dangor, "CDC's Six-Foot Social Distancing Rule Was 'Arbitrary', Says Former FDA Commissioner," *Forbes* (Sept. 19, 2021), https://www.forbes.com/sites/graisondangor/2021/09/19/cdcs-six-foot-social-distancing-rule-was-arbitrary-says-former-fda-commissioner/?sh=238cdd32e8e6

12   Face The Nation, "Gottlieb calls CDC's six-foot distancing recommendation 'arbitrary'," *Twitter* (Sept. 19, 2021), https://twitter.com/FaceTheNation/status/1439582587173941248

13   Spectator TV, "WHO Special Envoy on COVID - Dr David Nabarro on Lockdowns (Oct. 8, 2020)," *YouTube ...of Record*, 00:09:39, https://www.youtube.com/watch?v=YdSpiLBQWV0

14   Centers for Disease Control and Prevention, Association of State-Issued Mask Mandates and Allowing On-Premises Restaurant Dining with County-Level COVID-19 Case and Death Growth Rates — United States, March 1–December 31, 2020, *MMWR*, (Mar 12, 2021). https://www.cdc.gov/mmwr/volumes/70/wr/mm7010e3.htm

15   Donald G. McNeil, Jr., "How Much Herd Immunity Is Enough?," *New York Times* (Dec. 24, 2020), https://www.nytimes.com/2020/12/24/health/herd-immunity-covid-coronavirus.html?searchResultPosition=2

16   Wen Shi Li, et al, Antibody-dependent enhancement and SARS-CoV-2 vaccines and therapies, Nature Microbiology, (Sep 9, 2020). https://sci-hubtw.hkvisa.net/10.1038/s41564-020-00789-5

17   Nouara Yahi et al, Infection-enhancing anti-SARS-CoV-2 antibodies recognize both the original

Wuhan/D614G strain and Delta variants. A potential risk for mass vaccination?, *Journal of Infection*, (Aug 9, 2021). https://www.journalofinfection.com/article/S0163-4453(21)00392-3/fulltext

18   Jordan Lancaster, Fauci Doesn't Have An Answer To Why Those Who Recovered From Covid Are Required To Take Vaccine, The Daily Caller, (Sep 10, 2021). https://dailycaller.com/2021/09/10/fauci-doesnt-answer-recovered-covid-required-take-vaccine/

19   Disclose.tv, Dr. Fauci calls for unvaccinated Americans to be banned from air travel and to mandate #COVID19 vaccines for school children, *Twitter*, (Sep 13, 2021). https://twitter.com/disclosetv/status/1437382113548980232

20   Nieman Foundation, "Dr. Anthony Fauci: 'You've Got to Evolve with the Science. Science Is a Self-Correcting Process'," *Nieman Reports* (Jun. 30, 2021), https://niemanreports.org/articles/dr-anthony-fauci-youve-got-to-evolve-with-the-science/

21   Alan Reynolds, "How One Model Simulated 2.2 Million U.S. Deaths from COVID-19," Cato Institute (Apr. 21, 2020, 3:05 p.m.), https://www.cato.org/blog/how-one-model-simulated-22-million-us-deaths-covid-19

22   Ryan Chatelain, "U.S. COVID deaths in 2021 surpass 2020 total," Spectrum News NY1 (Oct. 06, 2021), https://www.ny1.com/nyc/all-boroughs/health/2021/10/06/u-s--covid-deaths-in-2021-top-2020-total *Models predicted 2.2 million deaths, 352,000 deaths were reported in 2020, the difference was 1.848 million deaths or a 525% overestimation of deaths related to COVID-19 based on models used to determine lockdowns.

23   BMGF Committed Grants, "Imperial College London 2016-2020," *BMGF Grants Database,* https://www.gatesfoundation.org/about/committed-grants?q=imperial%20college&yearAwardedEnd=2020&yearAwardedStart=2016

24   Centers for Control and Prevention, Weekly Updates by Select Demographic and Geographic Characteristics, (Oct 6, 2021). https://www.cdc.gov/nchs/nvss/vsrr/covid_weekly/index.htm?fbclid=IwAR3-wrg3tTKK5-9tOHPGAHWFVO3DfslkJ0KsDEPQpWmPbKtp6EsoVV2Qs1Q

25   Vincent Racaniello & Rich Condit, "COVID-19 with Dr. Anthony Fauci-Episode 641" *TWiV* (Jul. 17, 2020), https://www.youtube.com/watch?v=a_Vy6fgaBPE&t=27s

26   Daniel Payne, In little noticed July interview, Fauci warned that widely used COVID tests may pick up 'dead' virus, Just the News, (Dec 10, 2020). https://justthenews.com/politics-policy/coronavirus/newly-surfaced-video-july-fauci-tests-dead-virus

27   Michelle Rogers, "Fact check: Hospitals get paid more if patients listed as COVID-19, on ventilators- Senator Scott Jensen video," *USA TODAY*, (Apr. 24, 2020), https://www.usatoday.com/story/news/factcheck/2020/04/24/fact-check-medicare-hospitals-paid-more-covid-19-patients-coronavirus/3000638001/

28   Ibid. *USA Today*

29   Ross Lazarus et al, Medicare paid hospitals $39,000 per deaths from treating COVID-19, The Agency for Healthcare Research and Quality (AHRQ), (2010). https://digital.ahrq.gov/sites/default/files/docs/publication/r18hs017045-lazarus-final-report-2011.pdf

30   Jonathan Rothwell, "U.S. Adults' Estimates of COVID-19 Hospitalization Risk," *GALLUP*, (Sep 27, 2021). https://news.gallup.com/opinion/gallup/354938/adults-estimates-covid-hospitalization-risk.aspx

31   Elizabeth Fernandez, "Smoking Nearly Doubles the Rate of COVID-19 Progression," *UCSF* (May 12, 2020), https://www.ucsf.edu/news/2020/05/417411/smoking-nearly-doubles-rate-covid-19-progression

32   Becky McCall, "Vitamin D Deficiency in COVID-19 Quadrupled Death Rate," Medscape (Dec. 11, 2020), https://www.medscape.com/viewarticle/942497

33   Berkeley Lovelace, Jr., "CDC study finds about 78% of people hospitalized for Covid were overweight or obese," *CNBC* (Mar. 8, 2021), https://www.cnbc.com/2021/03/08/covid-cdc-study-finds-roughly-78percent-of-people-hospitalized-were-overweight-or-obese.html

34   Carolyn Crist, "Study: In U.S., Lockdowns Added 2 Pounds per Month," *WebMD,* (Mar. 23, 2021) https://www.webmd.com/lung/news/20210323/lockdown-weight-gain-study

35   M. Science, J. Johnstone, et al, "Zinc for the treatment of the common cold: a systematic review and meta-analysis of randomized controlled trials," CMAJ. 2012;184(10):E551-E561. doi:10.1503/cmaj.111990 https://www.ncbi.nlm.nih.gov/pmc/articles/PMC3394849/

36   College of Physicians and Surgeons, Statement on Public Health Information, (Apr 30, 2021). https://www.cpso.on.ca/News/Key-Updates/Key-Updates/COVID-misinformation

37   Emma Talkoff & Ashley Fetters Maloy, "The anti-vax wellness influencers," *Post Reports-Podcast* (Oct. 1, 2021) 00:10:11, https://www.washingtonpost.com/podcasts/post-reports/the-antivax-wellness-influencers/ https://open.spotify.com/episode/2N9m96jw8o9QtrMHPoI5Aa (00:10:25)

38   Gabby Landsverk, "Long-term quarantines can weaken your immune system due to loneliness and stress. Here's how to cope," Insider (May 12, 2020), https://www.insider.com/staying-inside-could-weaken-the-immune-system-from-stress-loneliness-2020-5

39   "Stress Weakens the Immune System," APA (Feb. 26, 2006), https://www.apa.org/research/action/immune

40   Yale, "Harvey Risch, MD, PhD Biography," *Yale School of Public Health* (Jul. 15, 2019), https://ysph.yale.edu/profile/harvey_risch/

41   Peter A. McCullough, et al., The Pathophysiologic Basic and Clinical Rationale for Early Ambulatory

Treatment of COVID-19, V:134 I:1 The American Journal of Medicine P16-22, (Jan 1, 2021,) DOI:https://doi.org/10.1016/j.amjmed.2020.07.003. https://www.amjmed.com/article/S0002-9343(20)30673-2/fulltext

42   Peter A. McCullough, et al., Multifaceted highly targeted sequential multidrug treatment of early ambulatory high-risk SARS-CoV-2 infection (COVID-19), 30;21(4):517-530 Rev Cardiovasc Med. (Dec. 2020), doi: 10.31083/j.rcm.2020.04.264, https://pubmed.ncbi.nlm.nih.gov/33387997/

43   FLCCC, "Dr. Pierre Kory, M.P.A., M.D. curriculum vitae," *FLCCC Alliance,* https://covid19criticalcare.com/wp-content/uploads/2021/01/FLCCC-Alliance-member-CV-Kory.pdf

44   Joel Rose, U.S. Field Hospitals Stand Down, Most Without Treating Any COVID-19 Patients, *NPR,* (May 7, 2020). https://www.npr.org/2020/05/07/851712311/u-s-field-hospitals-stand-down-most-without-treating-any-covid-19-patients

45   New York Times staff, Nearly One-Third of U.S. Coronavirus Deaths Are Linked to Nursing Homes, (Jun 1, 2021). *New York Times*, https://www.nytimes.com/interactive/2020/us/coronavirus-nursing-homes.html

46   *New York Times* staff, Nearly One-Third of U.S. Coronavirus Deaths Are Linked to Nursing Homes, (Jun 1, 2021). New York Times, https://www.nytimes.com/interactive/2020/us/coronavirus-nursing-homes.html

47   The FLCCC Alliance, FLCCC (2021), https://covid19criticalcare.com/network-support/the-flccc-alliance/

48   Paul G. Harch, Hyperbaric oxygen treatment of novel coronavirus (COVID-19) respiratory failure, *Medical Gas Research*, ()pr-Jun, 2020). https://www.ncbi.nlm.nih.gov/labs/pmc/articles/PMC7885706/

49   Scott A Gorenstein et al, Hyperbaric oxygen therapy for COVID-19 patients with respiratory distress: treated cases versus propensity-matched controls, Undersea & Hyperbaric Medical Society, (Third-quarter, 2020). https://pubmed.ncbi.nlm.nih.gov/32931666/

50   Saikiran Kannan, Did nasal rinsing help ASEAN control pandemic or swift action?, India Today, (jul 24, 2020). https://www.indiatoday.in/news-analysis/story/did-nasal-rinsing-help-asean-control-pandemic-or-swift-action-1703801-2020-07-24

51   Peter A. McCullough et al., The Pathophysiologic Basic and Clinical Rationale for Early Ambulatory Treatment of COVID-19, V:134 I:1 The American Journal of Medicine P16-22, (Jan 1, 2021,) DOI:https://doi.org/10.1016/j.amjmed.2020.07.003. https://www.amjmed.com/article/S0002-9343(20)30673-2/fulltext

52   Jianjun Gao et al, Breakthrough: Chloroquine phosphate has shown apparent efficacy in treatment of COVID-19 associated pneumonia in clinical studies, *BioScience Trends,* (Mar 26, 2020). Breakthrough: Chloroquine phosphate has shown apparent efficacy in treatment of COVID-19 associated pneumonia in clinical studies

53   National Health Commission & State Administration of Traditional Chinese Medicine, Diagnosis and Treatment Protocol for Novel Coronavirus Pneumonia, (Mar 3, 2020). https://www.chinalawtranslate.com/wp-content/uploads/2020/03/Who-translation.pdf

54   Aimee Quin, Interview with Robert F. Kennedy, Jr., (Oct. 1, 2021).

55   AAPS, "Physician List & Guide to Home-Based COVID Treatment," AAPS (Aug. 28, 2021),https://aapsonline.org/covidpatientguide/

56   The WHO Rapid Evidence Appraisal for COVID-19 Therapies (REACT) Working Group, Association Between Administration of Systemic Corticosteroids and Mortality Among Critically Ill Patients With COVID-19 A Meta-analysis, (Sep 22, 2020). https://jamanetwork.com/journals/jama/fullarticle/2770279

57   Univesity of Oxford, Common asthma treatment reduces need for hospitalisation in COVID-19 patients, study suggests, (Feb 9, 2021). Common asthma treatment reduces need for hospitalisation in COVID-19 patients, study suggests

58   Ly-Mee Yu, DPhil et al, Inhaled budesonide for COVID-19 in people at high risk of complications in the community in the UK (PRINCIPLE): a randomised, controlled, open-label, adaptive platform trial, The Lancet, (Aug 10, 2021). https://www.thelancet.com/journals/lancet/article/PIIS0140-6736(21)01744-X/fulltext

59   Aaron Z. Reyes et al, Anti-inflammatory therapy for COVID-19 infection: the case for colchicine, Annals of the Rheumatic Diseases, (May, 2021). https://pubmed.ncbi.nlm.nih.gov/33293273/

60   Peter A. McCullough et al, Multifaceted highly targeted sequential multidrug treatment of early ambulatory high-risk SARS-CoV-2 infection (COVID-19), Reviews in Cardiovascular Medicine, (2020). https://rcm.imrpress.com/article/2020/2153-8174/RCM2020264.shtml

61   R. Derwanda, M. Scholzb, & V. Zelenko, "COVID-19 outpatients: early risk-stratified treatment with zinc plus low-dose hydroxychloroquine and azithromycin: a retrospective case series study," International Journal of Antimicrobial Agents, Volume 56, Issue 6, Dec 2020, 106214, https://doi.org/10.1016/j.ijantimicag.2020.106214, https://www.sciencedirect.com/science/article/pii/S0924857920304258

62   McMaster University, Antidepressant fluvoxamine can save COVID-19 patients, McMaster-led research shows, (Aug 17, 2021). https://brighterworld.mcmaster.ca/articles/antidepressant-fluvoxamine-can-save-covid-19-patients-mcmaster-led-research-shows/

63   Gilmar Reis, et al., "Effect of Early Treatment With Hydroxychloroquine or Lopinavir and

Ritonavir on Risk of Hospitalization Among Patients With COVID-19: The TOGETHER Randomized Clinical Trial," *JAMA Netw. Open*, 4(4):e216468 (Apr. 22, 2021), doi:10.1001/jamanetworkopen.2021.6468. https://jamanetwork.com/journals/jamanetworkopen/fullarticle/2779044

64   Paul E. Alexander, et al., Early multidrug treatment of SARS-CoV-2 infection (COVID-19) and reduced mortality among nursing home (or outpatient/ambulatory) residents, V: 153 ISSN: 0306-9877 (Aug. 2021), https://doi.org/10.1016/j.mehy.2021.110622.

65   Juan IgnacioMorán Blanco, et al., Antihistamines and azithromycin as a treatment for COVID-19 on primary health care – A retrospective observational study in elderly patients, V: 67 ISSN 1094-5539 (Apr. 2021), https://doi.org/10.1016/j.pupt.2021.101989.

66   Peter A. McCullough, et al., Multifaceted highly targeted sequential multidrug treatment of early ambulatory high-risk SARS-CoV-2 infection (COVID-19), 30;21(4):517-530 Rev Cardiovasc Med. (Dec. 2020), doi: 10.31083/j.rcm.2020.04.264, https://pubmed.ncbi.nlm.nih.gov/33387997/

67   The New York Times, "See How the Vaccine Rollout Is Going in Your State," *NYT* (Jan. 31, 2021), https://web.archive.org/web/20210201124039/https://www.nytimes.com/interactive/2020/us/covid-19-vaccine-doses.html

Endnotes - Killing Hydroxychloroquine

1   Bill & Melinda Gates Foundation, 2018 Form 990, Attachment C, page 4, (2018). https://www.causeiq.com/organizations/view_990/911663695/011be856013f829aa3c948c35c2aa163

2   U.S. Food & Drug Administration, *Emergency Use Authorization of Medical Products and Related Authorities*, (Jan 2017). https://www.fda.gov/regulatory-information/search-fda-guidance-documents/emergency-use-authorization-medical-products-and-related-authorities#preeua

3   Jonathan Saltzman, *US Government Has Invested $6 Billion In Moderna's COVID-19 Vaccine*, STAT News, (April 30th 2021). https://www.statnews.com/2021/04/30/u-s-government-has-invested-6-billion-in-modernas-covid-19-vaccine/https://www.statnews.com/2021/04/30/u-s-government-has-invested-6-billion-in-modernas-covid-19-vaccine/

4   Bob Herman, NIH-Moderna Confidential Disclosure Agreement, AXIOS, (Nov 6, 2015) https://www.documentcloud.org/documents/6935295-NIH-Moderna-Confidential-Agreements.html

5   Zain Rizvi, *The NIH Vaccine*, PUBLIC CITIZEN, (Jun 25, 2020). https://www.citizen.org/article/the-nih-vaccine/#_ftn2

6   Legal Information Institute, 15 U.S. Code § 3710c - Distribution of royalties received by Federal agencies, CORNELL LAW SCHOOL, https://www.law.cornell.edu/uscode/text/15/3710c

7   Marisa Taylor and Aram Roston, Pressed by Trump, U.S. pushed unproven coronavirus treatment guidance, REUTERS, (Apr 4, 2020). https://www.reuters.com/article/us-health-coronavirus-usa-guidance-exclu/exclusive-pressed-by-trump-u-s-pushed-unproven-coronavirus-treatment-guidance-idUSKBN21M0R2

8   World Health Organization, Unedited Report of the 18th Expert Committee on the Selection and Use of Essential Medicines, (Mar 21-25, 2011). https://www.who.int/selection_medicines/Complete_UNEDITED_TRS_18th.pdf?ua=1

9   Concordia Pharmaceuticals Inc, PLAQUENIL® HYDROXYCHLOROQUINE SULFATE TABLETS, USP, (Jan, 2017). https://www.accessdata.fda.gov/drugsatfda_docs/label/2017/009768s037s045s047lbl.pdf

10   STELLA IMMANUEL, MD, LET AMERICA LIVE: EXPOSING THE HIDDEN AGENDA BEHIND THE 2020 PANDEMIC: MY JOURNEY, CH 1 (Charisma House 2021)

11   Centers for Disease Control and Prevention, Medicines for the Prevention of Malaria While Traveling Hydroxychloroquine (Plaquenil™), https://www.cdc.gov/malaria/resources/pdf/fsp/drugs/hydroxychloroquine.pdf

12   Peter A. McCullough, MD, Why doctors and researchers need access to hydroxychloroquine, THE HILL, (Aug 7, 2020). https://thehill.com/opinion/healthcare/510700-why-doctors-and-researchers-need-access-to-hydroxychloroquine

13   Els Keyaerts et al, In vitro inhibition of severe acute respiratory syndrome coronavirus by chloroquine, BIOCHEMICAL AND BIOPHYSICAL RESEARCH COMMUNICATIONS, Oct 8, 2004).

14   Keyaerts, Els et al., *In vitro inhibition of severe acute respiratory syndrome coronavirus by chloroquine Biochemical and biophysical research communications* vol. 323,1 (2004): 264-8. doi:10.1016/j.bbrc.2004.08.085, https://www.ncbi.nlm.nih.gov/pmc/articles/PMC7092815/pdf/main.pdf

15   Martin J. Vincent et al, Chloroquine is a potent inhibitor of SARS coronavirus infection and spread, (Aug 22, 2005). https://www.ncbi.nlm.nih.gov/pmc/articles/PMC1232869/

16   Julie Dyall, et al, *Repurposing of clinically developed drugs for treatment of Middle East respiratory syndrome coronavirus infection*, Volume 58 Issue 8, ANTIMICROBIAL AGENTS AND CHEMOTHERAPY, 4885-93 (2014), https://www.ncbi.nlm.nih.gov/pmc/articles/PMC4136000/

17   Adriaan H. de Wilde et al., *Screening of an FDA-Approved Compound Library Identifies Four Small-Molecule Inhibitors of Middle East Respiratory Syndrome Coronavirus Replication in Cell Culture*, VOL 58 ISSUE 8, ASM JOURNALS ANTIMICROBIAL AGENTS AND CHEMOTHERAPY, 4875-4884 (2014), https://doi.org/10.1128/AAC.03011-14

18   Yi Su et al, Efficacy of early hydroxychloroquine treatment in preventing COVID-19 pneumonia aggravation, the experience from Shanghai, China, BIOSCIENCE TRENDS, (Jan 23, 2021). https://pubmed.ncbi.nlm.nih.gov/33342929/

19   Jean-Christophe Lagier et al, Outcomes of 3,737 COVID-19 patients treated with hydroxychloroquine/azithromycin and other regimens in Marseille, France: A retrospective analysis, TRAVEL MEDICINE AND INFECTIOUS DISEASE, (Jul-Aug ,2020). https://www.ncbi.nlm.nih.gov/pmc/articles/PMC7315163/

20   Saja H. Almazrou et al, Comparing the impact of Hydroxychloroquine based regimens and standard treatment on COVID-19 patient outcomes: A retrospective cohort study, SAUDI PHARMACEUTICAL JOURNAL, (Dec 2020). https://www.ncbi.nlm.nih.gov/pmc/articles/PMC7527306/

21   Majid Mokhtaria et al, Clinical outcomes of patients with mild COVID-19 following treatment with hydroxychloroquine in an outpatient setting, INTERNATIONAL IMMUNOPHARMACOLOGY, (July, 2021). https://www.sciencedirect.com/science/article/pii/S1567576921002721

22   The COVID-19 RISK and Treatments (CORIST) Collaboration, Use of hydroxychloroquine in hospitalised COVID-19 patients is associated with reduced mortality: Findings from the observational multicentre Italian CORIST study, EUROPEAN JOURNAL OF INTERNAL MEDICINE, (Dec 1, 2020). https://www.ejinme.com/article/S0953-6205(20)30335-6/fulltext

23   Awadhesh Kumar Singh et al, Chloroquine and hydroxychloroquine in the treatment of COVID-19 with or without diabetes: A systematic search and a narrative review with a special reference to India and other developing countries, DIABETES & METABOLIC SYNDROME: CLINICAL RESEARCH AND REVIEWS, (May-Jun, 2020). https://www.sciencedirect.com/science/article/abs/pii/S1871402120300515#!

24   Alyssa Paolicelli, *Drug Combo with Hydroxychloroquine Promising: NYU Study*, SPECTRUM NEWS NY1 (May 20, 2021, 7:18 AM), https://www.ny1.com/nyc/all-boroughs/news/2020/05/12/nyu-study-looks-at-hydroxychloroquine-zinc-azithromycin-combo-on-decreasing-covid-19-deaths

25   Roland Derwand , Martin Scholz and Vladimir Zelenko, COVID-19 outpatients: early risk-stratified treatment with zinc plus low-dose hydroxychloroquine and azithromycin: a retrospective case series study, INTERNATIONAL JOURNAL OF ANTIMICROBIAL AGENTS, (Dec 2020). https://pubmed.ncbi.nlm.nih.gov/33122096/

26   Samia Arshad, Treatment with hydroxychloroquine, azithromycin, and combination in patients hospitalized with COVID-19, INTERNATIONAL JOURNAL OF INFECTIOUS DISEASES, (Aug 2020). https://www.ncbi.nlm.nih.gov/pmc/articles/PMC7330574/

27   Fabricio Souza Neves, Correlation of the rise and fall in COVID-19 cases with the social isolation index and early outpatient treatment with hydroxychloroquine and chloroquine in the state of Santa Catarina, southern Brazil: A retrospective analysis, TRAVEL MEDICINE AND INFECTIOUS DISEASE, (May-Jun, 2021). https://pubmed.ncbi.nlm.nih.gov/33667717/

28   Didier Raoult et al, Hydroxychloroquine and azithromycin as a treatment of COVID-19: results of an open-label non-randomized clinical trial, INTERNATIONAL JOURNAL OF ANTIMICROBIAL AGENTS, (Jul 2020). https://www.ncbi.nlm.nih.gov/pmc/articles/PMC7102549/

29   Roland Derwand , Martin Scholz and Vladimir Zelenko, COVID-19 outpatients: early risk-stratified treatment with zinc plus low-dose hydroxychloroquine and azithromycin: a retrospective case series study, INTERNATIONAL JOURNAL OF ANTIMICROBIAL AGENTS, (Dec 2020). https://pubmed.ncbi.nlm.nih.gov/33122096/

30   Harvey A. Risch, Early Outpatient Treatment of Symptomatic, High-Risk COVID-19 Patients That Should Be Ramped Up Immediately as Key to the Pandemic Crisis, AMERICAN JOURNAL OF EPIDEMIOLOGY, (Nov 2, 2020). https://pubmed.ncbi.nlm.nih.gov/32458969/

31   Wesley H. Self et al, *Effect of Hydroxychloroquine on Clinical Status at 14 Days in Hospitalized Patients With COVID-19*, JAMA, (Nov 9, 2020). https://jamanetwork.com/journals/jama/fullarticle/2772922?utm_campaign=articlePDF&utm_medium=articlePDFlink&utm_source=articlePDF&utm_content=jama.2020.22240

32   Harvey A. Risch, *Early Outpatient Treatment of Symptomatic, High-Risk COVID-19 Patients That Should Be Ramped Up Immediately as Key to the Pandemic Crisis*, AMERICAN JOURNAL OF EPIDEMIOLOGY, (Nov 2, 2020). https://pubmed.ncbi.nlm.nih.gov/32458969/

33   A Anglemyer, HT Horvath, L. Bero, "Healthcare outcomes assessed with observational study designs compared with those assessed in randomized trials," *Cochrane Database of Systematic Reviews 2014*, Issue 4. Art. No.: MR000034. DOI: 10.1002/14651858.MR000034.pub2. Accessed 11 October 2021, https://www.cochranelibrary.com/cdsr/doi/10.1002/14651858.MR000034.pub2/full

34   Peter A. McCullough et al, *Multifaceted highly targeted sequential multidrug treatment of early ambulatory high-risk SARS-CoV-2 infection (COVID-19)*, REVIEWS IN CARDIOVASCULAR MEDICINE, (2020). https://rcm.imrpress.com/article/2020/2153-8174/RCM2020264.shtml

35   Joseph A. Ladapo, et al, *Randomized Controlled Trials of Early Ambulatory Hydroxychloroquine in the Prevention of COVID-19 Infection, Hospitalization, and Death: Meta-Analysis*, MEDRXIV (Sep. 30, 2020), https://doi.org/10.1101/2020.09.30.20204693

36   Jia Liu, et al, Hydroxychloroquine, a less toxic derivative of chloroquine, is effective in inhibiting

SARS-CoV-2 infection in vitro, NATURE, (Mar 18, 2020). https://www.nature.com/articles/s41421-020-0156-0#citeas

37 Zhaowei Chen et al, *Efficacy of hydroxychloroquine in patients with COVID-19: results of a randomized clinical trial*, MEDRXIV, (Apr 10, 2020). https://www.medrxiv.org/content/10.1101/2020.03.22.20040758v3

38 Wei Tang et al, Hydroxychloroquine in patients with mainly mild to moderate coronavirus disease 2019: open label, randomised controlled trial, BMJ, (May 14, 2020). https://www.bmj.com/content/369/bmj.m1849

39 Lotta Ulander, et al., Hydroxychloroquine reduces interleukin-6 levels after myocardial infarction: The randomized, double-blind, placebo-controlled OXI pilot trial, 15;337:21-27 Int J Cardiol (May 4, 2021), doi: 10.1016/j.ijcard.2021.04.062. https://pubmed.ncbi.nlm.nih.gov/33961943/

40 Abdulrhman Mohana et al, Hydroxychloroquine safety outcome with an approved therapeutic protocol for COVID-19 outpatients in Saudi Arabia, INTERNATIONAL JOURNAL OF INFECTIOUS DISEASE, (2021). https://www.ijidonline.com/article/S1201-9712(20)32235-9/pdf

41 Drugs.com, *Hydroxychloroquine Prices, Coupons and Patient Assistance Programs*, DRUGS.COM (Sept. 1, 2021), https://www.drugs.com/price-guide/hydroxychloroquine

42 Allison Inserro, *Gilead Sciences Sets US Price for COVID-19 Drug at $2340 to $3120 Based on Insurance*, AJMC (Jun. 29, 2020), https://www.ajmc.com/view/gilead-sciences-sets-us-price-for-covid19-drug-at-2340-to-3120-based-on-insurance

43 and Nathalie Boudet-Gizardin, *Can Hydroxychloroquine Be Legally Prescribed in France for Patients with COVID-19?*, Ginestié Magellan Paley-Vincent, (July 4, 2020). https://www.ginestie.com/en/covid-19-can-hydroxychloroquine-be-legally-prescribed-in-France-for-patients-with-covid-19/

44 Id.

45 Public Health Agency of Canada, *Chloroquine and hydroxychloroquine can have serious side effects. These drugs should be used only under the supervision of a physician*, (Apr 25, 2020). https://healthycanadians.gc.ca/recall-alert-rappel-avis/hc-sc/2020/72885a-eng.php

46 Barry Bateman, *Illegally imported chloroquine destroyed*, ENCA.COM, (Sep 23, 2020). https://www.enca.com/news/illegally-imported-chloroquine-destroyed

47 Bethany Rodgers, *Feds want a Utah pharmacist who illegally imported malaria drugs to destroy them. He's pitching a different plan.*, THE SALT LAKE TRIBUNE, (Apr 24, 2021). https://www.sltrib.com/news/2021/04/24/feds-want-utah-pharmacist/

48 Joan Donovan, et al, Trading Up the Chain: The Hydroxychloroquine Rumor, MEDIA MANIPULATION, (March, 2020). https://mediamanipulation.org/case-studies/trading-chain-hydroxychloroquine-rumor

49 James Todaro, MD, HCQ Tweet, TWITTER, (Mar 13, 2020). https://twitter.com/JamesTodaroMD/status/1238553266369318914?s=20

50 Ben Hirschler, *Google venture arm backs UK universal flu vaccine company, Reuters* (Jan. 14, 2018), https://www.reuters.com/article/us-health-flu-vaccine/google-venture-arm-backs-uk-universal-flu-vaccine-company-idUSKBN1F400H

51 Project Vaseline by Verily, Building a research community of engaged participants to aid scientists in the fight against COVID-19, https://verily.com/solutions/covid-19-research/

52 Ben Hirschler, *GSK and Google parent forge $715 million bioelectronic medicines firm*, REUTERS, (Aug 1, 2016). https://www.reuters.com/article/us-gsk-alphabet/gsk-and-google-parent-forge-715-million-bioelectronic-medicines-firm-idUSKCN10C1K8

53 Christina Farr, *Alphabet's Verily builds its own lab to speed up coronavirus test results*, CNBC, https://www.cnbc.com/2020/08/11/alphabets-verily-builds-its-own-coronavirus-testing-lab.html

54 Kayla Epstein, *Trump keeps touting a decades-old malaria pill as a coronavirus 'game changer', undercutting his top infectious-disease expert*, BUSINESS INSIDER, (Mar 21, 2020). https://www.businessinsider.com/trump-undercuts-fauci-on-whether-chloroquine-can-treat-coronavirus-2020-3

55 Scott Sayare, He was a Science Star. Then He Promoted a Questionable Cure for COVID-19., NEW YORK TIMES MAGAZINE, (May 21, 2020). https://www.nytimes.com/2020/05/12/magazine/didier-raoult-hydroxychloroquine.html

56 Elizabeth Cohen & Wesley Bruer, US stockpile stuck with 63 million doses of hydroxychloroquine, CNN (Jun. 17, 2020), https://www.cnn.com/2020/06/17/health/hydroxychloroquine-national-stockpile/index.html

57 BBC News, 'Ousted' US vaccine expert Rick Bright to file whistleblower complaint, BBC NEWS, (Apr 24, 2020). https://www.bbc.com/news/world-us-canada-52400721

58 FDA News Release, Coronavirus (COVID-19) Update: FDA Revokes Emergency Use Authorization for Chloroquine and Hydroxychloroquine, FDA (Jun. 15, 2020), https://www.fda.gov/news-events/press-announcements/coronavirus-covid-19-update-fda-revokes-emergency-use-authorization-chloroquine-and

59 U.S. Food and Drug Administration, Coronavirus (COVID-19) *Update: FDA Revokes Emergency Use Authorization for Chloroquine and Hydroxychloroquine*, (Jun 15, 2020). https://www.fda.gov/news-events/press-announcements/coronavirus-covid-19-update-fda-revokes-emergency-use-authorization-chloroquine-and

60 World Health Organization, *WHO discontinues hydroxychloroquine and lopinavir/ritonavir treatment*

    arms for COVID-19, (July 4, 2020). https://www.who.int/news/item/04-07-2020-who-discontinues-hydroxychloroquine-and-lopinavir-ritonavir-treatment-arms-for-covid-19

61   Meryl Nass, MD, *Covid-19 Has Turned Public Health Into a Lethal, Patient-Killing Experimental Endeavor*, ALLIANCE FOR HUMAN RESEARCH PROTECTION, (Jun 20, 2020). https://ahrp.org/covid-19-has-turned-public-health-into-a-lethal-patient-killing-experimental-endeavor/

62   *Id.*

63   Meryl Nass, *WHO and UK trials use potentially lethal hydroxychloroquine dose--according to WHO consultant*, ANTHRAX VACCINE-POSTS BY MERYL NASS, M.D. (Jun. 14, 2020), http://anthraxvaccine.blogspot.com/2020/06/who-trial-using-potentially-fatal.html

64   RECOVERY: Randomized Evaluation of COVID-19 Therapy, (2021). https://www.recoverytrial.net/

65   Transparency data: List of participants of SAGE and related sub-groups, GOV.UK (Jun. 18, 2021), https://www.gov.uk/government/publications/scientific-advisory-group-for-emergencies-sage-coronavirus-covid-19-response-membership/list-of-participants-of-sage-and-related-sub-groups

66   *Member Bios Update*, NERVTAG (Nov. 2018), https://assets.publishing.service.gov.uk/government/uploads/system/uploads/attachment_data/file/756256/NERVTAG_Member_Bios_update_November2018.pdf

67   Tess de la Mare, Zoonotic disease needs to be stopped at source, top scientist warns, (Jun 11, 2021). https://www.belfasttelegraph.co.uk/news/uk/zoonotic-disease-needs-to-be-stopped-at-source-top-scientist-warns-40529238.html

68   World Health Organization, *WHO R&D Blueprint COVID-19 Informal consultation on the dose of chloroquine and hydroxychloroquine for the SOLIDARITY Clinical Trial*, 5-6 (Apr. 8, 2020). https://www.who.int/docs/default-source/documents/r-d-blueprint-meetings/rd-blueprint-expert-group-on-cq-dose-call-apr8.pdf

69   World Health Organization, *WHO R&D Blueprint COVID-1 Informal consultation on the potential role of chloroquine in the clinical management of COVID 19 infection*, (Mar 13, 2020). https://www.who.int/blueprint/priority-diseases/key-action/RD-Blueprint-expert-group-on-CQ-call-Mar-13-2020.pdf

70   Mayla G.S. Borba, et al. *Effect of High vs Low Doses of Chloroquine Diphosphate as Adjunctive Therapy for Patients Hospitalized With Severe Acute Respiratory Syndrome Coronavirus 2 (SARS-CoV-2) Infection: A Randomized Clinical Trial*, 3(4):e208857 JAMA (Apr. 24, 2020), https://jamanetwork.com/journals/jamanetworkopen/fullarticle/2765499

71   *Id.*

72   Marcus Vinicius Guimarães de Lacerda, ESCAVADOR, https://www.escavador.com/sobre/3931344/marcus-vinicius-guimaraes-de-lacerda

73   Ruanne V. Barnabas et al, *Hydroxychloroquine as Postexposure Prophylaxis to Prevent Severe Acute Respiratory Syndrome Coronavirus 2 Infection*, ANNALS OF INTERNAL MEDICINE, (Dec 8, 2020). https://www.ncbi.nlm.nih.gov/pmc/articles/PMC7732017/

74   Christine Johnston et al, *Hydroxychloroquine with or without azithromycin for treatment of early SARS-CoV-2 infection among high-risk outpatient adults: A randomized clinical trial*, ECLINICAL MEDICINE (published by THE LANCET). (Feb 26, 2021). https://pubmed.ncbi.nlm.nih.gov/33681731/

75   *Researchers Overdosing COVID-19 Patients on Hydroxychloroquine, States Association of American Physicians & Surgeons (AAPS)*, CISION PR NEWSWIRE (Jun. 17, 2020), https://www.prnewswire.com/news-releases/researchers-overdosing-covid-19-patients-on-hydroxychloroquine-states-association-of-american-physicians--surgeons-aaps-301078986.html

76   Lindzi Wessel, *'It's a nightmare.' How Brazilian scientists became ensnared in chloroquine politics*, SCIENCE, (Jun 22, 2020). https://www.science.org/news/2020/06/it-s-nightmare-how-brazilian-scientists-became-ensnared-chloroquine-politics

77   Bill Gates, *Opinion: Bill Gates: Here's how to make up for lost time on covid-19*, WASHINGTON POST, (Mar 31, 2020). https://www.washingtonpost.com/opinions/bill-gates-heres-how-to-make-up-for-lost-time-on-covid-19/2020/03/31/ab5c3cf2-738c-11ea-85cb-8670579b863d_story.html

78   Ibid.

79   Elizabeth Cohen & Wesley Bruer, *US stockpile stuck with 63 million doses of hydroxychloroquine*, CNN (Jun. 17, 2020), https://www.cnn.com/2020/06/17/health/hydroxychloroquine-national-stockpile/index.html

80   America's Front Line Doctors, *This Virus Has a Cure*, (Jul 27 2020). https://www.dailymotion.com/video/x7vbtzf

81   Max Zahn, *Bill Gates: Spread of 'Outrageous" Coronavirus Video Shows Flaw in Social Media Platforms*, YAHOO, (Jul 29, 2020). https://www.yahoo.com/now/bill-gates-spread-of-outrageous-coronavirus-video-shows-flaw-in-social-media-platforms-124710983.html

82   Dina Bass and Candy Cheng, Bill Gates on Covid Vaccine Timing, Hydroxychloroquine, and That 5G Conspiracy Theory, BLOOMBERG, (Aug 13, 2020). https://www.bloomberg.com/features/2020-bill-gates-covid-vaccine/

83   Bill & Melinda Gates Foundation, 2018 Form 990, Attachment C, page 4, (2018). https://www.causeiq.com/organizations/view_990/911663695/011be856013f829aa3c948c35c2aa163

84   Dina Bass and Candy Cheng, Bill Gates on Covid Vaccine Timing, Hydroxychloroquine, and That 5G Conspiracy Theory, BLOOMBERG, (Aug 13, 2020). https://www.bloomberg.com/features/2020-bill-gates-covid-vaccine/

85   Mandeep Mehra, et al., RETRACTED: Hydroxychloroquine or chloroquine with or without a macrolide for treatment of COVID-19: a multinational registry analysis, THE LANCET (May 22, 2020), https://www.thelancet.com/journals/lancet/article/PIIS0140-6736(20)31180-6/fulltext

86   Mandeep Mehra, et al, RETRACTED: Cardiovascular Disease, Drug Therapy, and Mortality in Covid-19, 382:e102 NEJM (May 2020), https://www.nejm.org/doi/full/10.1056/NEJMoa2007621

87   Melissa Davey and Stephanie Kirchgaessner, *Surgisphere: governments and WHO changed Covid-19 policy based on suspect data from tiny US company*, THE GUARDIAN, (Jun 3, 2020). https://www.theguardian.com/world/2020/jun/03/covid-19-surgisphere-who-world-health-organization-hydroxychloroquine

88   Denise M. Hinton, Letter revoking EUA for chloroquine phosphate and hydroxychloroquine sulfate, U.S. FOOD AND DRUG ADMINISTRATION, (Jun 15, 2020). https://www.fda.gov/media/138945/download

89   BBC, *Coronavirus: WHO halts trials of hydroxychloroquine over safety fears*, (May 25, 2020). https://www.bbc.com/news/health-52799120

90   Jacqueline Howard, *UK Covid-19 trial ends hydroxychloroquine study because there's no evidence the drug benefits patients*,

91   Matthias Blamont, Alistair Smout and Emilio Parodi, EU governments ban malaria drug for COVID-19, trial paused as safety fears grow, REUTERS, (May 27, 2020). https://www.reuters.com/article/health-coronavirus-hydroxychloroquine-fr/eu-governments-ban-malaria-drug-for-covid-19-trial-paused-as-safety-fears-grow-idUSKBN2340A6

92   Kelly Servick & Martin Enserink, A mysterious company's coronavirus papers in top medical journals may be unraveling, SCIENCE (Jun. 2, 2020), https://www.science.org/news/2020/06/mysterious-company-s-coronavirus-papers-top-medical-journals-may-be-unraveling

93   Roni Caryn Rabin, Scientists Question Validity of Major Hydroxychloroquine Study, NYT (May 29, 2020), https://www.nytimes.com/2020/05/29/health/coronavirus-hydroxychloroquine.html

94   James Heathers, The Lancet has made one of the biggest retractions in modern history. How could this happen?, THE GUARDIAN (Jun. 5, 2020), https://www.theguardian.com/commentisfree/2020/jun/05/lancet-had-to-do-one-of-the-biggest-retractions-in-modern-history-how-could-this-happen

95   WHO Solidarity Consortium, Repurposed Antiviral Drugs for Covid-19 — Interim WHO Solidarity Trial Results, 384:497-511 NEJM (Dec. 2, 2020), DOI: 10.1056/NEJMoa2023184. https://www.nejm.org/doi/full/10.1056/nejmoa2023184

96   The RECOVERY Collaborative Group, Effect of Hydroxychloroquine in Hospitalized Patients with Covid-19, 383:2030-2040 NEJM (Oct. 8, 2020), DOI: 10.1056/NEJMoa2022926. https://www.nejm.org/doi/full/10.1056/NEJMoa2022926

97   Richard Smith, Medical journals are an extension of the marketing arm of pharmaceutical companies, 2.5 PLOS MEDICINE e138 (May 2005). doi:10.1371/journal.pmed.0020138. https://www.ncbi.nlm.nih.gov/pmc/articles/PMC1140949/

98   *Id.*

99   *Id.*

100   Horton, R. [2004], The dawn of McScience. *New York Rev Books* 51(4): 7–9. Angell, M. (2005), *The Truth About Drug Companies: How They Deceive Us And What To Do About It*. New York: Random House.

101   Devan Cole, Fauci: Science shows hydroxychloroquine is not effective as a coronavirus treatment, CNN (May 27, 2020), https://www.cnn.com/2020/05/27/politics/anthony-fauci-hydroxychloroquine-trump-cnntv/index.html

102   Christopher M. Wittich, et al., Ten common questions (and their answers) about off-label drug use, 87(10), MAYO CLINIC PROCEEDINGS, 982–990, https://doi.org/10.1016/j.mayocp.2012.04.017

103   FDA, *FDA cautions against use of hydroxychloroquine or chloroquine for COVID-19 outside of the hospital setting or a clinical trial due to risk of heart rhythm problems,* FDA (Jul. 1, 2020), https://www.fda.gov/drugs/drug-safety-and-availability/fda-cautions-against-use-hydroxychloroquine-or-chloroquine-covid-19-outside-hospital-setting-or

104   George C. Fareed, MD et al, *Open letter to Dr. Anthony Fauci regarding the use of hydroxychloroquine for treating COVID-19*, THE DESERT REVIEW, (Aug 13, 2020), https://www.thedesertreview.com/opinion/columnists/open-letter-to-dr-anthony-fauci-regarding-the-use-of-hydroxychloroquine-for-treating-covid-19/article_31d37842-dd8f-11ea-80b5-bf80983bc072.html

105   *Id.*

106   Samia Arshad, et al., Treatment with hydroxychloroquine, azithromycin, and combination in patients hospitalized with COVID-19, 97 IJID (Aug. 2020) 396-403, doi:10.1016/j.ijid.2020.06.099. https://www.ncbi.nlm.nih.gov/pmc/articles/PMC7330574/

107   Beth LeBlanc, Fauci: Henry Ford Health's hydroxychloroquine study 'flawed', THE DETROIT NEWS (Aug. 1, 2020), https://www.detroitnews.com/story/news/local/michigan/2020/07/31/anthony-fauci-henry-ford-health-hydroxychloroquine-study-flawed/5559367002/

108   Boards of pharmacy and other actions relating to COVID-19 prescribing, AMA (2020), https://www.ama-assn.org/system/files/2020-04/board-of-pharmacy-covid-19-prescribing.pdf

109   Valerie Richardson, *Rudy Giuliani urges Andrew Cuomo to lift hydroxychloroquine restrictions for coronavirus*, THE WASHINGTON TIMES, (Apr 5, 2020). https://www.washingtontimes.com/news/2020/apr/5/rudy-giuliani-urges-andrew-cuomo-lift-hydroxychlor/

110   News 4 and FOX 11 Digital Team, *Gov. Sisolak signs emergency regulation restricting drug distribution during COVID-19*, (Mar 24, 2020). https://mynews4.com/news/local/gov-sisolak-signs-emergency-regulation-restricting-drug-distribution-during-covid-19

111   American Medical Association, *Boards of pharmacy and other actions relating to COVID-19 prescribing*, AMA ADVOCACY RESOURCE CENTER, (Apr 27, 2020). https://www.ama-assn.org/system/files/2020-04/board-of-pharmacy-covid-19-prescribing.pdf

112   Free Republic, *Is HCQ banned in your state? (probably not)*, (Aug 18, 2020). https://freerepublic.com/focus/f-news/3875427/posts?q=1&;page=1

113   FDA FAQ, *Frequently Asked Questions on the Revocation of the Emergency Use Authorization for Hydroxychloroquine Sulfate and Chloroquine Phosphate,* FDA (Jun. 15, 2021), https://www.fda.gov/media/138946/download

114   National Institute of Allergy and Infectious Diseases, *BULLETIN—NIH Clinical Trial Evaluating Hydroxychloroquine and Azithromycin for COVID-19 Closes Early*, (Jun 20, 2020). https://www.niaid.nih.gov/news-events/bulletin-nih-clinical-trial-evaluating-hydroxychloroquine-and-azithromycin-covid-19

115   Denise M. Hinton, *Letter revoking EUA for chloroquine phosphate and hydroxychloroquine sulfate,* U.S. FOOD AND DRUG ADMINISTRATION, (Jun 15, 2020). https://www.fda.gov/media/138945/download

116   World Health Organization, *Coronavirus disease (COVID-19): Solidarity Trial and hydroxychloroquine*, (Jun 19, 2020). https://www.who.int/news-room/q-a-detail/coronavirus-disease-covid-19-hydroxychloroquine

117   Natasha Donn, Covid-19: Portugal becomes 4th country to suspend use of hydroxychloroquine, PORTUGAL RESIDENT, (May 29, 2020). https://www.portugalresident.com/covid-19-portugal-becomes-4th-country-to-suspend-use-of-hydroxychloroquine/

118   Michel Jullian and Xavier Azalbert, Covid-19: hydroxychloroquine works, a proof ?, FRANCE SOIR, (Jul 13, 2020). https://www.francesoir.fr/societe-sante/covid-19-hydroxychloroquine-works-irrefutable-proof

119   Catherine Perea, *Doctor Sánchez Cárdenas believes that the use of hydroxychloroquine in Panama has yielded results*, TELEMETRO, (Aug 26, 2020). https://www.telemetro.com/nacionales/2020/08/26/doctor-sanchez-cardenas-dice-hidroxicloroquina/3169570.html

120   Association of American Physicians & Surgeons (AAPS), *AAPS vs. Food & Drug Administration et al,* (Jun 2, 2020). https://aapsonline.org/judicial/aaps-v-fda-hcq-6-2-2020.pdf

121   Association of American Physicians & Surgeons, *More Evidence Presented for Why Hydroxychloroquine Should be Made Available, in a New Court Filing by AAPS*, AAPS ONLINE, (Jul 22, 2020). https://aapsonline.org/more-evidence-presented-for-why-hydroxychloroquine-should-be-made-available-in-a-new-court-filing-by-aaps/

122   Jo Nova, Countries that use Hydroxychloroquine may have 80% lower Covid death rates, JONOVA (Aug. 8, 2020), https://joannenova.com.au/2020/08/countries-that-use-hydroxychloroquine-may-have-80-lower-covid-death-rates/

123   C19Study.com, *COVID-19 treatment studies for Hydroxychloroquine*, (Sep 20, 2020). https://c19hcq.com/

124   C19Study.com, *Countries based on current HCQ/CQ usage*, https://c19hcq.com/countries.html,

125   Statista, Coronavirus (COVID-19) deaths worldwide per one million population as of September 21, 2021, by country, (Sep 21, 2121) https://www.statista.com/statistics/1104709/coronavirus-deaths-worldwide-per-million-inhabitants/

126   George C. Fareed, MD et al, *Open letter to Dr. Anthony Fauci regarding the use of hydroxychloroquine for treating COVID-19*, THE DESERT REVIEW, (Aug 13, 2020), https://www.thedesertreview.com/opinion/columnists/open-letter-to-dr-anthony-fauci-regarding-the-use-of-hydroxychloroquine-for-treating-covid-19/article_31d37842-dd8f-11ea-80b5-bf80983bc072.html

127   Statista, *Coronavirus (COVID-19) deaths worldwide per one million population as of September 24, 2021, by country*, (Sep 24, 2021). https://www.statista.com/statistics/1104709/coronavirus-deaths-worldwide-per-million-inhabitants/

128   *Id.*

129   *Id.*

130   *Id.*

131   *Reported Cases and Deaths by Country or Territory*, WORLDOMETER (Sept. 24, 2021), https://www.worldometers.info/coronavirus/

132   The Global Economy, Netherlands: Covid new deaths per million, (2021).https://www.theglobaleconomy.com/Netherlands/covid_new_deaths_per_million/

133   Marine Strauss, *Belgium, once hard-hit, reports zero coronavirus deaths for first time since March,*

REUTERS, (Jul 14, 2020). https://www.reuters.com/article/us-health-coronavirus-belgium/belgium-once-hard-hit-reports-zero-coronavirus-deaths-for-first-time-since-march-idUSKCN24F0U7

134   Statista, *Coronavirus (COVID-19) deaths worldwide per one million population as of September 24, 2021, by country*, (Sep 24, 2021). https://www.statista.com/statistics/1104709/coronavirus-deaths-worldwide-per-million-inhabitants/

135   *More Evidence Presented for Why Hydroxychloroquine Should be Made Available, in a New Court Filing by the Association of American Physicians & Surgeons (AAPS)*, CISION PR NEWSWIRE (Jul. 22, 2020), https://www.prnewswire.com/news-releases/more-evidence-presented-for-why-hydroxychloroquine-should-be-made-available-in-a-new-court-filing-by-the-association-of-american-physicians--surgeons-aaps-301098030.html

136   Harvey Risch, *The Key to Defeating COVID-19 Already Exists. We Need to Start Using It*, NEWSWEEK (Jul. 23, 2020), https://www.newsweek.com/key-defeating-covid-19-already-exists-we-need-start-using-it-opinion-1519535

137   Meryl Nass, *How a False Hydroxychloroquine Narrative was created, and more*, ANTHRAX VACCINE-POST BY MERYL NASS, M.D. (Jun. 27, 2020), https://anthraxvaccine.blogspot.com/2020/06/how-false-hydroxychloroquine-narrative.html

138   *Id.*

139   *Id.*

140   *Id.*

141   *Id.*

142   Ronald L. Trowbridge, Anthony Fauci: A Product of His Training, THE INDEPENDENT (Apr. 27, 2020), https://www.independent.org/news/article.asp?id=13129

143   NIH ends key COVID-19 studies of hydroxychloroquine, BIOPHARMADIVE (Jun. 22, 2020), https://www.biopharmadive.com/news/nih-hydroxychloroquine-coronavirus-trial-halt/580270/

144   Meryl Nass, *Fauci the Hypocrite. Do NIAID royalties cloud his thinking?*, TRUTH IN THE AGE OF COVID, (Apr 19, 2020). https://merylnassmd.com/fauci-hypocrite-do-niaid-royalties/

145   Darryl Falzarano, Emmie de Wit, et al., *Treatment with interferon-a2b and ribavirin improves outcome in MERS-CoV-infected rhesus macaques*, 19(10) NAT MED 1313-1317, (2013), doi:10.1038/nm.3362 https://pubmed.ncbi.nlm.nih.gov/24013700/

146   Liz Szabo, Scientists fight deadly new coronavirus, USA TODAY (Apr. 18, 2013), https://www.usatoday.com/story/news/nation/2013/04/18/scientists-develop-possible-treatment-for-new-sars-like-virus/2091443/

147   *Id.*

148   *Id.*

149   *Id.*

150   U.S. Food and Drug Administration, Package Insert, PEG-IntronTM (Peginterferon alfa-2b) Powder for Injection, Schering Corporation, (Oct 3, 2001). https://www.accessdata.fda.gov/drugsatfda_docs/label/2001/pegsche080701LB.htm#warn

151   Pfizer and BioNTech, FDA Briefing Document Pfizer-BioNTech COVID-19 Vaccine, Vaccines and Related Biological Products Advisory Committee Meeting, (Dec 10, 2020). https://www.fda.gov/media/144245/download

152   ModernaTX, Inc., FDA Briefing Document Moderna COVID-19 Vaccine, Vaccines and Related Biological Products Advisory Committee Meeting, (Dec 17. 2020). https://www.fda.gov/media/144434/download

153   Janssen, Ad26.COV2.S Vaccine for the Prevention of COVID-19, Vaccines and Related Biological Products Advisory Committee Meeting, (Feb 26, 2021). https://www.fda.gov/media/146217/download

154   Christopher J. Boggs, *Insurance Implications of Required Vaccines*, INSURANCE JOURNAL, (Feb 22, 2021). https://www.insurancejournal.com/blogs/big-i-insights/2021/02/22/602123.htm

155   Kevin J. Hickey, The PREP Act and COVID-19: Limiting Liability for Medical Countermeasures, CONGRESSIONAL RESEARCH SERVICE (Updated Mar. 19, 2021), https://crsreports.congress.gov/product/pdf/LSB/LSB10443

156   Children's Health Defense Team, *Here's why Bill Gates wants indemnity… Are you willing to take the risk?*, CHILDREN'S HEALTH DEFENSE, (Apr. 11, 2020), https://childrenshealthdefense.org/news/heres-why-bill-gates-wants-indemnity-are-you-willing-to-take-the-risk/

157   *Dr. Anthony Fauci: No placebo-controlled trial has shown hydroxychloroquine is effective*, 1:17 YOUTUBE CNBC (Jul. 31, 2020), https://www.youtube.com/watch?v=xDjVwXM8ESE

158   WHO, *WHO recommends against the use of remdesivir in COVID-19 patients*, WHO NEWSROOM (Nov. 20, 2020) https://www.who.int/news-room/feature-stories/detail/who-recommends-against-the-use-of-remdesivir-in-covid-19-patients

159   Charan, Jaykaran et al. *Rapid review of suspected adverse drug events due to remdesivir in the WHO database; findings and implications*, 14,1 EXPERT REVIEW OF CLINICAL PHARMACOLOGY 95-103 (2021), PMID: 33252992 doi:10.1080/17512433.2021.1856655

160   Real-World Evidence, *Real-world data (RWD) and real-world evidence (RWE) are playing an increasing role in health care decisions*, FDA (Jul. 16, 2021), https://www.fda.gov/science-research/science-and-research-special-topics/real-world-evidence

161   FDA, 21st Century Cures Act, US FOOD & DRUG ADMINISTRATION (Jan. 31, 2020), https://www.fda.gov/regulatory-information/selected-amendments-fdc-act/21st-century-cures-act

162   Michael Thau, *COVID19: Top Yale Professor Says 'Decisions Not Based On Science'*, PRINCIPIA SCIENTIFIC, (Oct 14, 2020). https://principia-scientific.com/covid19-top-yale-professor-says-decisions-not-based-on-science/

163   *Id.*

Endnotes - Ivermectin

1   The Nobel Prize in Physiology or Medicine 2015, The Nobel Prize Press Release (Oct. 5, 2015), https://www.nobelprize.org/prizes/medicine/2015/press-release/

2   World Health Organization, "Executive summary: the selection and use of essential medicines 2021: report of the 23rd WHO Expert Committee on the selection and use of essential medicines," WHO (Sep. 30, 2021), https://www.who.int/publications/i/item/WHO-MHP-HPS-EML-2021.01

3   World Health Organization Model List of Essential Medicines—22nd List, WHO (2021), https://apps.who.int/iris/bitstream/handle/10665/345533/WHO-MHP-HPS-EML-2021.02-eng.pdf

4   Andy Crump and Satoshi Ōmura, "Ivermectin, 'Wonder drug' from Japan: the human use perspective," Proceedings of the Japan Academy, Series B, (Feb 10, 2011), https://www.ncbi.nlm.nih.gov/labs/pmc/articles/PMC3043740/

5   Fatemeh Heidary and Reza Gharebaghi, *Ivermectin: a systematic review from antiviral effects to COVID-19 complementary regimen*, The Journal of Antibiotics (June 12, 2020), https://www.nature.com/articles/s41429-020-0336-z

6   "Lab experiments show anti-parasitic drug, Ivermectin, eliminates SARS-CoV-2 in cells in 48 hours." Monash Biomedicine Discovery Institute (April 3, 2020), https://www.monash.edu/discovery-institute/news-and-events/news/2020-articles/Lab-experiments-show-anti-parasitic-drug,-Ivermectin,-eliminates-SARS-CoV-2-in-cells-in-48-hours

7   Ibid.

8   Juan Chamie-Quintero et al., "Ivermectin for COVID in Peru: 14-fold reduction in nationwide excess deaths, p<0.002 for effect by state, then 13-fold increase after ivermectin use restricted," *OSF Preprints* (March 8, 2021), https://osf.io/9egh4/

9   Hector Carvallo, et al., "Study of the Efficacy and Safety of Topical Ivermectin + IotaCarrageenan in the Prophylaxis against COVID-19 in Health Personnel," *Journal of Biomedical Research and Clinical Investigation* V. 2 I 1 at 1007 (Nov. 17, 2020), https://www.medicalpressopenaccess.com/single_article.php?refid=82

10   Alam et al., "Ivermectin as Pre-exposure Prophylaxis for COVID-19 among Healthcare Providers in a Selected Tertiary Hospital in Dhaka—An Observational Study," *European Journal of Medical & Health Sciences*, Vol. 2 No. 6 (Published December 15, 2020).

11   AD Santin, et al., "Ivermectin: a multifaceted drug of Nobel prize-honoured distinction with indicated efficacy against a new global scourge," 43:100924 New Microbes New Infect (Aug. 3, 2021) doi: 10.1016/j.nmni.2021.100924, https://pubmed.ncbi.nlm.nih.gov/34466270/

12   Statement to U.S. Committee on Homeland Security and Governmental Affairs: "Early Outpatient Treatment: An Essential Part of the COVID-19 Solution, Part II." THE ICON STUDY: Dr. Jean-Jacques Rajter Reviews Ivermectin, *Hope Pressworks* (January 18, 2021), https://hopepressworks.org/f/the-icon-study-dr-jean-jacques-rajter-reviews-ivermectin

13   Alam et al., "Ivermectin as Pre-exposure Prophylaxis for COVID-19 among Healthcare Providers in a Selected Tertiary Hospital in Dhaka—An Observational Study," Alam et al., *European Journal of Medical & Health Sciences*, Vol. 2 No. 6 (Published December 15, 2020), https://www.ejmed.org/index.php/ejmed/article/view/599

14   Peter A. McCullough et al., "Multifaceted highly targeted sequential multidrug treatment of early ambulatory high-risk SARS-CoV-2 infection (COVID-19)," *Reviews in Cardiovascular Medicine*, 2020, Vol. 21 Issue (4): 517-530 https://rcm.imrpress.com/EN/10.31083/j.rcm.2020.04.264

15   Ivermectineforcovid.com, "A Letter to NIH and Dr. Anthony Fauci. Is Anybody Home?," *Ivermectin for COVID* (Mar. 4, 2021), https://ivermectinforcovid.com/a-letter-to-nih-and-dr-anthony-fauci-is-anybody-home/

16   C. Bernigaud et al., "Bénéfice de l'ivermectine: de la gale à la COVID-19, un exemple de sérendipité," *Annales de Dermatologie et de Vénéréologie* (December, 2020), https://www.sciencedirect.com/science/article/abs/pii/S0151963820030627X?via%3Dihub

17   C. Bernigaud et al., "Ivermectin benefit: from scabies to COVID-19, an example of serendipity," *Annals of Dermatology and Venereology* (December, 2020), https://c19ivermectin.com/bernigaud.html

18   Choudhury, Abhigyan et al.,"Exploring the Binding Efficacy of Ivermectin against the Key Proteins of SARS-CoV-2 Pathogenesis: An in Silico Approach," *Future Virology* (March 25, 2021), https://www.ncbi.nlm.nih.gov/pmc/articles/PMC7996102/

19   Peter A. McCullough et al., "Early Ambulatory Multidrug Therapy Reduces Hospitalization and Death in High-Risk Patients with SARS-CoV-2 (COVID-19)," Vol. 6 No. 03 (March, 2021) | Page No.: 219–221

20   "FLCCC Alliance Invited to the National Institutes of Health (NIH) COVID-19 Treatment Guidelines Panel to Present Latest Data on Ivermectin," Press Release, Front Line COVID-19 Critical Care Alliance (FLCCC) (January 7, 2021), https://covid19criticalcare.com/wp-content/uploads/2021/01/FLCCC-PressRelease-NIH-C19-Panel-FollowUp-Jan7-2021.pdf

21   Dr. Andrew Hill, "Covid-19: WHO-sponsored preliminary review indicates Ivermectin effectiveness," Dr. Andrew Hill, *Swiss Policy Research* (December 31, 2020), https://swprs.org/who-preliminary-review-confirms-Ivermectin-effectiveness/

22   Carlos Chaccour, "The effect of early treatment with ivermectin on viral load, symptoms and humoral response in patients with non-severe COVID-19: A pilot, double-blind, placebo-controlled, randomized clinical trial," EClinical Medicine Published by the *Lancet*, (Jan 19, 2021), https://www.thelancet.com/journals/eclinm/article/PIIS2589-5370(20)30464-8/fulltext

23   Pierre Kory, et al., "Review of the Emerging Evidence Demonstrating the Efficacy of Ivermectin in the Prophylaxis and Treatment of COVID-19," *Front Line COVID-10 Critical Care Alliance* (Jan 16, 2021), https://covid19criticalcare.com/wp-content/uploads/2020/11/FLCCC-Ivermectin-in-the-prophylaxis-and-treatment-of-COVID-19.pdf

24   "Dr. Pierre Kory, president of the FLCCC Alliance testifies before Senate Committee on Homeland Security and Governmental Affairs looking into early outpatient COVID-19 treatment," Press release, *Front Line COVID-19 Critical Care Alliance (FLCCC)* (December 8, 2020), https://www.newswise.com/coronavirus/dr-pierre-kory-president-of-the-flccc-alliance-testifies-before-senate-committee-on-homeland-security-and-governmental-affairs-looking-into-early-outpatient-covid-19-treatment

25   Pierre Kory et al., "Clinical and Scientific Rationale for the "MATH+" Hospital Treatment Protocol for COVID-19," *Journal of Intensive Care Medicine* (December 15, 2020), https://journals.sagepub.com/doi/10.1177/0885066620973585

26   "One Page Summary of the Clinical Trials Evidence for Ivermectin in COVID-19," *Front Line COVID-19 Critical Care Alliance (FLCCC)* (January 11, 2021), https://covid19criticalcare.com/wp-content/uploads/2020/12/One-Page-Summary-of-the-Clinical-Trials-Evidence-for-Ivermectin-in-COVID-19.pdf

27   Testimony by Dr. Pierre Kory to the U.S. Senate Committee on Homeland Security and Governmental Affairs (2020), *Front Line COVID-19 Critical Care Alliance (FLCCC)*, https://covid19criticalcare.com/senate-testimony/

28   "FLCCC Alliance Invited to the National Institutes of Health (NIH) COVID-19 Treatment Guidelines Panel to Present Latest Data on Ivermectin," Press Release, Front Line COVID-19 Critical Care Alliance (FLCCC) (January 7, 2021), https://covid19criticalcare.com/wp-content/uploads/2021/01/FLCCC-PressRelease-NIH-C19-Panel-FollowUp-Jan7-2021.pdf

29   Donato Paolo Mancini, "Cheap antiparasitic could cut chance of Covid-19 deaths by up to 75%," *The Financial Times* (January 19, 2021), https://www.ft.com/content/e7cb76fc-da98-4a31-9c1f-926c58349c84

30   "NIH Updates its Position on Ivermectin," *COVEXIT: COVID-19 News & Policy Analysis* (January 14, 2021), https://covexit.com/nih-updates-its-position-on-ivermectin/

31   "Ivermectin is not approved by the FDA for the treatment of any viral infection," *National Institutes of Health COVID-19 Treatment Guidelines* (Last Updated: February 11, 2021), https://www.covid19treatmentguidelines.nih.gov/therapies/antiviral-therapy/ivermectin/

32   Peter Yim, "Federal Judge Allows NIH Ivermectin Deception Case to Proceed." *Trial Site News* (August 28, 2021), https://trialsitenews.com/federal-judge-allows-nih-ivermectin-deception-case-to-proceed/

33   National Institutes of Health email from Safia Kuruiakose, PharmD to partially redacted list of recipients (January 5, 2021). Email was obtained via Freedom of Information Act. https://drive.google.com/file/d/1qh9U0FvrF-lNnNyuZ9_ljPS8PCXpAdzR/view?usp=sharing

34   National Institutes of Health email from Andy Pavia to Dr. Susanna Naggie, M.D., and Cc: Safia Kuruiakose, PharmD and partially redacted list of recipients (January 6, 2021). Email was obtained via the Freedom of Information Act. https://drive.google.com/file/d/18OfEy2byeYQBxY_xRuQjgmwtRS8m23zu/view?usp=sharing

35   "Large clinical trial to study repurposed drugs to treat COVID-19 symptoms," National Institutes of Health news release (April 19, 2021), https://www.nih.gov/news-events/news-releases/large-clinical-trial-study-repurposed-drugs-treat-covid-19-symptoms

36   *TrialSite News* staff, "Chairman of Tokyo Metropolitan Medical Association Declares During Surge, Time for Ivermectin is Now," *TrialSite News* (August 16, 2021), https://trialsitenews.com/chairman-of-tokyo-metropolitan-medical-association-declares-during-surge-time-for-ivermectin-is-now/

37   *TrialSite News* staff, "Ivermectin Authorized in Indonesia as Pharma Issued License for Production to Battle COVID-19," *TrialSite News* (June 24, 2021), https://trialsitenews.com/ivermectin-authorized-in-indonesia-as-pharma-issued-license-for-production-to-battle-covid-19/

38   *World Tribune* staff, "Media mum on India's Ivermectin success story; El Salvador offers it free to citizens," *World Tribune* (Sept. 19, 2021), https://www.worldtribune.com/media-mum-on-indias-ivermectin-success-story-el-salvador-offers-it-free-to-citizens/

39   Nick Corbishley, "As US Prepares to Ban Ivermectin for Covid-19, More Countries in Asia Begin Using It," *Naked Capitalism* (September 7, 2021), https://www.nakedcapitalism.com/2021/09/as-us-prepares-to-ban-ivermectin-for-covid-19-more-countries-in-asia-begin-using-it.html?utm_source=dlvr.it&utm_medium=twitter

40   Rodrigo Guerrero et al., "COVID-19: The Ivermectin African Enigma," Europe PMC, doi: 10.25100/cm.v51i4.4613 (December 30, 2020), https://europepmc.org/article/PMC/PMC7968425

41   Justus R. Hope, M.D., "Ivermectin obliterates 97 percent of Delhi cases," *The Desert Review* (Jun 1, 2021 Updated Jun 7, 2021), https://www.thedesertreview.com/news/national/ivermectin-obliterates-97-percent-of-delhi-cases/article_6a3be6b2-c31f-11eb-836d-2722d2325a08.html

42   Justus R. Hope, M.D., "Ivermectin crushes Delhi cases," *The Desert Review* (May 18, 2021; updated September 8, 2021), https://www.thedesertreview.com/opinion/letters_to_editor/ivermectin-crushes-delhi-cases/article_31f3afcc-b7fa-11eb-9585-0f6a290ee105.html

43   See video at 6:58. However, the video is unavailable as of 9/25/2021. https://youtu.be/pko4LIdUQCI

44   Bhuma Shrivastava and Malavika Kaur Makol, "Forecaster who predicted India's COVID peak sees new wave coming," *Boston Globe* (Aug 1, 2021), https://www.bostonglobe.com/2021/08/01/business/forecaster-who-predicted-indias-covid-peak-sees-new-wave-coming/

45   The Hindu Net Desk, "Coronavirus updates | January 1, 2021," *The Hindu* (Jan 1, 2021), https://www.thehindu.com/news/national/coronavirus-updates-january-1-2021/article33469047.ece

46   Maria Caspani and Peter Szekely, "U.S. tops 21 million COVID-19 cases with record hospitalizations as states ramp up vaccinations," *Reuters* (Jan 6, 2021), https://ph.news.yahoo.com/u-sets-covid-19-hospitalization-161111470.html

47   Maulshree Seth, "Uttar Pradesh government says early use of Ivermectin helped to keep positivity, deaths low," *The Indian Express* (May 12, 2021), https://indianexpress.com/article/cities/lucknow/uttar-pradesh-government-says-ivermectin-helped-to-keep-deaths-low-7311786/

48   "Uttar Pradesh government says early use of Ivermectin helped to keep positivity, deaths low," MSN (May 12, 2021), https://www.msn.com/en-in/news/other/uttar-pradesh-government-says-early-use-of-ivermectin-helped-to-keep-positivity-deaths-low/ar-BB1gDp5U

49   "India State of 241 MILLION People Declared COVID-Free After Government Promotes Ivermectin," Infowars.com (September 18, 2021), https://www.infowars.com/posts/india-province-of-241-million-people-declared-covid-free-after-government-promotes-ivermectin/

50   "33 districts in Uttar Pradesh are now Covid-free: State govt," *Hindustan Times*, New Delhi (September 10, 2021), https://www.hindustantimes.com/cities/lucknow-news/33-districts-in-uttar-pradesh-are-now-covid-free-state-govt-101631267966925.html

51   Justin R. Hope, "Ivermectin A Success In Indian States Where It Is Used," *Peckford 42* (Aug 22, 2021), https://peckford42.wordpress.com/2021/08/22/ivermectin-a-success-in-indian-states-where-it-is-used/

52   Ibid.

53   Stefan J. Bos, "India's Largest State Nearly COVID-Free After Using Alternative Drug," *Worthy News* (Sept 13, 2012), https://www.worthynews.com/61607-indias-largest-state-nearly-covid-free-after-using-alternative-drug

54   Justus R. Hope, "India's Ivermectin Blackout: Part II," The Desert Review (Aug 13, 2021), https://www.thedesertreview.com/news/national/indias-ivermectin-blackout-part-ii/article_a0b6c378-fc78-11eb-83c0-93166952f425.html

55   Justus R. Hope, M.D., "Tamil Nadu leads India in new infections, denies citizens ivermectin," *The Desert Review* (May 21, 2021/Updated Sep 8, 2021), https://www.thedesertreview.com/opinion/letters_to_editor/tamil-nadu-leads-india-in-new-infections-denies-citizens-Ivermectin/article_32634012-ba66-11eb-9211-ab378d521f9a.html

56   https://journals.lww.com/americantherapeutics/fulltext/2021/08000/ivermectin_for_prevention_and_treatment_of.7.aspx

57   Justus R. Hope, M.D., "Dr. Tess Lawrie: The Conscience of Medicine." *The Desert Review* (May 4, 2021), https://www.thedesertreview.com/opinion/letters_to_editor/dr-tess-lawrie-the-conscience-of-medicine/article_ff673eca-ac2d-11eb-adaa-ab952b1d2661.html

58   Luke Andrews, "Cheap hair lice drug may cut the risk of hospitalized Covid patients dying by up to 80%, study finds," *Daily Mail* (January 4, 2021), https://www.dailymail.co.uk/news/article-9110301/Cheap-hair-lice-drug-cut-risk-hospitalised-Covid-patients-dying-80-study-finds.html

59   Dr. Tess Lawrie, "Who are the BIRD Group?," *BIRD* (Jan. 2021), https://bird-group.org/who-are-bird/

60   *TrialSite News* staff, "Are the Independent Actually Dependent? Is Cochrane's Ivermectin Analysis Biased?" *TrialSite News* (May 6, 2021), https://trialsitenews.com/are-the-independent-actually-dependent-is-cochranes-ivermectin-analysis-biased/

61   National Institutes of Health, NIH Awards by Location and Organization, https://report.nih.gov/award/index.cfm?ot=&fy=2021&state=&ic=&fm=&orgid=4997001&distr=&rfa=&om=y&pid=&view=state

62   Unitaid Press Release, "Unitaid hails new US$ 50 million contribution from the Bill & Melinda Gates Foundation," *Unitaid* (Dec. 7, 2017), https://unitaid.org/news-blog/unitaid-hails-new-us-50-million-contribution-bill-melinda-gates-foundation/#en

63   "Early Outpatient Treatment: An Essential Part of a COVID-19 Solution, Part II." Testimony by Dr. Pierre Kory to U.S. Senate Homeland Security and Governmental Affairs Committee (December 8, 2020), https://www.hsgac.senate.gov/early-outpatient-treatment-an-essential-part-of-a-covid-19-solution-part-ii

64   U.S. Sen. Ron Johnson, "Google/YouTube censor Senate hearing on ivermectin and early treatment of Covid-19," *Wall Street Journal* (February 4, 2021), https://anthraxvaccine.blogspot.com/2021/02/googleyoutube-censor-senate-hearing-on.html

65   "Currently available data do not show ivermectin is effective against COVID-19," "Why You Should Not Use Ivermectin to Treat or Prevent COVID-19." *U.S. Food and Drug Administration* (September 3, 2021), https://www.fda.gov/consumers/consumer-updates/why-you-should-not-use-ivermectin-treat-or-prevent-covid-19

66   ClinicalTrials.gov, "Ivermectin COVID-19 Search Results," *ClinicalTrials.gov*, https://www.clinicaltrials.gov/ct2/results?cond=COVID-19&term=ivermectin&cntry=&state=&city=&dist=&Search=Search

67   David R. Henderson and Charles L. Hooper, "Why Is the FDA Attacking a Safe, Effective Drug?" *Wall Street Journal* (Jul 28, 2021)

68   Rapid Increase in Ivermectin Prescriptions and Reports of Severe Illness Associated with Use of Products Containing Ivermectin to Prevent or Treat COVID-19, *Centers for Disease Control and Prevention* (Aug 26, 2021), https://emergency.cdc.gov/han/2021/han00449.asp

69   "AMA, APhA, ASHP statement on ending use of ivermectin to treat COVID-19." Press Release, *American Medical Association* (September 1, 2021). https://www.ama-assn.org/press-center/press-releases/ama-apha-ashp-statement-ending-use-ivermectin-treat-covid-19

70   "Psychiatric Drug Facts: What your doctor may not know," COVID-19 and Coronavirus Resource Center, with Peter R. Breggin MD, https://breggin.com/coronavirus-resource-center/

71   Merck, "Merck Statement on Ivermectin use During the COVID-19 Pandemic," *Merck* (Feb. 4, 2021), https://www.merck.com/news/merck-statement-on-ivermectin-use-during-the-covid-19-pandemic/

72   Special to *The New York Times*, "Merck Offers Free Distribution of New River Blindness Drug," *New York Times* (Oct. 22, 1987), https://www.nytimes.com/1987/10/22/world/merck-offers-free-distribution-of-new-river-blindness-drug.html

73   Rick Speare and David Durrheim, "Mass treatment with ivermectin: an underutilized public health strategy," *Bulletin of the World Health Organization* (Aug. 2004), https://www.who.int/bulletin/volumes/82/8/562.pdf

74   Staff reporter, "Merck and AstraZeneca to start COVID-19 drug trials in Japan," *The Japan Times* (Mar. 13, 2021), https://www.japantimes.co.jp/news/2021/03/13/national/science-health/merck-astrazeneca-clinical-trials/

75   Annalisa Merelli, "Merck's new Covid-19 drug could be one of the most lucrative drugs ever," *Quartz* (Oct 1, 2021), https://qz.com/2068247/merck-could-make-up-to-7-billion-from-its-covid-19-drugs-in-2021/

76   Marc Lipsitch, "Why Do Exceptionally Dangerous Gain-of-Function Experiments in Influenza?" https://pubmed.ncbi.nlm.nih.gov/30151594/

77   Jon Cohen and Charles Piller, "Emails Offer Look Into Whistleblower Charges of Cronyism Behind Potential COVID-19 Drug," *Science Insider* (May 13, 2020), https://www.science.org/news/2020/05/emails-offer-look-whistleblower-charges-cronyism-behind-potential-covid-19-drug

78   U.S. Government Defense Threat Reduction Agency (DTRA), https://www.dtra.mil/

79   Sharon Lerner, "Merck Sells Federally Financed COVID Pill to U.S. for 40 Times What it Costs to Make," *The Intercept* (Oct 5, 2021), https://theintercept.com/2021/10/05/covid-pill-drug-pricing-merck-ridgeback/

80   U.S. Department of Health and Human Services, "Biden Administration announces U.S. government procurement of Merck's investigational antiviral medicine for COVID-19 treatment" (June 9, 2021), https://www.hhs.gov/about/news/2021/06/09/biden-administration-announces-us-government-procurement-mercks-investigational-antiviral-medicine-covid-19-treatment.html

81   Ibid.

82   Manojna Maddipatla and Amruta Khandekar, Bengaluru, "Pfizer begins study of oral drug for prevention of COVID-19," *Reuters* (September 27, 2021), https://www.reuters.com/business/healthcare-pharmaceuticals/pfizer-begins-study-covid-19-antiviral-drug-2021-09-27/

83   Tyler Durden, "Pfizer Launches Final Study For COVID Drug That's Suspiciously Similar to 'Horse Paste,'" *Zero Hedge* (September 28, 2021), https://www.zerohedge.com/covid-19/pfizer-launches-final-study-covid-drug-thats-suspiciously-similar-ivermectin

84   "You are not a horse. You are not a cow. Seriously, y'all. Stop it." U.S. FDA @US_FDA Twitter post (August 21, 2021 at 6:57 a.m.), https://twitter.com/us_fda/status/1429050070243192839

85   "Intestinal Parasite Guidance: Summary of Recommendations. All Middle Eastern, Asian, North African, Latin American, and Caribbean refugees should receive presumptive therapy with: Albendazole, single dose of 400 mg (200 mg for children 12-23 months) AND Ivermectin, two

doses 200 mcg/Kg orally once a day for 2 days before departure to the United States." CDC Immigrant, Refugee, and Migrant Health website (as of September 26, 2021), https://www.cdc.gov/immigrantrefugeehealth/guidelines/overseas-guidelines.html

86 Jake Tapper, "'Don't do it': Dr. Fauci warns against taking Ivermectin to fight Covid-19," CNN video (August 29, 2021), https://www.cnn.com/videos/health/2021/08/29/dr-anthony-fauci-ivermectin-covid-19-sotu-vpx.cnn

87 Leah Willingham, "Livestock medicine doesn't work against COVID, doctors warn," AP News (August 25, 2021), https://apnews.com/article/health-coronavirus-pandemic-69c5f6d4476ca9b25bc7038e99a4a075

88 Vanessa Romo, "Joe Rogan Says He Has COVID-19 And Has Taken The Drug Ivermectin," NPR (September 1, 2021), https://www.npr.org/2021/09/01/1033485152/joe-rogan-covid-ivermectin

89 Tyler Durden, "*Rolling Stone* Issues 'Update' After Horse Dewormer Hit-Piece Debunked," *Zero Hedge* (September 5, 2021), https://www.zerohedge.com/covid-19/rolling-stone-horse-dewormer-hit-piece-debunked-after-hospital-says-no-ivermectin

90 Ken Miller, "Dozens line up at Oklahoma City church for COVID vaccine," Associated Press (January 26, 2021), https://apnews.com/article/race-and-ethnicity-baptist-oklahoma-city-oklahoma-coronavirus-pandemic-ec6293de7e118079141e90b8874533a8

91 Peter Wade, "One Hospital Denies Oklahoma Doctor's Story of Ivermectin Overdoses Causing ER Delays for Gunshot Victims: The hospital says it hasn't experienced any care backlog due to patients overdosing on a drug that's been falsely peddled as a covid cure," *Rolling Stone* (Updated September 5, 2021, 8:55 pm ET), https://www.rollingstone.com/politics/politics-news/gunshot-victims-horse-dewormer-ivermectin-oklahoma-hospitals-covid-1220608/

92 "Video: Livestock feed stores are reporting shortages of ivermectin as many Americans buy up the anti-parasite drug and misuse it to treat Covid," *The Daily Mail* (September 17, 2021), https://www.dailymail.co.uk/video/health/video-2504931/Ivermectin-Explained-Anti-Parasite-Drug-Emerged.html

93 Jon Jackson, "Patients Overdosing on Ivermectin Are Clogging Oklahoma ERs: Doctor," *Newsweek* (September 2, 2021), https://www.newsweek.com/patients-overdosing-ivermectin-are-clogging-oklahoma-ers-doctor-1625631

94 Martin Pengelly and agencies, "Ivermectin misuse adding to Covid pressures at Oklahoma hospitals, doctor says: Jason McElyea says people overdosing on anti-parasitic drug that some people believe without evidence can cure or treat Covid," *The Guardian* (September 13, 2021), https://www.theguardian.com/world/2021/sep/04/oklahoma-doctor-ivermectin-covid-coronavirus

95 Graig Graziosi, "Doctor says gunshot victims forced to wait for treatment as Oklahoma hospitals overwhelmed by coronavirus patients. (First paragraph: "People poisoning themselves with a drug most commonly used as a horse dewormer are turning up hospitals in rural Oklahoma, a doctor has claimed"), *The Independent* (September 4, 2021), https://www.independent.co.uk/news/world/americas/us-politics/gunshot-oklahoma-hospitals-ivermectin-overdose-b1914322.html.

96 "Patients overdosing on ivermectin backing up rural Oklahoma hospitals, ambulances": "'The scariest one I've heard of and seen is people coming in with vision loss,' he said." Twitter repost by Rachel Maddow, MSNBC @maddow (September 2, 2021), https://twitter.com/maddow/status/1433521336282976256

97 Tyler Durden, "*Rolling Stone* Issues 'Update' After Horse Dewormer Hit-Piece Debunked," Zero Hedge (September 5, 2021), https://www.zerohedge.com/covid-19/rolling-stone-horse-dewormer-hit-piece-debunked-after-hospital-says-no-ivermectin

98 Martin Pengelly, "Oklahoma hospitals deluged by ivermectin overdoses, doctor says," *Guardian* (Sep. 4, 2021), https://archive.is/iIYB7

99 U.S. Food & Drug Administration, "Why You Should Not Use Ivermectin to Treat or Prevent Covid-19," https://www.fda.gov/consumers/consumer-updates/why-you-should-not-use-ivermectin-treat-or-prevent-covid-19

100 Centers for Disease Control and Prevention, Rapid Increase in Ivermectin Prescriptions and Reports of Severe Illness Associated with Use of Products Containing Ivermectin to Prevent or Treat COVID-19, (Aug 26, 2021). https://emergency.cdc.gov/han/2021/han00449.asp

101 Erin Woo, "How Covid Misinformation Created a Run on Animal Medicine," *The New York Times* (September 28, 2021), https://www.nytimes.com/2021/09/28/technology/ivermectin-animal-medicine-shortage.html?searchResultPosition=1

102 Dr. Eddy Bettermann, MD, "Massive 'horse' lies about Nobel prize winning treatment," (September 27, 2021), https://dreddymd.com/2021/09/27/massive-horse-lies-about-nobel-prize-winning-treatment/

Endnotes - Remdesivir

1 Sydney Lupkin, Remdesivir Priced At More Than $3,100 For A Course Of Treatment, NPR, (May 8, 2020). https://www.npr.org/sections/health-shots/2020/05/08/851632704/putting-a-price-on-covid-19-treatment-remdesivir

2 Kathryn Ardizzone, Role of the Federal Government in the Development of Remdesivir, KEI

Briefing Note 2020:1., (March 20, 2020). https://www.keionline.org/wp-content/uploads/KEI-Briefing-Note-2020_1GS-5734-Remdesivir.pdf

3   Bill & Melinda Gates Foundation, Form 990-PF, 2016. https://docs.gatesfoundation.org/Documents/A-1_2016_Form_990-PF_Signed.pdf

4   United States Securities and Exchange Commission, Gilead Sciences, Inc.. Form 10-K https://www.sec.gov/Archives/edgar/data/882095/000119312506045128/d10k.htm

5   Allison DeAngelis, Gilead, Gates Foundation join Watertown biotech's $55M financing round, *Boston Business Journal*, (2018). https://www.bizjournals.com/boston/news/2019/01/29/gilead-gates-foundation-join-watertown-biotechs.html

6   Bill & Melinda Gates Foundation, University of California San Francisco, (Sept. 2016). https://www.gatesfoundation.org/about/committed-grants/2016/09/opp1159068

7   Christopher Rowland, Taxpayers paid to develop remdesivir but will have no say when Gilead sets the price, Washington Post, https://www.washingtonpost.com/business/2020/05/26/remdesivir-coronavirus-taxpayers/

8   Gilead Sciences, GlobalData offers view on Gilead's suspension of Covid-19 trials in China, News-Medical Life Sciences, (April 29, 2020), https://www.globaldata.com/gileads-suspension-of-covid-19-trials-in-china-should-serve-as-bellwether-for-studies-in-other-countries-says-globaldata/https://www.news-medical.net/news/20200429/GlobalData-offers-view-on-Gileade28099s-suspension-of-Covid-19-trials-in-China.aspx

9   ClinicalTrials.gov, Investigational Therapeutics for the Treatment of People With Ebola Virus Disease, (October 28, 2018). https://clinicaltrials.gov/ct2/show/NCT03719586

10   Ibid.

11   Erika Check Hayden, "Experimental drugs poised for use in Ebola outbreak," Nature (May 18, 2018), https://www.nature.com/articles/d41586-018-05205-x

12   National Institutes of Health, Investigational Drugs Reduce Risk of Death from Ebola Virus Disease, NIH News Release, (November 27, 2019). https://www.nih.gov/news-events/news-releases/investigational-drugs-reduce-risk-death-ebola-virus-disease

13   Sabue Mulangu, "A Randomized, Controlled Trial of Ebola Virus Disease Therapeutics, *New England Journal of Medicine*," (December 12, 2019). https://www.nejm.org/doi/full/10.1056/NEJMoa1910993

14   Ibid. at Table 2. Comparison of Death at 28 Days According to Treatment Group.

15   NIH Press Release, "NIH clinical trial of remdesivir to treat COVID-19 begins," *NIH*, (Feb. 25, 2020), https://www.nih.gov/news-events/news-releases/nih-clinical-trial-remdesivir-treat-covid-19-begins

16   Ibid. NIH Press Release

17   Ibid. NIH Press Release

18   Andres Hill, et al, "Minimum costs to manufacture new treatments for COVID-19," *ScienceDirect* Journal of Virus Eradication Volume 6, Issue 2, April 2020, Pages 61-69, https://www.sciencedirect.com/science/article/pii/S2055664020300182

19   Angus Liu, "Fair price for Gilead's COVID-19 med remdesivir? $4,460, cost watchdog says," *FIERCE PHARMA,* (May 4, 2020) https://www.fiercepharma.com/marketing/gilead-s-covid-19-therapy-remdesivir-worth-4-460-per-course-says-pricing-watchdog

20   Matthew Herper, Gilead announces long-awaited price for Covid-19 drug remdesivir, STAT (Jun. 29, 2020), https://www.statnews.com/2020/06/29/gilead-announces-remdesivir-price-covid-19/

21   Angus Liu, Gilead banks on blockbuster remdesivir with sunnier 2020 outlook, FiercePharma, (July 21, 2020). https://www.fiercepharma.com/pharma/gilead-buoyed-by-potential-remdesivir-covid-19-sales-elevates-2020-outlook-despite-weak-base

22   Ibid.

23   "Gillings School researchers receive $6M+ grant to fight infectious diseases," UNC Gillings School of Global Public Health (Aug. 31, 2017), https://sph.unc.edu/sph-news/gillings-researchers-receive-6m-grant-to-fight-infectious-disease/

24   NIH Funding: Awards by Location & Organization, Ralph S. Baric: summary for Fiscal Year 2017, https://report.nih.gov/award/index.cfm?ot=&fy=2017&state=NC,46&ic=&fm=&orgid=578206&distr=&rfa=&pid=1885536&om=n#tab5

25   Reality Check Team, "Coronavirus: Was US money used to fund risky research in China?," BBC (Aug. 2, 2021), https://www.bbc.com/news/57932699

26   Understanding The Risk Of Bat Coronavirus Emergence, Award No. R01AI110964, US Dept. HHS, https://taggs.hhs.gov/Detail/AwardDetail?arg_AwardNum=R01AI110964&arg_ProgOfficeCode=104

27   Jon Cohen, "Quarantined at home now, U.S. scientist describes his visit to China's hot zone," SCIENCE (Mar. 6, 2020), https://www.science.org/news/2020/03/quarantined-scientist-reveals-what-it-s-be-china-s-hot-zone

28   Zhang Yan, David Stanway, "China lab seeks patent on use of Gilead's coronavirus treatment," Reuters (Feb. 4, 2020), https://www.reuters.com/article/us-china-health-patent/china-lab-seeks-patent-on-use-of-gileads-coronavirus-treatment-idUSKBN1ZZ0RL

29   Ibid.

30   BMGF Press Release, "Bill & Melinda Gates Foundation, Wellcome, and Mastercard Launch Initiative to Speed Development and Access to Therapies for COVID-19," *BMGF* (Mar. 10, 2020), https://www.gatesfoundation.org/Ideas/Media-Center/Press-Releases/2020/03/COVID-19-Therapeutics-Accelerator

31   Matt Krantz, "Bill Gates' Coronavirus Manifesto Reveals 5 Forecasts For Investors," *Investor's Business Daily* (Apr. 24, 2020 01:48 PM ET) https://www.investors.com/etfs-and-funds/sectors/bill-gates-coronavirus-manifesto-reveals-forecasts-investors/

32   Associated Press, Fauci calls Henry Ford study on hydroxychloroquine and COVID-19 'flawed', (July 31, 2020). https://www.wxyz.com/news/coronavirus/fauci-calls-henry-ford-study-on-hydroxychloroquine-and-covid-19-flawed

33   CNBC, Dr. Anthony Fauci: No placebo-controlled trial has shown hydroxychloroquine is effective, (July 31, 2020). https://www.youtube.com/watch?v=xDjVwXM8ESE

34   CNN, Dr. Fauci warns against taking Ivermectin to fight Covid-19, (August 29, 2021). https://www.cnn.com/videos/health/2021/08/29/dr-anthony-fauci-ivermectin-covid-19-sotu-vpx.cnn

35   Jason D. Goldman, et al. Remdesivir for 5 or 10 Days in Patients with Severe Covid-19. NEJM, (May 27, 2020). https://www.nejm.org/doi/10.1056/NEJMoa2015301

36   European Medicines, Assessment Report: Vaklury, (June 25, 2020). https://www.ema.europa.eu/en/documents/assessment-report/veklury-epar-public-assessment-report_en.pdf

37   NIAID, "A Multicenter, Adaptive, Randomized Blinded Controlled Trial of the Safety and Efficacy of Investigational Therapeutics for the Treatment of COVID-19 in Hospitalized Adults," *NIH Clinical Trials .GOV* (Apr. 2, 2020), https://clinicaltrials.gov/ct2/show/NCT04280705 (https://clinicaltrials.gov/ProvidedDocs/05/NCT04280705/Prot_001.pdf)

38   John H. Beigel, M.D., et al, "Remdesivir for the Treatment of Covid-19 — Final Report," N Engl J Med 2020; 383:1813-1826, DOI: 10.1056/NEJMoa2007764, https://www.nejm.org/doi/full/10.1056/NEJMoa2007764

39   Health News Review, "Primary Outcome Measures: Screenshot of ClinicalTrials.gov," *Health NewsReview* (Apr. 30, 2020, 3:18 PM), https://www.healthnewsreview.org/wp-content/uploads/2020/04/Screen-Shot-2020-04-30-at-3.18.42-PM.png https://www.healthnewsreview.org/2020/04/what-the-public-didnt-hear-about-the-nih-remdesivir-trial/

40   COVID-19 Treatment Guidelines Panel Members, (Last Updated: August 4, 2021), https://www.covid19treatmentguidelines.nih.gov/about-the-guidelines/panel-roster/

41   Jon Cohen, "Quarantined at home now, U.S. scientist describes his visit to China's hot zone," SCIENCE (Mar. 6, 2020), https://www.science.org/news/2020/03/quarantined-scientist-reveals-what-it-s-be-china-s-hot-zone

42   COVID-19 Treatment Guidelines Panel. Coronavirus Disease 2019 (COVID-19) Treatment Guidelines. National Institutes of Health. Available athttps://www.covid19treatmentguidelines.nih.gov/about-the-guidelines/panel-financial-disclosure/Accessed October 1, 2021.

43   AHRP, "Brazilian Scientists Speak Out Against Corrupted "Science" & the Use of Inhumane Study Methods,"*AHRP* (May 27, 2020), https://ahrp.org/brazilian-scientists-speak-out-against-corrupted-science-the-use-of-inhumane-study-methods/

44   AHRP, "Dissolving Illusions–"Authoritative" Medical Information Sources Are Corrupted,"*AHRP* (Nov. 26, 2017), https://ahrp.org/dissolving-illusions-medical-information-sources-are-corrupted/

45   AHRP, "Nobody Should Volunteer for Clinical Trials As Long As Research Data Is Secret," *AHRP* (Apr. 12, 2012), https://ahrp.org/nobody-should-volunteer-for-clinical-trials-as-long-as-research-data-is-secret/

46   AHRP, "Medical Journals Complicit in Corruption of Medicine," *AHRP* (Nov. 13, 2010), https://ahrp.org/medical-journals-complicit-in-corruption-of-medicine/

47   AHRP, "BMJ Demands Raw Data From Pharma Clinical Trials," *AHRP* (Dec. 10, 2009), https://ahrp.org/bmj-demands-raw-data-from-pharma-clinical-trials/

48   Yeming Wang, MD, et al., Remdesivir in adults with severe COVID-19: a randomised, double-blind, placebo-controlled, multicentre trial, VOL 395, I 10236, P1569-1578, (MAY 16, 2020), https://doi.org/10.1016/S0140-6736(20)31022-9

49   Ibid.

50   Phil Taylor, "Gilead slides as second Chinese trial of remdesivir is stopped," PHARMAPHORUM (Apr. 15, 2020), https://pharmaphorum.com/news/gilead-slides-as-two-chinese-trials-of-remdesivir-are-stopped/

51   Erika Edwards, Remdesivir shows promising results for coronavirus, Fauci says, NBC, (April 29, 2020). https://www.nbcnews.com/health/health-news/coronavirus-drug-remdesivir-shows-promise-large-trial-n1195171

52   Gina Kolata, Peter Baker and Noah Weiland, "Remdesivir Shows Modest Benefits in Coronavirus Trial," *New York Times,* (April 29, 2020, Updated Oct. 29, 2020), https://www.nytimes.com/2020/04/29/health/gilead-remdesivir-coronavirus.html

53   FDA, "EUA Letter of Approval Veklury (remdesivir)," *RADM Denise M. Hinton-FDA,* (May 1, 2020 revised Oct. 22, 2020), https://www.fda.gov/media/137564/download

54   Veklury (remdesivir), "Information for US Healthcare Professionals", https://www.vekluryhcp.com/

55   Sarah Boseley, US secures world stock of key Covid-19 drug remdesivir, *The Guardian,* (June 30,

2020). https://www.theguardian.com/us-news/2020/jun/30/us-buys-up-world-stock-of-key-covid-19-drug

56   Francesco Guarascio, EU makes 1 billion-euro bet on Gilead's COVID drug before trial results, Reuters, (Oct 13, 2020.) https://www.reuters.com/article/us-health-coronavirus-eu-remdesivir/eu-makes-1-billion-euro-bet-on-gileads-covid-drug-before-trial-results-idUSKBN26Y25K

57   Gillings School News, Remdesivir, developed through a UNC-Chapel Hill partnership, proves effective against COVID-19 in NIAID human clinical trials, UNC Gillings School of Global Public Health, (April 29, 2020). https://sph.unc.edu/sph-news/remdesivir-developed-at-unc-chapel-hill-proves-effective-against-covid-19-in-niaid-human-clinical-trials/

58   Ibid.

59   AHRP, "Fauci's Promotional Hype Catapults Gilead's Remdesivir," *AHRP* (May 6, 2020), https://ahrp.org/faucis-promotional-hype-catapults-gileads-remdesivir/

60   Public Citizen, "The Public Already Has Paid for Remdesivir," *Public Citizen,* (May 7, 2020), https://www.citizen.org/news/the-public-already-has-paid-for-remdesivir/

61   Umair Irfan, "The FDA approved remdesivir to treat Covid-19. Scientists are questioning the evidence.," Vox (Oct. 24, 2020), https://www.vox.com/21530401/remdesivir-approved-by-fda-covid-19-fda-gilead-veklury

62   Jeremy Hsu, "Covid-19: What now for remdesivir?," 371:m4457 BMJ (Nov. 20, 2020), https://www.bmj.com/content/371/bmj.m4457

63   Jon Cohen & Kai Kupferschmidt, "The 'very, very bad look' of remdesivir, the first FDA-approved COVID-19 drug," Science (Oct. 28, 2020), https://www.science.org/news/2020/10/very-very-bad-look-remdesivir-first-fda-approved-covid-19-drug

64   Umair Irfan, "The FDA approved remdesivir to treat Covid-19. Scientists are questioning the evidence," Vox (Oct. 24, 2020) https://www.vox.com/21530401/remdesivir-approved-by-fda-covid-19-fda-gilead-veklury

65   Jon Cohen & Kai Kupferschmidt, "The 'very, very bad look' of remdesivir, the first FDA-approved COVID-19 drug," SCIENCE (Oct. 28, 2020), https://www.science.org/news/2020/10/very-very-bad-look-remdesivir-first-fda-approved-covid-19-drug

66   Science Media Centre, expert reaction to a study about compassionate use of remdesivir for patients with severe COVID-19, (April 11, 2020). https://www.sciencemediacentre.org/expert-reaction-to-a-study-about-compassionate-use-of-remdesivir-for-patients-with-severe-covid-19/

67   Ibid.

68   Steven Levy, Bill Gates on Covid: Most US Tests Are 'Completely Garbage. *Wired,* (August 7, 2020). https://www.wired.com/story/bill-gates-on-covid-most-us-tests-are-completely-garbage/

69   Owen Dyer, Covid-19: Remdesivir has little or no impact on survival, WHO trial shows. *BMJ* 2020;371:m4057 (Published October 19, 2020). https://www.bmj.com/content/371/bmj.m4057?ijk ey=f9301c3ac0aca0af6e1df53f1c90e72b9dc769cd&keytype2=tf_ipsecsha

70   FDA News Release, "FDA Approves First Treatment for COVID-19," *FDA* (Oct.22, 2020), https://www.fda.gov/news-events/press-announcements/fda-approves-first-treatment-covid-19

71   Kari Oakes, "WHO backs off on remdesivir as FDA issues another EUA." *Regulatory Focus* (November 19, 2020). https://www.raps.org/news-and-articles/news-articles/2020/11/who-backs-off-on-remdesivir-as-fda-issues-another

72   Michael E. Ohl, et al., "Association of Remdesivir Treatment With Survival and Length of Hospital Stay Among US Veterans Hospitalized With COVID-19," JAMA 4(7):e2114741 (2021) doi:10.1001/jamanetworkopen.2021.14741, https://jamanetwork.com/journals/jamanetworkopen/fullarticle/2781959

73   PRAC reviews a signal with Veklury, PRAC (Oct. 2, 2020), https://www.ema.europa.eu/en/news/meeting-highlights-pharmacovigilance-risk-assessment-committee-prac-28-september-1-october-2020

74   "EU medicines agency studies effect of COVID-19 drug on kidneys," Euronews (Oct. 2, 2020), https://www.euronews.com/2020/10/02/eu-medicines-agency-studies-effect-of-covid-19-drug-on-kidneys

75   CDC, "Interim Clinical Guidance for Management of Patients with Confirmed Coronavirus Disease (COVID-19): Clinical Management and Treatment" *Centers for Disease Control and Prevention* (Feb. 16, 2120), https://www.cdc.gov/coronavirus/2019-ncov/hcp/clinical-guidance-management-patients.html

76   NIH, "COVID-19 Treatment Guidelines:Therapeutic Management of Hospitalized Adults With COVID-19" *NIH* (Aug 25, 2021), https://www.covid19treatmentguidelines.nih.gov/management/clinical-management/hospitalized-adults--therapeutic-management/

77   NIH, "COVID-19 Treatment Guidelines: General Management of Nonhospitalized Patients With Acute COVID-19" NIH (Jul. 8, 2021), https://www.covid19treatmentguidelines.nih.gov/management/clinical-management/nonhospitalized-adults--therapeutic-management/

78   Sanna Gevers, et al, 'Remdesivir in COVID-19 Patients with Impaired Renal Function,' *JASN* (Feb. 2021) 32 (2) 518-519; DOI: https://doi.org/10.1681/ASN.2020101535

79   Jonathan Grein, M.D., et al, "Compassionate Use of Remdesivir for Patients with Severe Covid-19," *N Engl J Med* 2020; 382:2327-2336, DOI: 10.1056/NEJMoa2007016 https://www.nejm.org/doi/10.1056/NEJMoa2007016

80   Tavares, Remdesivir is approved for testing on coronavirus patients in Brazil and Europe, https://
       www.tavaresoffice.com.br/en/remdesivir-is-approved-for-testing-on-coronavirus-patients-in-brazil-
       and-europe/

81   Lisandra Paraguassu, Ricardo Brito, After record COVID-19 deaths, Bolsonaro tells Brazilians to stop
       'whining', *Reuters*, (March 5, 2021). https://www.reuters.com/article/us-health-coronavirus-brazil/
       after-record-covid-19-deaths-bolsonaro-tells-brazilians-to-stop-whining-idUSKBN2AX114

82   "Kidney problems more prevalent in NYC COVID-19 patients," Columbia (Jun. 8, 2020), https://
       www.cuimc.columbia.edu/news/kidney-problems-more-prevalent-nyc-covid-19-patients

Endnotes - Final Solution: Vaccines or Bust

1   Robert Kuznia, "The timetable for a coronavirus vaccine is 18 months. Experts say that's risky,"
       CNN (Apr. 1, 2020), https://www.cnn.com/2020/03/31/us/coronavirus-vaccine-timetable-concerns-
       experts-invs/index.html

2   Stuart A. Thompson, "How Long Will a Vaccine Really Take?," NYT (Apr. 30, 2020), https://www.
       nytimes.com/interactive/2020/04/30/opinion/coronavirus-covid-vaccine.html

3   Luke Andrews, 'It can prevent pneumonia': Oxford professor running coronavirus vaccine trial comes
       out in its defence after all of the monkeys given the treatment catch the disease, *The Daily Mail*,
       (May 22, 2020). https://www.dailymail.co.uk/sciencetech/article-8347391/It-prevents-pneumonia-
       Oxford-professor-defences-coronavirus-vaccine.html

4   Sara G. Miller, 'The looming question': Fauci says studies suggest vaccines slow virus spread, NBC
       News, (Feb 17, 2021). https://www.nbcnews.com/health/health-news/looming-question-fauci-says-
       studies-suggest-vaccines-slow-virus-spread-n1258142

5   Veronika Kyrylenko, Nobel Prize Winner Warns Vaccines Facilitate Development of Deadlier
       COVID Variants, Urges Public to Reject Jabs, The New American, (May 20, 2021). https://
       thenewamerican.com/french-nobel-prize-winner-warns-vaccines-facilitate-development-of-deadlier-
       covid-variants-urges-the-public-to-reject-jabs/

6   Zane Rizvi, "The NIH Vaccine," The Public Citizen (Jun. 25, 2020), https://www.citizen.org/article/
       the-nih-vaccine/#_ftn29

7   Alex Newman, Dr Peter McCullough on Vaccine Death Rate, Raccoon Medicine, (May 29, 2021).
       https://raccoonmedicine.com/wp/2021/05/29/dr-peter-mccullough-vaccine-death-rate-ignored/

8   Associated Press, "COVID-19 Vaccine Boosters Could Mean Billions for Drugmakers," USNews
       (Sept. 25, 2021), https://www.usnews.com/news/business/articles/2021-09-25/covid-19-vaccine-
       boosters-could-mean-billions-for-drugmakers

9   Peter Hotez March 5, 2020 testimony before the House Science, Space and Technology Committee
       on Coronavirus. Hotez speaks at around the 25:00 mark. https://www.c-span.org/video/?470035-1/
       house-science-space-technology-committee-hearing-coronavirus

10   Peter Hotez March 5, 2020 testimony before the House Science, Space and Technology Committee
       on Coronavirus. Hotez speaks at around the 25:00 mark. https://www.c-span.org/video/?470035-1/
       house-science-space-technology-committee-hearing-coronaviru

11   The Reality About Coronavirus Vaccine (W/Dr. Paul Offit), Dr. Paul Offit interview with Dr. Zubin
       "ZDogg" Damania, M.D. (Apr. 5, 2020). https://zdoggmd.com/paul-offit-2/

12   Ibid.

13   Dr. Anthony Fauci, 00:57:48, "White House coronavirus briefing (March 26, 2020)," *YouTube*,
       https://www.youtube.com/watch?v=uOruI8Rs0pg&t=3465s

14   ClinicalTrials.gov, "Study to Describe the Safety, Tolerability, Immunogenicity, and Efficacy of
       RNA Vaccine Candidates Against COVID-19 in Healthy Individuals," (April 30, 2020), https://
       clinicaltrials.gov/ct2/show/NCT04368728?term=NCT04368728&draw=2&rank=1

15   Ross Lazarus, "Electronic Support for Public Health–Vaccine Adverse Event Reporting System"
       (ESP:VAERS),(Sep 30, 2010), https://digital.ahrq.gov/sites/default/files/docs/publication/
       r18hs017045-lazarus-final-report-2011.pdf

16   Jerry Dunleavy, "Republicans press Facebook for documents on COVID-19 origins 'censorship' and
       Fauci emails," Yahoo News (Jun. 9, 2021), https://www.yahoo.com/now/republicans-press-facebook-
       documents-covid-230200549.html

17   Leopold, NIH FOIA: Anthony Fauci Emails, pp. 2065-2068, https://s3.documentcloud.org/
       documents/20793561/leopold-nih-foia-anthony-fauci-emails.pdf

18   Who Pays for Politifact? *Feb, 2021). https://www.politifact.com/who-pays-for-politifact/

19   Our Funding, FactCheck.org (2021), https://www.factcheck.org/our-funding/

20   Facebook Fact-Checkers Secretly Funded by Johnson and Johnson," Vision News (May 6, 2021),
       https://www.visionnews.online/post/facebook-fact-checkers-secretly-funded-by-johnson-and-johnson

21   @RWMalone Twitter (Oct. 7, 2021, 9:08 AM), https://twitter.com/RWMaloneMD/
       status/1446100124267057160

22   Children's Health Defense Team, "Home Run King Hank Aaron Dies of 'Undisclosed Cause' 18
       Days After Receiving Moderna Vaccine," The Defender-Children's Health Defense (Jan.22, 2021),
       https://childrenshealthdefense.org/defender/hank-aaron-dies-days-after-receiving-moderna-vaccine/

23   Robert F. Kennedy, Jr., "National Media Pushes Vaccine Misinformation — Coroner's Office

Never Saw Hank Aaron's Body," The Defender-Children's Health Defense (Feb. 12, 2021), https://childrenshealthdefense.org/defender/hank-aaron-dies-days-after-receiving-moderna-vaccine/

24 "'Die Lymphozyten laufen Amok'—Pathologen untersuchen Todesfälle nach COVID-19-Impfung." *RT Question More* (September 21, 2021). https://de.rt.com/inland/124390-lymphozyten-laufen-amok-pathologen-untersuchen-todesfaelle-nach-impfung/

25 National Institutes of Health, NIH Awards by Location and Organization, https://report.nih.gov/award/index.cfm

26 Bill & Melinda Gates Foundation Committed Grants Database, https://www.gatesfoundation.org/about/committed-grants

27 Frontline, Dr. Anthony Fauci: Risks From Vaccines Are "Almost Nonmeasurable", Frontline, (Mar 23, 2015). https://www.pbs.org/wgbh/frontline/article/anthony-fauci-risks-from-vaccines-are-almost-nonmeasurable/

28 Katie Rogers & Sheryl Gay Stolberg, "Biden Mandates Vaccines for Workers, Saying, 'Our Patience Is Wearing Thin'," New York Times, https://www.nytimes.com/2021/09/09/us/politics/biden-mandates-vaccines.html

29 Sivan Gazit, Roei Shlezinger, et al, "Comparing SARS-CoV-2 natural immunity to vaccine-induced immunity: reinfections versus breakthrough infections" medRxiv 2021.08.24.21262415; doi: https://doi.org/10.1101/2021.08.24.21262415

30 Brownstone Institute, "Natural Immunity and Covid-19: Twenty-Nine Scientific Studies to Share with Employers, Health Officials, and Politicians," Brownstone Institute (Oct. 10, 2021), https://brownstone.org/articles/natural-immunity-and-covid-19-twenty-nine-scientific-studies-to-share-with-employers-health-officials-and-politicians/

31 "COVID-19 vaccine doses administered by manufacturer, United States," Our World in Data, https://ourworldindata.org/grapher/covid-vaccine-doses-by-manufacturer?country=~USA

32 Stephen J. Thomas et al., Six Month Safety and Efficacy of the BNT162b2 mRNA COVID-19 Vaccine, medRxiv preprint (July 28, 2021). https://www.medrxiv.org/content/10.1101/2021.07.28.21261159v1.full.pdf

33 Ronald B. Brown, Outcome Reporting Bias in COVID-19 mRNA Vaccine Clinical Trials, Medicina, Medicina, (Feb. 26, 2021). https://pubmed.ncbi.nlm.nih.gov/33652582/

34 Stephen J. Thomas et al., Six Month Safety and Efficacy of the BNT162b2 mRNA COVID-19 Vaccine, medRxiv preprint (July 28, 2021). https://www.medrxiv.org/content/10.1101/2021.07.28.21261159v1.full.pdf

35 U.S. Food & Drug Administration, Good Review Practice: Clinical Review of Investigational New Drug Applications, (Dec, 2013). https://www.fda.gov/media/87621/download

36 Senate Testimony, Dr. Pierre Kory, FLCCC (Dec. 8, 2020) https://covid19criticalcare.com/senate-testimony/

37 Gérard Delépine, High Recorded Mortality in Countries Categorized as "Covid-19 Vaccine Champions". Increased Hospitalization, Freedom of Speech, (Oct 1, 2021). https://fos-sa.org/2021/10/01/high-recorded-mortality-in-countries-categorized-as-covid-19-vaccine-champions-increased-hospitalization/

38 Ibid.

39 Ibid.

40 "COVID-19 in Iceland: Vaccination Has Not Led to Herd Immunity, Says Chief Epidemiologist," Jelena Ćirić, Iceland Review (August 3, 2021), https://www.icelandreview.com/society/covid-19-in-iceland-vaccination-has-not-led-to-herd-immunity-says-chief-epidemiologist/

41 Delépine.

42 Ibid.

43 Ibid.

44 Ibid.

45 Holly Ellyatt, "Here's why herd immunity from Covid is 'mythical' with the delta variant," CNBC (Aug. 12, 2021), https://www.cnbc.com/2021/08/12/herd-immunity-is-mythical-with-the-covid-delta-variant-experts-say.html

46 "Vaccine Effectiveness Drops Further in the Over-40s, To as Low as Minus 53%, New PHE Report Shows – And That's a Fact." Will Jones, *The Daily Sceptic* (September 24, 2021). https://dailysceptic.org/2021/09/24/vaccine-effectiveness-drops-further-in-the-over-40s-as-low-as-minus-53-new-phe-report-shows-and-thats-a-fact/

47 Delépine.

48 Ibid.

49 Ibid.

50 "Population Wide Epidemiological Geography Demonstrates Vaccination Doesn't Correlate to Reduction in SARS-CoV-2 Infection." *TrialSite News* (October 3, 2021). https://trialsitenews.com/population-wide-epidemiological-geography-demonstrates-vaccination-doesnt-correlate-to-reduction-in-sars-cov-2-infection/

51 S.V. Subramanian, A. Kumar, "Increases in COVID-19 are unrelated to levels of vaccination across 68 countries and 2947 counties in the United States," *European Journal of Epidemiology* (September 30, 2021), https://link.springer.com/article/10.1007/s10654-021-00808-7

52 A comparison of age adjusted all-cause mortality rates in England between vaccinated and

unvaccinated. Norman Fenton and Martin Neil, *Probability and Risk* (September 23, 2021). http://probabilityandlaw.blogspot.com/2021/09/all-cause-mortality-rates-in-england.html

53   SARS-CoV-2 variants of concern and variants under investigation in England. Technical briefing 23, Public Health England (September 17, 2021). https://assets.publishing.service.gov.uk/government/uploads/system/uploads/attachment_data/file/1018547/Technical_Briefing_23_21_09_16.pdf

54   Ibid.

55   "Breakthrough Cases Surge: Vaccinated Individuals Accounted for 87% of Covid Hospitalizations Over the Past Week in Wales UK; 99% of All New Cases Were Under 60 Years Old." Julian Conradson, *Gateway Pundit* (September 30, 2021). https://www.thegatewaypundit.com/2021/09/ready-breakthrough-cases-surge-vaccinated-individuals-accounted-87-covid-hospitalizations-past-week-wales-uk-99-new-cases-60-years-old/

56   Senedd Research, COVID-19 vaccination data, (Jul 10, 2021). https://research.senedd.wales/research-articles/covid-19-vaccination-data/

57   Nosocomial outbreak caused by the SARS-CoV-2 Delta variant in a highly vaccinated population, Israel, July 2021. Pnina Shitrit et al., *Eurosurveillance* (Volume 26, Issue 39, 30/Sep/2021). https://www.eurosurveillance.org/content/10.2807/1560-7917.ES.2021.26.39.2100822

58   Centers for Diseases Control and Prevention, Outbreak of SARS-CoV-2 Infections, Including COVID-19 Vaccine Breakthrough Infections, Associated with Large Public Gatherings — Barnstable County, Massachusetts, July 2021, MMWR, (Aug 6, 2021

59   Committed Grants Johns Hopkins University, Bill & Melinda Gates Foundation, https://www.gatesfoundation.org/about/committed-grants?q=johns%20hopkins%20university

60   National Institutes of Health, NIH Awards by Location and Organization, Johns Hopkins University, https://report.nih.gov/award/indexcfm?ot=&fy=2020&state=&ic=&fm=&orgid=4134401&distr=&rfa=&om=y&pid=&view=state

61   World Snapshot, PANDA Pandemics Data & Analytics, https://rb.gy/vsyfnv

62   Shari Roan, Swine flu 'debacle' of 1976 is recalled, Los Angeles Times, (Apr 27, 2009). https://www.latimes.com/archives/la-xpm-2009-apr-27-sci-swine-history27-story.html

63   Found 16,310 cases where Vaccine is COVID19 and Patient Died, NVIC (From Oct. 1, 2021 release of VAERS data), https://www.medalerts.org/vaersdb/findfield.php?TABLE=ON&GROUP1=AGE&EVENTS=ON&VAX=COVID19&DIED=Yes

64   Found 778,685 cases where Vaccine is COVID19, NVIC (From Oct. 1, 2021 release of VAERS data), https://www.medalerts.org/vaersdb/findfield.php?TABLE=ON&GROUP1=CAT&EVENTS=ON&VAX=COVID19

65   Med Alerts, Found 14,925 cases where Vaccine is COVID19 and Patient Died (September 10, 2021). https://medalerts.org/vaersdb/findfield.php?TABLE=ON&GROUP1=O2D&EVENTS=ON&VAX=COVID19&DIED=Yes

66   Id.

67   COVID-19 Vaccinations 98 Times More Deadly Than Flu Vaccines (According to VAERS Reports), TrialSite News, (August 28, 2021). Archived at: https://web.archive.org/web/20210913051243/https://trialsitenews.com/covid-19-vaccinations-98-times-more-deadly-than-flu-vaccines-according-to-vaers-reports/

68   Ronald Kostoff, Why Are We Vaccinating Children Against COVID19? Toxicology Reports, Vol 8 2021, pages 1165-1684, https://www.sciencedirect.com/science/article/pii/S221475002100161X

69   Steve Kirsch, Vaccine Safety Evidence, (July 20, 2021). http://www.skirsch.com/covid/Vaccine.pdf

70   U.S. Food and Drug Administration, Vaccines and Related Biological Products Advisory Committee Meeting, Steve Kirsch Segment, (Sep 17, 2021). YouTube, 04:20:16, https://youtu.be/WFph7-6t34M

71   U.S. Food and Drug Administration, Vaccines and Related Biological Products Advisory Committee Meeting, Steve Kirsch Segment, (Sep 17, 2021). YouTube, 04:21:33, https://youtu.be/WFph7-6t34M

72   Whistleblower Lawsuit! Government Medicare Data Shows 48,465 DEAD Following COVID Shots—Remdesivir Drug has 25% Death Rate! Brian Shilhavy, Health Impact News (September 28, 2021). https://medicalkidnap.com/2021/09/28/whistleblower-lawsuit-government-medicare-data-shows-48465-dead-following-covid-shots-remdesivir-drug-has-25-death-rate/

73   Attorney Files Lawsuit Against CDC Based on "Sworn Declaration" from Whistleblower Claiming 45,000 Deaths are Reported to VAERS—All Within 3 Days of COVID-19 Shots. Health Impact News (2021). https://healthimpactnews.com/2021/attorney-files-lawsuit-against-cdc-based-on-sworn-declaration-from-whistleblower-claiming-45000-deaths-are-reported-to-vaers-all-within-3-days-of-covid-19-shots/

74   "How CDC Manipulated Data to Create 'Pandemic of the Unvaxxed' Narrative." Dr. Joseph Mercola, The Defender (August 16, 2021). https://childrenshealthdefense.org/defender/cdc-manipulated-data-create-pandemic-unvaxxed-narrative/

75   Public health investigations of COVID-19 vaccine breakthrough cases, CDC, https://www.cdc.gov/vaccines/covid-19/downloads/COVID-vaccine-breakthrough-case-investigations-Protocol.pdf

76   Virginia Langmaid, "Data on hospitalizations and deaths in the unvaccinated do not reflect Delta variant, CDC director says," CNN (Aug. 5, 2021), https://www.cnn.com/us/live-news/coronavirus-pandemic-vaccine-updates-08-05-21/h_82f976bb0f238323e3e0482af5d2d563

77   Our World in Data, "COVID-19 vaccination doses administered per 100 people, Jan 1, 2021," Our World in Data, (Jan. 1, 2021), https://ourworldindata.org/explorers/coronavirus-data-explorer?zoomToSelection=true&time=2021-01-01&facet=none&pickerSort=asc&pickerMetric=location&Metric=People+vaccinated+%28by+dose%29&Interval=7-day+rolling+average&Relative+to+Population=true&Align+outbreaks=false&country=ARE~PRT~ESP~SGP~URY~DNK~CHL~IRL~CAN~FIN~CHN~IND~USA~IDN~PAK~BRA~NGA~BGD~RUS~MEX~JPN~ETH~PHL~EGY~VNM~TUR~IRN~DEU~THA~GBR~FRA~TZA~ITA~ZAF~KEN~OWID_WRL

78   Our World in Data, "COVID-19 vaccination doses administered per 100 people, Apr 15, 2021," Our World in Data, (Apr. 15, 2021), https://ourworldindata.org/explorers/coronavirus-data-explorer?zoomToSelection=true&time=2021-04-15&facet=none&pickerSort=asc&pickerMetric=location&Metric=People+vaccinated+%28by+dose%29&Interval=7-day+rolling+average&Relative+to+Population=true&Align+outbreaks=false&country=ARE~PRT~ESP~SGP~URY~DNK~CHL~IRL~CAN~FIN~CHN~IND~USA~IDN~PAK~BRA~NGA~BGD~RUS~MEX~JPN~ETH~PHL~EGY~VNM~TUR~IRN~DEU~THA~GBR~FRA~TZA~ITA~ZAF~KEN~OWID_WRL

79   Our World in Data, "COVID-19 vaccination doses administered per 100 people, June 15, 2021," Our World in Data, (Jun. 15, 2021), https://ourworldindata.org/explorers/coronavirus-data-explorer?zoomToSelection=true&time=2021-06-15&facet=none&pickerSort=asc&pickerMetric=location&Metric=People+vaccinated+%28by+dose%29&Interval=7-day+rolling+average&Relative+to+Population=true&Align+outbreaks=false&country=ARE~PRT~ESP~SGP~URY~DNK~CHL~IRL~CAN~FIN~CHN~IND~USA~IDN~PAK~BRA~NGA~BGD~RUS~MEX~JPN~ETH~PHL~EGY~VNM~TUR~IRN~DEU~THA~GBR~FRA~TZA~ITA~ZAF~KEN~OWID_WRL

80   Victoria Male, "Menstrual changes after covid-19 vaccination" BMJ (2021) 374 doi: https://doi.org/10.1136/bmj.n2211

81   MedAlerts, "5,990 cases where Location is U.S., Territories, or Unknown and Vaccine is COVID19 and Symptom is Amenorrhoea or Dysmenorrhoea or Menopausal disorder or Menopausal symptoms or Menopause or Menopause delayed or Menstrual discomfort or Menstrual disorder or Menstruation delayed or Menstruation irregular," MedAlerts/VAERS, (Oct. 1, 2021), https://medalerts.org/vaersdb/findfield.php?TABLE=ON&GROUP1=AGE&EVENTS=ON&SYMPTOMS[]=Amenorrhoea+%2810001928%29&SYMPTOMS[]=Dysmenorrhoea+%2810013935%29&SYMPTOMS[]=Menopausal+disorder+%2810058825%29&SYMPTOMS[]=Menopausal+symptoms+%2810027304%29&SYMPTOMS[]=Menopause+%2810027308%29&SYMPTOMS[]=Menopause+delayed+%2810027310%29&SYMPTOMS[]=Menstrual+discomfort+%2810056344%29&SYMPTOMS[]=Menstrual+disorder+%2810027327%29&SYMPTOMS[]=Menstruation+delayed+%2810027336%29&SYMPTOMS[]=Menstruation+irregular+%2810027339%29&VAX=COVID19&STATE=NOTFR

82   Sue Hughes, "CVST After COVID-19 Vaccine: New Data Confirm High Mortality Rate," MEDSCAPE, (September 30, 2021) https://www.medscape.com/viewarticle/959992#vp_3

83   Ruud Lensen et al., Hepatitis C Virus Reactivation Following COVID-19 Vaccination—A Case Report. Int Med Case Rep J 2021; 14: 573–576, https://www.ncbi.nlm.nih.gov/pmc/articles/PMC8412816/

84   Centers for Disease Control and Prevention, Severe Outcomes Among Patients with Coronavirus Disease 2019 (COVID-19) — United States, February 12–March 16, 2020, MMWR, (Mar 27, 2020). https://www.cdc.gov/mmwr/volumes/69/wr/mm6912e2.htm

85   MedAlerts, "Found 7,537 cases where Vaccine is COVID19 and Symptom is Myocarditis or Myopericarditis or Pericarditis,"MedAlert/VAERS, (Oct. 1, 2021), https://medalerts.org/vaersdb/findfield.php?TABLE=ON&GROUP1=AGE&EVENTS=ON&SYMPTOMS[]=Myocarditis+%2810028606%29&SYMPTOMS[]=Myopericarditis+%2810028650%29&SYMPTOMS[]=Pericarditis+%2810034484%29&VAX=COVID19

86   MedAlerts, "Found 5,602 cases where Vaccine is COVID19 and Manufacturer is PFIZER/BIONTECH and Symptom is Myocarditis or Pericarditis," MedAlert/VAERS, (Oct. 1, 2021), https://medalerts.org/vaersdb/findfield.php?TABLE=ON&GROUP1=AGE&EVENTS=ON&SYMPTOMS[]=Myocarditis+%2810028606%29&SYMPTOMS[]=Pericarditis+%2810034484%29&VAX=COVID19&VAXMAN=PFIZER/BIONTECH

87   MedAlerts, "Found 476 cases where Age is 12-or-more-and-under-18 and Vaccine is COVID19 and Manufacturer is PFIZER/BIONTECH and Symptom is Myocarditis or Pericarditis,"https://medalerts.org/vaersdb/findfield.php?TABLE=ON&GROUP1=AGE&EVENTS=ON&SYMPTOMS[]=Myocarditis+%2810028606%29&SYMPTOMS[]=Pericarditis+%2810034484%29&VAX=COVID19&VAXMAN=PFIZER/BIONTECH&WhichAge=range&LOWAGE=12&HIGHAGE=18

88   A Report on Myocarditis Adverse Events in the U.S. Vaccine Adverse Events Reporting System (VAERS) in Association with COVID-19 Injectable Biological Products. Jessica Rose PhD, MSc, BSc1 and Peter A. McCullough MD, MPH, Current Problems in Cardiology. In Press, Journal Pre-proof (October 1, 2021). https://www.sciencedirect.com/science/article/abs/pii/S0146280621002267#!

89   Michael Kang, Viral Myocarditis, StatPearls (Updated 2021 Aug 11), https://www.ncbi.nlm.nih.gov/books/NBK459259/

90   Megan Redshaw, "Sweden, Denmark Pause Moderna's COVID Vaccine for Younger Age Groups

Citing Reports of Myocarditis," The Defender (October 6, 2021), https://childrenshealthdefense.org/defender/sweden-denmark-pause-moderna-covid-vaccine-myocarditis/

91   Essi Lehto, "Finland joins Sweden and Denmark in limiting Moderna COVID-19 vaccine," Reuters (October 7, 2021), https://www.reuters.com/world/europe/finland-pauses-use-moderna-covid-19-vaccine-young-men-2021-10-07/

92   Letter to Amit Patel, Pfizer Inc., BioNTech Manufacturing GmbH from U.S. Food & Drug Administration (August 23, 2021). https://www.fda.gov/media/151710/download

93   Mehul Suthar et al., "Durability of immune responses to the BNT162b2 mRNA vaccine," BioRxiv preprint (October 2021), https://www.biorxiv.org/content/10.1101/2021.09.30.462488v1

94   Maggie Fox, "Studies confirm waning immunity from Pfizer's Covid-19 vaccine," CNN (October 7, 2021), https://www.cnn.com/2021/10/06/health/pfizer-vaccine-waning-immunity/index.html

95   Sara Y. Tartof, Ph.D. et al., "Effectiveness of mRNA BNT162b2 COVID-19 vaccine up to 6 months in a large integrated health system in the USA: a retrospective cohort study." The Lancet (October 04, 2021), https://www.thelancet.com/journals/lancet/article/PIIS0140-6736(21)02183-8/fulltext

96   Will Jones, New Lancet Study Confirms Plummeting Vaccine Effectiveness, The Daily Sceptic (October 7, 2021), https://dailysceptic.org/2021/10/06/new-lancet-study-confirms-plummeting-vaccine-effectiveness/

97   Sara Y Tartof, et al, Effectiveness of mRNA BNT162b2 COVID-19 vaccine up to 6 months in a large integrated health system in the USA: a retrospective cohort study, (Oct 4, 2021). https://www.thelancet.com/journals/lancet/article/PIIS0140-6736(21)02183-8/fulltext

98   Ibid.

99   Sunil S Bhopal, et al, Children and young people remain at low risk of COVID-19 mortality. (May 1, 2021). https://www.thelancet.com/journals/lanchi/article/PIIS2352-4642(21)00066-3/fulltext

100   "Investigation: Deaths among Teenagers have increased by 47% in the UK since they started getting the Covid-19 Vaccine according to official ONS data." Will Jones, The Exposé (October 2, 2021). https://theexpose.uk/2021/09/30/deaths-among-teenagers-have-increased-by-47-percent-since-covid-vaccination-began/

101   Ibid.

102   Sunil S Bhopal et al., Children and young people remain at low risk of COVID-19 mortality, The Lancet, (May 2021).https://www.thelancet.com/action/showPdf?pii=S2352-4642%2821%2900066-3

103   Facebook page for WXYZ-TV Channel 7, accessed at September 20, 2021: https://www.facebook.com/wxyzdetroit/posts/10158207967261135

104   Nguyen Van Vinh Chau et al, Transmission of SARS-CoV-2 Delta Variant Among Vaccinated Healthcare Workers, Vietnam, preprints with The Lancet (Aug 10, 2021). https://papers.ssrn.com/sol3/papers.cfm?abstract_id=3897733

105   Lindsey Tanner, Mike Stobbe and Philip Marcelo, "Study: Vaccinated people can carry as much virus as others," AP News (July 30, 2021) https://apnews.com/article/science-health-coronavirus-pandemic-d9504519a8ae081f785ca012b5ef84d1

106   Apoorva Mandavilli, C.D.C. Internal Report Calls Delta Variant as Contagious as Chickenpox." The New York Times (July 30, 2021, Updates September 1, 2021), https://www.nytimes.com/2021/07/30/health/covid-cdc-delta-masks.html

107   Pnina Shitrit, et al, Nosocomial outbreak caused by the SARS-CoV-2 Delta variant in a highly vaccinated population, Israel, July 2021 (Sep 30, 2021). https://www.eurosurveillance.org/content/10.2807/1560-7917.ES.2021.26.39.2100822#html_fulltext

108   Ibid.

ChildrensHealthDefense.org/fauci-book
childrenshd.org/fauci-book

For updates, new citations and references, and new information about topics in this chapter:

CHAPTER 2

PHARMA PROFITS OVER PUBLIC HEALTH

"Of all tyrannies, a tyranny sincerely exercised for the good of its victims may be the most oppressive. It would be better to live under robber barons than under omnipotent moral busybodies. The robber baron's cruelty may sometimes sleep, his cupidity may at some point be satiated; but those who torment us for our own good will torment us without end for they do so with the approval of their own conscience."

—C. S. Lewis

For five decades, Dr. Anthony Fauci has wielded formidable power to fortify the pharmaceutical industry's explosive growth and its corrosive influence over our government regulatory agencies and public health policy. During his fifty-year career, Dr. Fauci has nurtured a complex web of financial entanglements among pharmaceutical companies and the National Institute of Allergy and Infectious Diseases (NIAID) and its employees that has transformed NIAID into a seamless subsidiary of the pharmaceutical industry. Dr. Fauci unabashedly promotes his sweetheart relationship with Pharma as a "public-private partnership."[1]

From his perch at NIAID, Dr. Fauci has used his $6 billion annual budget[2] to achieve dominance and control over a long list of agencies and governing bodies, including the Centers for Disease Control and Prevention (CDC), the Food and Drug Administration (FDA), Health and Human Services (HHS) agencies, the National Institutes of Health (NIH), the Pentagon, the White House, the World Health Organization (WHO), the United Nations (UN) organizations, and into the deep pockets of the Clinton and Gates Foundations, and Britain's The Wellcome Trust.

A leviathan yearly grant budget gives Dr. Fauci power to make and break careers, enrich—or punish—university research centers, manipulate scientific journals, and to dictate not just the subject matter and study protocols, but also the outcome of scientific research across the globe. Since 2005, the Defense Advanced Research Projects Agency (DARPA) has funneled an additional $1.7 billion[3] into Dr. Fauci's annual discretionary budget to launder sketchy funding for biological weapons research, often of dubious legality. This Pentagon funding brings the annual total of grants that Dr. Fauci dispenses to an astonishing $7.7 billion—almost twice the annual donations of the Bill & Melinda Gates Foundation. Working in close collaboration with pharmaceutical companies and other large grant makers, including Bill Gates—the biggest funder of vaccines in the world—Dr. Fauci has consistently used his awesome power to defund, bully, silence, de-license, and ruin scientists whose research threatens the pharmaceutical paradigm, and to reward those scientists who support him. Dr. Fauci rewards loyalty with prestigious sinecures on key HHS committees when they continue to advance his interests. When the so-called "independent" expert panels license and recommend new pharmaceuticals, Dr. Fauci's control over these panels gives him the power to fast-track his pet drugs and vaccines through the regulatory hurdles, often skipping key milestones like animal testing or functional human safety studies.

Dr. Fauci's funding strategies evince a bias for developing and promoting patented medicines and vaccines, and for sabotaging and discrediting off-patent therapeutic drugs, nutrition, vitamins, and natural, functional, and integrative medicines. Under his watch, drug companies engineered the opioid crisis and made American citizens the globe's most over-medicated population.[4] During his half-century as America's Health Czar, Dr. Fauci has played a central role in crafting a world where Americans pay the highest prices for medicine[5] and suffer worse health outcomes compared to other wealthy countries.[6] Adverse drug reactions are among the nation's top four leading causes of death, after cancer and heart attacks.[7,8] Dr. Fauci's impressive longevity at NIAID is largely due to his enthusiasm for promoting this Pharma-centric agenda.

NIAID: A Pharma Subsidiary

Under Dr. Fauci's management, NIAID has become the center of a web of corrupting financial ties with the pharmaceutical industry. Dr. Fauci's NIAID looks much more like a drug company than any sort of agency to advance science.

"I've been interviewing scientists for a long time in this country, and let me tell you something. There are two kinds: Those who are serfs of Anthony Fauci and those who are genuine scientists. The serf class will refract whatever the latest Lysenkoism is from Fauci and NIAID. They are protecting their grants," says Celia Farber, whose 2006 Harper's article, "Out of Control: AIDS and the Destruction of Medical Science," laid bare the culture of squalor, corruption, and violence at the vendetta-driven Division of AIDS (DAIDS). "The latter [genuine scientists] are the minority. They look, sound, and behave like scientists. And to varying degrees, they all live in a climate of both economic and reputational persecution. Peter Duesberg is one very famous example but there are others. Fauci's vendetta system has many ways of crushing the natural scientific impulse—to question and to demand proof. Breathtakingly, because of Fauci's impact since 1984, this tradition has been all but snuffed out in the US. 'Everybody is afraid.' How many times have I heard that line?"

By all accounts, Anthony Fauci has implemented a system of dysfunctional conflicts and a transactional culture that have made NIAID a seamless appendage of Big Pharma. There is simply no daylight between NIAID and the drugmakers. It's impossible to say where Pharma ends and NIAID begins. "It's like *Ozark*," says Farber.

Researchers in NIAID's labs supplement their income with honoraria they earn by attending Pharma seminars and briefing pharmaceutical company personnel with inside information about research progress on new drugs in NIAID's pipeline.[9] Dr. Fauci's underlings routinely perform private projects for drug companies in their NIAID labs and take contract work running clinical trials for Pharma's new drugs. Journalist and author Bruce Nussbaum reports that it is standard practice for Dr. Fauci's employees to pocket enough gravy from the deal flow to add 10–20 percent to their NIAID salaries from this sort of work. NIAID officials justify this controversial practice arguing that the influx of pharmaceutical dollars strengthens NIAID's labs and allows the agency to retain talented staff. NIAID also deducts 40, 50, or 60 percent off the top of these contracts for "overhead," cementing the agency's partnership with the industry.[10] It's no surprise that a 2004 Office of Government Ethics investigation chided Dr. Fauci for failing to control the corrupting entanglements between his staffers and pharmaceutical companies.[11,12] That report cited NIAID for failing to review and resolve possible ethical conflicts affecting two-thirds of NIAID's workers who were moonlighting in private industry.

The investigators also found[13] that NIAID had failed to obtain approval for a full 66 percent of "outside activities" the institute had undertaken over the review period. Outside activities, according to the NIH,[14] are undertakings that "generally involv[e] providing a service to or a function for an outside organization, with or without pay or other compensation." That could include generating income from a pharmaceutical patent from a drug company, consulting for industry, obtaining silent or equity involvement with biotech firms, or conducting paid lectures and seminars. Dr. Fauci's management style thrives on creating many such opportunities for his agency and its employees to participate in profitable ventures with pharmaceutical companies.

Dr. Fauci's drug development enterprise is rife with other corrupting conflicts. Most Americans would be surprised to learn, for example, that pharmaceutical companies routinely pay extravagant royalties to Dr. Fauci and his employees and to NIAID itself. Here's how the royalty system works: Instead of researching the causes of the mushrooming epidemics of allergic and autoimmune diseases—the function for which US taxpayers pay his salary—Dr. Fauci funnels the bulk of his $6 billion budget to the research and development of new drugs. He often begins the process by funding initial mechanistic studies of promising molecules in NIAID's own laboratories before farming the clinical trials out to an old boys' network of some 1,300 academic "principal investigators" (PIs) who conduct human trials at university-affiliated research centers and training hospitals, as well as foreign research sites. After these NIAID-funded researchers develop a potential new drug, NIAID transfers some or all of its share of the intellectual property to private pharmaceutical companies, through HHS's Office of Technology Transfer. The University and its PIs can also claim their share of patent and royalty rights, cementing the loyalty of academic medicine to Dr. Fauci.

Once the product gets to market, the pharmaceutical company pays royalties—a form of legalized kickbacks—through an informal scheme that allows Pharma to funnel its profits from drug sales to NIAID and to the NIAID officials who worked on the product. Under a secretive, unpromulgated HHS policy, Dr. Fauci and his NIAID underlings may personally pocket up to $150,000 annually from drugs they helped develop at taxpayers' expense.[15,16,17]

The United States Department of Health and Human Services (HHS) is the named owner of at least 4,400 patents. On October 22, 2020, the United States Government Accountability Office (GAO) published a report titled: *BIOMEDICAL RESEARCH: NIH Should Publicly Report More Information about the Licensing of Its Intellectual Property*. In this document, the authors reported that the NIH has received, "up to $2 billion in royalty revenue for NIH since 1991, when FDA approved the first of these drugs. Three licenses generated more than $100 million each for the agency."[18]

However, Dr. David Martin has reported that the NIH Office of Technology Transfer licensing records[19] suggest that NIH was less than transparent with the GAO investigators. Conspicuously absent from the GAO report are over 130 NIH patents associated with active compounds generating billions of dollars in revenue.

NIAID grants have resulted in 2,655 patents and patent applications, of which only 95 include an assignment to the Department of Health and Human Services as an owner.[20] Dr. Fauci assigned most of these patents to universities, thereby making the ultimate commercial beneficiaries entirely opaque while binding the invaluable loyalty of American medical schools and the nation's most influential physicians to Dr. Fauci and his policies.

Somewhat fishily, one of the largest holders of NIAID-generated patents is SIGA Technologies (NASDAQ: SIGA).[21] SIGA publicly acknowledges a close affiliation with NIAID, but the GAO omits all mention of SIGA in its report. SIGA's CEO, Dr. Phillip L. Gomez, spent nine years working for Dr. Fauci at NIAID developing Dr. Fauci's signature vaccine programs for HIV, SARS, Ebola, West Nile Virus, and Influenza before exiting to commercial ventures. While NIAID clearly developed SIGA's technology, the company reports revenue from NIAID but no royalty or commercial payments to NIH or any of its programs.

Eight US patents list Dr. Anthony Fauci as an inventor. However, NIAID, NIH, and GAO do not list any of them in their reports of active licensing despite the fact that Dr. Fauci has acknowledged collecting patent royalties on his interleukin-2 "invention."[22]

Furthermore, GAO reported none of NIAID's patents despite clear evidence that Gilead Sciences and Janssen Pharmaceuticals (a division of Johnson & Johnson) have generated over $2 billion annually from sales directly resulting from NIAID-funded technologies.[23] Missing from the GAO report are two patents for Janssen's Velcade® that have generated sales in excess of $2.18 billion annually for many years. The GAO report also omits any mention of the patents for Yescarta®, Lumoxiti®, or Kepivance® in violation of 37 USC §410.10 and 35 USC §202(a). At least thirteen of the twenty-one patents in the GAO report, including Dr. Fauci's Moderna vaccine, illegally fail to disclose government interest despite their indisputable NIH pedigrees.

How big is Dr. Fauci's drug development enterprise? Since Dr. Fauci arrived at NIH, the agency has spent approximately $856.90 billion.[24,25] Between 2010 and 2016, every single drug that won approval from the FDA—210 different pharmaceuticals—originated, at least in part, from research funded by the NIH.[26]

Following drug approval, Dr. Fauci continues to collaborate with his pharmaceutical partners on promoting and pricing and profiting from their new product. Over the decades since Dr. Fauci took over NIAID, the agency has formalized an elaborate process of negotiating against US taxpayers to allow Pharma to extract maximum profits back from NIAID's germinated drugs. With NIAID's help, the lucky pharmaceutical company walks the new drug through accelerated FDA approval. The CDC then sets obscene retail prices for these collaborative products in secretive negotiations. Such sweetheart deals—at taxpayer and consumer expense—and accelerated approvals can yield direct financial benefits to NIAID, to Dr. Fauci's favored employees, and even to Dr. Fauci himself.[27]

Dr. Fauci launched his career by allowing Burroughs Wellcome (now GlaxoSmith-Kline) to charge $10,000 annually[28] for azidothymidine (AZT), an antiretroviral medication developed exclusively by NIH and tested and approved by Dr. Fauci himself. Dr. Fauci knew that the product cost Burroughs Wellcome a mere $5/dose to manufacture.[29]

Higher profit for industry "partners" often means more extravagant royalty payments for his NIAID and NIH cronies.

Another antiviral drug developed by Dr. Fauci's shop, remdesivir, provides a recent example of a similar Pharma money-making scheme facilitated by NIAID/NIH. While remdesivir proved worthless against COVID, Dr. Fauci altered the study protocols to give his pet drug the illusion of efficacy.[30, 31] Despite opposition from FDA and WHO, Dr. Fauci declared from the White House that remdesivir "will be the standard of care" for COVID, guaranteeing the company a massive global market. Dr. Fauci then overlooked Gilead's price gouging; the company sold remdesivir for $3,300–$5,000 per dose, during the COVID pandemic. The raw materials to make remdesivir cost Gilead under $10. Medicaid must, by law, cover all FDA-approved drugs, so taxpayers again foot the bill. Through these boondoggles, Anthony Fauci has made himself the leading angel investor of the pharmaceutical industry.

The disparate treatment of patented versus less expensive off-patent COVID-19 drug treatments by federal health agencies clearly exposes Dr. Fauci's historic bias for high-ticket patent medicines that favor extravagant pharmaceutical industry profits over public health.[32]

A 2017 study in the *Emory Corporate Governance and Accountability Review* summarizes how compromised federal public health officials like Dr. Fauci have transformed NIAID, NIH, CDC, and FDA into pharmaceutical marketing machines.[33] The Emory researchers paint drug and vaccine makers as "thick as thieves," with HHS officials acting not as regulators, but as "enablers, or perhaps worse still, [they are] complicit in questionable or ethically unsound activity as a result of being driven by self-serving motives . . . " According to Dr. Michael Carome, a former HHS official and a director of the advocacy group Public Citizen, "Instead of a regulator and a regulated industry, we now have a partnership. . . . That relationship has tilted the agency [HHS] away from a public health perspective to an industry friendly perspective."[34] Dr. Fauci is the human face of this corrupt dynamic.

Under Dr. Fauci's leadership, the commercial features of this partnership have eclipsed his agency's mission to advance science. At NIAID, the Pharma tail now wags the public health dog. Dr. Fauci has done almost nothing to advance NIAID's core obligation of researching the causes of the devastating explosions in epidemics of chronic allergic and autoimmune diseases that, under his tenure, have mushroomed to afflict 54 percent of children,[35] up from 12.8 percent when he took charge of NIAID in 1984.[36] While ignoring the explosion of allergic conditions, Dr. Fauci has instead reshaped NIAID into the leading incubator for new pharmaceutical products, many of which, ironically, profit from the cascading chronic disease pandemic.

Over the last fifty years at NIH, Dr. Fauci has played a leading role in Big Pharma's engineered demolition of American health and democracy, working hand in glove with pharmaceutical companies to overcome federal regulatory obstacles and transform the NIH and NIAID into a single-minded vehicle for development, promotion, and marketing of patented pharmaceutical products, including vaccines and vaccine-like products.

Most of us would like "America's Doctor" to properly diagnose our illnesses using the best science, and then instruct us on how to get healthy. What if, instead of spending their entire budgets developing profitable pharmaceutical products, Dr. Fauci and the heads of other NIH institutes deployed researchers to explore the links between glyphosate in food and the explosion of gluten allergies, the link between pesticide residues and the epidemic of neurological diseases and cancers, the causal connections between aluminum and Alzheimer's disease, between mercury from coal plants and escalating autism rates, and the association of airborne particulates with the asthma epidemic? What if NIH financed research to explore the association between childhood vaccines and the explosion of juvenile diabetes, asthma, and rheumatoid arthritis, and the links between aluminum vaccine adjuvants and the epidemics of food allergies and allergic rhinitis? What if they studied the impacts of sugar and soft drinks on obesity and diabetes, and the association between endocrine disruptors, processed foods, factory farms, and GMOs on the dramatic decline in public health? What would Americans look like if, for fifty years, we had a public health advocate running one of our top health agencies—instead of a Pharma shill? What would have happened if we'd spent that hundreds of billions dollars on real science, instead of drug development? Dr. Fauci seems willing only to give us diagnoses and cures that benefit Big Pharma—instead of public health—and to cover his trail with artifice.

His critics have compared Dr. Fauci to a similarly long-lived federal agency bureaucrat, J. Edgar Hoover, who used his five-decade dictatorial control of the FBI to transform the agency into a vehicle for shielding organized crime, fortifying his corrupt political partners, oppressing Black Americans, surveilling his political enemies, suppressing free speech and dissent, and as a platform for building a cult of personality around his own inflated ego. More recently, Dr. Fauci's perennial biographer, Charles Ortleb, analogized Dr. Fauci's career and pathological mendacity to the sociopathic con men Bernie Madoff and Charles Ponzi.[37] Another critic, author J. B. Handley, labeled Dr. Fauci "a snake oil salesman" and a "bigger medical charlatan than Rasputin."[38] Economist and author Peter Navarro, former Director of Trade and Manufacturing Policy, observed during a national network television interview in April 2021 that "Fauci is a sociopath and a liar."[39]

His white lab coat, his official title, and his groaning bookshelves crowded with awards from his medical cartel collaborators allow Dr. Fauci to masquerade as a neutral, disinterested scientist and selfless public servant driven by a relentless commitment to public health. But Dr. Fauci doesn't really do public health. By every metric, his fifty-year regime has been a catastrophe for American health. But as a businessman, his success has been boundless.

In 2010, Dr. Fauci told adoring *New Yorker* writer Michael Specter that his go-to political playbook is Mario Puzo's novel *The Godfather*.[40] He spontaneously recited his favorite line from Puzo's epic: "It's nothing personal, it's strictly business."

Endnotes

1. Coronavirus Response, 116th United States Congress, May 12, 2020, Testimony of Dr. Anthony Fauci, https://www.rev.com/blog/transcripts/dr-anthony-fauci-cdc-director-senate-testimony-transcript-may-12 at 34:06
2. "NIAID Budget Data Comparisons," NIH/NIAID (2021), https://www.niaid.nih.gov/grants-contracts/niaid-budget-data-comparisons
3. Dr. David E. Martin, *The Fauci/COVID-19 Dossier* (Jan. 18, 2021) https://f.hubspotusercontent10.net/hubfs/8079569/The%20FauciCOVID-19%20Dossier.pdf
4. Teresa Carr, "Too Many Meds? America's Love Affair with Prescription Medication," *Consumer Reports,* Aug. 3, 2017, https://www.consumerreports.org/prescription-drugs/too-many-meds-americas-love-affair-with-prescription-medication/#nation
5. M. Jackson Wilkinson, "Lies, Damn Lies, and Prescriptions," MJACKSONW.COM, Nov. 6, 2015, https://mjacksonw.com/lies-damn-lies-and-prescriptions-f86fca4d05c
6. Maggie Fox, "United States Comes in Last Again on Health, Compared to Other Countries," *NBC News*, Nov. 16, 2016, https://www.nbcnews.com/health/health-care/united-states-comes-last-again-health-compared-other-countries-n684851
7. Peter C Gøtzsche, "Prescription drugs are the third leading cause of death," THE BMJ OPINION, June 16, 2016, https://blogs.bmj.com/bmj/2016/06/16/peter-c-gotzsche-prescription-drugs-are-the-third-leading-cause-of-death/
8. "Preventable Adverse Drug Reactions: A Focus on Drug Interactions," U.S. Food and Drug Administration, March 6, 2018, https://www.fda.gov/drugs/drug-interactions-labeling/preventable-adverse-drug-reactions-focus-drug-interactions
9. Bruce Nussbaum, *Good Intentions: How Big Business and the Medical Establishment are Corrupting the Fight Against AIDS* (Atlantic Monthly Press, 1990), 162
10. Nussbaum, op. cit., 162–163
11. Daniel Payne, John Solomon, "Fauci Files: Celebrated doc's career dotted with ethics, safety controversies inside NIH," *Just the News* (July 23, 2020), https://justthenews.com/accountability/political-ethics/fauci-says-americans-should-trust-doctors-himself-his-career
12. Marilyn L. Glynn, Letter to Edgar M. Swindell, Jul. 26, 2004, https://justthenews.com/sites/default/files/2020-07/OGE-2004NIHEthicsReview_0.pdf
13. Ibid.
14. National Institutes of Health, Visiting Scientists, Outside Activity, https://www.ors.od.nih.gov/pes/dis/VisitingScientists/Pages/OutsideActivityJ-1.aspx
15. J. Solomon, "Researchers mum on financial interests," CBS News (Associated Press), Jan 10, 2005; https://www.nbcnews.com/health/health-news/report-researchers-mumon-financial-interests-flna1c9475821
16. *Information for NIH Inventors, Inventor Royalties*, NIH Office of Technology Transfer, https://www.ott.nih.gov/royalty/information-nih-inventors
17. J. H. Tanne, "Royalty payments to staff researchers cause new NIH troubles," BMJ, Jan 22, 2005, https://www.ncbi.nlm.nih.gov/pmc/articles/PMC545012/
18. "Biomedical Research: NIH Should Publicly Report More Information about the Licensing of Its Intellectual Property," US Government Accountability Office, October 22, 2020, https://www.gao.gov/products/GAO-21-52
19. Martin, op. cit.
20. Ibid.
21. Ibid.
22. Tanne, op. cit.
23. Martin, op. cit.
24. "Proposed 1969 Budget Asks $1,196.6 Million Appropriation for NIH," *NIH Record*, Feb. 6, 1968, https://nihrecord.nih.gov/sites/recordNIH/files/pdf/1968/NIH-Record-1968-02-06.pdf
25. Appropriations History by Institute/Center (1938 to Present), NIH Office of Budget, https://officeofbudget.od.nih.gov/approp_hist.html
26. Ekaterina Galkina Cleary et al., "Contribution of NIH funding to new drug approvals 2010-2016," PNAS, Mar 6, 2018; first published Feb 12, 2018; https://doi.org/10.1073/pnas.1715368115
27. Tanne, op. cit.
28. Mark H. Furstenberg, "AZT the First AIDS Drug," *Washington Post*, Sept 15, 1987, https://www.washingtonpost.com/archive/lifestyle/wellness/1987/09/15/azt-the-first-aids-drug/f38de60b-1332-4bb3-a49f-277036b1baf2/
29. Philip J. Hiltz, "AIDS Drug Maker Cuts Price by 20%," *New York Times*, Sept 19, 1989, https://www.nytimes.com/1989/09/19/us/aids-drug-s-maker-cuts-price-by-20.html
30. Meryl Nass, "Faking results: Fauci's NIAID-paid Remdesivir Study changed its Outcome Measures Twice, in order to show even a whiff of benefit,"Anthrax Vaccine-Posts by Meryl Nass, M.D., May, 2, 2020, https://anthraxvaccine.blogspot.com/2020/05/faking-results-faucis-niaid-paid.html
31. Changes (Side-by-Side) for Study: NCT04280705 March 20, 2020 (v10), April 23, 2020 (v16), NIH, https://clinicaltrials.gov/ct2/history/NCT04280705?A=10&B=16&C=Side-by-Side#StudyPageTop

32. Elizabeth L. Vliet, "A Tale of Two Drugs: Money vs. Medical Wisdom," American Association of Physicians and Surgeons, May 7, 2020, https://aapsonline.org/a-tale-of-two-drugs-money-vs-medical-wisdom/

33. Leslie E. Sekerka & Lauren Benishek, "Thick as Thieves? Big Pharma Wields Its Power with the Help of Government Regulation," *Emory Law Scholarly Commons* Vol. 5, Issue 2 (2018), https://scholarlycommons.law.emory.edu/ecgar/vol5/iss2/4/

34. Caroline Chen, "FDA Repays Industry by Rushing Risky Drugs to Market," PROPUBLICA (June 26, 2018) https://www.propublica.org/article/fda-repays-industry-by-rushing-risky-drugs-to-market

35. Christina D. Bethell, Michael D. Kogan, et al., "A National and State Profile of Leading Health Problems and Health Care Quality for US Children: Key Insurance Disparities and Across-State Variations," *Academic Pediatrics*, (May–June 2011), https://doi.org/10.1016/j.acap.2010.08.011

36. Jeanne Van Cleave, Steven L. Gortmaker, James M. Perrin, "Dynamics of Obesity and Chronic Health Conditions Among Children and Youth," *JAMA*, (Feb. 17, 2010), doi:10.1001/jama.2010.104

37. Charles Ortleb, *Fauci: The Bernie Madoff of Science and the HIV Ponzi Scheme That Concealed the Chronic Fatigue Syndrome Epidemic*, (HHV-6 University Press, 2020), 27, 39, 41

38. Robert F. Kennedy Jr., "'TRUTH' with Robert F. Kennedy, Jr.–Episode 7," Interview with J.B. Handley, *Children's Health Defense*, July 9, 2020. https://childrenshealthdefense.org/news/truth-with-robert-f-kennedy-jr-episode-7/

39. Sinéad Baker, "Trump advisor Peter Navarro went on a wild rant on Fox News, calling Fauci the 'father' of the coronavirus," *Business Insider*, (Mar 31, 2021), https://www.businessinsider.com/peter-navarro-trump-advisor-calls-fauci-father-of-coronavirus-fox-news-rant-2021-3

40. Cory Steig, "Dr. Fauci uses this line from 'The Godfather' to help deal with stress and politicians," *CNBC*, (Oct. 21, 2020). https://www.cnbc.com/2020/10/01/dr-anthony-fauci-on-lesson-from-the-godfather-book-.html

ChildrensHealthDefense.org/fauci-book
childrenshd.org/fauci-book

For updates, new citations and references, and new information about topics in this chapter:

CHAPTER 3
THE HIV PANDEMIC TEMPLATE FOR PHARMA PROFITEERING

"Guys like Fauci get up there and start talking and you know he doesn't know anything really about anything, and I'd say that to his face. Nothing. The man thinks you can take a blood sample and stick it in an electron microscope and if it's got a virus in there, you'll know it… He doesn't understand electron microscopy and he doesn't understand medicine. And he should not be in the position like he's in. Most of those guys up there on the top are just total administrative people and they don't know anything about what's going on at the bottom. Those guys have got an agenda, which is not what we'd like them to have, being that we pay them to take care of our health in some way. They've got a personal kind of agenda. They make up their own rules as they go, they change them when they want to, and they smugly, like Tony Fauci, do not mind going on television, in front of the people that pay his salary, and lie directly into the camera."

—Dr. Kary Mullis, winner of the 1993 Nobel Prize for Chemistry for his invention of the Polymerase Chain Reaction (PCR) technique, from interview with Gary Null, 1993.

"Of course! I will always give you truth. Just ask the question and I'll give you the truth. At least to the extent, that I think it is, right [laughs]."

—Dr. Fauci, *Der Spiegel*, September 2020

"Scientifically," he [Harvey Bialy] says, "cancer is still an interesting question. AIDS has not been an interesting question for fifteen years."
"Why do you say that?"
"Because it's been a closed book for fifteen years. It has been clear for fifteen years that this is a non-infectious condition that has its cause in a whole variety of chemicals."
His voice rises. "Doesn't the book demonstrate very clearly that scientifically, nothing happened between 1994 and 2003? Zero. Absolutely nothing except one wrong epidemiological prediction after another, one failed poisonous drug after another. 0.000.000 cured. No vaccine, or even a fake vaccine. It's a total failure. We've turned virology inside out and upside down to accommodate this bullshit hypothesis for seventeen years now. It's enough."

—From *Serious Adverse Events: An Uncensored History of AIDS*, by Celia Farber

Prior to 1987, Peter Duesberg never had a single grant proposal rejected by the NIH. Since 1987, he has written a total of thirty research proposals; every single one has been rejected. He has submitted several proposals on aneuploidy, as recently as last year—they too have been rejected.
"They just took him out," says Richard Strohman, a retired UC Berkeley biologist. "Took him right out."
"The system works," says Dave Rasnick. "It's as good as a bullet to the head."

—From *Serious Adverse Events: An Uncensored History of AIDS*, by Celia Farber

Beginnings

Anthony Stephen Fauci was born in Brooklyn's Dyker Heights neighborhood on December 24, 1940. Three of his grandparents were native Italians; his maternal grandfather was born in the Italian-speaking region of Switzerland. All four came to the United States at the end of the nineteenth century. Both his parents were born in New York City. His father, Stephen Fauci, graduated from the College of Pharmacy, Columbia University. His mother, Eugenia, went to Brooklyn College and Hunter College. They married at eighteen years old. It's tempting to link his emergence as the modern champion of the pharmaceutical paradigm to the fact that Dr. Fauci's parents owned a drugstore. His father, a pharmacist, filled prescriptions; his mother worked the cash register, and young Tony apprenticed on his Schwinn bicycle for a lifelong career delivering drugs.

Anthony attended Our Lady of Guadeloupe Grammar School in Brooklyn and Regis High School, an elite Jesuit academy, where his tenacity distinguished him in the classroom and on the basketball court. Regis heavily weighted its curriculum toward the classics: "We took four years of Greek, four years of Latin, three years of French, ancient history, theology, etc.," he told an NIH oral historian in 1989. He was a good athlete in a borough of stickball aces. An early Yankees fan, he preferred the reliable champions to the hometown heroes and describes himself as "somewhat of a sports outcast among my friends, who were all Brooklyn Dodgers fans."[1] The underdog Dodgers lost eight of eleven World Series encounters against the Bronx Bombers. Tony's idols were Joe DiMaggio, Mickey Mantle, and Mets/Dodgers/Giants great Duke Snider. His appetite for total victory and domination made him a ferocious contender. Despite his diminutive size—he is 5′7″—he played basketball and football and was a star point guard and captain of Regis's 1958 basketball squad. Tony scored an impressive ten points per game, according to his yearbook. It wasn't enough; the Raiders ended the season with a dismaying 2-16 record. A teammate, Bob Burns, recalls that "he was ready to drive through whoever was in his way." Another class-mate, John Zeman, told *Wall Street Journal* reporter Ben Cohen, "He was just a ball of fire. He would literally dribble through a brick wall."[2]

Dr. Fauci went to Holy Cross College in 1958, studying philosophy, French, Greek, and Latin and graduating in 1962 with a BA. "I still am very interested in the classics," he said in a 1989 interview with Dr. Victoria Harden, director of the NIH Historical Office.[3] Dr. Fauci grew up Roman Catholic: "I credit very much the Jesuit training in precision of thought and economy of expression in solving and expressing a problem and the presentation of a solution in a very succinct, accurate way. This has had a major, positive influence on the fact that I enjoy very much and am fairly good at being able to communicate scientific principles or principles of basic and clinical research without getting very profuse and off on tangents."[4] Perhaps reason became the enemy of his faith—or, perhaps, Jesuit discipline robbed the catechisms of their fun. Today, Dr. Fauci brushes off queries about his Catholicism, describing himself as a humanist.[5]

Dr. Fauci never doubted that he wanted to be a doctor, commenting that in high school, "[T]here really was no question that I was going to be a physician. I think there was subliminal stimulation from my mother, who, right from the very begin-ning when I was born, wanted me to be a physician."[6]

Dr. Fauci earned his medical degree from Cornell in 1966, graduating first in his class. Like his wife, immunologist and NIH's Bioethics Department Director Dr. Christine Grady, Dr. Fauci is a lifelong germaphobe, but he confesses that he went into virology and immunology not so much to kill bugs as to avoid combat service in Vietnam: "I left Cornell and went into my internship and residency in 1966. That was at the exponential phase of the Vietnam War, and every single physician went into military service. I can remember very clearly when we were gathered in the auditorium at Cornell early in our fourth year of medical school. The recruiter from the Armed Forces came there and said, 'Believe it or not, when you graduate from medical school at the end of the year, except for the two women, everyone in this room is going to be either in the Army, the Air Force, the Navy, or the Public Health Service. So, you're going to have to take your choice. Sign up and give your preferences.' So I put down Public Health Service as my first choice and then the Navy. Essentially, I came down to the NIH because I didn't have any choice."[7]

The US Public Health Service was a heavily militarized public health agency led by its uniformed officer corps, including the surgeon general, which had grown out of military hospitals operated by the early Navy. NIH was its research arm created during World War II to support soldiers' health during the war. As infectious disease mortalities in the US declined precipitously in the mid-1950s, NIH maintained its relevance by declaring war on cancer.[8,9]

"I was very lucky because I knew that it was a phenomenal scientific opportunity. I wanted to learn some basic cellular immunology with the ultimate aim of going into what has been my theme for the past twenty-one years—human immunobiology and the regulation of the human immune system."[10]

After completing his residency at Cornell Medical Center, Dr. Fauci joined NIH in 1968 as a clinical associate at the NIAID, one of two dozen of NIH's sub-agencies. In 1977, he became deputy clinical director of NIAID. Oddly, his specialty was applied research in immune-mediated illness—a subject of increasingly grave national concern. He would spend the next fifty years largely ignoring the exploding incidence[11] of autoimmunity and allergic diseases, except to the extent they created profitable markets for new pharmaceuticals. Dr. Fauci became NIAID's director on November 2, 1984, just as the AIDS crisis was spiraling out of control.

NIAID: A Sleepy, Irrelevant Agency

When Dr. Fauci assumed leadership of NIAID, the agency was a backwater. Allergic and autoimmune disorders were hardly a factor in American life. Peanut allergies, asthma, and autoimmune diseases (e.g., diabetes and rheumatoid arthritis) were still so rare that their occasional occurrences in schoolchildren were novelties. Most Americans had never seen a child with autism; only a tiny handful would recognize the term until the 1988 film *Rain Man* introduced it into the vernacular. Cancer was the disease Americans increasingly feared, with nearly all the attention at NIH and the bulk of federal health funding going to the National Cancer Institute (NCI).

Worst of all, by the era of Dr. Fauci's ascendance as an ambitious bureaucrat at NIAID, infectious diseases were no longer a significant cause of death in America. Dramatic improvements in nutrition, sanitation, and hygiene had largely abolished the frightening mortalities from mumps, diphtheria, smallpox, cholera, rubella, measles, pertussis, puerperal fever, influenza, tuberculosis, and scarlet fever.[12] The devastating lethality from these former scourges that decimated earlier generations of Americans had dwindled. From 1900, when one-third of all deaths were linked to infectious diseases (e.g., pneumonia, tuberculosis, and diarrhea and enteritis), through 1950, infectious disease mortality decreased dramatically (except for the 1918 Spanish flu), leveling off in the 1950s to what we see today, about 5 percent of all US deaths.[13]

Annual deaths from communicable disease dropped in the 1980s to around 50 per hundred thousand population, from 800 per hundred thousand in 1900.[14] By the twentieth century, more people were dying of old age and heart attacks than from contagious illnesses.[15]

At NIAID and at its sister agency, CDC, the bug hunters were sliding into irrelevance. NIAID's heyday was a distant memory; it had served at the forefront of the war against deadly pestilence. NIH had mobilized scientists to track the epidemics of cholera, Rocky Mountain spotted fever, and the 1918 Spanish flu contagion that infected and killed millions globally.

Today CDC and NIAID promote the popular orthodoxy: that intrepid public health regulators, armed with innovative vaccines, played the key role in abolishing mortalities from these contagious illnesses. Both science and history dismiss this self-serving mythology as baseless. As it turns out, the pills, potions, powders, surgeries, and syringes of modern medicine played only a minor role in the historic abolition of infectious disease mortalities.

An exhaustive 2000 study by CDC and Johns Hopkins scientists published in *Pediatrics,* the official journal of the American Academy of Pediatrics, concluded, "Thus vaccination does not account for the impressive declines in [infectious disease] mortality seen in the first half of the [20th] century . . . nearly 90 percent of the decline in infectious disease mortality among US children occurred before 1940, when few antibiotics or vaccines were available."[16]

Similarly, a comprehensive 1977 study by McKinlay and McKinlay, formerly required reading in almost all American medical schools, found that all medical interventions, including vaccines, surgeries, and antibiotics, contributed only about 1 percent of the decline and at most 3.5 percent.[17] Both CDC and the McKinlays attributed the disappearance of infectious disease mortalities not to doctors and health officials, but to improved nutrition and sanitation—the latter credited to strict regulation of food preparation, electric refrigerators, sewage treatment, and chlorinated water. The McKinlays joined Harvard's iconic infectious disease pioneer, Edward Kass, in warning that a self-serving medical cartel would one day try to claim credit for these public health improvements as a pretense for imposing unwarranted medical interventions (e.g., vaccines) on the American public.

As the McKinlays and Kass[18] had predicted, vaccinologists successfully hijacked the astonishing success story—the dramatic 74 percent decline in infectious disease mortalities of the first half of the twentieth century—and deployed it to claim for themselves, and particularly for vaccines, a revered and sanctified—and scientifically undeserving—prestige beyond criticism, questioning, or debate.

An Agency Without a Mission

In 1955, as deaths from epidemic disease declined, NIAID's forerunner organization at NIH, the National Microbiological Institute (NMI), became part of the NIAID,[19] to reflect the diminished national significance of infectious diseases and the unexplained increases in allergic and immune system diseases. Congress ordered NIAID to support "innovative scientific approaches to address the causes of these diseases and find better ways to prevent and treat them."

Food allergies and asthma were still rare enough to be considered remarkable. Eczema was practically unknown, as were most autoimmune diseases, including diabetes, rheumatoid arthritis, lupus, Graves' disease, Crohn's disease, and myelitis.[20,21]

As early as 1949, Congressional bills to abolish CDC because of the remarkable decline in infectious disease mortalities twice won by impressive majorities.[22] From the mid-1970s, CDC was seeking to justify its existence by assisting state health departments to track down small outbreaks of rabies and a mouse disease called hantavirus, and by linking itself to the military's bioweapons projects. Looking back from 1994, Red Cross officer Paul Cummings told the *San Francisco Chronicle* that "The CDC increasingly needed a major epidemic" to justify its existence.[23] According to Peter Duesberg, author of *Inventing the AIDS Virus,* the HIV/AIDS theory was salvation for American epidemic authorities.[24]

James Curran, the Chief of the CDC's Sexually Transmitted Diseases unit, described the desperation among the public health corps in the early 1980s: "There was double-digit inflation, very high unemployment, a rapid military buildup and a threat to decrease all domestic programs, and this led to workforce cuts at the Public Health Service, and particularly CDC."[25] Nobel Laureate Kary Mullis similarly recalled the institutional desperation during the Reagan administration era. He said of the CDC: "They were hoping for a new plague. Polio was over. There were memos going around the agency saying, 'We need to find the new plague'; 'We need to find something to scare the American people so they will give us more money.'"[26] NIH scientist Dr. Robert Gallo—who would become Dr. Fauci's partner, coconspirator, and confidant—offered a similar assessment: "The CDC in Atlanta was under threat for reductions and even theoretically for closure."[27]

Drumming up public fear of periodic pandemics was a natural way for NIAID and CDC bureaucrats to keep their agencies relevant. Dr. Fauci's immediate boss and predecessor as NIAID Director, Richard M. Krause, helped pioneer this new strategy in 1976, during Dr. Fauci's first year at the agency. Krause was a champion of what he called "The Return of the Microbes" strategy,[28] which sought to reinstate microbes to their former status as the feared progenitors of deadly diseases. That year, federal regulators concocted a fake swine flu epidemic that temporarily raised hopes around CDC for the resurrection of its reputation as a life-saving superhero.[29]

Even in that idealistic era, regulators were allowing Pharma to craft public health policy behind closed doors. Director Krause, whom Dr. Fauci would shortly succeed, invited Merck executives to sit in on internal planning meetings as collaborators.[30] Working with Merck, NIAID[31] used taxpayer funds to subsidize development and distribution of vaccines,[32] and to rush untested products to market.[33] But the swine flu pandemic was a dud, and HHS's response was a global embarrassment. Only one casualty—a soldier at Fort Dix[34]—succumbed to the "pandemic," and Merck's experimental vaccine triggered a national epidemic of Guillain-Barré syndrome, a devastating form of paralysis resembling polio, before regulators recalled the jab.[35] The four vaccine manufacturers—Merck & Co., Merrell, Wyeth, and Parke-Davis—had refused to sell the vaccines to the government unless they were guaranteed profits and indemnity. They were sued for $19 million within months of the vaccination campaign. The Department of Justice handled the lawsuits.[36]

Prior to 1997, the FDA forbade pharmaceutical advertising on television, and the drug companies had not yet transformed television reporters into pharmaceutical reps. Journalists, in short, were still permitted to do journalism. *Sixty Minutes* aired a scathing segment in which Mike Wallace mercilessly exposed the corruption, incompetence, and cover-ups at HHS that led to the phony swine flu pandemic and the wave of casualties from NIH's experimental vaccine.[37] The scandal forced the resignation of CDC Director David Sencer for his role in concocting the phony pandemic and pushing the dangerous vaccine.[38] NIAID chief Richard Krause quietly resigned in 1984, deeding his seat to his faithful deputy, Tony Fauci.[39]

In a poignant emblem of the ascending power of the pharmaceutical paradigm under Dr. Fauci's stewardship, the *Sixty Minutes* report on the 1976 pandemic scandal is now largely scrubbed from the Internet. You can still view it on the Children's Health Defense website.

HIV/AIDS

Despite those catastrophic outcomes, Dr. Fauci's takeaway from the 1976 swine flu crisis seems to have been the revelation that pandemics were opportunities of convenience for expanding agency power and visibility, and for cementing advantageous partnerships with pharmaceutical behemoths and for career advancement. Four years later, the AIDS pandemic proved a redemptive juncture for NIAID and the launch pad for Dr. Fauci's stellar rise. The lessons he learned from orchestrating regulatory responses to the AIDS crisis would become familiar templates for managing subsequent pandemics.

Tony Fauci spent the next half-century crafting public responses to a series of real and concocted viral outbreaks[40,41]—HIV/AIDS[42] in 1983; SARS[43] in 2003; MERS[44,45,46] in 2014; bird flu[47,48] in 2005; swine flu ("novel H1N1")[49] in 2009; dengue[50,51] in 2012; Ebola[52] in 2014–2016; Zika[53] in 2015–2016; and COVID-19[54] in 2020. When authentic epidemics failed to materialize, Dr. Fauci became skilled at exaggerating the severity of contagions to scare the public and further his career.

Even all those years ago, Anthony Fauci had already perfected his special style of *ad-fear-tising*, using remote, unlikely, farfetched and improbable possibilities to frighten people. Fauci helped terrify millions into wrongly believing they were at risk of getting AIDS when they were not; emphasis in his statement is added to highlight the caveats and conditional language:

> The long incubation period of this disease, we *may be* starting to see, as we're seeing *virtually*, as the months go by, other groups that *can* be involved, and seeing it in

children is really quite disturbing. *If* the close contact of the child is a household contact, *perhaps* there will be a *certain number* of individuals who are just living with and in close contact with someone with AIDS *or at risk of* AIDS who *does not necessarily* have to have intimate sexual contact or share a needle, but just the ordinary close contact that one sees in normal interpersonal relationships. Now that *may be* farfetched *in a sense* that there have been no cases recognized *as yet* in which individuals have had merely casual contact, close *or albeit* with an individual with AIDS who *for example* have gotten AIDS. *For example*, there have been no cases *yet* reported of hospital personnel, who have *fairly* close contact with patients with AIDS. There have been no case reports of them getting AIDS; but the *jury is still out* on that because the situation is constantly *evolving* and the incubation period is so long, as you know. It's a *mean of about* fourteen months, *ranging from* six to eighteen months. So what medical researchers and public health service officials *will be*—are *concerned with* is what *we felt* were the confines of transmissibility now going to be *loosening up and broadening up* so that *something less than truly* intimate contact *can* give transmission of this disease.

The message people took away from those 250 rambling and obfuscating words: "Something less than truly intimate contact can give you this disease."

Translated into English, however, it's just twelve words of truth: ***There have been zero cases of AIDS spread by ordinary close contact.***

Dr. Fauci's most vocal critics complain that, from his earliest days running NIAID, he was neither a competent manager nor a particularly skilled or devoted scientist. His gifts were his aptitude for bureaucratic infighting; a fiery temper; an inclination for flattering and soft-soaping powerful superiors; a vindictive and domineering nature toward subordinates and rivals who dissented; his ravenous appetite for the spotlight; and finally, his silver tongue and skilled tailor. He won his initial beachhead by wresting jurisdiction over the AIDS crisis from NIH's Big Kahuna, the National Cancer Institute (NCI).[55]

In 1981, the CDC first recognized the emergence of a new disease that health officials dubbed Acquired Immune Deficiency Syndrome (AIDS) among about fifty gay men in Los Angeles, San Francisco, and New York. The AIDS crisis initially landed at NCI because the condition's most pronounced signal was Kaposi's sarcoma, which was then considered a deadly skin cancer associated with immune suppression.

A decade earlier, in 1971, President Nixon had launched the "War on Cancer."[56] The medical establishment promised a cancer cure by 1976.[57] Instead, Pharma quickly transformed NCI into its cash cow as captured regulators funneled hundreds of billions of dollars into single-purpose patented cancer remedies and wonder-drug production that the agency developed with pharmaceutical company partners. The money enriched Pharma, researchers, doctors, and universities, but yielded little net public health benefit. Fifty years and $150 billion dollars later,[58,59] soft tissue and non-smoking cancers have increased dramatically.[60] NCI, ever-sensitive to offending Big Pharma, Big Food, Big Ag, and Big Chemical, had spent almost nothing to address public exposures to carcinogens from medicines, vaccines, meats, processed foods, sugar, and chemical-laden agriculture. Mainstream cancer research suggests that one-third of all cancers could be eliminated through lifestyle changes. But according to cancer expert Samuel Epstein, NCI spent "Just 1 million—that is 0.02

percent of its $4.7 billion budget in 2005—on education, press releases, and public relations to encourage" better eating habits to prevent cancer.[61]

Under NIH's regulatory rubric, the only exposures that are permissible targets of criticism and research in that universal bugaboo are Big Tobacco and the sun, which doesn't pay lobbyists. NIH's unbridled criticism of UV light has made sunscreen lotions another booming profit center for Big Pharma.

For Pharma and its NCI regulators and enablers, the AIDS crisis looked like another ATM machine. But in 1984, NIH scientist Robert Gallo linked AIDS to his virus, HTLV-III, which in time would be renamed the "human immunodeficiency virus" (HIV). Dr. Fauci then moved aggressively to capture that revenue stream for his agency. In a dramatic confrontation with NCI's Sam Broder that year, Dr. Fauci persuasively argued that, since AIDS was an infectious disease, NIAID must have jurisdiction. His victory over NCI in that tip-off placed Dr. Fauci in position to capture the sudden flood of congressional AIDS appropriations flowing to NIH through the adept lobbying of a well-organized AIDS community then besieging the Capitol for resources to study and treat the "gay plague."

In 1982, congressional AIDS funding was a pitiful $297,000.[62] By 1986, that number jumped to $63 million.[63] The following year, it was $146 million.[64] By 1990, NIAID's annual AIDS budget was $3 billion. But Gallo's HIV/AIDS hypothesis proved a PR windfall for Dr. Fauci, as well. "The most dangerous place in America is between Tony and a microphone," recalls Dr. Fauci's perennial Boswell, Charles Ortleb, the former publisher of the *New York Native*, the gay newspaper that chronicled the early AIDS epidemic. "Once people recognized that this was caused by a virus," recalled CDC's James Curran, "media attention went from no news coverage to the most-covered news story in history. People went from neglecting it, to fear and panic."[65]

The expanded flow of cash spelled opportunity for Dr. Fauci. "AIDS was his big chance," wrote historian and journalist Bruce Nussbaum, who penned the definitive history of early AIDS research, *Good Intentions: How Big Business and the Medical Establishment are Corrupting the Fight Against AIDS*.[66] "He wasn't well known as a brilliant scientist, and he had little background in managing a big bureaucracy; but Fauci did have ambition and drive to spare. This lackluster scientist was about to find his true vocation—empire building."[67]

"Teflon Tony"

The AIDS crisis's best-known activist—and the most vocal critic of the NIH response—playwright Larry Kramer, may have been the first to make the cold assessment about Dr. Fauci's winning capacity for combining charm and flattery with evasion, misdirection, and misinformation to bedazzle the media into suspending skepticism and overlooking his reliable incompetence. "The main reason that Fauci has gotten away with so much," Kramer observed in 1987, "is that he's attractive and handsome and dapper and extremely well spoken and he never answers your question."[68] Historians Torsten Engelbrecht, author of *Virus Mania*, and Konstantin Demeter call Fauci "Dr. Baron of Lies."[69]

Asked to offer thoughts on Fauci, veteran AIDS "war" reporter Celia Farber pulls back and takes a broad view. She said:

People understand the Arendt concept of the "banality of evil."

You have set yourself the formidable task of deconstructing him. Why is he "evil"? (Which he is.)

It's not because he is so "banal," so bureaucratic, so *boring*. That's the drag costume.

In fact, he is a revolutionary—a very dangerous one, who slipped behind the gates when nobody understood what he was bringing in.

What was he bringing in? He was bringing in—as a trained Jesuit and committed Globalist—a new potion that would achieve any and all aims for Pharma and the powers he served. The potion was then known as Political Correctness—now called "woke."

Fauci switched the entire linguistic system of American science, from classical "speak," to woke "speak." He brought in Cancel Culture, essentially, before anybody could imagine what it was. It was too perverse for genuine scientists to conceive of such a thing mixing with science, they could not believe it, or grasp it. Like a rape. It was incredibly confusing. That's what I documented, on the ground, that horror and confusion among real scientists, as American science changes so radically before their eyes, to accommodate HIV.

Farber went on:

Let me elaborate a bit. Fauci's reign begins in 1984, a year of total change. Everything changes, all of a sudden. Gallo is deployed with Margaret Heckler to make the declaration by US Government fiat that the "probable cause of AIDS" had been "found" and that it was some kind of trans-Atlantic fusion that looked "virus like" on the big screen, but was really neither a cogent virus nor a pathogen. The reason it "flew" to use [Nature Bio/Technology founding editor] Harvey Bialy's word, was because everything had already changed. It was understood, without overt commands, that the "gay cancer" that had everybody in such a panic could not be assessed as complex toxic illness with a complex cause. The entire US media understood what to say and not say, and not only because of the allegiance to the shadow government, but because the era of classical science had ended. It ended that day. It would henceforth be a crime against decency to, for example, address anything that could be making gay men sick other than "the virus."

That's not "bad science." That's perfectly executed political correctness. And they are diametrically at odds, in the Biblical sense of good and evil.

What Fauci did was he made political correctness the new currency, of his funding empire. Peter Duesberg was not "wrong" about HIV and AIDS, he was politically incorrect about it and that was *how* Fauci banished him—sentenced him to funding and reputation death, as though he had done something *really bad* by dissenting against HIV theory. Stop and think how insane this is. An elite cancer virologist brought over from Germany's Max Planck Institute whose credentials are *so* outstanding, who was well on his way to solving cancer's genetics . . . felled suddenly by a fatwah, issued by this . . . Mufti? Who was *he* to issue a fatwah against America's top cancer virologist? Well, he did. He blocked every federal research dollar to Duesberg after 1987, because Duesberg repudiated the woke ideology Fauci's HIV empire, in a few paragraphs of a scientific paper that was about something else. He sustained the economic and reputational attack/vendetta for the next 3 decades. Without blinking. It's really an unbelievable story. It would make Americans' blood boil if they knew about it—because almost all have lost somebody in their family to cancer.

Fauci had, by 1987, when Duesberg wrote the *Cancer Research* paper that sealed his scientific fate, an apparatus that included mass media, psychological operations, public health—this octopus that just straight-up throttled the entire scientific tradition of Western civilization. Evidence based science and the discourse culture that goes with it—gone. That's what he did. It's no small feat. He destroyed American science by snuffing out its spirit, the spirit of open inquiry, proof and *standards*.

The reason so many outstanding scientists lent their names to opposing Fauci's vendetta on Duesberg was not that they cared, necessarily, about the cause of AIDS; This was, for them, a battle over the very soul of science. Kary Mullis [PCR inventor] broke down crying in an interview I did with him in 1994, talking about it—talking about what Fauci did to Peter Duesberg and what it meant.

The *real* scientists were horrified. Suddenly a guillotine was present. A new and strange terror. People were "guilty," of thought-crimes like "HIV denialism." Fauci had made political correctness the new revolutionary language, see? And that meant if you were "bad," if you didn't push agenda driven science, *everything* was taken away from you. And the media cheered. And anybody who didn't was destroyed, vilified, harassed, fired, in a word, canceled.

His gifts for deflection, misdirection, and obfuscation, and perhaps his boyish charm, give Dr. Fauci a Teflon quality—which he shared with President Ronald Reagan, under whom he initially came to power. Something about Dr. Fauci allows him to escape responsibility for (or even mild questioning about) his steady parade of sketchy decisions, his confident claims unsupported by scientific evidence, his relentless cascade of lies and failed predictions, and his miserable track record for keeping Americans healthy.

As the nation's newly appointed AIDS czar, Dr. Fauci was now a gatekeeper for almost all AIDS research. NCI already had long experience and robust infrastructure for conducting clinical trials on new drugs. NIAID had neither. Nevertheless, parroting NCI's vows to cure cancer, Dr. Fauci promised Congress that he would quickly produce drugs and vaccines to banish AIDS. In his 1990 book, Nussbaum concludes that Fauci's triumph over NCI cost many thousands of Americans their lives during the AIDS crisis.[70] Myriad contemporary critics concurred with that assessment.

The PIs: The Pharma/Fauci Mercenary Army

NIAID's lack of in-house drug development capacity allowed Dr. Fauci to build his new program by farming out drug research to a network of so-called "principal investigators," or PIs, effectively controlled by pharmaceutical companies. Today, when people refer to the "Medical Cartel," they are principally speaking of pharmaceutical companies, hospital systems, HMOs and insurers, the medical journals, and public health regulators. But the glue that holds all these institutions together, and allows them to march in lockstep, is the army of PIs who act as lobbyists, spokespersons, liaisons, and enforcers. Tony Fauci played a key historic role in elevating this cohort to dominate public health policy.

PIs are powerful academic physicians and researchers who use federal grants and pharmaceutical industry contracts to build feudal empires at universities and research hospitals that mainly conduct clinical trials—a key stage in the licensing process—for new pharmaceutical products. Thanks to NIH's largesse, and to NIAID in particular, a relatively tiny network of PIs—a few hundred—determines the content and direction of virtually all America's biomedical research.

In 1987, some $4.6 billion of NIH's $6.1 billion budget went to these off-campus researchers.[71] By 1992, NIH's budget had expanded to $8.9 billion,[72] with $5 billion going to outside scientists at 1,300 universities, laboratories, and other elite institutions.[73,74] Today, Dr. Fauci's NIAID alone controls $7.6 billion in annual discretionary expenditures that he distributes mainly to PIs around the globe.[75]

PIs are pharmaceutical industry surrogates who play key roles promoting the pharmaceutical paradigm and functioning as high priests of all its orthodoxies, which they proselytize with missionary zeal. They use their seats on medical boards and chairmanships of university departments to propagate dogma and root out heresy. They enforce message discipline, silence criticism, censor contrary opinions, and punish dissent. They populate the Data and Safety Monitoring Boards (DSMBs) that influence the design of clinical trial protocols and guide the interpretation of clinical trial outcomes and conclusions; the external advisory FDA panel, Vaccines and Related Biological Products Advisory Committee (VRBPAC), that guides determination of whether new vaccines are "safe and effective" and merit licensure (marketing); and the CDC panel, The Advisory Committee on Immunization Practices (ACIP), that essentially mandates vaccines to children. They are the credentialed and trusted medical experts who prognosticate on television networks—now helplessly reliant on pharmaceutical ad revenue—to push out Pharma content. These "experts"—Paul Offit, Peter Hotez, Stanley Plotkin, Ian Lipkin, William Schaffner, Kathleen Edwards, Arthur Caplan, Stanley Katz, Greg Poland, and Andrew Pollard—appear between Pharma ads on network and cable news shows to promote the annual flu shots and measles scares, to drum up fears about COVID, and to rail against "anti-vaxxers." They write the steady stream of editorials that appear in local and national newspapers to reinforce the hackneyed orthodoxies of the pharmaceutical paradigms—"all vaccines are safe and effective," etc. They root out heresy by sitting on the state medical boards—the "Inquisition" courts—that censure and de-license dissident doctors. They control the medical journals and peer-review journal literature to fortify Pharma's agenda. They teach on medical school faculties, populate journal editorial boards, and chair university departments. They supervise hospitals and chair hospital departments. They act as expert witnesses for pharmaceutical companies in civil court and the federal vaccine court. They present awards to one another.

The 2006 meeting of CDC's ACIP provides an illustrative blueprint for how Tony Fauci and his Pharma partners use their PIs to control the key FDA and CDC panels that license and "recommend" new vaccines for addition to the childhood schedule. That 2006 ACIP panel recommended two new blockbuster Merck shots: the Gardasil HPV vaccine for all girls ages nine through twenty-six,[76] and three doses of a Merck rotavirus vaccine, Rotateq, for infants at ages two, four, and six months.[77] Both Bill Gates[78] and Tony Fauci (via NIAID)[79] had provided seed and clinical trial funding for the development of both Gardasil and the rotavirus vaccine.[80,81] Merck maintained it had not tested either vaccine against an inert placebo in pre-approval trials, so no one could scientifically predict if the vaccines would avert more injuries or cancers than they would cause. Nevertheless, the sister FDA panel, VRBPAC, approved Gardasil—to prevent cervical cancer—without requiring proof that the vaccine prevented any sort of cancer, and despite strong evidence from Merck's clinical trial that Gardasil could dramatically raise risks of cancer and autoimmunity in some girls.[82] ACIP, nevertheless, effectively mandated both jabs. Gardasil would be the most expensive vaccine in history, costing patients $420 for the three-jab series and generating revenues of over $1 billion annually for Merck.[83]

That year, nine of the thirteen ACIP panel members and their institutions collectively received over $1.6 billion of grant money from NIH and NIAID.

Systemic Conflicts of Interest

Pharma and Dr. Fauci similarly rig virtually all the critical drug approval panels using this strategy of populating them with PIs who, bound by financial fealty to Pharma and NIAID funders, reliably approve virtually every new drug upon which they deliberate—with or without safety studies.

From 1999 to 2000, Government Oversight Committee (GOC) Chairman Republican Congressman Dan Burton investigated the systemic corruption of these panels during two years of intense investigations and hearings. According to Burton, "CDC routinely allows scientists with blatant conflicts of interest to serve on influential advisory committees that make recommendations on new vaccines . . . while these same scientists have financial ties, academic affiliations and other . . . interests in the products and companies for which they are supposed to be providing unbiased oversight."[84, 85]

Paul Offit: Voting Himself Rich

The notorious "Television Doctor" Paul Offit was the codeveloper of the rotavirus vaccine that ACIP approved in that 2006 session. Offit is one of Dr. Fauci's most prominent PIs and an exemplar of the kind of power, influence, and lucre available to PIs whose entrepreneurial energies are unobstructed by scruples. Offit is the darling of both mainstream and social media. He is a perennial guest on CBS, NBC, ABC, and CNN, on cable shows such as *The Daily Show,* and a former guest on *The Colbert Report.* He is the *New York Times*'s guest expert and provides regular editorials for the *Times*'s op-ed pages. He is a frequently quoted expert on evening news broadcasts and a regular contributor to online media outlets including *HuffPost, Politico,* and *The Daily Beast.*[86] Media platforms uniformly identify Offit as a "vaccine expert" from the University of Pennsylvania and the Children's Hospital of Philadelphia (CHOP). With Offit's encouragement, they seldom, if ever, disclose his pervasive financial entanglements with Dr. Fauci and the pharmaceutical companies. In 2011, for example, while presenting at NIH for the Great Teachers Lecture Series, he unabashedly declared, "I'm sorry, I have no financial conflicts of interest."[87] Given his voluminous conflicts, the brashness of that claim indicates his shameless arrogance. Dr. Offit, in fact, is a vaccine developer who has made millions monetizing his relationships with vaccine companies. He occupies the "Hilleman Chair" at CHOP (Children's Hospital of Philadelphia), which Merck funded with a $1.5 million donation and named in honor of the company's heavyweight vaccinologist.[88]

Offit and his university and hospital affiliates have flourished largely based on hundreds of millions in grant monies from Dr. Fauci's agency and from virtually all the big vaccine companies. In 2006 alone, his institution, CHOP, received $13 million from NIAID and $80 million from NIH. Offit's biennial propaganda books—including titles like *Vaccines: What Every Parent Should Know* and *Autism's False Prophets: Bad Science, Risky Medicine, and the Search for a Cure*—are unabashed paeans to Big Pharma, and scourges to industry detractors and natural health. Offit uses these plugola tomes to exalt a wide range of "miracle" pharma products, to vilify vaccine hesitancy, and gaslight and bully the mothers of vaccine-injured children. Merck launders hundreds of thousands of dollars in personal payments to Offit through bulk purchases of these propaganda broadsides, which the company then distributes to pediatricians across the country.[89]

Offit is the most visible spokesperson for Pharma, its allied industries, and the chemical paradigm in general. He represents himself as an authoritative source of reliable information, but he is actually a font of wild industry ballyhoo, prevarication, and outright fraud. He brazenly claims, against all scientific evidence, that vaccine injuries are a myth—that all vaccines are safe and effective, that children can safely receive ten thousand vaccines at once,[90] and that aluminum is safe in vaccines for babies because it is a "vital nutrient."[91] (There is no scientific study suggesting that aluminum is safe or that it has any nutritional value.) Offit says that mercury in vaccines is harmless and is quickly excreted from the body.[92] (Published science demonstrates decisively that mercury is a cataclysmically harmful and persistent toxin, and it is well known that both ethyl and methylmercury bioaccumulate.) Dr. Offit vocally supports GMO foods[93] and chemical pesticides and is an obstreperous foe of vitamins, nutrition, and integrative medicine.[94] He warns against the fallacy of going "GMO free," and takes the radical position that dichlorodiphenyltrichloroethane (DDT) is harmless. He bitterly demonizes Rachel Carson for killing millions of people by hatching the plot against Monsanto's DDT.[95]

Dr. Offit counsels his fellow PIs that lying is part of their job. He justifies any whopper that maximizes vaccine uptake. In 2017, Offit coached a group of fellow PIs, "You can never really say that MMR doesn't cause autism but frankly when you get in front of the media you better get used to saying it because otherwise people hear a door being left open when a door shouldn't be left open."[96] In his 2008 book, *Autism's False Prophets*, Offit fabricated a conversation claiming that a vaccine safety advocate, J. B. Handley—a prominent Portland, Oregon, businessman with a severely autistic son—threatened one of Offit's acolytes. Handley sued Offit for libel,[97] forcing him to retract the statement, to publicly apologize for the fabrication, and to make a humiliating $5,000 donation to Jenny McCarthy's autism charity.[98] Despite such embarrassments, the mainstream media treat Offit's most outlandish statements as gospel. Physicians rely upon the veracity of his pronouncements in making treatment decisions. Dr. Offit serves on the board of various pharma front groups[99] and astroturf organizations[100] and commands a vast network of bloggers and trolls, each of them directly or indirectly paid by the pharmaceutical companies to stifle debate, propagate lies, bully and intimidate the mothers of intellectually disabled children, silence scientific and medical dissent, and root out heresy.

In 1998, Offit sat on the CDC's ACIP Committee and participated in the debate that added rotavirus vaccine to the mandatory schedule for the first time, neither from the debate nor the vote, despite the fact that he had his own rotavirus vaccine then in development. He voted that year to add Wyeth-Ayerst Pharmaceuticals's rotavirus vaccine, RotaShield, to the mandatory schedule despite the absence of functional safety studies. Offit knew that ACIP's positive vote on Wyeth's rotavirus jab would virtually guarantee a similar approval for his own rotavirus vaccine during an upcoming ACIP session.[101]

Before arriving at ACIP, every vaccine must first get reviewed by FDA's sister "independent panel" called VRBPAC (which is also populated with Dr. Fauci's and Big Pharma's PIs), then licensed as "safe and effective" by the FDA. According to the findings of that 2000 Congressional investigation,[102] four of the five FDA VRBPAC committee members who voted to license the Wyeth rotavirus vaccine that year had financial conflicts with the four pharmaceutical companies, Sanofi, Merck, Wyeth, and Glaxo, that were developing versions of the vaccine.

Once the FDA committee gave RotaShield its blessing, the vaccine moved to ACIP to vie for a CDC "recommendation," which effectively mandates the vaccine for 3.8 million school children annually, guaranteeing the manufacturer a trapped market worth hundreds of millions.

During the 1998 ACIP session, Dr. Offit sat as one of five full voting members. (There were five additional nonvoting members.) His Rotateq codeveloper, Stanley Plotkin, also sat on the committee. The ACIP Committee unanimously recommended Wyeth's RotaShield vaccine.

The August 2000 Congressional investigation found that the majority of ACIP members were conflicted in that vote.[103] That report found that seven out of ten ACIP working group committee members who voted to approve the rotavirus vaccine in June 1998 had financial ties to the pharmaceutical companies that were developing different versions of the vaccine.

According to the Congressional Report:

- The Chairman served on Merck's Immunization Advisory Board.
- One member was under contract with the Merck vaccine division, received funds from various vaccine manufacturers, including Pasteur (now Sanofi), and was under contract as a principal investigator from SmithKline (now GSK).
- Another member (of that same ACIP panel) received a salary from Merck as well as other payments from Merck.
- Another member was participating in vaccine studies with Merck, Wyeth (now Pfizer), and SmithKline (now GSK).
- Another member received grants from Merck and SmithKline (now GSK).
- Another member shared a patent on his own rotavirus vaccine funded by a $350,000 grant for Merck to develop this vaccine and was a paid consultant to Merck.

The last of these bullet points referred to Paul "I Have No Conflicts" Offit. Dr. Fauci's and Pharma's corrupt control of those two panels allowed Wyeth to obtain both an FDA license and a CDC "recommendation" without having to genuinely safety test this product, a process that would have revealed terrible risks. Even the truncated trials of Wyeth's RotaShield, conducted with no placebo, revealed serious side effects in babies, including "failure to thrive," fevers high enough to cause brain injury, and a condition called intussusception, wherein a child's intestines telescope into themselves, causing an agonizing blockage that, in some instances, results in death. The intussusception figures alone were statistically significant—cited as one in two thousand of the children who received the vaccine.[104] At this time, there were around 3.8 million children in the target age group living in the United States; this translated to around 1,890 statistically likely cases of intussusception.[105]

Nevertheless, VRBPAC, under Fauci's and Pharma's tight control, approved the vaccine, and ACIP put it on the mandatory schedule. Less than a year after Dr. Offit and his confederates on ACIP voted to mandate RotaShield with no authentic safety testing, Offit again sat on the ACIP committee that revoked this earlier recommendation. ACIP pulled RotaShield from the market in October 1999 due to the many children who, predictably, suffered intussusception.[106] VAERS, the Vaccine Adverse Event Reporting System, contains fifty reports of vaccine-related intussusception for the year 1999.[107] Paul Offit's shrewd maneuvering through this sequence of events opened an unobstructed path to approval and enormous riches for his own rotavirus vaccine, RotaTeq.

Since its approval, Dr. Offit's rotavirus vaccine has caused a wave of catastrophic illnesses and agonizing deaths in babies from intussusception.[108]

From 1985 to 1991, prior to the introduction of the rotavirus vaccine, the rotavirus disease caused only 20–60 deaths per year nationwide, mainly due to dehydration associated with diarrhea.[109,110] Since dehydration is easily treated, virtually all deaths from rotavirus are avoidable with timely and appropriate medical care.

Reported adverse reactions from Dr. Offit's RotaTeq vaccine range from 953 to 1,689 per year. These included fever, diarrhea, vomiting, irritability, intussusception, SIDS, severe combined immunodeficiency, otitis media, nasopharyngitis, bronchospasm, urinary tract infection, hematochezia, seizures, Kawasaki disease, bronchiolitis, urticaria, angioedema, gastroenteritis, pneumonia, and death.[111]

The best evidence indicates that Dr. Offit's rotavirus vaccine causes negative net public health impacts; in other words, Dr. Offit's vaccine almost certainly kills and injures more children in the United States than the rotavirus disease killed and injured prior to the vaccine's introduction.

Finally, in 2010, after its introduction, NIH learned that Offit's vaccine, RotaTeq, also contained the porcine retrovirus that causes an HIV-like syndrome called "wasting disease" in pigs.[112] Neither Dr. Fauci nor any other agency has ever funded a study to establish the safety of injecting their dangerous pig retroviruses into babies. Millions of American children have now been inoculated with the virus, thanks to Offit.

In 2006, ACIP added Offit's vaccine to the schedule, allowing Offit and his business partners to sell his patent rights for the formulation to Merck for $186 million.

Offit made a declared profit of over $20 million as a result of this series of transactions. Offit reported, in a gushing 2008 *Newsweek* story, that the millions he made from his rotavirus caper was "like winning the lottery."[113] In a less-adoring assessment of the scam, UPI journalist Dan Olmsted and coauthor Mark Blaxill accused Offit of "voting himself rich."[114]

The disturbing saga of Paul Offit and his rotavirus vaccine illustrates how Tony Fauci's PIs stuff the sausages at HHS.

How PIs Control Public Marketing

Dr. Fauci's choice to transfer virtually all of NIAID's budget to pharmaceutical PIs for drug development was an abdication of the agency's duty to find the source and eliminate the explosive epidemics of allergic and autoimmune disease that began under his watch around 1989.[115,116] Refereed science, surveillance data, and manufacturers' inserts all implicate the very drugs and vaccines that Tony Fauci largely helped develop as culprits in those new epidemics. NIAID money effectively became a giant subsidy to the blossoming pharmaceutical industry to incubate a pipeline of profitable new drugs targeted to treat the symptoms of those very diseases.

While NIH remains a massive funding source for PIs, rich contracts from big drug companies and royalty payments from drug products often dwarf their government funding. Pharma money is the PIs' bread and butter, commanding their loyalties and dictating their priorities. They and their clinics and research institutions are, effectively, arms of the pharmaceutical industry. Their empires rely on Pharma for their growth and survival.

Moreover, PIs typically function in quasi-feudal fiefdoms: loyal to a single pharmaceutical company. Each drug company—Glaxo, Pfizer, Merck, Sanofi, Johnson & Johnson, and Gilead—cultivates a cadre of its own reliable PIs whom it funds to conduct clinical trials and drug research. Unwritten protocols dictate that a Merck PI will not customarily perform research for a Merck competitor. Typically, the drug company contracts with the reliable PI's medical school, attending hospital, or research institution to run clinical trials. The company makes payments ranging from a few hundred dollars to $10,000 (depending on the trial phase, complexity, and the company) for each patient enrolled in the drug trial,[117] with the university skimming one-half to two-thirds of those funds for "academic overhead."[118] Those payments from the pharmaceutical company secure long-term loyalty from the institution and its board. Moreover, both the researcher and the university customarily share patent interests in any product the PI helps develop, collecting rich royalties when it hits the market. Additional money from the Pharma sponsor supports the PI's assistants and laboratory costs. The drug company also pays "legalized bribes" to the PI grantee through honoraria, expert witness fees, speaking gigs, and first-class travel to exclusive resorts for conferences. All these perquisites tend to fortify loyalty and incentivize the favorable research results necessary to securing FDA drug approvals. On all sides of these transactions, each stakeholder understands that positive reviews of the subject drug promise future work.

According to Nussbaum, "PIs do their own kind of science and, more often than not, their experiments have little to do with either health or the public. They test drugs by private pharmaceutical companies for personal gain, for money that goes to their universities, and for power."[119]

The system allows pharmaceutical companies to systematically divert federal monies—the initial NIAID grant—to serve their own private profit priorities. Naturally, the system is hostile to drugs with expired patents or those that emerge from companies that are not paying the PI's research expenses. This bias explains Dr. Fauci's signature animosity toward non-pharmaceutical, unpatentable, or patent-expired and generic remedies.

In his unpublished history of the HIV era, *Down the Rabbit Hole*, author and historian Terry Michael offers a similar description of Dr. Fauci's abrogation of his scientific role to the army of Pharma PIs: "But NIH has other clients, including thousands of grant-seeking medical science Ph.D.'s produced by American universities after World War II. NIAID funds much of the pharmaceutical industry's research and clinical trials. In fact, Big Pharma has become a client of the NIH and especially its NIAID."[120]

This powerful army, garrisoned at hospitals and universities in every large American community, allows Pharma and Dr. Fauci to control the public health narrative around the country. Before I understood its structure, I encountered the pervasive power of the combination.

Between 1990 and 2020, I served as president of an influential environmental group, Waterkeepers, with 350 affiliates around the county and the globe. Waterkeepers is the world's largest water protection group. I published regularly in the *New York Times* and all the major papers: *Boston Globe, Houston Chronicle, Chicago Sun-Times, Los Angeles Times, Miami Herald*, and *San Francisco Chronicle*; in magazines including *Esquire, Rolling Stone*, and *The Atlantic*; and in online publications, most often in *HuffPost*. I delivered over 220 speeches each year, including sixty paid speaking engagements to large audiences at universities and corporate events. I earned a substantial income from those appearances. All that changed in 2005, after I published an article, "Deadly Immunity," about corruption in CDC's vaccine branch, simultaneously in *Rolling Stone* and *Salon*.

Newspapers thereafter generally refused to publish my articles on vaccine safety and ultimately banned me from publishing on any issues. In 2008, without consulting me or citing a specific reason, *Salon* retracted and removed my 2005 article. *Salon*'s founder, David Talbot, faulted *Salon* for caving in to Pharma. *Rolling Stone* finally removed the article without explanation in February 2021, and *HuffPost* purged all half-dozen of my vaccine articles. The editors of those online journals had thoroughly fact-checked my pieces prior to publication. They removed them without notice to me, and without ever explaining their decisions. It was the beginning of the mass censorship of any vaccine information that departs from official narratives. That year, universities and corporate hosts and municipal speakers' forums suddenly cancelled my scheduled speeches in droves. My bookings dropped from sixty paid speeches per year down to one or two. My speakers' bureau told me that floods of telephone calls from powerful members of the medical community had prompted the cancellations. They deluged the offices of presidents and board members of the colleges, businesses, and community groups that were hosting me, protesting my appearances. The callers were public health officials and leading doctors from local hospitals, university medical schools, and influential research centers in those locales. Using similar language, they offered dire warnings that I was anti-vaccine, anti-science, a "baby killer," and that my appearance would jeopardize public health and vital funding to university medical school programs.

The threat to interrupt money flows to the university PIs invariably trumps the traditions of speech freedom revered—in theory—by university administrators. Starting in 2019, PIs at NYU attempted to force the ouster of popular historian and propaganda expert Professor Mark Crispin Miller from its faculty roster and law professor Mary Holland from its law school faculty because they dared question reigning vaccine orthodoxies.

Terry Michaels summarized how Dr. Fauci exploited the strategic landscapes of the HIV pandemic to launch his career on a trajectory toward the unimaginable power that would allow him to dictate official orthodoxies, control the press, set international health policies, and even to shut down the global economy: "Dr. Anthony Fauci seized an opportunity to create a multi-billion dollar bureaucracy, distributing thousands of grants to seekers of federally funded research largesse, with a disproportionate (to other diseases) number going to HIV-AIDS researchers."[121]

Tony Fauci did not create the PI system, but his inexperience both as a scientist and as an administrator meant that he relied upon it and was, at first, at its mercy. Later, he took command of those troops and organized them into a powerful juggernaut that journalist John Lauritsen calls "the Medical Industrial Complex."[122]

Endnotes

1   Scott Allen, "Dr. Fauci Is Looking Forward to Seeing the Nationals Play Again," *Washington Post*, April 15, 2020, https://www.washingtonpost.com/sports/2020/04/15/dr-fauci-is-looking-forward-seeing-nationals-play-again/

2   Ben Cohen, "Fauci Was Basketball Captain. Now He's America's Point Guard," *Wall Street Journal*, March 29, 2020, https://www.wsj.com/articles/dr-fauci-was-a-basketball-captain-now-hes-americas-point-guard-11585479601

3   Dr. Victoria Harden, "Dr. Anthony S. Fauci Oral History 1989," National Institutes of Health, Office of NIH History and Stetten Museum, Mar 7, 1989, https://history.nih.gov/display/history/Fauci%2C+Anthony+S.+1989

4   Ibid.

5   "Q & A With Dr. Anthony Fauci," *C-SPAN*, Jan 8, 2015, https://www.c-span.org/video/?c4873572/user-clip-fauci-humanist

6   Harden, "Dr. Anthony S. Fauci Oral History 1989"

7   Ibid.

8   "Commissioned Corps of the U.S. Public Health Service: Our History," U.S. Department of Health and Human Services; https://www.usphs.gov/history

9   "A Short History of the National Institutes of Health," National Institutes of Health, Office of NIH History and Stetten Museum, U.S. Department of Health and Human Services; https://history.nih.gov/display/history/A+Short+History+of+the+National+Institutes+of+Health

10   Harden, "Dr. Anthony S. Fauci Oral History 1989"

11   Cezmi A. Adkis, "Does the epithelial barrier hypothesis explain the increase in allergy, autoimmunity and other chronic conditions?" *Nature Reviews Immunology* (2021). https://doi.org/10.1038/s41577-021-00538-7s

12   McKeown, Thomas. *The Role of Medicine: Dream, Mirage, or Nemesis?* (Princeton Univ. Press 2014)

13   Victoria Hansen et al., "Infectious Disease Mortality Trends in the United States, 1980-2014," *JAMA* 2016;316(20):2149-2151. https://jamanetwork.com/journals/jama/fullarticle/2585966

14   Gregory L. Armstrong et al., "Trends in Infectious Disease Mortality in the United States During the 20th Century," *JAMA* 281(1), 61-66 (1999). https://jamanetwork.com/journals/jama/fullarticle/768249

15   "Achievements in Public Health, 1900-1999: Control of Infectious Diseases," *CDC MMWR* (July 30, 1999), 48(29): 621-629. https://www.cdc.gov/mmwr/preview/mmwrhtml/mm4829a1.htm#fig1

16   Bernard Guyer et al., "Annual Summary of Vital Statistics: Trends in the Health of Americans During the 20th Century," *Pediatrics* (December 2000), 106 (6) 1307-1317, at 1314, 1315. https://www.factchecker.gr/wp-content/uploads/2017/10/PediatricsDec.2000-VOl-106No.6.pdf

17   John B. McKinlay and Sonja M. McKinlay, "The Questionable Contribution of Medical Measures to the Decline of Mortality in the United States in the Twentieth Century," Milbank Memorial Fund (Summer, 1977), https://www.milbank.org/wp-content/uploads/mq/volume-55/issue-03/55-3-The-Questionable-Contribution-of-Medical-Measures-to-the-Decline-of-Mortality-in-the-United-States-in-the-Twentieth-Century.pdf

18   Edward H. Kass, "Infectious Diseases and Social Change," *The Journal of Infectious Diseases* (123, no. 1 1971), 110-14. Accessed May 11, 2021. http://www.jstor.org/stable/30108855

19   *Supra* note 32, at "New Institutes"

20   Jean-François Bach, "The Effect of Infections on Susceptibility to Autoimmune and Allergic Diseases," *New England Journal of Medicine* (2002), 347:911-920; https://www.nejm.org/doi/pdf/10.1056/NEJMra020100?articleTools=true

21   Megan Scudellari, "Cleaning up the hygiene hypothesis," *PNAS* (Feb 14, 2017), 114(17), 1433-1436; https://www.pnas.org/content/pnas/114/7/1433.full.pdf

22   Peter H. Duesberg, *Inventing the AIDS Virus* (Regnery Publishing 1996), 134

23   Ibid., 146

24   Ibid., 146

25   James Curran, MD, 00:08:15 - "House of Numbers: Anatomy of an Epidemic," YouTube Video (2009), https://www.youtube.com/watch?v=lvDqjXTByF4

26   Kary Mullis, PhD, 00:08:03 - "House of Numbers: Anatomy of an Epidemic," YouTube Video (2009), https://www.youtube.com/watch?v=lvDqjXTByF4

27   Robert C. Gallo, MD, 00:08:36 "House of Numbers: Anatomy of an Epidemic," YouTube Video (2009), https://www.youtube.com/watch?v=lvDqjXTByF4

28   Richard M. Krause MD, National Institute of Allergy and Infectious Diseases (Aug 13, 2019), https://www.niaid.nih.gov/about/richard-m-krause-md

29   Robert F. Kennedy Jr., "'60 Minutes' - Swine Flu 1976 Vaccine Warning," Video, April 1, 2020, https://childrenshealthdefense.org/video/60-minutes-swine-flu-1976-vaccine-warning/

30   Richard Krause, "The Swine Flu Episode and the Fog of Epidemics," *Emerging Infectious Diseases* (Jan, 2006), https://dx.doi.org/10.3201%2Feid1201.051132

31   U.S. Comptroller General Elmer B. Staats, Letter to the U.S. House of Representative Committee on Interstate and Foreign Commerce (Feb 6, 1979) https://www.gao.gov/assets/hrd-79-47.pdf

32   Richard E. Neustadt and Harvey V. Fineberg, "The Swine Flu Affair: Decision Making on a Slippery Disease," *National Academies Press* (1978). https://www.nap.edu/catalog/12660/the-swine-flu-affair-decision-making-on-a-slippery-disease

33   Rick Perlstein, "Gerald Ford Rushed out a Vaccine. It Was a Fiasco," *New York Times* (Sep 2, 2020) https://www.nytimes.com/2020/09/02/opinion/coronavirus-vaccine-trump.html

34   Patrick Di Justo, *The "Great" Swine Flu Epidemic of 1976*, REAL CLEAR POLITICS, (Apr 28, 2009) https://www.realclearpolitics.com/2009/04/28/the_great_swine_flu_epidemic_of_1976_212900.html

35   Rebecca Kreston, 'The Public Health Legacy Of The 1976 Swine Flu Outbreak," *DISCOVER MAGAZINE*, (Sep 30, 2013) https://www.discovermagazine.com/health/the-public-health-legacy-of-the-1976-swine-flu-outbreak

36   Thomas O'Toole, "4 Swine Flu Vaccine Firms Sued for Total $19 Million," *Washington Post* (WP Company, January 27, 1977), https://www.washingtonpost.com/archive/local/1977/01/27/4-swine-flu-vaccine-firms-sued-for-total-19-million/f8e17851-0eb7-4991-85ba-245a4cd5d8d0/.

37   Mike Wallace, "Swine Flu," *60 Minutes*, YouTube Video (Nov 4, 1979) https://www.youtube.com/watch?v=4bOHYZhL0WQ

38   Mike Stobbe, "David Sencer; lost job for urging swine flu vaccinations in '76," boston.com (May 3, 2011) http://archive.boston.com/bostonglobe/obituaries/articles/2011/05/03/david_sencer_lost_job_for_urging_swine_flu_vaccinations_in_76/

39   "NIAID Director Resigns to Be Emory Med. Dean," U.S. Department of Health and Human Services, *The NIH Record* (July 3, 1984) https://nihrecord.nih.gov/sites/recordNIH/files/pdf/1984/NIH-Record-1984-07-03.pdf

40   Anthony S. Fauci, M.D.; National Institute of Allergy and Infectious Disease, National Institutes of Health (Director's page; content last reviewed by NIH on March 14, 2021); https://www.niaid.nih.gov/about/anthony-s-fauci-md-bio

41   Anthony S. Fauci, M.D. 1988–1994; Office of AIDS Research, National Institutes of Health (Director's Corner; content last reviewed by NIH on July 11, 2018); https://www.oar.nih.gov/about/directors-corner/fauci

42   Anthony S. Fauci, "The Acquired Immune Deficiency Syndrome The Ever-Broadening Clinical Spectrum," *JAMA* (1983), 249(17):2375-2376; https://jamanetwork.com/journals/jama/article-abstract/386561

43   John R La Montagne et al., "Severe acute respiratory syndrome: developing a research response," *Journal of Infectious Diseases* (Feb 15, 2004) 189(4):634-41; https://pubmed.ncbi.nlm.nih.gov/14767816/

44   Shauna Milne-Price et al., "The Emergence of the Middle East Respiratory Syndrome Coronavirus," *PMC* (Jul 2014), 71(2): 121-136; https://www.ncbi.nlm.nih.gov/pmc/articles/PMC4106996/

45   "First Confirmed Cases of Middle East Respiratory Syndrome Coronavirus (MERS-CoV) Infection in the United States, Updated Information on the Epidemiology of MERS-CoV Infection, and Guidance for the Public, Clinicians, and Public Health Authorities — May 2014," *MMWR* (May 2014) 63: 431-436; https://www.cdc.gov/mmwr/preview/mmwrhtml/mm6319a4.htm

46   "NIH: We're working on MERS vaccine," interview with Dr. Anthony Fauci, *CNN*, 2014. https://www.cnn.com/videos/health/2014/05/15/lead-intv-fauci-mers-nih.cnn

47   "NIAID Initiates Trial of Experimental Avian Flu Vaccine," *Science Daily*, March 25, 2005; https://www.sciencedaily.com/releases/2005/03/050325100848.htm

48   Robert Roos, "YEAR-END REVIEW: Avian flu emerged as high-profile issue in 2005." *CIDRAP News*, Jan 5, 2006. https://www.cidrap.umn.edu/news-perspective/2006/01/year-end-review-avian-flu-emerged-high-profile-issue-2005

49   Sundar S Shrestha et al., "Estimating the burden of 2009 pandemic influenza A (H1N1) in the United States (April 2009–April 2010)," *Clinical Infectious Diseases,* (Jan 1, 2001). https://academic.oup.com/cid/article/52/suppl_1/S75/499147

50   Nidhi Bouri et al., "Return of Epidemic Dengue in the United States: Implications for the Public Health Practitioner," *Public Health Reports,* (May-Jun, 2012). https://www.ncbi.nlm.nih.gov/pmc/articles/PMC3314069/

51   "Jordan Report—Accelerated Development of Vaccines 2012," National Institute of Allergy and Infectious Disease, National Institutes of Health, at 95 "Vaccine Updates"; https://www.niaid.nih.gov/research/jordan-report-accelerated-development-vaccines-2012

52   "Researching Ebola in Africa," National Institute of Allergy and Infectious Diseases, National Institutes of Health (Content last reviewed by NIH on September 23, 2019); https://www.niaid.nih.gov/diseases-conditions/researching-ebola-africa

53   "Why NIAID Is Researching Zika Virus," National Institute of Allergy and Infectious Disease, National Institutes of Health (Content last reviewed by NIH on September 8, 2017); https://www.niaid.nih.gov/diseases-conditions/why-niaid-researches-zika

54   "COVID-19 Clinical Research," National Institute of Allergy and Infectious Disease, National Institutes of Health (Content last reviewed by NIH on November 2, 2020); https://www.niaid.nih.gov/diseases-conditions/covid-19-clinical-research

55   Nussbaum, op. cit., 127–129

56   Anna D Barker and Hamilton Jordan, "Legislative History of the National Cancer Program," in Donald W Kufe et al., *Holland-Frei Cancer Medicine, 6th ed.*, 2003 (Chapter 81); https://www.ncbi.nlm.nih.gov/books/NBK13873/

57   Torsten Engelbrecht, Claus Köhnlein, et al., "Virus Mania: How the Medical Industry Continually Invents Epidemics, Making Billions at our Expense" (Books on Demand 3rd ed, 2021), 98

58   Eliot Marshall, "Cancer Research and the $90 Billion Metaphor," *Science* (Mar. 11, 2011). https://science.sciencemag.org/content/331/6024/1540.1

59   National Cancer Institute, NCI Budget and Appropriations, (Jan. 13, 2021). https://www.cancer.gov/about-nci/budget

60   National Cancer Institute, Cancer Stat Facts: Soft Tissue including Heart Cancer, https://seer.cancer.gov/statfacts/html/soft.html

61   Engelbrecht et al., op. cit., 24.

62   Cristine Russell, "Anthony S. Fauci," *Washington Post*, November 3, 1986, https://www.washingtonpost.com/archive/politics/1986/11/03/anthony-s-fauci/8d270beb-e95d-46e5-808b-d670630223f4/

63   Ibid.

64   Ibid.

65   James Curran, MD, 00:08:54 -"House of Numbers: Anatomy of an Epidemic," YouTube Video (2009), https://www.youtube.com/watch?v=lvDqjXTByF4

66   Nussbaum, op. cit., 126.

67   Ibid., 126–127

68   Ibid., 122.

69   Torsten Engelbrecht and Konstantin Demeter, Anthony Fauci: 40 Years of Lies From AZT to Remdesivir, OFF-GUARDIAN, (Oct. 27, 2020). https://off-guardian.org/2020/10/27/anthony-fauci-40-years-of-lies-from-azt-to-remdesivir/

70   Nussbaum, op. cit., 127.

71   Ibid., 131

72   "History of Congressional Appropriations 1990-1999," National Institutes of Health, Office of Budget,: https://officeofbudget.od.nih.gov/pdfs/FY08/FY08%20COMPLETED/appic3806%20-%20transposed%20%2090%20-%2099.pdf

73   "NIH Awards by Location & Organization," National Institutes of Health, https://report.nih.gov/award/index.cfm?ot=&fy=1992&state=&ic=&fm=&orgid=&distr=&rfa=&om=n&pid=

74   "Trends in NIH Funding (1992-2013)," Blue Ridge Institute for Medical Research, , http://www. brimr.org/NIH_Awards/Trends/

75   "Budget Updates From January 2021 NIAID Advisory Council Meeting," National Institute of Allergy and Infectious Diseases, (Feb 3, 2021). https://www.niaid.nih.gov/grants-contracts/budget-updates-january-2021

76   Elissa Meites et al., "Human Papillomavirus Vaccination for Adults: Updated Recommendations of the Advisory Committee on Immunization Practices," Morbidity and Mortality Weekly Report, *CDC-MMWR* (Aug 16, 2019) https://www.cdc.gov/mmwr/volumes/68/wr/mm6832a3.htm

77   "Advisory Committee on Immunization Practices, Rotavirus vaccination coverage and adherence to the Advisory Committee on Immunization Practices (ACIP)-recommended vaccination schedule--United States, February 2006-May 2007," Morbidity and Mortality Weekly Report, *CDC-MMWR* (Apr 18, 2008) https://pubmed.ncbi.nlm.nih.gov/18418345/

78   "Summary of Bill & Melinda Gates Foundation-supported HPV Vaccine Partner Activities," World Health Organization, , (2006) https://www.who.int/immunization/sage/HPV_partner_info_gates.pdf

79   "Safety of and Immune Response to a Novel Human Papillomavirus Vaccine in HIV Infected Children," ClinicalTrials.gov (Jun, 2006). https://clinicaltrials.gov/ct2/show/NCT00339040

80   "Bill and Melinda Gates Announce a $100 Million Gift to Establish the Bill and Melinda Gates Children's Vaccine Program," Bill and Melinda Gates Foundation, (Dec, 1998) https://www.gatesfoundation.org/ideas/media-center/press-releases/1998/12/bill-and-melinda-gates-childrens-vaccine-program

81   National Institute of Allergy and Infectious Disease, NIAID 60th Anniversary Timeline, (Apr 1, 2008) https://www.niaid.nih.gov/about/niaid-60th-anniversary

82   VRBPAC Background Document, "Gardasil™ HPV Quadrivalent Vaccine May 18, 2006 VRBPAC Meeting," at 13 (Table 17, Study 013), archived at http://wayback.archive-it.org/7993/20180126170205/https://www.fda.gov/ohrms/dockets/ac/06/briefing/2006-4222B3.pdf.

83   "News Release: Merck Announces Fourth-Quarter and Full-Year 2020 Financial Results," Merck (Feb. 4, 2021). https://www.merck.com/news/merck-announces-fourth-quarter-and-full-year-2020-financial-results/

84   Mark Benjamin, "UPI Investigates: The vaccine conflict," *UPI*, (Jul 21, 2003) https://www.upi.com/Odd_News/2003/07/21/UPI-Investigates-The-vaccine-conflict/44221058841736/

85   "Close Ties and Financial Entanglements: The CDC-Guaranteed Vaccine Market," The Children's Health Defense Team, *Children's Health Defense* (Jun 6, 2019). https://childrenshealthdefense.org/news/close-ties-and-financial-entanglements-the-cdc-guaranteed-vaccine-market/

86   Suzanne Humphries, "Herd Immunity: Flawed Science and Mass Vaccination Failures," *Waking Times*, (Jul 17, 2012). http://www.wakingtimes.com/herd-immunity-flawed-science-mass-vaccination-failures/

87   Paul Offit, M.D., "Communicating Vaccine Science to the Public," Contemporary Clinical Medicine: Great Teachers Lecture Series, 3:19 NIH (Dec. 14, 2011), https://videocast.nih.gov/summary.asp?live=10824

88   Sharyl Attkisson, "How Independent Are Vaccine Defenders?" *CBS News* (Jul 25, 2008) https://www.cbsnews.com/news/how-independent-are-vaccine-defenders/

89   Mark Benjamin, "UPI Investigates: The vaccine conflict," *UPI* (Jul 21, 2003) https://www.upi.com/Odd_News/2003/07/21/UPI-Investigates-The-vaccine-conflict/44221058841736/

90   Sharyl Attkisson, "How Independent Are Vaccine Defenders?" *CBS NEWS* (Jul 25, 2008) https://www.cbsnews.com/news/how-independent-are-vaccine-defenders/

91   "The Children's Hospital of Philadelphia, Vaccine Education Center: Vaccines and Aluminum, April 2013," Archived at: https://web.archive.org/web/20130528091723/http:/www.chop.edu:80/service/vaccine-education-center/vaccine-safety/vaccine-ingredients/aluminum.html/

92   Paul A. Offit, "Is There Mercury in Vaccines?" Children's Hospital of Philadelphia (Aug 11, 2015) https://www.chop.edu/centers-programs/vaccine-education-center/video/there-mercury-vaccines

93   Karen Iris Tucker, "'Bad Advice' by Penn scientist Paul A. Offit: Take back the science conversation in our culture, " *The Philadelphia Enquirer* (Aug 31, 2008), https://www.inquirer.com/philly/entertainment/arts/bad-advice-by-penn-scientist-paul-a-offit-take-back-the-science-conversation-in-our-culture-20180831.html

94   Paul A. Offit, MD, *Do You Believe in Magic? Vitamins, Supplements, and All Things Natural: A Look Behind the Curtain* (Harper Collins 2014)

95   Dan McQuade, "CHOP Professor: *Silent Spring* Author's Mistake Killed Millions," *Philadelphia Magazine* (Feb 6, 2017), https://www.phillymag.com/city/2017/02/06/silent-spring-paul-offit-rachel-carson/

96   "Paul Offit Accidentally Speaks The Truth About MMR and Autism," YouTube video, (May 19, 2017), https://www.youtube.com/watch?v=c2cHZa8t98w&t=0s

97   "Columbia University Press and Dr. Paul Offit Sued for Autism's False Prophets," *The Age of Autism* (Feb 10, 2009), https://www.ageofautism.com/2009/02/columbia-university-press-and-dr-paul-offit-sued-for-autisms-false-prophets.html

98   J.B. Handley, "Dr. Paul Offit, The Autism Expert. Doesn't See Patients with Autism?" *The Age of Autism* (Oct 26, 2009), https://www.ageofautism.com/2009/10/dr-paul-offit-the-autism-expert-doesnt-see-patients-with-autism.html

99 Vaccinate Your Family, Board of Directors, (Apr 20, 2020), https://vaccinateyourfamily.org/about-us/staff-boards/

100 Immunization Action Coalition, Advisory Board, (Sep 8, 2020), https://www.immunize.org/aboutus/advisoryboard.asp

101 Mark Benjamin, "UPI Investigates: The vaccine conflict," *UPI* (Jul 21, 2003), https://www.upi.com/Odd_News/2003/07/21/UPI-Investigates-The-vaccine-conflict/44221058841736/

102 "Conflicts of Interest in Vaccine Policy Making Majority Staff Report," U.S. House of Representatives, Government Reform Committee, (Aug 21, 2000), https://childrenshealthdefense.org/wp-content/uploads/conflicts-of-interest-in-vaccine-policy-making-majority-staff-report-us-house-of-representatives-3.5.pdf

103 Ibid.

104 Stephan Foster, "Rotavirus Vaccine and Intussusception," *Journal of Pediatric Pharmacology* (Jan-Mar, 2007), https://www.ncbi.nlm.nih.gov/pmc/articles/PMC3462159/

105 "Conflicts of Interest in Vaccine Policy Making, Majority Staff Report," U.S. House of Representatives, Committee on Government Reform, (Jun 15, 2000), https://www.nvic.org/nvic-archives/conflicts-of-interest.aspx

106 "Suspension of Rotavirus Vaccine After Reports of Intussusception—United States, 1999," Centers for Disease Control and Prevention (Sep 3, 2004), https://www.cdc.gov/mmwr/preview/mmwrhtml/mm5334a3.htm

107 "Found 50 Cases where Symptom is Intussusption and Vaccine Dates from 1990-01-01," National Vaccine Information Center, MedAlerts.org, https://medalerts.org/vaersdb/findfield.php?TABLE=ON&GROUP1=AGE&EVENTS=ON&SYMPTOMS=Intussusception+%2810022863%29&VAX_YEAR_LOW=1999&VAX_MONTH_LOW=01&VAX_YEAR_HIGH=1999&VAX_MONTH_HIGH=12

108 Mark Benjamin, "UPI Investigates: The vaccine conflict," *UPI* (Jul 21, 2003), https://www.upi.com/Odd_News/2003/07/21/UPI-Investigates-The-vaccine-conflict/44221058841736/

109 "Delayed Onset and Diminished Magnitude of Rotavirus Activity — United States, November 2007–May 2008," Centers for Disease Control and Prevention, Morbidity and Mortality Weekly Report, *CDC-MMWR* (Jun 27, 2008), https://www.cdc.gov/mmwr/preview/mmwrhtml/mm57e625a1.htm

110 P. E. Kilgore, et al., "Trends of Diarrheal Disease--Associated Mortality in US Children, 1969 through 1991," JAMA, (Oct 11, 1995). https://jamanetwork.com/journals/jama/article-abstract/389779

111 U.S. Food and Drug Administration, Package Insert - RotaTeq, (2006), https://www.fda.gov/media/75718/download

112 Shasta D. McClenahan, et al., "Molecular and infectivity studies of porcine circovirus in vaccines," *VACCINE* (May 11, 2001), https://sci-hub.do/10.1016/j.vaccine.2011.04.087

113 Claudia Kalb, "Dr. Paul Offit: Debunking the Vaccine-Autism Link," Newsweek (Oct 24, 2008), https://www.newsweek.com/dr-paul-offit-debunking-vaccine-autism-link-91933

114 Dan Olmstead and Mark Blaxill, "Voting Himself Rich: CDC Vaccine Adviser Made $29 Million Or More After Using Role to Create Market," *Age of Autism* (Feb 16, 2009), https://www.ageofautism.com/2009/02/voting-himself-rich-cdc-vaccine-adviser-made-29-million-or-more-after-using-role-to-create-market.html

115 Christina D. Bethell, Michael D. Kogen, et al., "A National and State Profile of Leading Health Problems and Health Care Quality for US Children: Key Insurance Disparities and Across-State Variations," Academic Pediatrics (May-June, 2011). https://www.academicpedsjnl.net/article/S1876-2859(09)00010-2/fulltext

116 "Chronic Illness in Children—Who is Sounding the Alarm?" The Children's Health Defense Team, (Sep 12, 2019), https://childrenshealthdefense.org/news/chronically-ill-children-who-is-sounding-the-alarm/

117 Maria Trimarchi, "How much do pharmaceutical test subjects get paid?" *How Stuff Works*, (2021), https://money.howstuffworks.com/how-much-do-pharmaceutical-test-subjects-get-paid.htm

118 Jane Redecki, "University budget Models and Indirect Costs," Ithaca S&R, February 25, 2021, https://sr.ithaka.org/publications/university-budget-models-and-indirect-costs/

119 Nussbaum, op. cit., 330

120 Terry Michaels, Down the Rabbit Hole: How US Medical Bureaucrats, Pharma Crony Capitalists, and Science Literate Journalists Created and Sustain the HIV-AIDS Fraud (Unpublished Manuscript), 10

121 Ibid.

122 John Lauritsen, *The AIDS War: Propaganda, Profiteering and Genocide from the Medical-Industrial Complex* (Asklepios, 1993), 322

ChildrensHealthDefense.org/fauci-book

childrenshd.org/fauci-book

For updates, new citations and references, and
new information about topics in this chapter:

CHAPTER 4

THE PANDEMIC TEMPLATE: AIDS AND AZT

"Doctors need three qualifications: to be able to lie and not get caught, to pretend to be honest, and to cause death without remorse."

—Jean Froissart 1337–1405

The AZT approval process was a shakedown cruise for Tony Fauci. As he ran AZT around the regulatory traps, Dr. Fauci pioneered and perfected the retinue of corrupt, deceitful, and bullying practices and strategies that he would replicate again and again over the next thirty-three years, to transform NIAID into a drug development dynamo.

When Dr. Fauci entered the principal investigator (PI) drug-testing universe, only one pharmaceutical company, Burroughs Wellcome (predecessor to GlaxoSmithKline), had a drug candidate teed up to test as an AIDS remedy—a toxic concoction, azidothymidine, known popularly as "AZT."

US government–financed researchers developed AZT in 1964[1] as a leukemia chemotherapy. AZT is a "DNA chain terminator," randomly destroying DNA synthesis in reproducing cells. AZT's developer, Jerome Horwitz, theorized that the molecule might inject itself into cells and interfere with tumor replication. FDA abandoned the toxic chemotherapy compound after it proved ineffective against cancer and breathtakingly lethal in mice.[2] Government researchers deemed it too toxic even for short-regimen cancer chemotherapy. Horwitz recounted that the drug's "extreme toxicity made it 'so worthless' that he 'didn't think it was worth patenting.'" Former *BusinessWeek* journalist Bruce Nussbaum recounted that Horwitz "dumped it on the junk pile" and "didn't [even] keep the notebooks."[3]

Soon after NIH's team identified HIV as the probable cause of AIDS in 1983, Samuel Broder, head of the National Cancer Institute (NCI)—another sub-agency of the NIH—launched a project to screen antiviral agents from around the world as potential treatments. In 1985, his team, along with colleagues at Duke University, found that AZT killed HIV in test tubes.[4]

NCI's study inspired Burroughs Wellcome to retrieve AZT from Horwitz's scrap heap and patent it as an AIDS remedy. Recognizing financial opportunity in the desperate terror of young AIDS patients facing certain death, the drug company set the price at up to $10,000/year per patient—making AZT one of the most expensive drugs in pharmaceutical history.[5] Since Burroughs Wellcome could manufacture AZT for pennies per dose, the company anticipated a bonanza.

In order to justify these exorbitant prices for an existing drug, wrote Dr. Marcia Angell, the longtime editor of the *New England Journal of Medicine* in her 2004 book, *The Truth About Drug Companies*, "the company claimed far more credit than it deserved."[6] After Burroughs Wellcome's CEO sent a self-congratulatory letter to the *New York Times* rationalizing AZT's exorbitant sticker price with the standard Pharma embroidery about the high risks and extravagant costs of early drug development,

Broder and four colleagues from the NCI and Duke responded angrily, reciting the seminal contributions Burroughs Wellcome did *not* make:

> The company specifically did not develop or provide the first application of the technology for determining whether a drug like AZT can suppress live AIDS virus in human cells, nor did it develop the technology to determine at what concentration such an effect might be achieved in humans. Moreover, it was not first to administer AZT to a human being with AIDS, nor did it perform the first clinical pharmacology studies in patients. It also did not perform the immunological and virological studies necessary to infer that the drug might work and was therefore worth pursuing in further studies. All of these were accomplished by the staff of the National Cancer Institute working with staff at Duke University.[7]

The NCI scientists pointedly added that the company's squeamishness about handling the HIV pathogens made it impossible for Burroughs Wellcome to perform any meaningful research: "Indeed one of the key obstacles to the development of AZT was that Burroughs Wellcome did not work with live AIDS virus nor wish to receive samples from AIDS patients."[8]

When Fauci appropriated the HIV program from the National Cancer Institute, NIAID inherited AZT, which was then further down the clinical trial path than any other drug.[9]

AZT proved to be an irresistible opportunity for Fauci. After all, Burroughs Wellcome not only had a head start in the AIDS drug program, the company also had its own army of veteran "principal investigators" (PIs) with plenty of expertise at running the complex regulatory hurdles—which Dr. Fauci had not yet mastered. Dr. Fauci needed a visible success to jump-start his program and anoint his new regime with the patina of competence. Nussbaum described how the British pharmaceutical company manipulated its leverage over Dr. Fauci to gain monopoly control over the government's HIV response: "Wellcome's PIs came to dominate NIAID's clinical trial system. They formed a web linking Wellcome, the drug AZT, and the NIH. They came to sit on the institute's key drug selection committee, and they voted on whether to give high or low priority to the testing of each anti-AIDS drug, including those that might possibly compete with AZT in the marketplace. The PIs were a power unto themselves. They were, in fact, out of control."[10]

Dr. Fauci would later mimic this successful model to populate key drug and vaccine approval committees in FDA, CDC, and at the Institute of Medicine (IOM) with his Pharma PIs, giving him, and his Pharma partners, complete, vertically integrated control over the drug approval process from molecule to market.

But all did not go smoothly. Even with Burroughs Wellcome holding the reins, progress at NIAID was glacial. AZT's horrendous toxicity hobbled researchers struggling to design study protocols that would make it appear either safe or effective. With AZT devouring his bandwidth, Dr. Fauci failed to populate clinical trials for any competing drug. After three years and hundreds of millions spent, NIAID had not produced a single new approved treatment.

Meanwhile, bustling networks of community-based AIDS doctors mushrooming in cities like San Francisco, Los Angeles, New York, and Dallas had become specialists in treating the symptoms of AIDS. As Dr. Fauci swung for the fences—the miraculous new antiviral "cure" for AIDS—these community doctors were achieving promising results with off-label therapeutic drugs that seemed effective against the constellation of symptoms that actually killed and tormented people with AIDS. These included off-the-shelf remedies like ribavirin, alpha interferon, DHPG, Peptide D, and Foscarnet for retinal herpes; and Bactrim, Septra, and aerosol pentamidine for AIDS-related pneumonias. Despite years of pleading by the HIV community, Dr. Fauci refused to test any of those repurposed drugs, which had older or expired patents and no Pharma patrons.[11] One of the most promising of these "street drugs" was AL 721, an antiviral that was far less toxic than AZT. Two of Dr. Fauci's top scientists, Robert Gallo and Jeffrey Laurence from NCI, had found AL 721 effective in reducing HIV viral loads—but, under pressure from his phalanx of Burroughs Wellcome PIs, Dr. Fauci refused to follow up.[12] Big Pharma and its PIs were loath to test any drug with patents they didn't control. None of the big pharmaceutical companies were interested in cultivating rivals for their high-margin blockbusters like AZT.

Dr. Fauci's failure to move these remedies through the NIAID system spawned a burgeoning sub-rosa market where people with AIDS and community doctors purchased remedies from underground "buyers' clubs."[13]

One of NCI's top virologists, Dr. Frank Ruscetti, who worked directly under Robert Gallo, recalls of that era, "We could have saved millions of lives with repurposed and therapeutic drugs. But there's no profit in it. It's all got to be about newly patented antivirals and their mischievous vaccines."[14]

The PIs made sure that Pharma's AZT was the only arrow in NIAID's clinical trial quiver. Because of Dr. Fauci's inexperience and perhaps deliberate sandbagging, he and his PIs had only managed to fill 5–10 percent of the slots in his clinical trials for other promising drugs that would compete with AZT. According to Nussbaum, "In time, the clinical trials network Fauci set up would come to be known as the 'HUD of the nineties.' Money was spent, but trials went under-enrolled, drug treatments never seemed to emerge, and people with AIDS continued to get sick and die."[15]

At the mercy of Burroughs Wellcome, Dr. Fauci cut the company PIs every courtesy to accelerate AZT's approval. FDA and NIH waived long-term primate studies that would be a high-risk gambit on a compound of such well-known toxicity. (Dr. Fauci would take the same shortcut thirty-six years later to accelerate approvals of his pet drug, remdesivir, and Moderna's coronavirus vaccine.) Dr. Fauci endorsed Burroughs Wellcome's scheme to price AZT at a sumptuous $10,000 per patient per year by agreeing to pay the top-shelf sticker price for the pills used in NIAID's clinical trials.[16]

According to Nussbaum, "Tony Fauci's managerial incompetence," which put him utterly at the mercy of Burroughs Wellcome and its AZT and AZT-only agenda, "had exacted a staggering cost. By 1987, more than a million Americans were infected by the AIDS virus. Not a single drug treatment had come out of the government's enormous biomedical research system."[17]

Nussbaum chronicles the escalating frustration among AIDS activists who were winning vast Congressional appropriations for NIAID, with nothing to show. By 1988, Nussbaum recounts, "several hundred million tax dollars had somehow disappeared into the nation's biomedical establishment and not one new drug had been produced." Tony Fauci's incompetence was frustrating the national response to the pandemic. "Where was Tony Fauci at this time?" Nussbaum asks. "Nowhere. . . . He wasn't, after all, a 'details' man. He was busy being a 'hit-the-front-pages-every-day' kind of guy."[18]

AIDS activists and public health officials were wondering, "Where did all the grant money go? Did NIAID keep the money? Who benefited? Certainly not the tens of thousands of people with AIDS who grew angrier and angrier with each wasted, passing day."[19] Activists complained that Dr. Fauci was not being forthcoming about the status and enrollment of his clinical trials. He was stonewalling inquiries and had veiled the entire process in secrecy.

Despite pleas from patients, their doctors, and advocates, despite the vast financial windfalls flowing to his agency from the HIV community's adept lobbying, Dr. Fauci refused to meet with the AIDS community leadership during his first three years as America's "AIDS Czar." That reticence further soured Dr. Fauci's already difficult relationships with the community he was responsible to serve.

It was a hardwired reflex at NIAID to exaggerate public fears of pandemics, and Dr. Fauci's first instinct as national AIDS czar had been to stoke contagion terror. He made himself a villain among AIDS activists with a fear-mongering 1983 article in the *Journal of the American Medical Association* warning that AIDS could spread by casual contact.[20] At the time, AIDS was almost exclusive to intravenous drug users and males who had sex with other males, but Dr. Fauci incorrectly warned of "the possibility that routine close contact, as within a family household, can spread the disease." Given that "nonsexual, non-blood-borne transmission is possible," Fauci wrote, "the scope of the syndrome may be enormous." In his history of the AIDS crisis, *And the Band Played On*, author Randy Shilts reports that the world's leading AIDS expert, Arye Rubinstein, was "astounded" at Fauci's "stupidity" because his statement did not reflect the contemporary scientific knowledge.[21] The best scientific evidence suggested the infectivity of HIV, even in intimate contact, to be so negligible as to be incapable of sustaining a general epidemic.

Nevertheless, Dr. Fauci's reflexive response was to amplify the widespread panic of dreaded pestilence that would naturally magnify his power, elevate his profile, and expand his influence. Amplifying terror of infectious disease was already an ingrained knee-jerk institutional response at NIAID.

In 1987, the *Wall Street Journal* won a Pulitzer Prize for its investigation of an HHS scheme its writers characterized as a deliberate campaign by officials to misrepresent AIDS as a general pandemic to secure greater public funding and financial support.[22]

The flimflam worked. Terror of pestilence, it turns out, is a potent impulse, and Fauci was adept at weaponizing it—and he quickly learned that other "respected authorities" would follow his lead. Following Dr. Fauci's fear-mongering prophecy, Theresa Crenshaw of the President's AIDS Commission made the astonishing forecast that within fourteen years, double the number of people then on the planet would be dying from lethal infections: "If the spread of AIDS continues at this rate, in 1996 there could be one billion people infected; five years later, hypothetically ten billion." Crenshaw asked, "Could we be facing the threat of extinction during our lifetime?"[23] Crenshaw's dire soothsaying never materialized. In 2007, WHO estimated only 33.2 *million* people worldwide were HIV-positive.[24] The HIV prevalence curves based on CDC's own data show that at least in the US, HIV has not spread at all since testing was first available, stubbornly remaining at the same levels relative to population.

The *Oprah* show broadcast Crenshaw's subsequent prognostication that "By 1990, one in five heterosexuals may be dead of AIDS."[25] Thankfully, this prognosis was also hyperbolic. According to CDC data, about one in 250 Americans tests HIV-positive, and outside the risk groups this number drops to about one in *five thousand*—about 1/1,000th Crenshaw's bodement.[26] The hysteria following Fauci's dystopian prediction prompted *Der Spiegel* to warn that AIDS infections would entirely exterminate the German population by 1992.[27] The following year (1985), the magazine *Bild Der Wissenschaft* also forecast the prompt extinction of the Teutonic race.[28]

A slightly less exuberant 1986 prophecy by *Newsweek* had five to ten million Americans lethally infected by 1991.[29] *Newsweek*'s auguring was off by ten times; US authorities have since identified only one million HIV infections.[30]

Dr. Fauci's embellishments quickly made HIV-positives the modern equivalent of lepers. Paranoia of AIDS from nonsexual contact persisted for years. In New York in 1985, for instance, 85 percent of schoolchildren at one public elementary school stayed home during opening week, while hundreds of parents demanded the school system bar any HIV-positive children from attending classes.[31] The Reagan administration made it unlawful for persons with AIDS to enter the United States. The Cuban government quarantined AIDS victims in modern leper colonies. AIDS activists charged Dr. Fauci with causing the "irrational, punitive" response that followed his hysterical statements.[32]

A year later, growing furor over his assertion forced Dr. Fauci to acknowledge that health officials had never detected a case of the disease spread through "casual contact."[33]

Finally, AIDS activists further complained that Dr. Fauci lacked sensitivity and human compassion toward people suffering from the disease. His laser focus on a single magic bullet antiviral left Dr. Fauci reluctant to study drugs that treated the constellations of grim infections that tortured and killed people with AIDS; patient care—which typically involved off-the-shelf drugs—was incompatible with NIAID's mushrooming mercantile obsession with high-price patented antivirals. Dr. Fauci's narrow focus on AZT over off-patent therapeutic medications prompted the AIDS plague's most vocal activist, Larry Kramer, to call Dr. Fauci a "damned bungler"[34] and "Public Enemy Number One."[35]

Melisa Wallack and Craig Borten, who received Oscar nominations for their script, *Dallas Buyers Club*, intensively researched NIAID's institutional hostility to patient care and repurposed drugs during the 1986 AIDS crisis. Dr. Fauci's campaign to sabotage therapeutic remedies played a key role in precipitating the emergence of the organized underground medical network. So-called "Buyers Clubs" filled the vacuum by providing treatments that community doctors and their patients considered effective against AIDS, but that FDA refused to approve. "Dr. Fauci was a liar," recalls Wallack, who researched Dr. Fauci intensively for her film. "He was utterly beholden to pharmaceutical companies and was hostile to any product that would compete with AZT. He was the real villain of this era. He cost a lot of people their lives."[36]

By 1987, thousands of AIDS activists from organizations like amfAR and ACT UP—many of them dressed in burial frocks—began mounting mass protests against Dr. Fauci at NIH's Bethesda, Maryland, research complex and demanding that he, at last, meet with them. Carrying signs that read, "Red Tape Kills Us," and "NIH—Negligence, Incompetence and Horror," protesters were met by a line of police officers in riot gear.[37] The protestors objected to Dr. Fauci's narrow focus on Wellcome's single patented antiviral and wanted more attention for existing therapeutic drugs that seemed to reduce the worst of AIDS's most agonizing and deadly symptoms.

As the clamoring crowds multiplied on NIH's expansive Bethesda campus, Congressman Henry Waxman intervened to force Dr. Fauci to finally sit down with activists in the spring of 1987. It was his first meeting with AIDS advocates since he became AIDS Commissar three years earlier. "The arrogance was simply part of NIH culture," wrote Nussbaum. "No one thought that people with AIDS and their local doctors had anything to recommend in terms of their own treatment. The same was true of people with cancer. They were all 'patients' or 'victims' to be pitied and helped by white-coated scientist-heroes."[38]

Larry Kramer, Nathan Kolodner, Dr. Barry Gingell, and singer/songwriter and pioneering AIDS activist Michael Callen finally took their seats across a broad table from Dr. Fauci and fifteen of his selected scientists from FDA and NIH. Throughout that meeting, the advocates found Dr. Fauci both manipulative and "dismissive" of their concerns. According to Nussbaum, these leaders "had said time and again that NIAID was obsessed with AZT, that most of the trials and people with AIDS involved in the trials were on just that one drug."[39] They began by confronting Dr. Fauci with the fact that his own most trusted scientists, Dr. Laurence and Dr. Gallo, had found AL 721 effective in reducing viral loads;[40] Dr. Fauci responded with a barrage of misdirection and obfuscation. He cherry-picked a single assay from an obscure laboratory that had found AL 721 ineffective and refused to discuss or acknowledge the two studies by his own agency that supported its use.

They next questioned him about his sandbagging on aerosol pentamidine. According to Nussbaum, ". . . dozens of community doctors and thousands of PWAs [people with AIDS] already knew: that Aerosol Pentamidine prevented AIDS' most lethal symptom—*pneumocystis carinii* pneumonia (PCP)."[41] Doctors had also found early intervention with Bactrim and Septra to be effective prophylaxis against PCP. The activists presented Dr. Fauci with a modest request: that NIAID agree to make guidelines for physicians who wanted to use Bactrim to treat people with AIDS preventively, or even a statement supporting consideration of the use. An official declaration by NIH that doctors consider these treatments "standard of care" would require insurance companies to cover their costs, making them available to AIDS victims, many of whom were destitute. Dr. Fauci met both requests with refusal. He said he simply could not recommend a drug until he saw "randomized, blinded, placebo-controlled trial" results. That was the "gold standard," he said. It would be that, or nothing. When they asked him, "Why not?" he shouted, "There's no data!"[42] He told them that the treatment experiences and voluminous case study reports of dozens of community AIDS doctors was not real science. The activists were aware of this increasingly lethal irony: It had been NIAID's decision to not fund any randomized trials on these unpatented drugs. Dr. Fauci himself had constructed this dead end. This pattern of resourceful stonewalling to obstruct repurposed off-patent drugs with lifesaving potential would become a pattern familiar to Dr. Fauci's critics during the COVID crisis.

According to Callen, "We asked him—no, we begged him—to issue interim guidelines urging physicians to prophylax those patients deemed at high risk for PCP (pneumonia) [with Bactrim or aerosol pentamidine]. Although it would not have cost the government much to have done so, he steadfastly refused to issue such guidelines. His reason: no data. So, the Catch-22 was complete and many people died of PCP who didn't have to."[43]

When the activists asked Dr. Fauci to at least add AL 721, Peptide D, DHPG, and aerosolized pentamidine to his clinical trials, Dr. Fauci's refusal was loud: "I can't do that!" he shouted. "I can't convene a consensus conference."[44] The choice, he explained, of which compounds would enter NIAID's clinical trial pipeline was made, not by public agreement, but by a panel of "independent scientists." Dr. Fauci did not mention that virtually all the members of his "independent panel" were pharmaceutical PIs, with ties to NIAID and Burroughs Wellcome.

Following that meeting, a group of frustrated community doctors raised money from their own AIDS patients to collect data for a randomized trial on Bactrim. It took them two years, and their results strongly supported Bactrim's effectiveness against pneumonia. AIDS activists lamented that two years of stalling by Fauci on aerosol pentamidine and Bactrim had cost seventeen thousand people their lives.[45]

Following the NIH parley, the fury of the AIDS patient advocates against Dr. Fauci mounted. In their view, the community doctors were generating plenty of good science. Those treatment experiences—often published—had as much validity as case studies upon which scientists routinely rely. As Nussbaum points out, "There was plenty of data, if only Fauci and the rest of NIH were willing to look at real people in real communities instead of the endless bottoms of their test tubes."[46]

Michael Callen told Nussbaum that Dr. Fauci's single-minded concern seemed to be avoiding the mortification of acknowledging success by doctors outside his agency. "He would not be humiliated even if 'Fauci's decision cost the lives of tens of thousands of people with AIDS.'"[47]

Michael Callen, Larry Kramer, and the other AIDS activists left the NIH sit-down in a fierce rage. In June 1987 at a postmortem at ACT UP's circus-like New York City headquarters auditorium (where I often spoke on environmental issues during that era), Kramer lambasted Dr. Fauci for his Pharma bias:

"Where are the drugs the government promised?" he asked. "After we got them millions of dollars for their experiments, what do we get? A ten-thousand-dollar drug! What about all the other drugs out there?"[48]

Congressional Confrontation April 28, 1988

Dr. Fauci had given Kramer and the other activists the bum's rush. He could not do the same with his congressional patrons. For years, my uncle, Senator Ted Kennedy, the chair of the Senate Health Committee, and Senator Lowell Weicker, who chaired Senate Appropriations, along with their allies in the House, California Congressman Henry Waxman and Manhattan Congressman Ted Weiss, had fought hand-to-hand combat with Ronald Reagan's tight-fisted budget director, David Stockman, to free up money for AIDS research.

In 1980, Teddy became the first presidential candidate to actively campaign for gay rights. I stumped with him in San Francisco's Castro District when he shattered political taboos by barnstorming the gay bars, shaking hands, and snapping photos. When the AIDS epidemic broke a year later, Teddy defied convention by hiring Terry Beirn, the first openly gay/HIV-infected Senate aide, to stage-manage the legislative battle against AIDS. Beirn became the leading national advocate for the community-based clinical trials for remedies like Bactrim and aerosolized pentamidine, to which Dr. Fauci had shown such hostility.[49] Beirn had hatched the idea for a community research initiative (CRI) with Teddy's close friend Mathilde Krimm, of the activist group of amfAR, and Martin Delaney of Project Inform. Their proposal was to create a "parallel track" approval system that would allow community AIDS doctors to conduct clinical studies on the off-the-shelf drugs that neither Pharma nor NIAID wanted to test. Delaney, who did not have AIDS but made his bones in the movement smuggling ribavirin from Mexico for the Buyers Clubs, described the parallel track program as "medically supervised guerilla drug trials."[50] Appealing to his friend Senator Orrin Hatch's Mormon sense of compassion toward the ill, Senator Kennedy had recruited the Utah conservative Republican to cochampion the AIDS issue. Independent-minded Connecticut Senator Lowell Weicker was another key ally. Those three most powerful senators from three different political perspectives worked in tandem and with Waxman and Weiss in the House. Their coordinated bipartisan efforts freed up hundreds of millions of dollars from the White House bean counters, over the objections of powerful Christian conservatives who framed AIDS as God's just punishment for the homosexual lifestyle.

For two years, Senator Kennedy and Beirn vainly urged Dr. Fauci to create a "parallel track." Kennedy was frustrated by Dr. Fauci's reticence to listen to the HIV community. He considered it petty, cruel, and irresponsible that Dr. Fauci would not allow testing of the buyers' club drugs.

In a September 2007 interview, Dr. Fauci recalled the urgency that Teddy brought to the topic. He said that Kennedy urged him, "We've got to have a clinical trial process that reaches out to the community. He was really the one who pushed very hard for the community program for clinical research on AIDS. That was one of his big agenda items. He wanted to get community access to clinical trials at the community level, not just limited to the trials run by drug companies and NIAID."[51]

By 1987, Dr. Fauci's political partners from all parties realized that Dr. Fauci's program was "in shambles."[52] Despite the millions from Congress, not a single AIDS drug had emerged from NIAID's pipeline. Senator Kennedy was beginning to suspect that Dr. Fauci was either inept or "in the tank" with Pharma. Ronald Reagan was pushing to transfer the entire AIDS effort to "more efficient" private pharmaceutical companies. Dr. Fauci's failed predictions, organizational inadequacies, and obfuscations had steamed his Capitol Hill allies past their boiling points.

In the spring of 1988, Dr. Fauci's congressional sponsors turned on him during a dramatic Capitol Hill confrontation. The April 28 hearing began with Rep. Weiss—perhaps Dr. Fauci's most loyal sponsor—demanding that the NIAID chief explain his snail's progress. Dr. Fauci responded by whining that he had no budget to purchase lab space, computers, desks, and office supplies, or to hire new workers.[53]

The stunned Upper West Side congressman reminded Dr. Fauci that he had accepted $374 million from Congress for AIDS research. It seemed astonishing that those sums were insufficient to purchase clerical supplies and furniture. Oblivious that his lame excuses were only stoking his benefactor's scorching rage, Fauci moaned that his office items required separate budget columns not provided for in the massive congressional appropriation. In a barely controlled fury, Rep. Waxman coldly asked Dr. Fauci why he never informed his congressional mentors of this logjam. That question provoked a cavalcade of vague and dissembling bellyaching during which Dr. Fauci suggested, obliquely, that he had feared antagonizing the Reagan White House—which might have frowned on his cozy bonhomie with congressional Dems.

Dr. Fauci's fuzzy equivocation prompted Rep. Waxman to darken visibly. "He was furious," recounts Nussbaum. "He practically levitated out of his chair."[54]

California Congresswoman Nancy Pelosi complained of Dr. Fauci's lackluster performance that "from our perspective, we have a burning building behind us and we're coming to you all for water and we're finding out that there's not somebody there to turn on the faucet."[55]

Pelosi next delivered the "coup de grâce," as Nussbaum chronicled the explosive exchange. Rep. Pelosi asked Dr. Fauci to assume that he had AIDS and found himself dying of pneumonia: "You know the theory behind aerosol pentamidine to prevent pneumonia is strong. You know that the aerosol pentamidine was evaluated by the NIH as highly promising. You know that many studies in San Francisco recommend it routinely and that it is available. . . . Would you take aerosol pentamidine or would you wait for a study?"[56]

For three years, Dr. Fauci had done everything in his power to deny aerosol pentamidine and its companion drug, Bactrim, to AIDS sufferers. But here's what he told the panel in 1988: "If I were an individual patient, I would probably take aerosolized pentamidine if I already had a bout of *Pneumocystis*. In fact, I might try, even before then, taking prophylactic Bactrim."[57] These were two promising remedies that everyone on the panel and in the audience knew that Fauci had refused to either test or recommend. At that very moment, Dr. Fauci was denying tens of thousands of AIDS patients access to these lifesaving remedies.

Nussbaum describes the scene that followed: "Silence. There was dead silence in room 2154 of the Rayburn House Office Building. People at the hearing just stared at Fauci and at one another. Here was the head of the NIH effort against AIDS publicly admitting that he personally would not follow the government's own guidelines and recommendations. Here was a top government scientist basically admitting that the government effort should be circumvented by the millions of people with AIDS. Here was Tony Fauci openly calling for the prophylaxis of *Pneumocystis carinii* pneumonia while his own clinical trials system did not have a single preventative drug in trial. It was a truly mind-wrenching admission. Fauci himself was calling into question the very foundation of the government's entire research effort against AIDS."[58]

Thirty-two years later, Dr. Fauci performed an encore of this kabuki dance during the early COVID crisis. On March 24, 2020, he answered a question from a journalist by admitting that, if he became ill with COVID, he would take hydroxychloroquine as his remedy.[59] Shortly thereafter, Dr. Fauci launched his aggressive campaign to deny HCQ—and all early treatments—to the rest of humanity.

Dr. Fauci's 1988 Capitol Hill performance left all his former friends wanting a piece of him. "Fauci was in deep trouble. These were his supporters, his financial mentors, his political protectors from an administration that was so aligned against the gay community and so ideologically antagonistic to the very existence of the NIH that it wanted Pharma to privatize the whole shebang. Now, Weiss and Waxman were clearly gunning for him. Fauci realized that the entire hearing was a setup to show his personal shortcomings."[60]

Larry Kramer was thunderstruck: "When he read about the NIH delays, the ineptitude and perhaps the moral cowardice behind them, Kramer lost control."[61]

On May 31, 1988, Kramer wrote his famous "Open Letter to Tony Fauci" in the *Village Voice*. Kramer's diatribe compared NIAID to the fraternity of miscreants, delinquents, and dimwitted knuckleheads in the comedy film *Animal House*. He called Dr. Fauci an "idiot" and a "murderer." He described Fauci sweating and squirming under Representative Ted Weiss's questioning: "You were pummeled into admitting publicly what some have been claiming since you took over some three years ago. You have admitted that you are an incompetent idiot."[62]

Said Kramer, "You expect us to buy this bullshit and feel sorry for you? YOU FUCKING SON OF A BITCH OF A DUMB IDIOT, YOU HAVE HAD $374 MILLION AND YOU EXPECT US TO BUY THIS GARBAGE OF EXCUSES!"[63]

Kramer accused Fauci of keeping his mouth shut for thirty-six months to pander to the Reagan White House. He asked Fauci, "WHY DID YOU KEEP QUIET FOR SO LONG?" while people perished in the pandemic. It reminded him, he said, of Hitler's "good lieutenant": Adolf Eichmann. He accused Fauci of being too cowardly and self-involved to speak up until forced to by a Congressional committee: "We lie down and die and our bodies pile up higher and higher in hospitals and homes and hospices and streets and doorways."[64]

Referring to aerosol pentamidine, Kramer pointed out, "[W]e know and hear what is working on some of us somewhere. You couldn't care less about what we say. You won't answer our phone calls or letters, or listen to anyone in our stricken community. What tragic pomposity!"[65]

"How many years ago did we tell you about aerosol pentamidine, Tony? That this stuff saves lives. And WE discovered it ourselves. We came to you, bearing this great news on a silver platter as a gift, begging you: Can we get it officially tested, can we get it approved by you so that insurance companies and Medicaid will pay for it (as well as other drugs we beg you to test) as a routine treatment, and our patients going broke for medicine can get it cheaper? You monster."[66]

"We tell you what the good drugs are, you don't test them, and YOU TELL US TO GET THEM ON THE STREETS! You continue to pass down word from On High that you don't like this drug or that drug—WHEN YOU HAVEN'T EVEN TESTED THEM!"[67]

"There are more AIDS patients dead because you didn't test drugs on them," Kramer said, "than because you did."[68]

After the Congressional hearing, everyone realized that the little Emperor had no clothes; Dr. Fauci recognized that his political life was dangling by a thread. He had spent hundreds of millions of dollars building a drug-testing network that didn't work. The Congress he had always been able to charm, double-talk, and bamboozle had finally called fraud! His only hope for reputation and career salvation was a dramatic and unexpected change.

"He had been tarred with an 'incompetence' brush by the very people who were his major supporters in the past. Only a complete change of strategy could resuscitate Tony Fauci's career. If he was to continue receiving financial support for AIDS research from Congress, if he was to continue being the head of NIAID, he had to reinvent himself."[69]

Dr. Fauci's Strategic Pivot

Anthony Fauci needed a makeover, and this master of bureaucratic survival responded to his existential crisis with a breathtaking pivot. Suddenly, Dr. Fauci turned to embrace the AIDS activists he had previously reviled. In the summer of 1989, he accosted Larry Kramer on a Montreal street during an international AIDS conference, took him for a walk, effectively begged forgiveness, and proposed a working partnership.[70] He began testing AIDS community drugs in parallel trials, as Senator Kennedy and amfAR had long requested.

Dr. Fauci partnered with the AIDS doctors—the contemporary equivalents of Front Line COVID-19 healers Dr. Pierre Kory, Dr. Peter McCullough, Dr. Richard Urso, and Dr. Ryan Cole (among others)—giving them authority and millions of dollars to launch local Community Research Initiative (CRI) programs that allowed community AIDS clinics to test promising drugs outside the formal clinical trial programs dominated by Dr. Fauci's Pharma PIs, and to quickly win federal approvals. "Fauci himself was now trying to build a system that consisted of greater access to drugs at a much earlier stage in the testing game,"[71] said Nussbaum. In a gesture of reconciliation with his biggest critics, Dr. Fauci named the parallel track program after Senator Kennedy's aide, Terry Beirn, and he gave Larry Kramer a seat at the table.

Most ironically, in light of his successful campaign to sabotage hydroxychloroquine and ivermectin during the COVID crisis, Dr. Fauci suddenly dropped his knee-jerk insistence that every drug needed randomized placebo-controlled testing prior to approval. In an extraordinary volte-face, he fiercely argued that if a drug looked promising for alleviating potentially lethal illness during pandemics, patients ought to be able to get access to it, even if it hadn't been through a double-blind placebo trial. In a brassy display of chutzpah and brazenly hypocritical misdirection, he questioned the ethics of FDA regulators who insisted on placebo testing of beneficial drugs during a global pandemic when people were dying. He seemed to have forgotten that this was precisely his posture until just a few weeks before. Dr. Fauci accused the FDA of foot-dragging and overmanaging drug development. He openly attacked NIAID's sister agency for its cruel and rigid insistence on randomized double-blind placebo testing for DHPG, a promising remedy for retinal herpes. In order to quiet the AIDS community, Tony Fauci even put AL 721 into trials. Dr. Fauci became a vocal cheerleader of "parallel track" approval of the retinue of popular buyers' club drugs: "It doesn't make any sense to deprive those people of the choice of whether or not they want to take a chance on a drug that has proven to be effective, as long as it doesn't interfere with clinical trials. As a scientist, I think it's an appropriate thing to do."[72]

"Fauci transformed himself in the summer of 1989. He became an aggressive advocate for speeding up testing and drug approval for all life-threatening diseases, not just AIDS," recalls Nussbaum. "Fauci adopted virtually the entire ACT UP program at once and as a whole. It was the kind of flip-flop that comes with a true religious conversion. It was so startling that it appeared as if Fauci had found the light, had an epiphany, and transformed himself into another being."[73] This sudden flip-flop presaged Dr. Fauci's 2021 neck-wrenching switcheroo when he suddenly demanded an investigation of the Wuhan lab after energetically forestalling that inquiry for over a year.

By the end of 1989, his insurrection against his own old orthodoxies and his merciless attacks on the beleaguered satraps at FDA had made Dr. Anthony Fauci into something of a hero to some in the HIV community.

Industry to the Rescue

Not everyone was happy. Dr. Fauci's U-turn had infuriated his industry PIs. Big Pharma's front-line troopers were in open revolt against his ballyhooed reforms. The CRI system was proving a disaster for the industry. The AIDS community's network of two hundred CRI doctors was testing anti-AIDS drugs in "parallel track" programs with low cost and quick enrollments. The community doctors, Nussbaum explained in 1990, "know more about treatment than do [Dr. Fauci's] ivory-tower PIs hidden away from the realities of life and driven by careers that don't reward them for furthering the public health."[74] So many AIDS patients were flocking to participate in CRI trials with caring doctors they knew and trusted that Dr. Fauci's traditional

Pharma PIs were having trouble recruiting volunteers to their clinical trials. The CRI was so successful that it began challenging the primacy of NIAID's traditional top-down university- and hospital-based research. The PI network that formerly enjoyed an unchallenged monopoly on drug trials balked as the gay community's upstart doctors threatened their exclusive position at NIAID's billion-dollar research funding teat.

Big Pharma's PIs were to Dr. Fauci what the Praetorian Guard was to the Roman emperors: Fauci was at once their commander and their hostage. Ultimately, they exercised life-or-death power over him. It's worth recalling that the vast majority of Roman emperors died at the hands of their subordinates, with either assistance or acquiescence in their murders by their "loyal" Praetorians.

His fifty years at NIH are resounding proof of Dr. Fauci's unerring survival skills. The political instincts that have made him history's longest-lived—and highest-paid—public health apparatchik must have informed him that antagonizing his Praetorians would eventually be fatal. He needed to make peace.

Whether Dr. Fauci's brief conversion was ever heartfelt, it was necessarily short-lived. Fauci's managerial style and his deep reliance on his network of Pharma PIs doomed parallel track from the outset. Nussbaum always doubted Dr. Fauci's authenticity: "Fauci's conversion," he concluded, "smacked of opportunism." Subsequent history, including the history we are living today, supported Nussbaum's cynical assessment.[75]

AIDS activists afterward learned that at the same time Dr. Fauci was telling them and Senator Kennedy's office that he was finally testing AL 721, Teflon Tony was confiding to his Pharma PIs that he had rigged the AL 721 studies to fail: "I wanted to debunk it," he reassured them.[76] Just as he would do with hydroxychloroquine during the COVID crisis thirty years later, he designed his AL 721 clinical trials in a way that would ensure their failure and thus discredit the unpatentable medicine. Dr. Fauci told the Burroughs Wellcome PIs who dominated his "independent" committee, "Let's put the thing into trial and get it over with once and for all."[77]

Nussbaum's verdict: "If there was any chance for a fair test for AL 721, it wasn't going to come from Tony Fauci's clinical trials system."[78]

At first, his devious plan backfired. Instead of debunking AL 721, the NIAID study confirmed that AL 721 stopped viral replication. When those promising results began emerging, Dr. Fauci and his PIs cancelled the trial, making sure that AL 721 never went to Phase 2. Dr. Fauci told skeptical activists that he could not get any volunteers to enroll in the study. (In 2021, he would invoke the same bunko to kill NIAID's ivermectin trials.)

Around the same time, activists realized that Dr. Fauci's vows to test aerosol pentamidine—which he admitted before Congress was effective—were a subterfuge. Dr. Fauci opened clinical trials for aerosol pentamidine but again claimed, disingenuously, that he couldn't populate them. Dr. Fauci's sandbagging finally prompted frustrated HIV activists to finance and conduct their own trial of aerosol pentamidine. Completed in 1990, that study demonstrated the drug's clear effectiveness against PCP. "The data had not been generated out of Tony Fauci's multimillion-dollar drug-testing system," Callen recalled. "That [Fauci's] system has not been able to enroll a single person in its trials of aerosol pentamidine. The HIV community and community doctors generated the data. A private company, LyphoMed, funded the study." Said Nussbaum, "The community has rolled up its sleeves and done an end run around federal incompetence and indifference."[79]

Nussbaum points out that even at the height of Dr. Fauci's "conversion," NIAID continued to ignore hundreds of other effective drugs for opportunistic diseases because "PIs have their own scientific agenda, which is not necessarily the same as the country's."[80]

Dr. Fauci's whole charade ended the moment the FDA approved AZT.

By then, Dr. Fauci had rigged the key committees that controlled drug approvals at NIH and FDA by stacking them with academic and industry scientists and doctors from his PI system: "Scientists who . . . made their entire careers in AZT . . . sat on committees voting on potential commercial competitors. Scientists who have had financial dealings with Burroughs Wellcome or other pharmaceutical companies have come to dominate the government's entire clinical trials network."[81]

While they actively stymied clinical trials for aerosolized pentamidine and AL 721, Dr. Fauci's insider's cabal greased the skids, allowing Burroughs Wellcome to skip animal testing and to proceed directly to human trials. This omission was unprecedented in the history of chemotherapy drugs, but again foreshadowed the decision to allow the Pfizer/BioNTech COVID-19 vaccine to proceed to human testing without completing the usual panel of safety testing in animal models.[82] Government researchers had thoroughly assessed AZT's frightening toxicity, including its lethal effects on rodents after short-term exposures with minuscule doses. Neither NIAID nor Burroughs Wellcome ever completed any long-term animal study. Burroughs Wellcome financed Dr. Fauci's fast-tracked human trials, fragmenting their study groups in twelve cities into small cohorts, making safety signals difficult to detect.

In 1987, Dr. Fauci's team declared the human study a success and terminated it after four months of a proposed six-month study—a record-setting speed for chemotherapy approval. That four-month observation period was far too short for researchers to detect side effects that would occur in patients taking AZT for years, or even a lifetime. But Dr. Fauci argued that his decision to abort the study was the only ethical choice: after sixteen weeks, nineteen trial subjects in the inactive placebo group and only one participant from the AZT group had died—an outcome that could be hailed as an extraordinary 95-percent efficacy! Dr. Fauci said that those results proved AZT safe and effective against AIDS. Even more importantly for Burroughs Wellcome shareholders, Dr. Fauci cleared AZT for use on healthy HIV-positive people, meaning people with no symptoms. Following those brief clinical trials, FDA granted AZT fast-tracked Emergency Use Approval in March 1987.

A Moment of Triumph

For Dr. Fauci, the FDA licensure was a moment for exultation. After years of humiliation and failure with his critics pounding him against the ropes, he finally had something to show: a double-blind, placebo-controlled study of 3,200 people, which allegedly showed that AIDS patients receiving AZT survived at rates exponentially higher than those denied the treatment. Dr. Fauci now had a product that validated his clinical trial system. At this first whiff of AZT's success, even before his AZT study was published, the young technocrat seized the moment to do what he always did best. He called a press conference.

Two years later, Dr. Fauci would reminisce about those halcyon days: "When I first got involved in AIDS research, I was reluctant to deal with the press. I thought it was not dignified."[83] There is, in fact, little evidence of that reticence in the public record. From the outset, Tony Fauci seemed almost desperate for such indignities.

Dr. Fauci launched his media blitz with an unprecedented action: At ten o'clock in the morning following his evening receipt of the initial study results, Dr. Fauci began personally calling key journalists to announce his triumph. "No director of an NIH institute had ever contacted the press like that," says Nussbaum.[84] Traditionally, the NIH director himself made major announcements, but Dr. Fauci was apparently unwilling to share the glory with his nominal boss, NIH Director James Wyngaarden, or with HHS Secretary Otis Bowen. In making his proclamation, Dr. Fauci employed the gimmick that he watched Robert Gallo pioneer during his premature announcement of Gallo's study linking HIV to AIDS. That announcement had shattered another tradition: Historically, agencies didn't announce the results of clinical trials until the data were peer-reviewed and published so that journalists—and the scientific community—could read the study and reach their own conclusion about what the science said. Gallo had trailblazed the technique of "science by press release" four years earlier, when he had staged an HHS press event to announce that the probable cause of AIDS had been found, a retrovirus that would later be named the "Human Immunodeficiency Virus" or "HIV." The press reported Gallo's discovery as scientific fact, even though Gallo had not published a peer-reviewed paper supporting his enormously consequential assertion. Here was a useful innovation that allowed regulatory officials to craft and control the public narrative from inception. The science was what the regulators declared it to be. There could be no opportunity for journalists to read the ambiguous data, consider contrary expert opinion, or second-guess official pronouncements.

Dr. Fauci made himself the virtuoso of this technique, displaying it, at its apogee, during his April 28, 2020, announcement of remdesivir's miraculous performance during NIAID's rigged and fraud-tainted clinical trials, while seated on an Oval Office couch beside President Trump. He had no peer-reviewed or published study, no authentic placebo trial, no data, and not even a handout for the press. With this vague hearsay claim, he forced through Emergency Use Authorization for his darling drug and sold Gilead's entire inventory to the president without publishing a word or ever leaving the sofa.

Under Dr. Fauci's leadership, this practice would become a routine vehicle for extreme abuse in the COVID-19 era, when vaccine companies habitually disclosed cherry-picked highlights of their clinical trials in press releases weeks before publishing far less bullish study results. Those tactics drew criticism as "pump and dump" enterprises with company executives simultaneously unloading stock timed with deceptive announcements that drove up share prices. At least one case—Dr. Fauci's Moderna vaccine—prompted a federal securities investigation.[85]

Using the same extravagant language he would later apply to remdesivir, Dr. Fauci boasted to reporters that his trial had produced "clear-cut evidence" that AZT "saved lives." Any reporter who wanted to cover the story for the evening news had to take his word for it. And then, as now, some people simply couldn't conceive that Anthony Fauci would lie or exaggerate. Dr. Fauci giddily declared that his agency would recommend AZT not only for individuals with full-blown AIDS, but for asymptomatic people who had tested positive for HIV but showed no sign of AIDS. He never mentioned that AZT cost $10,000 for annual treatment—only that Burroughs Wellcome would sell it for $500/bottle. The FDA approval meant the taxpayers would subsidize AZT's costs.

Burroughs Wellcome's shares soared 45 percent on Dr. Fauci's announcement, adding 1.4 billion pounds to the company's UK stock market value in one day.[86] The company's CEO predicted that AZT profits would bring in over $2 billion per year.[87]

The PIs had handed NIAID its first successful drug trial. Dr. Fauci was now in the clear and he knew that the PIs had pulled his chestnuts from the fire. Not only had they given him a blockbuster AIDS drug, they had also built him a tried-and-tested system for producing future drug approvals. He no longer needed to pander to the CRI doctors. Dr. Fauci wasted no time in putting an end to his parallel-track charade.

When Dr. Fauci abandoned the CRI system, NIAID just as quickly lost its brief interest in patient care or in testing repurposed new drugs against the opportunistic infections that killed people with AIDS. NIAID went back to its comfortable niche nurturing pharmaceutical blockbusters. "It was the same old story," recounts Nussbaum. "Nothing had changed for years."[88]

There was only one problem: Dr. Fauci's entire clinical trial for AZT had been an elaborate fraud.

A Moment of Truth, Uncovering the Fraud

In July 1987, the *New England Journal of Medicine* (*NEJM*) finally published Burroughs Wellcome's official report on the Phase II AZT trials—the so-called "Fischl study"—which was the basis of the FDA's approval of AZT.[89] Outside scientists finally had the chance to scrutinize the study's details for the first time. Many had earlier expressed shock at its abbreviated duration, but now they began to uncover evidence of fatal methodological flaws—some attributable to confirmation bias, but others clearly the product of corruption, and deliberate falsification. Within days, reporters, researchers, and scientists began lobbing aspersions on Dr. Fauci's Pollyannaish and self-serving interpretation of the data. European scientists complained that NIAID's raw data showed no benefit of reducing symptoms, a finding that threatened Glaxo's biggest anticipated profit pool. The Swiss newspaper *Weltwoche* termed his AZT trials a "gigantic botch-up."[90,91]

Investigative journalist and market research analyst John Lauritsen, who had covered the AIDS crisis since 1985, became the first intrepid journalist to critically analyze the details of the AZT trials. When he saw the *NEJM* reports, he quickly realized that the research was invalid. In his first AZT article, "AZT on Trial" (19 October 1987), he wrote: "The description of methodology was incomplete and incoherent. Not a single table was acceptable according to statistical standards—indeed, not a single table made sense. In particular, the first report, on 'efficacy,' was marred by contradictions, ill-logic, and special pleading."[92] He telephoned the nominal authors of the report, Dr. Margaret Fischl and Douglas Richman, and spoke to each for half an hour: "Neither one of them could explain the tables in the reports that they themselves had allegedly written." They could only say that he should call Burroughs Wellcome for answers to his questions.

The *New York Native* published Lauritsen's reports beginning in 1987. These reports later appeared in two books, *Poison by Prescription: The AZT Story* (*Poison*) (1990) and *The AIDS War: Propaganda, Profiteering and Genocide from the Medical-Industrial Complex* (*TAW*) (1993).[93]

Eighteen months after AZT's approval, FDA conducted its own investigation of the study. For many months, the FDA, cowering before Fauci's bullying, kept its damning reports secret. The most shocking revelations about Dr. Fauci's systemic conduct would emerge after Lauritsen finally obtained some five hundred pages from the FDA investigators' trove of documents, using the Freedom of Information Act. Those papers clearly demonstrated that the Fauci/Burroughs Wellcome research teams had engaged in widespread data tampering, which some have viewed rose to the level of homicidal criminality.

These documents showed that the "double-blind, placebo-controlled" trials had become unblinded almost immediately, which alone rendered them invalid. Internal FDA communications with the research team revealed rampant falsification of data, sloppiness, and departure from accepted procedures.[94]

In one of the Freedom of Information Act (FOIA) documents, Harvey Chernov, the FDA analyst who reviewed the pharmacology data, recommended that AZT should not be approved. Chernov noted many serious toxicities of AZT, especially its effect on the blood: "Although the dose varied, anemia was noted in all species (including man) in which the drug has been tested." Chernov further noted that AZT is likely to cause cancer: "[AZT] induces a positive response in the cell transformation" assay and is therefore "presumed to be a potential carcinogen."[95]

The Phase II trials were supposed to last for twenty-four weeks, but Wellcome and Dr. Fauci aborted them at the halfway point. The investigators claimed that AZT was miraculously prolonging the lives of those taking it. Lauritsen analyzed the mortality data and concluded that they were certainly false. Although few patients finished the full twenty-four weeks of treatment, and two dozen lasted less than four weeks on the drug, the investigators analyzed the skimpy data anyway, using bizarre statistical projections to forecast the probability of a patient's experiencing various opportunistic infections if the trials had continued as planned. Lauritsen scathingly comments: "This is analogous to estimating the probability of developing arthritis by the age of seventy, using a sample in which only a few people had reached this age, and in which some were still teenagers."

Most seriously, FDA investigators found a great many instances of cheating in the

Boston center where they began their review. Dr. Fauci's decision to terminate the trials prevented the inspectors from investigating the other eleven centers, which were, presumably, just as dreadful as Boston. After agonizing over whether to exclude data from the delinquent Boston center or from patients with protocol violations, the FDA decided to exclude nothing: "False data were retained. Garbage was thrown in with the good stuff." The FDA argued that if all the false data were excluded, there would be an insufficient number of patients left to complete the trials. Lauritsen pointed out that FDA's knowing use of false data constituted fraud.[96]

In 1991, four years later, Lauritsen filed a Freedom of Information request asking for various FDA documents pertaining to the Phase II AZT trials—most importantly, the "Establishment Inspection Report" on the Boston center, written by FDA investigator Patricia Spitzig. After months of lies, evasions, and obstructions from the FDA, a courageous female FDA whistleblower breached all the stonewalling and saw to it that Lauritsen got the Spitzig Report.[97] It was a bombshell:

> As it turned out, the Boston Principal Investigators (PIs) cheated on almost every patient. The Burroughs Wellcome PIs had quickly realized that AZT was so reliably deadly that they were hard-pressed to keep the trial recruits alive for the full six-month study. The Boston team solved this dilemma by lying about the length of time patients were in the trials. The company incentivized this sort of fraud by paying its PIs according to how many months they kept the AZT trial subjects alive. "Simply put," says Lauritsen, "the doctors received a great deal more money," from longer-term enrollments.

Pharma PIs know that their careers and paychecks depend on their ability to consistently produce study outcomes that will win FDA approval for the subject drug. Such perverse incentives naturally drive research bias, confirmation bias, data tampering, strategic laziness, and deliberate falsification and cheating. PIs routinely covered up adverse events, violated protocols, falsely reported AZT patients as being placebo patients, and lost control of the test product.

FDA based its AZT approval on Case Report Forms (CRFs) filed by Burroughs Wellcome PIs, who each had compelling financial and career inducements to downplay injuries to achieve a successful trial. However, there were also reams of shocking information in the medical records of private physicians, hospitals, and the diaries of patients that contradicted the crisis. In virtually every patient, the FDA's Spitzig found serious discrepancies between the medical records and what the PIs had entered on their CRFs.

The rules of the trials clearly stated that the PIs must record all adverse reactions on their CRFs and report immediately to the FDA. The Boston PIs did neither.

The FDA documents showed that the PIs knew very well which patients were on AZT and which on placebo, that they were skewing safety results in AZT's favor to give advantage to the AZT participants. Researchers began by placing the sickest patients in the placebo group. The researchers then bent over backward to coddle the group that took AZT, giving them more supportive medical services than the placebo subjects. For example, individuals taking AZT during the four-month study received six times more blood transfusions than the placebo group.

Of those who got AZT, all suffered from its unspeakable toxicity. "A number of them . . . would very definitely have died from anemia," had the PIs not given the blood transfusions to keep them alive, says Lauritsen. AZT causes anemia in every animal species ever studied, including human beings. In his book, *Poisoned by Prescription*, Lauritsen explains how "[p]atients taking AZT became anemic, and suffered low white blood cell counts accompanied by vomiting." FDA's documents showed that everyone in the AZT group suffered severe toxicities and anemia, yet NIAID's official report listed no adverse effects among AZT recipients.

Some of the AZT patients suffered adverse reactions so deadly that they needed multiple blood transfusions just to keep them alive. Dr. Fauci's crooked researchers pumped these individuals with regular blood transfusions and then neglected to record their multiplicity of health problems. In the AZT group, thirty patients—over half the total—clung to life until the end of the study only with help from multiple blood transfusions. In each case the Boston PIs checked "no adverse reactions" on the CRFs. Some 20 percent received multiple transfusions. In the placebo group, on the other hand, only five patients received transfusions.

"What happens when you get a blood transfusion?" asks noted AIDS researcher and author Dr. Robert E. Willner, MD, PhD. "You look better, you feel better, and you live a little bit longer. But the most important question and lesson from all of this, you must ask the question: Why do those on AZT need six times more transfusions in a four-month period than the individuals on the placebo? Because you're dealing with a killer drug. . . ."[98]

"Many of the patients would have died from the toxicities of AZT if they had not been given emergency blood transfusions," reports Lauritsen. "This is a serious adverse effect. That means literally that they would have died from the poison. And yet the case report forms that showed up eventually would report no adverse effects. I mean, this is a type of dishonesty. It's hard to go any further than that."

Dr. Willner, who died in 1995, accused Dr. Fauci of using transfusions and other artifices to systematically conceal AZT's horrendous toxicity. "What do we have to say about the National Institutes of Health, when a private, independent laboratory, found AZT to be 1,000 times more toxic than the laboratory of the NIH? We can understand a 5 percent error in a laboratory, even a 10 percent error, but a 10,000 percent error or a 100,000 percent error? That's fraud."[99]

One typically appalling item in Spitzig's report concerned Patient #1009, who was already taking AZT and was therefore ineligible to participate in the clinical trial. The Boston PIs nevertheless illegally entered him in the study and assigned him to the placebo group, although he never stopped taking AZT. He suffered typical AZT toxicities including severe headaches and anemia, dropped out of the study after less than a month, and died two months later. The PIs counted him as a death in the placebo group. Lauritsen wrote: "Further comment would be superfluous. If this is not fraud, the word has no meaning."[100,101]

Even in that innocent era, the United States mainstream media heavily censored journalistic criticism of Dr. Fauci and the corruption in the AZT studies. Most Americans were therefore unaware of any dissent from the AIDS orthodoxy. This was less true in Europe and the UK. On February 12, 1992, Channel 4 Television in London broadcast a documentary, "AZT: Cause for Concern." Produced by Meditel, the film described the material from the FOIA documents, exposed the crooked AZT trials as rank fraud, and chronicled the terrible toxicities of the drug. The next day, the charity, Wellcome Foundation, divested itself of most of its stock in Wellcome Pharmaceuticals, the parent company of Burroughs Wellcome, the manufacturer of AZT. Burroughs Wellcome stocks plunged, and the company suffered a series of hostile takeovers by SmithKline Beecham and then by Glaxo. Millions around the world viewed the UK documentary, but neither it nor any of the Medical AIDS-critical documentaries have ever been broadcast in the US.[102]

AZT is the most toxic drug ever approved for long-term use. Molecular biologist Professor Peter Duesberg has explained AZT's mechanism of action: It is a random terminator of DNA synthesis, the life process itself. Dr. Joseph Sonnabend stated simply: "AZT is incompatible with life."[103]

On January 27, 1988, NBC News broke the censorship blockade to broadcast the first of reporter Perry Peltz's three-part exposé on the AZT Fischl trial.[104,105] Peltz reported additional evidence of widespread tampering with the rules and the pervasive cheating, which she discovered had started on day one. Peltz learned that Fauci's claim that the study was double-blind was a wholesale canard and reported that most volunteers knew who was on the drug and who wasn't. Since everyone was desperate for the "miracle drug," the volunteers on AZT admitted to sharing their drug with placebo group members. This practice assured that researchers would get no clean results from either cohort. Furthermore, Peltz learned both placebo and study subjects were taking other drug regimens they obtained by purchasing remedies from buyers' clubs. Peltz was practicing understatement when she branded NIAID's AZT experiments as "seriously flawed."

Dr. Fauci loves the media spotlight, but only when the pitcher is throwing softballs. Peltz closed her report with a pointed comment: "When preparing this report, we repeatedly tried to interview Dr. Anthony Fauci at the National Institutes of Health. But both Dr. Fauci and Food and Drug Administration Commissioner Frank Young declined our request for interviews."[106] When Lauritsen saw the NBC broadcast, he commented, "Welcome to the club, Perri!" Fauci also refused to speak to the BBC, Canadian Broadcasting Corporation Radio, Channel 4 Television (London), Italian television, *The New Scientist*, and Jack Anderson. All these outlets had expressed skepticism about the Fischl report.[107]

Of course, Dr. Fauci remained a constant presence on the more obeisant media outlets. Despite years of ineptitude and catastrophe, he has managed to survive by cultivating credulous journalists who do not ask critical questions and give him free rein to broadcast self-serving propaganda. Furthermore, he had already become a master at persuading media outlets against giving platforms to his critics, a technique that served him well in 2020 and 2021.

By September of 2021, Dr. Fauci's power to muzzle his critics had achieved a mastery over free expression unprecedented in human history. That month, with a single phrase, Dr. Fauci silenced pop icon Nicki Minaj after she questioned whether COVID vaccines might be causing problems involving testicular swelling. When CNN's Jake Tapper asked him about Minaj's claim, Dr. Fauci simply declared, "The answer to that, Jake, is a resounding no."[108] As usual, he cited no study to support this assertion. The vaccine manufacturers acknowledge that the products are not tested for effects on fertility.[109] [110] Nevertheless, based upon Dr. Fauci's word alone, Twitter immediately evicted Minaj from its platform, censoring her communication with her 22 million followers. Pharma's obedient attack dogs CNN, CBS, and NBC rushed on to the dog pile to defame and discredit the rapper and to assure the public that Minaj was wrong. Dr. Fauci, after all, had spoken!

On February 19, 1988, Dr. Fauci appeared with hosts Charles Gibson and Joan Lunden on ABC's flagship television program, *Good Morning America*. His appearance was part of a propaganda blitz of the friendly media platforms to resurrect himself and AZT from the all-out assault by scientists and independent reporters like Lauritsen and Peltz.[111] Initially, *GMA* invited Dr. Fauci's most vocal and credible nemesis, perhaps the world's leading virologist, Berkeley professor Dr. Peter Duesberg, to appear on its show. Duesberg, who had at that date received more NIH grants than any other scientist, was enraging his benefactor agency by claiming that AZT was not just worthless, it was killing more people than AIDS. Duesberg had flown across the country to appear. On the evening before his scheduled appearance, *GMA*'s producer called Dr. Duesberg in his Manhattan hotel room to inform him that the show had been cancelled. The following morning, Duesberg awoke to watch Dr. Fauci promoting AZT and defending his study on *GMA*, unchallenged. This was, by then, a common motif for Dr. Fauci—his gift at strong-arming obsequious, slavish, credulous reporters to silence critics and to shield him from debate. The fawning *GMA* hosts asked Dr. Fauci why only one drug, AZT, had been made available. He replied: "The reason why only one drug has been made available—AZT—is because it's the only drug that has been shown in scientifically controlled trials to be safe and effective."[112] The sycophantic *GMA* team, characteristically, accepted Dr. Fauci's statement as gospel. Almost all of Dr. Fauci's claims in that broadcast were lies.[113]

Lauritsen points out that "this brief statement contains several outstanding falsehoods": "First, there have been no 'scientifically controlled trials' of AZT; to refer to the FDA-conducted AZT trials as 'scientifically controlled' is equivalent to referring to garbage as la haute cuisine. Second, AZT is not 'safe': it is a highly toxic drug—the FDA analyst who reviewed the toxicology data on AZT recommended that it should not be approved. Third, AZT is not known objectively to be 'effective' for anything, except perhaps for destroying bone marrow."[114]

Only thirty-three years later did Dr. Fauci finally concede that AZT's performance in his ballyhooed clinical trials—ostensibly saving lives at a 19-1 ratio—was actually less than stellar. Ironically, his delayed confession arrived just as Dr Fauci was minting a new whopper. In May 2020, during the White House meeting where he pronounced the miraculous efficacy of Gilead's antiviral remdesivir—another beneficiary of Dr. Fauci's manipulations—he admitted, "The first randomized placebo-controlled trial with AZT . . . turned out to give an effect that was modest."[115]

That's not what he said at the time. In 1987, he claimed that AZT was 95 percent effective; nineteen had died in the placebo group and only one in the AZT group.[116] In 2020, based on equally flimsy and contrived evidence, he made similar claims for his lethal remedy, remdesivir, and his dubious Moderna vaccine.

The media's reportage of AZT in the late 1980s almost universally lamented the cruelty of AZT's astronomical costs that ranged between $8,000 and $12,000, not counting the cost of the required blood transfusions when patients' platelets plummeted. Anthony Fauci solved this problem by making AZT "standard of care" for otherwise-healthy people with no AIDS symptoms who nevertheless were diagnosed with HIV via PCR tests. In 1989, when Dr. Fauci recommended universal testing, the *LA Times* dutifully gushed that AZT could "benefit about 600,000"[117] of the estimated 1.5 million HIV-positive people in the country. Dr. Fauci promised these healthy Americans that taking AZT could delay their inevitable death sentences and would "have the broadest impact of any of the therapeutic advances shown in recent years to prolong the lives of patients with AIDS or HIV infection."[118] The *New York Times's* Philip J. Hilts uncritically reported that everyone should now get tested: "Dr. Fauci, the director of the National Institute of Allergy and Infectious Diseases . . . said that now people who are at risk for AIDS, even if they have 'absolutely no symptoms,' it behooves them to get themselves tested."[119] The resultant flood of additional customers clamoring for the drug significantly expanded the AZT market, allowing Burroughs Wellcome (now GlaxoSmithKline) to lower per-unit costs.

No mainstream media outlet told the public about the behind-closed-doors meetings, where FDA green-lighted Dr. Fauci's sketchy new initiative. The meetings' transcripts reveal the deep anxieties of the FDA panelists, who worried that they had no idea if AZT might actually help healthy people, or whether it may, perhaps, kill them. Among all the American journalists covering the AIDS beat, only Celia Farber showed curiosity about the particulars of this milestone debate. In 1989, she quoted from the FDA transcript in an article titled "Sins of Omission," in *SPIN*:

> Everybody was worried about this one. To approve AZT, said Ellen Cooper, an FDA director, would represent a "significant and potentially dangerous departure from our normal toxicology requirements." One doctor on the panel, Calvin Kunin, summed up their dilemma. "On the one hand," he said, "to deny a drug which decreases mortality in a population such as this would be inappropriate. On the other hand, to use this drug widely, for areas where efficacy has not been demonstrated, with a potentially toxic agent, might be disastrous."
>
> "We do not know what will happen a year from now," said panel chairman Dr. Itzhak Brook. "The data is just too premature, and the statistics are not really well done. The drug could actually be detrimental." A little later, he said he was also "struck by the facts that AZT does not stop deaths. Even those who were switched to AZT still kept dying."
>
> "I agree with you," answered another panel member, "There are so many unknowns. Once a drug is approved there is no telling how it could be abused. There's no going back."[120]

By invoking the "people are dying argument" to rush through AZT's licensing for healthy Americans, the FDA's drug approval process was decimated. Farber told me "the idea that complying with the normal safeguards of the regulatory process and taking time to prudently study a drug for safety or efficacy was artfully conflated with murder." In that sense, an unbroken devolution of FDA's regulatory function leads from AZT to the fraud-fueled "Emergency use approvals" of remdesivir and the Moderna mRNA vaccine during the COVID pandemic.

"The death blow to FDA's safety function was AZT," says Farber. "After that, any potentially deadly disease became an excuse for curtailing clinical trials. Death by medication was normalized as an inherent part of progress." All those poisoned Americans were just unfortunate casualties in Little Napoleon's noble war against the germs.[121]

Dr. Fauci's fraud persuaded hundreds of thousands of people to take AZT. For many of them, it was a lethal choice. In 1987, AZT became the AIDS "therapy" even though in the recommended dosage of 1,500 mg/day, it was absolutely fatal.[122] Throughout the 1980s, the average lifespan of a patient on AZT was four years. The life expectancy only began to increase in 1990, when the FDA lowered the recommended dosages from 1,200 mg/day to 600.[123] The quality of life on AZT was universally pretty miserable. Many credible scientists argued that AZT was killing more people than AIDS. Lauritsen estimated that AZT killed 330,000 gay men between 1987 and 2019.[124] Many of the dead were perfectly healthy before beginning the AIDS regimen. Absent AZT, Lauritsen says, the vast majority of those men would not have died.

Fast-Track Template

AZT's record-setting race to approval did not stand for long. By 1991, Dr. Fauci had effectively abandoned testing low-profit repurposed drugs in the parallel track CRI program. But he used a parallel track to open a loophole in the FDA drug approval system, a loophole large enough to drive through truckloads of Pharma's new high-profit patented antivirals. Using CRI's relaxed rules, Dr. Fauci and his Pharma partners shattered a series of new speed records at FDA. Still smarting from the public roasting Dr. Fauci had administered to them, bedraggled and bullied FDA officials lowered agency standards to green-light Dr. Fauci's dark pharmacopoeia of deadly chemotherapy drugs with minimal safety testing. That year, exploiting the regulatory breach he had created with CRI's fast-track system, Dr. Fauci waved through another DNA chain antiretroviral terminator drug to quick approval, allowing it to skip the double-blind placebo testing he had previously declared indispensable. NIH had developed and patented didanosine (ddI) before licensing it to Bristol Myers Squibb.[125] Didanosine won FDA approval without even a pretense of a placebo-controlled study. The drug had so many debilitating and lethal side effects that FDA, in an uncharacteristic act of civil disobedience against NIAID's diminutive dictator, issued a black box warning. Nevertheless, desperate HIV-infected Americans rushed like doomed lemmings to take the drug. In 2010, FDA issued a statement that ddI can cause potentially a fatal liver disease called non-cirrhotic portal hypertension.[126] Even with its demonstrated toxicity, Dr. Fauci used CRI parallel-track process to bypass the usual controls, to win approval for use of ddI in pregnant mothers who test positive for HIV. A 2019 study [Hleyhel et al., *Environ Mol Mutagen* (2019)[127]] found that ddI accounted for 16 percent of prescriptions for infected mothers and 30 percent of the cancers in their children.

In 1996, Dr. Fauci used his expedited fast track to break another record by winning FDA approval for Merck's HIV antiviral Crixivan; this time it took only six weeks.[128] Dr. Fauci achieved that feat by allowing Merck to run Crixivan through a skeleton CRI process on a tiny cohort of ninety-seven volunteers in three groups, thereby winning the swiftest approval in history: forty-two days. That approval prompted open revolt by the AIDS community, which felt betrayed when Merck hiked up the price of the drug. Activists led by the Treatment Action Group condemned Merck's misuse of the CRI exemptions to secure approval for its deadly and ineffective drug.

In 2016, Dr. Fauci boasted that his efforts had led to the approval of some thirty new drugs to treat HIV/AIDS.[129] Dr. Fauci called this "extraordinary" accomplishment "one of the most important transformative discoveries in biological sciences."

These drugs generated billions of dollars in revenue for drugmakers: in 2000, global revenue from AIDS remedies was $4 billion; by 2004, it jumped to $6.6 billion. In 2010, AIDS drugs cracked the $9 billion mark[130] for pharmaceutical giants and topped $30 billion in 2020.[131]

"On the surface of AIDS, what the public sees, is a benevolent exterior, devoted to 'saving lives,' of originally mostly gay men in the west, then, since they shifted the narrative, primarily Africans. A global apparatus now worth over $2 trillion and composed of more NGOs, more organizations than anybody could count, obliterates all dissent, all real language, history and truth," says Celia Farber, author of *Serious Adverse Events: An Uncensored History of AIDS*. "It's a Beast system, and Fauci created it. It's not 'capitalism,' at all. It detests merit, standards, and all the values of Western Civilization. It uses the violence of the 'woke' economy to re-cast lies as truth, and to proudly crush and block any and all dissenting voices. It does this always in the name of 'saving lives.' Only now, with COVID, are Americans able to see Fauci's cold, ruthless face behind the mask. Americans have tried to follow what that man has said for a year and a half now, and we who have been dealing with him for so long, we feel like: *Welcome to our nightmare.* Nothing he says makes sense, yet nobody stands over him, to reign him in. Tower of Babble. Americans are trying to make him into a benign figure, but more and more, they feel a sinking feeling. Is he a madman? Why can't we understand what he is actually *saying*, what he means? This is very unsettling, when people are as afraid as people are now, since Covid."

Aftermath

A key and enduring legacy of the AZT battle was Dr. Fauci's emergence as the alpha wolf of HHS. His enormous budget, and multiplying contacts on Capitol Hill, the White House, and the medical industry, thereafter allowed him to influence or ignore a succession of politically appointed HHS directors and to bully, manipulate, and dominate HHS's other sister agencies, most notably FDA.

In his biography of Dr. Fauci, author Terry Michael described the drug approval system that NIAID nurtured post-AZT: "What has evolved into the HIV-AIDS industry is supported by a knowledge monopoly, comprised of federal government bureaucratic authorities led by Dr. Anthony Fauci, who hands out billions of dollars in research grants, who collude with crony capitalists from international pharmaceutical cartels, who distribute billions to AIDS advocacy non-profits, and whose official stories are communicated to the public by a science-illiterate mass media. With few exceptions, it is a media populated by journalists who don't even attempt to understand the science. These journalistic interpreters of those they label scientists are pawns in the hands of authorities in long-sleeved, white laboratory coats. That chief authority about HIV-AIDS, Dr. Fauci, has tightly held the purse strings on all HIV-AIDS research since he was appointed head of the NIAID in November 1984."[132]

As Michael suggests, the unique skill sets that allowed Dr. Fauci's extraordinary longevity and continuing public credibility—despite his miserable record of preventing and managing chronic and infectious disease—were his gifts for weaponizing media relationships, magically deploying journalists to promote his self-serving narratives, and relentlessly silence dissent.

"Dissent was effectively shut down in mass-mediated public discourse," Michaels observes. "And it was scrubbed from peer reviewed science and medical journals, which reap significant revenue from drug company advertisements for anti-retroviral drugs. Journal revenue is also derived from expensive annual subscriptions, purchased with funds from tens of thousands of HIV-AIDS related grants, funded by US taxpayers—if approved by Anthony Fauci."[133]

Template

His success at using the AIDS crisis to bring a deadly, toxic, and ineffective AIDS drug to market taught Dr. Fauci some key career lessons that he would faithfully repeat again and again and again throughout his long regime.

During his battle to win FDA approval for AZT, Dr. Fauci pioneered the strategies upon which he would build his career and then showcase for the world during the COVID epidemic. These include:

- pumping up pandemic fears to lay the groundwork for larger budgets and greater powers,
- incriminating an elusive pathogen,
- fanning hysteria by exaggerating disease transmissibility,
- periodically stoking waning fear levels by warning of mutant super-strains and future surges,
- suggesting substantial changes in how people live, ostensibly to save their lives,
- keeping the public and politicians engaged through confusing and contradictory pronouncements,
- using faulty PCR and antibody tests and manipulating epidemiology to inflate non-verifiable case and death numbers, to maximize the perception of an imminent calamity,
- ignoring and dismissing effective off-the-shelf therapeutic remedies,
- directing energy and money toward profitable new patented drugs and vaccines,
- championing dangerous and ineffective drugs originating in government laboratories as the only winning solution to end the pandemic,
- funding and orchestrating confirmation-biased research to validate his chosen remedy,

- partnering with large pharmaceutical companies and giving his partners advantages in the race for approval,
- allowing preferred companies to skip key testing metrics,
- curtailing clinical trials to conceal severe safety and efficacy problems,
- sabotaging, discrediting, and sweeping aside more effective therapies, antiretrovirals, off-the-shelf remedies, and non-patentable medicines that might compete with his new patented antiretrovirals and vaccines,
- subjecting competitive products to efficacy and safety studies that are designed to fail,
- allowing thousands of sick patients to suffer and die by denying them access to demonstrably effective competitive remedies, by publicly protesting the existing remedies were not subject to "randomized placebo testing,"
- controlling the key "independent" committees (DSMB, VRBPAC, ACIP) that approve and mandate new drugs by populating them with his own hand-picked PIs,
- presenting these agencies as "independent" and trustworthy experts,
- using the Emergency Use Authorization to fast-track the concoctions through a rigged approval process to market,
- using official government propaganda to market his concoctions,
- employing "Science by Press Release" to control narratives,
- making exaggerated claims for the efficacy of his products,
- using pervious and ineffective post-marketing surveillance systems to conceal mass injuries and deaths from the public,
- papering over all these testing deficiencies by crafting and promoting enduring narratives about the benefits, safety, and efficacy,
- citing "leading experts" to promote hypotheses that are practically never scientifically verified with peer-reviewed studies or appropriate controls,
- allowing pharmaceutical companies to charge Medicare, government programs, and insurance companies inflated prices bearing no relationships to cost,
- ensuring that research funding is restricted to projects supporting the dogma, excluding research into alternative hypotheses,
- preventing debate and censoring dissenting voices in popular media, social media, and scientific publications, and
- promising ultimate salvation with vaccines.

In addition, Dr. Fauci honed the skill of always speaking with authority—even when making contradictory assertions with no scientific basis—to rapidly reshape all government pronouncements into dogma, efficiently perpetuated in a quasi-religious manner by the media.

By repeatedly using these formulas for fifty years, Fauci directed his agency away from its core responsibility—basic research on infectious, allergic, and autoimmune diseases that have become epidemic since he took over NIAID—and transformed his agency into a profit-making appendage for itself and for Big Pharma.

Mark Twain once observed that "It's easier to fool people than to convince them that they have been fooled." AIDS activist Christine Maggiore lamented this feature of human gullibility when she assessed the mendacious fifty-year travesty of corrupted public health research that Tony Fauci put in motion during the 1984 AIDS crisis: "Commercial interests are definitely part of the problem here, and it's also our collective inability or challenge to say, 'All this time, all these years, all these lives, all these billions and billions of dollars. Can we just stop a second and go back to the very beginning and make sure we got this right?' I mean, that is so hard to do. People don't even know it's a lie. It's not so much a lie as business as usual."[134]

Endnotes

1   "The First AIDS Drugs," National Cancer Institute *Center for Cancer Research – Landmarks* (2017, P 18), https://ccr.cancer.gov/sites/default/files/landmarks_2017_web-508.pdf
2   Peter H. Duesberg, *Inventing the AIDS Virus* (Regnery Publishing Inc. 1996), 315.
3   Ibid, 309.
4   Simon Garfield, "The Rise and Fall of AZT: It was the drug that had to work. It brought hope to people with HIV and AIDS, and millions for the company that developed it. It had to work. There was nothing else. But for many who used AZT—it didn't." *Independent* (Oct. 23, 2001, 02:44), https://www.independent.co.uk/arts-entertainment/rise-and-fall-azt-it-was-drug-had-work-it-brought-hope-people-hiv-and-aids-and-millions-company-developed-it-it-had-work-there-was-nothing-else-many-who-used-azt-it-didn-t-2320491.html
5   Marlene Cimons, "Who Pays for Experimental Drugs? AIDS Drug AZT May Cost Patients $10,000 a Year," *Los Angeles Times* (Mar 11, 1987), https://www.latimes.com/archives/la-xpm-1987-03-11-mn-10107-story.html
6   Marcia Angell, *The Truth About Drug Companies: How They Deceive Us and What to Do About It* (Random House Trade Paperbacks, 2005), 26.
7   "Orphan Drug Amendments of 1991: Hearing Before the Committee on Labor and Human Resources," United States Senate, 102nd Congress, Second Session, on S. 2060, Tuesday, March 3, 1992, Volume 4, p 181, https://books.google.com/books?id=EV2X1EZ_fT0C&printsec=frontcover#v=onepage&q&f=false
8   Angell, op. cit., 26
9   Antiretroviral Drug Discovery and Development, NIAID, November 26, 2018, https://www.niaid.nih.gov/diseases-conditions/antiretroviral-drug-development
10   Bruce Nussbaum, *Good Intentions: How Big Business and the Medical Establishment are Corrupting the Fight Against AIDS* (The Atlantic Monthly Press, 1990), 133
11   Ibid., 120
12   Ibid.
13   Charles Linebarger, "Guerrilla Clinics and Buyers' Clubs Search for Alternative AIDS Treatments," *San Francisco Sentinel*, February 5, 1988, Vol 16 No 6, https://digitalassets.lib.berkeley.edu/sfbagals/Sentinal/1988_SFS_Vol16_No06_Feb_07.pdf
14   Robert F. Kennedy Jr., Interview with Dr. Frank Ruscetti (November 2020).
15   Nussbaum, op. cit., 139
16   Ibid., 176
17   Ibid., 146
18   Ibid., 126
19   Ibid., 149
20   Anthony S. Fauci, "The Acquired Immune Deficiency Syndrome, The Ever-Broadening Clinical Spectrum," *JAMA*. (1983) 249(17), 2375–2376, doi:10.1001/jama.1983.03330410061029
21   Randy Shilts, *And the Band Played On: Politics, People and the AIDS Epidemic* (Penguin Books, 1988), 300. https://archive.org/details/andbandplayedon00shil/page/300/mode/2up?q=fauci
22   "Evolution of an Epidemic: 25 Years of HIV/AIDS Media Campaigns in the U.S.," The Henry J. Kaiser Family Foundation (June, 2006), https://www.kff.org/hivaids/report/evolution-of-an-epidemic-25-years-of/
23   "Condom Advertising and AIDS," Hearing Before the Subcommittee on Health and the Environment of the Committee on Energy and Commerce, House of Representatives, 100th Congress, First Session, February 10, 1987, p. 79, https://files.eric.ed.gov/fulltext/ED289099.pdf
24   "AIDS Epidemic Update: December 2007," WHO (November 20, 2007), https://www.who.int/hiv/pub/epidemiology/epiupdate2007/en/
25   Rebecca Culshaw, *Science Sold Out:Does HIV really cause AIDS?* (North Atlantic Books, 2007), p. 4
26   Ibid.
27   Ibid., 91–92
28   Ibid., 92
29   Torsten Engelbrecht, Claus Köhnlein, et al., *Virus Mania: How the Medical Industry Continually Invents Epidemics, Making Billions at our Expense* (Books on Demand, 2021), p. 101
30   David Rasnick, "One Million HIV+ in USA 1986-2019. ~ Constant 1 million HIV positive in U.S. 1986-2019," davidrasnick.com, http://www.davidrasnick.com/aids/constant-one-million-hiv.html
31   Evan Thomas, "The New Untouchables: Anxiety over AIDS is verging on hysteria in some parts of the country," *TIME* (September 23, 1985), http://content.time.com/time/subscriber/article/0,33009,959944,00.html
32   Gregg Gonsalves and Peter Staley, "Panic, Paranoia, and Public Health—The AIDS Epidemic's Lessons

for Ebola," *New England Journal of Medicine* (December 18, 2014), https://www.nejm.org/doi/full/10.1056/NEJMp1413425

33   The National Institutes of Health, Dr. Anthony S. Fauci Oral History 1986 A (1986). https://history.nih.gov/display/history/Fauci%2C+Anthony+S.+1986

34   Larry Kramer, " 'We are in the middle of a plague and you behave like this!,' " AIDS Forum NYC - 1991 (Oct. 7, 1991), https://speakola.com/ideas/larry-kramer-aids-forum-1991.

35   Tribute to Anthony Fauci, Congressional Record (Dec. 19, 2007), https://www.govinfo.gov/content/pkg/CREC-2007-12-19/html/CREC-2007-12-19-pt1-PgS15994.htm

36   Robert F. Kennedy, Jr., Interview with Melisa Wallack (Aug. 2020)

37   Diane Bernard, "Three decades before coronavirus, Anthony Fauci took heat from AIDS protesters," *Washington Post* (May 20, 2020), https://www.washingtonpost.com/history/2020/05/20/fauci-aids-nih-coronavirus/

38   Nussbaum, op. cit., 119

39   Ibid., 299

40   Ibid., 120

41   Ibid., 121

42   Nussbaum, op. cit, 121

43   Ibid., 123

44   Ibid., 121

45   "Americans Need COVID Treatment NOW, States the Association of American Physicians & Surgeons *(AAPS),*" *Intrado GlobeNewswire*, October 20, 2020, http://www.globenewswire.com/en/news-release/2020/10/20/2111364/22503/en/Americans-Need-COVID-Treatment-NOW-States-the-Association-of-American-Physicians-Surgeons-AAPS.html

46   Nussbaum, op. cit.,121

47   Nussbaum, op. cit., 122

48   Nussbaum, op. cit.,191

49   "Pentamidine Wars Continue," ACT UP New York, ACT UP historical archive (Sep. 12, 1988), https://actupny.org/documents/FDAhandbook5.html

50   Dennis Hevesi, "Martin Delaney, 63, AIDS Activist, Dies," *New York Times* (Jan. 26, 2009), https://www.nytimes.com/2009/01/27/us/27delaney.html

51   "Anthony S. Fauci Oral History, AIDS Researcher," Interview with Dr. Anthony Fauci, The Miller Center Foundation and the Edward M. Kennedy Institute for the United States Senate (Sept. 10, 2007), https://www.emkinstitute.org/resources/anthony-s-fauci-oral-history-aids-researcher

52   Nussbaum, op. cit., 143

53   Nussbaum, op. cit., 270–271

54   Nussbaum, op. cit., 270

55   "Therapeutic Drugs for AIDS: Development, Testing and Availability," U.S. House of Representatives, Hearings Before a Subcommittee of the Committee on Government Operations (Apr. 28 and 29, 1988), https://play.google.com/books/reader?id=GLfLAxm5i3MC&pg=GBS.PP1&hl=en

56   Nussbaum, op. cit., 272

57   Ibid., 272

58   Ibid., 272

59   Larry O'Connor, "Fauci Would Prescribe Chloroquine to Patient Suffering From COVID-19," *Townhall* (March 25, 2020), https://townhall.com/columnists/larryoconnor/2020/03/25/fauci-would-prescribe-chloroquine-to-patient-suffering-from-covid19-n2565678

60   Nussbaum, op. cit., 271

61   Ibid., 272

62   Ibid., 272–274

63   Larry Kramer, "An Open Letter to Dr. Anthony Fauci," *The Village Voice*, May 31, 1988, 18, 20, https://www.villagevoice.com/2020/05/28/an-open-letter-to-dr-anthony-fauci/

64   Ibid., 20

65   Ibid.

66   Ibid.

67   Ibid.

68   Ibid.

69   Bruce Nussbaum, *Good Intentions: How Big Business and the Medical Establishment are Corrupting the Fight Against AIDS* (Penguin Books, 1990), 279

70   Ibid., 284

71   Ibid., 282

72   Ibid., 289

73   Ibid., 282

74   Ibid., 332

75   Ibid., 283

76   Ibid., 260

77   Ibid., 260

78   Ibid., 260

79   Ibid., 123

80   Ibid., 331

81   Ibid., 331

82   Japanese regulatory common technical document discovered by Byram Bridle et al., https://trialsitenews.com/did-pfizer-fail-to-perform-industry-standard-animal-testing-prior-to-initiation-of-mrna-clinical-trials/

83   "Will Fauci screw up Zika as much as he has screwed up HHV-6, AIDS and Chronic Fatigue Syndrome?" HHV-6 University (Aug. 21, 2016), http://hhv6.blogspot.com/2016/08/will-fauci-screw-up-zika-as-much-as-he.html

84   Nussbaum, op. cit., 314

85   Stephen Gandel, "Watchdog Urges SEC to Investigate Vaccine Maker Moderna," *CBS News* (June 3, 2020 4:05 PM), https://www.cbsnews.com/news/insider-trading-allegations-moderna-accountable-us-securities-exchange-commission/

86   Nussbaum, op. cit., 317

87   Ibid.

88   Nussbaum, op. cit., 310

89   Margaret A Fischl, MD, "The Efficacy of Azidothymidine (AZT) in the Treatment of Patients with AIDS and AIDS-Related Complex"; and Douglas D. Richman, MD, "The Toxicity of Azidothymidine (AZT) in the Treatment of Patients with AIDS and AIDS-Related Complex"; *New England Journal of Medicine*, 23 July 1987. https://europepmc.org/article/med/3299089

90   Engelbrecht, 134

91   Roger Müller, "Skepsis gegenüber einem Medikament [AZT], das krank macht," *Weltwoche,* 25 June 1992, 55–56

92   John Lauritsen, "AZT On Trial," *New York Native* (Oct. 19, 1987), https://www.duesberg.com/articles/jltrial.html

93   John Lauritson, *The AIDS War: Propaganda, Profiteering, and Genocide from the Medical Industrial Complex* (Asklepios, 1993), in its fourth printing, is still in print. *Poison By Prescription* is out of print but available as a free PDF book: https://www.paganpressbooks.com/jpl/POISON.PDF

94   Ellen C. Cooper, MD, M.P.H., "Medical Officer Review of NDA19-655"

95   Harvey I. Chernov, "Review & Evaluation of Pharmacology & Toxicology Data," NDA 19-655, 26 December 1986

96   John Lauritsen, *Poison by Prescription: The AZT Story* (Asklepios, 1990), 15, https://www.paganpressbooks.com/jpl/POISON.PDF

97   Lauritsen, *The AIDS War*, Chapter 29

98   Robert E. Willner, MD, PhD, "Deadly Deception-Presentation," YouTube (00:28:12 – 00:28:42), https://youtu.be/y2Q0rpnXq7Q

99   Ibid., YouTube (15:17–15:43)

100   John Lauritsen, "FDA Documents Show Fraud in AZT Trials," *New York Native* (March 30, 1992), https://www.duesberg.com/articles/jlfraud.html

101   John Lauritsen, *The AIDS War: Propaganda, Profiteering and Genocide from the Medical-Industrial Complex* (1993), Chapter 29: "FDA Documents Show Fraud in AZT Trials"

102   Previously two AIDS-critical Meditel documentaries were broadcast over Channel Four: "AIDS: The Unheard Voices" (1987) and "The AIDS Catch" (1990). Producer Joan Shenton has put all the Meditel documentaries, and many other videos, on the website of the Immunity Resource Foundation: immunity.org.uk

103   Robert E. Willner, MD, PhD, "Deadly Deception-Presentation," YouTube (00:28:12 – 00:28:42), https://youtu.be/y2Q0rpnXq7Q

104   Anayansi Vanderberg, "Fauci the Fraud: 40 Years of Lies from AZT TO Remdesivir," *Data Scientist* (Jan 11, 2020), https://edcdeveloper.wordpress.com/2020/11/01/faucithefraud-anthony-fauci-40-years-of-lies-from-azt-to-remdesivir/

105   Lauritsen, *The AIDS War* op. cit., 73–74

106   Lauritsen, *The AIDS War*

107   Lauritsen, *The AIDS War*

108   Annabelle Timsit and Marisa Iati, "Fauci Debunks Coronavirus Vaccine Infertility Conspiracies after Nicki Minaj Tweets," *Washington Post* (Sept. 15, 2021). https://www.washingtonpost.com/nation/2021/09/15/fauci-responds-nicki-minaj-covid/

109   U.S. Food and Drug Administration, *Prescription Highlights: COMIRNATY® (COVID-19 Vaccine, mRNA) suspension for injection, for intramuscular use* (2021), https://www.fda.gov/media/151707/download

110   U.S. Food and Drug Administration, *Fact Sheet for Healthcare Providers Administering Vaccine (Vaccination Providers) Emergency Use Authorization (EUA) of the Moderna COVID-19 Vaccine to Prevent Coronavirus Disease 2019 (COVID-19)* (Aug 27, 2021), https://www.fda.gov/media/144637/download

111   Lauritsen, *The AIDS War*

112   ABC News, Anthony Fauci Interview, YouTube video, 00:01-00:09, 1988, https://youtu.be/fKWwEkBZjhM .

113   Lauritsen, op. cit., 78

114   Engelbrecht et al., op. cit., 151
115   Ibid.
116   Lauritsen, *The AIDS War*, op. cit., 77
117   Marlene Cimons, "AZT Found to Delay AIDS in Those Free of Symptoms," *Los Angeles Times*, Aug. 18, 1989, https://www.latimes.com/archives/la-xpm-1989-08-18-mn-561-story.html
118   Ibid.
119   Philip J. Hilts, "Drug Said to Help AIDS Cases With Virus but No Symptoms," *NYT*, (Aug. 18, 1989), https://www.nytimes.com/1989/08/18/us/drug-said-to-help-aids-cases-with-virus-but-no-symptoms.html
120   Celia Farber,"AIDS and the AZT Scandal: Spin's 1989 Feature, 'Sins of Omission,'" *SPIN*, Oct. 5, 2015, 3:47 PM, https://www.spin.com/featured/aids-and-the-azt-scandal-spin-1989-feature-sins-of-omission/
121   RFK Jr. Interview with Celia Farber, September 10, 2021
122   Engelbrecht et al., op. cit., 142, 153
123   Lauritsen, *Poison by Prescription*, 114
124   HIV & AIDS: Fauci's First Fraud Documentary, YouTube Video, Sept 6, 2020, (1:30:48), https://youtu.be/YVjcq3m3JNo. *John Lauritsen, in a tape-recorded interview, discusses his estimate that a third of a million gay men were killed by AZT poisoning. He notes that "here one has to guess," as he followed CDC numbers closely for years, but at a certain point, CDC stopped reporting numbers for "certain things."
125   NIH Office of Technology Transfer, "Videx® Expanding Possibilities: A Case Study" (Sep. 2003), https://www.ott.nih.gov/sites/default/files/documents/pdfs/VidexCS.pdf
126   I-base, FDA safety announcement about ddI and non-cirrhotic portal hypertension, (Apr 2, 2010). https://i-base.info/htb/10215
127   Mira Hleyhel et al., "Risk of cancer in children exposed to antiretroviral nucleoside analogues in utero: The french experience," *Environ Mol Mutagen* (2019), https://doi.org/10.1002/em.22162
128   Philip J. Hilts, "With Record Speed, F.D.A. Approves a New AIDS Drug," *New York Times*, Mar 15, 1996, https://www.nytimes.com/1996/03/15/us/with-record-speed-fda-approves-a-new-aids-drug.html
129   Anthony S. Fauci, "Ending the HIV/AIDS Pandemic: Follow the Science," YouTube video, Mar 21, 2016, https://www.youtube.com/watch?v=HMSBtuCWk_M
130   Engelbrecht et al., op. cit., 128
131   Laura Wood, "HIV Drugs Market Worth $30.5 Billion in 2020 due to the Increase in Demand of HIV Drugs for the Treatment of COVID-19 Patients - ResearchAndMarkets.com," *Business Wire*, May 5, 2020, https://www.businesswire.com/news/home/20200505005791/en/HIV-Drugs-Market-Worth-30.5-Billion-in-2020-due-to-the-Increase-in-Demand-of-HIV-Drugs-for-the-Treatment-of-COVID-19-Patients---ResearchAndMarkets.com
132   Terry Michaels, *Down the Rabbit Hole: How US Medical Bureaucrats, Pharma Crony Capitalists, and Science Literate Journalists Created and Sustain the HIV-AIDS Fraud* (Unpublished Manuscript), 7
133   Ibid.
134   *HIV & AIDS: Fauci's First Fraud Documentary,* YouTube Video, Sept 6, 2020, (40:03), https://youtu.be/YVjcq3m3JNo

ChildrensHealthDefense.org/fauci-book
childrenshd.org/fauci-book

For updates, new citations and references, and new information about topics in this chapter:

CHAPTER 5

THE HIV HERESIES

"A man living outside the circle of delusion which imprisons most men has a question of eve-
ryone he meets, usually asked silently, 'Can you get outside of yourself for even a split second to
hear something you have never heard before?' Those who learn to hear will enter a new world."
—Khalil Gibran

I hesitated to include this chapter because any questioning of the orthodoxy that HIV is the sole cause of AIDS remains an unforgivable—even dangerous—heresy among our reigning medical cartel and its media allies. But one cannot write a complete book about Tony Fauci without touching on the abiding—and fascinating—scientific controversy over what he characterizes as his "greatest accomplishment" and his "life's work."

From the outset, I want to make clear that I take no position on the relationship between HIV and AIDS. I include this history because it provides an important case study illustrating how—some four hundred years after Galileo—politics and power continue to dictate "scientific consensus," rather than empiricism, critical thinking, or the established steps of the scientific method. It is a hazard to both democracy and public health when a kind of religious faith in authoritative pronouncements supplants disciplined observation, rigorous proofs, and reproducible results as the source of "truth" in the medical field.

While consensus may be an admirable political objective, it is the enemy of science and truth. The term "settled science" is an oxymoron. The admonishment that we should "trust the experts" is a trope of authoritarianism. Science is disruptive, irreverent, dynamic, rebellious, and democratic. Consensus and appeals to authority (be it CDC, WHO, Bill Gates, Anthony Fauci, or the Vatican) are features of religion, not science. Science is tumult. Empirical truth generally arises from the tilled, agitated, and upturned soils of debate. Doubt, skepticism, questioning, and dissent are its fertilizers. Every great scientific advance in history, every transformative idea, from evolution to heliocentrism to relativity, met initial ridicule from the panjandrums of "scientific consensus." As novelist and physician Dr. Michael Crichton observed,

> Consensus is the business of politics. Science, on the contrary, requires only one investigator who happens to be right, which means that he or she has results that are verifiable by reference to the real world. In science consensus is irrelevant. The greatest scientists in history are great precisely because they broke with the consensus. There is no such thing as consensus science. If it's consensus, it isn't science. If it's science, it isn't consensus. Period.[1]

Specifically, the original hypothesis on AIDS is an illustration of how vested interests (in this case, Dr. Anthony Fauci), using money, power, position, and influence, can engineer consensus on incomplete theories, and then ruthlessly suppress dissent.

The many thoughtful critics of Dr. Fauci's central canon offer various plausible, but wildly divergent, alternatives to the official orthodoxy that HIV alone causes AIDS. There is one issue upon which they all agree: During the thirty-six years since

Dr. Fauci and his colleague, Dr. Robert Gallo, first claimed that HIV is the sole cause of AIDS, no one has been able to point to a study that demonstrates their hypothesis using accepted scientific proofs. The fact that Dr. Fauci has obstinately refused to describe a convincing scientific basis for his proposition, or to debate the topic with any qualified critics, including the many Nobel laureates who have expressed skepticism, makes it even more important to give air and daylight to dissenting voices.

Even today, incoherence, knowledge gaps, contradictions, and inconsistencies continue to bedevil the official dogma. The unified chorus demanding blind adherence to that official dogma drowned out the lively public disputes of earlier years and ignored the clamor for scientific proof. An obsequious national media had consecrated the orthodoxy and anointed Anthony Fauci with an infallibility formerly reserved for popes. In the February 28, 1994, issue of *New York Native*, Neenyah Ostrom wrote an editorial titled "The Canonization of Anthony Fauci": "Anthony Fauci, the man who has so mangled and misdirected US 'AIDS' research that 13 years into the epidemic there is no clear idea of its pathogenesis and no effective treatment, was recently raised to near sainthood, once again, by the *New York Times*."[2,3]

Instead of responding to critics by answering common-sense inquiries, Dr. Fauci has cultivated a theology that denounces questioning of his orthodoxy as irresponsible, uninformed, and dangerous heresy. It's axiomatic that American democracy thrives on the free flow of information and abhors censorship, so Dr. Fauci's extraordinary capacity to ruthlessly silence, censor, ridicule, defund, and ruin prominent dissidents seems more congruent with the Spanish Inquisition or with Soviet and other totalitarian systems. Today, "The First Amendment simply does not apply to Tony Fauci," says Charles Ortleb. "Any scientist who disputes his official cosmology or any of the canons that promote the orthodoxy that HIV is the one and only cause of AIDS is dead in terms of the rewards and sustenance of science."

Finally, many of the tactics Dr. Fauci has pioneered to dodge debate—bedazzling and bamboozling the press into ignoring legitimate inquiry of the credo, and undermining, gaslighting, punishing, bullying, intimidating, marginalizing, vilifying, and muzzling critics—have become his mainstays for derailing skepticism about his mismanagement of subsequent pandemics, including COVID. So without attempting to draw conclusions about the underlying HIV/AIDS disputes, it is worth reviewing the weapons Dr. Fauci honed during his natal struggle to construct and fortify a "scientific" theology.

The loudest, most influential, and persistent challenge to the thesis that HIV might not be the only cause of AIDS came from Dr. Peter Duesberg, who in 1987 enjoyed a reputation as the world's most accomplished and insightful retrovirologist. Specifically, Dr. Duesberg accuses Dr. Fauci of committing mass murder with AZT, the deadly chemical concoction that according to Duesberg causes—and never cures—the constellations of immune suppression that we now call "AIDS." But Duesberg's critique goes deeper than his revulsion for AZT. Duesberg argues that HIV does not cause AIDS but is simply a "free rider" common to high-risk populations who suffer immune suppression due to environmental exposures. While HIV may be sexually transmittable, Duesberg argues, AIDS is not. Duesberg famously offered to inject himself with HIV-tainted blood "so long as it doesn't come from Gallo's lab."[4] For starters, Duesberg points out that HIV is seen in millions of healthy individuals who never develop AIDS. Conversely, there are thousands of known AIDS cases in patients who are not demonstrably infected with HIV. Dr. Fauci has never been able to explain these phenomena, which are inconsistent with the pathogenesis of any other infectious disease.

Many other prominent and thoughtful scientists have offered a variety of well-reasoned hypotheses to explain these baffling fissures in the HIV orthodoxy. Most of these alternative conjectures accept that HIV plays a role in the onset of AIDS but argue that there must be other cofactors, a qualifier that Dr. Fauci and a handful of his diehard PIs stubbornly deny.

Prior to advancing his own theory for the etiology of AIDS, Duesberg methodically laid out the logical flaws in Dr. Fauci's HIV/AIDS hypothesis in a groundbreaking 1987 article in *Cancer Research*.[5] Dr. Fauci has never answered Duesberg's common-sense questions.

In his subsequent book, *Inventing the AIDS Virus*, Duesberg, in 724 riveting pages, expands his dissection of the hypothesis's flaws and outlines his own explanation for the etiology of AIDS.[6]

For those subsumed in the theology that HIV is the sole cause of AIDS, Dr. Duesberg's critiques seem so outlandish that they automatically debase anyone who even considers them. It's telling, then, to discover how much traction his arguments have among the world's most thoughtful and brilliant scientists, including many Nobel laureates, perhaps most notably Luc Montagnier, who first isolated HIV. To date, Dr. Fauci has been able to silence but not to answer or to refute Duesberg's thesis.

I restate that I take no side in this dispute. It seems undeniable to me that the dissidents have raised legitimate queries that should be researched, debated, and explored. I believe public health officials have a duty to answer these sorts of questions, and I yearn to hear those arguments in an energized debate; Dr. Fauci's aggressive censorship campaign and his refusal to debate arouse my suspicion and my ire. It brings to mind George R. R. Martin's observation that entrenched powers remove men's tongues not to prevent them from telling lies, but to stop them from speaking the truth.

If any of Dr. Duesberg's revelations are solid, his story has momentous relevance today—as the removal of his tongue illustrates the capacity of the pharmaceutical cartel, in league with self-interested technocrats, to exaggerate and exploit viral pandemics, to foist toxic and dangerous remedies onto a credulous public, and promote self-serving agendas—even those with terrible outcomes—with the complicity of a fawning and scientifically illiterate media. Duesberg and others charge that by stifling debate and dissent, Dr. Fauci milled public fear into multi-billion-dollar profits for his Pharma partners while expanding his own powers and authoritarian control. The resulting policies, they say, have caused calamity to global economies and public health, and vastly expanded the pool of human suffering.

The first time that someone—Dr. Tom Cowan, a physician from Northern California—suggested to me that HIV was not the sole cause of AIDS, I dismissed the comment as ridiculous. I had watched many HIV-positive friends die of AIDS during the 1980s and 1990s. I personally knew two of the celebrities—Arthur Ashe and Rudolf Nureyev—whose pioneering deaths from "AIDS" shocked the world at the epidemic's dawn. It seemed self-evident that HIV was the culprit. I had no idea that the supposition was controversial. I have since learned that today, a disturbing number of virologists quietly doubt the theory that HIV is the sole cause of AIDS.

To understand the skepticism by many of the world's leading scientific minds, we need to venture back through history and briefly down a very deep rabbit hole. That journey pulls the curtain back on a shockingly corrupt NIH culture distinguished by lacunae that most Americans associate with politics, not science: cutthroat ambition, backstabbing duplicity, and moral bankruptcy.

In July 1981, CDC reported a unique outbreak of immune deficiency–related health problems in a group of highly promiscuous gay men in Los Angeles, New York, and San Francisco. A May 1983 *Science* article by French Institut Pasteur virologist Luc Montagnier first identified a retrovirus that would later earn the name HIV.[7] Montagnier believed he had detected signals of HIV in the lymph nodes of some of the AIDS victims he had sampled. After hearing a lecture by Montagnier, Dr. Robert Gallo, a blustering, ambitious National Cancer Institute (NCI) researcher, entrepreneur, and homophobe, persuaded the Frenchman to send him a sample of the newly discovered retrovirus, promising to use his considerable influence with the journal *Science* to get Montagnier's work published expeditiously. Instead, Dr. Gallo stalled the publication to give himself time to cultivate and steal Montagnier's virus. With the help of other HHS officials, Gallo then claimed Montagnier's pilfered virus as his own discovery and used an imaginative and cunning retinue of subterfuges and intricate frauds to obscure his larceny. In his book, *Science Fictions: A Scientific Mystery, a Massive Cover-up and the Dark Legacy of Robert Gallo*, Pulitzer Prize–winning *Chicago Tribune* reporter John Crewdson meticulously documents Gallo's brazen flimflam, perhaps the boldest, most outrageous, and most consequential con operation in the history of science. The book exposes Gallo as a mountebank who built his career poaching discoveries from other scientists and claiming them as his own.[8]

Scientists who worked for Gallo described his NIH lab, where he presided over some fifty scientists and a budget of $13 million, as a "den of thieves."[9] One of Gallo's scientists told Crewdson, "It's hard to be an honest person in this place." She said she knew three employees who committed suicide.[10] Gallo confided to a henchman that he liked to hire foreigners "because if they don't do what he wants, he can deport them." Gallo's former mistress and lab employee, Flossie Wong-Staal, reported that Gallo voiced his craven need for the Nobel Prize and his bitterness at being denied the honor so frequently that it was practically a "rhetorical device."[11]

It was natural that Gallo found a powerful and reliable ally in Tony Fauci. Gallo's "proof" that the cause of AIDS was a virus—as opposed to toxic exposures—provided the critical foundation stone of Dr. Fauci's career. This claim allowed Dr. Fauci to capture the AIDS program and its attendant cash flows from the National Cancer Institute (NCI) and launch the project of building NIAID into the world's leading drug-production empire.

On April 23, 1984, Gallo recruited his boss, HHS Secretary Margaret Heckler, to lend credibility and weight to his dramatic announcement. Heckler took the stage before a packed scrum of international press. "Good afternoon," she told the world, "Ladies and gentlemen, first, the probable cause of AIDS has been found—a variant of a known human cancer virus." She pointedly added, "Today we add a new miracle to the long honor roll of American medicine and science."[12]

Heckler's participation at Gallo's press event was important stagecraft because it gave the imprimatur of NIH's institutional gravitas to a theory that had not been subject to peer review.

Only later did the public learn that NIH allowed Gallo to delay the announcement until he had personally patented an antibody kit that he claimed capable of detecting HIV. He had developed the test at taxpayer expense.

Crewdson writes that Gallo conspired with a CDC official, James Curran, to improperly certify Gallo's test as equivalent in quality to a far better test developed by Montagnier. Gallo would make himself a millionaire from his innovation while fanning fears of the presumably deadly virus, which coincidentally drove sales. A subsequent lawsuit over Gallo's swindle by the French government ultimately forced Gallo to disgorge half his proceeds.

Gallo's premature announcement pioneered a new strategy of "Science by Press Release" that would become a familiar mainstay in Dr. Fauci's arsenal of narrative control, culminating in the COVID-19 pandemic. The journal *Science* did not publish Gallo's paper until over a week after his spectacular TV press conference. At the time, Gallo's tactic marked a severe breach of professional scientific etiquette. This gimmick assured that nobody could review Gallo's work prior to his proclamation.

Both Dr. Gallo and Dr. Montagnier, who had devoted their careers to studying retroviruses, were cancer researchers. Before the appearance of AIDS, both men had vainly strived to implicate retroviruses as the culprit in leukemia. In 1975, before he ever published a paper on the subject, Gallo gained national headlines when he publicly announced his discovery of a human retrovirus HL-23 that he claimed caused leukemia.[13] He told colleagues he expected to win the Nobel Prize for his detection of HL-23 in human leukemia cells.[14] He didn't.

Major labs around the country were intensely interested in HL-23, but when they requested samples from Gallo, he ordered subordinates to damage the infected cells, before sending them out, to make them useless for research by others.[15] Leukemia incidence was exploding at the time, but ethical elasticity apparently insulated Gallo against qualms about purposefully delaying vital research during a global pandemic. Other scientists complained that they could not reproduce Gallo's success. Subsequently, two groups of US researchers literally made a monkey out of Gallo's discovery—if not Gallo—by proving his HL-23 virus was actually a humiliating laboratory contamination consisting of a mélange of three viruses from a gibbon, a woolly monkey, and a baboon.[16] Instead of a Nobel laureate, Gallo became a laughingstock.

Undeterred by mortification, Gallo declared that a so-called HTLV virus, which he also claimed to have discovered (he had stolen the work of Japanese researchers, according to Crewdson), was the cause of AIDS.[17] Puzzled that he could not reproduce Gallo's results, another AIDS researcher, working with gay patients, asked Gallo if the discrepancy was because Gallo might be studying a different risk group. "Was your patient a Haitian? A hemophiliac?" the scientist queried. "It was a fucking fag," replied Gallo.[18]

When asked to address Duesberg's announcements about the HIV/AIDS hypothesis, Gallo often dismissed Duesberg's objections because, Gallo suggested, Duesberg was gay and/or mentally disturbed (Duesberg is straight, and sane): "[Duesberg] comes to meetings with guys with leather jackets and the hair and so on in the middle. I mean, that's a little bit odd. Doesn't it speak of something funny?"[19] These were the sorts of petty defamations that Gallo generously offered, instead of argument, to defend his work.

But Gallo's failure to demonstrate that he could find HTLV in the blood of men suffering from AIDS threatened to put the final nail into his naked Nobel ambitions. At the height of that personal crisis, Gallo learned of Montagnier's success. Unwilling to accept defeat by the French, he gulled the credulous virologist into sending him a sample, which he cultured on a substrate that, according to Crewdson, he stole from yet another scientist. When he succeeded in finding signs of Montagnier's virus in the blood of gay men suffering from immune system collapse, Gallo rebranded it HTLV and claimed it to be the same virus he had lately "discovered."[20] Gallo's lab notes, obtained by the *Chicago Tribune*, show that Gallo renamed the French virus repeatedly, apparently to further obscure its pedigree.

The following spring, *Science* published the four papers from Gallo's lab, upon which Gallo's celebrity as the "Superman of AIDS" entirely rests. The first paper reported Gallo's isolation of a so-called "new" virus from AIDS patients. (Gallo's lab had apparently cultivated and rechristened the French virus.) The second paper declared that the new virus had been "isolated from a total of forty-eight subjects," a finding that would go far toward proving that the virus caused the disease.[21] Examination of Gallo's lab notes by the *Chicago Tribune* found no traces of these forty-eight isolates.[22]

American and French governments skirmished over which scientist "discovered" HIV, until the combatants agreed in 1987 to call it a "co-discovery." The WHO delayed its response for two years as Gallo employed a series of artifices to pretend that there were two different viruses. By delaying the announcement of the French scientist's earlier discoveries, Gallo stalled the introduction of a widely available blood test for the AIDS virus by about a year. During that 1983–1984 interregnum, thousands of hospital patients and hemophiliacs received tainted blood from blood banks and became infected with HIV, and many of the already infected unwittingly spread the virus.[23,24]

The Nobel committee awarded Montagnier its prize in 2008, conspicuously snubbing Gallo, whose notorious ethical lapses were, by then, abundantly documented. Gallo's unsupported claims and sketchy conduct resulted in two US government inquiries into his professional ethics (NIH and congressional).[25,26] Pulitzer Prize–winner John Crewdson's 55,000-word exposé in the *Chicago Tribune* documenting Gallo's theft provided a withering portrait of Gallo as a sociopath and pathological liar who employed thieving felons to run his lab, a pirate enterprise engaged in pilfering money from the federal government and swiping discoveries from other scientists.[27]

The Sturm und Drang around the competing claims obscured the fact that both cancer researchers produced scientific papers that did nothing more than *suggest* their retrovirus might cause AIDS. Montagnier always moderated his own claims that HIV was proven the sole cause of AIDS and would eventually disavow the theory.

Recalling how public revelations about Bob Gallo's acrobatic chicanery during his efforts to link leukemia to HIV had nearly destroyed Gallo's career, Nobel Laureate Kary Mullis—who, unfortunately, died in August 2019, just before the COVID-19 pandemic—noted, "HIV didn't suddenly pop out of the rain forest or Haiti. It just popped into Bob Gallo's hands at a time when he needed a new career."[28] Duesberg later said, "He stole the fake diamonds from Luc Montagnier."[29]

Pouring Concrete on Confirmation Bias

But, like Dr. Fauci, Gallo had both the PIs and press in his pocket. NIH's mythical prestige lent Heckler's statement a near-religious authority. The medical establishment quickly embraced Gallo's scientific hypothesis. Suspending traditional skepticism toward government pronouncements, the press ordained Gallo's theory as indisputable doctrine and beatified Gallo as a saint.

Says journalist and editor Mark Gabrish Conlan of Gallo's big press event, "The Conference was held before any of Robert Gallo's papers were published. Therefore, before any other scientists had a chance to review them and look at the evidence and ask, has he got it right or wrong?"[30]

Gallo's announcement was a windfall for Anthony Fauci. Pinning the AIDS epidemic on a virus allowed him to divert the cascading river of AIDS money from the National Cancer Institute into NIAID's overflowing coffers.

Dr. Fauci opened the floodgates of NIAID cash to develop new antivirals against HIV. He unleashed his kennel of grant-hungry PIs to concoct and test new drugs that would kill the virus. Remarkably, Dr. Fauci never funded to completion a single grant to explore whether HIV actually caused AIDS.

Federal law requires that NIH's grant-review committee be composed of true peers—independent outside scientists knowledgeable about a given proposal's subject matter—to assess the application on its scientific merit. Ignoring those laws, Dr. Fauci began populating these committees with his own PIs. Researchers who reliably supported Dr. Fauci's orthodoxy watched their applications sail through the approval process. But scientists seeking to research ideas that departed from official doctrine encountered impenetrable obstacles. In 1988, a veteran NIH awardee, Seymour Grufferman, had his first experience with the new regime. Grufferman, the former chairman of NIH's Review Committee, had submitted a proposal to study the phenomenon of Chronic Fatigue Syndrome—a touchy subject potentially threatening to the dominant cosmology, since many of Dr. Fauci's critics believe that CFS is non-HIV AIDS. "I never got scores like that before," Grufferman told Hillary Johnson, author of *Osler's Web*. "My data sheets were *ATROCIOUS*." When he protested to Dr. Fauci, he recounted, Dr. Fauci was "nasty."[31]

Dr. Fauci's tsunami of research money poured the concrete of confirmation bias onto Gallo's hypothesis. NIAID's PI army welcomed the fierce new bug hunt around this novel medical mystery. "Thousands of health science PhDs seeking government grants rushed to study the virus," historian Terry Michael recounts.[32] Dr. Fauci's PIs became the fierce guard dogs of the pervasive HIV orthodoxy.

Nobel Laureate Kary Mullis knew the effect of NIH funding on cementing official dogma. "All the old virus hunters from the National Cancer Institute put new signs on their doors and became AIDS researchers. [US President Ronald] Reagan sent up about a billion dollars just for starters," noted Mullis, who in 1993 won the Nobel Prize in Chemistry for his invention of the Polymerase Chain Reaction (PCR) technique. "And suddenly everybody who could claim to be any kind of medical scientist and who hadn't had anything much to do lately was fully employed."[33]

The End of Science

According to Mark Gabrish Conlan, "The Department of Health and Human Services decided from now on we are only going to fund AIDS research that assumes that Robert Gallo's virus is the cause. Dr. Fauci will not fund research into any other possibilities. Therefore, those scientists who might have wanted to critique Gallo's papers would not be able to do so, at least not with anything supported by the federal government, which is virtually all science in this country today, from that moment on."[34]

For thirty-six years, Fauci targeted all federal grants toward the single pathogen theory of AIDS. The "little emperor" made NIAID the go-to agency for AIDS research grants and spent lavishly so long as grant writers toed the official line about the purported viral cause of AIDS, the only hypothesis for which NIAID would provide funding. He used his awesome leverage to discourage inquiry into any multi-factorial hypothesis. The PIs that he funded became his ideological commissars; the growing enterprise became the launch platform for his career as the most successful medical science bureaucrat in American history.

One of the inevitable outcomes of this "confirmation-biased" research was the rapidly expanding definition of "AIDS." Dr. Fauci's battalion of scientists implemented a wide-ranging HIV testing program using indiscriminate PCR tests capable of amplifying tiny strands of long-dead genetic debris billions of times. The PCR test could not identify active HIV infection. Mullis, who invented the tests, pointed out that the PCR was capable of finding HIV signals in large segments of the population who suffered no threat from HIV and had no live HIV virus in their bodies. Researchers naturally found harmless HIV DNA detritus in people with a constellation of other diseases. All those unrelated ailments soon became incorporated beneath the umbrella definition of AIDS. Individuals with Candida or Kaposi's sarcoma and a positive PCR test had AIDS. Those same individuals with a negative PCR would have Kaposi's sarcoma or Candida. Under this rubric, the AIDS definition rapidly metastasized to encompass a galaxy of some thirty separate well-known diseases, including Kaposi's sarcoma (KS), Hodgkin's disease, herpes zoster (shingles), *Pneumocystis carinii pneumonia* (PCP), Burkitt's lymphoma, isosporiasis, Salmonella septicemia, and tuberculosis, all of which also occur in individuals who had no HIV infection.[35,36]

"Most people consider it blasphemous when you point out AIDS is not a disease, it's a syndrome,"[37] Paul Philpott, MS, Editor, *Rethinking AIDS*, explained. "It's a collection of diseases and those diseases get called AIDS if they occur in a patient that the doctor somehow concludes is HIV-positive."[38] "All of the diseases in the category called AIDS occur to people who are HIV-negative. None of them are exclusive to people who test HIV-positive. And all of them have causes and treatments that are well-known; they're completely unrelated to HIV. So any of the diseases, when they happen to somebody who tested HIV-negative, are called by their old name; but when they occur in someone who tested HIV-positive, then they're called AIDS."[39]

In the hands of Dr. Fauci's opportunistic PIs, AIDS became an amorphous malady subject to ever-changing definitions, encompassing a multitude of old diseases in hosts who test positive for HIV.

Asked to define AIDS in a 2009 documentary, Fauci said, "When your CD4 count falls below a certain arbitrary level, by definition you have AIDS."[40] But how do we explain the many individuals who have low CD4 counts and no HIV?

The growth of the AIDS pandemic was predictably explosive. Using PCR and expanded diagnosis, WHO estimates that HIV has infected 78 million people and caused 39 million deaths. Today, 35 million people live with HIV with over 2 million new infections each year.[41,42]

This loose diagnostic system and the gravy train of financial incentives for finding AIDS everywhere guaranteed riches for institutions and individuals who signed on to Dr. Fauci's gold rush. The pharmaceutical multinationals, like GlaxoSmith-Kline, minting enormous profits marketing antivirals to kill HIV, had little incentive to challenge Dr. Fauci's orthodoxy.

Africa's AIDS Bonanza

With grants from Tony Fauci, intrepid researchers quickly found that the contagion had somehow reached Africa and infected up to 25 million Africans, with no one having taken notice. Researchers, extrapolating from small cohorts with positive PCR results, used murky statistical models to report HIV had infected nearly half the adult population in some nations—and forecast widespread depopulation of the African continent. None of the shrilly predicted depopulation has ever occurred, and most HIV-infected Africans showed no sign of illness. In those who were sick, the infirmities looked very much like the illnesses that doctors had previously diagnosed as malaria, pneumonia, malnutrition, leprosy, bilharzia, anemia, tuberculosis, dysentery, or infection with a grim inventory of pathogens and parasites familiar to doctors in Africa.

Because HIV antibody tests are too costly for widespread use in Africa, the World Health Organization has since 1985 used the "Bangui definition"[43,44] to diagnose AIDS, based on clinical symptoms. WHO's enthusiasm for this loose, all-encompassing definition may reflect the early revelation that the AIDS plague loosened purse strings like no other crisis on Africa's beleaguered landscapes.

The statistical picture of AIDS in Africa, consequently, is a sketchy projection based on very rough computer-generated estimates from the World Health Organization (WHO), built on a highly questionable data pool, dubious assumptions, and grotesque exaggeration. Uncertainty prevails, even in those extremely rare cases when doctors actually performed HIV tests on Africans; many diseases that are endemic to Africa, such as malaria, TB, flu, and simple fevers, trigger false positives. Duesberg and many other critics accused Dr. Fauci, and an opportunistic pharmaceutical industry, of taking this long inventory of ancient afflictions and recasting them as AIDS.

It's undeniable that African AIDS is an entirely different disease from Western AIDS. Whereas AIDS in Western countries continued to be a disease of drug addicts and homosexuals—with women reporting only 19 percent of US and European AIDS cases—in Africa, 59 percent of AIDS cases are in women, with 85 percent of cases occurring in heterosexuals, and the remaining 15 percent in children. No one has ever explained how a disease largely confined to male homosexuals in the West is a female heterosexual disease in Africa.

"AIDS in Africa looks nothing like AIDS in North America or Europe," observed Duesberg to me. "Africans were rarely tested with expensive PCR tests, so every unexplained death became 'AIDS.'"

The clinical symptoms of African AIDS are high fever, a persistent cough, loose stools for thirty days, and a 10 percent loss of body weight over a two-month period. By that definition, a large percentage of Western tourists have AIDS while in Africa. The simple cure is to get on a plane back to New York, where no doctor would dream of bestowing an AIDS diagnosis based on that symptomology alone.

After 1993, WHO added tuberculosis to the definition. Duesberg told me, "It became a garbage pail definition applied to anyone sick with an uncertain diagnosis."

"Due to compelling financial drivers, in Africa, AIDS is nearly always a presumptive diagnosis, applied without any 'positive' reaction to HIV tests," science journalist Celia Farber told me. "Big Pharma, researchers, clinics, international health agencies beginning with WHO, and local governments conspire to keep this stunningly broad and generic clinical definition of AIDS in Africa," she explains. "From the beginning it was a signal for funding. They are all in on the joke, because they are all helping themselves by skimming the unprecedented international funding streams that flow to African AIDS relief."

"AIDS is huge business, possibly the biggest in Africa," says James Shikwati in a 2005 interview with *Der Spiegel*. Shikwati is founder of the Inter Region Economic Network, a society for economic promotion in Nairobi (Kenya). "Nothing else gets people to fork out money like shocking AIDS figures. AIDS is a political disease here: we should be very skeptical."[45]

Former epidemiological director of WHO, Professor James Chin, in his 2006 book, *The AIDS Pandemic: The Collision of Epidemiology and Political Correctness*, admits unambiguously that the AIDS case figures for developing countries were massively manipulated in order to maintain the flow of billions of dollars.[46]

Dr. Rebecca Culshaw, PhD, a former HIV researcher and professor of Mathematical Biology and Population Dynamics at the University of Texas at Tyler, admits that "The paradox of how a disease could cause both vastly different epidemiologies and symptomatic progressions in the First and Third World"[47] was one of the irreconcilable problems that sowed her initial disillusionment with the HIV/AIDS orthodoxy: "The

African epidemic looks suspiciously nothing like the American and European epidemic, and closer inspection reveals it likely that this African epidemic is pure fabrication."[48]

The questions about widely divergent symptomology of this mysterious disease only amplify when we consider that WHO maintains twelve different descriptions of AIDS, depending on national boundaries. In 2003, AIDS activist Christine Maggiore told documentarians:

> In 1993, in this country, we adopted a definition that caused the number of AIDS cases to double overnight. And part of that reason was for the first time we'd began counting people as AIDS victims who were not ill and who did not have any symptoms. They had a low T-cell count and that's [all]. And T-cells are something that can fluctuate a 100 percent in a given day. So based on a low T-cell count that year, the number of AIDS cases doubled overnight. And with that definition, there have been 182,000 Americans who are not ill diagnosed with AIDS, who would not have AIDS if they moved to Canada. Because in Canada, they don't recognize that T-cell definition as a criteria for having an AIDS diagnosis.[49]

Many US AIDS sufferers can become "cured" by crossing the border into Canada. No other disease is so subject to this sort of nationalism.

Correlation Is Not Causation

In May 1984, a month after his momentous press conference, Robert Gallo finally published his paper claiming to have "discovered" the HIV virus, in *Science*.[50] He also explained in detail his rationale for linking HIV to the AIDS disease by reporting that he had found evidence of the virus in several afflicted gay men. Gallo reported a "frequent detection and isolation" of [HIV] from patients with AIDS and at risk for AIDS.[51] Scientists were shocked to learn for the first time that Gallo had found faint traces of HIV in only twenty-six of the seventy-two AIDS patients whose blood he examined. That weak conclusion was Gallo's only basis for claiming that HIV might *cause* AIDS. It's axiomatic that correlation does not prove causation. There were many other viruses, including herpes simplex, cytomegaloviruses, and a range of predatory herpes viruses found with a far higher frequency in AIDS patients upon which Gallo could have just as easily blamed AIDS.

A year earlier, Dr. Luc Montagnier also had only suggested—in his May 1983 paper in *Science*—that his claimed virus "may be *involved* in several pathological syndromes, *including* AIDS."[52] Montagnier, a brilliant scientist known for his integrity, had found evidence of HIV in the lymph nodes of 72 percent of the forty-four AIDS patients he tested. Montagnier always remained tentative about claiming the weak correlation as proof. As early as 1992, Montagnier told *Nature* that "HIV is a necessary but not, without the cofactor, a sufficient cause of AIDS."[53] As we shall see, Montagnier's later statements indicate that his doubts about HIV's role in the etiology of AIDS continued to grow thereafter.[54] Based upon Gallo and Montagnier's slender scientific reeds, these seminal papers introduced the idea that a single, discrete virus was causing the AIDS pandemic.

Dr. Fauci has since routinely claimed that HIV was "proven definitively to be the cause of AIDS by Bob Gallo here when he was at NIH."[55] But critics argue that evidence in Gallo's article is far too anemic to support Dr. Fauci's characterization. Neither Gallo nor Dr. Fauci has ever demonstrated, using any of the conventional scientific proofs, that the HIV virus alone actually causes AIDS. Rather than allowing his HIV hypothesis to triumph in the marketplace of ideas, Dr. Fauci sent clear signals to the American press that debate on this theory could no longer be tolerated.

In September 1989, Dr. Fauci broadcast an angry threat about journalists who dared to give a platform to Peter Duesberg. He ended with this warning: "And they should realize that their accuracy is noted by the scientific community. Journalists who have made too many mistakes or who are sloppy are going to find that their access to scientists may diminish."[56]

Dr. Fauci Leveraged Uncertain Tests to Paint AIDS as a Widespread Viral Plague

Instead of using traditional methods for diagnosing disease based on symptoms, Dr. Fauci encouraged doctors to perform blood tests on both healthy and unhealthy individuals to diagnose AIDS. Since none of the available tests are particularly accurate, Dr. Fauci must have understood that his reliance on blood tests alone was likely to yield highly dubious results capable of dramatically overstating the spread of HIV.[57]

In the decade preceding the AIDS crisis, a wave of new technologies, including PCR and super powerful electron microscopes, had opened windows on teeming new worlds containing millions of species of previously unknown viruses to scientists. Molecular genetics not only revolutionized biological science, but also made that science fabulously profitable. The lure of fame and fortune ignited a chaotic revolution in virology as ambitious young PhDs scrambled to inculpate newly discovered microbes as the cause of old malignancies. Making such connections could be a profitable pursuit for enterprising young biologists and pharmaceutical companies.

Under this new rubric, every theoretical breakthrough, every find, became potentially the basis for a new generation of drugs. The opportunity to capitalize on the transfer of information transformed researchers into entrepreneurs and their discoveries into "inventions." Science became big business.

All this new equipment made science expensive—too expensive to perform without financial support from Big Pharma and Big Government. Researchers increasingly relied on Tony Fauci and drug makers to furnish and support their laboratories. Long-term funding became the first requirement of any new research. The researcher got his financing, and Dr. Fauci and the pharmaceutical company got proprietary rights on new discoveries. The self-interest of the researcher, the research institution, and the biotech company converged.

Finance dictated the direction of research and—too often—warped its conclusions. Armies of scientists fresh from graduate schools joined the gold rush as Dr. Fauci and Big Pharma grub-staked brigades of young PhDs to prospect for novel viruses in the diseased tissues of sick patients.

It was often unclear that the new viruses they found in ailing tissues were actually causing the diseases, whether the tiny microbes were free riders colonizing decayed tissue, or altogether innocent bystanders. Harvard's Jim Watson, who won the Nobel Prize in 1962 for discovering the molecular structure of DNA, fretted that the "gold rush" mentality was likely to "scare off the sensible and leave the field to a combination of charlatans and fools."[58] In 2001, alarmed by the precipitous decline in scientific discipline, fourteen renowned virologists of the "old guard" published an appeal to the young high-technology–focused generation of researchers in *Science*. The graybeards warned the young scientists against attributing culpability to a microbe based upon correlation without first understanding how a newly discovered virus actually causes the disease:

> Modern methods like PCR, with which small genetic sequences are multiplied and detected, are marvelous [but they] tell little or nothing about how a virus multiplies, which animals carry it, how it makes people sick. It is like trying to say whether somebody has bad breath by looking at his fingerprint.[59]

Moreover, the evidence linking specific viruses to probable diseases was often subjective and not reproducible. The specific tests that researchers used to detect HIV had their own manner of additional deficiencies.

The most significant diagnostic tools that doctors use to determine if someone is infected with HIV or not, and therefore, whether they have AIDS are:

1. HIV antibody tests
2. PCR viral load tests
3. Helper cell counts (T-cells, or rather the T-cell subgroup CD4)

Antibody Test

Gallo used an "antibody" test of his own invention to detect the presence of the HIV virus in several gay men. But what did his test actually prove?

Gallo based his test on an antigen-antibody theory, which assumes the immune system fights against foreign viruses, by generating targeted antibodies specific to that virus. In order to calibrate a test to recognize that specific antibody, the inventor must isolate the target virus and expose it to human cells in a petri dish, which then generate the specific antibodies responsive to that virus. However, since it is unclear whether Gallo or any other researcher was ever able to isolate HIV,[60] he took from his AIDS patients a sample of antibodies that he found in great abundance in their blood and made a leap of faith that they were HIV antibodies. Geneticists have pointed out that these antibodies may have been associated with tuberculosis or herpes, or any of the many other pathogenic illnesses that multiply in collapsing immune systems.[61] Indeed, Gallo's HIV antibody test also reacts to people with fever, pregnant women, and individuals who have overcome a tuberculosis infection.[62] Therefore, it is unclear if the antibodies detected by his kit are really HIV antibodies.[63] Neither Gallo's test nor any of the later-developed antibody tests have ever proven that these proteins they identify as HIV antibodies have anything to do with HIV, or any other retrovirus.

The antibody test manufacturers recognize this deficiency with a caveat on their inserts: "There is no recognized standard for establishing the presence or absence of antibodies to HIV-1 and HIV-2 in human blood."[64]

The same also holds true for the quantitative PCR-based HIV diagnostic test. "It's not even a test for HIV," protested Kary Mullis, who invented the DNA amplification technique commonly used to diagnose AIDS infection. "Quantitative PCR is an oxymoron. PCR is intended to identify substances qualitatively, but by its very nature is unsuited for estimating numbers. Although there is a common misimpression that the viral load tests actually count the number of viruses in the blood, these tests cannot detect free, infectious viruses at all; they can only detect proteins that are believed, in some cases wrongly, to be unique to HIV. The tests can detect genetic sequences of the virus, but not the viruses themselves."[65]

In 1986, Thomas Zuck of the FDA warned that the HIV antibody tests were not actually designed specially to detect HIV. "Rather, numerous other germs or contaminants, including TB, pregnancy, or simple flu, also produce false positives." Zuck made that admission at a World Health Organization meeting but conceded that stopping the use of these HIV tests was "simply not practical." He explained that "Now that the medical community has identified HIV as an infectious sexually transmitted virus, public pressure for an HIV test was just too strong."[66]

Finally, and most importantly, critics point out that Gallo's HIV antibody tests flipped traditional immunology on its head. Throughout all of medical history, a high antibody level indicated that a person had already successfully battled against an infectious pathogen and was now protected from the disease. With all other viral diseases, the presence of antibodies signals a welcomed immunity from the disease. But Gallo and Dr. Fauci's PIs suddenly began informing people that the positive antibody test was a death sentence. How could this be so? Dr. Fauci has never explained this inexplicable paradox.

It gets even weirder when one contemplates Dr. Fauci's $15 billion-dollar HIV vaccine enterprise.[67] Usually, regulators measure a vaccine's success by its ability to produce robust and durable antibodies. Now, for the first time in history, Dr. Fauci and Bob Gallo were asking the world to believe that antibodies were a sign of active, deadly disease. This begs the question, "What is the HIV vaccine supposed to do?"

Mulling this conundrum, Reinhard Kurth, former director of the Robert Koch Institute, shrugged his shoulders in bewilderment during a 2004 interview with *Der Spiegel*: "To tell the truth, we really don't know exactly what has to happen in a vaccine so that it protects from AIDS." Perhaps that is the dilemma that has frustrated Dr. Fauci's AIDS vaccine project for thirty-six years.

PCR Testing Deficiencies

The Polymerase Chain Reaction (PCR) technique does not measure the actual, live virus in the body, but the amplified fragments of DNA that are thought to be similar to HIV.[68] But even if those fragments are amplified from the authentic HIV DNA, they could be from an old exposure—from a long-dead virus genetically similar to HIV, left over from an infection that has been suppressed by antibodies, perhaps decades earlier.

"The HIV test has never been validated," said Kary Mullis. "It doesn't show infection; it shows viral particles that may exist in millions of people." In the late 1980s, the biting and sardonic Mullis became Gallo and Fauci's most fierce critic—in fact, ridiculer. Mullis added, "With the PCR method, mind you, not a complete virus, but only very fine traces of genes (DNA, RNA) may be detected, but whether they come from a [certain] virus, or from some other contamination, remains unclear."[69]

Heinz Ludwig Sanger, professor of molecular biology and 1978 winner of the renowned Robert Koch prize, stated that "HIV has never been isolated, for which reason its nucleic acids cannot be used in PCR virus load test as the standard for giving evidence of HIV"[70] ("Misdiagnosis of HIV infections by HIV-1 viral load testing: a case series," a 1999 paper published in the *Annals of Internal Medicine*).[71]

Knowing the above, it's not surprising that every PCR kit includes a manufacturer's warning, "Do not use this kit as the sole basis for detecting HIV infection" or similar labeling.

Gallo's leap from correlation to causation troubled Mullis from the outset: "PCR made it easier to see that certain people are infected with HIV, and some of those people came down with symptoms of AIDS, but that doesn't *begin*, even, to answer the question: Does HIV cause it? Human beings are *full* of retroviruses."[72]

CD4 Tests

Similar deficiencies plague tests that count CD4+ "helper T cells." AIDS doctors look at low CD4 cell counts as the key marker for AIDS diagnoses. However, not a single study confirms this most important principle of the HIV only theory: that HIV destroys CD4 cells by means of an infection. Furthermore, even the most significant of all AIDS studies, the 1994 Concorde study,[73] questions using helper cell counts as a diagnostic test for AIDS. The problem is the use of a surrogate endpoint, which is notoriously im-precise. Many studies corroborate the skepticism. One of these is the 1996 paper "Surrogate Endpoints in Clinical Studies: Are We Being Misled?"[74] Published in the *Annals of Internal Medicine*, the paper concludes that CD4 T cell count in the HIV setting is as uninformative as "a toss of a coin"—in other words, not at all.[75]

Mullis added, "Now, is there a test that can definitively tell you if you're infected with the virus? What is that test?"[76]

The Party Line—At All Costs—Or Else

Critics of the HIV/AIDS hypothesis invariably cite Koch's Postulates as the most profound embarrassment for Gallo's theory. In 1884, Nobel Laureate Robert Koch, the father of bacteriology, first outlined the classical methodologies for proving causation between a pathogen and a disease. Summarizing Koch's postulates for *The Journal of Investigative Dermatology*, Julia A. Segre wrote:

> As originally stated, the four criteria are: (1) The microorganism must be found in diseased but not healthy individuals; (2) The microorganism must be cultured from the diseased individual; (3) Inoculation of a healthy individual with the cultured microorganism must recapitulate the disease; and finally (4) The microorganism must be re-isolated from

the inoculated, diseased individual and matched to the original microorganism. Koch's postulates have been critically important in establishing the criteria whereby the scientific community agrees that a microorganism causes a disease.[77]

Virologists—and every trial lawyer and judge—consider Koch's four criteria the gold standard for proof that a particular microorganism causes a particular malignancy.

The Problem of AIDS without HIV

Koch's first postulate requires that a truly pathogenic virus can be found in large quantities in every patient suffering from the disease. The failure of the HIV/AIDS hypothesis to meet this critical threshold remains one of Dr. Fauci's most exasperating dilemmas. For starters, Gallo claimed that he found HIV virus in fewer than half of the ailing AIDS patients from whom he drew blood.[78,79] Furthermore, every one of the thirty discrete illnesses we now call AIDS occurs also in persons uninfected by HIV.

In fact, AIDS commonly occurs in people who test HIV negative. If HIV is truly the only cause of AIDS, this should not be possible.

Soon after Robert Gallo's historic announcement, doctors around the country and CDC officials started seeing patients with low CD4 counts and signature AIDS diseases like PCP and immune system dysfunction, but who tested negative for HIV. Many of the victims were white heterosexual women. Dr. Fauci and the CDC kept this awkward information secret. Fauci-funded AIDS researchers—Dr. Fauci's PIs— also kept mum when they encountered such patients.

By 1992, media science writers also knew about these HIV-free AIDS cases, but they dutifully self-censored while awaiting signals from Dr. Fauci and the medical cartel. Lawrence Altman, the chief medical writer for the *New York Times*, confessed to *Science* magazine that he did not break the story because he didn't think it was his paper's place to announce something without the CDC's go-ahead.[80]

Then, in the first days of the 1992 Amsterdam AIDS Conference, a naive young *Newsweek* reporter, Geoffrey Cowley, innocently reported a cascade of cases of non-HIV AIDS that he uncovered during quiet confessional conversations with Dr. Fauci's AIDS researchers. Several scientists confided to Cowley their bewildered alarm at the large number of AIDS patients who were uninfected with HIV. Cowley's report almost precipitated the collapse of Dr. Fauci's entire carefully fortified HIV-only theology.

"The patients are sick or dying, and most of them have risk factors," Cowley reported in *Newsweek*.[81] He described a dozen such cases of non-HIV patients with AIDS-like symptoms, including brain lesions, corresponding cognitive deficits, chronic aggravation of herpes viruses, depleted C4 cells, PCP pneumonia, and immune system collapse. "What they don't have is HIV."[82]

The *Newsweek* article shattered the taboo. Conferees took the public disclosure as a signal that they could now discuss the previously verboten subject of AIDS patients without HIV. Dr. Fauci's researchers, gathered in Amsterdam an ocean away from his heavy hand, suddenly began sharing their own stories of AIDS without HIV across the United States and Europe.

With the floodgates opened by *Newsweek* threatening to sweep away Dr. Fauci's official orthodoxy, Dr. Fauci raced out to Andrews Air Force Base with CDC AIDS

Task Force Director James Curran and flew to the Netherlands on Air Force 2 on a mission to quell the uprising.[83] (Curran, the head of the CDC's AIDS division, had famously conspired with Gallo to take the antibody patent from the French.) But by the time the two bureaucrats arrived, the horse had left the stable. Dr. Fauci and Curran had to sit through a series of rollicking conference sessions as mobs of reporters, mutinous scientists, and enraged activists besieged them with case studies and unanswerable questions. Public health regulators, physicians, and researchers expressed indignation that Dr. Fauci hadn't come clean with them. Many physicians caring for AIDS patients were furious that the government agency had not informed them about the non-HIV AIDS cases. Curran confessed that the CDC had known about these cases for years.

He feebly protested, "These are not cases of AIDS," reasoning, with circular gymnastics, that they couldn't be AIDS since the definition of AIDS requires the presence of HIV.[84] Dr. Fauci weakly reassured the gathering that he would soon resolve the crisis. The *New York Native* reported that Dr. Fauci, "the little man with the compensatory ego . . . looked like he was going to have a nervous breakdown in Amsterdam. We kept waiting to see him curled up in a fetal position and crying hysterically—desperate for forgiveness, desperate to create a smokescreen to make everyone forget how he has elbowed every critical question about HIV out of the way." Dr. Fauci was trying to sell himself as an open-minded scientist. He was telling people, "Don't panic, don't panic."[85]

In the weeks following the Amsterdam conference, the number of cases identified in the United States alone continued to grow, almost daily. Within a few weeks, the escalating cascade forced CDC to admit to eighty-two certified cases in fifteen states. It was a pitiful underestimate. Duesberg sent a letter to *Science*, offering to provide "a list of references to more than 800 HIV-free immunodeficiencies and AIDS-defining diseases in all major American and European risk groups," along with references to "more than 2,200 HIV-free African AIDS cases."[86] Duesberg afterward identified more than four thousand documented AIDS cases in the peer-reviewed scientific literature in which there is no trace of HIV or HIV antibodies.[87] This number is impressive because Dr. Fauci had cultivated strong institutional deterrents to such descriptions, and because formal scientific papers never described the vast majority of AIDS cases.

In an editorial for the *Los Angeles Times,* Steve Heimoff allowed that reports of "AIDS without HIV" would "appear to signal at least partial, temporary vindication" of Duesberg.[88] Describing Duesberg as "the unofficial leader of the revisionists," "an international star of virology long before anyone heard of AIDS," and "not just another conspiratorialist," Heimoff observed that Duesberg's arguments "have the ring of common sense."[89]

"If there is even a remote chance that Duesberg is correct—and the latest reports increase that possibility—then the powers that be must leap into action."[90]

New York Native publisher Charles Ortleb commented, "It should have been the end of the HIV theory and absolute proof that the CDC had gotten the definition and cause of AIDS wrong. The fact that HIV-negative AIDS was also occurring in Chronic Fatigue Syndrome (CFS) patients fortified suspicions of many virus experts that AIDS and CFS were part of the same neuroimmunological epidemic."[91]

A large contingent of HIV/AIDS critics (although not Peter Duesberg) had been clamoring that CFS and AIDS were a single disease—neither caused by HIV. To derail this lethal heresy, Dr. Fauci had set the compass for the medical community's reprehensible dismissal of CFS as a "psychosomatic illness."[92] Following Dr. Fauci's lead, doctors dubbed CFS as "Yuppie Flu," characterizing it as a neurotic affliction among women genetically unequipped for high-pressure corporate jobs that suddenly opened to them in the 1980s, coterminous with the lockstep pandemics of AIDS and CFS.[93]

A September 6, 1992 *Newsweek* article[94] by Geoffrey Cowley asked "AIDS or Chronic Fatigue?" Though Cowley took some heat for the article, he was merely voicing the quiet suspicion among many of Dr. Fauci's own PIs that "non-HIV AIDS" was actually CFS, and that CFS was simply another name for AIDS when it occurred in heterosexuals who tested negative for HIV. "As more cases come to light," Cowley observed, "it's becoming clear that the newly defined syndrome has as much in common with CFS as it does with AIDS."[95]

Tony Fauci moved quickly to silence this existential threat. Three weeks after the Amsterdam riot, the CDC sponsored a special meeting at its Atlanta headquarters, inviting the scientists reporting HIV-free AIDS cases. In attendance was a doleful Cowley, the *Newsweek* journalist, by now on a short leash with a choke collar.[96]

In a brazen move to explain away the anomaly of AIDS without HIV, Dr. Fauci declared that the unexplained AIDS cases represented a new disease. To avoid suspicion that his "new disease" was, after all, CFS, Dr. Fauci labeled his discovery "idiopathic CD4+ lymphocytopenia," or "ICL." In this tongue-twister, "idiopathic" means "of unknown source." It might also have been Dr. Fauci's ironic play on the word "idiot." But such was his wizardry that everyone just swallowed it without questions. The press meekly nodded at his circular reasoning like religious zealots jotting down the words of an infallible pope.

(For the record, I believe that HIV is a cause of AIDS, but Dr. Fauci's acknowledgment of non-HIV AIDS shows that causation is more complex than the official theology.)

Dr. Fauci had somehow resuscitated his theory from certain death by erecting an arbitrary wall between AIDS with and sans HIV. Because there was no evidence the mystery illness was contagious, Dr. Fauci hazarded a guess, to the tractable reporters, that the blood supply was probably safe. He offered no evidence to support this assurance, and the kowtowing media requested none. That was more than enough for Cowley. "Cowley, the *Newsweek* reporter, almost lost his career," Charles Ortleb told me. *Newsweek* published a remorseful article, and Cowley stopped reporting on AIDS cases without HIV, or even Dr. Fauci's new disease, ICL.

Then, on August 18, *New York Newsday* revealed that two of the "non-HIV AIDS" patients had Chronic Fatigue Syndrome, reigniting the dangerous controversy.[97]

Dr. Fauci rushed to appear on CNN's *Larry King Live* to reassure the general public that the new illness was not a threat to people outside the AIDS "risk groups."[98]

Writing in the *New York Native*, Neenyah Ostrom described Dr. Fauci's interview with King:

King began by asking Fauci to describe what he thought was happening in the "mysterious AIDS" cases in which patients develop severe immunodeficiency and types of infections suffered by "AIDS" patients—but are not infected with HIV. Fauci kept saying that between twenty and thirty such cases had been identified [Dr. Fauci knew that CDC had already confirmed eighty-two cases in fifteen states, and Duesberg had found thousands documented in *PubMed*: The NIH official peer-review archives] and because such a small number of people were affected, it really was nothing to worry about. Fauci said it wasn't clear that these cases represented a new type of "AIDS"; these patients' immunodeficiency could, he stressed, be caused by something other than an infectious agent. Fauci speculated that the cases might not even represent a new illness, but that increasingly sophisticated testing of people's immune systems was turning up what could be "background" immunodeficiencies (whatever that is).[99]

Ostrom described Dr. Fauci's awkward denial when one caller to the show asked whether the new mystery illnesses had "anything to do with Chronic Fatigue Syndrome." Fauci stated emphatically that it did not.

"Fauci was clearly uncomfortable talking about chronic fatigue syndrome," Ostrom reported, "and couldn't quite figure out where to look, so his eyes darted everywhere. . . . The show ended with an angry call from a physician in the Midwest who treats AIDS patients. He demanded to know why Fauci and other health officials had not informed physicians about the cases of non-HIV 'AIDS' before the information appeared in *Newsweek*. Shouldn't the doctors know about this before the mass media, the doctor asked sarcastically. Fauci became very defensive, asserting that it had only become clear in the last couple of weeks that the non-HIV 'AIDS' cases constituted a real phenomenon and, therefore, there had previously been nothing to inform the physicians of. He did not look happy at the show's end."[100]

Ostrom added this observation: "Fauci's good on television, as long as he's being touted as President George Bush's hero or patted on the back for rushing toxic drugs through the approval process without adequate safety testing. But when reporters start acting like reporters, as they have since the non-HIV cases came to light, Fauci's thin skin gets him into trouble; he becomes defensive, condescending and sarcastic."[101]

King initially scheduled Peter Duesberg to appear on the same show and apparently canceled Duesberg at Dr. Fauci's insistence.[102]

The Problem of HIV Without AIDS

Koch's first postulate also requires that the suspected pathogen should only be found in sick individuals, and *never* in healthy individuals.

It is therefore equally frustrating for HIV-only aficionados that widespread PCR use quickly revealed hundreds of thousands of individuals with HIV and no sign of illness. Dr. Fauci initially predicted that all of these individuals would die of AIDS within two years. Later he doubled their life expectancy to four years, and then to eight. Then he stopped talking about these upcoming tragedies altogether. Today, even Dr. Fauci's most loyal clergymen acknowledge that there are over 165,000 Americans and millions of individuals globally who carry the HIV virus without ill effect.[103] According to CDC estimates, approximately one-third of HIV-positives in the United States do not know their status.[104] If this is the case, Harvey Bialy points out, there should be a huge number of people dying suddenly of AIDS. This is not happening. In fact, the vast majority of those who test positive for HIV remain healthy for years. Duesberg and other critics argued that there is meager proof that people with HIV alone will not live a normal life span.[105]

Dr. Fauci has also taken energetic precautions to ensure that nobody study the prevalence of healthy HIV-infected people. In July 1996, *Newsday* reported that Dr. Fauci had suddenly aborted a $16 million, five-year study of the phenomenon midstream. According to journalist Laurie Garrett's July 11 story in *Newsday*, "Key HIV Contract Is Killed: Some See Retribution at Hands of NIH Official,"[106] was the largest study on HIV AIDS ever commissioned, involving research from over 100 scientists from leading institutions, including Harvard, the Aaron Diamond AIDS Research Center (in Manhattan), Northwestern University (Chicago), Duke University (North Carolina), and the University of Alabama. One of the study's central purposes was to examine the question that Dr. Fauci apparently didn't want answered, why some HIV infected individuals never succumb to AIDS. The five-year contract, which began in 1994, to fund this collaboration (formerly named the Correlates of HIV Immune Protections, or CHIPS), has no parallel in US "AIDS" research. Dr. Fauci's action effectively scuttled a year's worth of work by about 100 independent scientists. Aaron Diamond's Dr. David Ho told Garrett, "I'd like to see if Tony could find a contract anywhere in his portfolio that could match the productivity of this one."[107]

Newsday reported that the shocking cancellation was a retaliation against a group of younger scientists among this group who had signed a report (the "Levine Report") that criticized NIH's policy of only funding research that supported Dr. Fauci's HIV/AIDS orthodoxies. "This is payback time for Tony Fauci,"[108] said AIDS activist Gregg Gonsalves of Treatment Action Group, the offshoot of ACT UP formed to openly receive Pharma funding. He told *Newsday*, "It was an act of retribution by Tony Fauci, plain and simple."[109] In reporting the incident, the *New York Native* quoted NIAID insiders (*New York Native*, July 22, 1996), complaining that Dr. Fauci had fostered a reprisal culture at NIH. They said that their boss's favorite expression was "What goes around comes around."[110] Gonsalves called the cancellation a "vendetta" against the young scientists in the group who dared to ask for science-based funding strategies. It's just as likely that Dr. Fauci was searching for an excuse to terminate a study that threatened the entire HIV/AIDS paradigm.

The Problem with Isolating the Virus

Koch's second postulate is that the virus can be isolated from an ill individual and made to grow in pure culture. Highly respected scientists including Éttienne de Harven argued that HIV has never been isolated or grown in pure culture. Both Montagnier and Gallo have periodically acknowledged this deficiency.[111]

Instigating Disease with Cultured HIV

Koch's third postulate requires that the cultured microorganisms should cause disease when introduced into healthy individuals. Duesberg and others argue, till this day, that this proof is incomplete. In 1984, Montagnier acknowledged that: "The only way to prove that HIV causes AIDS is to show this on an animal model."[112]

No one has tried injecting HIV into a healthy human being, but scientists have stuck all kinds of mice and rats and monkeys and chimpanzees, and none of them got anything resembling human AIDS. No one has yet been able to induce AIDS by inoculating a healthy experimental animal with the cultured microorganism.

"There is no animal model for AIDS," agreed Nobel Laureate for Chemistry Walter Gilbert in 1989,[113] "and where there is no animal model, you cannot establish Koch's postulates." This failure, by itself, said Gilbert, left such a gaping hole in Gallo's theory that he "would not be surprised if there were another cause of AIDS and even that HIV is not involved."[114]

Evolutionary biologist James Lyons-Weiler argues that genetic sequencing of infected individuals proves sexual transmission of HIV. He also points to a 1991 judicial decision against a Florida dentist, Dr. David Acer,[115] who allegedly infected five patients with a contaminated drill, as definitive proof of Koch Postulate 3.[116] Subsequent investigations by 60 Minutes and others raised new doubts about the Acer verdict.[117]

Re-Isolating the Pathogen

Koch's fourth and final postulate is that the microorganism must be re-isolated from this inoculated experimental diseased host.

Duesberg argues that vigorous efforts by HIV/AIDS proponents to satisfy the postulates have all failed.[118] In Djamel Tahi's 1996 documentary *AIDS—The Doubt,* Professor Luc Montagnier admitted that after years of trying, no one had succeeded: "There is no scientific proof." Montagnier therefore concludes, "that HIV causes AIDS."[119] Koch's principles are still taught to every student of epidemiology, but his name is now a source of embarrassment rather than admiration and affection among AIDS researchers.

From cases I have litigated, I know that entire court cases hinge on the capacity of the attorneys and scientists to persuade a fact finder that the proponent of causation has satisfied Koch's postulates. It is the standard protocol for proving the causative relationship of a pathogen to a particular disease. Therefore, it came as a shocking revelation for me to learn that there remain possibly viable arguments that the HIV/AIDS hypothesis had consistently failed that standard. In the American judicial system, that evidence would normally be sufficient to close a case. I am not opining on the science here.

Viral Load Does Not Necessarily Correlate to Illness

Yet another acute embarrassment to Gallo's hypothesis is the problem of viral load. With most bacteriological and viral illnesses, increased viral load correlates with the progression of the disease and declines the patient's health. If HIV is the sole cause of AIDS, titers should be able to track an increase in viral loads as physical deterioration progresses. Traditional viruses such as herpes, influenza, smallpox, etc., only cause disease at very high titer—thousands or millions of infectious units per cubic millimeter of infected tissue. In contrast, HIV has proven barely to be found in AIDS patients even in the final throes of illness. HIV can be detected, but only with difficulty, because even the sickest AIDS patients simply don't have much virus to be found. And even more baffling, neither Dr. Fauci nor Gallo has ever credibly explained the fact that viral load from HIV is always at its greatest in the days immediately following infection. Logically, it would be during this period that the virus is most likely to cause devastating illness. And yet, the onset of AIDS symptoms almost always arrive decades later (an average twenty years following exposure)—when viral loads are at their lowest.

In 2006, a study published in the *Journal of the American Medical Association* (*JAMA*) once again shook the foundation of the past decade of AIDS science to its core and incited apoplexy among many HIV/AIDS advocates.[120] A US nationwide team of orthodox, mainstream AIDS researchers led by doctors Benigno Rodriguez and Michael Lederman of Case Western Reserve University in Cleveland strongly challenged the claimed legitimacy of viral load testing—the standard method since 1996 for assessing patient health, predicting disease progression, and winning grant approval for new AIDS drugs. Their study of 2,800 positively tested people concluded, in over 90 percent of cases, viral load measures failed to predict or explain immune status.[121]

Today, Rodriguez's Group stands by its conclusion that viral load is only able to predict progression to disease in 4 percent to 6 percent of (so-called) HIV positives studied, challenging much of the basis for current AIDS science and treatment policy.

The Lancet published a study showing that decreases in so-called "viral load" did not "translate into a decrease in mortality" for people taking these highly toxic AIDS drug combinations.[122] The multi-center study—the largest and longest of its kind—tracked the effects of Dr. Fauci's antivirals on some 22,000 previously treated HIV positives between 1995 and 2003 at twelve locations in Europe and the United States. The study refutes popular claims that HIV meds extend life and improve health.[123]

Can a Retrovirus as Elusive and Rare as HIV Cause Deadly Illness?

Equally mysterious is the question of how an elusive, rare, difficult-to-find virus could be causing so much carnage. Peter Duesberg told me if HIV was causing infections, "You would never need a PCR, a machine that multiplies HIV segments a billion-fold, to 'see' whether a person is 'infected.' Infection would be as obvious as it is with active flu or active polio. The body would be swarming with microbes."

Lauritsen argues, "The virus infects very, very few cells—as few as one in 100,000—and on top of that, it doesn't even kill the cells it infects."[124]

Since HIV typically infects so few cells,[125] that means Dr. Fauci's antiviral concoctions like AZT must kill many healthy T-cells in order to eliminate the few cells that are infected. It's worth considering that Dr. Fauci endorses administration of AZT and other chemotherapy concoctions for months on end or for as many years as AIDS patients manage to survive.

Furthermore, I haven't found any evidence that HIV ever actually kills a T-cell.[126,127] They seem to instead get along quite well. For this reason, critics argue the collapse of the immune system cannot be plausibly explained merely by the presence of HIV.

Duesberg is not surprised at the gaps in the evidence. After all, he says, how can a virus be so destructive when it first enters the body, then turns around and plays dead for 10, 20, 30 years?[128] Yet this is the orthodoxy. Dr. Jay A. Levy, MD, a leading UC AIDS researcher, posits: HIV is a kind of time-bomb virus that lies dormant in the body until—for some unknown and unexplainable reason—it modifies its own genetic structure and transforms into a fast-growing, virulent, deadly virus. Duesberg chuckles at this speculation: "What kind of virus one day, out of nowhere, springs into action to destroy a person's immune system with no provocation?"

Gallo and Dr. Fauci originally claimed that HIV causes immunodeficiency by killing CD4+ T-cells. But even the most faithful acolytes no longer believe that HIV kills T-cells in any way. Instead, they make what might seem to an outsider like a desperate pitch, that HIV primes T-cells to commit mass suicide at some later date. Dr. Fauci's followers have advanced this "Jim Jones" hypothesis to explain the lack of evidence for any cell-killing mechanism that can be attributed to HIV.

Duesberg laughs at this explanation: "No virus has ever behaved that way." "There are many shortcomings in the theory that HIV causes all signs of AIDS," admits Luc Montagnier.

Among the most outspoken dissidents of the HIV orthodoxy are biologist Eleni Papadopulos and physician Val Turner of the Australian Perth Group.[129] Papadopulos and Turner believe the particles Gallo identified as HIV are not even retroviruses, but rather are a class of cellular debris generated entirely from within the human body. Even Luc Montagnier admitted in an interview with the journal *Continuum* in 1997 that after "Roman effort," with electron micrographs of the cell culture, with which HIV was said to have been detected, no particles were visible with "morphology typical of retroviruses."[130]

A British-German research team in 2006 proudly reported that, finally, "the structure of the world's most deadly virus has been decoded" and that they had succeeded in photographing HIV in a "3-D quality never achieved before." But after independent scientists inspected the team's paper, they found that the images depicted appear to be a series of nondescript clumps of debris ranging wildly in sizes and shapes. The study was funded by Wellcome Trust, that has had from its inception, a collaborative relationship with the pharmaceutical industry, including Burroughs Wellcome, the pharmaceutical giant that makes multibillion-dollar revenues from AIDS medications like Combivir, Trizivir, and of course AZT.[131] The Wellcome Trust is a kind of hybridized British version of NIAID and the Bill and Melinda Gates Foundation. It largely funds studies that tend to promote profit taking by British pharmaceutical companies.

Does Fauci's Hypothesis Fail Farr's Law?

William Farr was the British microbiologist who designed the accepted method for predicting the spread of a new virus across a naive population. Farr declared that every "new" viral epidemic follows the same intractable laws, spreading exponentially within weeks at most, after the first infection—and then declining exponentially as it runs out of new uninfected persons. He declared that the rigid symmetrical rise and fall of death rates was so predictable as to be intractable law: "The death rate is a fact; anything beyond this is inference."

New infectious disease epidemics can virtually all be reliably plotted in a predictable bell curve resembling, in appearance, Farr's graph from London's 1849 cholera epidemic (below).

The Predictable Spread of Infectious Disease with Farr's Law

Scientists who accepted Dr. Fauci's hypothesis that HIV was a new virus were initially confident they could accurately predict a catastrophic spread in a naive human population. But all those predictions were wrong. At the end of each year, HIV's disappointing performance in imposing mortalities forced CDC to revise its estimates precipitously downward. Instead of a steep rise in infections, CDC's annual estimates of how many Americans are infected with HIV between 1986 and 2019 has remained fairly constant at approximately one million.[132] HIV did not spread or kill at anywhere near the rate expected of a newly introduced sexually transmitted virus.

The growth of HIV in Africa and the West does not follow the laws that have governed population-wide viral pandemic transmission throughout history. Since 1984, HIV has followed a steady monotonic point trajectory spreading from twenty-nine million in 1998 to forty-nine million in 2008. In Africa and elsewhere, the graph of AIDS has been a gradual steady slope following population growth almost perfectly country by country, without any of the widely predicted decreases in population.

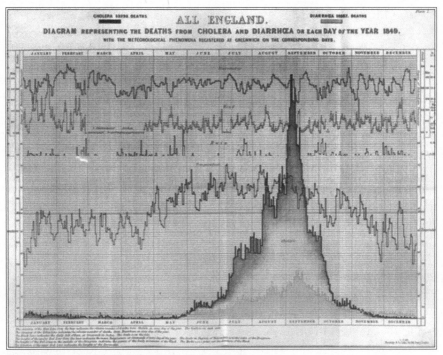

Report on the mortality of cholera in England, 1848-49.
Source: WellcomeCollection.org (Free to use under a CC-BY 4.0 license.)

The Spread of AIDS in the United States Post-1985

Dr. Rebecca Culshaw, a mathematical biologist and former AIDS researcher, went from unquestioning believer to converted heretic. The initial irony that captured her attention was the paradox of the preventive curve. It is, she observes, "indisputable fact that neither AIDS nor HIV have spread like they were predicted to. The predicted heterosexual AIDS explosion never happened, and even to mention this prediction now is almost taboo, as it is clearly an embarrassment to the AIDS establishment if HIV has not spread at all but rather it has remained constant in the population since its detection."[133]

In Western countries, AIDS has never broken away from its original core pool of homosexual men and drug addicts. That limit defies the pattern of every infectious and sexually transmitted disease throughout history. By definition, there can be no viral disease that does not break out of risk groups (poppers-consuming gays and those addicted to and frequently using hard drugs). This is especially true for HIV, because, as Dr. Fauci's acolytes claim, this is supposed to be "the most infectious virus that has ever existed." Assuming that is true, it is baffling that the virus did not frquently spread to women through sexual contact and did not affect all people all over the world equally.[134] It is especially baffling that AIDS does not spread to prostitutes except those who use intravenous drugs.[135,136]

The fact that AIDS does not obey the accepted rules that have reliably governed every other plague known to mankind is, Duesberg says, just more evidence that HIV is "an innocent bystander or a passenger virus."[137]

Enforced Consensus in a Sea of Dissenting Voices

The press long ago stopped reporting voices of dissent, but you now know, those voices are real. "It's like dying in outer space," Ortleb told me. "No one can hear you scream." But before questioning the orthodoxy became career suicide, some of the world's most prestigious scientists expressed such skepticism. It's worth revisiting some of these voices.

"We do not yet know how HIV causes AIDS," Dr. John Coffin of Tufts University, a member of the international committee that named the virus, told the delegates to the Sixth International Conference on AIDS in June, 1990.[138]

Dr. Shyh-Ching Lo, director of AIDS Pathology at the United States Armed Forces Institute of Pathology from 1986 through 2008, insisted that HIV could not be the sole cause of AIDS.[139]

In 2002, Dr. Bruce Evatt, CDC's director of the Division of Hematology, lamented that the CDC went to the public with statements for which there was "almost no evidence. We did not have proof it was a contagious agent."[140]

In September of 2004, Reinhard Kurth, former director of the Robert Koch Institute (one of the pillars of mainstream AIDS research), conceded in *Der Spiegel*: "We don't exactly know how HIV causes disease."[141]

In 1987, physiologist and MacArthur grant winner Robert Root-Bernstein told ABC correspondent John Hockenberry that he does not believe that HIV is necessarily the cause or the sole cause of "AIDS": "I've had people tell me bluntly that, 'I agree totally with your viewpoint that there are probably other things involved, that HIV can't cause AIDS by itself, that maybe you can get AIDS in the absence of HIV, but I'm not going to risk my million dollars of funding by saying that.'"[142]

Harvard Nobel Prize–winning molecular biologist Walter Gilbert told Hockenberry, "The major thing that concerns me, like calling HIV the cause of AIDS, is that we do not have a proof of causation. That's our major reason for being concerned."[143] Gilbert also said the problem with the HIV theory is the argument that "all cases of AIDS are associated with the virus and there is an inference made that all people with the virus will ultimately come down with AIDS. That's of course, not known to be a fact."

South Africa's pioneering AIDS researcher and physician, Dr. Joseph Sonnabend, chimed in: "The harm in the whole notion of the speculation being presented as fact is that if the speculation proved to be true, that means that research on whatever is truly going on has been neglected and this, of course, with a disease like AIDS, can be translated into the loss of tens of thousands of lives."[144]

Says prominent New York AIDS doctor Michael Lange, assistant head of Infectious Diseases and Epidemiology at St. Luke's Hospital: "We've lost . . . years in AIDS drug development . . . because of the Gallo/Essex/Haseltine axis boycotting other ideas."[145]

This chapter has outlined a meager skeletal description of just a few of the most common critiques of the hypothesis Dr. Fauci defends at all costs. Interested readers may find much more eloquent and thorough investigations in a number of books by various authors. Perhaps the best of these is mathematician Rebecca Culshaw's *Science Sold Out*. Culshaw was an AIDS researcher who slowly became disillusioned by the gaping chasms in the HIV/AIDS hypothesis, and by government corruption in maintaining the orthodoxy. Her book offers a sociological explanation as to how the theory was anointed by the media and scientific community. Other important books are Duesberg's *Inventing the AIDS Virus*, Lauritsen's book *The AIDS War*, *Osler's Web* by Hillary Johnson, and Harvey Bialy's *Oncogenes, Aneuploidy, and AIDS*. I also recommend "The Deconstruction of the AIDS" article by Yale mathematician Serge Lang, and an insightful chapter titled "Fear and Lawyers in Los Angeles" in Kary Mullis's *Dancing Naked in the Mind Field*.

Instead of civilly debating these dissidents and writers and common-sense questions posed by Duesberg and other critics, Dr. Fauci's strategy has been to exercise his frightening capacity to silence dissent and mangle reputations. History may credit him as the progenitor—even the inventor—of cancel culture.

My purpose here is not to take sides, much less to resolve disputes that have so far defied resolution for decades. Rather, I'm sharing something few people have been allowed to know: That there is a dispute, and that Tony Fauci has not allowed study that might resolve it. My hope is to chronicle Tony Fauci's role as high priest of an orthodoxy that today supports a multibillion-dollar global enterprise. Over the years, Dr. Fauci has deflected and evaded scientific debate and transformed theories into quasi-religious dogma, punishing and silencing dissent the way the Inquisition punished heresy. America's Doctor has never given the American taxpayers—or AIDS sufferers, 53 percent of whom are, in the United States, people of color—proof that AZT or its successive antivirals provide beneficial impacts on mortality. It seems fair, if not dangerous, to ask for that proof.

Endnotes

1   Mark J. Perry, "Michael Crichton explains why there is 'no such thing as consensus science,'" *AEI* (Dec. 15, 2019), https://www.aei.org/carpe-diem/michael-crichton-explains-why-there-is-no-such-thing-as-consensus-science/#:~:text=Science%2C%20on%20the%20contrary%2C%20requires,In%20science%20consensus%20is%20irrelevant.&text=There%20is%20no%20such%20thing%20as%20consensus%20science

2   Charles Ortleb, "The Chronic Fatigue Syndrome Epidemic Cover-Up," Rubicon Media, 2018, 356.

3   Natalie Angier, "SCIENTIST AT WORK: Anthony S. Fauci; Consummate Politician On the AIDS Front," NYT (Feb 15, 1994), https://www.nytimes.com/1994/02/15/science/scientist-at-work-anthony-s-fauci-consummate-politician-on-the-aids-front.html

4   Celia Farber, "The Passion of Peter Duesberg," *barnesworld* (Dec. 4, 2004), https://barnesworld.blogs.com/phdp.pdf

5   Peter H. Duesberg, "Retroviruses as Carcinogens and Pathogens: Expectations and Reality," *Cancer Research* (Mar 1, 1987), http://www.duesberg.com/papers/ch1.html

6   Peter H. Duesberg, *Inventing the AIDS Virus* (Regnery Publishing, Inc., 1996), 191

7   Luc Montagnier et al., "Isolation of a T-lymphotropic retrovirus from a patient at risk for acquired immune deficiency syndrome (AIDS)," *SCIENCE* (May, 1983), https://www.semanticscholar.org/paper/Isolation-of-a-T-lymphotropic-retrovirus-from-a-at-Barre%CC%81-Sinoussi-Chermann/7872732fd9c2e2bc5102408b477a8fd7adbe633f

8   John Crewdson, *Science Fictions: A Scientific Mystery, a Massive Cover-Up, and the Dark Legacy of Robert Gallo* (Little, Brown and Co., 2003)

9   Seth Roberts, "What AIDS Researcher Dr. Robert Gallo Did in Pursuit of the Nobel Prize," *SPY*, (June 1990), http://www.virusmyth.com/aids/hiv/srlabrat.htm

10   Ibid.

11   Ibid.

12   U.S. Department of Health and Human Services, Press Conference, Secretary Margaret Heckler (Apr 23, 1984), https://www.globalhealthchronicles.org/files/original/77248e9815176de4b40ffee97f4d855f.pdf

13   Crewdson, op. cit.,17

14   Ibid., 22

15   Roberts, op. cit. "What AIDS Researcher Dr. Robert Gallo Did"

16   David Remnick, "Robert Gallo Goes to War," *Washington Post,* Aug. 9, 1987, https://www.washingtonpost.com/archive/lifestyle/magazine/1987/08/09/robert-gallo-goes-to-war/224985a7-50f0-4526-acd5-91dedbbdc944/

17   Crewdson, op. cit., 29–32

18   Roberts, op. cit. "What AIDS Researcher Dr. Robert Gallo Did"

19   Anthony Liversidge, Interview Robert Gallo, *SPIN*, (Mar. 1989), https://www.virusmyth.com/aids/hiv/alinterviewrg2.htm

20   Roberts, op. cit. "What AIDS Researcher Dr. Robert Gallo Did"

21   Robert Gallo et al., "Frequent detection and isolation of cytopathic retroviruses (HTLV-III) from patients with AIDS and at risk for AIDS," *Science*, May 4, 1984, https://pubmed.ncbi.nlm.nih.gov/6200936/

22   Roberts, op. cit. "What AIDS Researcher Dr. Robert Gallo Did"

23   Ibid.

24   Peter D. Weinberg, Jennie Hounshell et al.,"Legal, Financial, and Public Health Consequences of HIV Contamination of Blood and Blood Products in the 1980s and 1990s," *Medicine and Public Issues* (Feb. 19, 2002), https://sci-hub.do/10.7326/0003-4819-136-4-200202190-00011

25   Jon Cohen, "HHS: Gallo Guilty of Misconduct," *Science* (Jan 8, 1993), https://www.sciencemag.org/site/feature/data/aids2002/pdfs/259-5092-168.pdf

26   Paul Kefalides, "Apathy, Outrage Accompany Leak of Unofficial Report On Gallo Case," *The Scientist* (Apr. 2, 1995), https://www.the-scientist.com/news/apathy-outrage-accompany-leak-of-unofficial-report-on-gallo-case-58571

27   John Crewdson, "Science Under the Microscope," *Chicago Tribune* (Nov. 19, 1989), https://www.chicagotribune.com/news/ct-xpm-1989-11-19-8903130823-story.html

28   Kerry Mullis, *Dancing Naked in the Mind Field* (Vintage Books, 1998), 178

29   Terry Michaels, *Down the HIV Rabbit Hole* (Unpublished Manuscript, 2017), 3

30   "HIV & AIDS Fauci's First Fraud," Documentary, YouTube video (Sept 6, 2020), 12:32, https://youtu.be/YVjcq3m3JNo

31   Hillary Johnson, *Osler's Web: Inside the Labyrinth of the Chronic Fatigue Syndrome Epidemic* (Crown Publishers, 1996), 250

32   Michael, op. cit. 2
33   Torsten Engelbrecht, Claus Köhnlein, et al., *Virus Mania: How the Medical Industry Continually Invents Epidemics, Making Billions at our Expense* (Books on Demand 3rd ed, 2021), 99
34   "HIV & AIDS Fauci's First Fraud," Documentary, YouTube video (Sept 6, 2020), 12:32, https://youtu.be/YVjcq3m3JNo
35   Engelbrecht et al., op. cit., 103
36   Centers for Disease Control and Prevention, HIV/AIDS Surveillance Report (1979), https://www.cdc.gov/hiv/pdf/library/reports/surveillance/cdc-hiv-surveillance-report-1997-vol-9-2.pdf
37   *HIV & AIDS Fauci's First Fraud Documentary,* 1:10.45
38   Paul Philpott, MS, Editor, *Rethinking AIDS* 1:10:43- 1:11:02
39   Christine Maggiore, Director Alive & Well AIDS Alternatives 1:11:03-1:11.30
40   Anthony Fauci, MD, 00:10:10 "House of Numbers: Anatomy of an Epidemic," YouTube Video (2009), https://www.youtube.com/watch?app=desktop&v=Vq8gT0xUcKY
41   *HIV & AIDS Fauci's First Fraud,* 45:04-45:21
42   "UNAIDS, FACT SHEET 2021 Preliminary UNAIDS 2021 epidemiological estimates" (2021) https://www.unaids.org/sites/default/files/media_asset/UNAIDS_FactSheet_en.pdf
43   F X Keou et al., "World Health Organization clinical case definition for AIDS in Africa: an analysis of evaluations," *East African Medical Journal* (Oct., 1992), https://pubmed.ncbi.nlm.nih.gov/1335410/
44   World Health Organization, *Workshop on AIDS in Central Africa- Bangui, Central African Republic,* WHO (22-25 Oct. 1985), https://www.who.int/hiv/strategic/en/bangui1985report.pdf
45   Engelbrecht et al., op. cit., 165
46   Ibid., 321
47   Rebecca Culshaw, *Science Sold Out: Does HIV Really Cause Aids?* (North Atlantic Books, The Terra Nova Series, 2007), 5
48   Ibid., 4
49   "HIV & AIDS Fauci's First Fraud" documentary, https://www.youtube.com/watch?v=YVjcq3m3JNo at 1:09.57
50   Mikulas Popovic, Robert Gallo, et al., "Detection, Isolation, and Continuous Production of Cytopathic Retroviruses (HTLV-111) from Patients with AIDS and Pre-AIDS," *Science* (May, 1984), https://sci-hub.do/10.1126/science.6200935
51   Robert Gallo, Mikulas Popovic, et al, "Frequent Detection and Isolation of Cytopathic Retroviruses (HTLV-III) from Patients with AIDS and at Risk for AIDS," *Science* (May, 1984), http://dx.doi.org/10.1126/science.6200936
52   Françoise Barré-Sinoussi, Luc Montagnier, et al., "Isolation of a T-Lymphotropic Retrovirus from a Patient at Risk for Acquired Immune Deficiency Syndrome (AIDS)," *Science* (May, 1983), https://sci-hub.do/10.1126/science.6189183
53   *Nature* 1992, 357:189—John Maddox, Editor, *Nature Magazine*
54   Progressive Radio Voices, "Is 'HIV' Really the Cause of AIDS? Are There Really Only 'a Few' Scientists Who Doubt This? (Part 1)," https://prn.fm/is-hiv-really-the-cause-of-aids-are-there-really-only-a-few-scientists-who-doubt-this-part-1/
55   *HIV & AIDS Fauci's First Fraud,* op. cit., 10:35
56   Charles Ortleb, *Fauci: The Bernie Madoff of Science and the HIV Ponzi Scheme that Concealed the Chronic Fatigue Syndrome Epidemic* (Rubicon Media, 2019), 26
57   Engelbrecht et al., op. cit., 107
58   Dr. James D. Watson Directors Report, Cold Spring Harbor Laboratory Annual Report (Dec. 31, 1971), 4, https://core.ac.uk/reader/158273440
59   Engelbrecht et al., op. cit., 96
60   The Perth Group, "The Perth Group revisits the existence of HIV," *The Perth Group* (May 29, 2008), http://www.theperthgroup.com/LATEST/PGRevisitHIVExistence.pdf
61   Engelbrecht et al., op. cit., 107
62   Ibid., 107
63   Ibid., 107
64   "Is the 'AIDS Test' Accurate?" (Abbott Laboratory's ELISA HIV antibody test kit pamphlet), *ALIVEANDWELL.ORG,* https://www.aliveandwell.org/html/questioning/questioningthetests.html
65   *HIV & AIDS Fauci's First Fraud,* Documentary YouTube video (Sept 6, 2020), Christine Maggiore, Director, Alive & Well AIDS Alternatives (01:03:03), https://youtu.be/YVjcq3m3JNo
66   Engelbrecht et al., op. cit., 109
67   Carolyn Johnson and Lenny Bernstein, "Decades of research on an HIV vaccine boost the bid for one against coronavirus," *Washington Post,* Jul 14, 2020, https://www.washingtonpost.com/health/decades-of-research-on-an-hiv-vaccine-boosts-the-bid-for-one-against-coronavirus/2020/07/13/3eb1a37a-c216-11ea-b4f6-cb39cd8940fb_story.html
68   Prescott L Deininger and Mark A Batzer,"Mammalian Retroelements" *Genome Research* (Oct. 12, 2002), https://pubmed.ncbi.nlm.nih.gov/12368238/
69   Engelbrecht et al., op. cit., 109
70   Ibid., 109

71   Josiah Rich et al., "Misdiagnosis of HIV Infection by HIV-1 Plasma Viral Load Testing: A Case Series," *Annals of Internal Medicine* (Jan. 5, 1999), https://sci-hub.do/10.7326/0003-4819-130-1-199901050-00007

72   Celia Farber, "Interview Kary Mullis: AIDS; Worlds from the Front," *Spin* (July 1994), http://virusmyth.com/aids/hiv/cfmullis.htm

73   Concorde Coordinating Committee, "Concorde: MRC/ANRS randomised double-blind controlled trial of immediate and deferred zidovudine in symptom-free HIV infection," *The Lancet* (Apr 9, 1994), https://sci-hub.do/10.1016/S0140-6736(94)90006-X

74   Thomas Fleming and David DeMets, "Surrogate End Points in Clinical Trials: Are We Being Misled?" *Annals of Internal Medicine* (Oct 1, 1996), https://sci-hub.do/10.7326/0003-4819-125-7-199610010-00011

75   Engelbrecht et al., op. cit., 110

76   *HIV & AIDS Fauci's First Fraud*, Documentary, YouTube video (Sep 6, 2020), 1:06, https://youtu.be/YVjcq3m3JNo

77   Segre, Julia A. "What does it take to satisfy Koch's postulates two centuries later? Microbial genomics and Propionibacteria acnes," *The Journal of investigative dermatology*, vol. 133,9 (2013): 2141-2. doi:10.1038/jid.2013.260 https://www.ncbi.nlm.nih.gov/pmc/articles/PMC3775492/

78   Peter Duesberg, *Human immunodeficiency virus and acquired immunodeficiency syndrome: correlation but not causation*, Proceedings of the National Academy of Sciences of the United States of America (Feb. 1989), https://www.ncbi.nlm.nih.gov/pmc/articles/PMC286556/

79   "Morbidity and Mortality Weekly Report (Supplement 1S)," *JAMA* (Sep. 4, 1987)

80   John Cohen, "Doing Science in the Spotlight's Glare," *Science* (Aug. 21, 1992), https://science.sciencemag.org/content/257/5073/1033

81   Steven Epstein, *AIDS, Activism and the Politics of Knowledge* (University of California Press 1996), 160, https://publishing.cdlib.org/ucpressebooks/view?docId=ft1s20045x&chunk.id=d0e5802&toc.depth=100&toc.id=d0e5247&brand=ucpress

82   Ibid., 160

83   Bryan Ellison, "AIDS: Words From The Front," *SPIN* (Dec., 1993), http://www.virusmyth.com/aids/hiv/beeis.htm

84   Epstein, op. cit., 161

85   Ortleb, op. cit., 288–289

86   Peter Duesberg, *Impure Science* (Univ. of CA Press, 1996), 161, http://ark.cdlib.org/ark:/13030/ft1s20045x/

87   Peter Duesberg, "The HIV Gap in National AIDS Statistics," *Nature Biotechnology* (Aug. 11, 1983), https://sci-hub.do/https://doi.org/10.1038/nbt0893-955

88   Steve Heimoff, "Test Ideas With Science, Not Scorn: Critics who insist that HIV doesn't cause AIDS may be wrong, but their argument deserves checking," *Los Angeles Times*, Jul. 28, 1992, https://www.latimes.com/archives/la-xpm-1992-07-28-me-4481-story.html

89   Ibid.

90   Ibid.

91   Steven Epstein, *Impure Science: AIDS, Activism, and the Politics of Knowledge*, Ch 4, ref 82 and 83 (Berkeley: University of California Press, c1996), http://ark.cdlib.org/ark:/13030/ft1s20045x/

92   Ortleb, op. cit., 32

93   Robert F. Kennedy Jr., "The Truth About Fauci," KENNEDY NEWS & VIEWS, (Apr. 20, 2020), https://childrenshealthdefense.org/news/the-truth-about-fauci-featuring-dr-judy-mikovits/

94   Newsweek staff, "AIDS or Chronic Fatigue?" *Newsweek* (Sep. 6, 1992), https://www.newsweek.com/aids-or-chronic-fatigue-198572

95   Ibid.

96   Bryan Ellison, "AIDS; Words From the Front," *SPIN* (Dec., 1993), http://www.virusmyth.com/aids/hiv/beeis.htm

97   Neenyah Ostrom, *50 Things You Should Know About the Chronic Fatigue Syndrome Epidemic* (St. Martin's Press, 1993), 131

98   Ibid., 130

99   Ortleb, op. cit., 292

100   Ibid., 293

101   Ibid., 293

102   Duesberg, 392–393

103   HIV.gov, "Too Many People Living with HIV in the U.S. Don't Know It," (Jun. 10, 2019), https://www.hiv.gov/blog/too-many-people-living-hiv-us-don-t-know-it

104   Lawrence K. Altman, "Many in U.S. With H.I.V. Don't Know It Or Seek Care," *New York Times* (Feb. 26, 2002), https://www.nytimes.com/2002/02/26/us/many-in-us-with-hiv-don-t-know-it-or-seek-care.html

105   Ortleb, op. cit., 228

106 Laurie Garrett, "Key HIV Contract Is Killed: Some see retribution at hands of NIH official," *Newsday* (Jul 11, 1996), A34

107 Neenyah Ostrom, "Fauci Kills Huge Cooperative 'AIDS' Research Contract: Revenge for Negative Report Cited as Motive," *New York Native*, July 12, 1996, https://4.bp.blogspot.com/-sEW2rQgOt6A/V8b6ORJsapI/AAAAAAAAD2w/7GfidS40DugoAUQscWwRv6rjI5YKWak5ACLcB/s1600/New%2BYork%2BNative%2BJuly%2B22%252C%2B1996%2Bpage%2B2.png

108 Garrett, op. cit., A34

109 Ibid.

110 Ostrom, op. cit., https://3.bp.blogspot.com/-0M9QLFlnZhg/V8b5ub6vJgI/AAAAAAAAD2s/8nUSo2ngOp4XRlog8gdTPJ_IVBalYd4ZACLcB/s1600/New%2BYork%2BNative%2B%2BJuly%2B22%252C%2B1996.png

111 Eleni Papadopulos-Eleopulos et al., "Has Gallo proven the role of HIV in AIDS?" *The Perth Group* (April 2, 1993), http://www.theperthgroup.com/SCIPAPERS/EPEGalloProveRoleHIVEmergMedOCR1993.pdf

112 Engelbrecht et al., op. cit., 106

113 Ron Rapoport, "AIDS: The Unanswered Questions," *Oakland Times*, May 22, 1989, at A1-A2

114 Engelbrecht et al., op. cit., 138

115 Rob Hiaasen, "Dr. Acer's Deadly Secret: How AIDS joined the lives of a dentist and his patients." Palm Beach Post, September 29, 1991. https://www.palmbeachpost.com/article/20160105/LIFESTYLE/812033993

116 D. Hillis, J. Huelsenbeck, "Support for dental HIV transmission," Nature 369, 24–25 (1994), https://www.nature.com/articles/369024a0

117 Carol A. Ciesielski, M.D. et al., "The 1990 Florida Dental Investigation: The Press and the Science," Annals of Internal Medicine (December 1, 1994), https://www.acpjournals.org/doi/10.7326/0003-4819-121-11-199412010-00011

118 Ibid., 135

119 Ibid., 106

120 Benigno Rodriguez and Michael Lederman, "Predictive Value of Plasma HIV RNA Level on Rate of CD4 T-Cell Decline in Untreated HIV Infection," *JAMA* (Sep. 27, 2006), https://sci-hub.do/10.1001/jama.296.12.1498

121 Engelbrecht et al., op. cit., 109

122 The Antiretroviral Therapy (ART) Cohort Collaboration, "HIV treatment response and prognosis in Europe and North America in the first decade of highly active antiretroviral therapy: a collaborative analysis," *The Lancet* (Aug. 5, 2006), https://www.thelancet.com/journals/lancet/article/PIIS0140-6736(06)69152-6/fulltext

123 Engelbrecht et al., op. cit., 145

124 Lauritsen, *The AIDS War*, 57

125 Peter H. Duesberg, *Inventing the AIDS Virus*, 358

126 Peter H. Duesberg, "Human immunodeficiency virus and acquired immunodeficiency syndrome: Correlation but not causation*," Proceedings of the National Academy of Science (Feb. 1989), https://www.ncbi.nlm.nih.gov/pmc/articles/PMC286556/

127 Peter H. Duesberg and Jody R. Schwartz, "Latent Viruses and Mutated Oncogenes: No Evidence for Pathogenicity," *Progress in Nucleic Acid Research and Molecular Biology* (1992), https://www.sciencedirect.com/science/article/pii/S0079660308610478?via%3Dihub

128 Peter H. Duesberg, "Human immunodeficiency virus and acquired immunodeficiency syndrome: Correlation but not causation*," PROCEEDINGS OF THE NATIONAL ACADEMY OF SCIENCE, 758 (Feb. 1989), https://www.ncbi.nlm.nih.gov/pmc/articles/PMC286556/

129 Ibid., 104

130 Engelbrecht et al., op. cit., 103

131 Ibid.

132 David Rasnick, "Constant HIV Rates in the U.S." (2019), http://www.davidrasnick.com/aids/constant-one-million-hiv.html. Click on year for references.

133 Culshaw, *Science Sold Out*, 4

134 Engelbrecht et al., op. cit., 321–322

135 Duesberg, *Inventing the AIDS Virus*, op. cit., 470, 543, 567

136 Peter Duesberg, "AIDS Acquired by Drug Consumption and other Noncontagious Risk Factors," *Pharmacology & Therapeutics* (1992), https://pubmed.ncbi.nlm.nih.gov/1492119/AIDS

137 Duesberg, *Inventing the AIDS Virus*, 190

138 Elinor Burkett, "Is HIV Guilty?" *The Miami Herald*, Dec. 23, 1990, https://www.virusmyth.com/aids/hiv/ebhiv.htm

139 Lawrence Broxmeyer, "Is AIDS really caused by a virus?" *Medical Hypotheses* (May 2003), https://citeseerx.ist.psu.edu/viewdoc/download?doi=10.1.1.424.5640&rep=rep1&type=pdf

140 Engelbrecht et al., op. cit., 70

141 Ibid., 105–106

142 Ortleb, *The Chronic Fatigue Syndrome Cover-Up,* 320

143 Ibid., 321
144 Ortleb, *The Chronic Fatigue Syndrome Cover-Up*, 321
145 Barry Werth, "The AIDS Windfall," *New England Monthly* (Jun., 1988), https://www.virusmyth.com/aids/hiv/bwwindfall.htm

ChildrensHealthDefense.org/fauci-book
childrenshd.org/fauci-book

For updates, new citations and references, and
new information about topics in this chapter:

CHAPTER 6
BURNING THE HIV HERETICS

"Why they did it," he says, "I cannot figure out. Nobody in their right mind would jump into this thing like they did. The secretary of health just announcing to the world like that that this man Robert Gallo, wearing those dark sunglasses, had found the cause of AIDS. It had nothing to do with any well-considered science. There were some people who had AIDS and some of them had HIV—not even all of them. So they had a correlation. So what?"
— Kary Mullis, PhD, Nobel Laureate, PCR Inventor[1]

In 1991, seven years after Robert Gallo's May 1984 article in *Science*, Harvard microbiologist Dr. Charles Thomas organized the éminences grises of virology and immunology to formally register their objections to Gallo's HIV hypothesis in an historical letter to *Nature*. The group was a Who's Who of international scientific doyens and Nobel laureates, among them Dr. Walter Gilbert of Harvard; PCR inventor Kary Mullis; Yale mathematician Serge Lang (a member, and watchdog, of the National Academy of Sciences); Dr. Harry Rubin, professor of Cell Biology at UC Berkeley; Dr. Harvey Bialy, cofounder of *Nature Biotechnology*; Bernard Forscher, PhD, ret. editor of *Proceedings of the National Academy of Sciences;* and many others.

The letter was only four sentences long:

> It is widely believed by the general public that a retrovirus called HIV causes a group of diseases called AIDS. Many biomedical scientists now question this hypothesis. We propose a thorough reappraisal of the existing evidence for and against this hypothesis, to be conducted by a suitable independent group. We further propose that the critical epidemiological studies be devised and undertaken.[2,3]

It seemed like a reasonable request. These esteemed researchers were only asking for the open debate and investigation about an extremely consequential scientific assertion that had, somehow, never occurred. But in an early display of Dr. Fauci's and Big Pharma's combined power to control the medical journals, *Nature* declined to publish the letter. Nor would *New England Journal of Medicine, JAMA*, or the *Lancet*. These journals rely on the pharmaceutical industry for upward of 90 percent of their revenues and seldom publish studies that threaten the Pharma paradigm. As *Lancet* editor Richard Horton has observed, "The journals have devolved into information laundering operations for the pharmaceutical industry."[4] Dr. Fauci exercises direct influence on the content that appears in their journals. Control of peer-reviewed publishing is a vital ingredient for constructing orthodoxies.[5]

When *Nature* rejected the letter, Thomas and Bialy subsequently organized a consortium, *The Group for Scientific Reappraisal of HIV/AIDS Hypothesis*, and in 1992, Thomas called it ". . . tantamount to criminal negligence"[6] for scientists to remain silent. "Of the fifty-three who had signed by June 1992, twelve had M.D.'s and twenty-five had Ph.D.'s. Twenty of the fifty-three gave academic affiliations with departments like physiology, biochemistry, medicine, pharmacology, toxicology, and physics."[7] Over 2,600 people, including three Nobel laureates, Walter Gilbert, Kary Mullis and two-time winner Linus Pauling, and 188 reputable PhDs, added their signatures. ("Rethinking-AIDS" website lists more than two thousand distinguished members: www.rethinkingaids.com.)[8], [9]

But the steady flow of money from NIAID was already annealing Gallo's viral hypothesis into ironbound orthodoxy, and those dissenting voices met the hardened steel of fortified institutional resistance. Tony Fauci's loosened purse strings had launched the HIV gold rush, and the government virologists and pharmaceutical PIs had circled their stagecoaches around Gallo's sketchy hypothesis and were lined up for handouts at the NIAID chuckwagon.

"They've got to hold onto HIV. Why?" observed Dr. Charles Thomas dolefully. "To hold on to their funding."[10]

Scratching his head, Kary Mullis commented, "There's something wrong here. It's got to be financial."[11] He explained, "The mystery of that damn virus," he says, "has been generated by the $2 billion a year they spend on it. You take any other virus, and spend $2 billion, and you can make up some great mysteries about it, too."[12]

Peter Duesberg

Among the scientists who added their name to the later version of the letter was an iconoclastic German-born prodigy with twinkling eyes, a biting wit, a boyish face, and a ready smile.

In the 1970s and 1980s, molecular biologist Professor Peter Duesberg (born December 2, 1936) was a demigod of molecular biology and among the world's best-known and highly respected scientists. The National Institutes of Health (NIH) generously supported his virology and cancer research. In 1986, NIH awarded Duesberg its special Cancer Fellowship, as well as the highly coveted Outstanding Investigator Grant, which the agency reserves for the top scientists in the country. NIH designed the seven-year grant to allow gifted scientists to push the boundaries of their specialties by removing the pressures of grant writing. The elite National Academy of Sciences inducted Duesberg into its Scientist Hall of Fame at the age of fifty, making him one of its youngest members ever.

At the University of California, Berkeley, Duesberg became the first to map the genetic structure of retroviruses like HIV, making him among the world's most renowned retrovirologists. A retrovirus is a primitive life form that has no capacity to replicate on its own, as is true of all viruses. The retrovirus injects its RNA into an existing cell, where an enzyme called reverse transcriptase converts viral RNA into DNA, which is then inserted (or spliced) into the host cell's DNA. Virologists generally believe that retroviruses are harmless, even beneficial, in a symbiotic relationship with humans during three billion years of evolution, providing mobile DNA blocks in the human genome. In fact, many of our genes first entered our genome as retroviruses.[13,14] Some 8–10 percent of human DNA is retroviral," says Dr. David Rasnick, "That's a hell of a lot."[15]

By 1970, at thirty-three, Duesberg won acclaim for having discovered the first cancer-causing gene. Duesberg and his fellow virologist Peter Vogt discovered the so-called "oncogene" inside a retrovirus that appeared to cause cancer. Duesberg's discovery gave rise to the "mutant gene theory" and unleashed a boom in a new discipline of cancer research. Colleagues expected Duesberg to win the Nobel Prize.

But Duesberg was the consummate scientist, believing researchers ought to experiment and reason from what they observe and ruthlessly question every orthodoxy, including their own. Duesberg therefore subjected his oncogene theory to more rigorous tests than had any of its critics. Before he got the magical call from Stockholm, Duesberg became convinced that his own momentous discovery had been a clinically irrelevant lab fluke. Publicly shrugging off his hypothesis, which had already electrified a new field—Duesberg himself debunked the theory, incinerating his Nobel prospects and his friendship with Peter Vogt. Harvey Bialy, Duesberg's biographer, reports Duesberg saying, "I would prefer to be honest even against my own interests."[16]

Duesberg was uncompromisingly committed to clean functional proof, at a time when electron microscopy and other technologies for detecting new viruses were making biology—particularly the study of viruses—increasingly murky. Fame and finance were driving the frenzy in viral research. With official and commercial encouragement, researchers were blaming newly discovered viruses as the culprits in an assortment of ancient diseases. NIAID and pharmaceutical companies readily funded this research, which often opened a straight path to patentable antivirals. A virologist who convincingly linked a "new" virus to an existing cancer or disease could enjoy relevance, rich financial remuneration, and professional glory. Pharmaceutical companies were minting profits from a pharmacopoeia of patented antivirals devised by isolating these viruses and identifying compounds that could kill them. Every research scientist was aware of the Nobel committee's bias toward breakthroughs that boosted Pharma's profit potentials.

From the outset, Duesberg had nagging doubts about Robert Gallo's findings. From an evolutionary standpoint, it didn't make sense that an ancient retrovirus would attack its human host. Retroviruses, in the form of incomplete strands of DNA inserted into human DNA, have no metabolism and no proven capacity to digest, reproduce, or evolve. They are not, by accepted definition, a life form. It would be a surprise if evolution had, through some unknown mechanism, transformed any of these into a cancerous or a killer cell.

Gallo's outspoken ambitions for the Nobel Prize were notorious: "What else would you expect from a person like Gallo who had studied retroviruses all his life—that he would say that it was a retrovirus causing AIDS. That seemed to be the first coincidence that made me wonder whether that was an authentic claim. But to me, it was not a surprise that he would say that. He said it before, that it would cause Alzheimer's or leukemia or neurological diseases and it failed. So I was not too impressed that this was going to be a winner."[17]

Following Gallo's announcement, Duesberg spent eighteen months studying every scientific publication on HIV and AIDS. He finally published his observations in the prominent journal *Cancer Research* in March 1987, in an explosive article with the banal title *Retroviruses as Carcinogens and Pathogens: Expectations and Reality.*[18]

Duesberg's article was a tour de force from the reigning father of retrovirology, calling for sobriety in the booming field that he saw spinning out of control. A young generation of virologists, armed with electron microscopes and other novel instruments and seeking wealth and career advancement, were pinning retroviruses as the culprits for every malignancy, with meager functional or empirical proof, or rigorous evidence-based science to explain the mechanism by which they caused disease. Duesberg exploded the idea that retroviruses cause leukemia, cancers in general, and finally AIDS (the cellular opposite of leukemia). He pointed out that, however one feels about the HIV hypothesis, it was a total reversal of the universal consensus about retroviruses before Gallo's April 1984 press conference. Duesberg reminded his colleagues that retroviruses—which have been a part of the human genome for as long as three billion years—are not "cytocidal" (cell killers). AIDS, Duesberg mused, is a disease of cell death, while leukemia is a disease of cell proliferation. By claiming initially that HIV caused leukemia, and later, AIDS, Gallo was accusing the bug of opposite reactions. Furthermore, Duesberg adds, "it would have been the first time that a retrovirus would have been pinned down as a cause of a human disease. Or even a disease in wild animals."[19]

Duesberg argued that HIV is capable of causing neither cancer nor AIDS. It is instead, he declared, a harmless passenger virus that has almost certainly coexisted in humans for thousands of generations without causing diseases. Duesberg concluded that the creature Gallo claimed to be a pandemic pathogen was simply one of many harmless passenger viruses, which innate and adaptive human immunity quickly hold at bay. "There are no slow viruses" causing AIDS, the acid-tongued Duesberg quipped, "only slow scientists."[20] HIV is not pathogenic, either in the industrialized world or the Third World.

Duesberg's *Cancer Research* paper was a lengthy, highly technical paper that raised a series of clear, compelling questions challenging point by point the basis of Gallo's HIV/AIDS hypothesis.

Duesberg's opus was a sweeping reality check against overblown claims for retroviruses, written by the man who at that point in history was thought to know them better than anybody. Many of his colleagues who studied Duesberg's research came to the same conclusion: something was terribly wrong with the war on AIDS.

In 1997, Berkeley's brilliant cell biologist, Dr. Richard Strohman, recalled the impact of Duesberg's elegantly structured arguments in the elite universe of cancer research: "It was a remarkable review and it raised the fundamental issues about virus as a cause of both cancer and immunosuppression—basic questions that haven't been really responded to in any meaningful way in the almost ten years since the date it was published."[21]

Do Retroviruses Cause Diseases?

Duesberg's skepticism about HIV/AIDS hypotheses quickly spread across the research community. The most fertile ground for incredulity was among researchers who knew the most about retroviruses. During the late 1990s, diverse teams of elite scientists began working on decoding the Human Genome. The idea of a cell-killing retrovirus made little sense to them from an evolutionary standpoint. Molecular biologist Harvey Bialy, scientific editor of *Nature Biotechnology*, remembers where he was when he first heard the news that NCI's Bob Gallo had found the cause of AIDS and that it was a retrovirus. "A colleague told me," says Bialy. "I was on my way to New York. It was January 1984. I remember laughing. 'A cytopathic retrovirus? This is just more Gallo bullshit,' I said. I said, 'it will never fly.'"

Bialy points out, "We all have tens of thousands of retroviruses in our germline and yet none of them has ever been demonstrated to be pathogenic."

Bialy told Celia Farber that Gallo, Dr. Fauci, and the thousands of researchers that Dr. Fauci funded to develop ways to kill HIV have never explained how Montagnier's virus could possibly be responsible for all the harms and diseases attributed to it: "It would have been the major single explanation that [Gallo's] hypothesis would have had to provide in order to be taken seriously. How do you account for the pathogenicity of this sleepy virus that has not a single pathogenic relative and in fact has 98,000 relatives quietly residing in the human germline? *Fuck.* 98,000 in the germline! Not in your body cells! In your *ovaries! Getting passed on from generation to generation for as long as human beings have been on this goddamn planet. Every single one of them is clearly not only not pathogenic but totally harmless.* This is the most powerful proof that what Peter has been saying for twenty years now is absolutely correct."

Nobel Laureate Kary Mullis expressed his astonishment at the credulousness of the scientific community. For him, it defied common sense that, after hundreds of years of scientific research, one medical scientist, Bob Gallo, had suddenly discovered the true cause of thirty ancient diseases in the United States and Europe, and a retinue of at least thirty more in Africa, and traced them all to a simple creature with a hundred thousand relations, none of them known to cause any disease. "Things don't happen that fast in science. You don't suddenly notice that one new organism is causing every problem. I mean, it was a bizarre thing that happened. It really was. It didn't really have any precedents in terms of medicine before that. Unless perhaps you could think of the 'possession by the devil' stuff, right? In that once you're possessed by the devil, anything that happens to you? [. . .] So it makes it easier for you to get tuberculosis, and it makes it easier for you to get uterine cancer. It makes it easier for you to get *candida albicans.* And so all those things can now be called AIDS; why would anybody do that? Why would any reasonable doctor start lumping together various symptoms into one pile and think all this is caused by HIV?"[22] Christine Maggiore adds, "We have a test, but it's not a test for AIDS; and it's called an HIV test, but it's not a test for HIV; and we have a series of problems that we are calling AIDS, but that doesn't elevate AIDS into a disease."[23]

Thirty years later, many, if not most, virologists have come to grudgingly accept—in some part, at least—Duesberg's skepticism of the Gallo/Fauci claim that HIV, alone, could cause AIDS. Most research scientists now—quietly—assume that AIDS must have a multifactorial etiology. Significantly, Dr. Robert Gallo and Dr. Luc Montagnier have placed themselves in this cohort. Dr. Tony Fauci is one of the few exceptions.

Other respected scientists took Duesberg's doubts even further than Duesberg. Led by Dr. Eleni Papadopulos and Dr. Val Turner, The Perth Group in Australia argues that Gallo's claim was altogether specious and that neither Gallo nor Montagnier had ever succeeded in even isolating a discrete HIV.

In my conversations with Turner and Papadopulos, and in my reading of their paper, I find their arguments clear and convincing. However, I recognize that there are some fifty thousand articles on AIDS in the scientific literature. A casual novitiate like myself has little chance of unraveling this baroque controversy in a vacuum. Without rigorous debate, the public and press must form opinions based upon appeals to authority—a feature of religion, not of democracy or science. Any debate on that battleground will always be won by self-interested government and industry officials who control the bullhorn and the media.

Rather than airing and openly debating such critiques, Tony Fauci and his PI army moved actively and effectively to snuff out the careers and silence the arguments of any scientist or journalist who questioned the official canons of the new state theology.

Punishing Duesberg

On their face, Duesberg's incendiary queries seemed to create an irresistible bulwark against Dr. Fauci's HIV-only hypothesis. Even today, Duesberg's rationales appear so clean, so elegantly crafted, and so compelling that, in reading them, it seems impossible that the entire hypothesis did not instantly collapse under the smothering weight of relentless logic. The scientific world waited to see how Drs. Gallo and Fauci could possibly answer Duesberg's devastating questions.

But the AIDS cartel never attempted a reply. Instead, Dr. Fauci met this existential assault by simply ignoring it and by castigating anyone who credited it. He set about making Duesberg an example to discourage future inquiries. Dr. Fauci made sure that, in Bialy's words, the article had "disastrous professional consequences" for Duesberg and "sealed his scientific fate for a dozen years."[24] Dr. Fauci orchestrated a fusillade of withering and venomous attacks that effectively ended Duesberg's illustrious career.

Dr. Fauci summoned the entire upper clergy of his HIV orthodoxy—and all of its lower acolytes and altar boys—to unleash a storm of fierce retribution on the Berkeley virologist and his followers. The dispute became one of the most sensational, vicious, and personalized battles in the history of science. Dr. Fauci had a strong stake in the controversy. Blaming AIDS on a virus was the gambit that allowed NIAID to claim the jurisdiction—and cash flow—away from NCI. Dr. Fauci's career depended on the universal belief that HIV alone causes AIDS. The dispute, for him, was existential. Led by Dr. Fauci's college of cardinals, the medical cartel—the emerging highly profitable drug, research, testing and nonprofit charitable HIV-AIDS enterprise—attacked Duesberg and the other dissidents as "flat-earthers"[25] and Holocaust-type "denialists,"[26] or, in Dr. Fauci's estimation, murderers.[27] The AIDS establishment, down to its lowliest doctor, publicly reviled Duesberg, NIH defunded him, and academia ostracized and exiled the brilliant Berkeley professor. The scientific press all but banished him. He became radioactive.

From his perch at HHS, Dr. Fauci controlled all the levers of power and public opinion. Shortly after Duesberg's *Cancer Research* paper's publication, the office of the secretary of Health and Human Services (HHS) sent out a memo under the heading

"MEDIA ALERT." HHS announced the imposition of message discipline harking back to the agency's military roots. The HHS directive rebuked the NIH for allowing Duesberg's paper to reach publication in the first place. "The article apparently went through the normal pre-publication process and should have been flagged at NIH," it read. "This obviously has the potential to raise a lot of controversy," it added ominously, "I have already asked NIH public affairs to start digging into this."[28]

By questioning the official government theology, and especially by clashing with HHS's reigning technocrat, Duesberg would soon see his generous stream of NIH research grants run dry. When Duesberg's seven-year Outstanding Investigator grant came up for renewal, it was D.O.A. As usual, Dr. Fauci had stacked the board. The NIH review committee included one AIDS researcher with deep financial ties to Glaxo, which manufactured AZT, a drug Duesberg ferociously criticized for its extreme toxicity; and another was Gallo's mistress, a scientist in his lab who had mothered his child.[29] Three reviewers never even read Duesberg's proposal. NIH pulled the grant and never again gave Duesberg a single research dollar.

Prior to 1987, NIH had never rejected a single one of Peter Duesberg's proposals. After 1987, Duesberg wrote over thirty research proposals; NIH refused every one.

"The US military industrial complex—HHS, NIH, NCI, DAIDS—all of it, is designed along military command structure because it is," says Celia Farber. "It is the military. It's not 'science' and it's not 'merit.' Fauci understands this and has mastered the elimination of both dissent and any mercy for the destroyed. It's a *sin* as he has now openly said, to question him—to question 'science.' He's so far gone that he has actually come out and said *he is science."*

"I would like Americans to learn who Peter Duesberg is," Farber continues, "what his achievements were, on cancer genetics, on aneuploidy, and what became of it. I want them to demand answers. Why did Anthony Fauci set out to defund, bully, censor and destroy America's premier cancer virologist? How do we feel about that? We know how the AIDS activists feel—but how do *we* feel about it? Most of us have lost at least one family member to cancer, and none to AIDS. Anthony Fauci should be brought before a criminal court and stand trial for destroying American science, and virology, and cancer science. A lot of the destruction was done through the wildly personal destruction of Peter Duesberg, and anybody who tried to 'take him seriously,' or even, for that matter, interview him. The true history is emerging now, and will emerge. Fauci will go down as a very dark figure. A travesty. He was obsessed with AIDS—why? America needed this obsession like a hole in the head. All it was was a money trough, a global apparatus of colonial parasitism. We buckled under to Fauci and a handful of shrieking activists. It's truly a tragedy."

"They just took him out," agrees Richard Strohman, a retired UC Berkeley biologist. "Took him right out."[30]

A frenzy of anti-Duesbergism swept the field like grass fire. Duesberg's name became so degraded that debasing him became a means of career advancement. Being seen with him was career suicide for aspiring scientists.

"The system works. It's as good as a bullet to the head," said Dave Rasnick.[31]

In a 1988 interview laced with poison and enraged profanity, Gallo denounced

Duesberg for questioning his HIV/AIDS hypothesis: "HIV kills like a truck!" he hollered. "HIV would kill Clark Kent!"[32]

Duesberg's riposte, at the time, was that he wouldn't mind being injected with HIV—so long as the sample didn't come from Gallo's lab.[33]

The scientifically illiterate mass media largely ignored Duesberg's evidence-based arguments as dangerous apostasies. Dr. Fauci showcased his easy capacity to control his servile media toadies and mobilize the public health cartel to punish skepticism and dissent. It was a tour de force and an extraordinary preview of his later censorship campaigns. This was a decade before FDA's 1997 consequential decision to allow pharmaceutical advertising on television, so Dr. Fauci's urgency in quickly summoning the media to obediently fall in line was all the more impressive. Subsumed in the received orthodoxy, fawning media outlets parroted the official caveat of the NIAID inquisition: to even acknowledge Duesberg's arguments was itself dangerous because it deflected valuable time from the business of "saving lives" and lent credence to deadly heresy. To mention Duesberg's name was irresponsible journalism.

AIDS organizations posted warnings about Duesberg and his fellow "denialists" on their websites. Project Inform's Martin Delaney, living fat, by then, on Dr. Fauci's payroll, conducted letter-writing and phone campaigns vowing to get every journalist who interviewed Duesberg fired. (Delaney would later come around to Duesberg's view that HIV could not solely cause AIDS.) It wasn't a particularly time-consuming project; very few journalists wanted to undertake the risk. As noted earlier, Anthony Fauci personally made sure Duesberg almost never appeared on national television. Dr. Fauci demonstrated his mastery at intimidating TV networks. In one case, *Good Morning America*[34] had already booked Duesberg and flown him to New York. On the night preceding his appearance, a *GMA* producer called to say the show was canceled. In the morning, he turned on his hotel TV and saw Anthony Fauci himself on the show. Similarly, Larry King[35] asked Duesberg for a televised interview in 1992 and then abruptly canceled the night before. Dr. Fauci took Duesberg's place at King's table. In 1987, when President Reagan invited Duesberg and Dr. Fauci to the White House for a friendly debate in front of the president, Dr. Fauci forced Reagan to cancel. A member of President Reagan's administration told Duesberg that "Anthony Fauci, far from reacting as . . . anticipated, threw a 'small fit' when he was invited, and demanded to know why the White House was interfering in scientific matters that belonged to the NIH and the Office of Science and Technology Assessment."[36]

Anthony Fauci's uninterrupted flow of millions of dollars to its labs and med school had by the 1980s transformed Berkeley—a mecca for free speech in the 1960s—into an omphalos of reaction and medical heterodoxy. In a pioneering template for "cancel culture," the university unceremoniously stripped Duesberg—then at the very top of his field—of everything: government funding, grad students, a proper lab, and invitations to conferences. Only his tenured position prevented Berkeley from ridding itself of the iconoclastic researcher altogether. The university refused to endorse Duesberg's appeal to the NIH of his grant revocation; without university support, he could not legally proceed. Duesberg has had to hire a lawyer to fight for his standard annual merit pay increase, which usually comes automatically to professors of his stature. UC Berkeley denied Duesberg his raise for over a decade, claiming his work was "not of high significance."[37]

Wary of ruining their careers, all his grad students abandoned Duesberg. The university warned them that working with Duesberg would make them pariahs. All scientific conferences disinvited him; prominent colleagues demonstrated their rectitude by publicly declaring that they would decline invitations to any conference that included Duesberg.

One of his Berkeley colleagues complimented Duesberg lavishly in a private interview with journalist Celia Farber.[38] The colleague praised his integrity, his genius, his kindness, and his intelligence. She protested his shoddy treatment by the university and the scientific establishment, but she insisted that she did not want to be identified in Farber's story, explaining that she feared retribution.

Another Berkeley colleague from the Donner Lab explained to Farber the general hesitancy about Duesberg among the faculty: "Peter may be right about HIV. But there's an industry now."[39]

The scientific press banished Duesberg from publishing. *Nature* editor John Maddox himself wrote a theatrical editorial stating that Duesberg, by his heresy, had forfeited the standard scientific publishing practice "Right of Reply."[40] Maddox invited Duesberg's colleagues to slander the virologist without fear of response. Anti-Duesberg ambuscades became pro forma in each new edition of *Nature*. Bialy's biography of Duesberg renders this written record in vivid, often hilarious detail.[41] Even the Proceeding of the National Academy of Sciences's (PNAS) journal, where members are always invited to publish, crushed a Duesberg paper on HIV after he spent over a year revising and resubmitting it to meet their various editing requests.

Colleagues reckless enough to defend Duesberg found themselves in *malodour*. The virologist Harry Rubin, himself a member of the Academy, suffered toxic vitriol and career injury after he intervened vainly with PNAS to get Duesberg's paper published. In 1992, Duesberg's paper became the second one in the PNAS's 128-year history to be blocked from publication.[42] (The other was written by Linus Pauling.)

"Duesberg's problem was one that transcended science: It was career protection to partake in his bullying and degradation," said Farber. "The Fauci serf scientists were driven by fear that if they did not denounce Duesberg in sufficiently disgusted tones, and very publicly, they would themselves soon be punished by Fauci, possibly de-funded, or worse."

The medical cartel dangled the prizes of redemption and reinstatement before Duesberg if he would only agree to reform. In 1994, a high-ranking NIH geneticist, Dr. Stephen O'Brien, called Duesberg and said he urgently needed to see him about a professional matter. O'Brien flew in from Bethesda the next day, and the two met at the opera in San Francisco. After some small talk about the good old days, O'Brien pulled a manuscript from the inside pocket of his tuxedo. Headlined "HIV Causes AIDS: Koch's Postulates Fulfilled," it had three very incongruous names at the bottom: Stephen O'Brien, William Blattner, and Peter Duesberg.[43]

Nature editor John Maddox had commissioned this apologia as inducement. If Duesberg would only sign the mea culpa, O'Brien implored, he could have everything back. He would be back at the top again, back in the safe bastion of Dr. Fauci's medical and science establishment.

Duesberg refused the bribe.[44]

In a 2009 documentary, Duesberg is somewhat empathetic, if not sympathetic, toward his detractors: "They are prostitutes, most of them, my colleagues—and to some degree, myself. You have to be a prostitute to get money for your research. You're trained a little bit to be a prostitute." He smiles and adds, "But some go all the way."[45]

Refusal to Debate

For several years, journalist John Lauritsen tried to get *any* scientist at NIH to answer the questions in Duesberg's article. But the orders had come from NIAID that no government scientist should respond. NIH officials repeatedly told Lauritsen that "none of the scientists for Robert Gallo in government were interested in discussing the etiology of AIDS." Lauritsen was therefore intrigued when the *New York Times* reported Tony Fauci's laconic official response to Duesberg's article: "The evidence that HIV causes AIDS is so overwhelming that it almost doesn't deserve any discussion anymore."[46] Lauritsen complained to me, "As a member of the press, I thought I should have been allowed to speak to Dr. Fauci, and ask him to reveal just one or two pieces of 'overwhelming evidence' that HIV is the cause of AIDS. How did he get away with this? His only strategy was to act as if the evidence was so overwhelming that no one should be allowed to question the assertion. Fauci adopted the posture that neither he nor his colleagues had any obligation to reply to Duesberg, or any of his other critics. It was the secular version of the doctrine of Papal infallibility; everyone must just accept the 'AIDS virus' theory as a matter of fact because the public health pope declares it."[47]

Harvey Bialy, founding scientific editor of *Nature Biotechnology*, said: "I am very tired of hearing AIDS establishment scientists tell me they are 'too busy saving lives' to sit down and refute Peter Duesberg's arguments although each one assures me they could 'do it in a minute if they had to.'"[48]

In 2006, Britain's preeminent epidemiologist, Gordon Stewart, voiced a similar frustration: "I have asked the health authorities, editors-in-chief and other experts concerned with HIV/AIDS, repeatedly for proof of their theses—and I've been waiting for an answer since 1984."[49]

Dr. Fauci's own refusal to debate his theories is just the tip of the iceberg. Dr. Fauci's control of his PI army gives him the ability to shut down all debate. When National Public Radio attempted to stage a conversation between Duesberg and a supporter of the HIV hypothesis, it could find no one willing to confront him. "Critiquing a dubious theory would take time away from more productive efforts," Anthony Fauci, head of NIAID, told NPR producers.[50]

When Bialy challenged Dr. John Moore of Cornell University to a debate on AIDS, Moore wrote in reply: "Participating in any public forum with the likes of Bialy would give him a credibility that he does not merit. The science community does not 'debate' with the AIDS denialists, it treats them with the utter contempt that they deserve and exposes them for the charlatans that they are. Kindly do not send me any further communications on this or any related matter."[51]

Such scathing rebuffs infuriated Nobel Laureate Kary Mullis. In 2004, he said, "All we have is Bob Gallo saying, 'Gentlemen, this is the cause of AIDS.' That's all we have. That's all we had. That's not enough. That is not sufficient to publish even a meager little scientific paper somewhere [much less a basis to spend] millions [or] billions of dollars a year and the cost of a lot of lives and anguish . . . lives have been totally ruined on the basis of some flimsy little statement made by a guy who's known to be a crook in lots of other ways. He lied about a whole lot of other stuff. Why are we trusting him? If he was a witness in a courtroom, we wouldn't trust his testimony. We've caught him in too many lies. [We] don't trust him anymore."[52]

Some twenty years after Gallo's announcement, circumstances finally forced Dr. Fauci to defend his thesis. In 2009, documentarian Brent Leung persuaded Dr. Fauci to submit to a sit-down interview for Leung's feature-length film on the history of AIDS, *House of Numbers: Anatomy of an Epidemic*. Leung asked an uncomfortable, chafing Dr. Fauci for his best evidence linking HIV to immune deficiency disease. With two decades and ten billion dollars to prepare his answer, Dr. Fauci's best explanation was the classic Fauci soft shoe. Contemporary Americans will recognize the familiar refrain of double-talking and dissembling that we all now recognize from the NIAID Director's COVID-19 interviews:

> When you put the combined findings of the initial characterization as a distinct retrovirus isolated by Montagnier and his group together with Gallo linking the virus to being the cause of AIDS, and they put those things together, that's how we have a confirmation of the causative agent of AIDS, namely HIV.[53]

"Translating all that into regular English," which Charles Ortleb remarked to me with a laugh, "takes just three words: Gallo says so. That's what Fauci calls 'a confirmation.'"

Among Dr. Fauci's skeptics were numerous Nobel laureates, including geneticist Barbara McClintock and chemist Walter Gilbert, who added their voices to the chorus complaining about the lack of scientific proof supporting the HIV/AIDS hypothesis, and the inability or unwillingness of health officials to answer fundamental questions. "It is good that the HIV hypothesis is being questioned," Gilbert told the *Oakland Tribune* in 1989. Gilbert acknowledged it "is absolutely correct . . . that no one has proven that AIDS is caused by the AIDS virus. And [Duesberg] is absolutely correct that the virus cultured in the laboratory may not be the cause of AIDS."[54]

Mullis, one of the most significant Nobel laureates of the twentieth century, died in 2019. "People keep asking me," he explained in 1994, "'You mean you don't believe that HIV causes AIDS?' And I say, 'Whether I believe it or not is irrelevant! I have no scientific evidence for it!'[55] If there is proof that HIV is the cause of AIDS, there should be scientific documents which either singly or collectively demonstrate that fact, at least with a high probability. There is no such document."[56]

Mullis observed in 1994 that the financial and career incentives for advancement to any researcher who could demonstrate a formal proof of Dr. Fauci's proposition are so monumentally enormous that the inability of anybody to produce this demonstration is itself compelling evidence that HIV alone does not cause AIDS: "If a postdoc were to write a review of the literature that showed without much doubt that HIV was the cause of AIDS, that guy would be famous. There are a hundred thousand guys out there who had the opportunity. Ten years have passed; we've been waiting for this star postdoctoral fellow to distinguish himself forever and get a lifelong grant from Tony Fauci but he hasn't shown up. No one has bothered to write a definitive review. Any journal would take it. That right there proves that HIV does not cause AIDS."[57]

Duesberg's most surprising convert was Luc Montagnier, the man who first discovered the virus.

At the San Francisco International AIDS Conference in 1990, Dr. Montagnier made a startling confession about HIV that was clearly against his own interest: "HIV might be benign."[58] Montagnier was the father of the AIDS theory. He is also a scientist of integrity. That was his surrender flag. Montagnier's discounting of the HIV/AIDS association should have been earthshaking. Instead, the conventioneers—content with the orthodoxy that was paying off handsomely for so many of them—ignored Montagnier's momentous confession and went right on talking about exciting new antiviral drug treatments.

Kary Mullis was astonished that Fauci's dogma had such a powerful hypnotic force that acolytes would ignore its public retraction by the genius who invented it. "Years from now, people looking back at us will find our acceptance of the HIV theory of AIDS as silly as we find the leaders who excommunicated Galileo, just because he insisted that the Earth was not the center of the universe," predicts Mullis. "It has been disappointing that so many scientists have absolutely refused to examine the available evidence in a neutral, dispassionate way, regarding whether HIV causes AIDS."[59]

All about the Money

Today, the presumption that HIV is the sole cause of AIDS is the central presumption of a multibillion-dollar industry. Everyone agrees that at least part of the explanation for its stupefying resilience is Dr. Fauci's relentless flow of cash. Charles Ortleb observed to me, "Science costs money and he who dispenses the money can control the science."

"Look, there's no sociological mystery here," observed Mullis. "It's just people's income and position being threatened by the things Peter Duesberg is saying. Their personal income and positions are being threatened and that's why they're so nasty. In the 1980s, a lot of people started being dependent on Tony Fauci and his friends for their livelihood. All these people really wanted success in the sense of lots of people working for them and lots of power."[60]

Bialy agrees: "First of all, there are tremendous financial and social interests involved. Billions of dollars in research funding, stock options, and activist budgets are predicated on the assumption that HIV causes AIDS. Entire industries of pharmaceutical drugs, diagnostic testing, and activist causes would have no reason to exist."

The 2004 documentary *The Other Side of AIDS* includes a remarkable scene in which Canadian PI, Mark Wainberg, MD, president of the International AIDS Society (the world's largest organization of AIDS researchers and clinicians), angrily calls for

Duesberg and others who "attempt to dispel the notion that HIV is the cause of AIDS" to be "brought up on trial." He considers HIV/AIDS skeptics "perpetrators of death."[61]

"I suggest to you that Peter Duesberg is the closest thing we have on this planet to a scientific psychopath."[62]

Then he declares the interview over, rips the microphone from his lapel, and storms off.

What happened next was revealing.

The audience erupted in laughter, which turned to boos as the screen flashed a list of Wainberg's patents and other financial ties to the HIV industry.

Other Causes

If HIV doesn't cause AIDS, one is bound to ask, then what does? Leading scientists have advanced multiple credible theories to account for AIDS's pathogenesis. I will examine three of the most compelling, beginning with Duesberg's theory, since his explanation arrived first chronologically and inspired the largest and most influential following. Subsequent theories—including hypotheses promoted, ironically, by Robert Gallo and Luc Montagnier—have equal persuasive power but enjoyed meager public interest or support. Duesberg's battle royal had demonstrated Dr. Fauci's sizable power to destroy careers, and no one after Duesberg had the courage and appetite to challenge the "Little Director" by advancing new theories.

Duesberg's Theory

Duesberg, Mullis, and their school of critics blame all the lethal symptomology known as AIDS on a multiplicity of environmental exposures that became ubiquitous in the 1980s. The HIV virus, this group insists, was a kind of free rider that was also associated with overlapping lifestyle exposures. Duesberg and many who have followed him offered evidence that heavy recreational drug use in gay men and drug addicts was the real cause of immune deficiency among the first generation of AIDS sufferers. They argued that the initial signals of AIDS, Kaposi's sarcoma and *Pneumocystis carinii* pneumonia (PCP), were both strongly linked to amyl nitrite—"poppers"—a popular drug among promiscuous gays.[63] Other common "wasting" symptoms were all associated with heavy drug use and lifestyle stressors. (Those interested in exploring the debate should read Chapter 3, Virus Hunting Takes Over, of Duesberg's riveting book *Inventing the AIDS Virus*.) Suffice it to say that Duesberg makes a compelling case, and his arguments deserve to be aired and civilly debated.

Dr. Duesberg observed that critical AIDS cases in the 1980s were among men engaged in behaviors then commonplace in the post-Stonewall, drug-charged gay party scene. Risk factors included promiscuous sex with multiple partners and cumulative toxic exposures from psychoactive drugs including methedrine, cocaine, heroin, LSD, and a cocktail of antibiotics prescribed to treat ubiquitous sexually transmitted diseases. On average, the early AIDS patients had been on at least three antibiotics courses in the year preceding diagnosis.[64]

Some 35 percent[65] of early AIDS cases were among IV drug users. In his paper "The Role of Drugs in the Origin of AIDS," Duesberg cites over a dozen medical references documenting AIDS-like immunodeficiency symptoms among drug addicts since 1900.[66] The medical literature attests to the ravaging effects of heroin, morphine, speed, cocaine, and other injected drugs on the immune system: "From as early as 1909 evidence has accumulated that addiction to psychoactive drugs leads to immune suppression (clinical autoimmunity), similar to AIDS."[67] Today, thousands of American junkies who are not infected with HIV are losing the same CD4+ T-cells and getting the same diseases as AIDS patients. STDs from promiscuous sex and blood-borne diseases like hepatitis A, B, and C added to the immune suppression among this cohort.

Duesberg's theory was by no means novel or outlandish. Dr. Fauci himself conceded in 1984 that drugs were a reasonable explanation for PCP and other signature symptoms of AIDS: "If I were to take drugs that would markedly immunosuppress me, there would be a reasonably good chance that I would get that pneumonia. That's what happens to the AIDS individuals."[68]

Poppers and Drugs

Prior to Gallo's "discovery" of HIV, the initial guess by government researchers and leading scientists was that recreational drugs were the prime suspects. Duke Medical School's renowned infectious disease expert, Professor David Durack, who served on NIH's Bioethics Committee, asked the (still relevant) question in his lead article in the December 1981 *NEJM*[69]: How can AIDS be so evidently new, when viruses and homosexuality are as old as history?[70] Recreational drugs, according to Durack, should be considered as causes: "They are widely used in the large cities where most of these cases have occurred. Perhaps as suggested one or more of these recreational drugs is an immunosuppressive agent." Durack observed that, other than drug-using homosexuals, the only patients with AIDS symptoms were "junkies."[71] In Duesberg's view, the highest risk addiction was the ubiquitous use of amyl nitrite poppers, which had well-established links with autoimmune disease.

The first AIDS cases were five gay men—all unknown to one another—diagnosed with a rare (PCP) pneumonia and Kaposi's sarcoma, a form of cancer that had previously afflicted only elderly men. Dr. Michael Gottlieb, a researcher searching California hospitals for new diseases with unusual symptomology, is credited with the initial discovery and characterization of the disease and its epidemiologic context. in Los Angeles in 1981, by Dr. Michael Gottlieb, a researcher searching California hospitals for new diseases with unusual symptomology. The men were all promiscuous party enthusiasts in the "fast lane" gay lifestyle. They were taking many different recreational drugs simultaneously and combining drugs in excess of patterns among straight drug users. They frequented bars, clubs, and bathhouses. They had daily multiple anonymous sexual partners—upward of a thousand per year—and contracted most of the common sexually transmitted diseases like syphilis, gonorrhea, and hepatitis B. They were, therefore, also functionally addicted to a pharmacopoeia of antibiotic prescription medications; "all of that created a situation where a handful of gay men," says Mark Gabrish Conlan "were burning the candle at both ends and putting a blowtorch to the middle. It's no wonder that after a while, their immune systems started to collapse and they started getting sick in these unusual ways that previously had only been seen in older people whose immune systems had deteriorated from age."[72]

John Lauritsen, a gay activist, was probably the longest-running AIDS journalist: "My first major AIDS article was in 1985. The very early AIDS cases were really quite sick, and there were very good reasons why they were sick."[73]

Lauritsen and many leading medical researchers and government health officials concluded early in the epidemic that poppers were the lead culprit. Chemists developed amyl nitrite as a vasodilator in the 1850s and began, in the 1960s, packaging it in glass ampules that doctors would pop open under the noses of unconscious patients to reanimate them. That same mechanism that prompted reanimation provided the relaxation of the anal musculature and a powerful rush that made poppers the reigning sex drug.

Poppers became a mainstay of the gay social scene in the late 1970s. Prior to 1987, every AIDS patient acknowledged heavy consumption of poppers.[74] Every porn shop, bar, and bathhouse locker room sold poppers.[75] Party gays huffed them continuously in dance clubs and during extreme sex. The saloons and dance halls reeked of their pungent chemical aroma. At the end of each evening, bartenders routinely announced, "Last call for alcohol," "Last call for Poppers."[76] Researchers believe poppers to be the direct cause of Kaposi's sarcoma, a rare form of skin cancer that afflicts the nose, throat, lungs, and skin.[77] Kaposi's sarcoma was the initial indicator disease of AIDS, but it was also common in gay men who were not infected with HIV.

Poppers can severely damage the immune system, genes, lungs, liver, heart, or the brain; they can produce neural damage similar to that of multiple sclerosis, can have carcinogenic effects, and can lead to "sudden sniffing death."[78]

"I discovered there was a very extensive medical literature on the volatile nitrites," Lauritsen explains. "The simplest thing is that they are very powerful oxidizing agents, which is part of AIDS causes; in fact, several types of anemia. Secondly, poppers are powerfully mutagenic and carcinogenic—meaning that they cause cellular changes and cancer. One of my informants, Filson—who was very active and outgoing in the People With AIDS Coalition—claimed that he had interviewed several hundred gay men with AIDS and he said that virtually all of them had been heavy users of drugs. They said without a single exception. They had all been poppers users."[79] A study published by Toby Eisenstein showed that nitrites found in poppers are radically immunosuppressive in rodents.[80]

Government researchers and regulatory officials supported the association. Prior to Gallo's announcement, CDC had targeted poppers as the likely culprit for AIDS. A year before Gallo's announcement, CDC's in-house AIDS expert Harry Haverkos analyzed three surveys of AIDS patients conducted by the CDC. He concluded that drugs like poppers played a key role in the disease onset. L. T. Sigell wrote in the *American Journal of Psychiatry* that the inhaled nitrites produced nitrosamine known for its carcinogenic effects—Thomas Haley of the Food and Drug Administration (FDA) issued the same warning.[81]

Following Gallo's 1984 press conference, Dr. Fauci launched a mission to quash all conversation about cofactors like poppers. The CDC quickly fell in line. The CDC shelved the Haverkos study and began parroting Dr. Fauci's hostility toward the drug connection. The CDC actively suppressed disagreeable data and published one of its signature junk science papers to "prove" poppers safe.[82] The CDC researchers assumed that gays used poppers as single-use reanimators and exposed laboratory mice to lifetime doses 1/1,000 of what a gay man would get in one evening on the party circuit. The study was "utterly fraudulent," remarks Lauritsen.[83] For a partial list of studies that tested the association of nitrites to AIDS, see *Oppenheimer, In the Eye of the Storm*, note 34, p. 295.[84]

Haverkos transferred to the FDA in 1984 to become AIDS coordinator there. His paper finally appeared in the journal *Sexually Transmitted Diseases* in 1985,[85] prompting the *Wall Street Journal* to pen an article arguing that substance abuse was so universal among AIDS patients that drug use, and not Dr. Fauci's virus, must be considered the primary cause of AIDS.[86]

According to Randy Shilts, writing in his classic history of the AIDS crisis, *And the Band Played On*, the poppers' starting point offers a "compelling" explanation for AIDS. "Everybody who got diseases seemed to snort poppers," writes Shilts.[87]

As I wrote this book, Children's Health Defense researcher Robyn Ross, Esq., alerted me to one of the unheralded ironies of this saga. As it turns out, Burroughs Wellcome holds the 1942 patent on the popper container and remained one of the largest manufacturers of poppers during the 1980s and '90s. As early as 1977, a *New York Daily News* article described Burroughs Wellcome strategies for dodging criticism of widespread health injuries from its booming popper sales. As we shall presently see, Burroughs Wellcome and other popper manufacturers were the principal sources of advertising revenues to the gay press during that epoch, and they used that leverage to force censorship of any journalist attempting to link amyl nitrite to immune system collapse. If Duesberg and others are correct about that association, it means that Burroughs Wellcome was profiting from both causing the AIDS epidemic and then from poisoning a generation of gay men with the AZT "Cure." Tony Fauci played traffic cop in this feedback loop. On the one hand, he was using his regulatory authority to promote AZT, and to kill its competition, effectively orchestrating Burroughs Wellcome's monopoly control over AIDS treatment. At the same time, he was suppressing the study of the toxicity of poppers and directing the blame for AIDS on the virus, thereby shielding Burroughs Wellcome from significant liability.

Kaposi's Sarcoma

In 1990, four leading scientists at the CDC suggested in the *Lancet* that Kaposi's sarcoma was common in young gay men who indisputably did not have HIV. They concluded that KS—the disease most central to the definition of AIDS—"may be caused by an as yet unidentified infectious agent, transmitted mainly by sexual contact."[88]

This was a stunning development, because KS was the initial and defining symptom of AIDS. Prior to 1981, KS was a disease limited to very old people. Its sudden appearance in young men was the identifying signal that launched the AIDS crisis. It was fundamental doctrine within the medical establishment that KS was the diagnostic signal of the AIDS pandemic. The very existence of AIDS was inextricably linked to KS. If HIV was not responsible for the outbreak of Kaposi's Sarcoma, then there had to be another culprit. That insurmountable logic raised the question of whether poppers might also be causing the other symptoms of AIDS— particularly the other major manifestation, immunosuppression, which science also linked to amyl nitrite.

While publicly cleaving to Dr. Fauci's official HIV/AIDS orthodoxy, Robert Gallo himself privately signaled doubts about his own theory that HIV alone can cause either Kaposi's sarcoma or AIDS. At a high-level meeting of US health authorities in 1994—titled "Do Nitrites Act as a CoFactor in Kaposi's Sarcoma?"—Gallo made some astonishing confessions to his trusted colleagues. HIV, he acknowledged, might only be a "catalytic factor" in Kaposi's: "There must be something else involved." Then he added a breathtaking concession, which could have been taken from the very research in Duesberg's article: "I don't know if I made this point clear, but I think that everybody here knows—we never found HIV DNA in tumor cells of KS. So this is not directly transforming. And in fact, we've never found HIV DNA in T cells although we've only looked at a few. So, in other words we've never seen the role of HIV as a transforming virus in any way."[89]

One attendee of that meeting was Harry Haverkos, by then director of the AIDS department at National Institute on Drug Abuse (NIDA). Haverkos observed to Gallo that not a single case of Kaposi's sarcoma had been reported among blood recipients where the donor had Kaposi's sarcoma. If blood transfusions couldn't spread the disease, Haverkos said, then semen exchanges could hardly be a plausible culprit. In response, Gallo allowed: "The nitrites (poppers) *could* be the primary factor."[90]

To fully appreciate the seismic implications of Gallo's statement, we must recall that, in wealthy nations like the United States and Germany, Kaposi's sarcoma was— along with PCP—the signature disease for diagnosing patients with "AIDS." In 1987, for example, *Der Spiegel* described AIDS patients as the "sarcoma-covered skeletons" from the "same-sex scene."[91]

By 1990, government regulators were already scrambling to drop Kaposi's sarcoma from the AIDS definitions. "At present, it is accepted [even by CDC scientists] that HIV plays no role, either directly or indirectly, in the causation of Kaposi's sarcoma," wrote Australian biologist and AIDS expert Eleni Papadopulos in 2004.[92] This was a momentous "bait and switch." Kaposi's was the AIDS-defining illness. "In the beginning," says Farber, "AIDS was Kaposi's sarcoma."[93] Because its association with AIDS was so well established, the official concession that the two conditions are distinct has never penetrated the reigning orthodoxy. Kaposi's sarcoma remains part of the official AIDS definition in industrialized countries (anyone with KS and a positive test result counts as an AIDS patient)—and, contrary to the facts, mainstream media outlets like the *New Yorker* still report that "Kaposi's sarcoma is a sign of AIDS" (i.e., HIV causes KS).[94]

AZT as Culprit

After 1987, Dr. Duesberg and his followers argue, the vast majority of "AIDS deaths" were actually caused by AZT—Dr. Fauci's radical "antiretroviral" chemotherapy purposefully concocted to kill human cells. Duesberg describes the syndrome as "AIDS by AZT." Ironically, he argued AZT, the highly toxic medication that Dr. Fauci was prescribing to treat AIDS patients, actually does what the virus cannot—that is, it causes AIDS itself.

In a rational universe populated by critical thinkers, Duesberg's suspicion that AZT causes immune collapse should never have seemed revelatory. The FDA, after all, had deemed AZT too toxic to use for even short-term cancer therapy. AZT is highly mutagenic, meaning that it destroys the genes themselves. It causes cancer in rodents. It targets the bone marrow where blood cells called lymphocytes are made. These are the very cells that an AIDS patient needs most for immunity. AZT randomly destroys bones, kidneys, livers, muscle tissue, the brain, and the central nervous system.

Cancer patients typically take chemo drugs for only two weeks. Thanks to Tony Fauci's Fischl study, doctors were now prescribing AZT for life! "Chemotherapy," says Duesberg, "is restricted to a few months. The hope is that the cancer dies before you die."[95]

Duesberg believes that AZT was not only causing AIDS, it was killing more people than had previously been dying from autoimmunity caused by recreational drugs. "AZT is causing AIDS and its defining diseases. It doesn't cause Kaposi's sarcoma. But it does cause immune deficiency. It was designed to do that. In fact, the manufacturer says specifically that it can cause 'AIDS-like diseases.'" Burroughs Wellcome's insert warns that it is "often difficult to distinguish adverse events possibly associated with administration of RETROVIR (AZT) from underlying signs of HIV disease or intercurrent illnesses."[96] In other words, even the company acknowledges that AZT causes the diseases that define AIDS.

"If you start taking any other chemotherapeutic agent for the rest of your life, it would be that agent probably to kill you,"[97] Kary Mullis observed. "When you give chemotherapy to somebody with cancer, you give them a round of it for maybe fourteen days or a few days. Hopefully, you're not going to kill the patient. You're going to kill the cancer. Patient's going to survive. But you don't keep giving it to him until he dies, because he certainly will."[98] Luc Montagnier makes this same point about HIV: "Any drug active on HIV will be toxic because it's not 100 percent specific of the HIV enzymes."[99]

If Duesberg is right, AIDS is an iatrogenic (doctor-caused) pandemic, and Dr. Fauci would be its author. It wouldn't be the first one. Historically, there are many examples of prescribed medicines causing worse injuries than their target disease. The notorious Tuskegee Experiment (1932–1973), which my uncle, Senator Ted Kennedy, exposed and ended in 1973, began as an effort by public health regulators to unravel which syphilis symptoms were from the spirochete bacterium and which were from the mercury "cure" that doctors had, by then, been prescribing for more than 500 years. As it turned out, the most deadly and debilitating symptom of syphilis—the lethal second-stage neuropathy—was actually acute mercury toxicity, not surprising since mercury is nature's most toxic substance.

"AIDS is a chronic long-term breakdown of the immune system that can be caused by multiple factors," says Mark Gabrish Conlan, gay historian, publisher, "generally more than one of them operating within any particular person with AIDS or with what has been described as AIDS. And at the top of that, in the west would be recreational drug use, also pharmaceutical drug use and repeated infections, including with diseases that are genuinely sexually transmitted, repeated antibiotic treatments for these: a lifestyle that involves a lot of partying, lack of nutrition, and in the less-developed world, AIDS is primarily disease of malnutrition, starvation, and the endemic infections that have been part of those environments for years."[100]

Drs. Duesberg, Willner, and others believed that AZT killed tens of thousands of Americans between 1986 and 1996 before less toxic chemotherapy drugs were introduced, causing far more fatalities than the immune deficiencies associated with the recreational drugs during the first wave of the AIDS pandemic. A scientific study in the *New England Journal of Medicine* article in late July 1987 headlined "The Toxicity of Azidothymidine (AZT) in the Treatment of Patients with AIDS and AIDS-Related Complex,"[101] and a comprehensive investigation by *The Independent* of London in May 1993, "The rise and fall of AZT,"[102] both supported Duesberg's theory that AZT was a deadly killer of dubious efficacy against amorphous AIDS.

Rudolf Nureyev and Arthur Ashe

Rudolf Nureyev, greatest ballet dancer of all time, was friends with my parents. He visited our family home in the 1960s and '70s. Against his doctor's advice, he began taking AZT. Nureyev was HIV-positive, but otherwise in robust health. His personal physician, Michel Canesi, recognized the deadly effects of AZT and warned Nureyev not to take the drug. But Nureyev insisted, "I want that drug!"[103] He became sick soon after commencing treatment and died in Paris in 1993 at age fifty-four.

That year, former Wimbledon champion Arthur Ashe also died at age forty-nine. Ashe was also a family friend and a regular fixture at our family home at Hickory Hill and Hyannis Port. A heterosexual, Ashe learned he was HIV-positive in 1988. His doctor prescribed an extremely high AZT dose.[104] In October 1992, Arthur wrote a column for the *Washington Post* voicing his extreme misgivings about AZT. "The confusion for AIDS patients like me is that there is a growing school of thought that HIV may not be the sole cause of AIDS, and that standard treatments such as AZT actually make matters worse," Ashe acknowledged, adding, "There may very well be unknown cofactors, but the medical establishment is too rigid to change the direction of basic research and/or clinical trials." Ashe wanted to stop taking AZT, but he didn't dare: "What will I tell my doctors?" he asked the *New York Daily News.*[105]

If Arthur Ashe's suspicions and Duesberg's suppositions are correct, Dr. Fauci would be the father of the AIDS pandemic and responsible for prolific deaths. So that story must never be found to be true.

Is AZT Mass Murder?

There is little question that the character of AIDS changed dramatically in the early 1990s with the proliferation of AZT. Kaposi's sarcoma uncoupled from the disease and AIDS cases began to look increasingly like AZT poisoning. "Then at a certain point, when really that sort of AIDS virtually ceased to exist, there came a new type of AIDS," says John Lauritsen.[106] "So they expanded the definition, and also they began giving the anti-HIV drugs to people who were in fact not even sick, but merely positive on the HIV test. And in that case, of course, when they finally became sick enough from the AIDS drugs they were called 'AIDS patients.' I would simply have to say that my main concern is the gay men, who have been murdered," Lauritsen observed. "I don't think 'murder' is too strong a word to use when you have a drug like AZT and all the nucleoside analogues that followed, more or less on its coat tail, approved on the basis of fraudulent research, and where, as you know, Joseph Sonnabend said, 'AZT is incompatible with life.' Well, if it's incompatible with life, it's a poison and if it's a poison that kills people, in context like that, it is murder."[107]

Concurring with Sonnaband's assessment, John Lauritsen accuses Dr. Fauci of conducting genocides against gay men and Black Africans. The evidence seems to indicate that the proliferation of AZT increased death rates from "AIDS" dramatically.[108]

The annual mortalities from so-called AIDS during the early years of the pandemic for 1983–1987—prior to AZT's approval—were lower, perhaps ten to fifteen thousand people in a country of 250 million.[109] It wasn't until the late 1980s, when Dr. Fauci's AZT came along, that the number of deaths attributed to AIDS shot up.

According to the CDC, in the fifth full year of AIDS, 1986, 12,205 people "with" AIDS died in the United States. At that time, CDC—in a now-familiar scheme to stoke pandemic fears—used deceptive protocols to inflate the body counts. The CDC's mortality numbers include anyone with an HIV positive antibody "status," even if the deceased had no "AIDS defining illness," and instead succumbed to suicide, a drug overdose, a car accident, or a heart attack.

The death rate climbed precipitously after the commercial introduction of AZT. In 1987, "AIDS" deaths rose by 46 percent with 16,469 people dying. In 1988, as more and more people received AZT, the death toll rose to 21,176, and then to 27,879 in 1989. Death rates rose to 31,694 in 1990, and 37,040 in 1991.[110] At the end of the 1980s, HHS's standard prescription for AZT was 1,500 mg a day. In 1988, the average survival time for patients taking AZT was four months.[111] Even mainstream medicine couldn't overlook the fact that the administration of higher doses led to much higher death rates.[112] At the beginning of the 1990s, health officials lowered the daily dose to 500 mg. The average lifespan of AZT patients rose to twenty-four months in 1997, as deaths attributed to AIDS plummeted. Afterwards. CDC changed its counting metrics to make it difficult to count annual AIDS deaths.[113]

In his history of the era, historian Terry Michaels wrote, ". . . the CDC, for the years between about 1986 and 1996, created the illusion that tens of thousands in America died from AIDS or HIV in that decade, rather than AZT and other 'monotherapy' nucleoside analog drugs."[114]

According to Dr. Claus Köhnlein, MD, a German internist and coauthor of *Virus Mania*, "Most of the deaths attributed to AIDS, or HIV disease as eventually it would be called, from the mid-1980s to the mid-1990s were the result of iatrogenic illness, resulting from prescription of high dose, toxic, DNA chain-terminating chemotherapy, specifically, azidothymidine (AZT) ending in premature death for scores of thousands of 'HIV positive' gay men, plus many hemophiliacs, IV drug users, Sub-Saharan Africans, and a few heterosexuals unlucky enough to have taken the specious HIV test, like the late tennis star, Arthur Ashe, who died in 1993." Köhnlein observes,

"The treatment causes a very similar condition we would expect from an AIDS patient. That's why nobody noticed that there was something wrong with the treatment."[115]

The HIV dissent movement, propagandistically rendered the HIV "denialist" movement by the AIDS research establishment and media, was somewhat less under siege in Europe than in the United States.

The HIV establishment was transnational, sparking and condemning as a unified globalist voice. However, in European countries where funding is less reliable upon Dr. Fauci's approval, dissenting professionals could generally keep working, without intelligence to the state apparatus.

Dr. Claus Köhnlein, an oncologist from Kiel, Germany, was less subject to the financial discipline by the state actors or the political hysteria that was censoring dissident scientists in the United States and was in some ways more of a threat to the HIV propaganda juggernaut than even Peter Duesberg, as he spoke from direct clinical experience. Köhnlein saw his first AIDS patients in 1990 and treated several hundred over the decades in his very conventional Kiel clinic. Ignoring "HIV," and instead treating each symptom, he got almost all of his patients out alive. "I lost maybe a handful," he said in an email, when contacted for this book.

His views on AZT were unequivocal. "We virtually killed a whole generation of AIDS patients without even noticing it because the symptoms of the AZT intoxication were almost indistinguishable from AIDS,"[116] he said in one interview. He elucidated during an RT interview in 2010, during a "Rethink" conference in Vienna, "When I worked at the University in Kiel, I witnessed the mass intoxication of the patients with AZT. AZT was the first recommended treatment, and we all know today that the dosage was much too high. We gave 1500 mg on a daily basis and that literally killed everybody that took this treatment. That is the reason why everybody believes that HIV is a deadly virus but there is still no proof of this assumption."[117]

The reporter was incredulous, so Köhnlein elaborated, "They were all over-treated at that time and the reason why doctors didn't notice it was easy to explain because the placebo control was stopped after four months," he replied. "It was said that for ethical reasons nobody can withhold AZT treatment. After these four months the mortality rose tremendously in both groups.[118]

"This mistreatment was the very reason why everybody believed HIV [to be] a deadly virus and that HIV positive tests put everybody at equal risk, which is completely nonsense. So, a healthy pregnant mother today, an HIV positive pregnant mother, is told she carries the same deadly virus as a hopeless . . . IV drug addict."[119]

In an email, Köhnlein pegged the evidence against both HIV theory and AZT to three studies:

> "Harm is usually underreported," he wrote. "To prove it you need three studies: The AZT licensing Fischl study,[120] the Hemophiliac study in *Nature* where [editor] John Maddox showed that the HIV positive hemophiliacs started dying only the very year AZT was introduced.[121] And lastly, the Concorde *Lancet* study[122] which showed: the more AZT, the more Death."

In his Oct. 30, 2020 exposé, "The Other Media Blackout," *Wall Street Journal* columnist and editorial board member Holman W. Jenkins Jr. complained that the medical community has failed "to acknowledge complicity in poisoning hundreds of thousands of human beings. The illness and death that resulted from high AZT doses administered in the 1980s and 1990s is irrefutable."[123]

"From my personal contacts with people in the field," says Dr. David Rasnick, PhD, an AIDS researcher, chemist, and designer of protease inhibitors, "I can tell you that I've found no evidence anywhere that people live longer, better lives who take these anti-HIV drugs, these protease inhibitors, either alone or in cocktails, as compared to a similar group of HIV-positive people who do not take these drugs. So I do not know where the evidence is for the claims that you see in *New York Times* or on CNN, or wherever you see it that people are living longer, better lives as a consequence of taking these drugs."[124]

Duesberg points out that the annual mortality rate of HIV-positives undergoing antiviral therapy is 7 to 9 percent—far higher than the mortality rate of all HIV-positives worldwide, at about 1 to 2 percent per year.[125] Furthermore, there is ample evidence that treated HIV-positives die much faster of liver failure or cardiac failure than both HIV-infected individuals and AIDS patients who do not take AZT.[126]

Gays Join Dr. Fauci

In marshaling institutional resistance to dissent from the growing cadre of prominent scientists and doctors, Dr. Fauci found an unlikely ally: the AIDS community.

Beginning after his 1987 reconciliation with Larry Kramer in Toronto, Dr. Fauci quickly moved to build financial bridges to gay leadership and quiet dissent from AIDS activists. That year, he began by funding ACT UP and amfAR and leading AIDS activists, like Kramer and Martin Delaney. NIAID funneled extravagant annual public education grants to advocacy groups. The funding effectively muted their criticisms of Dr. Fauci.

The AIDS establishment—hospitals, medical and research centers, and pharmaceutical companies—created opulently paid consulting contracts for important members of gay organizations.[127] The gay community thereby became powerful gatekeepers for the AIDS establishment.[128]

Other political, economic, and ideological rationales helped Dr. Fauci recruit gay community leaders to his campaign to build a cancel culture against Duesberg and drown out his voice in the liberal mainstream press. In an era when Christian conservatism was so powerful that it credibly claimed to have put Ronald Reagan in the White House, ideology and medical opinions attributing the "gay disease" to orgies and excess partying tended to feed anti-gay bigotry. The gay community, therefore, happily endorsed Dr. Fauci's one-bug theory.

There were compelling mercantile drivers, as well. During the 1970s, the principal financial supports of the gay press were ads for the $50,000,000-a-year popper industry[129] and for the bars that flourished on popper sales.

As Ian Young explains in "The Poppers Story: The Rise and Fall and Rise of The Gay Drug," in *Steam*, "During the seventies and early eighties, much of the gay press, including the most influential glossy publications, came to rely on poppers ads for a huge chunk of its revenue, and poppers became an accepted part of gay sex. There was even a comic strip called Poppers, by Jerry Mills. The unwritten agreement was almost never breached: poppers ads appeared only in gay publications."[130]

The gay press glossed over urgent medical warnings from scientists about the dangers of poppers. *The Advocate*, a popular US magazine for homosexuals, refused to print letters from dissident scientists like Duesberg while accepting parades of poppers advertisements from Great Lakes Products, the era's largest manufacturer of sex drugs. Those advertisements exonerated poppers from any connection to AIDS, openly declaring them harmless.[131] Pharmaceutical companies including Hoffmann-La Roche invested money in the gay community with innumerable advertisements for AIDS medications. Burroughs Wellcome ran an ad for poppers calling amyl nitrite (i.e., poppers) "the real thing." Gay publications and organizations continued to promote poppers and censure stories about their health risks.[132]

His historical cultivation of relationships with gay leaders was one of the factors that made Dr. Fauci a darling of liberals during the early COVID crisis. Numerous other historical and personal factors induced liberals to accept Dr. Fauci without scrutiny. Blind faith in Saint Anthony Fauci may go down in history as the fatal flaw of contemporary liberalism and the destructive force that subverted American democracy, our constitutional government, and global leadership.

Deadly Viruses and Mycoplasma

As the HIV/AIDS hypothesis came under attack for its many discrepancies and internal contradictions, scientists besides Duesberg were discovering bugs that provided more plausible culprits in the AIDS pandemic.

Among these competing hypotheses were two advanced, individually, by Robert Gallo and Luc Montagnier. It's probable that diplomacy, self-interest, and honed survival instincts prompted both men to introduce their pathogens as "cofactors" that might work alongside HIV to trigger AIDS. Critics pointed out that the new pathogens these scientists uncovered were so clearly deadly on their own that they hardly needed HIV; the discovery of these genuinely lethal germs made HIV superfluous and redundant to explaining the pandemic. But for these gentlemen, it was obligatory to genuflect to the inviolable orthodoxy that anointed HIV as AIDS's ultimate cause. They may, in fact, have seen their discoveries as salvatory of the original HIV hypothesis. It was becoming increasingly challenging to credibly claim that HIV, which remained dormant for decades within its host, could somehow suddenly become virulent—"the most deadly disease in history"—without some external provocation.

HHV6

In 1986, Robert Gallo announced the discovery of human herpesvirus (HHV6). This new pathogen was no benign retrovirus. It was instead a savage cell-killing DNA virus. Gallo's lab had found the murderous HHV6 "killer cells" in the blood of AIDS-infected men and in patients suffering from Chronic Fatigue Syndrome (CFS), an immune deficiency disease extremely similar to AIDS, that had appeared in heterosexuals on the exact same timeline as AIDS appeared in homosexuals. Many critics already suspected the two diseases were one and the same. Gallo's discovery seemed to fortify that supposition.[133]

In a 1995 article titled "Human herpesvirus 6 in AIDS,"[134] Gallo says, "HHV6 may act as an accelerating factor" in HIV infection because "HHV6 can also infect and kill CD8+ T-cells, natural killer cells, and mononuclear phagocytes," all major components of the immune system. Ironically, Gallo's discovery of HHV6 might have won him the Nobel Prize if he hadn't jumped the gun a decade earlier by stealing HIV from Montagnier. None of the embarrassing questions about how in the world the seemingly benign HIV retrovirus could cause deadly disease bedeviled his lethal new killing machine.

While HIV was never shown to be cytocidal, HHV6 had a murderous affinity for CD4 and T-cell "in potential effects on the immune system and brains." Gallo declared that HHV6 was a major source of disease progression in AIDS.[135,136]

On May 11, 1988, the *Miami Herald* reported Gallo's announcement: "A newly discovered highly contagious herpes virus might play a role in causing several types of cancer and could be a co-cofactor in wiping out the immune systems of AIDS patients, one of the nation's premier virologists [Robert Gallo] said Tuesday."[137] The *Herald* also wrote, "Since the AIDS virus kills only a small percentage of T-4 cells at a time, Gallo said the new herpes virus [HHV-6], if proven to be the co-cofactor, could explain the total annihilation of T-4 cells in AIDS patients. 'The virus kills cells after using them to replicate,' he said." The *Herald* quotes Gallo as saying, "So if a co-cofactor is involved in the development of AIDS, and I'm not convinced it's absolutely needed . . . then we want to consider this one strongly."[138]

Charles Ortleb told me that Gallo's study "struck me as being backwards. If Gallo's new DNA virus explains the 'total annihilation' of T-4 cells, why would it need a cofactor? The 'cofactor' in this mystery would have to be HIV, not HHV-6."

Some scientists had similar reservations. HHV6 didn't seem to need a retrovirus wingman. Duesberg remarked dryly that to the extent that Gallo's newly discovered pathogen was "partnering" with HIV, then HHV6 was the senior partner in the collaboration. I can't help wondering if it occurred then to Gallo that if only he had not impetuously stolen Montagnier's discovery four years earlier, he might have collected his long-sought Nobel for his own authentic discovery of a much more plausible AIDS virus. Alas, it was not to be. But Dr. Fauci had committed his agency to the HIV hypothesis. And Gallo had built his career on HIV—even if he stole it from Montagnier, says Charles Ortleb. "When Gallo began that battle with Fauci," says Ortleb, "the agency was already fully committed to the HIV theory and could not afford any signs of retreat." I asked, "Why would Gallo not pull rank?" Ortleb answered, "Gallo is a classic sociopath. He knows that his survival means acquiescing to Fauci."

Following Gallo's "natural killer cells" article, other researchers confirmed the links between HHV6 and AIDS. In 1996, Konstance Knox, PhD, and Donald R. Carrigan, PhD, published a study demonstrating that 100 percent of HIV-infected patients studied (ten out of ten) had active Human Herpes Virus 6A infections in their lymph nodes early in the course of their disease.[139] This finding led Knox and Carrigan to conclude that "active HHV-6 infections appear relatively early in the course of HIV disease and in vitro studies suggest that HHV-6 is capable of breaking HIV latency, with the potential for helping to catalyze the progression of HIV infection to AIDS."

In April 1986, Dr. Knox stated in an interview with the *New York Native*: "We're finding HHV-6 in the lymph nodes early-active infection; this virus is replicating. This is unheard of for any other opportunistic infection, even TB."[140] Knox said she believed that HIV kind of acts as a "wet nurse" to HHV-6A.

Knox and Carrigan found that *every* AIDS patient had active replication of HHV-6A in *every* stage of AIDS, from their diagnosis to their autopsies, with many having CD4+ cell counts over 700. With HHV-6A, there were none of the bewildering questions about how a seemingly benign retrovirus could possibly cause all that carnage. "It's also much more destructive. . . . It kills very well, and it destroys tissue very well. It can infect the brain, the lungs, the lymphoid organs, and the bone marrow." When *New York Native* interviewer Neenyah Ostrom asked Knox, "Can HHV-6A do everything that HIV can do?" Knox gave this chilling answer: "As far as immunologic damage? Oh, HHV-6A does it much more efficiently than HIV."[141] Citing data from multiple studies by diverse scientists, Knox added: "Where we have seen HHV-6A in tissue, we see dead tissue. And where you see . . . HIV alone . . . you don't see dead tissue. You don't see destroyed organs and scar formation, and that's what you see when you see HHV-6A. We find replacement of the normal architecture of the lymph nodes with scar tissue. HHV-6A kills it. It kills the lymph node tissue."[142]

Knox parroted the obligatory language that HHV-6 was acting *in concert* with HIV. That language would preserve her from reputational and financial suicide. "I think they're a team. And, when the two of them are present, they induce the production of more of each other. It's a mutually enhancing relationship. It's our feeling that if you could interrupt or limit or suppress the HHV-6A infection, the levels of HIV would go down tremendously, and HIV would become just a chronic viral infection. . . . We don't have any evidence, looking in the tissue, that HIV is responsible for any of the destruction. And, if you think about it, HIV infects patients for years—a decade or more—without progressing to AIDS. When you look in their tissues, you have to ask how you can have such a long-term viral infection and have no damage?"[143]

NIH quickly cut off funding for Knox and for anyone else who wanted to research HHV-6. When Neenyah Ostrom asked, "Why can't you get more funding for this research?" Knox replied, "Well, I don't know if you've been tracking the kinds of exposés that *Science* magazine and others have published, that 80 percent of AIDS research monies are retained within the federal government programs on AIDS research. I think the science is very inbred. And I think there's been a real resistance to entertaining hypotheses or directions of AIDS research that aren't looking specifically at HIV, and that is the basic problem. Our studies themselves have been enthusiastically received, but the funding hasn't followed. And that is funding through the federal agencies—like the NIH."[144]

That summer, Italian researcher Dario Diluca published his findings in the *Journal of Clinical Microbiology*, reporting HHV-6 in the lymph nodes of 22 percent of Chronic Fatigue Syndrome patients and only 4 percent of healthy people. This research raised the possibility that AIDS, which affects gay men, is the same disease as CFS, which became widespread in heterosexuals and in virtual lockstep with AIDS in the early 1980s.[145]

Surveys of CFS patient groups in thirty-five states show an exponential rise in cases produced each year since the 1970s. This curious temporal and case production relationship with the AIDS epidemic prompted many researchers to characterize CFS as an AIDS epiphenomenon. Gallo's discovery and Knox's revelations suggested that the new human herpesvirus HHV-6 might be a critical causative cofactor shared by both AIDS and CFS. In June 1989, CFS research pioneer Dr. Paul Cheney, PhD, MD, testified before Congress that CFS might have a relationship with the AIDS epidemic.[146] In January 1993, six months after the Amsterdam conference, Dr. Anthony Komaroff at Harvard University and his coworkers published a study that showed that brain lesions developed in CFS patients who had Human Herpes Virus-6 active in their bodies.[147]

Such revelations could only have terrified Tony Fauci. Ever since the 1992 Amsterdam meeting, Dr. Fauci had been insisting that CFS was a psychosomatic disease. The suggestion that it might be related to AIDS threatened the entire HIV paradigm.

In their 1988 "natural killer cells" paper, Lusso and Gallo had quietly disclosed that they had found HHV-6 was infecting and killing NK cells in both AIDS and CFS patients. "They identified the problem in both sets of patients," said Knox, "so it makes sense that HHV-6A would also be a problem in Chronic Fatigue Syndrome."[148] When Gallo and Lusso conducted a trial treating half their AIDS patients with acyclovir—a remedy against herpes—and half with AZT alone, they found a significant prolongation of life in the patients who had AZT and acyclovir, as opposed to AZT alone.[149]

Said Knox, "In laboratory testing, HHV-6A is sensitive to acyclovir. So we have a curiosity as well. I mean, that would be pretty dandy, because certainly acyclovir has less toxicity than [AZT], and if you're talking about treating healthy people in a clinical trial, you're looking for something that people can take orally."[150]

These kinds of findings threatened to derail and discredit Anthony Fauci's entire HIV/AIDS paradigm. What, after all, would be the implications if a mild, off-patent remedy like acyclovir could safely treat AIDS more effectively than Dr. Fauci's expensive pharmacopoeia of deadly chemotherapy poisons? He choked off any further funding for HHV6 research, despite Knox's potentially lifesaving discovery of the efficacy of acyclovir against AIDS.

Mycoplasma

Dr. Shyh-Ching Lo, the Chief Researcher in charge of AIDS programs for the Armed Forces Institute of Pathology, was one of the many researchers baffled by Anthony Fauci's unconventional claim that antibodies—heretofore the signal of a robust immune response—should, uniquely with HIV, instead be the signal for impending death. That was a bridge too far for Dr. Lo: he took the conventional position that the presence of the antibodies to HIV—far from being a sign of doom—is proof that the body has coped successfully with the virus.

"There is no good explanation for why and how the virus breaks out of the antibody protection," complained Dr. Lo.[151,152] "I'm not saying that HIV plays no role in AIDS—the data shows a clear correlation with disease." He recited the mandatory disclaimer: "But AIDS is much more complicated than HIV."

In 1986, Dr. Lo announced that he had detected a previously unknown organism in cells taken from AIDS patients. Dr. Lo said that he believed the new organism, a bacterium-like creature known as a mycoplasma, worked with HIV to cause AIDS. Dr. Lo could not find the organism in any healthy individuals. When he injected his mycoplasma into four silvered leaf monkeys, three quickly developed low-grade fevers. All four lost weight and then died between seven and nine months of infection.[153] During autopsy, Dr. Lo found mycoplasma in their brains, livers, and spleens. That does not happen with HIV.

Dr. Lo also found the mycoplasma, dubbed *mycoplasma incognitus*,[154] in the damaged tissue of six HIV-negative human beings—perhaps CFS sufferers—who had died with suppressed immune systems after suffering from suspiciously AIDS-like symptoms.

For nearly three years, mainstream medicine and the captive mainstream and science media dutifully ignored Dr. Lo's research. A dozen scientific journals turned down Shyh-Ching Lo's studies for publication before the *Journal of Tropical Medicine* agreed to print his findings.[155] Despite his impressive credentials and his prestigious post as a top military scientist, Dr. Lo's attempts to find funding failed. Dr. Lo's research posed a unique annoyance for Dr. Fauci. Because he was a top military doctor with his own laboratory, he could not be easily dismissed, bullied, or defunded.

Then, in December 1989, Dr. Fauci opted to meet this threat from the military with a direct frontal assault. NIAID dispatched a dozen of Dr. Fauci's most skeptical specialists to investigate Dr. Lo's data.[156,157] Dr. Fauci flew his leading experts in AIDS and other infectious diseases to San Antonio, Texas, for the confrontation, expecting to obliterate Dr. Lo and to discredit his theory. Dr. Fauci's panel members quizzed Dr. Lo mercilessly for three days before surrendering to the conclusion that Dr. Lo had made a momentous discovery.

"The documentation was absolutely solid," said Joseph Tully, head of mycoplasma programs for NIAID.[158,159] The newly converted NIAID participants formally recommended further study of the link between the mycoplasma and AIDS, and experiments with drugs that could kill the new microbe.

That recommendation apparently displeased Dr. Fauci. "We have not been pulled into the AIDS programs in any real way," Tully complained in 1990. Thirty-five years after Dr. Lo's initial announcement, NIAID has still funded no research on Dr. Lo's mycoplasma hypothesis.

At the June 1990 San Francisco AIDS conference, Luc Montagnier made his tectonic announcement that "The HIV virus is harmless and passive, a benign virus."[160,161] He added that he had discovered that HIV only becomes dangerous in the presence of a second organism. He described a tiny, bacteria-like bug called a mycoplasma. His laboratory had demonstrated that in culture with his new mycoplasma, HIV becomes a vicious killer. Montagnier declared that he now believes that HIV is "a peaceful virus" that becomes lethal only when combined with *mycoplasma infertans*.

As Montagnier spoke, Dr. Shyh-Ching Lo sat in the audience, basking in vindication. Dr. Lo's important new ally, Montagnier, the Nobel laureate of AIDS, had independently discovered the same mycoplasma and concluded, like Lo, that it was the primary cause of the immune system collapse known as AIDS. The two had not shared their data. Separately, they had made the same earthshaking discovery four months apart.

In April of that year, Montagnier published his findings in *Research in Virology*, reporting that HIV and the microscopic pathogen react together, causing the body's cells to burst.[162] Even more exciting, he had discovered that in his test tubes, tetracycline stopped the mycoplasma's destruction entirely in its tracks. Montagnier's findings had transformative implications for AIDS treatment. They suggested that AIDS could be effectively treated and demolished with common patent-expired antibiotics instead of deadly and expensive chemotherapy concoctions.

At the San Francisco conference, Dr. Lo was almost the only person in the room who was excited. Of the twelve thousand people who attended the conference, only two hundred attended Montagnier's talk, and almost half of them exited before he finished.[163,164] Characteristically, the multibillion-dollar international research and development establishment opted to ignore his discovery.

Peter Duesberg: "There was Montagnier, the Jesus of HIV, and they threw him out of the temple."[165,166]

"Who were these people who are so much wiser, so much smarter than Luc Montagnier?" asks Harry Rubin, the dean of American retrovirology. "He became an outlaw as soon as he started saying that HIV might not be the only cause of AIDS.[167, 168]

When asked for an interview concerning Dr. Lo's work, NIAID director Anthony Fauci said through spokesperson Mary Jane Walker that he "will not talk about mycoplasma or any other AIDS cofactor."[169,170]

In a film interview with Brent Leung in 2006, Tony Fauci said, "Cofactors are not necessary. The data that indicate that any different type of infection like mycoplasma or something like that is a necessary cofactor, I believe those theories have been debunked."[171] As usual, Dr. Fauci never cited the study that debunked the work of America's top military AIDS researcher, or the Nobel laureate who discovered HIV.

Thirty-four years later, with over half a trillion dollars spent on AIDS research,[172] Dr. Fauci has not budgeted one dollar to study the role of Lo's and Montagnier's mycoplasma or in Gallo's, and Knox's HHV-6 virus in the etiology of AIDS. Between 1981 and 2020, US taxpayers alone shelled out $640 billion for AIDS research[173, 174] focused almost exclusively on developing drugs to address Dr. Fauci's sketchy HIV hypothesis. Yet the growing list of medications hasn't demonstrably extended the life of a single patient, and the cure for AIDS is still nowhere in sight.[175]

"The minute someone suggests that the orthodoxy might be wrong, the establishment starts to call him crazy or a quack," Rubin continued. "One week you're a great scientist; the next week, you're a jerk. Science has become the new church of America and is closing off all room for creative, productive dissent."[176,177]

After suggesting in print, two years earlier, that HIV might need a co-cofactor to cause AIDS, Gallo went dark. Gallo today refuses to discuss the matter. The normally loquacious and combative Gallo refused my request to talk about HHV6.

AIDS and Fear

In a rational universe, or in a functioning democracy, combatants would duke out the incendiary HIV/AIDS dispute in an open public debate in the scientific literature between the foremost establishment scientists and the best-credentialed dissenting ones. But in Tony Fauci's authoritarian technocracy, the ruling medical cabal refuses to allow this sort of dialogue. Like Inquisition priests, HIV's high clergy stubbornly resist the possibility that they might be wrong. From the outset, the HIV/AIDS religion has seen its survival in moral absolutism, outright discrimination, and merciless suppression of doubt.

Dr. Harvey Bialy argues that the medical establishment's top concern is not public health, but their own reputations and perquisites. "The scientific and medical communities have a great deal of face to lose. It is not much of an exaggeration to state that when the HIV/AIDS hypothesis is finally recognized as wrong, the entire institution of science will lose the public's trust, and science itself will experience fundamental, profound, and long-lasting changes. The 'scientific community' has risked its credibility by standing by the HIV theory for so long. This is why doubting the HIV hypothesis is now tantamount to doubting science itself, and this is why dissidents face excommunication."

As Kary Mullis says in his book *Dancing Naked in the Mind Field*, "What people call science today is probably very similar to what was called science in 1634. Galileo was told to recant his beliefs or be excommunicated. People who refuse to accept the commandments of the AIDS establishment are basically told the same thing."[178]

The quasi-religious nature of the debate is evident in the loathing and pious moralizing expressed toward Duesberg by an unnamed Berkeley scientist, interviewed by Celia Farber for her 2006 book, *Serious Adverse Events: An Uncensored History of AIDS*: "He did it to himself, you know. You see, he wouldn't give up an idea. He went at it with a hammer. He may well be 3,000 percent right, but he upset an awful lot of people. . . . Nobody believed in him because what he was doing was overturning generally held views. They felt betrayed. . . . You don't just stand up and say everybody is wrong."[179]

In her book, *Science Sold Out: Does HIV Really Cause AIDS?*, Rebecca Culshaw writes, "The persistence of this intellectually bankrupt theory in the public mind is attributable entirely to the campaign of fear, discrimination, and terror that has been waged aggressively by a powerful group of people whose sole motivation was and is behavior control. Yes, the money and the vast interests of the pharmaceutical industry and government-funded scientists are important, but the seeds of the HIV/AIDS hypothesis are sowed with fear. If the fear were to end, the myth would end."[180]

Endnotes

1   Celia Farber, "Fatal Distraction," *SPIN* (June 1992), http://www.virusmyth.com/aids/hiv/cffatal.htm
2   Steven Epstein, *Impure Science: AIDS, Activism and the Politics of Knowledge* (University of California Press, 1996), 144
3   The Group for the Scientific Reappraisal of the HIV-AIDS Hypothesis, The Group (Jun 6, 1991) https://www.virusmyth.com/aids/group.htm
4   Elliott Ross, "How drug companies' PR tactics skew the presentation of medical research," *The Guardian* (May 20, 2001), https://www.theguardian.com/science/2011/may/20/drug-companies-ghost-writing-journalism

5   See also, https://www.duesberg.com/about/bribepd.html
6   Celia Farber, "Fatal Distraction," *SPIN* (June 1992), https://www.duesberg.com/media/cffatal.html
7   Steven Epstein, *Impure Science: AIDS, Activism and the Politics of Knowledge* (University of California Press, 1996), 144, https://publishing.cdlib.org/ucpressebooks/view?docId=ft1s20045x&chunk.id=d0e5242&toc.depth=1&toc.id=0&brand=ucpress
8   Rethinking AIDS, "Rethinking AIDS (RA) History," Rethinking AIDS, https://rethinkingaids.com/index.php/about/ra-history
9   Rethinking AIDS, "The AIDS Industry and Media Want You to Think There Are Only a Handful of Scientists Who Doubt the HIV–AIDS Theory," Rethinking AIDS, https://web.archive.org/web/20121227201717/https://rethinkingaids.com/quotes/rethinkers.htm
10 "HIV & AIDS Fauci's First Fraud," Documentary, Sep 6, 2020, YouTube video, 21:08, https://www.youtube.com/watch?v=wy3frBacd2k
11 Ibid., 1:11
12 Celia Farber, "AIDS Words from the Front," *Spin* Vol 10, No 4, 64 (July 1994), https://books.google.com/books?id=9Zg4PvPMtTcC&pg=PA5&dq=SPIN+-+July+1994
13 Another article in NOVA goes a bit further: "'The process requires an astonishingly rare set of circumstances be met,' Katzourakis says. 'Although endogenous retroviruses make up a pretty large proportion of our genome, in terms of the number of times they've infiltrated our genome over the past sixty or so million years, it only comes down to about 30 or 40 distinct occasions,' he says." Carrie Arnold, "The Viruses That Made Us Human," *NOVA*, Sept. 28, 2016, https://www.pbs.org/wgbh/nova/article/endogenous-retroviruses/
14 Carl Zimmer, "Ancient Viruses Are Buried in Your DNA," NYT, (Oct. 4, 2017), https://www.nytimes.com/2017/10/04/science/ancient-viruses-dna-genome.html
15 RFK Jr. Interview with Dr. David Rasnick (August 14, 2021)
16 Celia Farber, "The Passion of Peter Duesberg," *Barnes World Blogs* (April 24, 2004), https://barnesworld.blogs.com/phdp.pdf
17 Stephen Allen, "HIV=AIDS: Fact or Fraud?" (1996), 16:40, YouTube, https://www.youtube.com/watch?v=JTxvmKHYajQ
18 Peter Duesberg, "Retroviruses as Carcinogens and Pathogens: Expectations and Reality," *Cancer Research* (Mar, 1987), https://cancerres.aacrjournals.org/content/47/5/1199
19 "HIV & AIDS: Fauci's First Fraud," op. cit., 16:40
20 Charles Ortleb, *Peter Duesberg and the Duesbergians: How a Brave and Brilliant Group of Scientists Challenged the AIDS Establishment and Inadvertently Exposed the Chronic Fatigue Syndrome Epidemic* (Charles Ortleb/Rubicon Media, 2019), 14
21 "HIV & AIDS: Fauci's First Fraud," op. cit., 17:51
22 "HIV & AIDS: Fauci's First Fraud," op. cit., 1:11:45
23 "HIV & AIDS: Fauci's First Fraud," op. cit., 1:12:40
24 Harvey Bialy, *Oncogenes, Aneuploidy, and AIDS: A Scientific Life & Times of Peter H. Duesberg* (Inst. of Biotechnology, 1994), 3
25 Ibid., 156
26 Jeanne Lenzer, "AIDS 'Dissident' Seeks Redemption . . . and a Cure for Cancer," *DISCOVER* (May 14, 2008), https://www.discovermagazine.com/health/aids-dissident-seeks-redemption-and-a-cure-for-cancer
27 Michael Specter, "The Denialists," *The New Yorker* (Mar. 4, 2007), https://www.newyorker.com/magazine/2007/03/12/the-denialists
28 Chuck Kline, Media Alert, Department of Health and Human Services (Apr. 28, 1987), http://www.duesberg.com/about/hhsalert.html
29 Celia Farber, "Fatal Distraction," op. cit.
30 Celia Farber, "The Passion of Peter Duesberg," Barnes World Blogs (April 24, 2004), https://barnesworld.blogs.com/phdp.pdf
31 Ibid.
32 Ibid.
33 Ibid.
34 Duesberg, *Inventing the AIDS Virus*, 392
35 Duesberg, op. cit., 392–393
36 Bialy, op. cit., 83
37 Celia Farber, "The Passion of Peter Duesberg," *Barnesworldblogs* (Dec. 4, 2004), https://barnesworld.blogs.com/phdp.pdf
38 Ibid.
39 Ibid.
40 John Maddox, "Has Duesberg a Right of Reply?" *Nature* (May 13, 1993), https://sci-hub.se/https://doi.org/10.1038/363109a0
41 Bialy, op. cit., 84–88
42 Celia Farber, "The Passion of Peter Duesberg"
43 Ibid.
44 Ibid.

45 Peter Duesberg, PhD, "House of Numbers: Anatomy of an Epidemic," YouTube Video 00: 50:31, (2009), https://www.youtube.com/watch?v=lvDqjXTByF4

46 John Lauritsen, *The AIDS War: Propaganda, Profiteering, and the Genocide from the Medical-Industrial Complex* (Asklepsios, 1993), 61

47 RFK Jr. Interview (September 10, 2021)

48 Torsten Engelbrecht, Claus Köhnlein, et al., *Virus Mania: How the Medical Industry Continually Invents Epidemics, Making Billions at Our Expense* (Books on Demand, 3rd ed, 2021), 152

49 Ibid., 151

50 Katie Leishman, "The AIDS Debate That Isn't," *Wall Street Journal* (Feb. 26, 1988), https://www.functionalps.com/blog/2012/04/07/the-aids-debate-that-isnt/comment-page-1/

51 Rebecca Culshaw, *Science Sold Out: Does HIV Really Cause AIDS?* (North Atlantic Books, 2007), 68

52 "HIV & AIDS: Fauci's First Fraud," op. cit., 14:25

53 Anthony Fauci, MD, 00:50:56 "House of Numbers: Anatomy of an Epidemic," YouTube Video (2009), https://www.youtube.com/watch?v=lvDqjXTByF4

54 Engelbrecht et al., op. cit., 138

55 Ibid., 106

56 Ibid., 101

57 "HIV & AIDS: Fauci's First Fraud," op. cit., 01:02:03

58 Office of Medical and Scientific Justice, "The 'Berlin Patient' Demystifying AIDS?" (Apr. 13, 2011), https://www.omsj.org/issues/the-berlin-patient-demystifying-aids

59 Engelbrecht et al., op. cit., 30

60 Celia Farber, "The Passion of Peter Duesberg"

61 Robin Scovill, "The Other Side of AIDS,"2011, YouTube video, 1:04:18, https://www.youtube.com/watch?v=0dVYJp5dHf8

62 Robin Scovill, "The Other Side of AIDS," 2011, YouTube video, 1:06:34, https://www.youtube.com/watch?v=0dVYJp5dHf8

63 Engelbrecht et al., op. cit., 111–112)

64 Duesberg, Inventing the AIDS Virus, 282–283

65 Engelbrecht et al., op. cit., 117

66 Peter Duesberg, "The Role of Drugs in the Origin of AIDS," *Biomedicine and Pharmacotherapy* (1992), https://sci-hub.se/10.1016/0753-3322(92)90063-d

67 Peter Duesberg PhD, "The role of drugs in the origin of AIDS," *Biomed & Pharmacother* Vol. 46, 3-15, 1992, https://duesberg.com/papers/pdbiopharm.html

68 Anthony Fauci, "AIDS: Acquired Immunodeficiency Syndrome," NIH, 1984, YouTube video, 35:47-35:59, .https://www.youtube.com/watch?v=pzK3dg59TuY

69 David Durack, "Opportunistic infections and Kaposi's sarcoma in homosexual men," *New England Journal of Medicine* (Dec 10, 1981) https://www.nejm.org/doi/10.1056/NEJM198112103052408

70 Engelbrecht et al., op. cit., 116

71 Ibid., 117

72 "HIV & AIDS: Fauci's First Fraud," op. cit., 00:49:58

73 Ibid., 1:29:22

74 "HIV & AIDS: Fauci's First Fraud," op. cit., 59:59

75 Guy R. Newell et al., "Toxicity, Immunosuppressive Effects and Carcinogenic Potential of Volatile Nitrites: Possible Relationship to Kaposi's Sarcoma," *Pharmacotherapy* (Sep–Oct, 1984). https://accpjournals.onlinelibrary.wiley.com/doi/10.1002/j.1875-9114.1984.tb03376.x

76 "HIV & AIDS: Fauci's First Fraud," op. cit., 51:50

77 Michael Marmor et al., "Risk Factors for Kaposi's Syndrome in Homosexual Men, *Lancet* (May 15, 1982), https://www.thelancet.com/journals/lancet/article/PIIS0140-6736(82)92275-9/fulltext

78 Engelbrecht et al., op. cit., 114

79 "HIV & AIDS: Fauci's First Fraud, Documentary," op. cit., 00:58:20

80 Toby K. Eisenstein, "The Role of Opioid Receptors in Immune System Function," *Front Immunol* (Dec 2019), doi: 10.3389/fimmu.2019.02904

81 Engelbrecht et al., op. cit., 114

82 John Lauritsen, *Death Rush* (Pagan Press, 1986), 30

83 "HIV & AIDS: Fauci's First Fraud," op. cit., 1:00:14

84 Gerald M. Oppenheimer, *AIDS, The Burdens of History*, 295, note 34 (University of California Press, Jan. 1988), https://www.researchgate.net/publication/280939793_In_the_Eye_of_the_Storm_The_Epidemiological_Construction_of_AIDS]

85 H. W. Haverkos et al., "Disease manifestation among homosexual men with acquired immunodeficiency syndrome: a possible role of nitrites in Kaposi's sarcoma," *Sexually Transmitted Diseases*, (Oct–Dec, 1985), https://pubmed.ncbi.nlm.nih.gov/3878602/

86 Engelbrecht et al., op. cit., 121

87 Ibid., 122

88 V. Beral et al., "Kaposi's sarcoma among persons with AIDS: a sexually transmitted infection?" *The Lancet* (Jan. 20, 1990), https://www.thelancet.com/journals/lancet/article/PII0140-6736(90)90001-L/fulltext

89 Engelbrecht et al., op. cit., 124–125

90   John Lauritsen, "NIDA Meeting Calls for Research into The Poppers-Kaposi's Sarcoma Connection," *New York Native* (Jun. 13, 1994), https://paganpressbooks.com/jpl/NIDA-KS.HTM

91   Engelbrecht et al., op. cit., 125

92   Eleni Papadopulos et al., "A critique of the Montagnier evidence for the HIV/AIDS hypothesis," *Medical Hypotheses* (2004), https://www.sciencedirect.com/science/article/abs/pii/S0306987704002415?via%3Dihub

93   RFK Jr. Interview with Celia Farber

94   Engelbrecht et al., op. cit., 125

95   "HIV & AIDS Fauci's First Fraud," op. cit., 41:23

96   ViiV Healthcare ULC, PRODUCT MONOGRAPH INCLUDING PATIENT MEDICATION INFORMATION, PrRETROVIR (AZT), p 20, https://viivhealthcare.com/content/dam/cf-viiv/viiv-healthcare/en_CA/pdf/Retrovir.pdf

97   "HIV & AIDS: Fauci's First Fraud," op. cit., 41:29

98   Ibid.

99   Leung, op. cit., 01:11:41

100   "HIV & AIDS: Fauci's First Fraud," op. cit., 1:28:07-1:29:05

101   Douglas D. Richman et al., "The Toxicity of Azidothymidine (AZT) in the Treatment of Patients with AIDS and AIDS-Related Complex," *New England Journal of Medicine* (Jul. 23. 1987), https://www.nejm.org/doi/full/10.1056/nejm198707233170402

102   Simon Garfield, "The rise and fall of AZT: It was the drug that had to work. It brought hope to people with HIV and Aids, and millions for the company that developed it. It had to work. There was nothing else. But for many who used AZT - it didn't," *Independent* (Oct. 23, 2011), https://www.independent.co.uk/arts-entertainment/rise-and-fall-azt-it-was-drug-had-work-it-brought-hope-people-hiv-and-aids-and-millions-company-developed-it-it-had-work-there-was-nothing-else-many-who-used-azt-it-didn-t-2320491.html

103   John Lauritsen, "Petrushka was Poisoned: Did AZT contribute to Nureyev's untimely death?" *New York Native* (Feb 1, 1993), archived at https://www.duesberg.com/media/jlpetrushka.html

104   Engelbrecht et al., op. cit., 157

105   Ibid., 158

106   "HIV & AIDS: Fauci's First Fraud Documentary," op. cit., 1:29:42

107   "HIV & AIDS: Fauci's First Fraud" op. cit., 1:29:49-1:30:48

108   Celia Farber, "AIDS and the AZT Scandal: SPIN's 1989 Feature, 'Sins of Omission,'" *SPIN* (Oct 5, 2015), https://www.spin.com/featured/aids-and-the-azt-scandal-spin-1989-feature-sins-of-omission/

109   CDC, "HIV and AIDS—United States, 1981—2000," *MMWR Weekly* (June 1, 2001), https://www.cdc.gov/mmwr/preview/mmwrhtml/mm5021a2.htm

110   "AIDS Cases, Deaths, and Persons Living with AIDS 1985–2005—United States and Dependent Areas," CDC, https://childrenshealthdefense.org/wp-content/uploads/US_AIDS_cases_deaths_livingwithAIDS_1985-2005.pdf

111   Engelbrecht et al., op. cit., 142

112   Ibid.

113   "*U.S. HIV and AIDS Cases Reported Through December 1997,"* CDC HIV/AIDS Surveillance Report, Year-end edition Vol. 9, No. 2, p. 1, p. 25, figure 6, https://www.cdc.gov/hiv/pdf/library/reports/surveillance/cdc-hiv-surveillance-report-1997-vol-9-2.pdf

114   Terry Michaels, *Down the HIV Rabbit Hole* (Unpublished Manuscript, 2017), 38

115   "HIV & AIDS: Fauci's First Fraud," op. cit., 26:30

116   Claus Köhnlein, "The AZT Disaster," *Rethink Aids* (Feb. 11, 2006), http://www.whale.to/a/kohnlein2.html

117   Celia Farber, "Russia Today: Doctor Claus Köhnlein," Interview w/ Dr. Claus Köhnlein, July 21, 2010, YouTube video, 00:15-00:54, https://www.youtube.com/watch?v=xqBfJMsJjw8&t=3s

118   Ibid., 01:01-1:30

119   Celia Farber, "Russia Today: Dr. Claus Köhnlein," 01:47-02:18

120   M. A. Fischl et al., "The efficacy of azidothymidine (AZT) in the treatment of patients with AIDS and AIDS-related complex. A double-blind, placebo-controlled trial," *New England Journal of Medicine* (Jul 23, 1987), https://pubmed.ncbi.nlm.nih.gov/3299089/

121   John Maddox, "Study confirms AZT's lack of prophylactic effect," *NATURE* (April 14, 1994), doi:10.1038/368577b0 https://sci-hub.se/https://doi.org/10.1038/368577b0

122   Concorde Coordinating Committee, "MRC/ANRS randomised double-blind controlled trial of immediate and deferred zidovudine in symptom-free HIV infection," *The Lancet* (Apr 9, 1994), https://doi.org/10.1016/S0140-6736(94)90006-X

123   Holman W. Jenkins Jr., "The Other Media Blackout: How can Americans use good sense about an epidemic about which they are fed false information?" *Wall Street Journal* (Oct. 30, 2020), https://www.wsj.com/articles/the-other-media-blackout-11604094677

124   "HIV & AIDS: Fauci's First Fraud," op. cit., 37:45

125   Peter Duesberg et al., "The chemical bases of the various AIDS epidemics: recreational drugs, anti-viral chemotherapy and malnutrition," *Journal of Biosciences* (June 2003), doi: 10.1007/BF02705115

126   Culshaw, op. cit., 76

127   Engelbrecht et al., op. cit.,119

128   Ibid.
129   John Lauritsen and Hank Wilson, *Death Rush: Poppers & AIDS* (Pagan Press, 1986), 6
130   Ian Young, "The Poppers Story: The Rise and Fall and Rise of The Gay Drug," *Steam*, Vol 2, Issue 4, https://www.duesberg.com/articles/iypoppers.html
131   Engelbrecht et al., op. cit., 121
132   Ibid.
133   Health Watch, "Researchers Implicate HHV-6 Virus In Chronic Fatigue Syndrome, AIDS And Multiple Sclerosis,"*Prohealth.com* (Apr. 1, 1998), https://www.prohealth.com/library/researchers-implicate-hhv-6-virus-in-chronic-fatigue-syndrome-aids-and-multiple-sclerosis-11510
134   Paolo Lusso and Robert C. Gallo, "Human herpesvirus 6 in AIDS," *Immunology Today* (Feb 1995), doi: 10.1016/0167-5699(95)80090-5
135   Paolo Lusso, Robert C. Gallo, et al., "Productive dual infection of human CD4+ T lymphocytes by HIV-1 and HHV-6," *Nature* (Jan. 26, 1989), https://sci-hub.se/10.1038/337370a0
136   Robert C. Gallo, "A perspective on human herpes virus 6 (HHV-6)," *Journal of Clinical Virology* (2006), https://sci-hub.se/10.1016/S1386-6532(06)70003-8
137   Rosemary Goudreau, "Highly Contagious Herpes Virus Linked to Cancer, AIDS," *Miami Herald*, May 11, 1988, at 1A, https://miamiherald.newspapers.com/image/633340403/
138   Ibid., at 12A. /
139   Konstance Kehl Knox and Donald R. Carrigan, "Active HHV-6 Infection in the Lymph Nodes of HIV-Infected Patients: In Vitro Evidence That HHV-6 Can Break HIV Latency," *Journal of Acquired Immune Deficiency Syndromes and Human Retrovirology* (April 1, 1996), https://journals.lww.com/jaids/Fulltext/1996/04010/Active_HHV_6_Infection_in_the_Lymph_Nodes_of.7.aspx
140   Neenyah Ostrom, "Dr. Konstance Knox explains why HHV-6 may be the key to dealing with AIDS," *New York Native* (Apr 15, 1996), https://hhv6.blogspot.com/2021/03/dr-konstance-knox-explains-why-hhv-6.html
141   Ibid.
142   Ibid.
143   Ibid.
144   Ibid.
145   Dario DiLuca et al., "Human Herpesvirus 6 and Human Herpesvirus 7 in Chronic Fatigue Syndrome," *Journal of Clinical Microbiology* (June, 1995), https://journals.asm.org/doi/pdf/10.1128/jcm.33.6.1660-1661.1995
146   Charles Ortleb, *The Chronic Fatigue Syndrome Epidemic Cover-Up: How a Little Newspaper Solved the Scientific and Political Mystery of Our Time* (Rubicon Media, 2018), 166
147   Anthony L. Komaroff, Dedra Buchwald, et al., "A Chronic Illness Characterized by Fatigue, Neurologic and Immunologic Disorders, and Active Human Herpesvirus Type 6 Infection," *Annals of Internal Medicine* (Jan. 15, 1992), https://sci-hub.se/10.7326/0003-4819-116-2-103
148   Neenyah Ostrom, "Dr. Konstance Knox explains why HHV-6 may be the key to dealing with AIDS," *New York Native* (Apr 15, 1996), https://hhv6.blogspot.com/2021/03/dr-konstance-knox-explains-why-hhv-6.html
149   Ibid.
150   Ibid.
151   Elinor Burkett, "HIV: Not Guilty?" *Tropic Miami Herald* (Dec 23, 1990), 12, https://www.newspapers.com/image/635684455
152   Elinor Burkett, "Is HIV Guilty?" *Miami Herald* (Dec 23, 1990), http://virusmyth.com/aids/hiv/ebhiv.htm
153   Duesberg, op. cit., 233
154   Ibid.
155   Shyh-Ching Lo, "Virus-like Infectious Agent (VLIA) is a Novel Pathogenic Mycoplasma: Mycoplasma Incognitus," *Journal of Tropical Medicine* (Nov., 1989), https://www.ajtmh.org/view/journals/tpmd/41/5/article-p586.xml
156   Burkett, "HIV: Not Guilty?" at 12
157   Burkett, "Is HIV Guilty?"
158   Burkett, "HIV: Not Guilty?" at 12
159   Burkett, "Is HIV Guilty?"
160   Burkett, "HIV: Not Guilty?" at 12
161   Burkett, "Is HIV Guilty?"
162   Luc Montagnier et al., "Protective activity of tetracycline analogs against the cytopathic effect of the human immunodeficiency viruses in CEM cells," *RESEARCH IN VIROLOGY* (1990), https://sci-hub.se/10.1016/0923-2516(90)90052-K
163   Burkett, "HIV: Not Guilty?" at 12
164   Burkett, "Is HIV Guilty?"
165   Burkett, "HIV: Not Guilty?" at 12
166   Burkett, "Is HIV Guilty?"
167   Burkett, "HIV: Not Guilty?" at 12
168   Burkett, "Is HIV Guilty?"
169   Burkett, "HIV: Not Guilty?" at 12

170 Burkett, "Is HIV Guilty?"
171 Brett Leung, "Knowledge Matters, House of Numbers - Anatomy of an Epidemic," YouTube video, Apr. 19, 2009, 00:47:33, https://youtu.be/lvDqjXTByF4
172 Henry J. Kaiser Family Foundation, Annual AIDS Funding in USA, http://www.davidrasnick.com/ewExternalFiles/AIDS%20funding%20through%202019.pdf
173 Ibid.
174 HIV.gov, Federal Domestic HIV/AIDS Programs & Research Spending, https://www.hiv.gov/federal-response/funding/budget
175 Engelbrecht et al., op. cit., 31
176 Burkett, "HIV: Not Guilty?" at 12
177 Burkett, "Is HIV Guilty?"
178 Kary Mullis, *Dancing Naked in the Mind Field* (Vintage Books, 1998), 180
179 Celia Farber, *Serious Adverse Events: An Uncensored History of AIDS* (Melville House Press, 2006), 55
180 Rebecca Crenshaw, *Science Sold Out* (North Atlantic Books, 2007), 60

ChildrensHealthDefense.org/fauci-book
childrenshd.org/fauci-book

For updates, new citations and references, and new information about topics in this chapter:

CHAPTER 7
DR. FAUCI, MR. HYDE: NIAID'S BARBARIC AND ILLEGAL EXPERIMENTS ON CHILDREN

"The Nazi medical experiments are an example of this sadism, for in the use of concentration camp inmates and prisoners of war as human guinea pigs very little, if any, benefit to science was achieved. It is a tale of horrors of which the German medical profession cannot be proud. Although the 'experiments' were conducted by fewer than two hundred murderous quacks— albeit some of them held eminent posts in the medical world—their criminal work was known to thousands of leading physicians of the Reich, not a single one of whom, so far as the record shows, ever uttered the slightest public protest."
—William L. Shirer, *The Rise and Fall of the Third Reich*

"Science advances one funeral at a time."

—Max Planck

During the nearly four decades since Dr. Anthony Fauci took the agency's reins, the National Institute of Allergy and Infectious Diseases (NIAID) has often treated America's most vulnerable children as collateral damage in its director's single-minded pursuit of profitable pharmacological solutions for steadily declining public health. AZT's sketchy and corrupt path to regulatory approval in 1988 blazed a trail for a multibillion-dollar boom in new HIV drugs, and Dr. Fauci gave broad leeway to his pharmaceutical partners and their PIs to conduct unethical human experimentation that exposed both children and adults to toxic compounds.

The US Department of Health and Human Services (HHS) and its predecessor agency, the Public Health Service, already had a long history of morally repugnant experiments on vulnerable subjects, including imprisoned convicts, institutionalized adults with intellectual disabilities, and orphaned children in hellholes like Staten Island's Willowbrook and the Fernald School in Waltham, Massachusetts. In 1973, Dr. Stanley Plotkin penned a letter to the *New England Journal of Medicine* in which he justified his experiments on vulnerable intellectually disabled children, saying they "are humans in form but not in social potential."[1] Those sorts of prejudices did nothing to damage his lofty reputation among his colleagues. Vaccinologists consider the annual Stanley Plotkin Award the Nobel Prize of vaccinology. In 2019, the *British Medical Journal* called Plotkin "the Godfather of vaccines."[2] These homegrown American medical Mengeles most often targeted impoverished American Indians and Blacks in Africa, the Caribbean, and in the United States as their laboratory rats. I am proud that my uncle, Senator Edward Kennedy, played a key role in ending the government's forty-year Tuskegee Syphilis Experiment (begun in 1932), another notorious medical research assault on a vulnerable population, when he learned about it in 1972 from a CDC whistleblower.[3]

Government regulators and their pharmaceutical industry partners often combined racial discrimination with child abuse in HHS's drug and vaccine development campaigns. During the government/industry polio vaccine experiments of the 1950s–1960s, US vaccinologists like Hilary Koprowski and Stanley Plotkin worked with Belgian colonial authorities in the Congo to recruit millions of Black African child "volunteers" for dozens of mass-population trials with experimental vaccines that were perhaps considered to be too risky to test on white children. As late as 1989, the CDC conducted lethal experiments with a hazardous measles vaccine on Black children in Cameroon, Haiti, and South-Central Los Angeles, killing dozens of little girls before halting the program.[4] CDC did not tell "volunteers" that they were participating in an experiment. In 2014, another CDC whistleblower, the agency's senior vaccine safety scientist, Dr. William Thompson, disclosed that top CDC officials had forced him and four other senior researchers to lie to the public and destroy data that showed disproportionate vaccine injuries—including a 340 percent elevated risk for autism—in Black male infants who received the Measles, Mumps, Rubella (MMR) vaccine on schedule.[5] So it was only natural that Dr. Fauci and his Pharma partners employed Black and Hispanic foster children for cruel and barbaric treatments in their efforts to develop their second-generation antivirals and chimeric HIV vaccines that provided the initial stepping-stones for his career.

In 1989, Dr. Fauci declined President George H. W. Bush's offer to become NIH director, explaining, "I was training for the AIDS epidemic before it even happened. My being involved with it has been my passion and my life's work."[6, 7] Dr. Fauci's philanthropic demurrer might have been disingenuous. By then, his power as NIAID director dwarfed the authority wielded by his nominal boss at NIH. His successful early machinations during the AIDS drug boom had won NIAID a massive discretionary budget and global influence over scientific research and international health policy, including de facto control over its sister HHS agencies, FDA and CDC. The NIAID directorship also offered dizzying publicity opportunities and lucrative partnerships with pharmaceutical companies as NIAID became Pharma's chief incubator and collaborator in new drug development and promotion. Biocontainment handling expert and trainer Sean Kaufman, who designed and built mock biosafety level (BSL) laboratories for NIAID in the mid-2000s, is a longtime admirer of Dr. Fauci and trained hundreds of BSL workers in safety protocols for NIAID. Kaufman told me, "Everyone knows that Dr. Fauci runs the whole show at HHS. All the other agency heads are figureheads. Tony Fauci pulls all the strings."

Jonathan Fishbein, MD, who served as head of the Division of AIDS (DAIDS) Office of Policy in Clinical Research Operations from 2003–2005, told me that Fauci's expanding influence seemed to eclipse that of his boss, NIH Director Dr. Elias Zerhouni: "When Zerhouni could have taken the high road and righted the misconduct that I exposed in the Division of AIDS, he chose to stay uninvolved. Fauci is a master at marketing himself and his Institute and leveraged AIDS to generate huge appropriations from Congress to the NIH. Who would ever have stood up to him? Certainly not Zerhouni or his successors! NIAID money is spread throughout the major medical institutions in the United States and for that reason, he wields enormous influence in the medical community."

Dr. Fauci's corrupt collaboration with pharmaceutical companies that yielded NIAID's scandalous approval of AZT in 1987 consolidated his symbiotic relationship with the Pharma PIs and lowered NIAID's standards for product approvals. His relationships with his PIs and their Pharma patrons yielded a cascade of beneficial personal opportunities, and Dr. Fauci quickly learned to overlook Pharma's excesses. The 1980 Bayh–Dole Act[8] allowed NIAID—and Dr. Fauci personally—to file patents on the hundreds of new drugs that his agency-funded PIs were incubating, and then to license those drugs to pharmaceutical companies and collect royalties on their sales. NIAID's drug development enterprise quickly eclipsed HHS's regulatory function. Millions of dollars began flowing in from drug royalties to NIH and to NIAID's high-level personnel, including Dr. Fauci—further blurring the boundaries between public health and Pharma profits.

According to an exposé by the Associated Press, "In all, 916 current and former NIH researchers are receiving royalty payments for drugs and other inventions they developed while working for the government."[9] That investigation concluded that scientists and administrators at the National Institutes of Health flagrantly disregard ethical and legal requirements of financial disclosure.

Financial conflicts with pharmaceutical companies quickly became the defining feature of Dr. Fauci's governance style. As early as 1992, a Department of Health and Human Services Inspector General investigation concluded that NIAID failed to police conflicts of interest by his PIs in a vaccine clinical trial.[10]

All that new NIH and NIAID money made clinical trials a vast, booming industry. Holocaust survivor Vera Sharav spent her long career investigating abusive human experiments by NIAID and other agencies. Sharav told me, "Beginning around 1990, clinical trials became the profit center for the medical community. The insurance industry and HMOs were squeezing doctors so that it became hard to make big money practicing medicine. The most ambitious doctors left patient care and gravitated toward clinical trials. Everybody involved was making money except the subjects of the human experiments. At the center of everything was NIH and NIAID. While people were not paying attention, the agency quietly became the partner of the industry."

Pharma's ethics quickly pervaded and corrupted NIAID's culture. The agency routinely overlooked and often sanctioned and engaged in routine manipulation of science to "prove" efficacy of dangerous and ineffective drugs. Callous disregard toward suffering and deaths among clinical trial volunteers became a feature of NIAID's modus operandi.

According to the AP investigation, NIH scientists who violate ethical and legal requirements and use underhanded recruitment tactics pose a very real and present threat to public safety: "hundreds, perhaps thousands, of patients in NIH experiments made decisions to participate in experiments that often carry risks without full knowledge about the researchers' financial interests."

In 2004, investigative journalist Liam Scheff chronicled Dr. Fauci's secretive experiments on hundreds of HIV-positive foster children at Incarnation Children's Center (ICC) in New York City and numerous sister facilities in New York and six other states between 1988 and 2002.[11] Those experiments were the core of Dr. Fauci's career-defining effort to develop a second generation of profitable AIDS drugs as an encore to AZT.[12]

Scheff described how Dr. Fauci's NIAID and his Big Pharma partners turned

Black and Hispanic foster kids into lab rats, subjecting them to torture and abuse in a grim parade of unsupervised drug and vaccine studies: "This former convent houses a revolving stable of children who've been removed from their own homes by the Agency for Child Services [ACS]. These children are Black, Hispanic, and poor. Many of their mothers had a history of drug abuse and have died. Once taken into ICC, the children become subjects of drug trials sponsored by [Dr. Anthony Fauci's] NIAID (National Institute of Allergies and Infectious Disease, a division of the NIH), NICHD (the National Institute of Child Health and Human Development) in conjunction with some of the world's largest pharmaceutical companies—GlaxoSmithKline, Pfizer, Genentech, Chiron/Biocine and others."[13]

NIAID's Pharma partners remunerated Incarnation Children's Center (ICC) for supplying children for the tests. As usual, Dr. Fauci had the safety oversight board rigged with his loyal PIs, foremost of whom was Dr. Stephen Nicholas, a generously funded NIAID AIDS drug researcher. "Stephen Nicholas was not only director of the ICC until 2002; he also simultaneously sat on the Pediatric Medical Advisory Panel, which was supposed to oversee the tests—which signifies a serious conflict of interest," criticizes Vera Sharav, president of the Alliance for Human Research Protection (AHRP), a medical industry watchdog organization.[14]

Scheff continued, "The drugs being given to the children are toxic—they're known to cause genetic mutation, organ failure, bone marrow death, bodily deformations, brain damage, and fatal skin disorders.[15]

"If the children refuse the drugs, they're held down and force fed. If the children continue to resist, they're taken to Columbia Presbyterian hospital, where a surgeon puts a plastic tube through their abdominal wall into their stomachs. From then on, the drugs are injected directly into their intestines.[16]

"In 2003, two children, ages six and twelve, had debilitating strokes due to drug toxicities. The six-year-old went blind. They both died shortly after. Another fourteen-year-old died recently. An eight-year-old boy had two plastic surgeries to remove large, fatty, drug-induced lumps from his neck."[17]

"This isn't science fiction. This is AIDS research."[18]

Even the foster children who survived Fauci's experiments reported dire side effects, ranging from skin outbreaks and hives, nausea, and vomiting, to sharp drops in immune response and fevers—all common adverse reactions associated with the drugs he was targeting for development.

During one of his trials involving the drug Dapsone, at least ten children died. A May 2005 Associated Press investigation reported that those "children died from a variety of causes, including four from blood poisoning." Researchers complained they were unable to determine a safe, useful dosage. Their guessing game cost those children their lives.[19]

"An unexpected finding in our study," the researchers pitilessly observed, "was that overall mortality while receiving the study drug was significantly higher in the daily Dapsone group." NIAID researchers shrugged off the deaths as a mystery: "This finding remains unexplained."[20]

Vera Sharav spent years investigating Dr. Fauci's torture chambers as part of her lifelong mission to end cruel medical experimentation on children. Sharav told me, "Fauci just brushed all those dead babies under the rug. They were collateral damage in his career ambitions. They were throw-away children." Sharav said that at least eighty children died in Dr. Fauci's Manhattan concentration camp and accused NIAID and its partners of disposing of children's remains in mass graves.

BBC's heartbreaking 2004 documentary, *Guinea Pig Kids*,[21] chronicles the savage barbarity of Dr. Fauci's science projects from the perspective of the affected children. That year, BBC hired investigative reporter Celia Farber to conduct field research for the film, which exposes the dark underside of Big Pharma's stampede to develop lucrative new AIDS remedies. "I found the mass grave at Gate of Heaven cemetery in Hawthorne, New York," she told me. "I couldn't believe my eyes. It was a very large pit with AstroTurf thrown over it, which you could actually lift up. Under it one could see dozens of plain wooden coffins, haphazardly stacked. There may have been 100 of them. I learned there was more than one child's body in each. Around the pit was a semi-circle of several large tombstones on which upward of one thousand children's names had been engraved. I wrote down every name. I'm still wondering who the *rest* of those kids were. As far as I know, nobody has ever asked Dr. Fauci that haunting question.

"I remember the teddy bears and hearts in piles around the pit and I recall the flies buzzing around. The job of recording all those names took all day. NIAID, New York, and all the hospital PIs were stonewalling us. We couldn't get any accurate estimate of the number of children who died in the NIAID experiments, or who they were. I went to check the gravestone names against death certificates at the NYC Department of Health, which you could still do at that time. BBC wanted to match these coffins to the names of children who were known to have been at ICC. It was a very slow, byzantine project with tremendous institutional resistance, but we did turn up a few names. We learned the story of a father who had come out of prison looking for his son. He was told his son had died at ICC of AIDS and there were no medical records, as they'd all been 'lost in a fire.' He was devastated. This story ran in the *NY Post*, believe it or not. But one after the other, every media outlet that touched this story got cold feet. Even then, the medical cartel had this power to kill this kind of story. Dr. Fauci has built his career on that attitude. Nobody even asks him a follow-up question. NIAID's narrative, at that time, was that these children were among the doomed as they 'had AIDS,' so supposedly they were all going to die anyway. When people died, in large numbers, gruesome deaths, NIAID's medical researchers called it 'lessons learned.'"

Two years later, Farber would follow the trail of child casualties left by Dr. Fauci's AIDS branch, DAIDS, in Uganda, exposing the pattern of abusing African mothers and children.

After the BBC documentary aired, AP reporter John Solomon made his own efforts to calculate the number of children who died in Dr. Fauci's AIDS drug experiments. Solomon's May 2005 AP investigation revealed that at least 465 NYC foster children were subjects in NIAID's trials and that Dr. Fauci's agency provided fewer than one-third (142) of those children with an advocate—the minimum legally mandated protection.[22]

A March 2004 letter from Vera Sharav to Dr. David Horowitz, director of FDA's Office of Compliance, charged Dr. Fauci's HIV drug trials with numerous violations of federal law, including NIAID's failure to protect the rights and safety of foster children, particularly during the perilous Phase I stages in which drug companies determine toxicity effects by exploring maximum tolerance levels.[23] Sharav accused Dr. Fauci's team of illegally failing to provide state wards and orphans with independent guardians to represent their interests and protect their rights during brutal, dangerous, and often agonizingly painful experiments.

The 2004 FDA investigation of Dr. Fauci's AIDS research division urged the head of NIH to insist on better management from NIAID. "The overall management of this Division requires careful review," the report said.[24] A May 2005 Congressional hearing also concluded that NIAID's experiments had violated federal statutes.[25]

In testimony before Congress, NIAID and its local partner—New York City's Administration for Children's Services (ACS)—sought to justify the unethical research practices by claiming they were providing first-class, cutting-edge treatments to HIV-infected children who could otherwise not afford expensive medicines.[26]

However, AHRP's investigation revealed that many of the children NIAID subjected to Dr. Fauci's experiments were perfectly healthy and may not even have been HIV-infected.[27] Those investigations focused on thirty-six of the trials. For obvious reasons, clinical trials virtually always occur in hospital settings with trained medical personnel, doctors and nurses, in attendance. However, ICC was a non-medical facility. The decision to allow experiments with highly toxic drugs at an orphanage devoid of medical personnel was, itself, a stunning act of malpractice. Subsequent events suggest that the decision was deliberate, calculated to avoid scientific and ethical objections that might have put Pharma PIs at odds with trained medical professionals. Publicly, NIAID pretended it would permit pharmaceutical companies to conduct their dangerous dose tolerance experiments only on children who had terminal AIDS and were therefore likely to die anyhow. However, AHRP found that NIAID was quietly allowing its Pharma partners to experiment not only on children with laboratory-confirmed HIV infection, but also those "presumed" to be infected. In other words, NIAID required no proof that these children actually had HIV. AHRP accused NIAID of exposing children who might never have developed AIDS to lethal risks and the horrific adverse effects of highly toxic drugs for purposes that were not therapeutic, but purely experimental.[28]

On March 8, 2004, NIH rejected a Freedom of Information Act (FOIA) request for the adverse event reports from NIAID's trials conducted at ICC, citing FOIA's "trade secrets" and "privacy" exemptions.[29] AHRP then filed a complaint on March 10 with the FDA and the Office of Human Research Protections (OHRP), charging that NIAID was depriving foster children of legally mandated federal protections against research risks. Two subsequent investigations validated AHRP's complaint.[30, 31]

John Solomon's AP investigation finally brought Dr. Fauci's experiments to national prominence. AP identified at least forty-eight AIDS experiments NIAID conducted on foster children in seven states—mostly in violation of the federal requirement that NIAID provide those children an advocate. In addition to the Dapsone trial that killed at least ten children, NIAID sponsored another study testing a combination of adult antiretroviral drugs. AP reported that of the fifty-two children in the trial, there were twenty-six moderate to severe reactions—nearly all in infants. The side effects included rash, fever, and dangerous drops in infection-fighting white blood cells.[32]

Casualties in the HIV Vaccine Enterprise

From the outset, Dr. Fauci's experiments served his vain obsession to develop an HIV vaccine. (Despite these expenditures of tens of billions of dollars, he has failed—for forty years—to ever develop an HIV vaccine that was safe or effective for human use.) Medical records that NIAID ultimately and reluctantly released proved that Dr. Fauci's PIs were testing his dangerous vaccines on children from one month to eighteen years old. AP writer John Solomon confirmed that despite contrary requirements in official NIAID protocols, NIAID was knowingly allowing its Pharma partners to violate NIAID's written study protocols by conducting these experiments on children with and without proof of HIV infection.[33,34]

For example, published reports acknowledge that NIAID, Genentech, and Micro-Genesys cosponsored a vaccine trial code-named ACTG #218. The ACTG #218 protocol states, "Patients must have: Documented asymptomatic HIV infection," and the "Expected Total Enrollment" was seventy-two. However, an internal report acknowledges that NIAID was allowing the companies to openly violate those requirements: "125 immunized children proved to be HIV uninfected."[35] Another report stated: "A total of 126 children were not infected."[36] NIAID's final analysis acknowledged that ACTG #218 "showed no clinical benefit to vaccine recipients."[37]

Another HIV Phase I vaccine trial, ACTG #230, tested two experimental vaccines, one by Genentech, another by Chiron/Biocine. This time, the protocol openly declared: "Accepts Healthy Volunteers."[38] As Solomon discovered, the "volunteer" subjects of that unethical experiment were newborns aged three days or less.[39] NIAID randomized these infants to one of three doses of either experimental HIV vaccine or placebo. These reports validate AHRP's concerns that Dr. Fauci experimented on infants and children who were never at risk of AIDS, and that he exposed them to deadly risks and agonizing discomforts in a speculative drug and vaccine exercise that offered absolutely no potential benefit for them.

Dr. Fauci was certainly aware of the peril to which he was subjecting his gallant infant "volunteers." Most of the drugs that his PIs tested on these children were previously approved for adults with AIDS and carried Black Box warnings of potentially lethal side effects: Aldesleukin, Dapsone, Didanosine, Lamivudine, Nevirapine, Ritonavir, Stavudine, and Zidovudine.[40, 41]

Finally, even in cases when the children were genuinely ill, Dr. Fauci's pretense that his experiments were compassionate gestures to impoverished orphans was always a sham. NIAID's claim that their experiments were the only opportunity for those children to receive "life-saving" drugs was a canard from the outset. New York State law requires that physicians provide "life-saving" treatment to wards of the state, if need be, to provide treatment "off-label."

Furthermore, drug companies do not primarily design clinical trials to benefit the individual subjects. Their purpose is to gain safety and efficacy information that may prove helpful for subsequent patients and be profitable for their bottom line. Finally, not all subjects get the "most promising" drug in a trial; some get placebos.

Liam Scheff's January 2004 article, "The House that AIDS Built," ignited an outraged Internet controversy, prompting the *New York Press* to publish a follow-up article by Scheff, "Inside Incarnation."[42] Scheff's detailed descriptions are worth reading if only to understand the sacrifices that Dr. Fauci demanded from his venturesome "volunteer" babies for "the greater good."

Scheff's chronicle suggests that Dr. Fauci and his PIs purposefully took advantage of Incarnation Children's Center's status as a non-medical facility. The PIs had free rein to engage in conduct that experienced professional nurses and doctors would have flagged as unethical and illegal.

When children declined to take the toxic drugs, NIAID and its Pharma partners arranged to surgically implant feeding tubes in their bellies to force obedience. Scheff wrote, "When Mimi started at ICC, the tubes were used infrequently. 'But when the kids got older, a lot of them started to refuse the medication,' she recalled. 'Then they started coming in with the tubes more and more. Kids who refused too much, or threw up too much, they'd get a tube. First it was through the nose. But then it was more and more through the stomach. You'd see a certain child refusing over and over, and one day they'd come back from the hospital from surgery, and they had a tube coming right out of their stomach. If you asked why, the doctors said it was for "compliance"—the regimen. Got to keep up the regimen,' said Mimi. 'Those were the rules.'"[43]

Mimi describes how children suffered—and how some died:

One girl, a six-year-old, Shyanne—she came in for adherence. She was the most delicate little flower—beautiful, polite, full of life. Her family never gave her meds. So, Administration for Children's Services brought her into ICC . . . she came in and started the meds. And it was three months, maybe three months. And she had a stroke. She could not see. She was this normal girl, singing, jumping, playing. Then, poof, stroked out. Blind. We were freaked out. Then, in a few months, she was gone—dead.[44]

Between 1985 and 2005, NIAID and its Pharma partners conscripted at least 532 infants and children from foster care in New York City as human subjects of clinical trials testing NIAID's experimental AIDS drugs and vaccines.[45] ICC and the medical research centers that conducted the trials received substantial payments for hosting the experiments, from both the National Institutes of Health and the manufacturers of the drugs. Among those companies were Merck, Bristol Myers Squibb, MicroGenesys, Biocine, Glaxo, Wellcome, and Pfizer.[46]

The subsequent independent investigations—by the Associated Press,[47] by the federal Office of Human Research Protections,[48] and by the Vera Institute of Justice[49]—confirmed that most children did not have the protection of an independent advocate to give or refuse consent to experimental interventions, and that they were almost all children of color: predominantly African American (64 percent) and Latino

(30 percent), suggesting discriminatory policies consistent with HHS's long history of medical racism.

The Vera Institute, relying mostly on city ACS documents, confirmed eighty deaths and that many other children suffered serious harm: "The child welfare files contained information indicating that some children experienced serious toxicities, or side effects, from trial medications, such as reduced liver function or severe anemia. These toxicities were consistent with toxicities described in published articles about the trials."

"Fauci pooh-poohed all those deaths," recalls Vera Sharav. "The very best thing you could say about Dr. Fauci is that he failed to get involved when problems emerged on his management watch."

The Associated Press reported that the scope of Dr. Fauci's experiments was much wider, extending beyond New York to "at least seven states." Among them: Illinois, ·Louisiana, Maryland, New York, North Carolina, Colorado, and Texas. AP reported that more than four dozen different studies were involved. The foster children ranged from infants to late teens.

Investigation by the Federal Office of Human Research Protections (OHRP)

In 2006, following journalist John Solomon's AP report, the OHRP launched its own investigation of the problems at NIAID. That study found NIAID's toxic culture had normalized chronic violations of product safety science. OHRP confirmed the allegations by the AHRP—that drug companies, their PIs, and government officers failed to obtain proper consent from an independent advocate, failed to ensure "equitable" selection, and failed to ensure safeguards for the foster children who "are likely to be vulnerable to coercion or undue influence."[50]

Vera Institute Report

In 2005, the NYC Administration of Child Services (ACS) commissioned a four-year investigation by the Vera Institute, at a cost of $3 million.[51] The Vera Institute issued its Annual Report in 2009. The Report investigated a twenty-year period during which Dr. Fauci's NIAID experiments endangered predominantly African American and Latino children in foster care by subjecting them to toxic Phase I and Phase II AIDS drug and vaccine experiments–mostly without parental consent and without the protection of an independent advocate.

Among the findings in the Vera study:

- eighty of the 532 children who participated in clinical trials or observational studies died while in foster care;
- twenty-five of the children died while enrolled in a medication trial;
- sixty-four children participated in thirty medication trials that were NOT REVIEWED by a special medical advisory panel, as the city's policy required;
- and twenty-one children participated in trials that the panel had reviewed but had NOT RECOMMENDED.

- (In both cases, thirteen of the enrollments occurred before the children were placed in foster care.)

Vera Institute's director, Timothy Ross, complained that the report only contained a portion of NIAID's atrocities because NIAID allowed the hospitals to deny the Institute staff access to the children's primary records or the clinical trial records, which the culprits kept sealed under the pretense of confidentiality. These are the hospitals that conduct the lucrative clinical trials for NIAID and Pharma that Dr. Fauci's PIs supervise. NIAID basically funnels tens of millions to hundreds of millions of dollars to these hospitals specifically, to give Dr. Fauci unquestioned power over the policies.[52]

Thanks to NIAID's stonewalling, the Vera Institute had to rely on secondary child welfare files and Pediatric AIDS Unit (PAU) records, both of which are notoriously incomplete. The Vera Institute did not even have access to minutes from research review boards (IRBs) for the medical centers that conducted the trials.

2008 NIH Report

Even after this scandal exploded, there was no evidence that Dr. Fauci made any effort to reform NIAID. Six years later, two biomedical ethicists inside the NIH concluded in a January 2008 article in *Pediatrics* that the agency still did not have adequate protections for vulnerable foster children: "Enrolling wards of the state in research raises two major concerns: the possibility that an unfair share of the burdens of research might fall on wards, and the need to ensure interests of individual wards are accounted for. . . . Having special protections only for some categories is misguided. Furthermore, some of the existing protections ought to be strengthened."[53]

During the decades since Dr. Fauci took over NIAID, he has sanctioned drug companies to experiment on at least fourteen thousand children, many of them Black and Hispanic orphans living in foster homes. He permitted these companies to operate without oversight or accountability. Under Dr. Fauci's laissez faire rubric, these companies systematically abused and, occasionally, killed children.[54, 55]

Dr. Fauci presided over these atrocities, collaborating with pharmaceutical company researchers and winking at their loose definitions of "informed consent" and "volunteer." Instead of looking out for the best interests of children, Dr. Fauci gave outlaw drug makers[56] free rein to torture vulnerable children behind closed doors, with neither parental permission nor requisite oversight from child welfare authorities.

***

In 1965, my father kicked down the door of the Willowbrook State School on Staten Island, where pharmaceutical companies were conducting cruel and often-deadly vaccine experiments on incarcerated children.[57] Robert Kennedy declared Willowbrook a "snake pit" and promoted legislation to close the institution and end the exploitation of children. Fifty-five years later, national media and Democratic Party sachems have beatified a man who presided over similar atrocities, somehow elevating him to a kind of secular sainthood.

What dark flaw in Anthony Fauci's character allowed him to oversee—and then to cover up—the atrocities at Incarnation Children's Center? At very best, there must be some arrogance or imperiousness that enables Dr. Fauci to rationalize the suffering and deaths of children as acceptable collateral damage in what he sees as his noble search for new public health innovations. At worst, he is a sociopath who has pushed science into the realm of sadism. Recent disclosures support the latter interpretation.

Freedom of Information documents obtained in January 2021 by the White Coat Waste project show that Dr. Fauci approved a $424,000 NIAID grant in 2020 for experiments in which dogs were bitten to death by flies.[58] The insects carried a disease-carrying parasite that can affect humans. The researchers strapped capsules containing infected flies to the bare skin of twenty-eight healthy beagle puppies and kept them in agonizing suffering for 196 days before euthanizing them. NIAID acknowledged it subjected other animals, including mice, Mongolian gerbils, and rhesus monkeys to similar experiments.

That same year, Dr. Fauci's agency gave $400,000 to University of Pittsburgh scientists to graft the scalps of aborted fetuses onto living mice and rats.[59, 60] NIAID sought to develop rat and mouse "models" using "full-thickness fetal skin" to "provide a platform for studying human skin infections." Dr. Fauci's sidekick and putative boss, Francis Collins—who casts himself as a pious Catholic—kicked in a $1.1 million sweetener from NIH for this malignant project.

Of all the desperate public health needs in America, of all the pain that a well-spent $2 million might alleviate, Tony Fauci and his government confederates deemed these demented and inhumane experiments the most worthwhile expenditures of America's taxpayer dollars.

These disclosures beg many other questions: From what moral wilderness did the monsters who devised and condoned these experiments descend upon our idealistic country? How have they lately come to exercise such tyrannical power over our citizens? What sort of nation are we if we allow them to continue? Most trenchantly, does it not make sense that the malevolent minds, the elastic ethics, the appalling judgment, the arrogance, and savagery that sanctioned the barbaric brutalization of children at the Incarceration Convent House, and the torture of animals for industry profit, could also concoct a moral justification for suppressing lifesaving remedies and prolonging a deadly epidemic? Could these same dark alchemists justify a strategy of prioritizing their $48 billion vaccine project ahead of public health and human life? Did similar hubris—that deadly human impulse to play God—pave the lethal path to Wuhan and fuel the reckless decision to hack the codes of Creation and fabricate diabolical new forms of life—pandemic superbugs—in a ramshackle laboratory with scientists linked to the Chinese military?

On my birthday in January 1961, three days before I watched my uncle John F. Kennedy take his oath as president of the United States, outgoing President Dwight Eisenhower, in his farewell address, warned our country about the emergence of a Military Industrial Complex that would obliterate our democracy. In that speech, Eisenhower made an equally urgent—although less celebrated—warning against the emergence of a federal bureaucracy, which, he believed, posed an equally dire threat to America's Constitution and her values:

In this revolution, research has become central; it also becomes more formalized, complex, and costly. A steadily increasing share is conducted for, by, or at the direction of, the Federal government. Today, the solitary inventor, tinkering in his shop, has been overshadowed by task forces of scientists in laboratories and testing fields. In the same fashion, the free university, historically the fountainhead of free ideas and scientific discovery, has experienced a revolution in the conduct of research. Partly because of the huge costs involved, a government contract becomes virtually a substitute for intellectual curiosity. The prospect of domination of the nation's scholars by Federal employment, project allocations, and the power of money is ever present and is gravely to be regarded. . . . *[We] must . . . be alert to the danger that public policy could itself become the captive of a scientific technological elite.*

Eisenhower demanded that we guard against this insipid brand of tyranny, by entrusting our government to responsible officials ever-vigilant against the deadly gravities of technocratic power and industry money that would pull our nation away from democracy and humanity and into diabolical dystopian savagery:

It is the task of statesmanship to mold, to balance, and to integrate these and other forces, new and old, within the principles of our democratic system—ever aiming toward the supreme goals of our free society.

During his half-century as a government official, Dr. Fauci has utterly failed in this charge. As we shall see, he has used his control of billions of dollars to manipulate and control scientific research to promote his own, and NIAID's, institutional self-interest and private profits for his pharma partners to the detriment of America's values, her health, and her liberties. Of late, he has played a central role in undermining public health and subverting democracy and constitutional governance around the globe and in transitioning our civil governance toward medical totalitarianism. Just as President Eisenhower warned. Dr. Fauci's COVID-19 response has steadily deconstructed our democracy and elevated the powers of a tyrannical medical technocracy.

Endnotes

1   Stanley Plotkin, "Ethics of Human Experimentation," Letter to the editor, *New England Journal of Medicine* (1973), https://sci-hub.se/10.1056/NEJM197309132891123

2   Elisabeth Mahase, "Vaccine hesitancy: an interview with Stanley Plotkin, rubella vaccine developer," *BMJ* (December 23, 2019), https://www.bmj.com/content/367/bmj.l6926

3   Robert F. Kennedy, Jr., "CDC's Latest Tuskegee Experiment African American Autism and Vaccines," *Children's Health Defense* (Jun. 25, 2015), https://childrenshealthdefense.org/news/cdcs-latest-tuskegee-experiment-african-american-autism-vaccines/

4   Barbara Loe Fisher, "Measles Vaccine Experiment on Minority Children Turn Deadly," *The Vaccine Reaction* (Jun., 1996), https://www.nvic.org/nvic-archives/newsletter/vaccinereactionjune1996.aspx

5   Brian S. Hooker, Ph.D., PE, "Reanalysis of CDC Data on Autism Incidence and Time of First MMR Vaccination," *Journal of American Physicians and Surgeons*, Volume 23, Number 4; Winter, 2018. https://www.jpands.org/vol23no4/hooker.pdf

6   Molly Roberts, "Anthony Fauci built a truce. Trump is destroying it," *Washington Post*, The Opinions Essay (July 16, 2020), https://www.washingtonpost.com/opinions/2020/07/16/anthony-fauci-built-truce-trump-is-destroying-it/?arc404=true

7   Edited transcript of three PBS interviews with Dr. Anthony Fauci conducted Jan. 28, March 10, and June 5, 2005 (posted May 30, 2006), https://www.pbs.org/wgbh/pages/frontline/aids/interviews/fauci.html

8   U.S. Government Publishing Office, Title 15 Commerce and Trade, Chapter 63–Technology Innovation (2010) https://www.govinfo.gov/content/pkg/USCODE-2010-title15/html/USCODE-2010-title15-chap63.htm

9   "NIH Scientists Caught Concealing Millions in Royalties for Experimental Treatments," Alliance for

Human Research Protection (October 26, 2006), *Associated Press* article (January 11, 2005), https://ahrp.org/nih-scientists-caught-concealing-millions-in-royalties-for-experimental-treatments-ap/

10   Daniel Payne and John Solomon, "Fauci Files: Celebrated doc's career dotted with ethics, safety controversies inside NIH," *Just the News* (Jul 23, 2020), https://justthenews.com/accountability/political-ethics/fauci-says-americans-should-trust-doctors-himself-his-career

11   Liam Scheff, "The House That AIDS Built," altheal.org (January 2004), https://www.altheal.org/toxicity/house.htm

12   Douglas Montero, "Shocking Experiments: AIDS tots used as 'guinea pigs,'" Reporting by *New York Post, U.K. Guardian, A&U Magazine, Associated Press, Fox News* and AHRP on the Incarnation Children's Center drug trials (February 28, 2004), http://www.leftgatekeepers.com/articles/AIDsTotsUsedAsGuineaPigsByDouglasMontero.htm

13   Scheff, op. cit.

14   Englebrecht, 143

15   Scheff, *The House That AIDS Built*

16   Ibid.

17   Ibid.

18   Ibid.

19   John Solomon, "Government tested AIDS drugs on foster kids," Associated Press (May 4, 2005), https://www.nbcnews.com/health/health-news/government-tested-aids-drugs-foster-kids-flna1c9443062

20   Ibid.

21   Website for *Guinea Pig Kids*, a BBC documentary on New York City's abuse of HIV-positive children under its supervision as human test subjects for experimental AIDS drug trials (November 30, 2004), https://www.guineapigkids.com/

22   Scheff, op. cit.

23   Vera Sharav, "Complaint: Phase I AIDS Drug/Vaccine Experiment on Foster Children," Alliance for Human Research Protection (March 10, 2004), https://ahrp.org/complaint-phase-i-aids-drugvaccine-experiment-on-foster-children/

24   Payne, op. cit.

25   U.S. House of Representatives, Committee on Ways and Means, Subcommittee on Human Resources, "Protections for Foster Children Enrolled in Clinical Trials," Hearing before the Subcommittee on Human Resources, U.S. House of Representatives, 109th Congress, First Session (May 18, 2005), https://www.govinfo.gov/content/pkg/CHRG-109hhrg36660/html/CHRG-109hhrg36660.htm

26   Janny Scott and Leslie Kaufman, "Belated Charge Ignites Furor Over AIDS Drug Trial," *New York Times* (Jul. 17, 2005), https://www.nytimes.com/2005/07/17/nyregion/belated-charge-ignites-furor-over-aids-drug-trial.html

27   Vera Sharav, "NYS Hearing – AIDS Drug /Vaccine Experiments on Foster Children," Alliance for Human Research Protection, (Sep. 8, 2005), https://ahrp.org/nys-hearing-aids-drug-vaccine-experiments-on-foster-children/

28   Vera Sharav, "NYS Hearing" op. cit.

29   Ibid.

30   Complaint: Phase I AIDS Drug/Vaccine Experiment on Foster Children, Alliance for Human Research Protection (Mar. 10, 2004), https://ahrp.org/complaint-phase-i-aids-drugvaccine-experiment-on-foster-children/

31   Solomon, op. cit.

32   Ibid.

33   Solomon, op. cit.

34   Scheff, op. cit.

35   "AIDS Drug/Vaccine Experiments on Foster children," op. cit.

36   Ibid.

37   Ibid.

38   Ibid.

39   Ibid.

40   DHHS Panel on Antiretroviral Guidelines for Adults and Adolescents, *Guidelines for the Use of Antiretroviral Agents in Adults and Adolescents with HIV* (Aug 16, 2011), https://clinicalinfo.hiv.gov/sites/default/files/guidelines/documents/AdultandAdolescentGL.pdf

41   Vera Sharav, "Complaint: Phase I AIDS Drug/Vaccine Experiment on Foster Children," Alliance for Human Research Protection (Mar. 10, 2004), https://ahrp.org/complaint-phase-i-aids-drugvaccine-experiment-on-foster-children/

42   Liam Scheff, "Inside Incarnation," *New York Press* (Jul. 27, 2005), https://pearl-hifi.com/11_Spirited_Growth/10_Health_Neg/04_Pandemics/01_AIDS/Liam_Scheff__Inside_Incantation.pdf

43   Ibid., 2

44   Ibid., 3

45   Vera Hassner Sharav, Vera Institute of Justice Final Report (2009): "80 NYC Foster Children Died in AIDS Drug Trials," Alliance for Human Research Protection (Aug. 1, 2015), https://ahrp.org/2009-commissioned-investigation-absolves-nyc-of-80-foster-children-deaths-in-aids-trials/

46 SourceWatch, "Foster child drug trials," The Center for Media and Democracy (Dec. 11, 2015), https://www.sourcewatch.org/index.php/Foster_child_drug_trials

47 Solomon, op. cit.

48 Subcommittee on Human Resources of the Committee on Ways and Means, U.S. House of Representatives, "Protections for Foster Children Enrolled in Clinical Trials" (May 15, 2005), https://www.govinfo.gov/content/pkg/CHRG-109hhrg36660/pdf/CHRG-109hhrg36660.pdf

49 Timothy Ross and Anne Lifflander, "The Experiences of New York City Foster Children in HIV/AIDS Clinical Trials," Vera Institute of Justice (Jan., 2009), https://www.vera.org/downloads/Publications/the-experiences-of-new-york-city-foster-children-in-hiv-aids-clinical-trials/legacy_downloads/Experiences_of_NYC_Foster_Children_in_HIV-AIDS_Clinical_Trials_FINAL_appendices.pdf

50 Andy Soltis, "AIDS Guinea Pigs–Foster Kids Used in 'Illegal' Drug Trials," *New York Post* (Jun. 17, 2005), https://nypost.com/2005/06/17/aids-guinea-pigs-foster-kids-used-in-illegal-drug-trials/

51 Vera Sharav, "Vera Institute of Justice Final Report"

52 Ross et al., op. cit.

53 Sumeeta Varma and David Wendler, "Research Involving Wards of the State: Protecting Particularly Vulnerable Children," Pediatrics (Jan., 2008), https://www.ncbi.nlm.nih.gov/pmc/articles/PMC2201985/

54 Faith Dyson, "What Was Fauci's Role in Funding Tuskegee-Like AIDS Experiments on Foster Children In Seven U.S. States?" Facebook (June 20, 2020), https://www.facebook.com/notes/faith-dyson/what-was-faucis-role-in-funding-tuskgegee-like-aids-experiments-on-foster-childr/10157586697714677/

55 Mary Otto, "Drugs Tested on HIV-Positive Foster Children Hill Investigates Ethical Questions Raised by 1990s Trials in Md., Elsewhere," *Washington Post* (May 19, 2005), https://www.natap.org/2005/newsUpdates/062005_02.htm?fbclid=IwAR02tZzsBxqNweSrl38oA8d_O2qhOZXI7Nhok6HPT63R94MV4Ev9sDxCp-E

56 Good Jobs First, *Violation Tracker Industry Summary Page* (2020), https://violationtracker.goodjobsfirst.org/industry/pharmaceuticals

57 The Bryant Park Project, "Remembering an Infamous New York Institution," NPR (Mar. 7, 2008), https://www.npr.org/templates/story/story.php?storyId=87975196

58 Amanda Nieves, "WCW Exposé: Fauci Spent $424K on Beagle Experiments, Dogs Bitten to Death by Flies." White Coat Waste Project (July 30, 2021), https://blog.whitecoatwaste.org/2021/07/30/fauci-funding-wasteful-deadly-dog-tests/

59 Yash Agarwal et al., "Development of humanized mouse and rat models with full-thickness human skin and autologous immune cells," *Scientific Reports*, volume 10, article 14598 (September 3, 2020), https://www.nature.com/articles/s41598-020-71548-z

60 David Daleiden, "University of Pittsburgh Won't Explain its Planned Parenthood Ties," Center for Medical Progress, *Newsweek* opinion page (May 26, 2021), https://www.newsweek.com/university-pittsburgh-wont-explain-its-planned-parenthood-ties-opinion-1594564

ChildrensHealthDefense.org/fauci-book

childrenshd.org/fauci-book

For updates, new citations and references, and new information about topics in this chapter:

CHAPTER 8

WHITE MISCHIEF:
DR. FAUCI'S AFRICAN ATROCITIES

"They increased the number of diseases from two to nearly thirty that could be classified as AIDS, and after that they started a global testing program of 'vulnerable populations,' which just coincidentally happen to be people not in a position to defend themselves easily. They started to find AIDS everywhere, including in Africa, but including in the United States—and wouldn't you know, one of the communities they found was the African-American community, and they tested a lot of women and they found a lot of HIV-positive women, and they decided, well, let's go forward."

—Kary Mullis, winner of 1993 Nobel Prize for Chemistry

As Vera Sharav points out, racism is an abiding feature of medical authoritarianism and human experimentation. Molecular biologist Harvey Bialy, the editor of the *Nature Biotechnology* journal, observed that the subtle backdrop of racial and sexual bigotry and bullying are the distinguishing attributes of AIDS research: "The fearful fascination with the contagion was amplified by the official narrative that the disease originated in Africans doing weird things with monkeys, and spread to the voodoo kingdom of Haiti, and that the sexual depravity of homosexuals drove the disease into the United States." Dr. Fauci's critic, Charles Ortleb, the editor of *New York Native* and author of a biography of the NIAID director, recalls that the theme of unwanted minorities spreading contagion was a standardized soliloquy of totalitarianism, most notoriously Hitler's stoking of public fears of tuberculosis to incite bigotry toward Jews: "There was always this undertone of bigotry with AIDS. I don't think we can dismiss as coincidence that the population that they targeted for their toxic concoctions were gays, Blacks, Hispanics, and Africans."

And Dr. Fauci did not restrict his unethical experiments with AIDS drugs to American children. By June 2003, NIH and NIAID were running 10,906 clinical trials in ninety countries, and Dr. Fauci's pioneering AIDS Branch, newly christened DAIDS (Division of Acquired Immunodeficiency Syndrome), was testing new toxic antiviral concoctions in some four hundred clinical trials in the United States and globally.[1] Dr. Fauci's PIs targeted developing nations that lacked strong institutional structures for protecting impoverished citizens from the abusive practices of powerful pharmaceutical multinationals. According to Vera Sharav, Dr. Fauci had NIAID and its pharmaceutical company partners move his most controversial and risky studies offshore "because they can do stuff that they could never get away with in the United States."

Journalist Celia Farber concurs with Sharav's assessment: "The racism is cloaked inside carefully crafted philanthropic manipulations such as 'access' to drugs. It's never access to clean drinking water, education, sanitation, nutrition. It's a very blighting message for the US to constantly be browbeating Africans with our self-serving messaging that they are so sick, and *we* have just the drugs to 'save' their lives. When the opposite happens, it's swept away and hidden behind the false front of charity. I call it Pharma-Colonialism."[2]

Africa has been a Pharma colony for over a century. It is the venue of choice for companies seeking cooperative government officials, compliant populations, the lowest per-patient enrollment costs, and lax oversight by media and regulatory officials. Powerless, often illiterate, and, if necessary, disposable quasi volunteers allow Pharma's PIs to paper over even catastrophic side effects and mistakes. In 2005, FDA officials learned that Dr. Fauci's DAIDS team had concealed scores of deaths and hundreds of injuries during HIV drug trials in Africa with another of his toxic chemotherapy vanity products, Nevirapine.[3, 4, 5]

Dr. Fauci's fingerprints were all over DAIDS's sketchy African experiments. In October 1988, his success at getting approval for AZT won him the equivalent of a billion-dollar lottery for a career technocrat—a mention during then-Vice President George H. W. Bush's presidential debate:

> You've probably never heard of him. He's a very fine researcher, top doctor at National Institute of Health, working hard doing something, research on this disease of AIDS.

The accolade gained him an even larger prize—access to and the trust of the new president.

Two administrations later, Dr. Fauci warned President George W. Bush that HIV had gotten a toehold in Africa and was spreading like wildfire. He persuaded the president to demonstrate his bona fides as a "compassionate conservative" by redirecting the United States foreign aid spending into the heroic enterprise of eliminating African AIDS. Accordingly, on January 19, 2002, President Bush announced a $15 billion package to combat AIDS, including a $500 million program to purchase millions of doses of Nevirapine for distribution to African mothers and children.[6] Dr. Fauci told the President that Nevirapine would save millions of lives by preventing maternal transmission of HIV to unborn children. President Bush would later repeat this promise in his 2003 State of the Union address.

Dr. Fauci's artful 1988 achievement of winning FDA approvals for AZT had launched the AIDS drug gold rush. Nevirapine was German pharmaceutical giant Boehringer Ingelheim's beachhead in the race. Boehringer had apparently lifted Nevirapine from the same toxic junk pile from which Burroughs Wellcome had retrieved AZT. Canadian regulators rejected Nevirapine—in 1996 and 1998—due to its potent toxicity and dubious efficacy.[7] In December 2000, the *Journal of the American Medical Association* advised health care workers exposed to HIV to avoid prescribing Nevirapine after the drug caused life-threatening liver toxicity in patients. A 2001 FDA review reported twenty "serious adverse events" (meaning, death, hospitalization, "life-threatening," or permanently disabling) resulting from brief, prophylactic Nevirapine exposure.[8] Nevertheless, the German chemical company found a soft landing for its product at NIAID.

Another Drug Too Big to Fail

Dr. Fauci apparently neglected to tell President Bush that Nevirapine had never won FDA approval as a safe and effective drug. "Dr. Fauci had to know all about the safety problems, but he must have either omitted or whitewashed them when he sold the program to Bush," says Celia Farber, who researched the episode extensively for her 2006 article in *Harper's Magazine*. NIAID's powerful apparatchik didn't fret that FDA had already refused Nevirapine its official safety imprimatur. "Dr. Fauci seemed confident that he eventually could get FDA to give him anything he wanted," Farber told me.

In the early 1990s, Ugandan dictator Yoweri Museveni rolled out the red carpet for Pharma. Uganda became one of many African nations seeking to cash in on the lucrative business of farming out their citizens for the booming clinical trial business. In 1997, Uganda granted Dr. Fauci's Johns Hopkins–based PI, Brooks Jackson, permission to run clinical trials on Nevirapine in Kampala.

NIAID's AIDS division, DAIDS, was sole sponsor of a study to test the efficacy and safety of Nevirapine and AZT on preventing maternal transmission of HIV to newborns.[9] DAIDS code-named its Uganda clinical trial HIVNET 012. In 1999, Jackson and his team reported in the Medical Journal *Lancet*, "Nevirapine lowered the risk of HIV-1 transmission during the first 14–16 weeks of life by nearly 50 percent in a breastfeeding population. This simple and inexpensive regimen could decrease mother-to-child HIV-1 transmission in less developed countries."[10] Fauci acolytes hailed this success as NIAID's largest triumph against HIV to date. Congress voted a hefty hike to the NIH budget.

But the study's sunny conclusions concealed glaring methodological deficiencies. When they can get away with it, Pharma researchers commonly employ the highly unethical gimmick of eliminating the placebo control group in order to mask injuries in the study group. The absence of an inert placebo comparator group allows PhD grifters to dismiss all injuries and deaths in the study group as sad coincidences not associated with the drug they are testing. DAIDS's official Nevirapine clinical trial protocols required an inert placebo group, but once in Uganda, DAIDS's cowboy research team simply made the placebo group vanish. Instead of using a placebo, Jackson and his team ended up comparing the health outcomes in 626 pregnant women, half of whom took Dr. Fauci's horrendously dangerous chemotherapy concoction AZT, while the other half took Nevirapine.

Based on this study, Dr. Fauci was able to persuade the WHO in 2000 to grant Emergency Use Authorization Approval (EUA) to single-dose Nevirapine for preventing mother-to-child transmission of HIV as its official recommendation. WHO was already a sock puppet for Big Pharma. Dr. Fauci used the stopgap WHO approval to persuade President Bush to purchase millions of dollars of Nevirapine. Boehringer began shipping cartons of its deadly and ineffective drug to clinics and maternity wards in fifty-three developing nations.[11]

The Boehringer study enrolled 626 supposedly HIV-infected pregnant Ugandan women. Even at its best, HIV diagnosis in Africa is a casual affair seldom verified by blood tests, and NIAID's trial team had a particularly cavalier approach to determining HIV infection. It is therefore unclear how many of the agency's "recruits" were actually HIV positive. From day one, the researchers trampled virtually all the study's safety/efficacy protocols, including the most critical requirement in "dosing safety" studies—a genuine placebo control group.

The gimmick of equalizing the carnage in both AZT and Nevirapine study groups allowed the NIAID researchers to cobble together the sunny assessment of both drugs, which they published in the *Lancet* in summer of 1999.[12] Using the deceptive code words that are de rigueur in NIAID's official reports of its clinical trials, the researchers reported that "[T]he two regimens were well-tolerated." Their proof of this fraudulent assertion was that "[A]dverse events were similar in the two groups." Only the fine print of the *Lancet* study revealed that thirty-eight babies had died, sixteen in the Nevirapine group and twenty-two in the AZT group.

But as we shall see, that deceptive swindle was just the start of the mayhem. A NIAID project officer later complained to Farber that the Uganda trials were "out of control" and researchers were trampling safety and regulatory standards.

In July 2001, Boehringer Ingelheim filed a supplemental application to the FDA to market Nevirapine for preventing mother-to-child transmission (PMTCT) of HIV solely based upon NIAID's Uganda trials. However, stories were already trickling back to Washington that Dr. Fauci's Kampala trials that underpinned the *Lancet* paper were a three-ring circus of flimflam fraught with serious accuracy and ethical issues. It was at this time that the FDA, in keeping with standard procedure regarding planned inspections of a foreign site, announced that it was sending investigators to Uganda. That declaration apparently irked Dr. Fauci and his NIAID team and terrified his Boehringer partners. In January 2002, Boehringer dispatched an audit team to Kampala.[13] In exchange for FDA agreeing to delay its visit, Boehringer promised to share its inspection report with the US licensing agency. That report did little to assuage FDA's alarm. Boehringer's own investigators described grisly mayhem in Kampala; the NIAID study was in shambles, including "serious non-compliance with FDA regulations." In its efforts to win FDA approvals for the dangerous and ineffective concoction, DAIDS's team had violated virtually every good clinical practice, including the unlawful failure to employ the standardized informed consent procedure of disclosing serious risks to study participants.

Boehringer's damning inspection report only heightened concern at FDA. In hopes of forestalling the FDA inspection, NIAID, in February 2002, hired a private consultant group, Westat, to conduct an investigation and audit of the Kampala site.

It's fair to assume that Dr. Fauci's crew was hoping for a whitewash from Westat. But Westat used seasoned auditors whose backgrounds included inspections on behalf of the FDA,[14] and the Westat audit confirmed the long inventory of severe violations of Good Clinical Practice, including—most disturbingly—the convenient "loss of critical records."[15] The missing records included a vital logbook that appeared to have documented the study's worst atrocities before its mysterious disappearance. NIAID's Uganda team told the Westat researchers that they had lost the critical log that, among other things, recorded all the adverse events and deaths. The remaining records didn't report which mothers received which drugs or even whether they survived the study. The auditors reported a scene of pure chaos. "Drugs were given to the wrong babies, documents were altered, and there was infrequent follow-up, even though one third of the mothers were marked 'abnormal' in their charts at discharge. The infants who did receive follow-up care were, in many cases, small and alarmingly underweight. 'It was thought to be likely that some, perhaps many, of these infants had serious health problems.'"[16] When Westat chose a random sample of forty-three of those infants to examine, *all* of them had "adverse events" twelve months after the study terminated. Only eleven of them were HIV positive.[17]

When Westat confronted Dr. Jackson's researchers with study discrepancies, they admitted that they routinely applied more lenient standards for their Black Ugandan subjects than FDA rules required for US safety studies.[18] The PIs admitted to systematically downgrading standardized definitions of serious adverse events to adapt to "local standards." Injuries that researchers would score as "serious" or "deadly" if they happened to white Americans became "minor" injuries when Black Africans were the victims. Under their relaxed rubric, clinical trials staff scored "life-threatening" injuries as "not serious." When they reported them at all, NIAID classified mortalities among its African volunteers as "serious adverse events," rather than "death." NIAID's Ugandan team had entirely neglected to report thousands of adverse events and at least fourteen deaths.[19]

Dr. Fauci's PI, Dr. Brooks Jackson, acknowledged that he had avoided reporting "thousands" of AEs and SAEs (adverse and severe adverse events) by applying those diluted definitions of "serious" and "of severity."[20] Researchers specifically excluded from the reports all deaths that occurred more than a few months after the study ended. When Westat pushed for answers, the NIAID/Hopkins local team pleaded that no one had trained them in Good Clinical Practice and that they had "never attempted a Phase III trial."[21] Finally, the Westat auditors refused to sign off on Nevirapine because they could find no valid data suggesting that this highly toxic drug prevented HIV transmission.[22]

After receiving Westat's audit report, panic-stricken NIAID and Boehringer officials again feared FDA making its own planned site visit.[23] But Dr. Fauci had already pushed his beleaguered sister agency past its high tolerance for bureaucratic humiliation. FDA demanded to see the Westat report.[24] Dr. Jonathan Fishbein told me that when FDA regulators finally reviewed the Westat report, "They read the riot act to NIAID and Boehringer's officers." FDA instructed Boehringer to withdraw its application for Nevirapine's approval or face the mortification of a public FDA rejection.

In March 2002, Boehringer Ingelheim consequently pulled its supplemental FDA application for Nevirapine's approval, and the Johns Hopkins/NIAID team closed the scandal-ridden Uganda study site.[25] The decision to shorten the study occurred at a tense meeting between the FDA and the NIAID. Everyone knew the enormous implications of the audit findings. FDA's refusal to rubber-stamp Nevirapine's approval meant the collapse of the Bush administration's most visible foreign policy program. Dr. Fauci had persuaded the president to make the abolishment of African AIDS his moonshot project, his career legacy, and Nevirapine was the foundation stone of that project.

The severe embarrassment to the president and to the NIH would also engulf Uganda's Makerere University, Boehringer, the investigators and their employers (Johns Hopkins University), and Family Health International (FHI)—the organization responsible for monitoring the trial. It would antagonize the South African government, whose drug regulatory agency, the Medicines Control Council (MCC), permitted the distribution of Nevirapine under duress based solely on the fraudulent results of the Uganda study published in *Lancet* in 1999.[26]

Curiously, Dr. Fauci did not attend the meeting to take responsibility for his Institute's leading role in the catastrophe. He dispatched his underlings to absorb the spanking. "It was a great embarrassment to the Bush administration because that was their big initiative," recalls Farber. In any other circumstances, Nevirapine would have been D.O.A. in FDA's licensing process. But Nevirapine was Tony Fauci's baby. He had staked his credibility with the president on the success of this trial. Like AZT, Nevirapine was therefore too big to fail. Such desperate circumstances summoned Teflon Tony to perform his greatest magic act: resurrect the dead.

"Tony Fauci knew that Nevirapine had fundamental safety and efficacy deficits that went way beyond recordkeeping," says Farber. Those problems were existential: the drug didn't work, and it killed both mothers and children. According to Farber, "Two inspections had now declared HIVNET 012 to be a complete mess: Boehringer's own and Westat's, which had been performed in conjunction with NIAID. But the ways in which the various players were tethered together made it impossible for NIAID to allow the study to die without embarrassing Dr. Fauci in his relationship with President Bush and without implicating NIAID in the Uganda scandal." NIAID sprang into cover-up mode; Dr. Fauci, by now adept at manipulating both elected officials and a credulous press, had his PR team begin the resurrection project by refashioning the Uganda charnel house scandal as a simple misunderstanding based on minor clerical errors.

Blithely ignoring FDA's dire safety and efficacy signals and Boehringer's demeaning withdrawal, NIAID issued a press release characterizing its Ugandan atrocities as mere record-keeping glitches. NIAID's statement claimed that while "certain aspects of the collection of the primary data may not conform to FDA regulatory requirements," "no evidence has been found that the conclusions of HIVNET 012 are invalid or that any trial participants were placed at an increased risk of harm."[27] To the contrary, the communiqué assured the public, NIAID's trial had proven Nevirapine both safe and effective. Summoning his extensive web of loyal dependencies, Dr. Fauci lined up a host of organizations, including the Elizabeth Glaser Pediatric AIDS Foundation, Johns Hopkins, Boehringer Ingelheim, and others, to issue statements and press releases supporting NIAID's official narrative. NIAID portrayed the devastating Boehringer withdrawal as merely a temporary setback, which Dr. Fauci, in a perverse but inspired twist of Orwellian Newspeak, recast as an admirable demonstration of corporate responsibility.

In July 2002, DAIDS announced that it would reassess the Uganda Nevirapine study with its own in-house "remonitoring protocol"—a fancy construction for "whitewash"—managed by Fauci's top AIDS henchman, DAIDS Director Edmund Tramont. [28] However, in an uncharacteristic faux pas, Tramont included his hand-picked DAIDS in-house review team, which included the agency's Medical Officer, Dr. Betsy Smith, who was not down for the cover-up. During her document inspection, Dr. Smith took notice of the poor quality and the incompleteness of the safety data. Shoddy recordkeeping at the site revealed that the study did not comply with Good Clinical Practice (GCP) guidelines. GCP is a requirement for all NIH-funded clinical research and any studies conducted for the purpose of supporting the safety and effectiveness of investigational drugs.

Dr. Smith's draft safety report raised all kinds of noisome alarms: she noted that medical records such as clinical notes, which are source documents needed to validate study data, were missing, incomplete, and often unsigned or undated.[29] This made it difficult to validate the occurrence of adverse events. Poor quality clinical records were "below expected standards of clinical research," especially for a study of such great importance.

Smith and the Regulatory Branch Chief, Mary Anne Luzar, also uncovered serious health injuries in the chaotic Uganda safety records. Babies in the AZT arm were showing consistently elevated liver enzymes—injuries consistent with Nevirapine's long history of provoking lethal liver failure.

She found that the Uganda team had neglected to report numerous infant deaths and routinely failed to track patients who had abnormal lab values, clinical signs, and symptoms to determine how these problems resolved. Further complicating that problem, the study team did not interpret laboratory results using the standard toxicity grading scales that the protocol required but had spitballed their assessment using "less stringent grading scales and creating a team-defined, reporting algorithm for study with the goal to report fewer AEs and SAEs (adverse and serious adverse events)."[30] This was the delicate lingua franca that bureaucrats employ to accuse one another of fraud.

Dr. Jackson had not trained his study staff on how to report SAEs, and his team neither tracked nor reported AEs, including serious ones. Instead of treating these grave deficiencies with appropriate concern, FHI's research monitors, who had been visiting the site for years, made light of the problems.

"Their on-the-ground solution to the Nevirapine toxicity problem in Africa was to simply not monitor for safety," says Farber.

Dr. Smith realized the monumental implications of her findings, which jeopardized the mission-critical project to license Nevirapine to prevent maternal to child transmission of HIV. Dr. Smith therefore trod delicately. She concluded her report by stating "safety reporting did not follow DAIDS reporting requirements during the conduct of HIVNET 012. Safety conclusions from this trial should be very conservative."[31]

"Dr. Fauci had sold his Nevirapine enterprise as a heroic moment for American greatness," recalls Celia Farber. "Dr. Fauci said he was going to save African pregnant women and their babies. It turns out that this is an extremely dangerous drug with no demonstrated ability to save a single life. This isn't rocket science. Dr. Fauci knew all about the 'safety problems,' but for Fauci and his cult of HIV drug worship, no drug is ever 'unsafe.'" Farber researched the episode extensively for her 2006 article in *Harper's Magazine*, "Out of Control: AIDS and the Corruption of Medical Science."

Dr. Smith's conclusions in her safety report—if allowed to stand—would kill Nevirapine's chances of winning FDA approval for preventing maternal-to-child transmission.

Despite all these setbacks, NIAID's powerful apparatchik didn't seem to fret that FDA was now unlikely to grant Nevirapine its official safety imprimatur.

Dr. Fauci employed the same ploy that Bob Gallo used when he recruited Margaret Heckler as his "useful idiot" to convince the world that NIH's intrepid scientists had identified the viral culprit behind AIDS. By now, Dr. Fauci was acting on a much larger stage. On January 29, 2003, the new president took the podium at his State of the Union speech and announced Dr. Fauci's new program to the world, the President's Emergency Plan for AIDS Relief (PEPFAR):

> On the continent of Africa, nearly 30 million people have the AIDS virus. . . . Yet across that continent, only 50,000 AIDS victims—only 50,000—are receiving the medicine they need. . . . I ask the Congress to commit $15 billion over the next five years, including nearly $10 billion in new money, to turn the tide against AIDS in the most afflicted nations of Africa and the Caribbean.[32]

His power to deliver a $15 billion health program and to summon unprecedented accolades from a sitting president gave Dr. Fauci unchallengeable power over the entire US health bureaucracy unmatched in American history. He now enjoyed consolidated power over HHS and all its subsidiaries.

"After Bush's State of the Union, all of HHS fell in line behind Dr. Fauci's project to rewrite history," recalls Farber. "The political stakes were very high." To save the reputation of his boss and employer, and by extension everyone else implicated in this scandal, Ed Tramont rose to the task.

Tramont went to work by eliminating inconvenient facts recorded by Betsy Smith and top regulatory compliance officer in the NIH's AIDS division Mary Anne Luzar by "reorganizing" the disqualified Uganda data. When DAIDS released Tramont's edited version of the remonitoring report on March 30, 2003, Dr. Smith's safety review had vanished.[33] In its place was a safety section Tramont later admitted having ghostwritten. Tramont had begun with a straightforward revision of the safety review committee's conclusion, altering it from "unfavorable" to "favorable." Tramont's purged draft boldly concluded, "Single-dose Nevirapine is a safe and effective drug for preventing mother-to-infant transmission of HIV. This has been proven by multiple studies, including the HIVNET 012 study conducted in Uganda." Tramont began massaging data sets to conform the rest of the report to this adjusted outcome. Tramont dismissed concerns raised by Luzar about pediatric liver problems and forged in his own bleached conclusions that the drug was safe. In Dr. Fishbein's words, Tramont simply "rewrote the safety section, minimizing concerns about the toxicities, deaths, and record-keeping problems that had been highlighted by his medical safety expert." Tramont's editing skills produced a document that laid the foundation, in December 2002, for FDA's approval of the lethal concoction for global use on pregnant women.

The Presidential Seal of Approval

Then Teflon Tony played his trump card. Dr. Fauci's coup de grâce was a White House announcement that Bush would anoint the scandal-ridden Nevirapine project with a personal site visit. The presidential junket would serve as a kind of public purification ritual to purge away the scandal and anoint Nevirapine with legitimacy.

The special bravado that allowed Dr. Fauci to summon a president to a distant continent, and to make Dr. Fauci's personal agenda the centerpiece of White House foreign policy, was a demonstration of power that could only provoke the entire awe-struck public health bureaucracy to stand at attention and salute. What FDA bureaucrat would now have the courage to taint this prestigious HHS triumph with awkward questions about safety and efficacy?

Dr. Fauci "wanted the HIVNET site reopened for President Bush's visit," Dr. Fishbein told me. "That visit was such an embarrassment to all of us who knew the truth, but everyone fell into line." The US AIDS media even began to refer to Museveni suddenly as a "benevolent dictator." Farber remarks, "That Presidential junket was so transparently phony—a shameless exercise in colonial public relations and lies."

On July 11, 2003, President Bush toured the clinical trial site in Kampala,[34] which DAIDS had hurriedly reopened and populated with temporary health workers for the occasion. Dr. Fishbein explained to me, "NIAID officials rushed to reopen the site despite my concerns that it wasn't ready. But Tramont overruled me. He wanted the restriction lifted ASAP because in his words, 'the site is now the best in Africa run by Black Africans' and President Bush was scheduled to be there in four days." Said Farber, "NIAID officials rushed to reopen the site to paper over the disgrace and to impress and deceive the President." She added, "It was really straight up—Potemkin's Village, a vast PR campaign, with nothing behind the Hollywood façade, except death. Dead babies, dead mothers—we will never know their names."

Almost all of HHS was now behind Dr. Fauci's project to rewrite history. In July 2002, DAIDS announced that it would reassess the Uganda Nevirapine study.

Presenting awards to one another is a knee-jerk strategy by which vaccine experts paper over malefactions and atrocities. It is therefore not surprising that, to advance the cover-up and absolve the Uganda research team, Tramont recommended that Dr. Fauci get his putative NIH boss, Elias Zerhouni, to present Dr. Jackson and his Uganda project researchers who had supervised the African debacle with an NIH award. This strategy would co-opt the NIH Director into the cover-up and fortify institutional resistance to a full-blown investigation. Tramont assigned the task to his flunky, DAIDS's deputy director, John Kagan. But in a rare display of independent good judgment, Kagan protested that giving awards to the clowns who had killed all those Africans—probably with criminal negligence—was a bridge too far: "We cannot lose sight of the fact that they screwed up big time. And you bailed their asses out," he advised Tramont by email. "I'm all for forgiveness, etc. I'm not for punishing them. But it would be 'over the top' to me, to be proclaiming them as heroes. Something to think about before pushing this award thing . . ."

But the conspirators had a problem. NIH medical officers Betsy Smith and Mary Anne Luzar were not willing participants to the cover-up. To tie up the last loose ends, Kagan ordered NIAID's ethics officer, Dr. Jonathan Fishbein, to reprimand Luzar for insubordination. The admonishment would bring the agency's Ethics Division into the cover-up and create an insurance policy if Luzar blabbed to anyone about all those African kids with collapsed livers; an official reprimand from a supposedly "independent" ethics officer would allow NIAID to discredit Luzar as a "disgruntled employee."

But Dr. Fishbein's investigation convinced him that the whistleblower, Luzar, was a hero. He told Tramont that he could find no justification for Luzar's reprimand and advised him against issuing the official rebuke: "They were out to get her because she refused to compromise her integrity," Dr. Fishbein told me. Faced with Dr. Fishbein's resistance, Tramont backed down. In Dr. Fishbein, Dr. Fauci's team had run up against a public health official naive enough or conscientious enough to say "no." Meanwhile, Dr. Fishbein's investigation of Luzar gave him additional reasons to mistrust Kagan's judgment. Female employees reported to Dr. Fishbein that Kagan was sexually harassing them.

Tramont may have felt that Dr. Fishbein was purposefully goading him when, instead of rebuking Luzar, Dr. Fishbein filed a sexual harassment complaint against Tramont's sidekick and enforcer. "Kagan was Fauci's bagman," AP reporter John Solomon told me. "He was a career Army guy from Fort Detrick or Walter Reed." Dr. Fishbein concurred in this assessment. "He was a 'just following orders' kind of guy, brought in to put a layer of insulation between Fauci and all the institutionalized mismanagement in his HIV clinical trials." Dr. Fishbein adds that the corruption that had begun with AZT "then metastasized throughout the entire program." Dr. Fishbein adds, "The sexual harassment issues aside, Kagan was a miserable manager."

Boehringer Ingleheim never resubmitted its application to the FDA for preventing maternal-to-child transmission of HIV. Nevertheless, WHO—which, as we shall soon see, was by then under the control of Bill Gates and Anthony Fauci—began shipping this lethal concoction to developing nations globally to use on their pregnant women.[35] "It's a mystery why Nevirapine was ever developed, launched or marketed to the developing world the way it was," says journalist Celia Farber, "since it was rejected by every Western drug safety agency—every single time. Why was it then re-purposed and shipped to non-Westerners? The double standard is quite stark. We need to start calling it what it is." Says Dr. Fishbein: "The tragic irony here is that the Kampala Nevirapine research was performed to a level of standards that would be insufficient for supporting the drug's approval for use in the United States, but Fauci fervently defended the study as adequate to justify giving nevirapine to Black Africans. Frankly, it strikes me as racist." Reverend Jesse Jackson echoed Fishbein's sentiment: "This was not a thoughtful and reasonable decision, but a crime against humanity. Research standards and drug quality that are unacceptable in the US and other Western countries must never be pushed onto Africa."[36]

Profits to Die For

The pharmaceutical and the medical cartel's historical preference was to test dangerous drugs and medical procedures on people of color. But by the late 1990s, Black Americans were increasingly suspicious of medical authorities. President Clinton's belated 1996 official apology[37] to the victims of the Tuskegee syphilis experiments (1935–1973) had reminded Blacks of other historic atrocities, including the barbaric gynecological experiments on Black women by Dr. J. Marion Simms ("Father of Modern Gynecology").[38] In 1992, a *Los Angeles Times* exposé revealed that the CDC had been conducting unlicensed experiments with a deadly flu vaccine on Black children in Haiti and Cameroon, and on 1,500 Black children in South Central Los

Angeles beginning in 1986.[39] Blacks were therefore understandably reluctant to sign up for clinical trials. Despite vigorous efforts by pharmaceutical companies and regulators to recruit Blacks, fewer than 4 percent of clinical trial enrollees in America were Black.[40] Nevertheless, Dr. Fauci seemed to have a genius for finding Blacks, both American and African, to participate in his HIV chemotherapy drug experiments.

In 2003, an HIV-positive African American mother in Memphis, Tennessee, died during one of Dr. Fauci's Nevirapine drug trials.[41] In April of that year, Joyce Ann Hafford—four months pregnant and already the mother of a gifted thirteen-year-old—was shocked to learn she had tested positive following a routine HIV test recommended by her pediatrician. Believing her diagnosis to be a death sentence, Hafford enrolled in DAIDS's clinical trial at the University of Tennessee in hopes of saving her soon-to-be-born son from getting AIDS. Dr. Fauci's local PI, Dr. Edwin Thorpe, planned to recruit 440 pregnant women to determine the "treatment limiting toxicities" of four HIV drugs in pregnant women.[42] It's an embarrassment to me, to my family, and particularly to my deceased aunt and godmother that NIH's Eunice Kennedy Shriver National Institute of Child Health and Human Development was a collaborator in this fraud.

Hafford was healthy and symptom-free. None of her subsequent tests ever showed any clinical markers for AIDS, and Dr. Thorpe never told Hafford that the HIV test measured only for the presence of antibodies and was not a reliable indicator of HIV infection. Furthermore, pregnancy frequently triggers false positive results on HIV antibody tests, and Dr. Thorpe tested Hafford only once. To make matters worse, her family later found that Joyce never signed her consent form, suggesting that Dr. Thorpe never informed her of Nevirapine's risks.

Hafford's health took a steep nosedive after her first dose of Nevirapine. It only took a few days before Hafford was showing undeniable signs of dwindling liver function. Instead of taking her off the drugs that he knew could be deadly, Dr. Thorpe prescribed cortisone cream for her skin rashes. Within weeks, Hafford was presenting with alarming signals of hepatic collapse. Forty-one days after starting the trial, she was dead from liver failure—the same injury about which both FDA and *JAMA* had issued clear warnings. On July 29, doctors delivered her baby, Sterling, by C-section three days before Joyce died.

When the shattered family gathered around her body, Dr. Thorpe and his team told them, to their bewilderment, that Joyce had died of rapidly progressed AIDS. They were lying. The year after her passing, Associated Press reporter John Solomon gave Joyce's family a trove of DAIDS reports he had obtained from a Freedom of Information Act request.[43] In those internal memos, DAIDS officials openly acknowledged to one another that Nevirapine caused Joyce Hafford's liver to fail.

Dr. Thorpe and his colleagues kept Sterling on AZT for three months. Fifteen months later, Sterling tested negative for HIV. Despite their repeated requests, Dr. Thorpe and his hospital refused to release Sterling's medical records to the Haffords. Sterling's family believes that NIAID withheld those records because they would prove that neither Joyce nor Sterling ever had HIV; all babies born to mothers with HIV test positive at birth, and almost all babies shed the maternal antibodies by eighteen months.

Celia Farber, who focused her *Harper's* exposé on Joyce's death and the HIVNET coverup, is still angry. Farber, who grew close to the Hafford family during the months she spent researching Nevirapine, blames Dr. Fauci directly: "The death of Joyce Anne Hafford in Memphis was a methodical calculated homicide of a black woman by Fauci's henchmen," Farber told me. "They had to know they were killing her when they saw her go into jaundice and they just watched her liver crash. They wouldn't let her off the Nevirapine. It seemed like very clear medical murder at Dr. Fauci's doorstep. I'm still trying to recover from it."

At that time, Dr. Jonathan M. Fishbein, MD, was DAIDS's first director of the Office for Policy in Clinical Research Operations.[44] His job was to monitor and enforce compliance to federal research and ethical policy in DAIDS-sponsored studies. In the summer of 2003, he intervened in Hafford's case. According to Dr. Fishbein, DAIDS's medical staff always knew that Hafford died of Nevirapine toxicity. "Nevirapine's toxicity," Dr. Fishbein told me, "particularly its association with liver failure, was well documented and the PI certainly had that knowledge."

That August, Dr. Fishbein sent a memo to Dr. Fauci's AIDS Branch Director, Ed Tramont, informing him that Nevirapine caused Hafford's lethal liver failure.[45] Tramont wrote back, "Ouch. Not much wwe [sic] can do about dumd [sic] docs!"[46] Tramont's glib riposte seems to have been like a subtle signal to Dr. Fishbein to get in line with NIAID's strategic cover-up. Dr. Fishbein told me that acknowledging Nevirapine's role in Hafford's death would have jeopardized Nevirapine's FDA approval. Despite Tramont's crass cypher, Dr. Fishbein's regulatory team nevertheless informed the FDA about Hafford's drug-related death.

Hafford was not the only trial recruit to suffer. In the initial Phase I trial on twenty-one pregnant women, NIAID's investigators would later report that four of twenty-two infants died and twelve suffered "serious adverse events." Furthermore, the studies suggested that Nevirapine was ineffective. None of the women experienced reduction of viral loads. When Thorpe and his colleagues finally published the results of their Nevirapine study in 2004, they acknowledged that "the study was suspended because of greater than expected toxicity. . . ."[47]

Rooting Out Integrity in the Workplace

Dr. Fishbein didn't last long in his official capacity as the DAIDS official in charge of enforcing compliance with clinical research and ethical policy. His lethal misstep was his decision to follow a trail of irregularities affecting a NIAID drug trial called ESPRIT, which tested interleukin-2 (IL-2), a cancer chemotherapy and AIDS drug, known by its brand name, Proleukin. The ESPRIT study was investigating IL-2 clinical outcomes in individuals with asymptomatic HIV+.[48] In December 2003, the ESPRIT Medical Officer alerted Dr. Fishbein to troubling side effects in the Proleukin trial, namely, capillary leak and an unusual psychiatric side effect: suicidal ideation. The Medical Officer, Larry Fox, worried that NIAID was putting volunteers in danger by withholding the information about those hazards from the Investigator Brochure, as the law required.[49] This brochure is an FDA-mandated document containing updated information detailing, among other things, the side effects and risks of an investigational drug. It provides clinical trial investigators with safety information compiled across study sites to keep study subjects informed about emerging hazards. Furthermore, without an up-to-date document (NIAID had issued the last one in 2000), NIAID was not adequately warning potential clinical trial enrollees about these serious dangers. Recalls Dr. Fishbein, "The drug had grave risks for suicidal ideation, and capillary leaks. The study leadership was ignoring their legal duty to inform the study recruits and participants about these troubling signals."

By this time, NIAID had invested some $36 million dollars in ESPRIT and had thousands of subjects enrolled at two hundred international locations over nearly four years.[50] If these asymptomatic participants were to learn about the emerging risks, NIAID feared they would drop out. It would also be difficult to attract new volunteers. The failure to retain subjects or recruit additional volunteers would nullify the study, one of NIAID's most costly ever. (Ironically, after eight years and 4,150 subjects, ESPRIT concluded Proleukin offered "no benefits" to clinical outcome in HIV+ patients.)

It was now evident to key NIAID officials that Dr. Fishbein was becoming an all-around nuisance. He was professional, curious, incorruptible, and far too serious about performing his duties. "His big problem," says Farber, "is that he thought his job was legit. Dr. Fishbein's personal virtues were all fatal character flaws within the NIAID institutional culture." Dr. Fishbein's refusal to toe the line sent him stumbling into the terminal career cul-de-sac at NIAID.

Dr. Fishbein explained further about the Proleukin trial: "It was a serious violation of protocols and the researchers were ignoring their legal duty to report the signal. They omitted and whitewashed all these safety problems. You can't just focus on efficacy and ignore safety." Dr. Fishbein told AP reporter John Solomon, "The ones that were in the study, and those that wanted to get in the study, neither were being informed. NIAID feared that if they understood the risks, they would drop out."

Dr. Fishbein had entered a dangerous realm at NIAID. He was interfering with ongoing drug approvals. Tramont was angry that Dr. Fishbein was allowing concerns about patient safety to become an obstacle to the agency's central mission of getting new drugs through the approval process with positive reviews. Tramont warned Dr. Fishbein to slow down. "You are moving too fast. You need to get to know how this place works," Tramont told him. "We need to act more like a pharmaceutical company; we need to get patients, and get studies done."

In the course of his IL-2 investigation, Dr. Fishbein stumbled on another awkward fact: Anthony Fauci personally owned patents to IL-2 and stood to make millions in royalties if the treatment won FDA approval. Dr. Fishbein was shocked: "Dr. Fauci had a personal financial interest in the drug being tested! He was listed as a co-owner on the patent for Proleukin, and stood to earn royalties from it!" According to little-known HHS rules at that time, NIH employees could collect unlimited royalty payments from drugs they worked on during their agency tenures.[51] Dr. Fishbein found it stunning that Dr. Fauci stood to personally gain significant revenues, providing HHS green-lighted Proleukin.

Contemporaneous records obtained by the AP found that some fifty-one NIH scientists were then involved in testing products for which they secretly receive royalties; Dr. Fauci and his trusty longtime sidekick, Dr. H. Clifford Lane, "have received tens of thousands of dollars in royalties for an experimental AIDS treatment they invented [interleukin-2]. At the same time, their office has spent millions in tax dollars to test the treatment on patients across the globe."[52]

The AP story expressed understandable indignation about the circumstances under which the government has licensed the commercial rights to IL-2 to Chiron Corp: "Fauci's division subsequently has spent $36 million in taxpayer money testing the treatment on patients in one experiment alone. Known as the ESPRIT experiment, it is one of the largest AIDS research projects in NIH history, testing IL-2 on patients at more than two hundred sites in eighteen countries over the last five years."

On February 6, 2004, Dr. Fishbein wrote the Study Executive Committee, requesting issuance of the long-overdue updated version of the Investigator Brochure within sixty days, to include warnings of the newly discovered risks. Within days, Dr. Fishbein recalled, "I wrote a letter to the executive committee telling them to update the brochure. From that point on, the floor came out below me."

Even though his mandate was to enforce research policy, Dr. Fishbein had crossed the red line at NIAID. He was not just interfering with the drug-approval process: he was meddling with research in which Fauci had a peculiar interest.[53]

Dr. Fishbein's questioning about Dr. Fauci's patents tripped NIH into DEFCON 1. "All sorts of alarms went off," recalls Dr. Fishbein. "I came into government very naive. At the very least, I assumed that since Dr. Fauci wanted me to make sure studies were properly done, safety came first and that the participants were protected," he laughed. "I was wrong." He recounts that he had met Dr. Fauci only once—at the interview when Dr. Fauci hired him as NIAID's chief ethical and regulatory compliance officer. Dr. Fishbein recalls Dr. Fauci's earnestness: "This is an important job. If you come across any problems in the agency, I want to hear about them personally. I want you to come directly to me."[54] Dr. Fauci told Dr. Fishbein that his "door would always be open." But when Dr. Fishbein asked to meet with him about the IL-2 trials, Dr. Fauci went dark, and Dr. Fishbein felt the institution turning against him. "His guardians said he'd get back to me," recalls Fishbein. "He didn't." He adds, "He basically ran away."

In the course of his subsequent grievance procedures and litigation over his firing, Dr. Fishbein obtained emails and other documents that chronicled what happened behind the scenes. Dr. Fauci's principal strategy in discussions with his upper-level management was how to sack Dr. Fishbein while keeping NIAID's Director out of the splatter zone when things exploded. On February 24, 2004, Dr. Fauci met with Kagan and Tramont to plan strategies for ridding himself of Dr. Fishbein. The men hatched a plan by which Kagan and Tramont would orchestrate Dr. Fishbein's dismissal while making Dr. Fauci's fingerprints undetectable.

The challenge was daunting. All the players knew that Dr. Fauci was the only one with legal authority to fire Dr. Fishbein. NIAID human resources officers originally told Dr. Fishbein that Dr. Fauci had authorized his firing. Dr. Fauci later protested to various investigators from NIH and the US Congress that he had not ordered the firing. The NIH also denied that Dr. Fauci had ordered the firing. Dr. Fishbein calls this statement a lie: "I was a Title 42 special expert: paid outside the agency budget. Dr. Fauci was the only NIAID officer with authority to fire me."

Dr. Fishbein's reputation, his integrity, and his sterling work record presented additional obstacles. In November 2003, three months before his dismissal, Dr. Fauci presented Dr. Fishbein with a commendation for exceptional work at NIAID. Three months later, on February 9, 2004, Dr. Tramont also recognized Dr. Fishbein's outstanding job performance by recommending him to receive a $2,500 Service Recognition Award. Five days later, on February 13, 2004, Kagan blocked the processing of the award, canceled the $2,500 merit prize.

DAIDS officials followed these actions with an exchange of frantic emails discussing how to axe Dr. Fishbein without implicating Fauci. In a February 23, 2004, note to Kagan, Tramont said, "Jon, let's start working on this—Tony [Fauci] will not want anything to come back on us, so we are going to have to have iron-clad documentation, no sense of harassment or unfairness and, like other personnel actions, this is going to take some work. In Clausewitzian style, we must overwhelm with 'force.' We will prepare our paper work, then . . . go from there." Several of Dr. Fauci's other trusted subordinates joined the email chain with recommendations for how to blow up Dr. Fishbein's career while keeping Dr. Fauci's hands clean.

Said Farber, "Jonathan Fishbein [was] tarred and feathered for pointing out that the NIH flagship study on Nevirapine was a complete disaster. Fishbein's failure to fall into line, his failure to understand that Nevirapine was too important to fail, meant that the AIDS bureaucracy's neutralizing antibodies had to be activated to destroy him."

Between February 14–18, after Tramont notified Dr. Fishbein that he was now reporting to Kagan—the same man whom he had recently cited for disciplinary action—Dr. Fishbein exchanged emails with Tramont (then traveling in Thailand) requesting an explanation for this odd demotion that had him working for a lower-level employee who was a key target of his misconduct investigation. An elusive Tramont refused to explain the decision and answered with a vague remonstration reminiscent of Dr. Fauci's signature obfuscating gobbledygook:

> It has not been lost on me that the most complaints [about Kagan] I heard from our constituents when I arrived revolved around [complaints filed by Dr. Fishbein's branch] and since you have arrived, I have NOT heard a single complaint; and when I have inquired about that, the answer has been the charge brought by you.

On February 25, 2004, Kagan canned Dr. Fishbein. Kagan explained to Dr. Fishbein that he had failed in every aspect of his job and that his bosses saw no chance for improvement. Kagan advised Dr. Fishbein to leave DAIDS immediately. Dr. Fishbein opted to stay and fight his dismissal.

Dr. Fishbein first wrote to Tramont and Dr. Fauci requesting a meeting. He never received a reply. He next appealed to Dr. Fauci's ostensible boss, NIH Director Elias Zerhouni, who likewise refused to meet with him. NIH banned all employees from speaking about or to Dr. Fishbein. "Everyone was terrified of Fauci," says Dr. Fishbein. "He runs the agency like a vindictive dictator. Everyone is frightened of him; everyone knows that you never cross Fauci." In Farber's words, "Fishbein became a 'ghost.' Nobody addressed him in the corridors, in the elevators, in the cafeteria.

'There was an active campaign to humiliate me,' he recalls. 'It was as if I had AIDS in the early days. I was like Tom Hanks in *Philadelphia*. Nobody would come near me.'"

On February 26, 2004, Dr. Fishbein met with NIH's Office of Management Assessment (OMA) to complain about the actions against him. OMA also declined to investigate. On March 1, 2004, Dr. Fishbein brought his charges to the HHS Inspector General. The IG, similarly, refused to lift the carpet at NIH. Later that month, in desperation, Dr. Fishbein moved for whistleblower protection and sought a Congressional investigation of the wide-ranging corruption at NIAID.

On Capitol Hill, he at last found sympathetic ears. Dr. Fishbein told investigators for United States Senator Charles E. Grassley (R-IA) and Senator Max Baucus (D-MT), the chairman of the Senate Finance Committee and ranking minority member, respectively, that his sacking was retribution for his reports of wrongdoing in the Nevirapine and Proleukin trials. Both senators began clamoring for HHS to investigate Dr. Fauci's corruption charges against NIAID, and to answer the troubling questions Dr. Fishbein had raised about the homicidal studies in Tennessee and Uganda, and sexual harassment and mismanagement in NIAID's home office.

In a series of stern letters to NIH Director Zerhouni and his boss, HHS Secretary Michael Leavitt, Senators Arlen Specter and Herb Kohl joined Grassley and Baucus in rebuking NIH for inaction on Dr. Fishbein's complaints. Maryland Congresspersons Reps. Ben Cardin, Barbara Mikulski, and Steny Hoyer signed a similar letter. It's illustrative of Dr. Fauci's overwhelming power that he and his bosses decided to ignore and defy these remonstrances. After all, these three representatives were the royalty of NIH's home state delegation.

In May 2004, under pressure from lawmakers, NIH agreed to commission an Institute of Medicine (IOM) investigation of HIVNET 012. The Institute of Medicine, a branch of the National Academies of Sciences, is ostensibly Congress's independent and trustworthy advisor on scientific issues. IOM regularly assembles panels of top scientists to oversee and review agency science. The presumption is that while regulated industries easily capture and compromise federal agencies, the Institute of Medicine is incorruptible. IOM members do not work for either industry or the government. Congress expects to get the straight poop from IOM.

However, by that time, Dr. Fauci had already figured out how to control the IOM with invisible strings. The Capitol Hill lawmakers never realized that Dr. Fauci's PIs dominated the IOM panel that assembled to investigate his wrongdoing. Six of its nine members were NIAID grant recipients then conducting their own trials for Dr. Fauci, with annual grants ranging from $120,000 to $2 million. The IOM's study on Dr. Fishbein's charges was predictably, therefore, yet another whitewash. The IOM panel strategically adopted an extremely narrow scope of investigation that did not include NIAID's outrageous misconduct in Uganda or Tennessee. On April 7, 2004, the IOM panel reported its finding that the HIVNET 012 data should be considered valid.[55]

That same day, Dr. Fishbein received a letter of termination from Tramont. Dr. Fishbein sought and received an automatic postponement of his sacking as he argued his whistleblower case before the Equal Employment Opportunity Commission. Tramont's action, in the middle of a congressional investigation, was a naked gesture of defiance toward NIH's congressional overlords from both political parties. It signaled HHS's resolution to protect Dr. Fauci at any cost and to muzzle criticism by his principal detractor.

Teflon Tony had come a long way since 1987, when his public blistering by Congress had left him remorseful and terrified for his future. By 2004, he had the protection of his boss, a powerful Republican president, who—thanks to Dr. Fauci—was also implicated in the corrupt HIVNET trials and who cared little for the distempers of a Democrat-controlled Congress. Frustrated and angry at Dr. Fauci's insubordination, Grassley and Baucus fired off a letter dated June 30 to NIH Director Elias A. Zerhouni, demanding an explanation for Dr. Fishbein's firing and accusing NIAID of retaliating against Dr. Fishbein to silence his corruption charges against NIAID.[56] The letter noted that retaliation against an employee for reporting misconduct is "unacceptable, illegal, and violates the Whistleblower Protection Act."

Meanwhile, a secret internal NIH review of the Nevirapine trials was confirming Dr. Fishbein's worst accusations about Dr. Fauci and HIVNET. On August 9, 2004, Dr. Ruth Kirschstein, senior advisor to Zerhouni, sent the NIH director the results of her investigation. Kirschstein warned that Dr. Fauci's efforts to fire Fishbein at the very least gave the "appearance of reprisal." Kirschstein added that "It is clear that [Dr. Fauci's AIDS Branch] is a troubled organization" and that Dr. Fishbein's complaint "is clearly a sketch of a deeper issue."[57] Zerhouni kept quiet about these damning results from the agency's internal investigations. Defying the Senate, he fired Dr. Fishbein on July 4, 2005.

Following his dismissal, Dr. Fishbein brought his case before the Merit Systems Protection Board, asserting protection from any official retaliation under federal whistleblower laws. The MSPB reinstated Dr. Fishbein after determining his firing was "wrongful retribution." It was clear, however, that Dr. Fishbein had no future at NIH. He negotiated a termination deal. The terms of Dr. Fishbein's settlement agreement with NIAID are secret, and the deal forbids him from discussing its particulars.

Dr. Fishbein told me that despite his nominal victory, Dr. Fauci continued to punish him from afar with reverberations reaching far beyond NIAID. "I couldn't get a job in public health for five years," Dr. Fishbein says of Dr. Fauci's vendetta. "Everyone in science is terrified of crossing him. He's like a mafia kingpin. He controls everything and everyone in public health." Dr. Fishbein added, "He spreads so much money around and everyone knows he is vindictive. I had one friend tell me, 'I can't risk hiring you because I can't afford to anger Fauci.'" Says Dr. Fishbein, "This was my first exposure to the cancel culture."

He further reminisced: "I left the private sector and took the NIH job because I wanted to do public service. But I was very naive. I believed the government could find solutions, and that justice always prevailed. My experience at the Division of AIDS really opened my eyes about how the system really operated. The federal budget is a big trough to feed special interest groups. But if you become wise to it, open your mouth, and get on the wrong side of someone really powerful, they are out for blood. The government lawyers up, and they have unlimited resources to burn you. Truth may not be on their side, but they can throw every obstacle in your way to getting a fair hearing of your grievance. And you can't get justice because litigation will drain you to your last penny. The system isn't designed to help the aggrieved party. I couldn't coerce Fauci for a deposition. He was too busy doing interviews and accepting awards. There were never any consequences for the perpetrators. They continued merrily in their careers. I had to start all over again. If they are determined to ruin your life, they can do it."

Farber is also disenchanted. "They unleash such violence over your whole existence if you cross them. You never walk the same again. They make you feel like you are a dead person, totally devalued. They put a lot of money into these attack campaigns over my article. They went nuclear. Their crusade to discredit and destroy me had lasting impacts on my life. But you know what? I didn't get murdered. Joyce [Hafford] did. I think of her all the time.

"And the real losers in that battle," added Farber, "were the millions of African women and babies forced to take Nevirapine, a drug that does not prevent AIDS but sickens and kills people who take it." In the end, Dr. Fauci succeeded in rigging corrupt clinical trials, concealing catastrophic cheating, and deftly manipulating the politics to bring his dangerous and inefficacious drug, Nevirapine, to market.

In March 2005, Dr. Valendar Turner, a surgeon at the Department of Health in Perth, Australia, pointed out in a letter to *Nature*: "None of the available evidence for Nevirapine comes from a trial in which it was tested against a placebo. Yet, as the study's senior author has said, a placebo is the only way a scientist can assess a drug's effectiveness with scientific certainty."[58]

Dr. Turner observed that the transmission rate that HIVNET 012 reported for HIV—13.1 percent—was *above* the background transmission rate. "The HIVNET 012 outcome is higher than the 12 percent transmission rate reported in a prospective study of 561 African women given no antiretroviral treatment. This, in effect, is the placebo group." If anything, then, Dr. Fauci's pet drug has aggravated rather than prevented transmission in all those African babies he was pretending to save.

Farber argues that, under Dr. Fauci's leadership, the failure of researchers to properly control with a placebo group "is perhaps the outstanding characteristic of AIDS research in general." The statistical gimmick of getting rid of the inert placebo control group would become a tool wielded by Dr. Fauci to gain approvals for hundreds of new drugs and vaccines, from AIDS to COVID.

According to Farber, "As it was, there was no placebo group, so HIVNET's results are a statistical trick, a shadow play, in which success is measured against another drug and not against an inert placebo—the gold standard of clinical trials."

The single beneficial outcome of Dr. Fishbein's ordeal was that Congressional and press questions about Dr. Fauci's personal financial stakes in the IL-2 drug forced Dr. Fauci to pledge to donate his royalties from the scheme to charity. HHS thereafter changed its royalty policies—a little—limiting royalty payments to contract employees to $150,000 per year, per employee, per patent. In the thirty years since, no member of the media has ever asked Dr. Fauci how much money he made on IL-2, or to which charity, if any, he directed his donations. Nor has Dr. Fauci ever disclosed the extent of his personal stakes or the financial returns from his patents on other NIAID drugs, or the royalty amounts he has rewarded to loyal cronies and underlings at NIAID for the thousands of other new drugs the agency has developed.

Finally, during all of Dr. Fauci's tenure at NIH, Dr. Zeke Emmanuel was director of the Department of Bioethics (DOB), the ethical oversight board for all of NIH. Emmanuel's deputy was Tony Fauci's wife, Christine Grady. In 2012, Grady took over as director of DOB. That department oversees bioethics at clinical trials for all NIH subagencies, including responsibilities for overseeing ethical issues in clinical trials commissioned by her husband, like those for Nevirapine and Proleukin.

Grady acknowledged in an interview with *Vogue* that she was aware of Tony Fauci's reputation as a very scary person, upon their first meeting in 1983.[59] "Everyone was afraid of [him]. And when I first saw [him] I thought, 'What are they talking about?' He's young, he's handsome, and doesn't seem that scary."

"Dealing with Tony Fauci is like dealing with organized crime," says Dr. Fishbein. "He's like the godfather. He has connections everywhere. He's always got people that he's giving money to in powerful positions to make sure he gets his way—that he gets what he wants. These connections give him the ultimate power to fix everything, control every narrative, escape all consequence, and sweep all the dirt and all the bodies under the carpet and to terrorize and destroy anyone who crosses him."

Endnotes

1   Celia Farber, "Out of control: AIDS and the corruption of medical science," The Free Library, 2006 *Harper's Magazine* Foundation (Jul. 7, 2021), https://www.duesberg.com/articles/2006%20 Harper's,%20Farber%20on%20AIDS%20&%20cancer.pdf

2   Interview with Robert F. Kennedy, Jr., February 2021

3   John Solomon, "U.S. knew of AIDS drug concerns," Associated Press (Dec. 14, 2004), http://www. newmediaexplorer.org/sepp/2004/12/14/africa_nevirapine_for_aids_mothers_us_hid_research_ concerns.htm

4   Anthony Brink, *The Trouble with Nevirapine* (Open Books, 2008), 179

5   Centre for Research on Multinational Corporations (SOMO), "SOMO briefing paper on ethics in clinical trials #1: Examples of unethical trials" (Feb., 2008), https://www.somo.nl/wp-content/ uploads/2008/02/Examples-of-unethical-trials.pdf

6   Brink, op. cit.,188

7   Brink, op. cit., 17

8   "Serious Adverse Events Attributed to Nevirapine Regimens for Postexposure Prophylaxis After HIV Exposures—Worldwide, 1997–2000," Morbidity and Mortality Weekly Report (Jan 5, 2001), https://www.cdc.gov/mmwr//PDF/wk/mm4951.pdf

9   Brooks Jackson, Laura A. Guay, et al., "Intrapartum and neonatal single-dose nevirapine compared with zidovudine for prevention of mother-to-child transmission of HIV-1 in Kampala, Uganda: HIVNET 012 randomised trial," The Lancet (Sep 4, 1999), https://www.thelancet.com/journals/ lancet/article/PIIS0140-6736(99)80008-7/fulltext

10   Ibid.

11   Brink, op. cit., 114

12   J. B. Jackson et al, "Intrapartum and neonatal single-dose nevirapine compared with zidovudine for prevention of mother-to-child transmission of HIV-1 in Kampala, Uganda: HIVNET 012 randomised trial," *The Lancet*, (Sep 4, 1999), https://pubmed.ncbi.nlm.nih.gov/10485720/

13   Brink, 179

14   Westat Division of AIDS Clinical Research Operations and Monitoring Center, Site Visit Report, Pre-Inspection Audit (Mar. 8, 2002), https://justthenews.com/sites/default/files/2020-07/westat_ audit.pdf

15   Ibid.

16   Brink, op. cit., 184

17   Ibid.

18   John Solomon, "Orphans in hospice in Nairobi, Kenya, some of the nearly 2 million children under 15 living with HIV/AIDS in sub-Saharan Africa: NIH was warned on '02 on AIDS drug for Africa," Associated Press (Dec 14, 2004)

19   Ibid.

20   Ibid.

21   Brink, op. cit.,186

22   Westat Division of AIDS Clinical Research Operations and Monitoring Center, Site Visit Report,

Pre-Inspection Audit,(Mar. 8, 2002), https://justthenews.com/sites/default/files/2020-07/westat_audit.pdf

23 Farber, op. cit.

24 Brink, 179

25 Michael Carter, "Nevirapine: South African govt tries new ploy to avoid providing HIV drug to mothers," NAM AIDSMAP (Mar 28, 2002), https://www.aidsmap.com/news/mar-2002/nevirapine-south-african-govt-tries-new-ploy-avoid-providing-hiv-drug-mothers

26 Ibid.

27 E-drug, "Nevirapine /MTCT indication withdrawn in USA" (Mar 25, 2002), http://lists.healthnet.org/archive/html/e-drug/2002-03/msg00062.html

28 The National Academies, "Review of the HIVNET 012 Perinatal HIV Prevention Study" (2005), https://www.ncbi.nlm.nih.gov/books/NBK22287/

29 "NIH AIDS Research Chief Rewrote Safety Report-AP," AHRP (Oct. 26, 2006), https://ahrp.org/nih-aids-research-chief-rewrote-safety-report-ap/

30 Betsy Smith MD, "The HIVNET 012 Safety Review Panel: Primary Objectives, Approach, Findings and Summary to Date" (Jan. 24, 2003), https://childrenshealthdefense.org/wp-content/uploads/betsy-smith-un-redacted-report-01242003.pdf

31 Ibid.

32 George W. Bush's State of the Union Speech, CNN (Jan. 29, 2003), https://www.cnn.com/2003/ALLPOLITICS/01/28/sotu.transcript/

33 HIVNET 012 Monitoring report, DAIDS, NIAID (March 2003) (unpublished)

34 "Bush Touts AIDS Initiative During Uganda Visit–2003-07-12," Voice of America News (Oct 30, 2009), https://www.voanews.com/archive/bush-touts-aids-initiative-during-uganda-visit-2003-07-12

35 Keith Alcorn, "Boehringer cuts nevirapine price for developing world," NAM AIDSMAP (May 15, 2007), https://www.aidsmap.com/news/may-2007/boehringer-cuts-nevirapine-price-developing-world

36 Associated Press, "Docs Worry AIDS Drug Use May Be Halted," Fox News (Dec 20, 2004), https://www.foxnews.com/story/docs-worry-aids-drug-use-may-be-halted.amp

37 William J. Clinton Presidential Library, "Apology to survivors of the Tuskegee syphilis experiment," May 2, 2012, YouTube video, https://www.youtube.com/watch?v=F8Kr-0ZE1XY

38 Brynn Holland, "The 'Father of Modern Gynecology' Performed Shocking Experiments on Enslaved Women," HISTORY (Dec 4, 2018), https://www.history.com/news/the-father-of-modern-gynecology-performed-shocking-experiments-on-slaves

39 Barbara Loe Fisher, "Measles Vaccine Experiments on Minority Children Turn Deadly," National Vaccine Information Center (June, 1996), https://www.nvic.org/nvic-archives/newsletter/vaccinereactionjune1996.aspx

40 Joseph P. Williams, "From Tuskegee to a COVID Vaccine: Diversity and Racism Are Hurdles in Drug Trials," U.S. News and World Report (Nov 19, 2020), usnews.com/news/health-news/articles/2020-11-19/from-tuskegee-to-covid-diversity-racism-are-hurdles-in-drug-trials

41 Farber, op. cit.

42 Ibid.

43 Ibid.

44 Ibid.

45 John Solomon and Randy Herschaft, "AIDS drugs likely killed test subject," Associated Press/The Spokesman Review (Dec 16, 2004), https://www.spokesman.com/stories/2004/dec/16/aids-drugs-likely-killed-test-subject/

46 AP, "Woman died during AIDS drug study," NBC News (Dec. 16, 2004), https://www.nbcnews.com/health/health-news/woman-died-during-aids-drug-study-flna1c9448535

47 Edwin M. Thorpe et al., "Maternal toxicity with continuous nevirapine in pregnancy: results from PACTG 1022," Journal of Acquired Immune Deficiency Syndrome (Jul 1, 2004), pubmed.ncbi.nlm.nih.gov/15213559/

48 NIAID, "An International Study to Evaluate Recombinant Interleukin-2 in HIV Positive Patients Taking Anti-retroviral Therapy (ESPRIT)," Clinicaltrials.gov (Aug 31, 2001), https://clinicaltrials.gov/ct2/show/NCT00004978?term=ESPRIT%2C+IL-2&cond=HIV&draw=2&rank=1

49 Email exchange between Drs. Larry Fox and Jonathan Fishbein, December 12–15, 2003

50 John Solomon, Report: "Researchers mum on financial interests," Associated Press/NBC, (Jan 10, 2005), https://www.nbcnews.com/health/health-news/report-researchers-mumon-financial-interests-flna1c9475821

51 Ibid.

52 Solomon, op. cit.

53 Janice Hopkins Tanne, "Royalty payments to staff researchers cause new NIH troubles," BMJ (Jan. 22, 2005), https://www.ncbi.nlm.nih.gov/pmc/articles/PMC545012/

54 Ibid.

55 Institute of Medicine (US) Committee on Reviewing the HIVNET 012 Perinatal HIV Prevention Study. Review of the HIVNET 012 Perinatal HIV Prevention Study. Washington (DC): National Academies Press (US); 2005. 6, Response to the Charge to the Committee. Available from: https://www.ncbi.nlm.nih.gov/books/NBK22299/

56   Sean Moulton, "NIH AIDS Division Director Fired Possible Retaliation for Whistleblowing," Center for Effective Government (July 11, 2005), https://www.foreffectivegov.org/node/2495

57   Dr. Ruth Kirschstein, Senior Advisor to NIH Director Elias Zerhouni, to Deputy Director, NIH: Subject: Management Review. August 9, 2004

58   Valendar Turner, "HIV drug remains unproven without placebo trial," *NATURE*, (Mar 2005), https://ur.booksc.eu/book/10543591/51e6c0

59   Michelle Ruiz, "For Dr. Anthony Fauci and Dr. Christine Grady, Love Conquers All," *Vogue* (Feb. 11, 2021), https://www.vogue.com/article/dr-anthony-fauci-and-dr-christine-grady-love-story

ChildrensHealthDefense.org/fauci-book
childrenshd.org/fauci-book

For updates, new citations and references, and
new information about topics in this chapter:

CHAPTER 9
THE WHITE MAN'S BURDEN

Take up the White Man's burden—
The savage wars of peace—
Fill full the mouth of Famine
And bid the sickness cease.

—Rudyard Kipling, "The White Man's Burden" 1897

In 1984, following Dr. Robert Gallo's notorious press conference, Dr. Fauci promised the world an AIDS vaccine forthwith. Delivering a functioning AIDS immunization would, of course, be the most persuasive debunking of the Duesbergians and other critics of the HIV/AIDS hypothesis. "Finally," Dr. Fauci assured the global press, "given the fact that we now have the virus in our hands, it is quite possible, in fact, it's inevitable that we will develop a vaccine for AIDS."[1] Margaret Heckler told the media scrum, "We hope to have . . . a vaccine ready for testing in about two years."[2] Heckler was off the mark by a third of a century and counting. In the intervening decades, the federal government spent well over half of a trillion dollars on AIDS. Dr. Fauci has dedicated much of that moolah to his quest for an elusive vaccine to immunize people against HIV. Dr. Fauci pumped our money into nearly 100 vaccine candidates, with none of these coming even close to the finish line. All these disappointments never darkened Dr. Fauci's buoyant optimism that he will soon collar that retreating horizon.

For a decade, Oklahoma's US Senator Tom Coburn, MD, occupied front-row seats in Congressional and Senate Health Committees during Dr. Fauci's annual gallivants to Capitol Hill. By 2010, Coburn had wearied at the NIAID director's bootless promises of the imminent delivery of his quixotic jab. When, on May 18, Dr. Fauci returned to the Senate hearing room to tout "significant progress in HIV vaccine research," the normally taciturn Dr. Coburn finally exploded. He lambasted Dr. Fauci for deliberately deceiving lawmakers and accused his fellow physician of hoodwinking Congress into approving appropriations with no purpose beyond sustaining his bureaucracy: "Most scientists involved in AIDS research believe that an HIV vaccine is further away than ever."[3]

It had taken years of Dr. Fauci's ritualistic pilgrimages for Coburn to recognize, with indignant clarity, that attempted HIV vaccines are an ATM for NIH, whether they work or not. From an institutional standpoint, none of Dr. Fauci's failed experiments were, after all, failures. They each resulted in massive transfers of public lucre to Dr. Fauci's Pharma partners, and sustaining funding for NIAID's laboratories and PIs. The only true failure at NIAID would be a shrinking workforce.

This verity remains utterly obscure to the dewy-eyed press, which faithfully applauds each of Dr. Fauci's *Groundhog Day* encores. In 2019, nearly a decade after Coburn's remonstrance, and only a few months before the COVID-19 pandemic, Dr. Fauci made a surprise announcement: he finally had a working HIV vaccine. While the inoculation had demonstrated a bare-bones 30 percent efficacy in human trials in Thailand, data from the Phase III trial in South Africa looked promising, and NIAID was getting teed up to test the vaccine on Americans.[4] Dr. Fauci added some deflating caveats: While his new vaccine didn't prevent transmission of AIDS, the nimble technocrat jauntily predicted that intrepid souls who took the jab would find that when they did get AIDS, the symptoms would seem to be much reduced. So confident was Dr. Fauci of the media's slavish credulity that he assumed, correctly, that he'd never need to answer the many questions raised by this feverish gibberish. That entire odd proposition received zero critical press commentary. His success at slapping lipstick on this donkey and selling it to the world as a Thoroughbred may have emboldened his ruse—a year later—of placing similar cosmetics on the COVID vaccines that, likewise, neither prevent disease nor preclude transmission.

A PARADE OF HORRIBLES

The thirty-year decampment of journalistic scrutiny means that there is still no coherent public narrative chronicling Dr. Fauci's futile quest for his "inevitable" AIDS vaccine, much less accountability. Industry and government scientists have instead shrouded the scandalous saga in secrecy, subterfuge, and prevarication, obscuring a thousand calamities and a sea of tears deserving its own book. Every meager effort to research the debacle—on Google, PubMed, news sites, and published clinical trial data—yields only shocking new atrocities—a grim, repetitive parade of horribles: heartbreaking tragedies, entrenched institutional arrogance and racism, broken promises, vast expenditures of squandered treasure, and the recurring chicanery of Anthony Fauci, Bob Gallo, and Bill Gates. It's a darker version of *Groundhog Day*, devoid of humor, irony, wisdom, or redemption. It will be an easier read if I touch on only a few random lowlights of the painful saga.

Gallo Redux

In 1991, as part of a settlement ending years of litigation, Bob Gallo finally admitted that he had stolen the HIV virus from Montagnier. He was hardly chastened. On April 14, John Crewdson reported, in the *Chicago Tribune*, that one of Gallo's experiments with an HIV vaccine had killed three AIDS patients in Paris the previous year.[5] NIH had launched the project before handing it off to Gallo and his trusty henchman, Daniel Zagury, who tested the concoction on volunteers in France and, predictably, an African country, this time Zaire. His cronies at the National Cancer Institute had granted Gallo's experiments "expedited review, approval." How expedited? Just twenty-five days. The patients died after Gallo's team inoculated them with an HIV vaccine derived from cowpox. NIH scientists formulated the preparation from vaccinia—a virus that causes cowpox in bovines—into which the government scientists genetically inserted a fragment of the HIV virus. Apparently, the cowpox remained infectious, and three of their nineteen Paris volunteers immediately developed "vaccinia," a frequently fatal necrosis, which caused acute lesions and an expanse of hardened, swollen, purplish-red skin around the victims' injection sites as the disease devoured their flesh.

As is typical of AIDS vaccine research, the NIH scientists cached the atrocity. Neither Gallo nor Zagury reported the deaths. Instead, Gallo vaunted the trial as a great success in the *Lancet*'s July 21, 1990, edition, audaciously claiming that there had been "no deaths" and "no complications or discomfort" among any of those to whom he administered the preparation.[6]

One of Dr. Gallo's casualties was a forty-two-year-old classic literature professor—regarded as a brilliant Egyptologist—who succumbed March 5, 1990, more than four months before Gallo's article appeared. A second volunteer, a thirty-six-year-old Paris University librarian, died on July 6, weeks before Gallo published his article. Friends described Gallo's two victims as healthy and vibrant in the days and weeks immediately preceding their deaths. "It was unimaginable," a co-worker said of the robust professor, "that he could have died six weeks later."[7]

A longtime friend of Dr. Gallo's third victim, who died on October 1, 1990, asked Zagury's principal assistant, Dr. Odile Picard, whether the experimental vaccine may have caused the destructive lesions that the coroner detected on the victim's brain during autopsy. Picard assured him the vaccines were not at fault, adding, "We don't know what this is." A month after this conversation, Picard delivered another paper, also signed by Gallo and Zagury, at an international AIDS meeting in Paris, the Colloque des Cent Gardes. Here again Picard mentioned nothing about the three deaths, telling her colleagues that the vaccinia preparation had shown itself "safe in patients."[8] Perhaps she meant safe for those patients who survived.

André Boué, the distinguished French geneticist and secretary to France's National Committee on Medical Ethics, who approved the vaccine trials in 1987, complained that Gallo never informed his panel that any of the subjects had died. Officials of Assistance Publique, the municipal hospital system in Paris, grumbled that Gallo's team also neglected to tell them of the three fatalities. French officials only learned of the deaths from physicians who became suspicious at hospitals where Gallo's team had shipped their ailing recruits to die.[9]

NIH managers also protested that Gallo had not come clean about the deaths. One functionary called Gallo's omission "very troubling." NIH records show that neither Gallo nor any of his NIH confederates informed the Office of Protection from Research Risks about the body count. Federal law requires that OPRR approve human experimentation and that researchers report adverse events, including, of course, the most adverse event. In February, citing multiple violations of federal regulations by Gallo and his team on both sides of the Atlantic, the OPRR abruptly halted the experiment.[10]

Channeling his mentor's hallmark chutzpah, Zagury, after submitting the chipper *Lancet* article, applied for a patent on the deadly vaccine technology called "Methods of Inducing Immune Responses to the AIDS Virus," with Zagury listing himself as an "inventor" in the application.[11]

Once again, the omertà held. There was no investigation, no accountability, and no word of what sort of injuries the volunteers in the Zaire arm of the study may have suffered. Characteristically, Gallo was unembarrassed, unbowed, and undaunted by this latest setback. The bought-and-bullied virology community stayed silent about a scandal that would have implicated NIH and provoked unwanted scrutiny of the HIV orthodoxies.

Five years later, Gallo left NCI and established the Institute for Human Virology

(IHV) with his two longtime cronies, William Blattner, who served for 22 years under Gallo as Director of Viral Epidemiology at NCI, and Robert Redfield, a military doctor and researcher who shared Gallo's lifelong obsession with HIV and his ethical lacunae.

Dr. Robert Redfield

Many Americans will recognize Redfield as Donald Trump's CDC Director during the 2020 COVID pandemic. Dr. Redfield and his faithful sidekick, Dr. Deborah Birx, served with Dr. Fauci on Trump's Coronavirus Task Force.

Both Redfield and Birx were former Army medical officers who, in the 1980s and 1990s, led the military's AIDS research, a specialty that seems like a magnet for hucksters and quacks.

US military documents[12] show that in 1992 Redfield and Birx, his then-assistant—both serving at Walter Reed in Washington—published inaccurate data in the *New England Journal of Medicine*, claiming that an HIV vaccine they helped develop and tested on Walter Reed patients was effective.[13] They both must have known the vaccine was worthless.

In 1992, an Air Force medical office accused Redfield of engaging in "a systematic pattern of data manipulation, inappropriate statistical analyses and misleading data presentation in an apparent attempt to promote the usefulness of the GP160 AIDS vaccine."[14] A specially convened Air Force tribunal on scientific fraud and misconduct concluded that Redfield's "misleading or, possibly, deceptive" information "seriously threatens his credibility as a researcher and has the potential to negatively impact AIDS research funding for military institutions as a whole. His allegedly unethical behavior creates false hope and could result in premature deployment of the vaccine." [15] The tribunal recommended investigation by a "fully independent outside investigative body."[16] Under threat of court-martial, loss of his medical license, and possible imprisonment, Dr. Redfield confessed to angry DOD interrogators and to the tribunal that his analyses were faulty and deceptive. He agreed to correct them and to publicly admit the vaccine was worthless at an upcoming AIDS conference at which he was scheduled to speak in July 1992. Perhaps it was the grandeur of the hall, the microphone, and the audience that conspired to weaken his resolve. Instead of retracting his falsehood, he boldly repeated his fraudulent claims at this and two subsequent international HIV conferences.[17] As astonished prosecutors watched, he then brazenly parroted his debunked perjuries in testimony before Congress, swearing that his vaccine cured HIV.[18]

Redfield's bold gambit worked. Bamboozled by Redfield's brazen ballyhoo, Congress immediately appropriated $20 million to the military to support Redfield and Birx's research project.[19] Enraged military prosecutors wanted to court-martial Redfield. But as Public Citizen complained in a 1994 letter to the Congressional Committee's Chairman, Henry Waxman, the dedicated budget hikes promised by Congress prompted the Army to kill the investigation, silence its own prosecutors, and "whitewash" Redfield's misdeeds.[20]

By snatching triumph from the jaws of career-ending disaster, Redfield had pulled off the perfect crime. The bold flimflam catapulted Birx and Redfield into their stellar careers as top federal health officials. Whatever other lessons he learned from the episode, recklessness and mendacity continued to be Redfield's go-to strategies. Gallo's partnership with Redfield became a gold mine for both men. Dr. Gallo told me in a May 11, 2021, email that the Institute of Human Virology's (IHV) annual budget is in excess of $100 million: "A majority of this funding is from PEPFAR." President George W. Bush created PEPFAR in 2003, at Dr. Fauci's urging, to coordinate AIDS assistance from all the various federal government, civilian, and military sources. Since 2014, PEPFAR's administrator has been Deborah Birx, who simultaneously served on the board of the Bill Gates–backed Global Fund.

In 2017, the IHV's Annual Report boasted that these two quacksalvers had won over $600 million in grants—much of it from NIH and Bill Gates—since they cemented their lucrative partnership.[21] They seem to have spent the bulk of that loot experimenting with failed HIV drugs and vaccines on Black people, including 20,000 residents of Washington and Baltimore and 1.3 million misfortunates from Africa and the Caribbean.

Gallo and Redfield's IHV partnership was a good bet. They had an academic affiliation with the University of Maryland, their own nonprofit to launder grant money from their old NIH, NIAID, and NCI cronies, and a for-profit spinoff that would allow them to monetize their taxpayer-funded discoveries. Their former accomplices at NIH were pumping $200 million annually into the HIV vaccine boondoggle.[22] Moreover, Redfield had an inside track through Birx and through his military confederates to the vast Pentagon budgets for bioweapons and infectious disease. Those connections yielded plenty of federal dough to keep everyone in the chips. Furthermore, in 1998 a new HIV funder appeared—one with deep pockets and a shared obsession with vaccines.

That year, the William H. Gates Foundation announced a nine-year, $500 million plan to fund AIDS vaccine development through Gates's International AIDS Vaccine Initiative (IAVI)—the predecessor organization to the Global Alliance for Vaccines and Immunizations (GAVI).[23] IAVI's president, Seth Berkley—Gates's faithful and extravagantly compensated[24] minion—said the plan would fund multiple efficacy trials of AIDS vaccine candidates in developing countries. If any of the vaccines worked even reasonably well on sub-Saharan Africans, they could then presumably be tested in Western countries.

Despite Redfield's well-publicized history as a charlatan and pretender, President Donald Trump put him in charge of the CDC at a time when the agency's overarching mission was promoting COVID vaccines. Trump also elevated Birx, a lifelong protégée to both Redfield and Anthony Fauci and confidante to Bill Gates. These three vaccine mountebanks—Redfield, Birx, and Fauci—led the White House coronavirus task force and steered America's COVID response during the first year of the pandemic.

The trio—none of whom ever treated a COVID patient—adopted controversial strategies to confine the nation under house arrest, shut down the global economy, deny the public access to early treatment and lifesaving therapeutics like hydroxychloroquine and ivermectin, excite persistent public terror through the broadcasts of deliberately exaggerated death and case counts, and repeatedly tell the world that "the only path back to normal is a miraculous vaccine." With minimal scientific support, they imposed draconian quarantine, mask, and social-distancing mandates, purposefully or accidentally inducing a species of mass psychosis called "Stockholm syndrome" wherein hostages become grateful to their captors convinced that the only path to survival is unquestioning obedience.

The Gates/Fauci Bromance

Two years after Gates announced IAVI, he summoned Dr. Fauci to Seattle to propose a partnership that, two decades later, would have profound impacts on humanity. Dr. Fauci first met Bill and Melinda Gates during that Seattle trip. Ostensibly for a conversation about combating tuberculosis, the Microsoft billionaire had invited the NIAID chief to a muster of global health honchos at his 40,000-square-foot, $127 million mansion rising from forty wooded acres on the banks of Lake Washington. After dinner, Gates culled Fauci from the herd and corralled him into his spectacular blue-domed library overlooking the lake. Fauci remembered: "Melinda was showing everyone on a tour of the house. And he said, 'Can I have some time with you in my library,' this amazingly beautiful library. . . . And we sat down. And it was there that he said, 'Tony, you run the biggest infectious disease institute of the world. And I want to be sure the money I spend is well spent. Why don't we really get to know each other? Why don't we be partners?'"[25]

Over the next two decades, that partnership would metastasize to include pharmaceutical companies, military and intelligence planners, and international health agencies all collaborating to promote weaponized pandemics and vaccines and a new brand of corporate imperialism rooted in the ideology of biosecurity. That project would yield Mr. Gates and Dr. Fauci unprecedented bonanzas in wealth and power and have catastrophic consequences for democracy and humanity.

The Microsoft Monopoly

Influence peddling fueled Bill Gates's drive to power from the outset. Gates came from a wealthy family; his great-grandfather made a fortune in banking and left Bill a trust fund worth millions in today's dollars. After dropping out of Harvard in 1975, Gates leveraged his passion for software engineering to launch Microsoft in an era when most Americans still used typewriters. At the time, his mother, Mary Gates, a prominent Seattle businesswoman, sat on the United Way board alongside then-IBM chairman John Opel. In 1980, IBM was looking to recruit a software concern to develop an operating system for its personal computer. Mary Gates persuaded Opel to take a chance on her son. That intervention propelled Gates's fledgling firm into the big leagues and made Gates a billionaire within two decades.

Gates's closest boyhood friend and the Microsoft cofounder, Paul Allen, described Gates in his 2011 book (*Idea Man: A Memoir*) as a sarcastic bully who in 1982 schemed to oust him and steal his share of their company. Back at work following a bout with cancer, an anemic Allen, depleted by radiation and chemotherapy, overheard Gates conniving with Microsoft's new manager, Steve Ballmer, to dilute Allen's stake. Allen recalled bursting in and shouting: "This is unbelievable! It shows your true character, once and for all.”[26] Declining Gates's $5-a-share buyout offer, Allen left Microsoft with his 25 percent stake intact, becoming a billionaire when the company went public in 1986.[27]

In May 1998, the Department of Justice and twenty state attorneys general filed antitrust charges against Microsoft, accusing Gates's company of illegally thwarting efforts by consumers to install competing software on its Windows-based computers. The DOJ asked the federal trial court in Seattle to fine Gates a record million dollars a day for antitrust violations. Judge Thomas Penfield Jackson ruled that Microsoft had violated the 1890 Sherman Antitrust Act prohibitions outlawing monopolies and cartels, saying, “Microsoft placed an oppressive thumb on the scale of competitive fortune, thereby effectively guaranteeing its continued dominance in the relevant market.”[28]

Judge Jackson ordered Microsoft to divide itself in halves and divest either its operating system or its software arm. An appeals court overturned that decision. In a settlement, the DOJ abandoned its drive to break up the company, and Microsoft agreed to pay an anemic $800,000 fine and to share computing interfaces with competing firms.[29,30] Aside from the financial cost, the litigation had blighted Gates's reputation. Judge Jackson complained that Gates's testimony was “evasive and forgetful”[31] and observed that “[He] has a Napoleonic concept of himself and his company, an arrogance that derives from power and unalloyed success, with no leavening hard experience, no reverses.”[32] The public had seen enough of the trial—and of Gates's revealing depositions—to share Judge Jackson's revulsion.[33] An online group called SPOGGE gained widespread popularity. The acronym stands for “Society for Preventing Gates from Getting Everything.” Class action lawsuits filed in 2000 against the company for gross discrimination against African American workers and for including racially charged messages in its software further blighted Gates's pockmarked public image. Legendary plaintiffs' lawyer Willie Gary complained that Microsoft had “a ‘plantation mentality’ when it comes to treating African-American workers”[34] and observed that “there are glass ceilings and walls for African-American workers at Microsoft.”[35] Gary settled the case for $97 million.[36] Two years later, European regulators levied a $1.36 billion fine against Microsoft, the highest penalty in EU history.

Gates reacted to snowballing popular disgust by lobbying Congress to slash the Justice Department's budget and by hiring an army of PR firms to soften his image as a ruthless and duplicitous king-baby robber baron. As part of a concerted offensive to recast his public persona, Gates and his wife formed a charity, the Children's Vaccine Program, with an impressive $100 million donation.[37]

The Rockefeller-Gates Nexus

A century earlier, America's first billionaire, John D. Rockefeller, had blazed his own wildly successful exit ramp from public loathing, bad press, and antitrust prosecution by launching a medical philanthropy. John D. Rockefeller's consigliere, Frederick Taylor Gates, served as John D.'s chief business confidant and philanthropic adviser. Frederick Gates helped Rockefeller structure his foundation, advising the mogul that “judicious disposal of his fortune might also blunt further inquiry into its origins.”[38]

Practically from his nativity, Bill Gates began coordinating his own foundations' giving with the Rockefeller organization. In 2018, Bill Gates made the salient observation that "Everywhere our foundation went, we discovered the Rockefeller Foundation had been there first."

At the twentieth century's dawn, Rockefeller's sanguinary maneuvering—including bribery, price-fixing, corporate espionage, and creating shell companies to conduct illegal activities—had won his Standard Oil Company control of 90 percent of US oil production and made him the richest man in world history with a net worth of over half a trillion in today's dollars. Senator Robert Lafayette excoriated Rockefeller as "the greatest criminal of the age."[39] The oil magnate's father, William "Devil Bill" Rockefeller, was a marauding con artist who supported his family by posing as a doctor and hawking snake oil, opium elixirs, patent medicines, and other miracle cures.[40] In the early 1900s, as scientists discovered pharmaceutical uses for refinery by-products, John D. saw an opportunity to capitalize on the family's medical pedigree. At that time, nearly half the physicians and medical colleges in the United States practiced holistic or herbal medicine. Rockefeller and his friend Andrew Carnegie, the Big Steel robber baron, dispatched educator Abraham Flexner on a cross-country tour to catalog the status of America's 155 medical colleges and hospitals.

The Rockefeller Foundation's 1910 Flexner Report[41] recommended centralizing America's medical schooling, abolishing miasma theory, and reorienting these institutions according to "germ theory"—which held that germs alone caused disease—and the pharmaceutical paradigm that emphasized targeting particular germs with specific drugs rather than fortifying the immune system through healthy living, clean water, and good nutrition. With that narrative in hand, Rockefeller financed the campaign to consolidate mainstream medicine, co-opt the burgeoning pharmaceutical industry, and shutter its competition. Rockefeller's crusade caused the closure of more than half of American medical schools; fostered public and press scorn for homeopathy, osteopathy, chiropractic, nutritional, holistic, functional, integrative, and natural medicines; and led to the incarceration of many practicing physicians.

Miasma vs. Germ Theory

"Miasma theory" emphasizes preventing disease by fortifying the immune system through nutrition and by reducing exposures to environmental toxins and stresses. Miasma exponents posit that disease occurs where a weakened immune system provides germs an enfeebled target to exploit. They analogize the human immune system to the skin of an apple; with the skin intact, the fruit will last a week at room temperature and a month if refrigerated. But even a small injury to the skin triggers systemic rot within hours as the billions of opportunistic microbes—thronging on the skin of every living organism—colonize the injured terrain.

Germ theory aficionados, in contrast, blame disease on microscopic pathogens. Their approach to health is to identify the culpable germ and tailor a poison to kill it. Miasmists complain that those patented poisons may themselves further weaken the immune system, or simply open the damaged terrain to a competitive germ or cause chronic disease. They point out that the world is teeming with microbes—many of them beneficial—and nearly all of them harmless to a healthy, well-nourished immune system. Miasmists argue that malnutrition and inadequate access to clean water are the ultimate stressors that make infectious diseases lethal in impoverished locales. When a starving African child succumbs to measles, the miasmist attributes the death to malnutrition; germ theory proponents (a.k.a. virologists) blame the virus. The miasmist approach to public health is to boost individual immune response.

For better or worse, the champions of germ theory, Louis Pasteur and Robert Koch, proved victorious in their fierce decades-long battle with their miasmist rival Antoine Béchamp. Pulitzer Prize–winning historian Will Durant suggests that germ theory found popular purchase by mimicking the traditional explanation for disease—demon possession—giving it a leg up over miasma. The ubiquity of pasteurization and vaccinations are only two of the many indicators of the domineering ascendancy of germ theory as the cornerstone of contemporary public health policy. A $1 trillion pharmaceutical industry pushing patented pills, powders, pricks, potions, and poisons and the powerful professions of virology and vaccinology led by "Little Napoleon" himself, Anthony Fauci, fortifies the century-old predominance of germ theory. And so with the microbe theory, the "cornerstone was laid for modern biomedicine's basic formula with its monocausal-microbial starting-point and its search for magic bullets: one disease, one cause, one cure," writes American sociology professor Steven Epstein.[42]

As Dr. Claus Köhnlein and Torsten Engelbrecht observe in *Virus Mania*, "The idea that certain microbes—above all fungi, bacteria, and viruses—are our great opponents in battle, causing certain diseases that must be fought with special chemical bombs, has buried itself deep into the collective conscience."[43]

Imperialist ideologues find natural affinity with germ theory. A "War on Germs" rationalizes a militarized approach to public health and endless intervention in poor nations that bear heavy disease burdens. And just as the military-industrial complex prospers in war, the pharmaceutical cartel profits most from sick and malnourished populations.

On his deathbed, the victorious Pasteur is said to have recanted, "Béchamp was right," declaring, "the microbe is nothing. The terrain is everything."[44] Miasma theory survives in marginalized, yet vibrant, pockets among integrative and functional medicine practitioners. And burgeoning science documenting the critical role of the microbiome in human health and immunity tends to vindicate Béchamp, and particularly his teachings that microorganisms are beneficial to good health. Köhnlein and Engelbrecht observe that:

> [But] even for mainstream medicine, it is becoming increasingly clear that the biological terrain of our intestines—the intestinal flora, teeming with bacteria [or weighing up to 1 kg in a normal adult human, totaling 100 trillion cells.] is accorded a decisive role, because it is by far the body's biggest and most important immune system.[45]

A doctrinal canon of the germ theory credits vaccines for the dramatic declines of infectious disease mortalities in North America and Europe during the twentieth century. Anthony Fauci, for example, routinely proclaims that vaccines eliminated mortalities from the infectious diseases of the early twentieth century, saving millions of lives. On July 4, 2021, he commented to NBC's Chuck Todd, "You know, as the director of the National Institute of Allergy and Infectious Diseases, it was my responsibility to make sure that we did the science that got us to the vaccines that as we know now have already saved millions and millions of lives."[46] Most Americans accept this claim as dogma. It will therefore come as a surprise to learn that it is simply untrue. Science actually gives the honor of having vanquished infectious disease mortalities to nutrition and sanitation. A comprehensive study of this foundational assertion published in 2000 in the high-gravitas journal *Pediatrics* by CDC and Johns Hopkins scientists concluded, after reviewing a century of medical data, that "vaccination does not account for the impressive decline in mortality from infectious diseases . . . in the 20th century."[47] As noted earlier, another widely cited study, McKinlay and McKinlay—required reading in virtually every American medical school during the 1970s—found that all medical interventions including vaccines, surgeries, and antibiotics accounted for less than about 1 percent—and no more than 3.5 percent—of the dramatic mortality declines. The McKinlays presciently warned that profiteers among the medical establishment would seek to claim credit for the mortality declines for vaccines in order to justify government mandates for those pharmaceutical products.[48]

Seven years earlier, the world's foremost virologist, Harvard Medical School's Dr. Edward H. Kass, a founding member and first president of the Infectious Diseases Society of America and founding editor of the *Journal of Infectious Diseases,* rebuked his virology colleagues for trying to take credit for that dramatic decline, scolding them for allowing the proliferation of "half-truths . . . that medical research had stamped out the great killers of the past—tuberculosis, diphtheria, pneumonia, puerperal sepsis, etc.—and that medical research and our superior system of medical care were major factors extending life expectancy."[49] Kass recognized that the real heroes of public health were not the medical profession, but rather the engineers who brought us sewage treatment plants, railroads, roads, and highways for transporting food, electric refrigerators, and chlorinated water.[50]

The illustrations on the following page pose an indomitable challenge to germ theory's central dogma and stark support for miasma's approach to medicine. These graphs demonstrate that mortalities for virtually all the great killer diseases, infectious and otherwise, declined with advances in nutrition and sanitation. The most dramatic declines occurred prior to vaccine introduction.

Note the mortality declines occurred in both infectious and noninfectious diseases, irrespective of the availability of vaccines.

> *"When the tide is receding from the beach it is easy to have the illusion that one can empty the ocean by removing the water with a pail."*
>
> —René Dubos

As Drs. Engelbrecht and Köhnlein observe:

> Epidemics rarely occur in affluent societies, because these societies offer conditions (sufficient nutrition, clean drinking water, etc.) which allow many people to keep their immune systems so fit that microbes simply do not have a chance to multiply abnormally.[51]

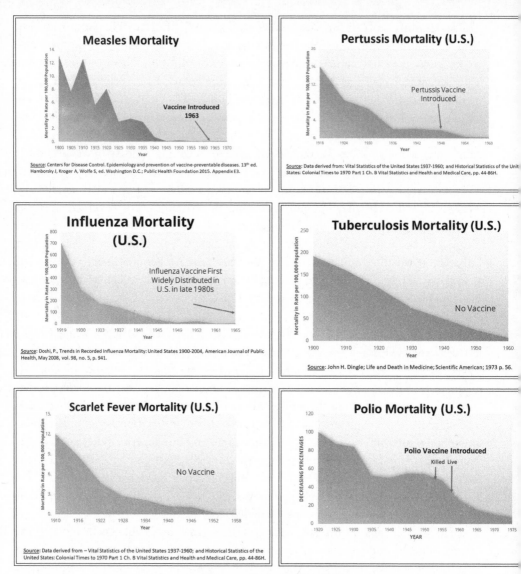

(Courtesy of Brian Hooker, PhD)

As a final side note, it seems to me that a mutually respectful science-based, evidence-based marriage incorporating the best of these two clashing dogmas would best serve public health and humankind.

Fauci and Gates; Germ Theory as Foreign Policy

The arcane conflict between germ and miasma theorists has important resonance for public health policy in the developing world, where many policy advocates fiercely protest that a dollar spent on food and clean water is far more effective than a dollar spent on vaccines. As we shall see, the Gates/Fauci militarized approach to medicine has precipitated an apocalyptic battle on the African and Asian continents between the two philosophies in a zero-sum game that pits nutrition and sanitation against vaccines in a life-and-death conflict for resources and legitimacy. The historic clash between these warring philosophies offers a useful framework for understanding Bill Gates's and Anthony Fauci's approach to public health. In order to assess the effectiveness of their mass-vaccination projects, we would need a disciplined accounting that compares health outcomes in vaccinated populations to similarly situated unvaccinated cohorts. This is the kind of accounting that neither of these men has been willing to provide. The facts suggest that it is the absence of such reliable metrics and science-based analysis that allows Gates and Fauci to get away with their dubious claims about the efficacy and safety of their prescriptions. Any even-handed examination of the role of immunizations in Africa must acknowledge that mass-vaccination programs may serve a larger agenda in which the priorities of power, wealth, and control can eclipse quaint preoccupations with public health. And, once again, it was the Rockefeller Foundation that pioneered germ theory as a foreign policy tool.

The Triumph of Germ Theory

In 1911, the Supreme Court ruled that Standard Oil constituted an "unreasonable monopoly" and splintered the behemoth into thirty-four companies that became Exxon, Mobil, Chevron, Amoco, Marathon, and others. Ironically, the breakup increased rather than diminished Rockefeller's personal wealth. Rockefeller donated an additional $100 million from that windfall to his philanthropic front group, the General Education Board, to cement the streamlining and homogenization of medical schools and hospitals. In accordance with the pharmaceutical paradigm, he simultaneously provided large grants to scientists for identifying the active chemicals in disease-curing plants utilized by the traditional doctors whom he had extirpated. Rockefeller chemists then synthesized and patented petrochemical versions of those molecules. The foundation's philosophy of "a pill for an ill" shaped how Americans came to view health care.[52]

In 1913, the patriarch founded the American Cancer Society and incorporated the Rockefeller Foundation. Philanthropic foundations were an innovation of the era, and detractors criticized, as "tax evasion," Rockefeller's scheme to take a $56 million deduction on his donation of 72,569 shares of Standard Oil to launch a foundation that would give him perpetual control of that "donated" wealth. A congressional investigation described the foundation as a self-serving artifice posing "a menace to the future political and economic welfare of the nation."[53] Congress repeatedly denied Rockefeller a charter. Attorney General George Wickersham denounced the foundation as a "scheme for perpetuating vast wealth" and "entirely inconsistent with the public interest."[54]

To reassure public, politicians, and press of its benign purposes, the Rockefeller Foundation declared its ambition to eliminate hookworm, malaria, and yellow fever. The Rockefeller Sanitary Commission for the Eradication of Hookworm Disease sent teams of doctors, inspectors, and lab technicians to administer deworming medication across eleven Southern states.[55] These ambassadors systematically exaggerated the medication's efficacy, glossed over its regular fatalities, and—through the graces of Rockefeller's mercenary army of journalists for hire—ignited enough favorable popular interest for the Foundation to justify the proposed expansion into the colonized world.

The Rockefeller Foundation launched a "public-private partnership" with pharmaceutical companies called the International Health Commission, which set about feverishly inoculating the hapless populations of the colonized tropics with a yellow fever jab.[56] The vaccine killed its beneficiaries in droves and failed to prevent yellow fever. The Rockefeller Foundation quietly dropped the useless vaccine after the foundation's star researcher, the yellow fever vaccine's inventor, Hideyo Noguchi, succumbed to the disease, likely contracted through careless laboratory exposure.[57] Noguchi's flexible scruples had greased his dicey experimentation on colonized "volunteers" and fueled his meteoric rise in the ethically barren landscapes of virology. At the time of his death, the New York district attorney was investigating Noguchi for illegally experimenting on New York City orphans with syphilis vaccines without the consent of their legal guardians.[58]

Despite such setbacks, the Rockefeller Foundation's yellow fever project caught the approbatory attention of army planners on the lookout for remedies against the tropical diseases that hamstrung the US military's expanding retinue of equatorial adventures. In 1916, the board's president made an early observation about the utility of biosecurity as a tool of imperialism: "For purposes of placating primitive and suspicious peoples, medicine has some advantages over machine guns."[59]

The Rockefeller Foundation's carefully heralded public health attainments eclipsed popular revulsion for the many abuses Americans associated with the Standard Oil petroleum empire. After World War I, its patronage of the League of Nations Health Organization gave the Rockefeller Foundation global reach and an impressive cortège of high-level contacts among the international elites. As the century progressed, the foundation became an exquisitely connected global enterprise with regional offices in Mexico City, Paris, New Delhi, and Cali. From 1913 to 1951, the Rockefeller Foundation's health division operated in more than eighty countries.[60] The Rockefeller Foundation was the world's de facto authority on how best to manage global diseases, with influence dwarfing all other nonprofits or government actors working in the field.[61] The Rockefeller Foundation provided almost half of the budget for the League of Nations Health Organization (LNHO) following its founding in 1922 and populated LNHO ranks with its veterans and favorites. The RF imbued the League with its philosophy, structure, values, precepts, and ideologies, all of which its successor body, the WHO, inherited at its inauguration in 1948.

By the time John D. Rockefeller disbanded the Rockefeller Foundation's International Health Division in 1951, it had spent the equivalent of billions of dollars on tropical disease campaigns in almost 100 countries and colonies. But these projects were window dressing for the Foundation's more venal preoccupations, according to a 2017 report, *U.S. Philanthrocapitalism and the Global Health Agenda*.[62] That idée fixe was opening developing world markets for US oil, mining, pharmaceutical, telecom, and banking multinationals in which the Foundation and the Rockefeller family were also invested. That white paper made the same complaints against the Rockefeller Foundation that contemporary critics level against the Bill & Melinda Gates Foundation:

> But the RF rarely addressed the most important causes of death, notably infantile diarrhea and tuberculosis, for which technical fixes were not then available and which demanded long-term, socially oriented investments, such as improved housing, clean water, and

sanitation systems. The RF avoided disease campaigns that might be costly, complex, or time-consuming (other than yellow fever, which imperiled [the military, and] commerce). Most campaigns were narrowly construed so that quantifiable targets (insecticide spraying or medication distribution, for example) could be set, met, and counted as successes, then presented in business-style quarterly reports. In the process, RF public health efforts stimulated economic productivity, expanded consumer markets, and prepared vast regions for foreign investment and incorporation into the expanding system of global capitalism.[63]

Here was a business model tailor-made for Bill Gates.

Philanthrocapitalism

Gates has dubbed his foundation's operational philosophy "philanthrocapitalism." Here is a stripped-down explanation of how philanthrocapitalism functions: Bill and Melinda Gates donated $36 billion of Microsoft stock to the BMGF between 1994 and 2020.[64] Very early on, Gates created a separate entity, Bill Gates Investments (BGI), which manages his personal wealth and his foundation's corpus. Renamed BMGI in January 2015 to include Melinda,[65] the company predominantly invests that loot in multinational food, agriculture, pharmaceutical, energy, telecom, and tech companies with global operations. Federal tax laws require the BMGF to give away 7 percent of its foundation assets annually to qualify for tax exemption. Gates strategically targets BMGF's charitable gifts to give him control of the international health and agricultural agencies and the media, allowing him to dictate global health and food policies so as to increase profitability of the large multinationals in which he and his foundation hold large investment positions. Following such tactics, the Gates Foundation has given away some $54.8 billion since 1994, but instead of depleting his wealth, those strategic gifts have magnified it.[66] Strategic philanthropizing increased the Gates Foundation's capital corpus to $49.8 billion by 2019. Moreover, Gates's personal net worth grew from $63 billion in 2000 to $133.6 billion today.[67] Gates's wealth expanded by $23 billion just during the 2020 lockdowns that he and Dr. Fauci played key roles in orchestrating.

In 2017, the *Huffington Post* observed that the Gates Foundation blurs "the boundaries between philanthropy, business and nonprofits" and cautions that calling Gates's investment strategy "philanthropy" was causing "the rapid deconstruction of the accepted term."[68]

Gates's pharmaceutical investments are particularly relevant to this chapter. Since shortly after its founding, his foundation has owned stakes in multiple drug companies. A recent investigation by *The Nation* revealed that the Gates Foundation currently holds corporate stocks and bonds in drug companies like Merck, GSK, Eli Lilly, Pfizer, Novartis, and Sanofi.[69] Gates also has heavy positions in Gilead, Biogen, AstraZeneca, Moderna, Novavax, and Inovio. The foundation's website candidly declares its mission to "seek more effective models of collaboration with major vaccine manufacturers to better identify and pursue mutually beneficial opportunities."[70]

Gates and Fauci: Colonizing the Dark Continent

After sealing their collaboration with a handshake, Gates and Dr. Fauci geared up their vaccine partnership quickly; by 2015, Gates was spending $400 million annually on AIDS drugs—mainly testing them on Africans.[71, 72] If he could prove that an AIDS remedy actually worked in Africa, the subsequent payoff from US and European customers would be astronomical.

For Gates, the immediate advantage of his new alliance with Dr. Fauci was clear. The imprimatur of his partnership with the US government's premier public health khedive anointed Gates's public health experiments with credibility and gravitas. Moreover, Dr. Fauci was an international power broker controlling a gargantuan bankroll and wielding Brobdingnagian political wallop across Africa. A trusted presidential confidant, Dr. Fauci had made himself the indispensable rainmaker for the river of HIV funding flooding the African continent. Dr. Fauci had, by then, persuaded a succession of US presidents to burnish their humanitarian bona fides by redirecting US foreign aid away from the causes of nutrition, sanitation, and economic development and toward solving Africa's HIV crisis with vaccines and drugs. His success in extracting a $15 billion commitment from George W. Bush in 2003 for AIDS drugs in Africa solidified Dr. Fauci's reputation as a global powerbroker capable of delivering US dollars to any African potentate who cooperated with his AIDS enterprise.[73] Despite his miserable track record at actually reducing illness over the next decade, he nevertheless persuaded President Bill Clinton, in May 1997, to set a new national goal for science by making the cure for African AIDS his JFK moonshot promise. In a speech he delivered at Morgan State University, Clinton said, "Today let us commit ourselves to developing an AIDS vaccine within the next decade."[74] Largely due to Tony Fauci's influence, Clinton would squander billions of taxpayer dollars on this fruitless crusade during his presidency and millions more of corporate and philanthropic contributions through the Clinton Foundation during his twilight years.[75]

George W. Bush similarly relied on Dr. Fauci's counsel, diverting $18 billion of the US government's relatively anemic foreign aid contributions to Dr. Fauci's pet global AIDS projects between 2004 and 2008 alone.[76]

In 2008, the *Journal of the European Molecular Biology Organization* published a peer-reviewed article examining how the Gates/Fauci partnership had skewed NIH funding to reflect Gates's priorities, "The Grand Impact of the Gates Foundation. Sixty Billion Dollars and One Famous Person Can Affect the Spending and Research Focus of Public Agencies." That article showed how, following the Gates/Fauci handshake, NIH had shifted $1 billion to Gates's global vaccine programs "at a time when overall NIH budget experienced little growth." The article outlines the technical details of the Gates NIH partnership; the Gates Foundation and the Wellcome Trust funneled their donations through the NIH Foundation, which administers the money while Gates determines how it is spent.[77] In this way, Gates has cloaked his pet projects with the imprimatur and credibility of the United States government. He has effectively purchased himself an agency directorate.

There is little objective evidence that all the treasure has extended or improved the lives of Africans, but every penny accrued to Fauci's reputation as Africa's foreign aid Golconda. When it came to public health policy in Africa, Dr. Fauci owned the keys to the kingdom. Gates needed Dr. Fauci to unlock the door.

Citing Ralph Waldo Emerson's observation that charity can be a "wicked dollar," sociology Professor Linsey McGoey explains that philanthropy can have evil effect when it "places its beneficiaries under a boot rather than recognizing their equal right to foster their own independence, to realize their individuality."[78] Professor McGoey is the author of the 2015 book *No Such Thing as a Free Gift: The Gates Foundation and the Profits of Philanthropy.*

Pharma had designs on Africa; Bwana Fauci and Bwana Gates donned pith helmets, grasped their machetes, shouldered their weaponized vaccines and toxic antivirals, and made themselves the twenty-first-century versions of the crusading European explorers Burton and Speke—bestowing the blessings of Western civilization upon the Dark Continent and requiring only obedience in return. "They are here to save the world," says McGoey of philanthrocapitalists, "as long as the world yields to their interests."[79] Thanks to their powerful collaboration, Pharma would emerge as, perhaps, Africa's cruelest and most deadly colonial overlord.

HIV provided Gates and Dr. Fauci a beachhead in Africa for their new brand of medical colonialism and a vehicle for the partners to build and maintain a powerful global network that came to include heads of state, health ministers, international health regulators, the WHO, the World Bank, the World Economic Forum, and key leaders from the financial industry and the military officials who served as command center of the burgeoning biosecurity apparatus. Their foot soldiers were the army of frontline virologists, vaccinologists, clinicians, and hospital administrators who relied on their largesse and acted as the community-based ideological commissars of this crusade.

Philanthrocapitalism's Global Imperium

In August 1941, President Franklin Roosevelt forced Winston Churchill to sign the Atlantic Charter as a condition for US support of the Allied effort in World War II. The Charter—a heartening emblem of American idealism—required the European allies to relinquish their colonies following the war. For two centuries, unimpeded access to the colonized world's rich national resources had been the principal source of European wealth. The Atlantic Charter and nationalist liberation movements in the 1950s and '60s dismantled the traditional colonial model in Africa. The continent, however, quickly reopened to "soft colonization" by multinational corporations and their state sponsors.

During the Cold War, the US military and intelligence agencies largely replaced Europe's colonial armies in those regions, supporting virtually any tinhorn dictator who proved his "anti-Communist" bona fides by rolling out red carpets for US multinationals. When the Berlin Wall fell, the United States already had 655 military bases (now 800)[80] across the developing world, and US companies had blank checks in host nations to extract agricultural, mineral, petroleum, and lumber resources, and large markets for finished goods including, notably, pharmaceuticals. After the Soviet bugaboo collapsed, Islamic terrorism and biosecurity supplanted communism as the rationale for a continued US military and corporate presence all over the developing world.

Pharma's acquisitive longing for Africa's natural resources and its teeming and compliant populations with their elevated disease burdens helped drive the rise of biosecurity as the spear-tip of corporate imperialism. Bill Gates and Dr. Fauci offered biosecurity as the underlying rationale for their medical neocolonialism project.

Paraphrasing the military's Cold War dogma, Gates and Dr. Fauci warned that if we didn't stop the germs in Africa, we'd end up fighting them in New York and Los Angeles. They echoed, also, the hackneyed crusaders' narrative that they were rescuing the continent from famine, pestilence, and ignorance with superior know-how and breakthrough technologies.

The combined Gates/Fauci power to rain foreign aid dollars on capital-starved African governments made them modern imperial viceroys on the continent. WHO became their colonial vassal, legitimizing and facilitating their campaigns to open African markets for drugmakers to dump unwanted products and to experiment with promising new cures.

AIDS Vaccines in Africa

In January 2003, as Gates and Dr. Fauci opened dozens of clinical trials for experimental AIDS vaccines across Africa, Dr. Fauci's perennial hagiographer, Michael Specter, writing in *The New Yorker*, raised trenchant questions about "the ethical problems associated with long-term vaccine trials in the developing world—funded by Western donors and designed, largely, by Western scientists." Specter asks, "Has the race to save Africa from AIDS put Western science at odds with Western ethics?" The article quotes African leaders asking why their continent needed to shoulder the burden of testing expensive vaccines and medicines that—if successful—would be primarily used in Western countries. They complained about pharmaceutical companies automatically lowering safety standards for clinical trials when they stepped onto the African continent. "Why us?" a prominent African journalist asked Specter. "It seems it's always us. For how many years does Uganda have to be the test case?"[81]

"I am very worried about these trials," said Peter Lurie, the deputy director of Public Citizen's Health Research Group. Lurie and his colleague, Sidney Wolfe, complained to Specter about the cavalier attitude of American researchers toward Third World subjects. "Instead of seeing themselves as activists for better care in Africa, scientists will use the poor quality of care to justify what they want to do anyway," Lurie said. "But you are not permitted simply to use subjects in order to collect data because it is useful to you. That is exploitation and abuse. That is what Tuskegee was." Lurie was referring to CDC's notorious decision to leave hundreds of Black Alabama sharecroppers with untreated syphilis for forty years beginning in 1932, in order to document the course of the disease. (I am proud that my uncle, Sen. Edward Kennedy, played a key role in exposing and ending the experiment in 1973.) Lurie added, "If we aren't careful, we could be in for the greatest injustice in the history of medicine."[82]

Later that year, Dr. Fauci's agency announced that NIAID's most recent AIDS vaccine experiment had failed. "Please don't say that I am pessimistic, because I am not," Anthony Fauci said in 2003, obliquely conceding that HIV and AIDS were not behaving the way his hypotheses predicted. "The best ways to vaccinate don't work with H.I.V. We need to come up with something new."[83]

Gates seemed to think that floods of new money could teach the virus to behave. In July 2006, the Bill & Melinda Gates Foundation announced sixteen grants totaling $287 million to create an international network of collaborative research consortia focused on accelerating the pace of HIV vaccine development by funding more than 165 PIs to conduct vaccine trials in nineteen countries.[84]

Two years later, on July 18, 2008, Dr. Fauci announced the cancellation of the largest human trial to date. It was NIAID's most promising HIV vaccine by far. Dr. Fauci contributed $140 million of taxpayer money to develop the Merck jab, and NIAID had already begun enrolling 8,500 US volunteers. It would be the first trial of an HIV jab on US citizens. Dr. Fauci said he intended the new trial to determine whether the vaccine could significantly lower the amount of HIV in the blood of those who become infected. Of course, Merck and NIAID had already by then tested the vaccine on 3,000 participants in nine African countries. The latest data were showing that the trial had not gone well. The *Times* reported coyly that "The PAVE trial had been postponed after a test of the Merck vaccine failed in its two main objectives: to prevent infection and to lower the amount of HIV in the blood among those who became infected."[85]

Buried near the end of the *New York Times* article were some key facts. It turned out that the vaccine was not only ineffective, but researchers reported alarming safety signals that caused a safety monitoring committee to halt the study. Furthermore, instead of preventing infection, the Merck/NIAID researchers reported data suggesting the vaccine actually *raised* the risk of contracting HIV![86]

Dr. Fauci said he reached his decision to abort the coming trial after meeting with scientists trying to understand why the Merck vaccine malfunctioned. Dr. Fauci's colleagues could offer no explanation for the vaccine's failure. Lawrence K. Altman of the *New York Times* reported that Dr. Fauci admitted that after a decade of effort, "scientists realized that they did not know enough about how HIV vaccines and the immune system interact." Dr. Fauci told the *Times* that it was becoming clearer that more fundamental research and animal testing would be needed before an HIV vaccine was ever marketed. These were stunning admissions, which seemed to validate the critiques by Duesberg and others who predicted the inevitable failure of a vaccine based on the defective HIV/AIDS hypothesis. Dr. Fauci said he had concluded that scientists must go a step at a time because they "did not yet know fundamental facts like which immune reactions are the most important in preventing the infection."[87] Cornell University scientist Kendall A. Smith made an even broader confession of error: "We really have not understood what actually constitutes a successful vaccine, despite the more than two centuries that have elapsed since Sir Edward Jenner described the first effective vaccine for smallpox virus in 1798. Consequently, all of the vaccines currently in use were developed empirically, and only within the past 50 years, without a comprehensive understanding as to how the immune system functions."[88]

"If Fauci's HIV/AIDS hypotheses were true, they should have been able to develop a vaccine," observes Dr. David Rasnick, a PhD biochemist who has worked for thirty years in the pharmaceutical biotech field. "Fauci's fundamental conundrum is that he has told everybody to diagnose AIDS based on the presence of HIV antibodies. With every other disease, the presence of antibodies is the signal that the patient has vanquished the disease. With AIDS, Fauci and Gallo, and now Gates, claim it's a sign you're about to die. Think about it; if the objective of an AIDS vaccine is to stimulate antibody production, then success would mean that every vaccinated person would also have an AIDS diagnosis. I mean, this is fodder for a comedy bit. It's like someone gave the Three Stooges an annual billion-dollar budget!"

On October 8, 2015, Gallo's Institute of Human Virology at the University of Maryland School of Medicine announced the launch of its Phase 1 human trials of Gallo's latest HIV vaccine candidate. A consortium led by the Bill & Melinda Gates Foundation gave $23.4 million to Gallo's research on this vaccine. Other money came from Redfield's pals in the US Military HIV Research Program.[89]

Gallo launched his clinical trial in collaboration with Profectus BioSciences, a biotech firm that he recently spun off from IHV to allow him to monetize the research he conducted with tax-deductible funding from Gates and taxpayer dollars from NIH and the military.[90, 91]

Gallo had already been testing his new HIV vaccines on animals, and "The results in monkeys are interesting, but they're not perfect." Undeterred by the vaccine's disappointing performance in macaques, Gallo was champing at the bit to test his concoction on some higher primates. "If we keep just using monkeys, we're never going anywhere. We need for humans to respond."[92] In May 2020, I asked Gallo what had ever happened to this experimental vaccine. Gallo claimed, in what I suspected to be an evasive nonanswer, that he was still (after six years) testing it for an immune response.[93]

By 2015, the BMGF was spending about $400 million a year on AIDS drug research. Gallo told me that his is only one of over 100 groups Gates has funded to find the elusive vaccine. Gates admitted publicly to Agence France-Presse that the quest for an AIDS vaccine has taken longer than expected, with many disappointments along the way.[94] Despite Gates and Fauci's impressive string of failures, Gates remained bullish. "A vaccine, that's a big area of funding for our foundation. But even in the best case that's five years away, and perhaps as long as 10," he jauntily predicted during a question-and-answer session with young people. "Probably the top priority is a vaccine. If we had a vaccine that can protect people, we can stop the epidemic."[95]

On February 3, 2020, Julie Steenhuysen of Reuters reported that NIAID had suddenly halted its clinical trial of its most promising HIV vaccine to date. NIAID was in the middle of Phase 3 trials on more than 5,000 South Africans when they realized that once again, the vaccine was *raising* the risk of AIDS in vaccinated individuals. Dr. Fauci issued another of his cheerful prognostications: "Research continues on other approaches to a safe and effective HIV vaccine, which I still believe can be achieved."[96]

Since 1984, undeterred by thirty-seven years of broken promises, failed clinical trials, billions of squandered dollars, and uncounted human carnage, Dr. Fauci and his old crony Bob Gallo continue to ride the AIDS vaccine gravy train. Neither man has advanced the search for a cure, but both have built impressive institutions. Existential questions about their scientific validity still bedeviled the two intertwined disciplines of virology and vaccinology for which those institutions form key nerve centers. Dr. Fauci's battle against AIDS is a religious crusade rooted in faith and appeals to authority rather than empiricism or rigorous scientific proof. Following the path of earlier colonial interventions in Africa, Dr. Fauci's evangelical campaign to impose the orthodoxies of germ warfare on Africans is an exercise in raw power, domination, and the ruthless extraction of profit.

Virology; A New Janissary Corps

As with the sultans, khans, czars, monarchs, and emperors of yore, Dr. Fauci's power derives from his capacity to fund, arm, pay, maintain, and effectively deploy a large and sprawling standing army. NIH alone controls an annual $42 billion budget mainly distributed in over 50,000 grants supporting over 300,000 positions globally in medical research.[97] The thousands of doctors, hospital administrators, health officials, and research virologists whose positions, careers, and salaries depend on AIDS dollars flowing from Dr. Fauci, Mr. Gates, and the Wellcome Trust (Great Britain's version of the Gates Foundation) are the officers and soldiers in a mercenary army that functions to defend all vaccines and Dr. Fauci's HIV/AIDS doxologies. The entire field of virology is Dr. Fauci's Janissary Corps—the elite warriors that he can rapidly muster to each new battlefield to achieve new conquests and ruthlessly suppress rebellion, dissent, and resistance.

In 2020, many of the Gates/Fauci HIV vaccine trials in Africa suddenly became COVID-19 vaccine trials, as the unprecedented tsunami of new COVID-19 plunder began flowing through Dr. Fauci to the same disciplined legions of the virology caste. At the outset of the pandemic, Dr. Fauci tapped his trusty procurator, Dr. Larry Corey, to launch the COVID-19 Prevention Network with the purpose of redeploying Dr. Fauci's most reliable and trusted PIs on a blitzkrieg campaign to win lightning vaccine approvals for his preferred jabs. Fauci accomplished this daunting mission by transforming his existing HIV trials, practically overnight, into Phase 3 COVID-19 vaccine trials. Without breaking stride, his PI army pivoted to march in lockstep into the new viral skirmish. Their exquisitely disciplined ranks also supplied the "independent experts" who populated the FDA and CDC committees that approved those shoddily tested COVID pokes, the doctors and "medical ethicists" who appeared on TV to run interference for every government-mandated COVID-19 countermeasure: masks, lockdowns, social distancing, and vaccination—including justifying the jab for children and pregnant women. (In any rational universe, giving these untested low-efficacy shots to children and pregnant women would constitute both medical malpractice and child abuse given the low risk for COVID and higher risk from the vaccine among these cohorts.) They penned editorials in the newspapers and articles in the scientific journals validating official orthodoxies and uniformly dismissing dissenters as screwballs, flakes, quacks, and "conspiracy theorists." From their ranks, Dr. Fauci and Bill Gates tapped the charlatans and biostitutes who conducted the fraudulent studies that torpedoed hydroxychloroquine and ivermectin and won approval for their pet blockbuster drug, remdesivir. When revelations that the COVID-19 virus was likely the product of genetic engineering threatened to discredit his empire, Tony Fauci dispatched the handpicked elite of virology's officer corps to draft and sign the consequential editorials published in *Nature*[98] and *The Lancet*[99] in February and March of 2020 assuring the world that the lab leak hypothesis was a "crackpot" conspiracy. The monolithic discipline of the virology caste and its capacity to rigorously enforce its omertà effectively silenced debate on COVID-19's origins for a year.

The saga of Fauci virologist Kristian Andersen, a PI who built his career on serial NIAID grants, offers a stark example of Fauci's system of payoffs. Andersen was the first grantee to alert Tony Fauci, in a 10:32 p.m. email on January 31, 2020, to the strong evidence that COVID-19 was lab-generated and that the experiment that created it may bear NIAID's fingerprints.[100,101] After midnight, Dr. Fauci warned his chief aide to keep his phone on and stand ready for some important work: To arrange a secretive emergency meeting to discuss damage control with eleven of the world's top virologists, including Andersen and five key researchers from the Wellcome Trust.[102] Dr. Fauci was the only US government official on this phone call. Four days later, Andersen, who less than a hundred hours earlier was convinced the virus did not come from nature, submitted a letter—secretly edited by Fauci—signed by five prominent virologists—all NIAID and Wellcome Trust PIs—ridiculing the suggestion that the circulating coronavirus could possibly have been lab generated.[103] One month later, Dr. Fauci—without disclosing his secretive involvement—cited that very letter at a White House press conference as proof that COVID-19 was naturally evolved.[104,105] In the months that followed, Andersen's employer, Scripps Research Institute, received an array of substantial grants from NIAID totaling $78 million for the calendar year.[106] The NIAID, by the end of 2020, had granted the employers of four of the five signatories on the paper a total of nearly $155,000,000.[107,108,109,110] That's how the game gets played. Dr. Fauci's disciples and soldiers understand that, as long as they support Dr. Fauci, they will continue to benefit from the endless stream of public health booty he controls—their spoils from the War on Germs and on skeptics.

HIV Vaccines: A New Lease on Life

In March 2020, Bill Gates stepped down from his position on the board of directors at Microsoft, explaining that he was "now spending the predominant amount of his time on the pandemic."[111]

Gates celebrated his Microsoft retirement by directing a river of money to build six manufacturing plants for different COVID vaccines and funding vaccine trials by companies like Inovio Pharmaceuticals, AstraZeneca, and Moderna Inc., all front-runners in the race to develop a COVID-19 jab. The Gates Foundation also invested $480 million in "a wide range of vaccine candidates and platform technologies" through the Coalition for Epidemic Preparedness (CEPI), which Gates founded—with Wellcome Trust director Jeremy Farrar.[112] Tony Fauci, meanwhile, took over managing the White House Coronavirus Task Force. The two men played tag team on the evening news and Sunday talk shows to promote remdesivir and to let their obsequious hosts and the American people know that the only way to end the global hostage crisis was compliance by 7 billion people with their new vaccines. No one ever questioned Gates's mantric pronouncement, which he repeated like a Gregorian chant: "Realistically, if we're going to return to normal, we need to develop a safe, effective vaccine. We need to make billions of doses, we need to get them out to every part of the world, and we need all of this to happen as quickly as possible."[113] He reiterated versions of this message ad nauseam, as he did again on CNN on March 22, 2020: "Things won't be back to truly normal until we have a vaccine that we've gotten out to basically the entire world."

Back to HIV

But despite all the excitement about COVID, neither of these partners forgot their first love, AIDS. On February 9, 2021, with his Operation Warp Speed vaccine rollout approaching the finish line, Dr. Tony Fauci took a well-earned breather to make an exciting announcement. He told his giddy media acolytes that NIH had just committed to a $200 million joint initiative with the Gates Foundation to fund the next generation of AIDS vaccines using NIAID's new mRNA technology: "This collaboration is an ambitious step forward, harnessing the most cutting-edge scientific tools and NIH's sizable global HIV research infrastructure to one day deliver a cure and end the global HIV pandemic."[114] Ignoring forty years of abysmal failure, NIH Director Francis S. Collins, MD, PhD, who plays Robin to Dr. Fauci's Batman, added, "We aim to go big or go home."[115] That thrilling announcement occurred almost exactly forty years after the first report of AIDS.[116] After four decades of cataclysmic outcomes, billions of squandered dollars, untold lives lost and failed promises, the press corps gave this most recent production the same rapt and credulous applause with which they greeted Teflon Tony's hundred other indistinguishable pronouncements over four decades. "He is the P. T. Barnum of public health," marvels journalist Celia Farber. "He cracks his whip and says 'Abracadabra,' and they all forget that they've seen the same trick so many times. It's really quite astonishing."[117]

By then, the Fauci/Gates arsenal of COVID pokes were topping the all-time charts for medical moneymaking by their pharma partners with Pfizer alone projecting $96 billion in COVID vaccine sales.[118]

Moderna followed up Dr. Fauci's appearance with a press release announcing new mRNA vaccines for Zika, Ebola, flu, cancer, and HIV. On July 25, 2021, Dr. Fauci expanded on this exciting communiqué by announcing a new multibillion-dollar government initiative to use taxpayer money and NIAID-patented mRNA technology to prepare distinct new vaccines for twenty families of viruses that might spark future pandemics. Dr. Fauci disclosed that he was already in discussions with the Biden White House about his electrifying proposal, which he said will cost "a few billion dollars" on top of NIAID's existing $6 billion budget. He said he planned to launch the project in 2022. Dr. Collins said he found Dr. Fauci's proposal "compelling," scolding that "As we begin to contemplate a successful end to the COVID-19 pandemic, we must not slip back into complacency."[119] On September 2, 2021, Joe Biden came through for Dr. Fauci again— announcing a $65 billion pandemic response effort. Dr. Fauci will be its point man.

Biden's announcement eclipsed some sad news. On August 31, 2021, Dr. Fauci acknowledged their premature termination of yet another of his African HIV vaccine experiments. A large trial on 2,600 African girls of a Johnson & Johnson HIV jab— funded jointly by NIAID and the Bill and Melinda Gates Foundation—had failed to demonstrate a beneficial effect.[120]

The Heartbreaking Legacy of Medical Colonialism

Rudyard Kipling originally coined the term "White Man's Burden" in his 1897 poem exhorting the moral imperative of the United States and England to impose Western civilizations and Christianity on tribal peoples for their own good.

Every student of African history is familiar with the recurring theme of well-intentioned white men visiting calamity on Africans. My interest in Africa began as a child. I have traveled the continent for six decades and met some of its most visionary leaders, including Tom Mboya, Jomo Kenyatta, Julius Nyerere, and Nelson Mandela. These anticolonial leaders understood that poverty is a complex conspiracy of social, historical, political, institutional, and technical maladies. It is most often best addressed through small-scale, locally tailored, trial-and-error experimentation. The optimal solutions are invariably homegrown with regular local input, disciplined self-assessment, accountability, frequent course corrections, and lots of humility by administrators, officials, and above all foreigners.

Gates's HIV vaccine and antiviral program is due to its continent-wide scale, arguably the worst in a long parade of paternalistic Western schemes by imperialists, avaricious adventurers, scammers, schemers, charlatans, double-dealing rogues, and well-meaning dolts who regularly pledge to end African suffering.

Instead of approaching Africa with humility, curiosity, open ears, and a willingness to support local initiatives, Gates leads with the same weisenheimer arrogance that Judge Penfield Jackson pegged as Gates's defining character trait in his 1998 ruling. At best, Gates and Dr. Fauci are just the latest in a long line of crusaders, con artists, and conquistadors who periodically appear on the continent armed with the conviction that they know what's best for Africans. Too often, these are self-serving, one-size-fits-all vanity projects that, in the end, only compound calamity and magnify suffering. At worst, in the words of Loffredo and Greenstein, Gates and his foundation function "as a trojan horse for Western corporations, which of course have no goal greater than an increased bottom line. . . . The foundation appears to see the Global South as both a dumping ground for drugs deemed too unsafe for the developed world and a testing ground for drugs not yet determined to be safe enough for the developed world."[121]

Magical vaccines are Gates's preeminent cookie-cutter solution for the ills of poverty, famine, drought, and disease. The absurdity of expensive shots as a remedy for indigence, a salve for malnutrition, or the dearth of potable water is obvious when one considers that three billion people live on less than two dollars per day. Eight hundred and forty million people don't have enough to eat. One billion lack clean water, or access to sanitation. One billion are illiterate. About a quarter of children in poor countries do not finish primary school.[122] Poverty is a target-rich environment, but the data suggest that Gates's vaccines miss the target altogether. Sociologist Linsey McGoey quotes a young health researcher based at the University of Cape Coast, on western Ghana: "From my point of view, it's more like [the Gates Foundation] are selling technology than solving problems. Most of their calls have to do with developing some new technology or vaccines."[123]

How Gates Controls the WHO

Worst, Gates has used his money strategically to infect the international aid agencies with his distorted self-serving priorities. The United States historically has been the largest direct donor to WHO with a contribution of $604.2 million in 2018–2019 (the last years for which numbers are available). That year BMGF gave $431.3 million and

GAVI gave $316.5 million.[124] Plus, Gates also routes funding to WHO through SAGE and UNICEF and Rotary International, bringing his cumulative total contributions to over $1 billion, making Gates the unofficial top sponsor of the WHO, even before the Trump administration's 2020 move to cut all his support to the organization.

Those $1 billion tax-deductible donations give Gates leverage and control over WHO's $5.6 billion budget and over international health policy, which he largely directs to serve the profit interest of his pharma partners. Pharmaceutical companies cement WHO's institutional bias toward vaccines with approximately $70 million of their own direct contributions. "Our priorities are your priorities," Gates declared in 2011.[125]

In 2012, WHO's then-Director General Margaret Chan complained that because the WHO's budget is highly earmarked, it is "driven by what [she calls] donor interests."[126] According to McGoey, "According to its charter, the WHO is meant to be accountable to member governments. The Gates Foundation, on the other hand, is accountable to no one other than its three trustees: Bill, Melinda, and Berkshire Hathaway CEO Warren Buffett. Many civil society organizations fear the WHO's independence is compromised when a significant portion of its budget comes from a private philanthropic organization with the power to stipulate exactly where and how the UN institution spends its money." McGoey observes that "Virtually every significant decision at WHO is first vetted by the Gates Foundation."[127] As the UK-based NGO Global Justice Now told Grayzone, "the Foundation's influence is so pervasive that many actors in international development which would otherwise critique the policy and practice of the Foundation are unable to speak out independently as a result of its funding and patronage."[128] (See also "The Perils of Philanthrocapitalism," Eric Franklin Amarante, *Maryland Law Review*, 2018.)

Gates's vaccine obsession has diverted WHO's giving away from poverty alleviation, nutrition, and clean water to make vaccine uptake its preeminent public health metric. And Gates is not afraid to throw his weight around. In 2011, Gates spoke at the WHO, ordering that "All 193 member states, you must make vaccines a central focus of your health systems."[129] The following year, the World Health Assembly, which sets the WHO agenda, adopted a "Global Vaccine Plan" that the Gates Foundation coauthored. Over half of WHO's total budget now goes to vaccines. That narrow focus on inoculations is deepening Africa's health crisis, according to global health experts and African officials.

Their control of several billion dollars in annual inputs gives Gates and Fauci effective control over not only WHO, but also the retinue of authoritative quasi-governmental agencies that Gates—often with Fauci's assistance and support—created and/or funded, including CEPI, GAVI, PATH, UNITAID, UNICEF, SAGE, the Global Development Program, the Global Fund, the Brighton Collaboration, and governmental health ministries in dozens of African nations that are largely dependent on the WHO and other global health partnerships. A 2017 analysis of the twenty-three global health partnerships revealed that seven were entirely dependent on Gates funding and another nine listed the foundation as its top donor. The Gates Foundation also controls the Strategic Advisory Group of Experts (SAGE), the principal advisory group to the WHO for vaccines. During a recent meeting, half of

SAGE's governing board of fifteen people listed conflicts of interest with the Gates Foundation.

The most powerful of these groups is GAVI, the second-largest non-state funder of the WHO. Gates created GAVI as a "public-private partnership" that facilitates bulk sales of vaccines from his pharma partners to poor countries.

GAVI is the template for Gates's impressive capacity to use his celebrity, credibility, and wealth to mesmerize key public officials and heads of state into giving Gates control over their foreign aid spending. Gates launched GAVI in 1999 with a $750 million donation. The BMGF occupies a permanent seat on the GAVI board.[130] Other organizations that Gates controls or can rely upon—WHO, UNICEF, the World Bank—and the pharmaceutical industry occupy additional seats, giving Gates dictatorial authority over GAVI's decision making. The BMGF has donated a total of $4.1 billion to GAVI to date.[131] But Gates has used that relatively trivial contribution—and his personal charm, I suppose—to attract over $16 billion from government and private donors,[132] including $1.16 billion annually from the US government, five times the amount that Gates donates to the WHO.[133]

When President Trump withdrew the United States from WHO in 2020, he continued the US contribution of $1.16 billion to GAVI.[134] The cumulative effect, therefore, of the withdrawal was to increase Gates's power over WHO and over global health policy. A recent assessment of GAVI by British Prime Minister Boris Johnson offers potent testimony of Gates's capacity to inspire the sort of obsequious adulation that has prompted Western leaders to hand over foreign policy and vast hordes of taxpayer dollars to Gates's discretion. In August 2021, Johnson declared that GAVI was the "new NATO."[135] Switzerland, which hosts GAVI's global headquarters in Geneva, has granted Gates's group full diplomatic immunity—a privilege Switzerland denies to many nations and their diplomats.

Additionally, the sheer magnitude of his foundation's financial contributions has made Bill Gates an unofficial—albeit unelected—leader of the WHO.

By 2017, Gates's power was so complete that he handpicked his deputy, Tedros Adhanom Ghebreyesus, as the WHO's new director general despite complaints that Tedros would be the first director general to the WHO without a medical degree and despite Tedros's dubious background. Critics credibly charge Tedros with running a terror group associated with extreme human rights violations including genocidal policies against a rival tribal group in Ethiopia.[136] As Ethiopia's foreign minister, Tedros aggressively suppressed freedom of speech, including arresting and jailing journalists who criticized his party's policies. Tedros's key qualification for the WHO gig was his loyalty to Gates. Tedros previously served on the boards of two organizations that Gates founded, funded, and controls: GAVI and the Global Fund, where Tedros was Gates's trusted chair of the board.[137,138]

GAVI is the most tangible outcome of the partnership Gates sealed with Fauci in early 2000. Under the terms of the partnership, Dr. Fauci greenhouses a pipeline of new vaccines in NIAID labs and farms them out for cultivation in clinical trials by his university PIs and the pharmaceutical multinationals in which Gates holds high investment stakes. Gates then builds out supply chains and creates innovative financial devices for guaranteeing those companies markets in Third World countries. A key feature of this scheme is Gates's capacity, through WHO, to pressure developing countries to expedite and purchase the vaccine, and to use GAVI as a bank through which wealthy countries cosign the debt. Western nations once funneled their foreign aid through traditional NGOs for food and economic development. Gates has captured those "deal flows" for GAVI and his pharma partners by pressuring Western countries to fork over their foreign aid to GAVI. Gates thereby hijacks the foreign assistance monies from wealthy governments, diverting it to drugmakers.

In May 2012, following two meetings with GAVI CEO Dr. Seth Berkley, Fauci candidly described the close relationship between GAVI and NIH.

"We, NIH, work on the upstream component of the fundamental research development. GAVI develop[s] a vaccine and get[s] it into the arms of people who need them." Dr. Fauci explained that while "NIH is way up in the upstream, and GAVI is way down in the downstream," there is no daylight between Gates's organization and his agency. ". . . there are areas of synergy and outright collaboration between us in setting the standard of what is needed and what kinds of research questions are important to answer. . . . We don't want to be putting resources particularly in the developing world if the research isn't going to be implemented, particularly with cold chain concerns. GAVI is much more of visible, coordinated force now, with a lot of resources, working in many, many countries." In contrast to some of the less reliable African governments, "It's an organization you can deal directly with."[139]

"Western nations originally conceived the World Health Organization and the United Nations to embody liberal ideologies implemented via a democratic structure of one nation, one vote," India's leading human rights activist, Dr. Vandana Shiva, told me. "Gates has single-handedly destroyed all that. He has hijacked the WHO and transformed it into an instrument of personal power that he wields for the cynical purpose of increasing pharmaceutical profits. He has single-handedly destroyed the infrastructure of public health globally. He has privatized our health systems and our food systems to serve his own purposes."[140]

As Jeremy Loffredo and Michele Greenstein concede in their July 2020 article, "The Gates Foundation has already effectively privatized the international body charged with creating health policy, transforming it into a vehicle for corporate dominance. It has facilitated the dumping of toxic products onto the people of the Global South, and even used the world's poor as guinea pigs for drug experiments. . . . The Gates Foundation's influence over public health policy is practically contingent on ensuring that safety regulations and other government functions are weak enough to be circumvented. It therefore operates against the independence of nation states and as a vehicle for Western capital."[141]

The Sanctity of Patents

A singular feature of Gates's vaccine caper—largely unnoticed until recently by the global press—is his ironclad commitment to protect pharma's intellectual property rights. Asked in a Sky News interview if sharing intellectual property and the recipe for vaccines would be helpful, Gates replied bluntly: "No."

"There's all sorts of issues around intellectual property having to do with medicines, but not in terms of how quickly we've been able to ramp up the volume here. . . . I do a regular phone call with the pharmaceutical CEOs to make sure that work is going at full speed."[142]

In April 2021, his unyielding allegiance to patent rights—and corporate profits—finally caused cracks to appear in the monolithic support for Gates among mainstream media and the public health establishment.

That month, the *New Republic* writer Alexander Zaitchik published a lengthy article, "Vaccine Monster,"[143] describing how Bill Gates had aggressively impeded global access to COVID vaccines by the world's poorest people in order to safeguard the profitable patent privileges of his pharmaceutical partners.

By March 2020, Indian and African nations anticipating severe vaccine shortages of COVID inoculations for their populations were clamoring for a waiver of patent rights that would allow local manufacturers to rapidly supply hundreds of millions of generic vaccines at prices that would provide access to the poor. Western nations joined the hullabaloo in the cause of patent exemptions recognizing that government innovation, vast flows of taxpayer subsidies, regulatory waivers, liability exemptions, coercive mandates, and licensing monopolies had given birth to the COVID vaccines with drug companies themselves playing relatively minor roles.

By August 2020, a global movement to waive patents for COVID-19 vaccines had gathered the momentum of a runaway locomotive. Proponents included much of the global research community, major NGOs with long experience in medicines development and access, and dozens of current and former world leaders and public health experts. In a May 2020 open letter, more than 140 political and civil society leaders called upon governments and companies to begin pooling their intellectual property. "Now is not the time . . . to leave this massive and moral task to market forces,"[144] they wrote. In early March 2021, the world's leading public health authorities launched a voluntary intellectual property pool inside the WHO to ensure that COVID-19 drugs and vaccines would be universally and cheaply available—the WHO COVID-19 Technology Access Pool, or C-TAP.

In May 2021, President Biden threw his weight behind the movement, calling for a temporary suspension of patent protections for COVID-19 vaccines to ensure coverage among poorer nations.[145] "We believe that intellectual property rights constitute a very substantial barrier to ensure equitable access," he said. "We believe that if we could have a limited, targeted waiver to ensure that we can ramp up production in various parts of the world, we would go a long way to ensure that we address not only the prevention but also the treatment of COVID-19." Biden's equity initiative forced Gates into the open. Gates's entire philanthrocapitalism business model rests on the sanctity of knowledge monopolies; and so, with the whole world watching, Gates revealed that patent integrity—the source of vaccine profits to his pharma partners—is the sine qua non of Gates's global health initiatives. When push turns to shove, patent protection eclipses his professed concerns for public health.

His ironclad control of WHO made Gates's opposition to C-Tap dispositive. The runaway train hit a granite mountain. Any pretense that democracy or equity should determine global health policy collapsed before the raw power and influence of Bill Gates. According to the *New Republic*, "Advocates for pooling and open science, who seemed ascendant and even unstoppable that winter, confronted the possibility that they'd been outmatched and outmaneuvered by the most powerful man in global public health."[146]

Gates derailed the C-Tap pool, replacing it with his own WHO program, the "COVID-19 Act-Accelerator," which consecrated industry patent rights and relegated developing world vaccination programs to the charitable impulses of pharmaceutical companies and Western donor nations fighting for their own share of the vaccines. As the predictable result of Gates's intervention, around 130 of the poorest of the world's 190 nations, 2.5 billion people, have had zero access to vaccines as of February 2021. As Zaitchik pointed out, the supply crisis was easily foreseeable: "Not only were the obstacles posed by intellectual property easily predictable a year ago, there was no lack of people making noise about the urgency of avoiding them." Gates had once again used his international reputation and money authority to shield corporate greed with a "halo effect."[147, 148] International health officials warned, for example, that despite all government expressions of concern about Africa, "Less than 2 percent of all doses administered globally have been in Africa. Just 1.5 percent of the continent's population are fully vaccinated." (Paradoxically, these nations happen to have lower COVID mortalities by orders of magnitude.)

"There has never been a point at which the Gates Foundation—before the pandemic, at the start of the pandemic, and now at the worst moment of the pandemic—is willing to surrender and look at IP as something that has to be managed differently to ensure that we're doing as much as possible," said Rohit Malpani, board member of the global health agency UNITAID.[149]

Gates opposed waiving some provisions of the World Trade Organization's Agreement on Trade-Related Aspects of Intellectual Property Rights, or TRIPS. A waiver would allow member nations to stop enforcing a set of COVID-19-related patents for the duration of the pandemic. "Bill Gates asked everyone to block the TRIPS waiver and trust a handful of companies hoarding IP and know-how," said James Love, director at Knowledge Ecology International.[150]

Gates's commitment to patent rights is existential and unyielding. Gates has ruthlessly defended intellectual property monopolies since his early battles with open-source hobbyists in Microsoft's natal days. Gates built both his fortune and his charitable model of philanthrocapitalism on the sanctity of intellectual property protections in software, food, and drugs.

Gates made his bones with his Big Pharma partners by triumphing over Nelson Mandela in hand-to-hand combat during the grim African AIDS crisis of the 1990s. South Africa was ground zero in the global AIDS epidemic, with HIV infection rates affecting one in every five adults. Mandela had made himself the paladin in a Third World crusade to allow generic drugmakers to give the global poor access to expensive AIDS drugs. Mandela's reputation as a kind of saint stymied the pharmaceutical companies, reluctant to defend a venal business model that—by their own estimation—was a death sentence for 29 million African children and adults. Cloaking himself in the moral authority as the world's largest charitable benefactor, Gates stepped forward as the industry champion, expounding the cause of intellectual property and knowledge monopolies over public health. That ghillie suit of selfless altruism successfully confused the press and public—especially the liberal establishment—about Gates's solipsistic motives for over two decades.

In December 1997, Mandela's administration pushed through a law allowing the health officials to import, produce, or purchase generic AIDS drugs that were out of reach for most Africans. Pharma was happy to test AIDS drugs on Africans but had priced the final product far out of their reach. Glaxo, for example, was still selling annual dosages of AZT for $10,000. Gates declared war on Mandela and his generic drug crusade by supporting a suit by thirty-nine pharma multinationals who sued South Africa to prevent poorer nations from accessing generic AIDS drugs for their people.[151] Once again, Gates put the halo on greed.

The *New Republic* chronicled the fight: "In Geneva, the lawsuit was reflected in a battle at the WHO, which was divided along a north-south fault line: on one side, the home countries of the Western drug companies; on the other, a coalition of most from the global south and dozens of leading public health groups including Médecins Sans Frontières and Oxfam joined the battle on behalf of Mandela."[152]

In the end, Gates and pharma won the legal case, and Gates helped push through enduring bullet-proof protection for pharmaceutical patents by his implacable support of the Trade-Related Aspects of Intellectual Property (TRIPS), an international agreement that outlawed the use of unsanctioned generics to combat AIDS and other diseases.

Today, leading public health officials agree that the primary drivers of the current artificial shortage of COVID-19 vaccines is Gates's defense of intellectual property rights to protect the profiteering by his pharma partners.

Zaitchik recounts how "battle-scarred" public health veterans saw clearly, for the first time, how Gates's addiction to proprietary science and market monopolies easily overrode his professed concern about the impacts of the pandemic and poor nations and the structural inequality in access to medicines: "COVID19 reveals the deep structural inequality in access to medicines globally, and a root cause is IP that sustains and dominates industry's interests at the cost of lives."[153]

Zaitchik offers a devastating indictment of Gates: "Gates is certain he knows better. But his failure to anticipate a crisis of supply, and his refusal to engage those who predicted it, have complicated the carefully maintained image of an all-knowing, saintly mega-philanthropist. COVAX presents a high-stakes demonstration of Gates's deepest ideological commitments, not just to intellectual property rights but also to the conflation of these rights with an imaginary free market in pharmaceuticals—an industry dominated by companies whose power derives from politically constructed and politically imposed monopolies."[154]

After describing how Gates pushed back ruthlessly "defending the status quo and running effective interference for those profiting by the billions from their control of COVID-19 vaccines," Zaitchik offers a glimmer of hope for humanity's most downtrodden third fighting for their lives against this "vaccine monster": "There are signs of overdue scrutiny of Gates's role in public health and lifelong commitment to exclusive intellectual property rights."[155]

Blacks to the Front of the Line

At the February 2021 press conference, Francis Collins said that NIH's new generation of HIV vaccines will specifically target Africans and African Americans, "to make sure everybody, everywhere, has the opportunity to be cured, not just those in high-income countries."[156] Such sympathies were a consistent preoccupation along the Gates/NIH nexus. Melinda Gates lamented on CNN, April 10, 2020, that she was "kept up at night" worrying about vulnerable populations in Africa.[157, 158] In June 2020, she told *Time Magazine* that, in the United States, Black people should get the COVID-19 vaccine first.[159] The idea that Blacks should be first in line for the vaccine—and official anxieties that many Blacks would resist this privilege—were persistent themes in pronouncements by the leading health agencies during the pandemic. As we shall see in Chapter 12, Gates, Fauci, and the intelligence agency and pharmaceutical company partners repeatedly wargamed strategies for overcoming anticipated Black resistance in many of the dozen pandemic simulations leading up to COVID-19. Once the pandemic was underway, HHS recruited Black preachers, HBCU college deans, civil rights leaders, and sports figures like Hank Aaron to soften jab hesitancy in the Black community. They staged press conferences and highly publicized celebrity vaccination confabs and extravagantly financed government advertising campaigns targeting Blacks in both the United States and Africa. In December 2020, Dr. Fauci scolded resistance in the Black community, saying, "The time is now to put skepticism aside." Without citing any studies demonstrating the vaccine was safe, he said that "The first thing that you might want to say to my African brothers and sisters is that the vaccine you're going to be taking was developed by an African-American woman—and that's just a fact."[160]

When Cicely Tyson, Marvin Hagler, and rapper Earl Simmons—a.k.a. DMX— all died soon after taking COVID vaccines, the medical community and CDC rushed in to assure the African American community that the deaths were not vaccine related. Social media and mainstream outlets censored or removed stories that suggested a vaccine association. Gates-funded "fact checker" organizations "debunked" any link. The desperation to discredit such talk inspired many "respectable" media outfits to simply lie. When home run king Hank Aaron, whom I knew, died seventeen days after receiving a vaccine at a staged press conference at Atlanta's Morehouse College, I wrote that his death was among a wave of deaths in older people following vaccination. (I never said the vaccine caused Aaron's death.) The *New York Times,* CNN, ABC, NBC, *Inside Edition*, and a hundred news organizations across the globe rushed to castigate me and rebuff my article as "vaccine misinformation," assuring the public that the Fulton County Coroner had declared Aaron's death "unrelated to the vaccine." When I called the Fulton County Coroner, the office informed me that they had never seen Hank Aaron's body and that Aaron's family had buried him without autopsy.[161] After I published this embarrassing fact, not a single news organization posted a retraction.

Federal law requires that every injury or death following vaccination during clinical trials—or, by logical extension, with emergency use products—must be attributed to the vaccine unless proven otherwise. Nevertheless, as of August 2021, the CDC officially took the Pollyannaish view that not one of the 13,000-plus deaths[162] reported to VAERS following vaccination as of August 20, 2021, was vaccine related.[163] Not one. As was the case with Hank Aaron, CDC apparently did nothing to actively investigate any of those deaths, exonerating the vaccines, instead, by fiat.

While unusual numbers of Black celebrities were dying postvaccination in America, an eyebrow-raising number of anti-vax political leaders were simultaneously expiring in Africa. The epidemic of untimely deaths among high-profile black African heads of state and key government ministers and physicians who opposed Bill Gates/COVAX policies provoked a wave of conspiracy theories suggesting that these men were murdered to silence dissent. The phenomenon was so striking during the first year of the pandemic that both Reuters and the *British Medical Journal* (*BMJ*) published articles seeking to explain the troubling trend. The Internet assassination speculations reached a boil following the strange murder of President Jovenel Moïse of Haiti by a team of elite, well-trained Colombian mercenaries with links to United States intelligence agencies. Moïse was a vocal opponent of the WHO vaccine program. The African leaders who died suddenly after criticizing WHO vaccination policy included President John Magufuli of Tanzania (March 17, 2021), Prime Minister Hamed Bakayoko of Ivory Coast (March 10, 2021), President Pierre Nkurunziza of Burundi (January 8, 2020), and Madagascar's popular, influential, and anti-vax ex-President Didier Ignace Ratsiraka (March 28, 2021). Kenya's beloved physician Stephen Karanja, the chairman of the Kenya Catholic Doctors Association—who had exposed the WHO sterilization program in 2014 and who criticized the agency's COVID rollout in 2020—also died, reportedly of COVID (April 29, 2021). A peer-reviewed article in the BMJ titled "Why have so many African leaders died of COVID" lists seventeen heads of state and leading government health ministers who passed in the twelve months between February of 2020 and February of 2021. The BMJ article states that almost all of these deaths resulted in dramatic shifts in national health policies from skepticism toward strong support for vaccination in their respective countries. The article points out that the overall death rates (1:33) among African elected leaders from COVID are seven times the rates for their sex and age and demographics of the general population during that time period.[164]

I do not endorse the theory that these men were murdered, nor do I dismiss such speculation out of hand. It is naive to believe that powerful men and women who threaten a trillion-dollar industry allied with Western military and intelligence agencies do so without risk. I document the keen interest by the Western intelligence community and militaries in the African vaccine enterprise in Chapter 12, "Germ Games." The historic involvement of Western intelligence agencies in coups and the murders of African leaders on behalf of their corporate clientele is well documented. I have a clear personal memory of the shocked reaction by my father and my uncle John Kennedy to the assassination of Congo's liberator Patrice Lumumba on my birthday, January 17, 1961, a week before my uncle John Kennedy's inauguration as US President. JFK regarded Lumumba as the "George Washington of the Congo." US and European mining companies had their eyes on the Congo's vast mineral wealth, and Lumumba—a beloved nationalist who led the Congo's liberation movement against Belgium—had sworn to deploy that wealth, instead, to benefit the Congolese people. We have since learned that the CIA and the Belgian intelligence agencies collaborated in Lumumba's murder. (In 2002, Belgium formally apologized for its role in the assassination.)[165] CIA Director Allen Dulles, who planned to kill Lumumba with poison toothpaste, knew that my uncle had enormous affection and admiration for Lumumba. Dulles feared that JFK would interfere with the CIA's plan to liquidate the charismatic leader. Among other mischief, the CIA overthrew governments in Ghana in 1966 and Chad in 1982.

Congressional investigations in the 1970s exposed the CIA's years of experimentation to develop untraceable poisons and secretive murder tools. CIA scientists, including NIH brain surgeon Maitland Baldwin, working under MKUltra's director Sidney Gottlieb at Ft. Detrick, concocted a diabolical arsenal of assassination weaponry including beamed radio frequency radiation, pathogenic microbes, and dissipating chemicals, all intended to mimic natural deaths. This armory of toxins gave the agency capacity to assassinate uncooperative foreign leaders while avoiding suspicion. Such shenanigans suggest that it is our duty as citizens to remain alert to the times democracy might lose control of rogue intelligence agencies.

Doctor Gates, I Presume!

Media recipients of pharma advertising dollars and Gates Foundation lucre like to characterize Gates as a "public health expert."[166] But six years after Gates summoned Dr. Fauci to his Seattle mansion, two *Los Angeles Times* investigative reporters, Charles Piller and Doug Smith, employed the term "White Man's Burden" to describe the catastrophic impact of Gates's medical meddling in Africa.[167] That title suggests that Gates's efforts to rescue the dark races from disease and famine mask all the familiar impulses for imperial control. The comprehensive study provides eloquent testimony to the lethal effect of Gates's natal arrogance on children.

Piller and Smith detail how Gates's systematic diversion of Africa's international medical spending to his high-tech, high-price, and often untested vaccines is killing babies across Africa. Gates's prioritization of vaccines has dried the stream of foreign assistance that once flowed to basic nutrition and that financed the cheap, functional medical devices that could prevent many deaths. The team at the *Los Angeles Times* documents how, in a single Lesotho hospital, one or two babies die from asphyxiation every day for lack of a $35 oxygen valve: "That life-saving valve is outside the purview of Gates's $400 million annual vaccine giving—almost all of which goes to HIV, polio, TB, and malaria vaccines." Gates's regimen has also deprioritized the off-patent malaria medicines like hydroxychloroquine that could prevent half of all malaria deaths at 12 cents per dose, as well as $4 mosquito nets that can spare a child from contracting malaria. It estimates that three dollars of food and conventional medicines to each new mother could prevent five million child deaths annually.[168]

The *Times* investigation found that Gates's programs, including those of the Global Fund and the GAVI Alliance, have had net negative consequences on public health. In fact, the *Times* found an inverse correlation between dollars spent by Gates's charities and declines in children's health. The nations that get the most Gates money see the worst health outcomes.[169]

By narrowing the focus of international relief aid to fund pharma solutions to a handful of celebrity diseases, Gates has not only reduced public expenditures on basic equipment and lifesaving food and water, he has pulled many of the best health-care workers and researchers away from lifesaving basic care.

The *LA Times* quotes leaders in half a dozen sub-Saharan African nations facing desperate shortages while doctors and nurses chase extravagant salaries that Gates's Global Fund pays to clinicians who provide antiretroviral drug therapy, known as ART, for HIV/AIDS patients: "The resulting staff shortages have abandoned many children of AIDS survivors to more common killers: birth sepsis, diarrhea, and asphyxia."[170]

In Rwanda, the *Los Angeles Times* reports, nurses earning $50 to $100 a month in local clinics work beside Gates-supported nurses earning $175 to $200 a month. "All over the country, people are furious about incentives for ART staff," said Rachel M. Cohen, who is Doctors Without Borders's Lesotho mission chief. Her organization staffs the government health clinics.[171]

The *Los Angeles Times* concludes that Gates's obsession with vaccine-preventable diseases has proportionally reduced assistance streams for nutrition, transportation, hygiene, and economic development, causing negative overall impacts on public health: "Many AIDS patients have so little food that they vomit their free AIDS pills. For lack of bus fare, others cannot get to clinics that offer lifesaving treatment."[172]

The Gates Foundation addresses these catastrophic impacts on broader health concerns by blocking Africans from talking about any problem that is not susceptible to a vaccine solution. According to the report, "Gates-funded vaccination programs have instructed caregivers to ignore—even discourage patients from discussing—ailments that the vaccinations cannot prevent. This is especially harmful in outposts where a visit to a clinic for a shot is the only contact some villagers have with health-care providers for years."[173]

WHO, GAVI, and the Global Fund effectively function as ideological commissars enforcing Gates's vanity priorities. The *Times* reporter found that their oversight has caused "key measures of societal health have stalled at appalling levels or worsened."[174]

Gates's claim that his vaccines have "saved several million lives" is a reflexive trope for which he offers no proof, no validation, and no accountability. Most of the preeminent decision makers and advisers in the Gates organization are former pharmaceutical industry moguls and regulators who not surprisingly share his pharma-centric worldview.

For example, Dr. Tadataka Yamada, an unsavory bully who served as president of the Gates Foundation's Global Health Program from 2005 to 2011, was the former research director for GlaxoSmithKline.[175] He left GSK just a few steps ahead of a US Senate Finance Committee seeking to question him about multiple accusations that he conducted an intimidation campaign to threaten and silence prominent doctors exploring the British drugmaker for knowingly killing some 83,000 Americans with its blockbuster diabetes drug, Avandia. Gates knew of Yamada's sordid conduct because the Senate Committee staffers sent his foundation a letter requesting Yamada submit to questioning. A 2007 article by one of these staffers, Alicia Mundy, describes how Yamada repeatedly lied to his interrogators.[176, 177, 178] Yamada's successor at BMGF, Trevor Mundel, was an executive at both Novartis and Pfizer. The foundation's chief communications officer, Kate James, worked at GSK for almost 10 years. Penny Heaton worked for Merck and Novartis before Gates named her as director of Vaccine Development at BMGF. So it's not surprising that Gates's success metrics rarely measure better health outcomes, but only the number of vaccines administered and the number of pills distributed and consumed.

"Many believe that [the Global Fund's] tight remit is increasingly becoming a straitjacket," complains a 2007 editorial on the Global Fund in the *Lancet Infectious Diseases*.[179] "The failure to support basic care as comprehensively as vaccines and research is a blind spot for the Gates Foundation," said Paul Farmer, recipient of a John D. and Catherine T. MacArthur Foundation fellowship and founder of Partners in Health, which has received Gates Foundation funding for research and training. "It doesn't surprise me that as someone who has made his fortune on developing a novel technology, Bill Gates would look for magic bullets" in vaccines and medicines, Farmer said. "But if we don't have a solid delivery system, this work will be thwarted." He added, "That's something that's going to be hard for the big foundations. They treat tuberculosis. They don't treat poverty."[180]

African public health leaders protest that Gates refuses to finance traditional medical supplies that spell life or death in African clinics.

Lesotho's Health Minister Mphu Ramatlapeng (now Executive Vice President of the Clinton Foundation) told the *Times* that a $7 million annual donation would allow her to raise the pay of every government health professional by two-thirds, sufficient to retain most of them. But this sort of banal need bores Mr. Gates. His Global Fund has poured $59 million into Lesotho to advance his priorities, which are the high-profit vaccines and drugs that enrich his pharma partners. Dr. Fauci and Gates's obsession with AIDS is great for companies like Merck and Glaxo, with which the two men partner, but it's been a lousy deal for Africans.[181]

Like Dr. Fauci, Gates raises expectations, yet takes no responsibility and offers no convincing proof that his schemes have had a beneficial impact on morbidities, public health, or quality of life. There are meager signs of tangible benefits to the poor.

Instead, every effort to measure the health outcomes of Gates's interventions has exposed them as cataclysmic for their beneficiaries. In 2017, the Danish Government commissioned a study of health outcomes among African children who received WHO's flagship DTP vaccine—the world's most popular inoculation. They found that vaccinated girls had ten times the death rate compared to unvaccinated girls.[182]

The investigation by the *Los Angeles Times* found that Botswana, a favored target of Gates's and his corporate amigos' largesse, has seen few tangible benefits from the attention. Botswana is a stable, well-governed democracy with a relatively high living standard and a small population, but one of the world's highest HIV infection rates.

In 2000, the Gates Foundation partnered with Merck to launch a $100 million pilot program in Botswana to showcase how mass AIDS treatment with vaccines, patented antivirals, and prevention could eliminate AIDS in Africa. The pilot's disastrous failure instead became a parable for how Gates's obsession with expensive pharmaceuticals is killing Africans. The project produced no reduction in HIV rates. By 2005, the virus had spread to a quarter of all adults.[183]

All those deadly retrovirals and vaccines from Tony Fauci's little shop of horrors exacted a fearsome toll on Botswana's mothers and infants. The rate of pregnancy-related maternal deaths nearly quadrupled, and child mortality rose dramatically.

Health economist Dean Jamison, formerly editor of the Gates Foundation–funded reference book, *Disease Control Priorities in Developing Countries*, acknowledged that the Gates Foundation's narrow obsession with AIDS drugs may have accelerated death and illness in Botswana by drawing the nation's top medical professionals away from primary care and child health. "They have an opportunity to double or triple their salaries by working on AIDS," Jamison said. "Maybe the health ministry replaces them [when they leave government service], maybe not."[184]

The Gates Foundation has poured billions into sub-Saharan Africa through the Global Fund, to finance vaccines and antivirals for AIDS and TB treatment for 3.9 million people. But one AIDS patient, Moleko, told the *Times*, "The clinics don't have what we need: food."[185]

Majubilee Mathibeli, the nurse at Queen II hospital who gives Moleko her pills, wept in frustration as she told the *Times* reporter that four out of five of her patients ate fewer than three meals a day.

"Most of them," she said, "are dying of hunger." In Lesotho and Rwanda, dozens of patients described hunger "so brutal that nausea prevented them from keeping their anti-AIDS pills down."[186]

Mathibeli said that Gates's Global Fund was out of touch. "They have their computers in nice offices and are comfortable," she said, nervous about speaking bluntly. But "they are not coming down to our level. We've got to tell the truth so something will be done."[187]

Dr. Jennifer Furin, the Lesotho director for Partners in Health, a Boston-based NGO, made a similar complaint. By giving African patients medicine without food, she said, "You're consigning that person to death because they are poor."[188]

Antipathy toward Locally Controlled Health Care Systems

Dr. Francis Omaswa, special adviser for human resources at the WHO, estimates that Gates's spending "could be five times more beneficial"[189] if he directed his philanthropy toward addressing poverty and supporting existing health systems. This is the most common critique among knowledgeable public health experts. According to Global Justice Now, the BMGF's "heavy focus on developing new vaccines detracts from other, more vital health priorities such as building resilient health systems."[190] Unfortunately, the idea of building local institutions to support democracy and public interest is inconsistent with Gates's technology-based approach to public health.

As Dr. David Legge explained to *The Gray Zone*, Gates "has got a mechanistic view of global health, in terms of looking for silver bullets. All of the things he supports are largely framed as silver bullets. . . . That means that major issues that have been identified in the World Health Assembly are not being addressed including, in particular, the social determinants of health and the development of health systems."[191]

University of Toronto public health Professor Anne Emanuelle Birn wrote in 2005 that the Gates Foundation had a "narrowly conceived understanding of health as the product of technical interventions divorced from economic, social, and political contexts."[192]

"The Gates Foundation has long championed private sector involvement in, and private sector profit-making from global health," Birn told *The Gray Zone*. One of GAVI's senior representatives even reported that Bill Gates often told him in private conversations that he is vehemently "against health systems" because it is a "complete waste of money."[193]

Katerini Storeng, researcher at Oslo's Centre for Development and Environment, writes that a GAVI staffer told her that the foundation was a "very loud, vocal voice, saying that we do not believe in the strengthening of health systems." "A former GAVI employee and HSS [health systems strengthening] proponent recounted how he and his colleagues used to 'roll down the HSS posters' when Bill Gates came to visit the GAVI headquarters in Geneva because he is known to 'hate this part' of GAVI's work, Gates's antipathy toward public health systems reflects a pathological—almost bigoted contempt for African institutions and science," Storeng's report also notes. Gates's patterns of funding reflect his bias toward white Western institutions and his hostility toward indigenous community-based African solutions.[194, 195]

Linsey McGoey argues that a commitment to "true equity should entail offering money directly to capable Health teams based in the global South, better resourcing of their universities, their access to scientific research, and their ability to publish more extensively leading journals."[196]

Gates seems impervious to the importance of cultivating local leadership, institutions, and talent. His giving patterns reinforce the colonial architecture that keeps the authority to "call the shots," outside Africa. Investigating the Gates Foundation's global health spending in 2009, British public health policy expert David McCoy found that of 659 grants BMGF awarded to nongovernmental or for-profit organizations, 560 went to organizations in high-income countries, mainly in the US. Only thirty-seven grants went to NGOs based in low- or middle-income countries. Similarly, of the 231 grants BMGF awarded to universities, only twelve went to universities based in developing regions. Linsey McGoey points out that the very limited direct funding to these countries automatically excludes scientists and program managers who best understand the problems from contributing creative solutions.[197]

In his book *The White Man's Burden*, economist William Easterly, who codirects the Development Research Institute at New York University, asks, "Who chose the human right of universal treatment of AIDS over other human rights?"[198] The answer to that question, of course, is Bill Gates.

Bill Gates's continent-wide experiment on the African population is a long tragic joke. *The Times* reporters deliver its devastating punchline: "2006 data, the most

313

recent available, show a paradoxical relationship between GAVI funding in Africa and child mortality. Overall, child mortality improved more often in nations that received smaller than average GAVI grants per capita. In seven nations that received greater-than-average funding, child mortality rates worsened."[199]

Neutralizing the Press

Piller and Smith's *Los Angeles Times* exposé on Gates's calamitous African adventure is an artifact of an expired era. Investigative journalism of this probative quality is a quaint relic of a time when editors and producers still permitted their reporters and correspondents to express skepticism toward Gates. Even before the open censorship of the COVID epoch, US media reports about Gates's charities operated in the narrow range between obsequious fawning and adulation. This is no accident. By 2006, the tsunami of advertising revenues from pharmaceutical firms—about $4.8 billion annually—had already drowned out most of the voices of vaccine dissent in mainstream media.[200] By 2020, those expenditures grew to $9.53 billion.[201]

After the devastating *Los Angeles Times* piece, Gates moved aggressively to neutralize the once-independent press with compromising grants that struggling news organizations couldn't refuse. An August 2020 expose by Tim Schwab in the *Columbia Journalism Review* showed how Gates dispensed at least $250 million in grants to media outlets including NPR, Public Television (PBS), *The Guardian, The Independent,* BBC, Al Jazeera, *Propublica, The Daily Telegraph, The Atlantic, The Texas Tribune,* Gannett, *Washington Monthly, Le Monde, The Financial Times, The National Journal,* Univision, Medium, and the *New York Times* to dampen journalistic appetites for—well—journalism.[202,203] In fact, the Bill and Melinda Gates Foundation finances *The Guardian's* entire "Global Development" section. That shrewd investment presumably earned the couple this February 14, 2017 *Guardian* headline: "How Bill and Melinda Gates helped save 122m lives—and what they want to solve next." *The Guardian* calls Gates and his partner Warren Buffett "Superman and Batman."[204]

The foundation has also invested millions in journalism training and in researching effective ways of crafting media narratives to support Gates's global ambitions. Gates, for example, gave grants totaling nearly $1.5 million from 2015 through 2019[205] to the Center for Investigative Reporting—apparently to discourage investigative reporting. According to the *Seattle Times,* "Experts coached in Gates-funded programs write columns that appear in media outlets from the *New York Times* to the *Huffington Post,* while digital portals blur the line between journalism and spin."[206]

The Gates Foundation frequently hosts "strategic media partners" meetings at its headquarters in Seattle. Representatives from the *New York Times, The Guardian,* NBC, NPR, and the *Seattle Times* all attended a 2013 convocation. The aim of the event, wrote Tom Paulson, a Seattle-based reporter, was to "improve the narrative" of media coverage of global aid and development, highlighting good news stories rather than tales of waste or corruption.[207] That same year, the BMGF gave marketing colossus Ogilvy & Mather, a global public relations firm, a $100,000 grant for a project titled "Aid is Working: Tell the World."[208]

Subsequent articles in *The Nation* reported that Gates had invested in a retinue of companies positioned to mint windfall profits from the COVID crisis and documented the reluctance of players in the philanthropic donor community and key charities to criticize their self-serving arrangements. Fearful of his prowess and reputation for vendetta, leading charities keep their mouths shut about Gates's recipe for leavening his altruism with profiteering. They call this omertà "the Bill Chill."[209]

Gates has also made large strategic investments in Poynter and the International Network of Fact Checking Organizations, which dutifully "debunks" virtually every public statement that seems critical of Gates, whether accurate or not.[210]

In 2008, the communications chief for *PBS NewsHour*, Rob Flynn, explained that "there are not a heck of a lot of things you could touch in global health these days that would not have some kind of Gates tentacle." This was around the time when the foundation gave *NewsHour* $3.5 million to establish a dedicated production unit to report on important global health issues.[211]

That kind of moolah purchased a lot of goodwill from the Fourth Estate. Jeff Bezos's *Washington Post* called Gates the "champion of science-backed solutions."[212] The *New York Times* gushes that he is "the most interesting man in the world."[213] *Time Magazine* dubbed him "Master of the Universe."[214] *Forbes* calls Gates "savior of the world" who "set the standard for a billionaire good citizen."[215] Looking on admiringly, editors of fashion magazine *Vogue* wondered, "Why Isn't Bill Gates Running the Coronavirus Task Force?"[216]

Ignoring the fact that Gates never graduated from college, much less medical school, mainstream media outlets unanimously parrot BBC's assessment that Gates is a "public health expert" and ridicule those who question whether the whole world should take his self-serving advice on lockdowns, masks, and vaccines. In just the month of April 2020, while the virus and lockdowns were severely impacting the United States, Gates and Fauci did tag-team appearances on CNN, CNBC, Fox, PBS, BBC, CBS, MSNBC, the *Daily Show*, and the *Ellen DeGeneres Show*, reinforcing their self-serving messages about lockdowns and vaccines. None of those reporters mentioned the fact that the quarantines that Gates was cheerleading on their networks have increased Gates's wealth by $22 billion over twelve months.

And Gates's efforts to promote his contrary narrative claims only exacerbate their limitations. Gates's emphasis on conditional lending, corporate partnerships, top-down control, high-tech cookie-cutter solutions, and patent privileges tends to favor wealthy nations and multinational corporations: "These are just a few of the ways in which current development policies are failing the global south."[217]

"If aid flows are working well," asks McGoey, "why do they need a masterful PR campaign to convey that message effectively? Many observers on the left and right suggest that the problem isn't a marketing failure; it's a failure with the underlying 'product.' Aid, they argue, is not working."[218]

Endnotes

1   Anthony Fauci, AIDS (Fauci), 55:24 NIH (1984), YouTube, youtube.com/watch?v=pzK3dg59TuY
2   Robert Bazell, "Finding the Way Again After Failed AIDS Vaccine," NBC News (Nov. 20, 2007, 3:22 AM), nbcnews.com/id/wbna21889662

3   HIV VACCINE AWARENESS DAY; Congressional Record Vol. 156, No. 75 (Senate, May 18, 2010), childrenshealthdefense.org/wp-content/uploads/03-26-20-Congressional-Record--Congress.gov--Library-of-Congress.pdf

4   Donald G. McNeil Jr., "AIDS Vaccine Trial Shows Only Slight Protection," *New York Times* (Oct. 20, 2009), nytimes.com/2009/10/21/health/research/21vaccine.html

5   John Crewdson, "3 Dead in AIDS Vaccine Tests," *Chicago Tribune* (Apr. 14, 1991), chicagotribune.com/news/ct-xpm-1991-04-14-9102030245-story.html

6   Robert C. Gallo et al., "AIDS Vaccine Therapy: Phase I Trial, *The Lancet* (1990), sci-hub.se/https://doi.org/10.1016/0140-6736(90)91699-B

7   Crewdson.

8   Ibid.

9   Ibid.

10   Ibid.

11   Methods of Inducing Immune Responses to the AIDS Virus, U.S. patent application (Jan. 7, 1991). patentscope.wipo.int/search/en/detail.jsf?docId=WO1992000098

12   Craig W. Hendrix, Maj, USAF, MC, and R. Neal Boswell, Col, USAF, MC, GP160 phase I Immunotherapy Data Presentation, Department of the Air Force (Oct. 21, 1992), CHD, childrenshealthdefense.org/uncategorized/hendricks-burke-letter-re-misleading-or-possibly-deceptive-presentations-by-dr-redfield-in-the-gp160-phase-i-data/

13   Robert R. Redfield, MD, Deborah L. Birx, MD, et al., "A Phase I Evaluation of the Safety and Immunogenicity of Vaccination with Recombinant gp160 in Patients with Early Human Immunodeficiency Virus Infection," *N Engl J Med* 1991; 324:1677–1684 (Jun. 13, 1991), nejm.org/doi/full/10.1056/NEJM199106133242401#article_references

14   Peter Lurie, MD, MPH, and Sidney M. Wolfe, MD, Letter to Rep. Henry Waxman, Public Citizen's Health Research Group (Jun. 7, 1994), CHD, childrenshealthdefense.org/uncategorized/public-citizen-letter-to-rep-henry-waxman/

15   SGS (Mrs. Whitaker), Minutes of the Institutional Review Committee (IRC) Subcommittee on Potential Scientific Misconduct in GP160 Phase I Immunotherapy Study, Department of the Air Force Wilford Hall USAF Medical Center (ATC) Lackland AFB, TX (Oct. 23, 1992), CHD, childrenshealthdefense.org/uncategorized/minutes-of-the-irc-subcommittee-on-potential-scientific-misconduct-in-gp160-phase-1-immunotherapy-study/

16   Hendrix and Boswell

17   Ibid.

18   Peter Lurie and Sidney M. Wolfe, *Public Citizen Letter to U.S. Representative Henry Waxman* (Jun. 7, 1994), https://khn.org/wp-content/uploads/sites/2/2018/03/940607plswtowaxman.pdf

19   Laurie Garrett, *Meet Trump's New, Homophobic Public Health Quack*, FOREIGN POLICY (Mar. 23, 2018), https://foreignpolicy.com/2018/03/23/meet-trumps-new-homophobic-public-health-quack/

20   Lurie and Wolfe

21   Institute of Human Virology, Annual Report, 2017, ihv.org/media/SOM/Microsites/IHV/documents/Annual-Report/2017-IHV-annual-report_singlepg-final-standard_Res.pdf

22   National Institutes of Health, Estimates of Funding for Various Research, Condition, and Disease Categories (RCDC) (Jun 25, 2021), report.nih.gov/funding/categorical-spending#/

23   Bill & Melinda Gates Foundation, New Plan To Speed AIDS Vaccine Development Released (Jun., 1998), gatesfoundation.org/ideas/media-center/press-releases/1998/06/international-aids-vaccine-initiative

24   Ian Birrell, "The Fattest Charity Fat Cat of Them All: Foreign Aid Boss Has Made MILLIONS out of the £1.5billion Handed to His Charity by British Taxpayers," *Daily Mail* (Dec. 31, 2016), dailymail.co.uk/news/article-4078904/The-fattest-charity-fat-cat-Foreign-aid-boss-MILLIONS-1-5billion-handed-charity-British-taxpayers.html

25   Michael Barbaro, "Can Bill Gates Vaccinate the World? How the Microsoft founder is changing the way the world is vaccinated and, potentially the course of the pandemic," *New York Times* podcast (Mar. 3, 2021), nytimes.com/2021/03/03/podcasts/the-daily/coroanvirus-vaccine-bill-gates-covax.html?showTranscript=1

26   Paul Allen, *Idea Man: A Memoir by the Cofounder of Microsoft* (Penguin Books, 2011), 165, archive.org/details/ideaman/page/n177/mode/2up?q=true+character

27   Amy Lamare, "How Much Did Paul Allen's Net Worth Grown From 1990 To His Death In 2018?," Celebrity Net Worth (Mar. 20, 2020), celebritynetworth.com/articles/billionaire-news/how-much-has-paul-allens-net-worth-grown-since-1990/

28   "The Breakup of Microsoft," *The Guardian* (Jun 8, 2000), theguardian.com/world/2000/jun/08/qanda

29   "U.S. Fines Bill Gates $800,000," CNN Money (May 3, 2004), money.cnn.com/2004/05/03/technology/gates_penalty/

30   Reuters, "EU Slaps Record Fine on Microsoft" (Feb. 27, 2008), cnbc.com/id/23365419

31   "MS Judge Rips Gates Again," *WIRED* (Jan. 8, 2001), https://www.wired.com/2001/01/ms-judge-rips-gates-again/

32   LA Times Archive, "Gates to Testify for the First Time in 4-Year Microsoft Antitrust Case," *Los Angeles Times* (Apr. 20, 2002), https://www.latimes.com/archives/la-xpm-2002-apr-20-fi-gates20-story.html

33   Joe John, "Best of Bill Gates Being Deposed by David Boies, United States v. Microsoft Corp., 1998–2001" (Nov. 2, 2018), youtube.com/watch?v=gRelVFm7iJE&t=495s

34   "Microsoft sued for racial discrimination," BBC News (Jan. 3, 2001), news.bbc.co.uk/2/hi/business/1098683.stm

35   "New Plaintiffs Join Microsoft Discrimination Suit," CNET (Jan 2, 2002), cnet.com/news/new-plaintiffs-join-microsoft-discrimination-suit/

36   The Rebecca Project for Justice, Depo-Provera Deadly Reproductive Violence Against Women, p. 20 (Jun. 25, 2013), rebeccaprojectjustice.org/wp-content/uploads/2019/12/depo-provera-deadly-reproductive-violence-rebecca-project-for-human-rights-2013-3.pdf

37   Bill & Melinda Gates Foundation, "Bill and Melinda Gates Announce a $100 Million Gift to Establish the Bill and Melinda Gates Children's Vaccine Program" (Dec., 1998), gatesfoundation.org/ideas/media-center/press-releases/1998/12/bill-and-melinda-gates-childrens-vaccine-program

38   Linsey McGoey, *No Such Thing as a Free Gift: The Gates Foundation and the Price of Philanthropy* (Verso, 2016), 55

39   Steven R. Weisman, "The Rockefellers," *New York Times* (Mar. 28, 1976), nytimes.com/1976/03/28/archives/the-rockefellers.html

40   Erin Blakemore, "Tycoon John D. Rockefeller Couldn't Hide His Father's Con Man Past," History (June 14, 2019), history.com/news/john-d-rockefeller-father-con-man-origins

41   Thomas P. Duffy, "The Flexner Report—100 Years Later," *Yale Journal of Biological Medicine* (Sep., 2011), ncbi.nlm.nih.gov/pmc/articles/PMC3178858/

42   Torsten Engelbrecht, Claus Köhnlein, et al., *Virus Mania: How the Medical Industry Continually Invents Epidemics, Making Billions at Our Expense* (Books on Demand 3rd ed, 2021), 27

43   Ibid., 66

44   Ibid., 35

45   Ibid., 28, 35

46   *Meet the Press with Chuck Todd* (Jul. 4, 2021), nbcnews.com/meet-the-press/meet-press-july-4-2021-n1273065

47   Bernard Guyer et al., Annual Summary of Vital Statistics: Trends in the Health of Americans During the 20th Century, *Pediatrics* (December 2000), DOI: doi.org/10.1542/peds.106.6.1307, pediatrics.aappublications.org/content/106/6/1307

48   J. B. McKinlay and S.M. McKinlay, "The Questionable Contribution of Medical Measures to the Decline of Mortality in the United States in the Twentieth Century" (Milbank Mem Fund Q Health Soc., 1977), columbia.edu/itc/hs/pubhealth/rosner/g8965/client_edit/readings/week_2/mckinlay.pdf

49   J. B. Handley, "The Impact of Vaccines on Mortality Decline Since 1900—According to Published Science," CHD (Mar. 12, 2019), childrenshealthdefense.org/news/the-impact-of-vaccines-on-mortality-decline-since-1900-according-to-published-science/

50   Mary M. Eichhorn Adams, "'The Journal of Infectious Diseases': Yesteryear and Today," *The Journal of Infectious Diseases* 138, no. 6 (1978): 709–11, accessed September 2, 2021. jstor.org/stable/30109031

51   Engelbrecht and Köhnlein et al., 70

52   Meridian Health Clinic, "How Rockefeller Created the Business of Western Medicine" (Dec. 27, 2019), meridianhealthclinic.com/how-rockefeller-created-the-business-of-western-medicine/

53   Gara LaMarche, "Is Philanthropy Bad for Democracy?" *The Atlantic* (Oct. 30, 2014), theatlantic.com/politics/archive/2014/10/is-philanthropy-good-for-democracy/381996/

54   Justin Fox, "Zuckerberg Charity Dust-Up Is Age-Old American Theme," *East Bay Times* (Dec. 7, 2015), eastbaytimes.com/2015/12/07/zuckerberg-charity-dust-up-is-age-old-american-theme/

55   E. Richard Brown, PhD, "Public Health in Imperialism:Early Rockefeller Programs at Home and Abroad," *AJPH* 66: 9 (September, 1976), ajph.aphapublications.org/doi/pdf/10.2105/AJPH.66.9.897

56   Ibid.

57   Siang Yong Tan, MD, and Jill Furubayashi, "Hideyo Noguchi (1876–1928): Distinguished Bacteriologist," *Singapore Medical Journal* (Oct. 2014), ncbi.nlm.nih.gov/pmc/articles/PMC4293967/

58   Susan Eyrich Lederer, "Hideyo Noguchi's Luetin Experiment and the Antivivisectionists," *Isis* (Mar., 1985), jstor.org/stable/232791?read-now=1&seq=7#page_scan_tab_contents

59   Brown.

60   McGoey, 150

61   Ibid.

62   Anne-Emanuelle Birn and Judith Richter, "U.S. Philanthrocapitalism and the Global Health Agenda" (2017), peah.it/2017/05/4019/

63   Howard Waitzkin, *Health Care under the Knife: Moving beyond Capitalism for Our Health* (Monthly Review Press, 2018), 159

64   Bill & Melinda Gates Foundation, Foundation FAQ (2020), gatesfoundation.org/about/foundation-faq

65   Anupreeta Das, "Bill Gates Investments Changes Its Name . . . Slightly," *Wall Street Journal* (Dec. 4, 2014), wsj.com/articles/BL-MBB-30516

66   Eva Mathews and Jonathan Stempel, "Warren Buffett Resigns from Gates Foundation, Has Donated

Half His Fortune," Reuters (Jun. 24, 2021), reuters.com/business/buffett-resigns-trustee-gates-foundation-2021-06-23/

67   *Forbes*, Real Time Billionaires (Aug. 5, 2021), forbes.com/real-time-billionaires/#6e8906283d78

68   Tom Watson, Huffington Post, "Exploding Philanthropy: What the Clinton Party Really Meant" (Dec. 6, 2017), huffpost.com/entry/exploding-philanthropy-wh_b_30381

69   Tim Schwab, "Bill Gates's Charity Paradox," *The Nation* (Mar. 17, 2020), thenation.com/article/society/bill-gates-foundation-philanthropy/

70   Pavan Kulkarni, "Malaria Vaccine Trials in Africa: Dark Saga of Outsourced Clinical Trials Continues," NEWSCLICK (Mar. 17, 2018), newsclick.in/malaria-vaccine-trials-africa-dark-saga-outsourced-clinical-trials-continues

71   Medical Xpress, "Bill Gates hopeful of AIDS vaccine in 10 years," (Jun 26, 2015). https://medicalxpress.com/news/2015-06-bill-gates-aids-vaccine-years.html

72   Charles Piller and Doug Smith, "Unintended Victims of Gates Foundation Generosity," *Los Angeles Times* (Dec. 16, 2007), latimes.com/nation/la-na-gates16dec16-story.html

73   The White House, Office of the Press Secretary, Fact Sheet: The President's Emergency Plan for AIDS Relief (Jan. 29, 2003), georgewbush-whitehouse.archives.gov/news/releases/2003/01/20030129-1.html

74   Commencement Address at Morgan State University in Baltimore, MD, The American Presidency Project (May 18, 1997), presidency.ucsb.edu/documents/commencement-address-morgan-state-university-baltimore-maryland

75   Michael Hamilton, "Taxpayers Funded Clinton Foundation's Bad AIDS Drugs," Heartland (Oct. 7, 2016), heartland.org/news-opinion/news/taxpayers-funded-clinton-foundations-bad-aids-drugs

76   McGoey, 154

77   Kristen R.W. Matthews and Vivian Ho, "The Grand Impact of the Gates Foundation: Sixty Billion Dollars and One Famous Person Can Affect the Spending and Research Focus of Public Agencies," *EMBO Rep* (2008) 9:409–412, embopress.org/doi/full/10.1038/embor.2008.52

78   McGoey, 245

79   Ibid., 243–244

80   Kim Hjelmgaard, "'A Reckoning Is Near': America Has a Vast Overseas Military Empire. Does It Still Need It?" USA Today, Feb. 25, 202, https://www.usatoday.com/in-depth/news/world/2021/02/25/us-military-budget-what-can-global-bases-do-vs-covid-cyber-attacks/6419013002/

81   Michael Specter, "The Vaccine," *New Yorker* (Jan. 26, 2003), michaelspecter.com/2003/02/the-vaccine/

82   Ibid.

83   Ibid.

84   Foundation Funds Major New Collaboration to Accelerate HIV Vaccine Development, BMGF Press Release (2006), gatesfoundation.org/ideas/media-center/press-releases/2006/07/hivaidspr060719

85   Lawrence K. Altman, "Trial for Vaccine Against HIV is Canceled," *New York Times* (Jul. 18, 2008), nytimes.com/2008/07/18/health/18vaccine.html

86   Ibid.

87   Ibid.

88   K. A. Smith, "The HIV Vaccine Saga," *Med Immunol* (2003) 2(1):1, doi:10.1186/1476-9433-2-1 ncbi.nlm.nih.gov/pmc/articles/PMC151598/

89   Andrea K. McDaniels, "HIV Vaccine to Be Tested on People," *Baltimore Sun* (Oct. 8, 2015), baltimoresun.com/health/bs-hs-aids-vaccine-20151007-story.html

90   U.S. Department of Health and Human Services, NIH Research Portfolio Online Reporting Tools, NIH Awards by Location & Organization, report.nih.gov/award/index.cfm?ot=&fy=2020&state=&ic=&fm=&orgid=10006808&distr=&rfa=&om=y&pid=&view=state

91   PitchBook, Profectus BioSciences Overview, pitchbook.com/profiles/company/53380-54#overview

92   Ibid.

93   McDaniels

94   Agence France-Presse, "Bill Gates Expects There Could Be an AIDS vaccine in 5 to 10 years," *Business Insider* (Jun 26, 2015), businessinsider.com/afp-bill-gates-hopeful-of-aids-vaccine-in-10-years-2015-6

95   Agence France-Presse

96   Julie Steenhuysen, "Trial of Promising HIV Vaccine Halted after Failing to Show Benefit," Reuters (Feb. 3, 2020), reuters.com/article/us-health-hiv-vaccine/trial-of-promising-hiv-vaccine-halted-after-failing-to-show-benefit-idUSKBN1ZX2QO

97   National Institutes of Health, Budget: Research for the People (Jun 29, 2020), nih.gov/about-nih/what-we-do/budget

98   Kristian G. Andersen et al., "The Proximal Origin of SARS-CoV-2," *Nat Med* 26, 450–452 (Mar. 17, 2020), doi.org/10.1038/s41591-020-0820-9

99   Charles Calisher et al., "Statement in Support of the Scientists, Public Health Professionals, and Medical Professionals of China Combatting COVID-19," *The Lancet*, vol. 395 iss.10026 E42–43 (Feb. 19, 2020), doi.org/10.1016/S0140-6736(20)30418-9

100   James Gorman and Carl Zimmer, "Scientist Opens Up About His Early Email to Fauci on Virus Origins," *New York Times* (Jun. 14, 2021), nytimes.com/2021/06/14/science/covid-lab-leak-fauci-kristian-andersen.html

101   LEOPOLD NIH FOIA, Anthony Fauci Emails, 3187, documentcloud.org/documents/20793561-leopold-nih-foia-anthony-fauci-emails

102   Ibid., 3197–3198

103   Andersen et al.

104   John Haltiwanger, "Dr. Fauci throws cold water on conspiracy theory that coronavirus was created in a Chinese lab," Insider (Apr. 18, 2020), https://www.businessinsider.com/fauci-throws-cold-water-conspiracy-theory-coronavirus-escaped-chinese-lab-2020-4

105   James S. Brady Press Briefing Room, Remarks by President Trump, Vice President Pence, and Members of the Coronavirus Task Force in Press Briefing (Apr. 17, 2020), https://web.archive.org/web/20210117030731/https://www.whitehouse.gov/briefings-statements/remarks-president-trump-vice-president-pence-members-coronavirus-task-force-press-briefing-april-17-2020/

106   NIH Awards by Location and Organization, Scripps Research Institute, NIAID Grants, 2020, https://report.nih.gov/award/index.cfm?ot=&fy=2020&state=&ic=NIAID&fm=&orgid=7375802&distr=&rfa=&om=n&pid=#tab2

107   Ibid.

108   NIH Awards by Location and Organization, Columbia University Health Sciences, NIAID Grants, 2020, https://report.nih.gov/award/index.cfm?ot=&fy=2020&state=&ic=NIAID&fm=&orgid=1833205&distr=&rfa=&om=n&pid=&view=state#tab2

109   NIH Awards by Location and Organization, University of Sydney, NIAID Grants, 2020, https://report.nih.gov/award/index.cfm?ot=&fy=2020&state=&ic=NIAID&fm=&orgid=8178701&distr=&rfa=&om=y&pid=&view=state

110   NIH Awards by Location and Organization, Tulane University of LA, NIAID Grants, 2020, https://report.nih.gov/award/index.cfm?ot=&fy=2020&state=&ic=NIAID&fm=&orgid=8424601&distr=&rfa=&om=y&pid=&view=state#tab2

111   Jay Greene, "The Billionaire Who Cried Pandemic," Washington Post (May 2, 2020), washingtonpost.com/technology/2020/05/02/bill-gates-coronavirus-science/

112   World Economic Forum, Responding with Strength to Pandemic Risk (May 22, 2020), weforum.org/our-impact/the-world-has-entered-a-new-era-of-pandemic-risk-we-re-responding-with-strength/

113   Bill Gates, "What You Need to Know about the COVID-19 Vaccine," Gates Notes (Apr. 20, 2020), gatesnotes.com/Health/What-you-need-to-know-about-the-COVID-19-vaccine?WT.mc_id=20200430165003_COVID-19-vaccine_BG-TW&WT.tsrc=BGTW&linkId=87665522

114   Kambiz Shekdar, PhD, "Dr. Fauci Moves to Cure AIDS," West View News (Feb. 9, 2021), westviewnews.org/2021/02/09/dr-fauci-moves-to-cure-aids/web-admin/

115   Ibid.

116   "CDC. Pneumocystis pneumonia—Los Angeles," MMWR 981;30:250--2, https://www.cdc.gov/mmwr/preview/mmwrhtml/june_5.htm

117   Interview with Robert F. Kennedy Jr.

118   Allana Aktar, "Pfizer Could Sell Nearly $100 billion Worth of COVID-19 vaccines in the Next Five Years, Morgan Stanley Estimates," Insider (May 10, 2021), businessinsider.com/pfizer-could-sell-96-billion-dollars-covid-vaccines-morgan-stanley-2021-5?op=1

119   Gina Kolata, "Fauci Wants to Make Vaccines for the Next Pandemic Before It Hits," Yahoo! (July 26, 2021), finance.yahoo.com/news/fauci-wants-vaccines-next-pandemic-120241706.html

120   John Lauerman, "J&J Halts Africa HIV Vaccine Trial as Prevention Insufficient," Bloomberg (Aug. 31, 2021), https://www.bloomberg.com/news/articles/2021-08-31/j-j-halts-africa-hiv-vaccine-trial-as-prevention-insufficient

121   Jeremy Loffredo and Michele Greenstein, "Why the Bill Gates Global Health Empire Promises More Empire and Less Public Health," The Gray Zone (July 8, 2020), thegrayzone.com/2020/07/08/bill-gates-global-health-policy/

122   William Easterly, The White Man's Burden (Penguin, 2008), 8

123   McGoey, 176

124   WHO Contributors, 2018-2019, (Updated until Q4 2019), http://open.who.int/2018-19/contributors/contributor

125   "Address by Mr Bill Gates to the Sixty-fourth World Health Assembly," apps.who.int/gb/ebwha/pdf_files/WHA64/A64_DIV6-en.pdf

126   Loffredo and Greenstein

127   McGoey, 8

128   Kevin Smith, "Why should Bill Gates get to set the agenda for international development?," Global Justice Now (Jan. 20, 2016), globaljustice.org.uk/blog/2016/01/why-should-bill-gates-get-set-agenda-international-development/

129   Prepared Remarks by Bill Gates, Co-chair and Trustee, World Health Assembly (May 17, 2011), gatesfoundation.org/ideas/speeches/2011/05/world-health-assembly

130   Bill and Melinda Gates Foundation, GAVI (Updated July 29, 2020), gavi.org/operating-model/gavis-partnership-model/bill-melinda-gates-foundation

131   Bill and Melinda Gates Foundation, GAVI (2021), gavi.org/investing-gavi/funding/donor-profiles/bill-melinda-gates-foundation

132   Funding Current Period 2021–2025, GAVI (2021), gavi.org/investing-gavi/funding/current-period-2021-2025

133   United States endorses Gavi with recommendation of US$ 1.16 billion, four-year commitment, GAVI (2021), gavi.org/news/media-room/united-states-endorses-gavi-recommendation-us-116-billion-four-year-commitment

134   Kris Nelson, "US Officially Withdrawals Funding for WHO, But Funds Billions to Gates' GAVI," Evolve Consciousness (Sept. 13, 2020), evolveconsciousness.org/us-officially-withdraws-funding-for-who-but-funds-billions-to-gates-gavi/

135   John Stone, "British Prime Minister Channels Churchill as He Surrenders to Gates and the Vaccine Cartel," Age of Autism, (Jun. 5, 2020, 6:02 AM), ageofautism.com/2020/06/british-prime-minister-channels-churchill-as-he-surrenders-to-gates-and-the-vaccine-cartel.html

136   "Ethiopia accuses WHO chief Tedros of supporting Tigray forces," *Daily Sabah* (Nov. 19, 2020), dailysabah.com/world/africa/ethiopia-accuses-who-chief-tedros-of-supporting-tigray-forces

137   "Gavi welcomes election of new WHO chief," GAVI (May 23, 2017), gavi.org/gavi-welcomes-election-of-new-who-chief

138   "Global Fund welcomes Dr. Tedros Adhanom Ghebreyesus as Director-General of WHO," The Global Fund (May 23, 2017), theglobalfund.org/en/news/2017-05-23-global-fund-welcomes-dr-tedros-adhanom-ghebreyesus-as-director-general-of-who/

139   "Fauci: forging closer ties with GAVI," GAVI (May 22, 2012), gavi.org/news/media-room/fauci-forging-closer-ties-gavi

140   Interview with Robert F. Kennedy Jr.

141   Loffredo and Greenstein

142   Sky News, "COVID-19: Bill Gates hopeful world 'completely back to normal' by end of 2022," (Apr 25, 2021), youtube.com/watch?v=0-Ic4EN0io4

143   Alexander Zaitchik, "How Bill Gates Impeded Global Access to Covid Vaccines," *New Republic* (Apr. 12, 2021), newrepublic.com/article/162000/bill-gates-impeded-global-access-covid-vaccines

144   " Uniting behind a people's vaccine against COVID-19," UNAIDS (May 14, 2020), unaids.org/en/resources/presscentre/featurestories/2020/may/20200514_covid19-vaccine-open-letter

145   Emma Bowman and Ashish Valentine, "President Biden threw his weight behind the movement calling for a temporary suspension of patent protections for COVID-19 vaccines," NPR (May 5, 2021, 6:43 PM), npr.org/sections/coronavirus-live-updates/2021/05/05/993998745/biden-backs-waiving-international-patent-protections-for-covid-19-vaccines

146   Alexander Zaitchik, "How Bill Gates Impeded Global Access to Covid Vaccines," *New Republic* (Apr. 12, 2021), newrepublic.com/article/162000/bill-gates-impeded-global-access-covid-vaccines

147   Zaichik

148   WHO, WHO Director-General's opening remarks at the media briefing on COVID-19—5 February 2021, https://www.who.int/director-general/speeches/detail/who-director-general-s-opening-remarks-at-the-media-briefing-on-covid-19-5-february-2021

149   Catherine Cheney, "Gates Foundation reverses course on COVID-19 vaccine patents," Devex (May 7, 2021), https://www.devex.com/news/gates-foundation-reverses-course-on-covid-19-vaccine-patents-99810

150   Ibid.

151   Zaitchik

152   Ibid.

153   Ibid.

154   Ibid.

155   Ibid.

156   Sony Salzman, NIH, Gates Foundation Hatch Plan to Develop Affordable Gene Therapy for HIV and Sickle Cell Disease, The Body Pro (Oct. 25, 2019), thebodypro.com/article/affordable-gene-therapy-for-hiv-sickle-cell-disease

157   Loffredo and Greenstein

158   Melinda Gates: This is what keeps me up at night, CNN (Apr. 10, 2020), 4:50–5:27, youtube.com/watch?v=qSVse07y2O4

159   Jamie Ducharme, "Melinda Gates Lays Out Her Biggest Concern for the Next Phase of the COVID-19 Pandemic," *TIME* (Jun. 4, 2020, 11:30 AM), time.com/5847483/melinda-gates-covid-19/

160   Zoe Christen Jones, "Fauci urges Black community to be confident in COVID-19 vaccine: 'The time is now to put skepticism aside,'" CBS News (Dec. 8, 2020), cbsnews.com/news/fauci-black-community-covid-19-vaccine/

161   Robert F. Kennedy, Jr., National Media Pushes Vaccine Misinformation—Coroner's Office Never Saw Hank Aaron's Body, *The Defender* (Feb. 12, 2021), childrenshealthdefense.org/defender/national-media-vaccine-misinformation-hank-aaron/

162   MedAlerts, "Cases Where Vaccine Targets COVID-19 and Patient Died," MedAlerts (current data),medalerts.org/vaersdb/findfield.php?TABLE=ON&GROUP1=AGE&EVENTS=ON&VAX=COVID19&VAXTYPES=COVID-19&DIED=Yes

163   Selected Adverse Events Reported after COVID-19 Vaccination, CDC (Aug. 2, 2021), https://www.cdc.gov/coronavirus/2019-ncov/vaccines/safety/adverse-events.html

164   Jean-Benoit Felisse et al., "Why have so many African Leaders died of COVID-19?," *BMJ Global Health* 2021;6:e005587, 10.1136/ bmjgh-2021-005587, https://gh.bmj.com/content/6/5/e005587

165   Daniel Boffey, "Reappearance of statue's missing hand reignites colonial row," *The Guardian* (Feb. 22, 2019), theguardian.com/world/2019/feb/22/statue-missing-hand-colonial-belgium-leopold-congo

166   BBC, Coronavirus: Bill Gates Interview @BBCBreakfast (Apr. 12, 2020), youtube.com/watch?v=ie6lRKAdvuY

167   Piller and Smith

168   Ibid.

169   Ibid.

170   Ibid.

171   Ibid.

172   Ibid.

173   Ibid.

174   Ibid.

175   Charlotte Schubert, "Tadataka Yamada, 1945–2021: Pioneer in drug development led global health at Gates Foundation," Geekwire (Apr. 5, 2021), geekwire.com/2021/tadataka-yamada-1945-2021-pioneer-drug-development-led-global-health-gates-foundation/

176   Paul D. Thacker, "I Never Trusted Bill Gates, Nor Should You," The Disinformation Chronicle (May 11, 2021), disinformationchronicle.substack.com/p/i-never-trusted-bill-gates-nor-should

177   Alician Mundy and Kristi Heim, "Senate committee turns attention to Gates Foundation official," *Seattle Times* (Nov. 20, 2007, 8:10 AM), seattletimes.com/seattle-news/senate-committee-turns-attention-to-gates-foundation-official/

178   Thacker

179   Piller and Smith

180   Ibid.

181   Piller and Smith

182   Soren Mogensen, Peter Aaby, et al., "The Introduction of Diphtheria-Tetanus-Pertussis and Oral Polio Vaccine Among Young Infants in an Urban African Community: A Natural Experiment," *Ebiomedicine*, 17, 192–198, (2017), doi.org/10.1016/j.ebiom.2017.01.041

183   Piller and Smith

184   Ibid.

185   Ibid.

186   Ibid.

187   Ibid.

188   Ibid.

189   Ibid.

190   Loffredo and Greenstein

191   Ibid.

192   Ibid.

193   Ibid.

194   Ibid.

195   Katerini Storeng, "The GAVI Alliance and the 'Gates approach' to health system strengthening," *Global Public Health* (Sep, 14, 2014), ncbi.nlm.nih.gov/pmc/articles/PMC4166931/

196   McGoey, 245

197   McGoey, 176

198   Easterly, 251

199   Piller and Smith

200   Dayna M. Porter, "Direct-to-Consumer Marketing: Impacts and Policy Implications," Grand Valley State University (2011), scholarworks.gvsu.edu/cgi/viewcontent.cgi?article=1013&context=spnharevi ew

201   Blake Droesch, "US Healthcare and Pharma Digital Aid Spending 2020," Emarketer, emarketer.com/content/us-healthcare-pharma-digital-ad-spending-2020

202   Tim Schwab, "Journalism's Gates Keepers," *CJR* (Aug. 21, 2020), cjr.org/criticism/gates-foundation-journalism-funding.php

203   Tim Schwab, "Bill Gates's Charity Paradox," *The Nation* (Mar. 17, 2020), thenation.com/article/society/bill-gates-foundation-philanthropy/

204   Sarah Boesley, "How Bill and Melinda Gates helped save 122 m lives—and what they want to solve next," *The Guardian* (Feb. 14, 2017), theguardian.com/world/2017/feb/14/bill-gates-philanthropy-warren-buffett-vaccines-infant-mortality

205   There are several grants listed to the *CIR*.
        2015—$300,612
        2016—$300,000
        2017— $253,300
        2019—$592,727
    "Center for Investigative Reporting 2015–2019," BMGF, gatesfoundation.org/about/committed-grants?q=Center%20for%20Investigative%20Reporting

206   Sandy Doughton and Kristi Heim, "Does Gates Funding of Media Taint Objectivity?," *Seattle Times* (Feb. 19, 2011), seattletimes.com/seattle-news/does-gates-funding-of-media-taint-objectivity/

207   McGoey, 202

208   Logan Christopher, "Non-Conspiracy Criticism of Gates Foundation" (Jun. 13, 2020), loganchristopher.com/non-conspiracy-criticism-of-gates-foundation/

209   Michael Barbaro et al., "Can Bill Gates Vaccinate the World?," *New York Times* (Mar 3, 2021), nytimes.com/2021/03/03/podcasts/the-daily/coroanvirus-vaccine-bill-gates-covax. html?showTranscript=1

210   Domina Petric, "Who Is Going to Fast Check the Fast Checkers?," *ResearchGate* (2010), researchgate. net/publication/343962629_Who_is_going_to_Fast_Check_the_Fast_Checkers

211   Loffredo and Greenstein

212   Greene

213   Timothy Egan, "Bill Gates Is the Most Interesting Man in the World," *New York Times* (May 22, 2020), nytimes.com/2020/05/22/opinion/bill-gates-coronavirus.html

214   Master of the Universe (Cover Page), *TIME Magazine* (June 5, 1995), content.time.com/time/ covers/0,16641,19950605,00.html

215   William Jeakle, "In Praise Of Bill Gates: Entrepreneur, Normal Guy, Potential Savior Of The World," *Forbes* (Mar. 19, 2020), forbes.com/sites/williamjeakle/2020/03/19/in-praise-of-bill-gates-entrepreneur-normal-guy-potential-savior-of-the-world/?sh=1116438a2d85

216   Stuart Emmrich, "Why Isn't Bill Gates Running the Coronavirus Task Force?," *Vogue* (Apr. 15, 2020), vogue.com/article/bill-gates-trump-who-coronavirus-task-force

217   McGoey, 203

218   Ibid.

ChildrensHealthDefense.org/fauci-book

childrenshd.org/fauci-book

For updates, new citations and references, and
new information about topics in this chapter:

CHAPTER 10

MORE HARM THAN GOOD

"It's like waking up in your house with a room full of smoke, opening the window to let the smoke out, and then going back to bed."

—Unnamed Medical Expert

In the last chapter, we heard global public health advocates accuse Bill Gates and Dr. Fauci of hijacking WHO's public health agenda away from the projects that are proven to curb infectious diseases (clean water, hygiene, nutrition, and economic development) and diverting international aid to wedge open emerging markets for their multinational partners and to serve their personal vaccine fetish. This chapter will examine Gates's underlying assertion that his African and Asian vaccines are yielding a net public health benefit.

Allergy to Placebo Testing

Most medicinal products cannot get licensed without first undergoing randomized placebo-controlled trials that compare health outcomes—including all-cause mortalities—in medicated versus unmedicated cohorts. Tellingly, in March 2017, I met with Dr. Fauci, Francis Collins, and a White House referee (and separately with Peter Marks from CBER at FDA) to complain that HHS was, by then, mandating 69 doses of sixteen vaccines[1] for America's children, none of which had ever been tested for safety against placebos prior to licensing. Dr. Fauci and Dr. Collins denied that this was true and insisted that those vaccines were safety tested. They were unable, however, after several weeks, to provide us a citation for a single clinical trial using an inert placebo against a vaccine. In October 2017, Del Bigtree and Aaron Siri—who both attended these meetings—joined me in suing HHS under the Freedom of Information Act to produce the long-promised safety studies.[2] Ten months after the meeting with Fauci and Collins, on the courthouse steps, HHS admitted that we were, in fact, correct: none of the mandated childhood vaccines had been tested for safety in pre-licensing inert placebo tests.[3] The best of Bill Gates's African vaccines are all on this list. But Bill Gates also uses a large retinue of much more dangerous and demonstrably ineffective vaccines in Africa—ones that Western countries have actually rejected because of dire safety signals.

That means that nobody knows the risks of these products and nobody can say, with specificity or certainty, that any of Bill Gates's flagship vaccines actually prevent more injuries and deaths than they cause. Furthermore, it means that all of Gates's African vaccines are experimental products. For Gates and his cronies, the continent is a mass human experiment—with no control groups and no functional data collection systems—for shoddily tested, high-risk medical interventions. His unwillingness to actually measure or prove the effectiveness of his prescriptions in reducing mortality and improving health suggests that Gates appreciates that his vaccines are not the human health miracle he proclaims.

Because Gates and Dr. Fauci suffer the same allergy to funding studies that examine the effectiveness of their vaccines in improving health and reducing mortality, neither man has ever offered empirical evidence to support their pivotal claim that their vaccines have "saved millions of lives." The meager published science examining this question indicates that virtually all of Gates's blockbuster African and Asian vaccines—polio, DTP, hepatitis B, malaria, meningitis, HPV, and Hib—cause far more injuries and deaths than they avert.

This chapter will offer a rough cost-benefit analysis of each of Bill Gates's flagship African and Indian vaccines.

Bill and Tony's African Safari

In the colonial era, Africa provided model precincts for testing new vaccines. In the 1950s, white colonial overlords rolled out the red carpet for pharmaceutical companies to perform vaccine experiments on compliant test subjects numbering in the millions. Drug companies spend some 90 percent of their drug development costs on Phase III human trials.[4] Every trial delay eats into the critical time period when the product enjoys patent protection. In the 1980s, Pharma therefore moved most of its clinical trials to poor nations where human guinea pigs are cheap and even the most severe injuries will rarely delay the study. Government complicity and anemic corporate liability laws allow vaccine makers to write off injuries as collateral damage with little consequence or accountability.

Today, Pharma still regards Africa as the beau ideal to test immunizations, and as a lucrative receptacle for dumping expired and defective stocks.[5] Bill Gates has played a key role in legitimizing this arrangement while collaborating with captive or corrupt WHO officials to scam Western donor nations into footing the bill, and guaranteeing rich profits for pharmaceutical companies in which, coincidentally, he holds hefty stock positions. Gates—the "biggest funder of vaccines in the world"[6]—is heavily invested in lucrative partnerships with almost all the world's largest vaccine companies.[7] Bill and Melinda Gates have continued the tradition of human experimentation in Africa with the WHO stepping neatly into the role of an enabling colonial vassal.

Following the colonial era, most of Africa's new nationalist governments considered healthcare a national priority, and many of them developed model health programs for their populations. During the 1970s, International Monetary Fund (IMF) austerity policies bankrupted the best of these programs and left African nations almost entirely dependent on the WHO to finance National Health Ministries and vital HIV programs.[8] Using its control of the flow of international assistance, WHO exerts discipline, rewards compliance, and punishes resistance to Pharma's African ambitions. WHO uses its funding power to bully African governments that slack on vaccine uptake. Gates's pervasive control over WHO has made Africa his fiefdom.[9] The continent's populations have become his guinea pigs. Vaccines, for Bill Gates, are a strategic philanthropy that feed his many vaccine-related businesses and give him dictatorial control over global health policies affecting millions of human lives.

DTP Vaccine: African Genocide

A wave of gruesome brain injuries and deaths followed the introduction of diphtheria, tetanus, and pertussis (DTP) vaccines in the United States and Europe in the 1970s.

As early as 1977, a study published by British physicians and researchers in *The Lancet* established that the risks of the whole-cell pertussis jab (used in the DTP vaccine) exceed the risks associated with wild pertussis.[10]

Six years later, a 1983 NIH-funded UCLA study found that Wyeth's DTP vaccine was killing or causing severe brain injury, including seizures and death, in 1 in every 300 vaccinated children.[11] The resultant lawsuits caused the collapse of insurance markets for vaccines and threatened to bankrupt the industry. Wyeth—now Pfizer—claimed to be losing $20 in downstream liability for every dollar it earned on vaccine sales, and induced Congress to pass the National Childhood Vaccine Injury Act in 1986 shielding vaccine makers from liability.[12]

In 1985, the Institute of Medicine (IOM) recommended the abandonment of the whole cell version of the pertussis vaccine—to avert the high incidence of encephalopathy and deaths.[13] In 1991, the United States, E.U. countries, and Japan switched to a far safer (but less effective) dead cell (attenuated) vaccine—DTaP—and discontinued use of the DTP jab.[14] While Western nations pulled the DTP, WHO gave pharma free rein and cash to dump its toxic inventories in Africa, Asia, and Central America, despite strong evidence of its deadly impacts.[15]

Its dangers aside, the old DTP is cheaper to manufacture and more lucrative for pharma, and so, after 2002, Gates and his surrogates, GAVI, WHO, and Global Fund made DTP the flagship for their African vaccine program and continued giving this neurotoxic and often lethal vaccine to some 156 million African children annually.[16, 17] WHO's use of DTP as its bellwether vaccine—to measure national compliance with WHO's vaccine schedule—has made DTP today the most popular vaccine on Earth.[18] Health ministries across the world must demonstrate specific uptake goals with the DTP recommendations in order to qualify for vital WHO assistance for HIV and other support.

Prior to 2017, neither HHS nor WHO performed the kind of study necessary to ascertain whether the DTP vaccine was actually yielding the beneficial health outcomes about which Gates frequently boasts. That year, the Danish government and the Scandinavian vaccine behemoths, Statens Serum Institut[19] and Novo Nordisk, commissioned prominent Scandinavian scientists Søren Mogensen and Peter Aaby—both vocal champions of the Africa vaccine program—to lead an illustrious team of international researchers to examine all-cause mortalities after the DTP inoculations.

That massive study put the lie to Gates's mantric incantation that his investment in the DTP vaccine has saved millions of lives. In June 2017, the team published a peer-reviewed study in *EBioMedicine*, a high-gravitas journal in Elsevier's publishing house armada. The article parsed data from a so-called "natural experiment" in Guinea Bissau, where half the children in certain age groups were vaccinated and the other half were not. The division was randomized.

That 2017 study (Mogensen et al., 2017)[20] shows that, following their DTP immunization at three months, vaccinated girls had tenfold higher mortality than unvaccinated children. The girls were dying of a wide range of diseases—pneumonia, anemia, malaria, dysentery—and for two decades no one noticed that the dying children were predominantly those who received the vaccine. The DTP vaccine—while protecting children against diphtheria, tetanus, and pertussis—had ruined their immune systems, making them vulnerable to a wide range of deadly nontarget infections. Mogensen's team arrived at that conclusion, as had the 1977 *Lancet* study researchers exactly forty years earlier: "DTP vaccine may kill more children from other causes than it saves from diphtheria, tetanus or pertussis."[21]

In other words, Gates's DTP vaccine—instead of saving 10 million lives, as he claims[22]—may have unnecessarily killed millions of African girls. At least seven other studies have confirmed DTP's association with high mortality in vaccinated girls compared to unvaccinated.[23] The idealistic Americans who donated to Gates's African vaccine project—believing they were saving African babies—were actually funding a continent-wide female genocide.[24]

After completing the study and verifying its shocking results, Peter Aaby—a virtual deity among African vaccine researchers—made an impassioned, and remorseful, plea to the WHO to reconsider the DTP vaccine. "I guess most of you think that we know what our vaccines are doing," he said. "We don't."[25]

Gates, WHO, and GAVI ignored Aaby's appeal and redoubled their efforts to expand DTP vaccinations and to shore up support for the girl-killing jab. *The Lancet* published a commentary by Gates Foundation plenipotentiary Chris Elias, Dr. Anthony Fauci, and three apparatchiks from lesser Gates-funded consortia, WHO's Margaret Chan, UNICEF's Director Anthony Lake, and Seth Berkley of GAVI, who portray their deadly African DTP program as a public health triumph. These charlatans proclaimed DTP as one of the "bright spots" in global well-being and gasconaded that "more children are being immunized worldwide than ever before with the highest level of routine coverage in history (as measured by coverage of three doses of the diphtheria-tetanus-pertussis (DTP)-containing vaccine)."[26] That project also involved reputationally demoting Aaby with a defamation campaign.

A subsequent expert review by the founder of the Cochrane Collaborative, Peter Gøtzsche, condemned the WHO's attempt to downplay the risks of DTP vaccine. The WHO, he observed, had been dismissive of studies finding detrimental nonspecific effects for the DTP vaccine while accepting studies finding beneficial nonspecific effects for the measles vaccine. The WHO is "inconsistent and biased toward positive effects of vaccines. When a result pleases the WHO, it can be accepted, but not when a result does not please the WHO."[27] Gøtzsche found the studies by Mogensen and Aaby "superior in every respect to the Gates-generated *Lancet* study."

Gates and his WHO vassals continue to bully African nations into taking their lethal DTP vaccines by threatening to withdraw financial aid to their health departments and HIV programs if the government fails to achieve national uptake targets (90 percent).

Mercury Rising

Many vaccines shipped to underdeveloped countries—including the hepatitis B, *haemophilus influenzae* type B, and DTP inoculations—contain bolus doses of the mercury-based preservative and adjuvant thimerosal.[28]

The immunity provisions of the 1986 Vaccine Act gave a blank check to US pharmaceutical companies to promote the most shoddily tested vaccines without consequences or cost. Pharma responded with a gold rush to add new lucrative vaccines to the schedule, and by 1991, mercury exposures to US children from the vaccine preservative thimerosal had more than doubled.[29] Parents, physicians, and researchers blamed a subsequent explosion of neurological and autoimmune disease on thimerosal.

Alarmed at the exploding epidemics of neurodevelopmental, allergic, and autoimmune diseases in children that began in 1986, CDC commenced in 1999 an in-house study of the vast repository of health and vaccination data from the ten largest HMOs stored in the Vaccine Safety Datalink (VSD). A specially assembled CDC research team led by Belgian epidemiologist Thomas Verstraeten compared health outcomes in hundreds of thousands of vaccinated versus unvaccinated children. The raw data from CDC's 1999 Verstraeten study showed that children who took thimerosal-containing hepatitis B vaccines in their first thirty days suffered an astonishing 1,135 percent higher rate of autism than children who did not.[30] Verstraeten also documented a grim inventory of other neurological injuries including ADD/ADHD, speech and language delays, tics, and sleep disorders in children exposed to thimerosal. Verstraeten reported that these shocking signals prompted him to review, for the first time, the published medical literature, where he confirmed the alarming toxicity of mercury (thimerosal) to cause these injuries was biologically plausible.

Overwhelming science—over 450 studies—by then attested to thimerosal's devastating toxicity.[31] Because testosterone amplifies the neurotoxicity of the mercury molecule, boys disproportionately suffered reduced IQ and a range of developmental disorders—ADD, ADHD, speech delay, tics, Tourette's syndrome, narcolepsy, ASD, and autism following exposure to ethylmercury in thimerosal. Numerous studies link thimerosal to miscarriage and Sudden Infant Death. There is simply no study ever published that demonstrates thimerosal's safety.

In 2017, Robert De Niro and I hosted a packed press conference at the National Press Club in Washington, DC. We offered a $100,000 reward to anyone who could point to such a study. A prestigious group of scientists, including UCLA Fielding School Emeritus Professor of Epidemiology and Statistics Dr. Sander Greenland, toxicologist and past director of the Environmental Toxicology Program at the National Institute of Environmental Health Sciences, Dr. George Lucier, and Dr. Bruce Lanphear of Simon Fraser University and British Columbia Children's Hospital, agreed to judge the study. There were no takers.

In 2001, the Institute of Medicine recommended thimerosal's removal from all pediatric vaccines. In accordance with the IOM recommendation, manufacturers removed thimerosal from childhood vaccines—Hib, hepatitis b, and DTP—except multi-dose flu vaccines in the United States beginning in 2001. Japan and the European governments had already dramatically reduced mercury levels in their vaccines as early as 1993.

The European and US bans left Pharma struggling to unload stocks and find new ways to monetize stranded assets—the hundreds of millions in production facilities committed to mercury-based vaccines. Bill Gates came to Pharma's rescue. Gates helped pharmaceutical companies unload their thimerosal inventories by dumping them in developing countries. Merck, with the help of Bill Gates and GAVI, brokered a deal to donate (dump) 1 million doses of their thimerosal-containing Recombivax HB hepatitis B vaccine to the Millennium Vaccine Initiative to African countries. The White House hailed Gates's corporate welfare initiative as an "unprecedented level of corporate support" in a press release issued March 3, 2000.[32]

Despite the discontinuance in Western nations, Bill Gates and WHO continue to use their power to force African children to submit to a battery of potentially dangerous mercury-laced pediatric vaccines. Strong evidence suggests that African boys with higher testosterone and chronic vitamin D deficiencies are far more vulnerable to vaccine and thimerosal injury than whites.[33, 34] When it comes to pharma profits, dead and brain-damaged African babies are merely collateral damage.

In 2012, Dr. Fauci waxed philosophical when a reporter asked him to describe an example of one of his useful collaborations with Gates. Perhaps, he speculated, NIAID would work with Gates and GAVI on a project to remove thimerosal from African vaccines. "What is used now is thimerosal, which is frowned upon because of concerns of mercury. So Seth [Berkley, Gates's GAVI Director] and I were talking about finding a preservative for these multi-dose vials without thimerosal so we no longer would have the baggage associated with it."[35] By "baggage," he apparently meant the millions of neurologically injured African children.[36] There is no evidence that this particular collaboration survived its stillbirth as a hypothetical reverie. Eight years later, Africans are still carrying that toxic baggage. It's a crushing—often mortal—load.[37]

Lethal Malaria Vaccine Experiments

Malaria claims some 655,000 lives annually, mostly African children aged under five.[38] In 2010, the Gates Foundation funded with $300 million a Phase III trial of GlaxoSmithKline's experimental malaria vaccine Mosquirix[39] in seven African countries, "aimed at young children because their immune system is still developing."[40] GlaxoSmithKline contributed $500 million, NIAID contributed tens of millions in a battery of grants. Lesser funders included USAID, CDC, and Wellcome Trust. Gates is heavily invested in GSK.[41] Apparently suspecting the vaccine might be lethal, Gates's team elected not to test it against a placebo. They used, instead, highly reactogenic meningitis and rabies vaccines that, themselves, were never tested against a placebo. The meningitis jab was famous for causing alarming numbers of injuries and deaths. The use of a reactogenic placebo—a so-called fauxcebo—is a deliberately fraudulent gimmick that unscrupulous vaccine companies deploy to mask injuries in the study cohort by purposefully inducing injuries among the placebo cohort. Clinical trials that omit true inert placebos marketing masquerading as science. Some 151 African infants died in the trial, and 1,048 of the 5,049 babies suffered serious adverse effects—in both control and study groups—including paralysis, seizure, and febrile convulsions.[42]

Eager to secure the WHO approval necessary to license GSK's vaccine for global distribution, the Bill and Melinda Gates Foundation brushed aside the lethal outcomes of these experiments, declaring the trial a mild disappointment but vowing to press on with the project, casualties be damned. "The efficacy came back lower than we had hoped, but developing a vaccine against a parasite is a very hard thing to do. The trial is continuing, and we look forward to getting more data to help determine whether and how to deploy this vaccine." He demonstrated his resolve by donating an additional $200 million[43] to finance more defective GSK research.

Even with Gates's generous grubstake, GSK's crooked clinical trial researchers could only muster a feeble claim of 30 percent efficacy for their infanticidal jab.[44] Undaunted, Gates rolled out Mosquirix in 2019 as the first malaria vaccine in sub-Saharan Africa. It turned out to be another "genocide-for-girls" project. According to the publication *Science*, "Mosquirix's efficacy and durability are mediocre. Four doses offer only 30 percent protection against severe malaria, for no more than 4 years. . . . The biggest concerns, however, are about the vaccine's safety." *BMJ*'s senior editor, Dr. Peter Doshi, points out, "These were a rate of meningitis in those receiving Mosquirix 10 times that of those who did not, increased cerebral malaria cases, and a doubling in the risk of death (from any cause) in girls." Dr. Doshi says WHO's Malaria Vaccine Study represents a "serious breach of international ethical standards."[45] The demonstrated risk worried WHO so much that it retreated from its plan to roll out the vaccine across Africa, in favor of smaller pilot programs in Malawi, Ghana, and Kenya that will administer the vaccine to hundreds of thousands of children instead of the 100 million that BMGF had hoped for.[46]

Virologists and academics around the world kept mum about Gates's Mosquirix deaths. Gates's plump purse, his impeccable connections, his power over the virology cartel, and the weakness and needs of African governments once again insulated him from the consequences of all these dead children—with the exception of Dr. Doshi.

Lethal Meningitis Vaccine Experiments

In 2010, Gates funded a MenAfriVac campaign in sub-Saharan Africa. Gates operatives forcibly vaccinated thousands of African children against meningitis, causing approximately 50 of 500 vaccinated children to develop paralysis.[47] Citing additional abuses, South African newspapers declared, "We are guinea pigs for the drug makers."[48] Professor Patrick Bond, a political economist who served in Nelson Mandela's South African government, describes Gates's unseemly business—philanthropic practices and the agenda of the Gates Foundation—as "ruthless and immoral."[49]

Population and Sterilization Vaccines

Early twentieth-century America saw the snowballing popularity of eugenics, a racist pseudoscience that aspired to eliminate human beings deemed "unfit" in favor of the Nordic stereotypes. Twenty-seven state governments enshrined elements of the philosophy as official policy by enacting forced sterilization and segregation laws and marriage restrictions. In 1909, California became the third state to adopt laws requiring sterilization of intellectually challenged Americans. Ultimately, eugenics practitioners coercively sterilized some 60,000 Americans."[50]

John D. Rockefeller, Jr.'s keen interest in eugenics colored his passion for population control. The oil baron scion joined the American Eugenics Society and served as trustee of the Bureau of Social Hygiene. The Rockefeller Foundation dispatched hefty donations in the 1920s and early 1930s to hundreds of German researchers, including those conducting Hitler's notorious "twins studies" at the Kaiser Wilhelm Institute for Anthropology, Human Heredity and Eugenics in Berlin.[51] The Rockefeller Foundation curtailed donations to Nazi Germany's medical institutions before Pearl Harbor, but Rockefeller's success promoting the eugenics movement had already captivated Adolf Hitler. "Now that we know the laws of heredity," Hitler told a fellow Nazi, "it is possible to a large extent to prevent unhealthy and severely handicapped beings from coming into the world. I have studied with interest the laws of several American states concerning prevention of reproduction by people whose progeny would, in all probability, be of no value or be injurious to the racial stock."[52]

In the early 1950s, the Rockefeller Foundation conducted fertility studies in India that historian Matthew Connolly characterizes as an example of "American social science at its most hubristic." In one of the collaborations with the Harvard School of Public Health and India's Ministry of Health, the Rockefeller Foundation studied 8,000 tribal people in seven villages in the Khanna section of Punjab to determine whether contraceptive tablets could dramatically reduce fertility rates.[53] According to Linsey McGoey, "The villagers were treated like lab specimens, subjected to monthly questioning but otherwise ignored."[54]

Rockefeller's researchers did not initially inform the Punjabis that their pills would prevent women from bearing children. McGoey describes the villagers as "shocked," "dismayed," and "resentful" to learn that the medication they credulously consumed was intended to render them infertile: "Some were incensed by the effort to limit their future progeny."[55]

Over the next two decades, the Rockefeller Foundation conducted frequent anti-fertility programs in India and elsewhere, earning the growing animosity of physicians, human rights activists, and poverty specialists who criticized the foundation for focusing on population growth while ignoring the realities of persistent poverty that makes large families so indispensable to Indian and African villagers.[56]

"Today," McGoey adds, "the Gates Foundation is pouring money into experimental medical trials that are facing criticism similar to those directed at the [Rockefeller Foundation's] Khanna study. Like earlier philanthropic foundations, The Gates Foundation has the financial and political clout to intervene in foreign nations with relative impunity, and to remain unfazed when the experiments it funds go awry."[57]

Gates's fetish for reducing population is a family pedigree. His father, Bill Gates Sr., was a prominent corporate lawyer and civic leader in Seattle with a lifelong obsession for "population control." Gates Sr. sat on the national board of Planned Parenthood, a neo-progressive organization founded in 1916 by the racist eugenicist Margaret Sanger to promote birth control and sterilization and to purge "human waste"[58] and "create a race of thoroughbreds."[59] Sanger said she hoped to purify the gene pool by "eliminating the unfit" persons with disabilities—preventing such persons from reproducing[60] by surgical sterilization or other means.

In 1939, Sanger created and directed the racist Negro Project, which strategically co-opted Black ministers in leadership roles to promote contraceptives to their congregations. Sanger stated in a letter to her eugenics colleague, Clarence Gamble (of Procter & Gamble), "We do not want word to go out that we want to exterminate the

Negro population and the minister is the man who can straighten out that idea if it ever occurs to any of their more rebellious members."[61]

"When I was growing up, my parents were always involved in various volunteer things," Gates told Bill Moyers in 2003. "My dad was head of Planned Parenthood. And it was very controversial to be involved with that."[62]

Overpopulation, Gates's father told *Salon* in a 2015 interview, was "an interest he's had since he was a kid."[63] In 1994, the elder Gates formed the William H. Gates Foundation (the family's first), focused on reproductive and child health in the developing world. Population control was an enduring preoccupation of his son's philanthropy from its inception.

Gates has made a long parade of public statements and investments that reflect his deep dread of overpopulation. He describes himself as an admirer and proponent of the population doomsayer Paul Ehrlich, author of *The Population Bomb*, whom Gates describes as "the world's most prominent environmental Cassandra," meaning a prophet who accurately predicts misfortune or disaster.

By the way, I share Gates's fear that if humanity persists in juxtaposing exponential population expansion atop linear resource growth, we will all land in a nightmarish Malthusian dystopia. I'm troubled, however, by his apparent comfort in using coercive and mendacious tactics to trick poor people into dangerous and unwanted contraceptive programs. The proven paths to zero population growth are the mitigation of poverty and empowerment of women. Women with alternative career opportunities seldom choose the heavy and hazardous burden of serial maternity. Virtually every nation with a stable middle class has fertility below replacement rates. But Gates's careless public statements and the programs that he habitually funds suggest that Gates has involved himself in sketchy stealth campaigns to sterilize dark-skinned and marginalized women without their informed consent—including by the deceptive use of dangerous sterility vaccines.

On February 20, 2010, less than one month after he famously committed $10 billion to the WHO, Bill Gates suggested in his "Innovating to Zero" TED Talk in Long Beach, California, that reducing world population growth could be done in part with "new vaccines":[64]

> The world today has 6.8 billion people. That's headed up to about 9 billion [here he is almost quoting Bryant et al.]. Now, if we do a really great job on new vaccines, health care, reproductive health services, we could lower that by, perhaps, 10 or 15 percent . . .[65, 66]

Gates's defenders—and the Gates-subsidized "Fact Checker" organizations—scoff at critics who interpret literally Gates's 2010 statement that he hoped to use vaccines to reduce population. They explain that Gates intended, by this inartful construct, to suggest that lifesaving vaccines will allow more infants to survive to adulthood, thereby reassuring impoverished parents that they need not have so many children. But this hypothesis rests on the sketchy premise that his vaccines reduce child mortality—a proposition that Gates has never demonstrated and that current science does not support. His peculiar choice of words naturally fueled speculation that he was engaging in a premeditated campaign to use vaccines to sterilize women. His questionable antics in promoting antifertility drugs and WHO's widespread use of stealth sterility vaccines credibly fuel such sentiments.

Depo-Provera: A Cruel Irony

Population control has been the central preoccupation of the Gates Foundation since its inception. In 1999, Gates's $2.2 billion commitment to the UN Population Fund doubled the size of the Gates Foundation.[67] The same year, he funded, with a $20 million contribution, the founding of the Johns Hopkins Center for Population.[68]

In 2017, the Gates Foundation adopted the goal of administering contraceptives to 214 million women in poor countries.[69] Gates's contraceptive of choice is the long-term infertility agent Depo-Provera. Population planners have administered Depo-Provera primarily to poor and Black women in the United States since its invention in 1967. In the United States, 84 percent of Depo-Provera users are Black, and 74 percent are low-income.[70] Depo-Provera's biggest promoter, Planned Parenthood, specifically targets Blacks[71] and Latinas[72] in its marketing campaigns. UN data demonstrate that Depo-Provera is seldom administered to White or affluent women or girls in the United States or Europe.

Depo-Provera is a powerful poison, with a devastating inventory of wretched side effects: Under federal law, the Depo-Provera label must bear FDA's most stringent Black Box warning—due to its potential to cause fatal bone loss. Furthermore, women have reported both missed periods and excessive bleeding; blood clots in arms, legs, lungs, and eyes; stroke; weight gain; ectopic pregnancy; depression; hair loss; decreased libido; and permanent infertility.[73] Some studies have associated Depo-Provera with dramatic increases (200 percent) in breast cancer risk.[74] The FDA warns women not to take Depo-Provera for longer than two years, but Gates's program prescribes at least a four-year course—or indefinitely—for African women and goes to great lengths to avoid warning Black women about the concoction's many drawbacks.[75]

Between 1994 and 2006, Bill & Melinda Gates teamed with the Rockefeller and Andrew W. Mellon Foundations, the Population Council, and USAID to fund a seminal family-planning experiment administering Depo-Provera to approximately 9,000 impoverished women in the town of Navrongo and districts of Ghana.[76] (Though USAID's stated underlying principles for family planning are "volunteerism and informed choice," it hasn't always worked out that way.)

A disturbing 2011 exposé of the collaboration by the Rebecca Project for Justice, "The Outsourcing of Tuskegee: Nonconsensual Research in Africa," documented how Gates's researchers lied to the Navrongo women, telling them that they were receiving "routine healthcare" and/or "social observations"—never informing them that they were part of a population control experiment.[77] Gates's researchers violated US research laws by failing to administer informed consent forms to the women they injected with Depo-Provera. Nor did they obtain institutional review board (IRB) approval for a human experiment that lasted an extraordinary six years. Under direction of Gates's PI, Dr. James Phillips, and his fellow Pfizer and Gates's PIs, deliberately fabricated and falsified research data to fraudulently "prove" Depo-Provera safe.[78] Based on such "proofs," in 2011, Gates expanded his project to fund Depo-Provera programs for some 12 million women across sub-Saharan Africa.[79,80]

That same year, 2011, a study by a another prestigious BMGF & NIH-funded research team from Gates's own Washington School of Public Health published an article in *Lancet Infectious Diseases*, Heffron et al. (2012), reporting that African women who used injectable Depo-Provera were much more likely to acquire HIV/AIDS compared to untreated women. Depo-Provera injections double a woman's risk of contracting and transmitting HIV.[81] This result was not an enormous surprise. For twenty-four years, diverse studies have shown that Depo-Provera thins the vaginal wall, easing transmission of HIV. Furthermore, the researchers found Depo-Provera exacerbates the rates of HIV/AIDS infections to a recipient's sexual partners. Despite her funding from Gates, the study's lead author, Dr. Renee Heffron, and her fellow researchers recommended informing HIV-infected women of Depo-Provera's grave risks and to use alternative non-progesterone-based contraceptives: "Women should be counseled about potentially increased risk of HIV-1 acquisition and transmission."[82] The confirmation of the risk by his own scientists posed an obvious conundrum for Gates since it pitted his passion for population control against his avowed commitment to end the spread of HIV in Africa. Population, it turns out, trumps HIV-prevention in Bill Gates's catechism.

Without offering any scientific research to substantiate their claims, Gates's deputies, a cabal of extreme population control advocates linked to Gates, worked with Pfizer intermediaries to viciously attack Heffron's research findings. The critics included BMGF, Planned Parenthood, the UN, Ronald Gray of the Gates-funded Johns Hopkins University, James Shelton of USAID's Office of Population, and others.

Under these fierce attacks by Gates's minions in the medical cartel, Dr. Heffron and her research team courageously stood their ground and retained their professional integrity. The *Lancet* published Heffron's withering response. Dr. Heffron pointed out that her attackers cited no convincing science and that the two recent studies—by Heffron and the WSPH team—capped a quarter-century of published research documenting increased HIV risk among women taking Depo-Provera.[83]

To combat this crisis, WHO—by then, wearing Bill Gates's boot on its neck—convened a group of handpicked experts, all sworn to secrecy, for a closed-door meeting in Geneva on January 31, 2012, to discuss damage control on the Heffron study and the mountain of HIV research that supported her. On February 16, 2012, WHO and its mysterious expert cabal—unsurprisingly—announced its preordained decision: Women living with HIV/AIDS or at high risk of HIV/AIDS can safely use Depo-Provera.[84]

Betsy Hartmann, a longtime reproductive rights advocate, ridiculed WHO's convenient new guidelines: "This reversal despite 25 years of studies citing an increased risk of HIV transmission among women using it raises question marks whether WHO abandoned caution due to 'outside encouragement' by special interest groups."[85] Hartmann was clearly referring to BMGF.

In the wake of WHO's self-serving declaration, Melinda Gates announced in July 2012 a billion-dollar contribution as BMGF's share of a four-billion-dollar collaboration with USAID, PATH, and Pfizer with the goal of promoting Pfizer's Depo-Provera across sub-Saharan Africa.[86] Pfizer and USAID committed the remaining $3 billion to African contraceptive projects.

Outcry and censure from dozens of international women's rights advocates and reproductive health groups greeted Melinda Gates's announcement.

According to a detailed report by Jacob Levich, "The Real Agenda of the Gates Foundation," "Mrs. Gates minimized the proven risk of acquiring HIV/AIDS with Depo-Provera by directing the public to a contrived eight-page 'Technical Statement' published by the Gates Foundation's supplicants at WHO, assuring the public that Depo-Provera is safe, and that all contrary scientific research that linked Depo-Provera to HIV infection was "inconclusive."[87]

To quell the growing uproar, Gates funded a WHO study to debunk the HIV association once and for all. This time he skipped over Heffron to fund a more "reliable" group of researchers (environmental lawyers call this sort "biostitutes"). On October 21, 2015, WHO released its investigation—which, not surprisingly, concluded that "There is no evidence of a causal association between DMPA use and an incidence in women's risk of HIV acquisition."[88] WHO then issued new guidelines that mirror precisely those recommended by Pfizer, Depo-Provera's manufacturer.

Some forty reproductive health groups demanded that WHO's director, Margaret Chan, sideline the new guidelines until Gates's study could survive a rigorous reevaluation process. WHO ignored those pleas.[89]

The centerpiece of the Gates $4 billion caper is a "self-injection" syringe—a plastic bubble attached to a needle—for administering Depo-Provera. Pfizer creates the gizmo, but Gates's Seattle-based legate, PATH, markets it under the new brand name "Sayana Press." PATH's former director, Chris Elias, was by then president of the BMGF. Through PATH, Gates will distribute these devices, costing $1 per three-month dose, to 120 million women in sixty-nine of the world's poorest countries.[90] With contributions that Gates plans to squeeze from those governments, these lucky ladies will pay little or none of the cost.

Pfizer, of course, will make a killing. According to the *Wall Street Journal*'s Market Watch, "Pfizer could potentially earn approximately $36 billion in sales resulting from an unprecedented Bill & Melinda Gates Foundation (BMGF) investment—$560 million from BMGF, totaling $4.3 billion with government contributions—that promotes Depo-Provera as the optimum contraceptive for women of color and low-income women."[91]

Levich explains that this scheme is a cunning dodge to evade US regulations that require Pfizer's label include its dire Black Box warning bearing the words: "FDA," "Black Box," "warning," and "osteoporosis," and that the administering clinician inform every recipient that the drug poses life-threatening harm. In the United States, pharmacists can never dispense Depo-Provera directly to a patient to self-inject, since the law requires that medical personnel counsel each patient about risks. Ignoring these safeguards in Africa would expose Pfizer to criminal prosecution and thousands of lawsuits under the Alien Tort Claims Act, which could allow aggrieved African women to sue negligent US drugmakers in US courts if they suffer injuries as the result of failure to warn.[92] Pfizer's apparent strategy for insulating itself from liability is to use PATH and BMGF as surrogates to market its contraception.

Furthermore, to promote Depo-Provera's uptake among Blacks, PATH makes a series of outlaw, off-label claims that Pfizer could not legally make about the product. PATH claims that Depo-Provera protects against endometrial cancer and uterine fibroids and reduces risks of sickle cell anemia and iron deficiency anemia—diseases that disparately injure Blacks. FDA has never approved Depo-Provera for cancer prevention or for any of these other uses. It is therefore illegal for Pfizer to promote these off-label claims. Presenting Gates and PATH as its intermediaries is apparently also Pfizer's strategy for evading US laws that prohibit off-label claims. Levich adds: "These statements taken in totality are contextually false and designed to specifically circumvent the FDA's Black Box warnings. If Depo-Provera is genuinely a safe and effective contraceptive, with only minimal side effects, why then are Gates, Hopkins, USAID, Planned Parenthood, and Pfizer's other intermediaries deliberately concealing the plain "Black-Letter" FDA Black-Box warnings in their effort to minimize and conceal Depo-Provera's life-threatening harm?"[93]

Put bluntly, Gates and his confederates are tricking African women into taking the contraceptive by deceiving them about its safety and lying about its efficacy against diseases that disproportionately harm Blacks—something Pfizer executives could go to jail for. Gates's willing partner in this fraud is USAID.

USAID's Director, Dr. Rajiv Shah, has been a serial coconspirator in Gates's many racist flim-flams. For a decade prior to his gig running USAID, Shah worked for Bill Gates's foundation (2001–2010) as the principal fundraiser for GAVI's World Immunization Programs. Shah candidly acknowledged that BMGF's and PATH's stamp of approval on Depo-Provera serves as a clever strategy for insulating Pfizer from criminal and civil prosecution for violating FDA regulations.[94] Gates's caper aims to artfully remove the FDA's jurisdiction by using PATH as its surrogate and by effectively transferring regulatory authority to the WHO.

The Rebecca Project for Justice characterizes Gates's African project as "A family planning strategy that unethically targets women of color to prohibit births of beautiful [Black] children, by not informing mothers of Depo-Provera's deadly risks as mandated under US law/regulations; thus, denying women of color their inalienable right to choose and access safe reproductive health."[95]

Depo-Provera came honestly to its notoriety as the tool of choice for racist eugenicists. Israel banned Depo-Provera in 2013 following a scandal in which government health workers seeking to radically reduce the number of Black births were targeting African Jews with Depo-Provera. Sharona Eliahu Chai, lawyer for the Association of Civil Rights in Israel (ACRI), condemned the government policy of preventing Black Israelis from reproducing: "Findings from investigations into the use of Depo-Provera are extremely worrisome, raising concerns of harmful health policies with racist implications in violation of medical ethics."[96]

In 2002, India banned this dangerous drug from all family welfare programs after a similar scandal: government officials were targeting lower-caste Indians.[97] Many other nations, including Bahrain, Israel, Jordan, Kuwait, Qatar, and Saudi

Arabia, prohibit the use of Depo-Provera on their nationals. European countries largely restrict the use of Depo-Provera and require full disclosure of risks for women and informed consent prior to its use. Gates and USAID have taken advantage of political disorganization in Pakistan to administer "self-inject" Depo-Provera to Muslim women. In contrast to its US counterpart, USAID, the Swedish International Development Authority (SIDA) does not fund, purchase, or provide Depo-Provera for Swedish-assisted projects in developing countries.[98]

Sterility Vaccines / Chemical Castration

Gates's defenders ridicule as "conspiracy theory" the suggestion that Gates, or any reputable public health authority, would use "life-saving vaccines" as a stealth vehicle for surreptitiously rendering women infertile. But one of Gates's earliest philanthropic undertakings was a 2002 project to administer tetanus vaccines to poor women in fifty-seven countries.[99] For reasons we are about to discover, critics credibly suggest that these vaccines may have been secretly laced with a formula the Rockefeller Foundation developed to sterilize women against their will.

On November 6, 2014, four years after Gates pledged at a TED Talk to use vaccines to lower birth rates, medical researchers and doctors associated with the Kenya Conference of Catholic Bishops (KCCB) and the Kenya Catholic Health Commission accused WHO, UNICEF, and GAVI of secretly conducting a mass sterilization program against Kenyan women, under the veil of eradicating tetanus disease.[100, 101] The *Washington Post* reported similar charges by the Kenya Catholic Doctors Association (KCDA).[102]

The Catholic doctors became suspicious due to WHO's glaring departures from the usual tetanus vaccine protocols. Normally a single tetanus vaccine provides a decade of immunity. Since men and women are equally susceptible, both sexes routinely get the vaccine. But WHO instructed Kenyan doctors to give the vaccine in five administrations, six months apart, and only to girls of childbearing years.

"The defense that the WHO intended only to target 'maternal and neonatal tetanus' seems odd in view of the fact that males are about as likely as females to be exposed to the bacterium which is found in the soil everywhere there are animals,"[103] observed a 2011 peer-reviewed study of the controversy. The Catholic doctors also noticed other unusual features of the campaign. For starters, WHO suspiciously initiated its jab campaign not from a hospital or medical center or any of the estimated 60 local vaccination facilities, but distributed shots from the luxurious New Stanley Hotel in Nairobi—an exclusive resort out of reach to most physicians or public health officials.[104] At considerable cost, a police escort accompanied the shots to vaccination sites, where police officers strictly supervised their handling by nursing staff and required clinicians to return each empty vial to WHO officials at Nairobi's only five-star hotel under the watchful eyes of armed officers.

Four years later, in October 2019, the Kenyan Catholic Doctors' Association accused UNICEF, GAVI, and the WHO of rendering millions of women and girls barren.[105] The doctors had by then produced chemical analyses of vaccines verifying their allegations. Three independent Nairobi accredited biochemistry laboratories tested samples of the WHO tetanus vaccine, finding human chorionic gonadotropin (hCG) where none should be present. In October 2014, Catholic doctors obtained six additional vials and tested them in six accredited laboratories, finding hCG in half of those samples.

In 2019, a group of independent researchers from Kenya and Great Britain led by University of British Columbia neurologist Dr. Christopher Shaw studied the charges and concluded that "the Kenya 'anti-tetanus' campaign was reasonably called into question by the Kenya Catholic Doctors Association as a front for population growth reduction." The medical researchers characterized the WHO program "an ethical breach of the obligation on the side of the WHO to obtain 'informed consent' from those Kenyan girls and women."[106]

Catholic medical personnel made similar accusations about WHO's tetanus projects in Tanzania, Nicaragua, Mexico, and the Philippines. Following indignant denials of all such accusations, and obligatory denunciations against its accusers, WHO grudgingly admitted it had been developing the sterility vaccines for decades. WHO nevertheless punished the Kenyan doctors and the community officials who reported the spiked vaccine by canceling contracts for future work.[107]

The Sordid History of Sterility Vaccines

It wasn't the first time that Catholic medical authorities accused the WHO of a stealth sterilization campaign against African women. As early as November 1993, Catholic publications charged that the WHO was spiking tetanus vaccines to neuter dark-skinned women globally with potent abortifacients.[108] WHO denied the explosive charges.

Shaw's research team showed that WHO and Rockefeller Foundation scientists began research on "anti-fertility" vaccines for "birth-control" as early as 1972, by lacing hCG with tetanus toxoid, which acts as a carrier for the hormone. That year, WHO researchers at a meeting of the US National Academy of Sciences[109, 110] reported their successful creation of a "birth-control" vaccine that diminishes the βhCG essential to a successful pregnancy and causes at least temporary "infertility." Subsequent experiments proved that repeated doses could extend infertility indefinitely.[111]

By 1976, WHO scientists had successfully conjugated a functional "birth-control" vaccine. The WHO researchers reported triumphantly that their formula could induce "abortions in females already pregnant and/or infertility in recipients not yet impregnated." They observed that "repeated inoculations prolong infertility."[112] More recently, in 2017, WHO researchers were working on more potent antifertility vaccines using recombinant DNA. WHO publications explain that the agency's long-range purpose is to reduce population growth in unstable "less developed countries."[113]

The Kenyan tetanus campaign occurred shortly after Gates made his pledge of $10 billion to the WHO with the stated purpose of reducing population with "new vaccines." Perhaps to emphasize his commitment to population control, Gates recruited his most influential vizier, Christopher Elias, as president of Global Development at the Gates Foundation the following year. Prior to that appointment, Dr. Elias was president/CEO of Gates's nonprofit PATH, which partners with pharmaceutical companies to distribute vaccines to poor countries by persuading rich and poor governments to fork over moolah to multinational drugmakers in which Gates is invested.

Elias ran PATH's innovative "Sayana Press" injectable Depo-Provera project designed to end-run US safety regulations while reducing fertility of Black African women. That brainchild earned Elias the Klaus Schwab Foundation's Social Entrepreneur of the Year award in 2005. The Gates Foundation provided numerous grants to PATH, including one in November 2020 (after Elias had moved over to BMGF) "to support clinical development of COVID-19 vaccines by Chinese manufacturers."[114]

Before PATH, Elias had been senior associate in the international Programs Division of the Population Council, with the responsibility of dampening fecundity throughout Southeast Asia. John D. Rockefeller III founded The Population Council in 1952 at a conference he convened for the high priesthood of population control, including the director of the new Planned Parenthood Federation of America and several well-known eugenicists. Lamenting that modern civilization had reduced the operation of natural selection by saving more "weak" lives and enabling them to reproduce, resulting in "a downward trend in . . . genetic quality," the group agreed to create an organization devoted to the "reduction of fertility." While Rockefeller formally launched the Council with a grant of $100,000 and served as the first president, the next two Council presidents were Frederick Osborn and Frank Notestein, both members of the American Eugenics Society. The NIH and USAID were among the "start-up" funders, and US and foreign governments soon became the Council's largest financial backers.[115]

The Council does research promoting the use of artificial birth control and abortion and biomedical research to discover and develop new contraceptive drugs and technologies. It collaborated with the Ford Foundation and International Planned Parenthood Foundation to develop large-scale IUD programs abroad, despite its own research doctors warning about acute adverse side effects. Later, the Council played a key role in developing the extremely dangerous hormonal contraceptive implant Norplant.[116]

Historian Donald T. Critchlow wrote that the Population Council "cultivated elite connections and avoided public controversy by identifying itself as a neutral, scientific organization."[117]

The US Agency for International Development (USAID) conducted a decades-long partnership with the Population Council and cultivated long-term alliances with the Rockefeller Foundation and the WHO researching the use of fertility controls to reduce world population, especially in sub-Saharan Africa.[118, 119] By 2014, Gates and Elias had a reliable collaborator at the federal program: USAID Director Rajiv Shah, who had, prior to winning that appointment, worked a decade for the Gates Foundation, running GAVI's immunization program for African children.

Dr. Shah joined the Gates Foundation in 2001 and oversaw its alliance with the Rockefeller Foundation in launching the Alliance for a Green Revolution in Africa. He directed the International Finance Facility for Immunization. The IFFI is a shady agency that finances Bill Gates's global vaccine enterprises in developing nations through a diabolically innovative bond issuance scheme that runs up huge debts in poor countries to finance Gates's self-serving vaccines. Using sleight of hand, IFFI enriches Gates's pharma partners with Western financial bonds by passing the costs to future generations in poor countries. Shah raised $5 billion through this swindle for GAVI. At USAID, his primary responsibility was reorganizing the agency to reflect its new biosecurity direction under Obama's 2009 executive order. Shah left USAID to become president of the Rockefeller Foundation in 2017. Shah has deep links to the intelligence agencies and the oil and chemical cartels. Shah serves on both the Trilateral Commission and the Council on Foreign Relations, two globalist organizations that the Rockefeller/Kissinger alliance largely defined. Shah is a board member of the International Rescue Committee, a nonprofit with long-standing CIA ties. In his 1991 book, *Covert Network: Progressives, the International Rescue Committee and the CIA*, University of Massachusetts economics professor Eric Thomas Chester exposes IRC as a CIA front. Bill Casey, a lifelong spy, who as Ronald Reagan's CIA Director helped manage the Iran-Contra affair in the 1980s, chaired IRC from 1970 to 1971. IRC operates in forty countries doing "humanitarian aid." According to its current CEO, David Miliband, the former UK foreign secretary, Shah's role on the high-level council is to "monitor political and non-health issues related to prevention and preparedness imperatives for a potential epidemic of global proportions."[120]

In 1974, USAID and WHO collaborated on the creation of a top-secret "Kissinger Report." Henry Kissinger—whose patron was Nelson Rockefeller and whose career was deeply enmeshed with the Rockefeller Foundation—drafted the classified White Paper,[121] which became official US policy under President Gerald Ford in 1975. That report, known as the US National Security Study Memorandum 200,[122, 123] outlined the geopolitical incentives for reducing population growth in "less developed countries" (LDCs) to near zero by "reducing fertility" so as to safeguard the economic interests of the United States and other industrialized nations in imported mineral resources.[124, 125]

Kissinger observed that the industrialized West was already having to import significant quantities of aluminum, copper, iron, lead, nickel, tin, uranium, zinc, chromium, vanadium, magnesium, phosphorous, potassium, cobalt, manganese, molybdenum, tungsten, titanium, sulphur, nitrogen, petroleum, and natural gas[126, 127] at high cost. The Kissinger Report anticipated rising prices as population growth triggered instability in African nations.[128]

The high-level US government commitment explains the WHO's monumental commitment to sterility vaccines. Shaw et al. found 150 research publications emanating from WHO on various infertility formulations between 1976 and 2016 with many thousands of citations.

In the years 1993 and 1994, WHO launched antifertility vaccination campaigns in Nicaragua, Mexico, the Philippines,[129] and Kenya in 1995.[130, 131] In each country, WHO and local government clinicians vaccinated women of childbearing age, telling them that the purpose of the WHO immunizations was to "eliminate maternal and neonatal tetanus."[132]

A subsequent WHO study of birth control policy, Bryant et al., acknowledged that WHO's family planning "services" had involved routinely deceiving the persons "served"[133] with "sterilization procedures being applied without full consent of the patient."[134] Similarly, a 1992 study titled "Fertility Regulating Vaccines" published by the UN and WHO Program of Research Training in Human Reproduction, reported "cases of abuse in family planning programs" dating from the 1970s including:

incentives . . . [Such as] women being sterilized without their knowledge . . . being enrolled in trials of oral contraceptives or injectables without . . . consent . . . [and] not [being] informed of possible side-effects of . . . the intrauterine device (IUD).[135]

The authors of that WHO report advised their partners against characterizing their work as "anti-fertility measures for population control," observing that milder descriptions like "family planning" and "planned parenthood" were more palatable for public appetites. Speaking on behalf of the WHO, Bryant et al. admitted, "It is perhaps more conducive to a rights-based approach to implement family planning programs in response to the welfare needs of people and communities rather than in response to international concern for global overpopulation."[136]

The targeted regions for the WHO tetanus campaigns are principally the same developing nations that the Kissinger Report targeted. For example, a 2015 news release by Associated Press announced "[tetanus] immunization campaigns to take place in Chad, Kenya, and South Sudan by the end of 2015 and contribute toward eliminating [maternal natal tetanus] in Pakistan and Sudan in 2016, saving the lives of countless mothers and their newborn babies."[137]

The Kenya schedule was identical to the one published for the WHO birth-control conjugate of tetanus toxoid linked to βhCG: five spaced doses of "TT" vaccine at six-month intervals, which, of course, strongly contrasts with the published schedule for authentic tetanus immunization schedules.[138]

Rajah Bill and his Indian Jabs

Polio Vaccine

Following his seminal meeting with Dr. Fauci in 2000, Gates launched a global polio vaccine campaign, pledging $450 million through BMGF of a $1.2 billion total and promising to eradicate polio by decade's end. Improved nutrition, disease management, and UNICEF's vaccine program had "vanquished" polio in India in 2011, meaning that the disease occurred in fewer than 300 people per year. Doctors diagnosed just over 200 new cases in 2012.[139] WHO declared the malady eradicated after its five-year near-absence in 2016. By that year, polio affected only about 2,000 sufferers globally. The last few hundred cases of an endemic disease are always the most difficult and expensive to prevent. But, apparently, the glory of claiming the triumph for its total obliteration appealed to Bill Gates as an irresistible challenge. He vowed, against sage advice, to eradicate polio and successfully exhorted rich and poor nations to finance his cause.

Even the high-end polio vaccines used in Western nations are linked to injuries and illnesses that dwarf historical harms from polio. A short list of these include the highly contagious SV-40 monkey virus[140] that scientists believe is responsible for the explosion of deadly soft tissue cancers in baby boomers and the Chimpanzee coryza agent that entered polio vaccines at the Walter Reed Hospital laboratories in 1955 and caused the devastating pandemic of respiratory syncitial virus (RSV) that the WHO estimates today causes 3 million hospitalizations annually and 60,000 deaths in children under five and 14,000 deaths among adults sixty-five years and older.[141]

In order to discourage public discussion of those embarrassing abcesses on its sacred cow, HHS in 1984—the year Anthony Fauci became director of NIAID—quietly pushed through an astonishing federal regulation that reflected the agency's institutional culture of paranoia, secrecy, and imperiousness but not America's democratic values or the US Constitution:

> Any possible doubts, whether or not well-founded, about the safety of the vaccine cannot be allowed to exist in view of the need to assure that the vaccines will continue to be used to the maximum extent consistent with the nation's "public health objectives."

—Fed Register Vol. 49 No 107

Most Americans are shocked to learn that today, this abominable regulation is the law of our land.

To complicate these problems, the low-rent polio vaccines Gates uses in Africa and Asia are dramatically different from those used in Western countries. The BMGF committed more than $1 billion pushing an oral polio vaccine (OPV) that contains a live polio virus across the global South. This live virus can replicate inside a child's gut and spread in regions with substandard sanitation and plumbing. That means people can contract the virus from the vaccine. Gates's program created windfall profits for pharmaceutical behemoths that could not market such dangerous products in Western countries.

Experts argued that Gates's attempts to exterminate polio would be counterproductive. Extirpating the final dwindling dead-end infections requires carpet-bombing entire regions with massive vaccination batteries, raising the paradoxical risk of vaccine-strain polio epidemics.

"I can't see myself how we can satisfactorily eliminate the vaccine-derived strains," said Prof. Donald Henderson, a distinguished scholar at the University of Pittsburgh Medical Center for Biosecurity. "I just don't think it can be done."[142] Henderson is the renowned WHO epidemiologist who led the successful campaign against smallpox during the 1960s.

Ignoring such advice, Gates declared war on polio in India and implemented a shock-and-awe strategy to exterminate those last few cases. Gates took control of India's vaccine oversight panel, the National Advisory Board (NAB), by stacking it with loyalists and friendly PIs. Under his control, the NAB mandated an astonishing barrage of fifty polio vaccines (up from five) for each child in several key Indian provinces before they reached the age of five.

As Henderson predicted, vaccine-derived poliovirus—a mutation of the virus contained in the oral vaccine—came back to bite Gates, and the unfortunate populations of the nations that submitted to his prescriptions. Indian doctors blame the Gates campaign for a devastating vaccine-strain epidemic of acute flaccid myelitis—a disease formerly classified as "polio"—that paralyzed 491,000 children in these provinces between 2000 and 2017, in direct proportion to the number of polio vaccines that Dr. Gates's minions administered in each area.[143]

Non-Polio Acute Flaccid Paralysis (NPAFP) is "clinically indistinguishable from polio but twice as deadly,"[144] according to Keith Van Haren, child neurologist at the

Stanford School of Medicine. Van Haren explains that Acute Flaccid Myelitis (AFM) is a polite term for polio: "It actually looks just like polio, but that term really freaks out the public-health people."[145]

In 2012, the *British Medical Journal* wryly noted that polio eradication in India "has been achieved by renaming the disease."[146]

That year, the disillusioned Indian government dialed back Gates's vaccine regimen and evicted Gates's cronies and PIs from the NAB. Polio paralysis rates dropped precipitously.[147] After squandering half of its total budget on the polio epidemic—at Gates's direction—the WHO reluctantly admitted that the global polio explosion is predominantly vaccine strain, meaning it is happening *because of* Gates's vaccine program. The most frightening epidemics in Congo, the Philippines, and Afghanistan are all linked to the vaccines he promoted. Polio had disappeared altogether from each of those nations until Gates reintroduced the dreaded disease with his vaccine.

In Syria, the Gates-backed GAVI committed $25 million for polio immunization in 2016.[148] The following year, the WHO reported that fifty-eight Syrian children had been paralyzed by the vaccine-derived form of the virus.[149]

Other vaccine-strain polio outbreaks occurred in China, Egypt, Haiti, and Malaysia. A study by Oxford's Clinical Infectious Diseases Periodical found that Gates's oral polio vaccine is not only giving kids polio, but also "seems to be ineffective in stopping polio transmission." By 2018, the WHO conceded, 70 percent of global polio cases came from Gates's vaccines.[150]

As the *British Medical Journal* reported in 2012, "the most recent mass polio vaccination programs [in India], fueled by the Bill and Melinda Gates Foundation, resulted in increased cases [of polio]."[151]

In an interview with NPR, professor of microbiology Raul Andino said, "It's actually an interesting conundrum. The very tool you are using for polio eradication is causing the problem."[152]

Dr. Henderson argued that Gates's futile campaign would strip money from other areas of need, forcing nations to prioritize polio immunization at the expense of other public health investments. Arthur Caplan, an eminent bioethicist and a polio vaccine fanatic who himself suffered from polio as a child, also criticized Gates's obsession with polio eradication, pointing out that "government budgets and resources in poor nations are diverted from other far more pressing local problems to try and capture the last marginal cases."[153]

Donald Henderson observes that only Western nations (and billionaires like Gates) consider eliminating the disease as a priority. Polio kills far fewer people in developing regions than scourges such as malaria, TB, malnutrition, and the greatest killer: dysentery from deficient water supplies. When Gates first floated his dream of eradicating polio, developing nations feared a diversion of resources towards an area where the money was least warranted.[154]

"When you're doing polio, you're not doing other things," Henderson says. "At least through 2011, in several countries—Nigeria, India, and Pakistan—they were giving polio vaccines."[155] "In 2012, there were only 223 reported cases of polio worldwide. . . . By any measure, polio is not one of the world's greatest killers. Road accidents, for example, kill about 1.25 million people each year. Measles kills about 150,000 children each year."[156] "A number of villagers say, 'What is polio? We've never seen it—why are we worried about it?'"[157]

Rather than provoking reevaluation, Henderson's concern seems to infuriate Gates. "I've got to get my D. A. Henderson response down better,"[158] Gates mumbled to one of his aides in 2011 after the *New York Times* editorial board interviewed him during his transglobal trek soliciting rich and poor governments to ratchet up their commitment to his polio enterprise. A reporter overheard and reported Gates's whispered comment. That response suggests that he is aware of the criticism by the man most knowledgeable about eradicating diseases. Instead of integrating Henderson's critique into his strategy or executing a mid-course correction, Gates treated Henderson's caveats as a marketing challenge and lumbered onward. His imperviousness to self-assessment allows him to treat the hundreds of thousands of casualties of his policies as acceptable collateral damage in his self-serving schemes for humanity.

Gates's strategic investments have made him immune to criticism by the media and the scientific community, and so, despite these atrocities, the Gates Foundation steers WHO like a rogue destroyer floundering forward full speed ahead through the mayhem, and the carnage of dead and paralyzed children whose ruined lives bob in their wake. In 2020, the BMGF boasted that the WHO is now providing "unprecedented levels of technical assistance" for polio vaccination campaigns in Nigeria, Pakistan, and Afghanistan.[159]

HPV Vaccine

In 2009 and 2012, the Gates Foundation funded tests of experimental HPV vaccines, developed by Gates's partners GSK and Merck, on 23,000 girls 11–14 years old in remote provinces of India. These experiments were part of Gates's effort to bolster those companies' sketchy claims that HPV vaccines protect women against cervical cancer that might develop in old age.[160] Gates and his foundation have large investments in both companies.[161] [162] Since deaths from cervical cancer occur on average at age 58 in the United States and affect only 1/40,000 women, and since virtually all these deaths are preventable with early detection by Pap smears, any vaccine given to young girls to prevent the low risk of preventable death half a century from now ought to be 100 percent safe—and this vaccine isn't even close.

Both Merck and Glaxo disclosed in their Shareholders Reports that profitable performances by their flagship HPV vaccines were top indicators of shareholder value. Gardasil has been a top seller for Merck, earning total global sales of $1.2 billion in 2011,[163] a windfall for the company floundering to recover from a $7 billion court settlement related to criminal charges that the company had knowingly killed between 100,000 and 500,000 Americans by defrauding customers about the safety of its blockbuster pain pill, Vioxx.[164] Merck's executives nicknamed the HPV vaccine "**H**elp **P**ay for **V**ioxx" and fast-tracked it to market after shoddy safety tests under pressure from Wall Street analysts itching to downgrade Merck's "buy" recommendations.

At least 1,200 of the girls in Gates's study—1 in 20—suffered severe side effects, including autoimmune and fertility disorders.[165] Seven died—about 10x the US death rates for cervical cancer, which almost never kills the young. India's Federal Ministry of Health suspended the trials and appointed an expert parliamentary committee to investigate the scandal. Indian government investigators found that Gates-funded researchers at PATH committed pervasive ethical violations: pressuring vulnerable village girls into the trial, bullying illiterate parents, and forging consent forms. Gates provided health insurance for his PATH staff but not to any participants in the trials, and refused medical care to the hundreds of injured girls.[166]

The PATH researchers targeted girls at ashram *paathshalas* (boarding schools for tribal children), to dodge the need to seek parental consent for the shots.[167] They gave the girls "HPV Immunization Cards" that were printed in English, which the girls couldn't read. They did not tell the girls that they were part of a clinical trial and instead hoodwinked them with the lie that these were "wellness shots" that would guarantee "lifelong protection" against cancer. That was not true. PATH conducted the trials in impoverished rural areas that lacked mechanisms for tracking the adverse effects and had no system for recording major adverse reactions to the vaccines, something legally mandated for large-scale clinical trials.[168]

In 2010, the Indian Council of Medical Ethics found that the Gates group had violated India's ethical protocols. In August 2013, a special parliamentary committee excoriated PATH, stating that the NGO's "sole aim has been to promote the commercial interests of HPV vaccine manufacturers who would have reaped windfall profits had PATH been successful in getting the HPV vaccine included in the UIP [universal immunization program] of the Country."[169] According to Dr. Colin Gonsalves, senior counsel of the Supreme Court of India,

> The Indian Parliament formed a committee, and it was to be a rather surprising move, because you generally don't often have such a high level inquiry into matters affecting poor people. And that was such an extraordinary report. I don't think the Indian Parliament has ever come out with such a scathing report. And the government officials came out and said, "We shouldn't have authorized this, we're sorry, and we're not going to allow them again"—and now they are back, doing their same old tricks again.[170]

In 2013, two separate groups of health activists and human rights advocates filed public interest litigation (PIL) petitions calling on India's Supreme Court to investigate the HPV trials and determine whether PATH and other stakeholders responsible for the trial should be held liable for financial damages in relation to the families of the seven deceased girls.[171]

One of the lead petitioners, Amar Jesani, a physician who directs the Centre for Studies in Ethics and Rights in Mumbai, told Professor McGoey that he regrets that he did not add the Gates Foundation as a defendant. "The ethical guidelines of the Indian Council for Medical Research talks about totality of responsibility. It defines the totality of responsibility in terms of everybody—that means sponsor . . . involved," Jesani said. "Under that principle, everyone should be held responsible. There is also no evidence at the moment that the Gates Foundation took any steps to discipline PATH for the research it carried out in India. . . . I think, to some extent, the Gates Foundation thinks PATH has done nothing wrong. And that is a concern. One needs to get a spotlight on the Gates Foundation."[172] The case is now before the country's Supreme Court.

CDC cited Merck's and Gates's cheery assessments of the grotesque Indian experiments to help justify its expanded recommendation for the Gardasil vaccine. Prior to COVID-19, Gardasil was the most dangerous vaccine ever licensed, accounting for some 22 percent of cumulative injuries from all adverse events reported to the US Vaccine Adverse Events Reporting System (VAERS). During clinical trials, Merck was unable to show that Gardasil was effective against cervical cancers.[173] Instead, the studies showed the vaccine actually increases cervical cancer by 46.3 percent in women exposed to HPV prior to vaccination—perhaps one-third of all women.[174] According to Merck's clinical trial reports, the vaccine was associated with autoimmune diseases in one out of every thirty-nine women.[175] Since introduction of that vaccine in 2006, thousands of girls have reported debilitating autoimmune diseases, and cancer rates have skyrocketed in young women.[176]

HPV Vaccines and Fertility

Gates's strong patronage of HPV vaccines (Gardasil and Cervarix) deepened suspicions that he was weaponizing vaccination against human fertility. Merck's clinical trials showed strong signals for reproductive harm from Gardasil.[177, 178] People in the study suffered reproductive problems including premature ovarian failure at ten times background rates. Female fertility has dropped precipitously beginning in 2006 in the United States, coterminous with Gardasil uptake.[179, 180] Historical drops in fecundity have occurred in every nation with high Gardasil uptake.[181]

Hepatitis B

The conspiracy by GAVI, WHO, and UNICEF to force India to mandate hepatitis B vaccines is yet another illustration of how, under Bill Gates's hegemony, vaccine industry profits trump public health. The WHO initially recommended hepatitis B vaccination only in countries with high incidence of hepatocellular carcinoma (HCC), the species of liver cancer that the vaccine promises to abolish. Since HCC is rare in India, the country did not qualify under WHO's initial criteria, which recommended the vaccine only in nations with significant HCC. WHO's policy meant the vaccine manufacturers would lose a market of 1.3 billion people.

Notwithstanding such concerns about the high costs and meager benefits of the vaccine, Gates, through his surrogates at GAVI, PATH, and WHO successfully arm-twisted the Indian government in 2007–8 into introducing the hepatitis B vaccines.

GAVI pushed WHO to change the official policy to a universal recommendation, meaning that even countries with low disease burdens would be required to vaccinate. GAVI hoped this would reopen the Indian markets. WHO obligingly changed its recommendation to include universal immunization with hepatitis B vaccine for *all* countries, even those where HCC was not a problem. The Indian government obediently adopted WHO's recommendation.

Indian academics and public health officials condemned the government's hepatitis B mandates, citing India's extremely low burden from HCC. The Indian Cancer

Registry (ICMR) shows the incidence of hepatocellular carcinoma due to hepatitis B infection is only 5,000 cases a year. Independent scientists and Indian physicians argued against immunizing 25 million babies each year to theoretically prevent 5,000 cases of HCC. Anticancer vaccines are poor performers, and there is not even meager proof that the vaccine can prevent *any* cancers. Dr. Jacob M. Puliyel, MD, Chair of the Department of Pediatrics, St. Stephen's Hospital, Delhi, told me that—even if the vaccine were 100 percent effective—the need to administer 15,000 vaccines to infants to prevent a single death from HCC that might occur decades later "intuitively seems an uneconomic way to spend scarce health resources."

In a July 17, 1999, commentary published in *BMJ*, Dr. Puliyel observed that the cheapest Indian hepatitis B vaccine costs 360 rupees ($5.00) for three doses. Dr. Puliyel points out that "a third of [India's] population earn less than 57 rupees (83p) per capita per month. The main causes of death in India are diarrhea, respiratory infections, and malnutrition." Puliyel says, "Should immunisation against hepatitis B take priority over provision of clean drinking water?"[182]

The study of Gates's forced introduction of hepatitis B vaccines in India showed that the vaccine did not reduce hepatitis B. The frequency of chronic carriers (HBsAg positivity) was similar in the unvaccinated as in the vaccinated. The study further suggested that maternal immunity was protecting newborn babies from infection at the time when they are most vulnerable to develop chronic carrier status and HCC, and that the vaccine program reduces this natural immunity. Paradoxically, therefore, there is a substantial likelihood that Gates's vaccine is increasing the incidence of HCC in the country. These findings demonstrated the absurd futility of hepatitis B vaccination in India. "No matter," says Puliyel, "Gates's opinion was the only thing that counted." WHO stood firm, taking the position that all countries must include hepatitis B vaccine in their immunization program, even if the vaccine was unnecessary.

Haemophilus Influenzae B (Hib)

WHO followed its hepatitis B debacle with a much weaker recommendation for vaccination against *Haemophilus influenzae* type b (Hib). WHO recommended Hib vaccines only in nations suffering a grave disease burden. In an editorial in the Bulletin of the WHO, Indian doctors questioned the need for Hib vaccine in Asia, where the incidence of invasive Hib disease was extremely low (Lau 1999).[183] In 2002, Dr. Thomas Cherian, who is now the WHO Coordinator of EPI, wrote that based on the available data, Hib vaccine should not be recommended for routine use in India.

To overcome such meddling from India's prying medical community, in 2005 Gates funded, through GAVI, a four-year, $37 million study of mass vaccination with Hib jabs in Bangladesh intending to showcase the vaccine's benefits.[184,185] GAVI's Bangladesh study backfired, showing no advantage from Hib vaccination. In response, a formidable coterie of superstar international health experts—all of them, coincidentally, from Gates-funded organizations WHO, GAVI, UNICEF, USAID, Johns Hopkins Bloomberg School of Public Health, the London School of Hygiene and Tropical Medicine, and CDC—issued a deceitful proclamation that fraudulently claimed that the Bangladesh study proved a Hib jab protects children from "significant burden of life-threatening pneumonia and meningitis."[186] Prominent Indian doctors responded with outraged commentaries in the *British Medical Journal* and the *Indian Journal of Medical Research*, describing the Gates-funded study as a devious artifice.[187,188] Based on Gates's orchestrated guile, WHO in 2006 took the official position that the "Hib vaccine should be included in all routine immunization programmes."[189] Once again, the Indian government caved in to Gates and mandated Hib vaccines in India, where Hib invasive disease was nearly nonexistent.

In self-congratulatory articles, GAVI boasted triumphantly of its role in rescuing the Hib vaccine project in India after the Bangladesh study proved the vaccine a worthless waste of money (GAVI 2007; Levine et al. 2010).[190,191] GAVI's article notes that, since there was little burden from Hib disease in India, it had been a great challenge to gin up support for WHO's recommendation. GAVI bragged—in technocratic argot—that it twisted WHO's arm to revise WHO's Hib vaccine policy from a weak permissive statement[192] to a firm recommendation calling for universal vaccine introduction in all countries.[193] WHO's volte-face dragooned reticent Indian health officials to recommend the useless vaccine. Dr. Puliyel complains that incident "highlights the influence GAVI and other vaccine manufacturer-funded organizations like the 'Hib Initiative' have on the WHO and how it impacts vaccine uptake internationally."[194]

Puliyel protests that the Gates Foundation has privatized and monetized international public health policy, transforming WHO recommendations into effective mandates and compelling poor countries to pay annual tribute to foreign Pharma overlords. Puliyel told me that India and other Asian nations are now effectively compelled to administer the vaccine and to increase Hib uptake targets, "irrespective of an individual country's disease burden, notwithstanding of natural immunity attained within the country against the disease, and not taking into account the rights of sovereign States to decide how they use their limited resources." He adds that "The mandate and wisdom of issuing such a directive, for a disease that has little potential of becoming a pandemic, needs to be questioned."

Dr. Puliyel's commentary in the *BMJ* denounced Gates and GAVI for pushing Hib vaccine in developing countries and for falsifying the characterization of the research data in their press release: "The directive has come after a number of failed attempts to convince the scientific community of the need for this vaccine in Asia." Puliyel described the HiB saga as "a case study on the visible and invisible pressures brought to bear on governments to deploy expensive new vaccines."[195]

Pentavalent Vaccine

Despite Gates's victory in winning recommendations for Hib and hepatitis B in Asia, actual uptake rates disappointed the pharma mikados. Defying the WHO and Indian Health Ministry recommendations, local physicians stonewalled the jab. Most Indians had never heard of either illness. Dr. Puliyel told me, "Indian doctors were not impressed by the need for either Hib or hepatitis B jabs and seldom recommended them to patients." Physician resistance stymied Indian health officials from meeting WHO's uptake metrics for the newly recommended shots. To overcome this problem, Pharma introduced a diabolically cunning strategy to euthanize three birds with one stone. The companies withdrew their flagging Hib and hepatitis B vaccines and reissued a new concoction that combined those immunizations with the DTP, which, despite its popularity, had become another sandbag on Big Pharma's profit ambitions.

By 2008, Pfizer's DTP patent was long expired, and there were sixty-three manufacturers making the vaccine in forty-two countries with large surpluses and very low margins. The Gates cabal solved these profiteering problems by brewing up a new (five diseases) vaccine by mixing the DTP, Hib, and hepatitis B formulas in a single syringe. That new combination became a "new vaccine." The Global Alliance for Vaccines and Immunizations (GAVI) and WHO christened the novel, untested, and unlicensed concoction the "Pentavalent Vaccine" and recommended its use in developing countries to replace the DTP vaccine. Compliant Indian health ministries then phased out the DTP, which had been popular with doctors. Now, if any physician or individual wanted DTP, their only choice would be the Pentavalent vaccine.

On its website, GAVI admitted that its underlying reason for this caper was to increase the uptake of the hepatitis B and Hib vaccines in these countries by piggybacking on the well-accepted DTP vaccine. It was an ingenious moneymaking connivance. Competition had driven down the cost of the DTP to 15.50 rupees—about 14 cents US. The hepatitis B vaccine retailed for 45 rupees and the Hib for 25. Therefore, the combined cost of all three vaccines if purchased separately was Rs 185. However, the new pentavalent vaccine—made by Gates's friend Cyrus Poonawalla, owner of the Serum Institute of India—costs Rs 550, a 1,440 percent increase in profits for every vaccine sold![196]

The Food and Drug Administration (FDA) has not licensed the combination vaccine for either safety or efficacy, and developed countries do not use it. A Cochrane meta-analysis showed that the combination is less effective than the vaccines given separately. Furthermore, the pentavalent vaccine is life-threatening to infants.

Before its Indian debut, Bhutan, Sri Lanka, Pakistan, and Vietnam previewed the pentavalent jab. In each of these countries, unexplained deaths followed immunization. Bhutan suspended the immunization program in October 2009 after five cases of encephalopathy/encephalitis occurred following the vaccine. WHO persuaded health officials to resume the program, insisting that viral meningoencephalitis caused the deaths. Bhutan obeyed and four infants died. Bhutan no longer uses the pentavalent vaccine. The director of Public Health, Dr. Ugen Dophu, observes that there have been no more cases of meningoencephalitis among infants after the vaccine was withdrawn.[197]

Sri Lanka unleashed the pentavalent vaccine in January 2008 and then suspended the program four months later after five babies died. Under pressure from WHO, Sri Lanka reintroduced the vaccine in 2010. Between 2010 and 2012, there were fourteen additional deaths following the vaccine, making the total number of deaths in Sri Lanka nineteen.[198]

Vietnam introduced the pentavalent jab in June 2010 and suspended the jab in May 2013, after twenty-seven infant deaths.[199]

The experience in Pakistan was similar, including at least three reported deaths.[200]

India introduced pentavalent vaccine in December 2011. Up to the first quarter of 2013, health officials reported eighty-three serious Adverse Events Following

Immunization (AEFI). Twenty-one babies have died in India following immunization with the pentavalent vaccine.[201]

Gates and WHO simply trivialize the deaths as sad coincidences or collateral damage. The vaccine has effectively reduced the incidence of Hib disease in India. However, there has been a proportionate increase in non-Hib strains of H. influenzae, including non-serotypable strains, causing invasive disease in the post-Hib vaccine era. As usual, there was no accounting.

This is only one of many examples of the Gates Foundation prioritizing the mandate for high-cost vaccines in the national immunization programs that Bill Gates effectively controls. Putting aside questions about net costs and benefits from these dangerous jabs, McGoey agrees with Puliyel that the diversion of sanitation and nutrition money is also deadly: "The problem is that by prioritizing the delivery of expensive vaccines, other proven interventions lose out."[202]

Real-world evidence, including his investments in pharmaceutical, petroleum, chemical, and GMO, processed, and synthetic food, suggests that Gates's obsession with vaccines does not evince any genuine commitment to healthy populations. According to Amy Goodman, Gates owns investments in sixty-nine of the world's worst-polluting companies.[203] His single-minded obsession with vaccines seems to serve his impulse to monetize his charity and to achieve monopoly control over global public health policy. His strategies and corporate alliances in the food, public health, and education sectors may also reflect messianic conviction that he is ordained to save the world with technology, top-down centralized cookie-cutter solutions to complex human problems, and a godlike willingness to experiment with the lives of lesser humans.

And Gates's vaccine cartel has amassed Midas-like riches. Early in 2021, a TV interviewer, Becky Quick, observed that Gates had spent $10 billion on vaccines over the past two decades and asked Gates, "You've figured out the return on investment for that and it kind of stunned me. Can you walk us through the math?" Gates responded: "We see a phenomenal track record . . . there's been over a 20-to-1 return. So if you just looked at the economic benefits, that's a pretty strong number." The interviewer pressed him: "If you had put that money into an S&P 500 and reinvested the dividends, you'd come up with something like $17 billion dollars, but you think it's $200 billion dollars." Gates continued: "Here, yeah," hastening to add that "helping young children live, get the right nutrition, contribute to their countries, that has a payback that goes beyond any typical financial return."[204]

The key to it all, he added, is "Having that big portfolio."

And the key to much of that portfolio is having Anthony Fauci.

Endnotes

1   NVIC, CDC Recommended Vaccine Schedule: 1983 vs 2017 (Jan. 2018), nvic.org/cmstemplates/nvic/pdf/downloads/1983-2017-vaccine-schedules.pdf

2   Informed Consent Action Network, Letter to the U.S. Department of Health and Human Services, (Oct. 12, 2017), childrenshealthdefense.org/wp-content/uploads/ican-notice-october-12-2017.pdf

3   U.S. Department of Health and Human Services, Letter to Informed Consent Action Network (Jan. 18, 2018), childrenshealthdefense.org/wp-content/uploads/hhs-response-january-29-2018.pdf

4   Avik Roy, "How the FDA Stifles New Cures, Part II: 90% of Clinical Trial Costs are Incurred in Phase III," Forbes (Apr. 25, 2012), forbes.com/sites/theapothecary/2012/04/25/how-the-fda-stifles-new-cures-part-ii-90-of-clinical-trial-costs-are-incurred-in-phase-iii/?sh=6c0e403c7b52

5   Lynsey Chutel, "Bad Medicine: Africa is the dumping ground for 40% of the world's reported fake medicines," *Quartz Africa* (Nov. 29, 2017), qz.com/africa/1140890/one-in-ten-medical-products-sent-to-developing-countries-are-falsified-or-below-standard-who/

6   Jeremy Loffredo and Michele Greenstein, "Why the Bill Gates Global Health Empire Promises More Empire and Less Public Health," *The Gray Zone* (July 8, 2020), thegrayzone.com/2020/07/08/bill-gates-global-health-policy/

7   Matthew Herper, "With Vaccines, Bill Gates Changes the World Again," *Forbes* (Nov. 2, 2011), forbes.com/sites/matthewherper/2011/11/02/the-second-coming-of-bill-gates/?sh=19b990e713fd

8   Linsey McGoey, *No Such Thing as a Free Gift: The Gates Foundatioin and the Price of Philanthropy* (Verso, 2016), 226

9   Julia Crawford, "Does Bill Gates have too much influence in the WHO?" (May 10, 2021), swissinfo.ch/eng/politics/does-bill-gates-have-too-much-influence-in-the-who-/46570526

10   Gordon T. Stewart, "Vaccination Against Whooping Cough: Efficacy versus Risks," *The Lancet* (Jan. 29, 1977), sci-hub.se/https://doi.org/10.1016/S0140-6736(77)91028-5

11   Larry J. Baraff et al., "Possible temporal association between diphtheria-tetanus toxoid-pertussis vaccination and sudden infant death syndrome," *Pediatric Infectious Disease Journal* (Jan. 1983), sci-hub.se/10.1097/00006454-198301000-00003

12   99th Congress (1985–1986), H.R.5546–National Childhood Vaccine Injury Act of 1986, (Oct 18, 1986). congress.gov/bill/99th-congress/house-bill/5546

13   Institute of Medicine, New Vaccine Development, *Establishing Priorities: Volume I: Diseases of Importance in the United States* (National Academies Press, 1985), 172–217, doi.org/10.17226/12085

14   Centers for Disease Control and Prevention, "Pertussis Vaccination: Use of Acellular Pertussis Vaccines Among Infants and Young Children Recommendations of the Advisory Committee on Immunization Practices (ACIP)," MMWR (Mar 28, 1997), cdc.gov/mmwr/preview/mmwrhtml/00048610.htm

15   Peter Aaby et al., "Evidence of increase in mortality after the introduction of diphtheria–tetanus–pertussis vaccine to children aged 6–35 months in Guinea-Bissau: a time for reflection?" *Frontiers in Public Health* (2018), ncbi.nlm.nih.gov/pmc/articles/PMC5868131/

16   Mariam Saleh, "Population of Africa in 2020, by age group, STATISTA" (Jun. 18, 2021), statista.com/statistics/1226211/population-of-africa-by-age-group/

17   Violaine Mitchell et al., Immunization in Developing Countries (2013), sciencedirect.com/topics/medicine-and-dentistry/vaccination-coverage

18   Anna N. Chard et al., "Routine Vaccination Coverage—Worldwide, 2019," *MMWR* (Nov. 13, 2020), cdc.gov/mmwr/volumes/69/wr/mm6945a7.htm#T2_down

19   Mogensen et al., "Evidence of Increase in Mortality After the Introduction of Diphtheria-Tetanus-Pertussis Vaccine to Children Aged 6-35 Months in Guinea-Bissau: A Time for Reflection?" *Front Public Health*, 2018;6:79 (March 19, 2018), ncbi.nlm.nih.gov/pmc/articles/PMC5868131/

20   Ibid.

21   Mogensen et al.

22   "Bill Gates Challenges Global Leaders at 64th World Health Assembly: Make Vaccines a Priority to Save 10 Million Lives," Bill & Melinda Gates Foundation (May, 2011), gatesfoundation.org/ideas/media-center/press-releases/2011/05/bill-gates-challenges-global-leaders-at-64th-world-health-assembly-make-vaccines-a-priority-to-save-10-million-lives

23   JPT Higgins et al., *Systematic Review of the Non-specific Effects of BCG, DTP and Measles Containing Vaccines* (March 13, 2014), who.int/immunization/sage/meetings/2014/april/3_NSE_Epidemiology_review_Report_to_SAGE_14_Mar_FINAL.pdf

24   Steve Stecklow, "Products Turn Red To Augment AIDS Fund," *Wall Street Journal* (Apr 13, 2006), wsj.com/articles/SB114488885543224665

25   *Medical Racism: The New Apartheid*, CHD Films, 2021, 41:52, medicalracism.childrenshealthdefense.org/medical-racism-the-new-apartheid/.

26   Margaret Chan, Chris Elias, Anthony Fauci, Anthony Lake, and Seth Berkley, "Reaching everyone, everywhere with life-saving vaccines," *The Lancet*, vol. 389, issue 10071 (2017): 777, doi.org/10.1016/S0140-6736(17)30554-8

27   Jeremy R. Hammond, "WHO Experimenting on African Children without Informed Consent," *Foreign Policy Journal* (Mar. 1, 2020), foreignpolicyjournal.com/2020/03/01/who-experimenting-on-african-children-without-informed-consent

28   Jose G. Dorea et al., "Neonate Exposure to Thimerosal Mercury from Hepatitis B Vaccines," *American Journal of Perinatology* (2009), sci-hub.se/10.1055/s-0029-1215431

29   Robert F. Kennedy Jr. and Mark Hyman, *Thimerosal: Let the Science Speak* (Skyhorse, 2014)

30   GENERATION ZERO, "Thomas Verstraeten's First Analyses of the Link Between Vaccine Mercury Exposure and the Risk of Diagnosis of Selected Neuro-Developmental Disorders Based on Data from the Vaccine Safety Datalink: November-December 1999," Safe Minds, November 2004, https://childrenshealthdefense.org/wp-content/uploads/safeminds-generation-zero.pdf

31   Kennedy and Hyman

32   "International Organizations, Drug Firms Convene at Vaccination Meeting," Reuters Health (March 3, 2000), childrenshealthdefense.org/wp-content/uploads/Merck-donates-TC-Hep-B-vaccine-to-3rd-world-March-2000.pdf

33   Children's Health Defense, "Summary of Key Facts: Vaccine Injury in the African American Community" (2020), childrenshealthdefense.org/wp-content/uploads/Vaccine-Injury-in-AA-community_BiFoldwChart_REVISED_r1.pdf

34   Carolyn M. Gallagher and Melody S. Goodman, "Hepatitis B Vaccination of Male Neonates and Autism Diagnosis, NHIS 1997–2002," *Journal of Toxicology and Environmental Health*, Part A (Apr 10, 2010), publichealth.stonybrookmedicine.edu/phpubfiles/Hep_B_and_autism.pdf

35   "Fauci: Forging closer ties with GAVI," GAVI Media Room (May 22, 2012), https://www.gavi.org/news/media-room/fauci-forging-closer-ties-gavi

36   Alan Challoner, "Rapid Response: How can vaccines cause damage?" BMJ (Feb 29, 2004), bmj.com/rapid-response/2011/10/30/how-can-vaccines-cause-damage

37   Mogensen et al.

38   "655,000 malaria deaths in 2010: WHO; Africa accounted for 91 percent of deaths," *New York Daily News* (Dec. 14, 2011, 10:48 AM), nydailynews.com/life-style/health/655-000-malaria-deaths-2010-africa-accounted-91-percent-deaths-article-1.991359

39   The RTS's Clinical Trials Partnership, "First Results of Phase 3 Trials of RTS,S/AS01 Malaria Vaccine in African Children," *N Engl J Med* 365:1863–1875 (Nov. 17, 2011), nejm.org/doi/full/10.1056/NEJMoa1102287

40   Ivana Kottasova, "World's first malaria vaccine, backed by Bill Gates, gets a green light," CNN Money (July 24, 2015, 12:05 PM), money.cnn.com/2015/07/24/news/malaria-vaccine-bill-gates/index.html

41   "New partnership between GSK and the Bill & Melinda Gates Foundation to accelerate research into vaccines for global health needs," GSK Press Release (Oct. 29, 2013), https://www.gsk.com/en-gb/media/press-releases/new-partnership-between-gsk-and-the-bill-melinda-gates-foundation-to-accelerate-research-into-vaccines-for-global-health-needs/

42   The RTS's Clinical Trials Partnership

43   PATH, "RTS'S malaria candidate vaccine reduces malaria by approximately one-third in African infants," PATH (Nov. 9, 2012), path.org/media-center/rtss-malaria-candidate-vaccine-reduces-malaria-by-approximately-one-third-in-african-infants/

44   Kate Kelland and Ben Hirschler, "Setback for first malaria vaccine in African trial," Reuters (Nov. 9, 2012, 1:05 AM), reuters.com/article/us-malaria-vaccine-gsk/setback-for-first-malaria-vaccine-in-african-trial-idUSBRE8A80I120121109

45   Peter Doshi, "WHO's malaria vaccine study represents a 'serious breach of international ethical standards,'" BMJ (Feb, 2020), bmj.com/content/368/bmj.m734/rapid-responses

46   Jop de Vrieze, "First malaria vaccine rolled out in Africa despite limited efficacy and nagging safety concerns," *Science* (Nov. 26, 2019, 3:04 PM), sciencemag.org/news/2019/11/first-malaria-vaccine-rolled-out-africa-despite-limited-efficacy-and-nagging-safety

47   Christina England, "Minimum of 40 Children Paralyzed After New Meningitis Vaccine," LaLeva (Jan 7, 2013), laleva.org/eng/2013/01/minimum_of_40_children_paralyzed_after_new_meningitis_vaccine.html

48   Sharmeen Ahmed, "Accountability of International NGOs: Human Rights Violations in Healthcare Provision in Developing Countries and the Effectiveness of Current Measures," Annual Survey of International and Comparative Law (2017), digitalcommons.law.ggu.edu/cgi/viewcontent.cgi?article=1205&context=annlsurvey

49   Clement Kpeklitsu, "Transhumanism: Dark Secrets You Need to Know," Ghanaian Times, ghanaiantimes.com.gh/transhumanism-dark-secrets-you-need-to-know

50   Edwin Black, "Eugenics and the Nazis," in *Beyond Bioethics* (University of California Press, 2018), 52, doi.org/10.1525/9780520961944-008

51   David Turner, "Foundations of Holocaust: American eugenics and the Nazi connection," *Jerusalem Post*, (Dec 30, 2012). https://www.jpost.com/blogs/the-jewish-problem---from-anti-judaism-to-anti-semitism/foundations-of-holocaust-american-eugenics-and-the-nazi-connection-364998

52   Edwin Black, "Hitler's Debt to America," an excerpt from: *War Against the Weak: Eugenics and America's Campaign to Create a Master Race, The Guardian* (Feb. 5, 2004 21.36 EST), theguardian.com/uk/2004/feb/06/race.usa

53   McGoey, 151

54   Ibid., 152

55   Ibid.

56   Ibid.

57   Ibid., 153

58   Margaret Sanger, *The Pivot of Civilization,* Project Gutenberg (2008), gutenberg.org/files/1689/1689-h/1689-h.htm#link2HCH0005

59   Tanya L. Green, "The NEGRO PROJECT: Margaret Sanger's EUGENIC Plan for Black America," Black Genocide (2012), blackgenocide.org/archived_articles/negro.html

60   Peter Engleman, *A History of the Birth Control Movement in America* (Praeger, 2011), 132–133

61   Neil Foster, "The Money, the Power and Insanity of Bill Gates—The Planned Parenthood Depopulation Agenda," Sovereign Independent UK (Jan. 31, 2013), adam.curry.com/art/1380297046_Au563ZWf.html

62   Bill Moyers, "A conversation with Bill Gates: Making a Healthier World for Children and Future Generations," *Moyers* (May, 9, 2003), billmoyers.com/content/conversation-bill-gates-making-healthier-world-children-future-generations-transcript/

63   Andrew Leonard, "Is Bill Gates a closet liberal? The money trail of his philanthropy suggests some clues to the political leanings of Microsoft's founder," *Salon* (Jan. 29, 1998 8:00 PM), salon.com/1998/01/29/feature_349/

64   Bill Gates, "Innovating to Zero!", TED Talk 4:33 (Feb. 2010), ted.com/talks/bill_gates_innovating_to_zero

65   Ibid. 4:25

66   Alexander Higgins, "Gates makes $10 billion vaccines pledge," Associated Press (Jan. 29, 2001), archive.boston.com/business/technology/articles/2010/01/29/gates_makes_10_billion_vaccines_pledge/

67   "UN Population Fund Hails Bill And Melinda Gates' $2.2 Billion Donation To Fund Population And Health Activities Worldwide," UN Press Release (Feb. 11, 1999), un.org/press/en/1999/19990211.pop704.html

68   "Gates Foundations Give Johns Hopkins $20 Million Gift to School of Public Health for Population, Reproductive Health Institute," Press Release: Bill & Melinda Gates Foundation (May 1999), gatesfoundation.org/ideas/media-center/press-releases/1999/05/johns-hopkins-university-school-of-public-health

69   Emma Batha, "Contraceptives are 'one of the greatest anti-poverty innovations': Melinda Gates," Reuters (July 11, 2017, 1:32 AM), reuters.com/article/us-global-contraception-summit/contraceptives-are-one-of-the-greatest-anti-poverty-innovations-melinda-gates-idUSKBN19W0PC

70   Kathryn Joyce, "The New War on Contraceptives: How the Christian Right is co-opting the women's rights movement to fight contraceptives in Africa," *Pacific Standard* (Updated: Sep 16, 2018, Original: Aug 17, 2017), psmag.com/magazine/new-war-on-birth-control

71   "Depo-Provera, the Birth Control Shot, as a Form of Birth Control," Planned Parenthood, YouTube (Feb. 4, 2010), youtube.com/watch?v=crw-8CPMPfE

72   "Que es el Provera?," YouTube (Nov 10, 2016), youtube.com/watch?v=xJ9wvjYtqTs

73   Pfizer Inc, Full Prescribing Information: Contents Depo-Provera, FDA, Revised Oct. 2010, accessdata.fda.gov/drugsatfda_docs/label/2010/020246s036lbl.pdf

74   Christopher I. Li et al., "Effect of Depo-Medroxyprogesterone Acetate on Breast Cancer Risk among Women 20 to 44 Years of Age," American Association for Cancer Research (2012), ncbi.nlm.nih.gov/pmc/articles/PMC3328650/

75   The Rebecca Project for Justice, 13

76   Ibid., 2

77   Ibid., 6

78   Ibid.

79   Keegan Hamilton, "Depo Danger: AIDS, Africa, and the UW researchers who rocked the medical world," Seattle Weekly (Dec 6, 2011), seattleweekly.com/news/depo-danger/

80   Jacob Puliyel, "Ethical questions that surround vaccine to reduce fertility," *Sunday Guardian Live* (May 28, 2018, 7:09 PM), sundayguardianlive.com/news/ethical-questions-surround-vaccine-reduce-fertility

81   Renee Heffron, MPH, et al., "Use of Hormonal Contraceptives and Risk of HIV-1 Transmission: a prospective cohort study," *The Lancet* (Oct 04, 2011), thelancet.com/journals/laninf/article/PIIS1473-3099(11)70247-X/fulltext

82   Renee Heffron et al., "Hormonal contraceptive use and risk of HIV-1 transmission: a prospective cohort analysis," *The Lancet Infectious Diseases* (Jan., 2012), sci-hub.se/10.1016/S1473-3099(11)70247-X

83   The Rebecca Project for Justice, 11–12

84   Ibid., 12

85   Lisa Correnti, APFLI, "Gates Foundation Suspected of Forcing Controversial Contraceptive on Africans," Physicians for Life (Nov. 15, 2015), physiciansforlife.org/gates-foundation-suspected-of-forcing-controversial-contraceptive-on-africans/

86   Sandi Doughton, "Gates Foundation: $1B for contraceptives in developing countries," Seattle Times (Jul 12, 2012), seattletimes.com/seattle-news/gates-foundation-1b-for-contraceptives-in-developing-countries/

87   The Rebecca Project for Justice, 11

88   Jonathan Abbamonte, "Depo-Provera contraceptive increases risk of HIV infection: new study," Lifesite News (Mar 8, 2016),lifesitenews.com/news/new-pri-study-shows-depo-provera-increases-risk-of-hiv-infection/

89   Correnti

90   PFIZER'S SAYANA® PRESS BECOMES FIRST INJECTABLE CONTRACEPTIVE IN THE UNITED KINGDOM AVAILABLE FOR ADMINISTRATION BY SELF-INJECTION, Pfizer Press Release (Sept. 23, 2015, 8:01pm), pfizer.com/news/press-release/press-release-detail/pfizer_s_sayana_press_becomes_first_injectable_contraceptive_in_the_united_kingdom_available_for_administration_by_self_injection.

91   The Rebecca Project for Justice, 1

92   Ibid., 11
93   Ibid., 13
94   Ibid., 9–10
95   Ibid., 1
96   Ibid.
97   Ibid., 10
98   Ibid., 6
99   Oller et al., "HCG Found in WHO Tetanus Vaccine in Kenya Raises Concern in the Developing World," *2017 Open Access Library Journal* 4(10):1–32 doi: 10.4236/oalib.1103937.
100   Steve Weatherbe, "'A mass sterilization exercise': Kenyan doctors find anti-fertility agent in UN tetanus vaccine," "This WHO campaign is not about eradicating neonatal tetanus but a well-coordinated forceful population control mass sterilization exercise using a proven fertility regulating vaccine," Lifesite News (November 6, 2014), lifesitenews.com/news/a-mass-sterilization-exercise-kenyan-doctors-find-anti-fertility-agent-in-u
101   Beth Griffin, "WHO, UNICEF deny Kenyan bishops' claim that they supplied sterility-causing tetanus vaccines," *National Catholic Reporter* (March 9, 2015), ncronline.org/news/world/who-unicef-deny-kenyan-bishops-claim-they-supplied-sterility-causing-tetanus-vaccines
102   Abby Ohlheiser, "The tense standoff between Catholic bishops and the Kenyan government over tetanus vaccines," *Washington Post*, (Nov 14, 2014), https://www.washingtonpost.com/news/worldviews/wp/2014/11/14/the-tense-standoff-between-catholic-bishops-and-the-kenyan-government-over-tetanus-vaccines/
103   Oller, Shaw, et al. (2017), "HCG Found in WHO Tetanus Vaccine in Kenya Raises Concern in the Developing World," *Open Access Library Journal*, 4, 1–32. doi: 10.4236/oalib.1103937. scirp.org/journal/paperinformation.aspx?paperid=81838
104   Kenya Catholic Doctors Association, "Catholic Church Warning: Neonatal Tetanus Vaccine by WHO Is DEADLY and Bad for Women Reproductivity," *Kenya Today*, kenya-today.com/news/catholic-warning-neonatal-tetanus-vaccine-wto-deadly-bad-women-reproductivity
105   Admin, Liberty Writers, "UNICEF is making women barren through polio vaccine, African nations should be alert: Kenyan doctors say" (October 10, 2019), branapress.com/2019/10/10/unicef-is-making-women-barren-through-polio-vaccine-african-nations-should-be-alert-kenyan-doctors-say/
106   Oller, Shaw, et al.
107   CHD Films, *Medical Racism: The New Apartheid*
108   Oller, Shaw, et al.
109   Ibid.
110   Talwar et al., "Isoimmunization against human chorionic gonadotropin with conjugates of processed beta-subunit of the hormone and tetanus toxoid," *Proceedings of the National Academy of Sciences*, Jan 1976, 73 (1): 218–222; DOI: 10.1073/pnas.73.1.218 https://www.pnas.org/content/73/1/218
111   Oller, Shaw, et al.
112   Oller, Shaw, et al.
113   Ibid.
114   Committed Grants, Bill & Melinda Gates Foundation (Nov. 2020), gatesfoundation.org/about/committed-grants/2020/11/inv023725
115   Population Council, Influence Watch (2021), influencewatch.org/non-profit/population-council/
116   Ibid.
117   Ibid.
118   Oller, Shaw, et al.
119   USAID, "Family Planning Program Timeline: Before 1965 to the Present" (Jun. 2, 2019), usaid.gov/global-health/health-areas/family-planning/usaid-family-planning-program-timeline-1965-present
120   David Miliband and Rajiv J. Shah, "Worldwide Leadership has been severely lacking over COVID-19—we need a better way," The Independent (May 25, 2001), independent.co.uk/voices/covid-vaccine-who-governments-un-b1853498.html
121   Oller, Shaw, et al.
122   Ibid.
123   USAID, U.S. National Security Study Memorandum 200, (Dec. 10, 1974), pdf.usaid.gov/pdf_docs/PCAAB500.pdf
124   Oller, Shaw, et al.
125   USAID, U.S. National Security Study Memorandum 200
126   Oller, Shaw, et al.
127   USAID, U.S. National Security Study Memorandum 200f
128   Ibid., 42
129   Oller, Shaw, et al.
130   Ibid., reference 3
131   Fredrick Nzwili, "Kenya's Catholic Bishops: Tetanus vaccine is birth control in disguise, *Washington Post* (Nov. 11, 2014), washingtonpost.com/national/religion/kenyas-catholic-bishops-tetanus-vaccine-is-birth-control-in-disguise/2014/11/11/3ece10ce-69ce-11e4-bafd-6598192a448d_story.html
132   Oller, Shaw, et al., 8
133   Bryant et al., "Climate change and family planning: least-developed countries define the

agenda," *Bulletin of the World Health Organization*, 852–853, 2009, who.int/bulletin/volumes/87/11/08-062562.pdf

134 Ibid., 852

135 WHO Special Programme of Research, Development and Research Training in Human Reproduction. (1993), "Fertility regulating vaccines: report of a meeting between women's health advocates and scientists to review the current status of the development of fertility regulating vaccines," Geneva, 17–18 August 1992, World Health Organization, p. 13, apps.who.int/iris/handle/10665/61301

136 Bryant et al., "Climate change and family planning: least-developed countries define the agenda," *Bulletin of the World Health Organization*, 853, 2009, who.int/bulletin/volumes/87/11/08-062562.pdf

137 Province Opinion, "David Morley and Dr. John Button: Just $2.18 can save a baby from an awful death," *The Province* (Apr. 25, 2015), theprovince.com/opinion/david-morley-and-dr-john-button-just-2-18-can-save-a-baby-from-an-awful-death

138 Oller, Shaw, et al.

139 McGoey, 155

140 Regis A. Vilchez et al, "Simian virus 40 in human cancers," The American Journal of Medicine, (Jun 1, 2003). https://www.amjmed.com/article/S0002-9343(03)00087-1/fulltext

141 Lyn Redwood, "From Chimpanzees to Children: The Origins of RSV—Respiratory Syncytial Virus," The Defender, (Aug 27, 2021). https://childrenshealthdefense.org/defender/chimpanzees-children-origins-respiratory-syncytial-virus/

142 Ibid., 159

143 Rachana Dhiman et al., "Correlation between Non-Polio Acute Flaccid Paralysis Rates with Pulse Polio Frequency in India," *International Journal of Environmental Research and Public Health* (Aug., 2018), ncbi.nlm.nih.gov/pmc/articles/PMC6121585/

144 Loffredo and Greenstein

145 Dan Hurley, "The Mysterious Polio-Like Disease Affecting American Kids," *The Atlantic* (Oct. 24, 2014), theatlantic.com/health/archive/2014/10/the-mysterious-polio-like-disease-affecting-american-kids/381869/

146 Viera Scheibner, "Rapid Response to Polio eradication: a complex end game," *BMJ* (Apr. 10, 2012), bmj.com/content/344/bmj.e2398/rr/578260

147 "Polio in India: Richard Gale, Dr. Gary Null, Ph.D., Neal Greenfield, Esq., The Myths about the Polio Vaccine's Safety and Efficacy," Progressive Radio Network (September 16, 2019), list.uvm.edu/cgi-bin/wa?A2=ind1909&L=SCIENCE-FOR-THE-PEOPLE&P=32923

148 Helen Branswell, "Global nonprofit says it will help deliver vaccines to Syrian children," Stat (Dec. 8, 2016), statnews.com/2016/12/08/syria-vaccines-children/

149 Helen Branswell, "Polio outbreak is reported in Syria, WHO says," Stat (June 8, 2017), statnews.com/2017/06/08/polio-outbreak-syria-who/

150 The Associated Press, "More Polio Cases Now Caused by Vaccine than by Wild Virus," ABC News (Nov. 25, 2019, 12:19 PM), abcnews.go.com/Health/wireStory/polio-cases-now-caused-vaccine-wild-virus-67287290

151 Scheibner

152 Jason Beaubien, "Rare but Real: Mutant Strains of Polio Vaccine Cause More Paralysis than Wild Polio," NPR (Jun 28, 2017, 3:22 PM), npr.org/sections/goatsandsoda/2017/06/28/534403083/mutant-strains-of-polio-vaccine-now-cause-more-paralysis-than-wild-polio

153 McGoey, 155.

154 Ibid., 157

155 Ibid., 158

156 Ibid., 160

157 Ibid., 158

158 Donald G. McNeil, Jr., "Can Polio Be Eradicated? A Skeptic Now Thinks So," *New York Times* (Feb. 14, 2011), nytimes.com/2011/02/15/health/15polio.html

159 Bill & Melinda Gates Foundation, Polio (2020), gatesfoundation.org/our-work/programs/global-development/polio

160 Christina Sarich, "Bill Gates Faces Trial in India for Illegally Testing Tribal Children with Vaccines," Natural Society (October 13, 2014), naturalsociety.com/bill-gates-faces-trial-india-illegally-testing-tribal-children-vaccines/

161 GlaxoSmithKline, "New partnership between GSK and the Bill & Melinda Gates Foundation to accelerate research into vaccines for global health needs" (Oct. 29, 2013), gsk.com/en-gb/media/press-releases/new-partnership-between-gsk-and-the-bill-melinda-gates-foundation-to-accelerate-research-into-vaccines-for-global-health-needs/

162 Bill & Melinda Gates Foundation, "The Bill & Melinda Gates Foundation, Merck & Co., Inc. and the Republic of Botswana Launch New HIV Initiative" (Jul, 2000), https://www.gatesfoundation.org/ideas/media-center/press-releases/2000/07/comprehensive-hivaids-partnership

163 Merck, "Merck Announces Full-Year and Fourth-Quarter 2011 Financial Results" (Feb. 2, 2012), merck.com/news/merck-announces-full-year-and-fourth-quarter-2011-financial-results/

164 McGoey, 161

165 John Spritzler, "Bill Gates Is Not a Benign Philanthropist, Quite the Contrary" (Aug. 22, 2018), newdemocracyworld.org/culture/gates.html

166 Ibid.

167 McGoey, 161

168 Ibid., 162

169 Ibid.

170 "Bill Gates: What You Were Not Told," BitChute, 00:07:50–00:08:33, https://www.bitchute.com/video/3SYybOqK9p5o/

171 Ibid., 164

172 Ibid., 166

173 Zachariah Otto v. Merck et al., Superior Court of the state of California (Sep. 16, 2020), https://dockets.justia.com/docket/california/cacdce/8:2021cv00899/820706

174 Ibid.

175 Ibid.

176 Ibid.

177 Merck. Gardasil package insert. (2011),merck.com/product/usa/pi_circulars/g/gardasil/gardasil_pi.pdf

178 Baum Hedlund Law, Merck Gardasil Clinical Trials (2021), baumhedlundlaw.com/prescription-drugs/gardasil-lawsuit/merck-gardasil-clinical-trials/

179 Gretchen Livingston, "Is U.S. fertility at an all-time low? Two of three measures point to yes," Pew Research Center (May 22, 2019), pewresearch.org/fact-tank/2019/05/22/u-s-fertility-rate-explained/

180 Robert F. Kennedy, Jr., "Is Gardasil Vaccine Linked to Record Birth Rate Declines?" CHD (Mar 5, 2021), childrenshealthdefense.org/defender/gardasil-vaccine-linked-birth-rate-declines/

181 Celeste McGovern, "Vaccine Boom, Population Bust: Study Queries the Links Between HPV Vaccine and Soaring Infertility," CHD (Nov. 28, 2018), childrenshealthdefense.org/news/vaccine-safety/vaccine-boom-population-bust-study-queries-the-link-between-hpv-vaccine-and-soaring-infertility/

182 Jacob Puliyel, "Should immunisation against hepatitis B take priority over provision of clean drinking water?," *BMJ Letters* (Jul. 17, 1999), ncbi.nlm.nih.gov/pmc/articles/PMC1116282/

183 Y. L. Lau, "*Haemophilus influenzae* type b diseases in Asia," *Bulletin of the World Health Organization* (1999), ncbi.nlm.nih.gov/pmc/articles/PMC2557750/

184 Gavi, "Hib Initiative: a GAVI success story," GAVI, gavi.org/news/media-room/hib-initiative-gavi-success-story

185 Abdullah H. Baqui et al., "Effectiveness of Haemophilus influenzae type B conjugate vaccine on prevention of pneumonia and meningitis in Bangladeshi children: a case-control study," *Pediatric Infectious Disease Journal* (Jul. 2007), pubmed.ncbi.nlm.nih.gov/17596795/

186 Johns Hopkins Bloomberg School of Public Health, "Hib Vaccine: A Critical Ally in Asia's Effort to Reduce Child Deaths" (Jun. 27, 2007), jhsph.edu/news/news-releases/2007/baqui-hib.html

187 Zubair Lone and Jacob M. Puliyel, "Introducing pentavalent vaccine in the EPI in India: A counsel for caution," *Indian Journal of Medical Research* (Jul., 2010), jacob.puliyel.com/download.php?id=212

188 Jacob Puliyel, "GAVI and WHO: Demanding Accountability," *BMJ Letters* (Aug. 7, 2010), jacob.puliyel.com/download.php?id=217

189 "WHO Position Paper on Haemophilus influenzae Vaccines" (Nov. 24, 2006), who.int/immunization/documents/HIB_references.pdf

190 GAVI, "Hib vaccine - critical in Asia's effort to reduce child deaths," GAVI (Jun. 28, 2007), gavi.org/news/media-room/hib-vaccine-critical-asias-effort-reduce-child-deaths

191 Orin S. Levine et al., "A policy framework for accelerating adoption of new vaccines," *Human Vaccines* (Dec., 2010), ncbi.nlm.nih.gov/pmc/articles/PMC3060382/

192 "Review Panel on Haemophilus Influenzae Type B (Hib) Disease Burden in Bangladesh, Indonesia and Other Asian Countries, Bangkok, January 28–29, 2004," *Weekly Epidemiological* Record 79, No.18 (Apr. 30, 2004): 173–174, https://pubmed.ncbi.nlm.nih.gov/15168565/

193 "WHO Position Paper on Haemophilus Influenzae Type B Conjugate Vaccines" (Replaces WHO Position Paper on Hib Vaccines Previously Published in the Weekly Epidemiological Record), *Weekly Epidemiological Record* 81, No. 47 (Nov. 24, 2006): 445–452

194 Imrana Qadeer and P. M. Arathi, *Universalising Healthcare in India: From Care to Coverage* (Aakaar Books, 2019), 291

195 J. Puliyel, "GAVI and WHO: Demanding Accountability," *BMJ*, 2010, doi: https://doi.org/10.1136/bmj.c4081

196 Jacob M. Puliyel, "AEFI and the pentavalent vaccine: looking for a composite picture," *Indian Journal of Medical Ethics* vol. 10, no. 3 (Jul.–Sep., 2013): 142–146, ijme.in/pdfs/213ed142.pdf

197 Gouri Rao Passi, "News In Brief," *Indian Pediatrics* (Dec. 2013), indianpediatrics.net/dec2013/dec-1165.htm

198 Ibid.

199 Ibid.

200 S. K. Mittal, "'Sudden Deaths' after Pentavalent vaccination: Is the vaccine really safe?," *BMJ* (Jul. 23,

2010), bmj.com/rapid-response/2011/11/02/sudden-deaths-after-pentavalent-vaccination-vaccine-really-safe

201   Global Advisory Committee on Vaccine Safety review of pentavalent vaccine safety concerns in four Asian countries (June 12, 2013), who.int/vaccine_safety/committee/topics/hpv/GACVSstatement_pentavalent_June2013.pdf

202   McGoey, 161

203   Amy Goodman, "Report: Gates Foundation Causing Harm with the Same Money It Uses to Do Good," *DEMOCRACY NOW!*, 46:39 (Jan. 9, 2007), democracynow.org/2007/1/9/report_gates_foundation_causing_harm_with

204   Mathew J. Belvedere, "Bill Gates: My 'best investment' turned $10 billion into $200 billion worth of economic benefit," CNBC (Jan. 23, 2019, 7:13 a.m.), cnbc.com/2019/01/23/bill-gates-turns-10-billion-into-200-billion-worth-of-economic-benefit.html

ChildrensHealthDefense.org/fauci-book
childrenshd.org/fauci-book

For updates, new citations and references, and
new information about topics in this chapter:

CHAPTER 11

HYPING PHONY EPIDEMICS: "CRYING WOLF"

"Governments do like epidemics, just the same way as they like war, really. It's a chance to impose their will on us and get us all scared so that we huddle together and do what we're told."
—Dr. Damien Downing, President, British Society of Ecological Medicine
(Al Jazeera, 2009)

"Fear is a market. To instill fear in people also has advantages. Not only in terms of drug use. Anxiety-driven people are easier to rule."
—Gerd Gogerenzer, Director Emeritus at the Max Planck Institute for
Educational Research (Torsten Engelbrecht, *Virus Mania*, 2021)

In 1906, infectious disease caused a third of all annual deaths in the United States, and 800–1000 of every 100,000 Americans died of infectious disease. By 1976, fewer than fifty Americans per hundred thousand died of infectious diseases, and CDC and NIAID were under extreme pressure to justify their budgets. Hyping pandemics became an institutional strategy in both agencies. Pharmaceutical companies and international health agencies, banking and military contractors soon found purchase in the ecosystem, and random pandemics discovered their own self-perpetuating rationale. Dr. Fauci's critics chide him for routinely exaggerating—and even concocting—global disease outbreaks to hype pandemic panic, elevate the biosecurity agenda, boost agency funding, promote profitable vaccines for his pharma partners, and magnify his own power. The historical record supports these charges.

1976 Swine Flu

As chief of the NIAID's Clinical Physiology Section of the Laboratory of Clinical Investigation, Dr. Fauci was, in 1976, a frontline spectator during the NIH's bogus swine flu pandemic. That year, a soldier at Fort Dix died of a lung ailment following a forced march. Army physicians sent some samples to CDC, which identified the malady as a swine flu. Dr. Fauci's NIAID boss, Richard Krause (who Dr. Fauci would shortly replace), labored with his CDC counterpart, David Sencer, to spread terror of a catastrophic pandemic and initiate public demand for a vaccine. The NIAID chief convened in-house strategy sessions with Merck's iconic vaccine developer Maurice Hilleman and other immunization industry nabobs.[1] Congressional investigators subsequently landed the notes from those consultations, in which Dr. Hilleman candidly confesses that the resulting vaccine "had nothing to do with science and everything to do with politics." In the August 2020 *Rolling Stone*, Gerald Posner, author of *Pharma: Greed, Lies, and the Poisoning of America*,[2] recounted how Merck and other manufacturers utilized their secret meeting with the regulators to hatch a scheme that would guarantee industry profits while shielding Pharma from liability.[3] This innovation—now a persistent feature of Big Pharma's business model—turned out to be carte blanche for negligent and even criminal behavior.

Pharma and NIAID told Congress, the White House, and the public that the Fort Dix swine flu was the same strain responsible for the 1918 Spanish flu pandemic, which, they warned, had killed 50 million people worldwide.[4] They were lying; scientists at Fort Dix, the CDC, and HHS knew that H1N1 was an ordinary pig virus posing no risk for humans.[5] Nevertheless, NIAID conducted a hard-sell campaign warning of one million deaths in the United States. Working with the pharmaceutical companies, NIAID, CDC, and Merck persuaded incoming president Gerald Ford to sign a bill appropriating $135 million for vaccine manufacturers to inoculate 140 million Americans against the pestilence.[6]

At the behest of federal regulators, Ford appeared on TV urging all Americans to get vaccinated. Ford's obligatory references to the 1918 Spanish flu mass fatalities inspired some 50 million US citizens to hotfoot it to their local health center for injections of hastily concocted, shoddily tested, zero-liability vaccines that HHS and Merck conspired to rush to market. CDC director David Sencer set up a swine flu "war room" to bolster public fear amongst an enthused media.[7] The government launched a full-scale promotional campaign, including terrifying TV commercials depicting remorseful patients who dodged their vaccination and suffered serious illness. A CDC press release claimed that popular TV star Mary Tyler Moore had taken the jab. Moore told *60 Minutes* she had avoided the shot due to her concerns about side effects. She said that she and her doctor were very happy she didn't get it.[8]

In the end, the actual number of pandemic swine flu casualties in 1976 was not 1 million, but one. Dr. Harvey Fineberg, who authored the government's 1978 comprehensive postmortem of NIAID's response to that fake pandemic, told the WHO *Bulletin*: "In '76, the virus was detected in a single military installation, at Fort Dix, New Jersey. In the ensuing weeks and months, not one related swine flu case was reported elsewhere in New Jersey, the USA or anywhere else in the world. . . . At the same time, political decision-makers consistently thought that the scientists were giving them no choice but to go ahead with a mass immunization programme."[9]

NIH's influenza and flu vaccine expert senior bacteriologist and virologist Dr. John Anthony Morris informed his HHS bosses that the flu scare was a farce and that NIAID's campaign was a boondoggle to promote a dangerous and ineffective flu vaccine for a greedy industry. Dr. Morris had worked for thirty-six years at federal public health agencies beginning in 1940. His office, at the time of the 1976 "outbreak," was a few doors down the hall from Tony Fauci's. Morris served as the government's chief vaccine officer and led research on the flu and flu vaccines for the Bureau of Biologics Standards (BBS) at NIH and later at FDA. Morris enjoyed a distinguished career researching viral respiratory diseases. When Dr. Morris protested the fraud, his direct superior ordered him to stand down, advising Morris "not to talk about this."[10] His NIH bosses threatened Dr. Morris with loss of employment and professional ruin if he failed to keep his mouth shut. When vaccine recipients began reporting adverse reactions, including Guillain-Barré Syndrome (GBS), Dr. Morris disobeyed orders. Publicly declaring that there was zero evidence that the Fort Dix swine flu was contagious to humans, he reiterated, the vaccine could induce neurological side effects.[11] In response, HHS officials confiscated Dr. Morris's research materials, changed the laboratory locks, moved him to a small room with no telephone, reassigned his laboratory staff, forbade him to see visitors except with permission, and blocked his efforts to publish his findings.[12] Finally, after months of threats and petty harassment, HHS fired Morris for insubordination, citing a long list of drummed-up charges, including failure to return library books on time.[13]

Over at CDC, scientist Dr. Michael Hatwick was also warning HHS bigwigs that the flu vaccine could cause widespread brain injuries.

The 1976 swine flu vaccine was so fraught with problems that HHS discontinued the jab after vaccinating 49 million Americans. According to news accounts, the incidence of flu was seven times greater among the vaccinated than the unvaccinated. Furthermore, the vaccine caused some 500 cases of the degenerative nerve disease Guillain-Barré Syndrome, 32 deaths,[14] more than 400 paralyzations, and as many as 4,000 other injuries.[15]

Public health officials pulled the vaccine. President Ford fired David Sencer.

American taxpayers ended up paying for the swine flu vaccine coming and going, through guaranteed profits for Merck at the front end and outlays for piles of lawsuits from vaccine injury victims on the other side.

The government paid $134 million for the swine flu vaccine program. Injured plaintiffs filed 1,604 lawsuits. By April 1985, the government had paid out $83,233,714 and spent tens of millions of dollars adjudicating and processing those claims.[16] In 1987, Dr. Morris testified before Congress, "These figures give some idea of the consequences resulting from a program in which the federal government assumes liability of a product known to produce, in an indeterminate number of recipients, serious damage to health. . . . When I left the FDA in 1976, there was no available technique to measure, reliably and consistently, neurotoxicity or potency of most of vaccines then in use, including DTP vaccines. Today [1987], 11 years later, the situation remains essentially the same."[17] Dr. Morris's research found that flu vaccines often induced fever in children and in pregnant women, and serious harm to the fetus. He worried that there were hidden risks for everyone because the vaccine was "literally loaded with extraneous bacteria."[18] According to Dr. Morris, "There is a great deal of evidence to prove that immunization of children does more harm than good."[19] In what serves as a concise epithet for his crosses, Dr. Morris stated, "There is a close tie between government scientists and manufacturing scientists. My results were hurting the market for flu vaccines."[20]

In 1977, Dr. Morris instituted a wrongful dismissal suit. The court overturned all NIH's charges against him. Subsequently a grievance committee unanimously found that his supervisors had harassed and wrongfully terminated Dr. Morris.[21] A group of former FDA and NIH scientists endorsed Dr. Morris's criticisms of the agency. The *New York Times* quoted a fellow scientist, B. G. Young, who characterized NIH's reprisals against honest scientists as "suppression, harassment, and censorship of individual investigators. . . . I finally came to realize that you either had to compromise yourself or leave. Morris and (Bernice) Eddy are the real heroes in that place because they stayed and fought. The others voted with their feet and left."[22]

Up until his death in July 2014, Dr. Morris remained an outspoken critic of CDC's annual flu shot program. In 1979, Dr. Morris told the *Washington Post*, "It's a medical rip-off. . . . I believe the public should have truthful information on the basis of which they can determine whether or not to take the vaccine. . . . I believe that given full information, they won't take the vaccine." Dr. Morris's 2014 *New York Times* obituary reported his statement, "The producers of these [influenza] vaccines know they are worthless, but they go on selling them anyway."[23]

Dr. B. G. Young told the *New York Times* that NIH's industry-dominated culture at the vaccine division had driven away all the honest regulators—those willing to stand up to pharma. Dr. Fauci, in contrast, is the rare scientist who lasted fifty years at HHS. He has done so largely by aligning himself with NIH's pharma overlords and carrying industry water.

The same weapons that NIH used to silence Dr. Morris—enforced isolation, disgrace, prohibiting him from publishing papers, presenting at conferences, or talking to the press, changing his laboratory locks to prevent further research—were already pieces of an established Soviet-style template for silencing dissident scientists at NIH. The agency first unsheathed those weapons in the 1950s to destroy the career of its award-winning virologist, Dr. Bernice Eddy, the discoverer of the poliomyelitis virus—who later found a cancer-causing monkey virus in the Salk and Sabin polio vaccines. When her research disclosed problems with vaccine safety, NIH officials banned Dr. Eddy from her lab, changed her office locks, and ordered her to refrain from interviews and speeches. After silencing Eddy, NIH gave the contaminated vaccine to 99 million baby boomers, who suffered a tenfold increase in soft tissue cancers, resulting in a public health disaster that dwarfs the harms of polio.[24,25] Dr. Fauci and government health regulators later used these same techniques to muzzle a parade of in-house scientists, including Dr. Judy Mikovits, NIH contract researcher Dr. Bart Classen, and CDC's varicella (chicken pox) vaccine researcher Dr. Gary Goldman, who dared to tell hard truths about vaccine safety and efficacy.

The 1976 swine flu event was the first time that the federal government agreed to serve as pharma's insurer. The episode taught the public an important lesson: tort immunity incentivizes dangerous and ineffective vaccines. Industry and the magisterial class learned an entirely different morale from the tragic episode. In 1986 they made swine flu vaccine template the model for the National Childhood Vaccine Injury Act, which shielded *all* mandated vaccines from liability.

At the dawn of Dr. Fauci's career, he learned that both pandemics and fake pandemics provide an opportunity to expand the bureaucracy's power and to multiply the wealth of its pharma partners.

2005 Bird Flu

In 2005, Dr. Fauci revived NIAID's script from the 1976 debacle. This time the villain was an avian flu, H5N1. Like an agitated Chicken Little, Dr. Fauci had been warning the world about the imminent bird flu pandemic since 2001. That year, in a paper, "Infectious Diseases: Considerations for the 21st Century," Dr. Fauci balefully forecast a bird-to-human transmission of an influenza scourge that would decimate global populations, beginning with Hong Kong.[26] He predicted unprecedented carnage from this "new strain of influenza A virus entering a population that is relatively naïve for the microbe in question."

In 2004, a Vietnam-based Oxford University Clinical Research Unit Director, Jeremy Farrar—who would later rise to both knighthood and to command of the powerful Wellcome Trust—and his Vietnamese colleague, Tran Tinh Hien, identified the reemergence of the deadly bird flu, or H5N1, in humans. "It was a little girl. She caught it from a pet duck that had died and she'd dug up and reburied," Farrar told the *Financial Times*.[27]

The Wellcome Trust heavily funded Oxford's Vietnam project. Drug developer Sir Henry Wellcome established Wellcome Trust with a donation of his stock in Burroughs Wellcome, the British pharmaceutical behemoth. In 1995, the Trust sold its stock to Burroughs Wellcome's chief competitor, GlaxoSmithKline,[28] to facilitate the merger of England's two pharmaceutical giants. Wellcome Trust's $30 billion endowment[29] makes it the world's fourth-largest foundation and the globe's most prodigious financier of biomedical research. Like the Gates Foundation, Wellcome targets its donations to promote the interests of the pharmaceutical industry.

In 2007, British medical journalist John Stone raised the issue of phony pandemics in a letter to BMJ online as part of the swine flu postmortem: "There always remains the issue of whether scares are being promoted because of sober assessment of risk or because they constitute another bonanza for the pharmaceutical industry. We need better institutional means to spot the difference, but so far pandemic flu has been disappointing for the horror merchants. . . . Does anyone recall the moral of the story of the little boy who cried wolf? Well, it is what the industry does all the time."[30]

In 2020, Farrar would partner with Bill Gates to fund modeler Neil Ferguson, the epidemiologist who produced the wildly exaggerated COVID-19 death forecasts that helped ratchet up the COVID-19 fear campaign and rationalize draconian lockdowns.[31] As Schwab mentions, Farrar was at the heart of the earlier fiasco involving avian flu, generated around the delusory fear that the virus would cross the species barrier.[32]

Ferguson is the modeling impresario at drumming up phony pandemics. His curriculum vitae includes:

In 2005, Ferguson predicted that up to 150 million people could be killed from bird flu.[33] In the end, only 282 people died worldwide from the disease between 2003 and 2009.[34]

In 2001, a published Imperial College projection by Ferguson sparked the mass culling of eleven million sheep and cattle during the 2001 outbreak of foot-and-mouth disease. In 2002, he projected human deaths of 136,000 in the UK from mad cow disease. The UK Government slaughtered millions of cows. The actual number of deaths was 177.[35]

In 2009, Ferguson's projected that the swine flu would kill 65,000 Brits. Swine flu killed 457 people in the UK.[36]

In 2020, Ferguson famously predicted up to 2.2 million COVID-19 deaths in the United States in 2020 alone.[37,38] Dr. Fauci, in many Western countries, used Ferguson's projection to justify lockdowns and other draconian mandates.[39]

Farrar played a key role in Dr. Fauci's campaign to cover up evidence of government involvement in the potential lab generation of COVID-19.[40, 41]

In 2005, Dr. Fauci crowed that his long-awaited bird flu had finally arrived.

Using data from Ferguson, he warned it would kill "millions of people" worldwide[42] unless he and his pharma partners could deploy a vaccine to derail the approaching holocaust. Political and medical establishment cheerleaders mobilized for the now-familiar drill boosting pandemic panic.

Parroting Dr. Fauci's bird anxieties, government ministries of countries like the United States, Canada, and France, and the World Health Organization bewailed that H5N1 was "highly contagious" and deadly. The World Health Organization and the World Bank screeched that the plague could cost the world $2 trillion![43] Anthony Fauci prophesied that H5N1 is "a time bomb waiting to go off." Klaus Stohr, then coordinator of the influenza program at the World Health Organization (WHO), amplified Dr. Fauci's augury, predicting that between 2–7 million people would die, and that billions would fall ill worldwide.[44] In September 2005, *Der Spiegel* quoted the United Nations's Chief Coordinator David Nabarro that the new flu pandemic "can kill up to a hundred and fifty million people."[45] The *New Yorker* offered over-wrought bodements of millions of deaths from "one of the greatest dangers facing the United States."[46] Pandemic expert Robert Webster invoked military vernacular that had become de rigueur for loosening public purse strings in the post-9/11 biosecurity era: "We have to prepare as if we were going to war—and the public needs to understand that clearly. This virus is playing its role as a natural bioterrorist."[47]

In response to Dr. Fauci's lathered forecasts, the White House unveiled a Christmas list for the Bush family's favorite medicine man, including $7.1 billion to protect Americans from his avian plague.[48, 49] President George W. Bush warned that "No country can afford to ignore the threat of avian flu."[50] Dr. Fauci trotted out his reliable old chestnut that the new version of bird flu could be as lethal as the 1918 Spanish flu epidemic that killed 50–100 million people.[51]

Dr. Fauci had reason to know that this weary bogeyman was a canard. In 2008, he coauthored a study for the *Journal of Infectious Disease* confessing that virtually all of the "influenza" casualties in 1918 did not actually die from flu but from bacterial pneumonia and bronchial meningitis, which are, today, easily treated with antibiotics unavailable in 1918.[52] The Spanish flu that government virologists have invoked to terrorize generations of Americans with vaccine compliance is, after all, a paper tiger.

Bush told the US Congress the country needed $1.2 billion for sufficient avian virus vaccine to inoculate 20 million Americans. Additionally, he added $3 billion for Dr. Fauci's new seasonal flu vaccines, and $1 billion for the storage of antiviral medications.[53] Bush also demanded that Congress pass the "Biodefense and Pandemic Vaccine and Drug Development Act of 2005" granting liability relief to vaccine manufacturers. The pharmaceutical firms told the White House that they would refuse to manufacture vaccines without an impervious shield from tort liability.[54] The act banned lawsuits against even the most negligent, reckless, and reprehensible behavior by vaccine makers, even for vaccinations administered by force. The immunity provision was a blank check to Big Pharma's greed and criminal profiteering. The National Vaccine Information Center called the scheme "a drug company stockholder's dream and a consumer's worst nightmare."[55] Dr. Fauci arranged for rich vaccine contracts to Sanofi and Chiron to shore up the fragile "vaccine enterprise."[56]

Once again, Dr. Fauci's pandemic was a no-show. By the time it was all over, the WHO estimated that by May 16, 2006, Dr. Fauci's bird flu had killed only 100 people worldwide.[57] As the investigative journalist and attorney Michael Fumento observed in his postmortem on Dr. Fauci's bird flu hoax: "Dr. Fauci's recurring disease 'nightmares' often don't materialize."[58] Fumento recounted in *Forbes* magazine, "Around the world nations heeded the warnings and spent vast sums developing vaccines and making other preparations."[59]

2009 Hong Kong Swine Flu

In 2009, Dr. Fauci once again hyped a fraudulent epidemic. This time it was the Hong Kong swine flu. That year, in a classic "bait and switch," which Dr. Fauci and the Wellcome Trust helped to mastermind, the WHO—by then under control of pharma and its emergent funder, Bill Gates—declared a swine flu pandemic. Three years earlier, Gates had appointed GlaxoSmithKline's director, Tachi Yamada, to run his foundation's Global Health Program. Yamada also sat on the board of Neil Ferguson's outfit, the Imperial College London, which ran the fraudulent modeling that grievously inflated projected death counts from the 2009 swine flu outbreak[60] (and more recently for COVID-19).[61] Gates is one of the largest funders for the Imperial College London's modeling center.[62] Neil Ferguson, the epidemiologist who produced the fraudulent projections, also sat on the Wellcome Trust staff with Jeremy Farrar. There was no sign of a pandemic. In May of that year, WHO had detected some excess cases of seasonal flu, but the symptoms were mild and death rates were very low—fewer than 145 people worldwide over eleven weeks since its first appearance.[63] Nevertheless, the agency decided, in secret meetings, to declare a global pandemic.

WHO's declaration activated $18 billion worth of sleeper contracts[64] that WHO—and Gates's other organizations—had pressured various African and European countries to sign with GlaxoSmithKline and other pharmaceutical companies.[65] These secretive agreements obliged signatory nations including Germany, Great Britain, Italy, and France to purchase 18 billion dollars of various experimental, untested fast-tracked zero-liability H1N1 flu vaccines, most notably Glaxo's product, Pandemrix, in the event that the WHO declared a Class 6 pandemic. Then, just in time to trigger the sleeper contracts, WHO—in a sleazy switcheroo—changed the definition of Class 6 "pandemic" deleting the words and the requirement for "mass deaths around the globe." "You could now have a pandemic with zero deaths,"[66] explained Michael Fumento in *Forbes* magazine.

Under hot pressure from apoplectic critics of the boondoggle, WHO denied and then sheepishly admitted that it had downgraded its definition in consultation with government and industry scientists. The names of these individuals, WHO explained, needed to remain top secret for reasons that WHO didn't explain. To date, WHO has refused to disclose the identities of its trusted confidants. There was widespread suspicion that most of those officials were PIs on the payroll of Glaxo and other vaccine makers. According to the *British Medical Journal*, the World Health Organization's handling of the swine flu pandemic was "deeply marred by secrecy and conflict of interest with drug companies."[67] The *BMJ* found that the experts who wrote WHO's guidelines on the use of antiviral drugs had received consulting fees from the top two manufacturers of these drugs, Roche and GlaxoSmithKline (GSK). Among the driving forces behind the pandemic declaration was Sir Roy Anderson,[68] a board member of GlaxoSmithKline and the rector of Imperial College London, which would play such a prominent role in concocting both the 2009 swine flu and the 2020 COVID-19 crises. WHO's pandemic declaration forced five European[69] and several African countries[70] to purchase millions of doses of Glaxo's dangerous pandemic vaccine, earning Glaxo a cool and fast $13 billion. Sanofi reported €1.95 billion profit on its swine flu vaccine revenue. According to a report on the episode by the London-based Bureau of Investigative Journalism, the WHO violated its own rules by not publicly disclosing the conflicts among its key advisers when it drew up the guidelines.[71, 72]

Contemporary news accounts identify Dr. Fauci as the chief proponent of the multibillion-dollar fast-track H1N1 flu vaccine given that year to millions of Americans. Dr. Fauci is "more responsible than any other single person for the fast-track development of this new flu vaccine," according to a contemporary report by National Public Radio's Richard Knox.[73, 74]

As usual, the fawning US media obediently spread fear and lies to promote Dr. Fauci's H1N1 jabs. NBC grimly forecasted that "Swine flu could strike up to 40 percent of Americans over the next two years and as many as several hundred thousand could die if a vaccine campaign and other measures aren't successful."[75]

Historian Dr. Russell Blaylock writes, "The Ministry of Fear (the CDC) was working overtime peddling doom and gloom, knowing that frightened people do not make rational decisions—nothing sells vaccines like panic."[76]

At a January 2019 conference hosted by the Gates Foundation–funded Centre on Global Health Security at London's Chatham House, Marc Van Ranst, a Belgian virologist and pharmaceutical industry insider financially and ideologically indentured to GSK, Sanofi-Pasteur, J&J, and Abbott, described his role during the swine flu hoax a decade earlier. Chatham House is an exclusive think tank for globalist and corporatist elites. Its deliberations are so closely guarded that its name is synonymous with secrecy.

In 2009, Van Ranst served as Belgium's flu commissioner, in charge of managing crisis communication. To audible and admiring guffaws, Van Ranst told his corps d'elite audience how to stage a pandemic: "You have one opportunity to do it right. You have to go for one voice, one message. . . . You have to be omnipresent that first day or days, so you attract media attention . . . and they're not going to search for alternative voices." He explained that "talking about fatalities is important because . . . people say wow, what do you mean, people die because of influenza? That was a necessary step to take. Then of course a couple of days later, you had the first H1N1 death in the country and the scene was set." He continued: "I misused the fact that the top football clubs in Belgium inappropriately and against all agreements made their soccer players priority people. I could use that, because if the population really believes that this vaccine is so desirable that even these soccer players would be dishonest to get their vaccine, okay I can play with that. So I made a big fuss about it. . . . It worked."[77]

In 2020, this kind of thinking earned Van Ranst appointments to the Belgian "Risk Assessment Group" (RAG) and to the "Scientific committee Coronavirus," which advises Belgian health authorities on combating the virus. He became the public face of Belgium's response to COVID-19.

By October 2009, many people were complaining of a wave of devastating illnesses from the flu shots. From the beginning of their concocted pandemic, Dr. Fauci and other trusted public health officials had stressed that pregnant women were at a special risk from the swine flu compared to the seasonal flu.[78] This was a lie, but terrified mothers queued up in droves to get the jab.

Many of them would regret their choice. Research by Goldman in 2013 documented an elevenfold increase in fetal loss reports following the 2009–2010 pandemic flu season when pregnant women received two seasonal flu vaccines during pregnancy, and the H1N1 vaccine.[79]

A 2017 CDC study links miscarriage to flu vaccines, particularly in the first trimester. Pregnant women vaccinated in the 2010/2011 and 2011/2012 flu seasons had two times greater odds of having a miscarriage within twenty-eight days of receiving the vaccine. In women who had received the H1N1 vaccine in the previous flu season, the odds of having a miscarriage within twenty-eight days were 7.7 times greater than in women who did not receive a flu shot during their pregnancy.[80]

To quiet the clamor, Dr. Fauci took to YouTube to reassure the global public that the flu shots were rigorously tested, perfectly safe and that the risks of serious adverse events for the influenza vaccine are "very, very, very small"[81] This statement was scientifically baseless. Heavy conflicts of interest marred the underlying studies, which received fast-tracked approval without any functional double-blind placebo-controls.[82] Dr. Fauci went on to explain, "The H1N1 pandemic flu vaccine is made exactly the same way by the same manufacturers with the same processing, the same materials, as we make seasonal flu vaccine, which has an extraordinarily good safety record."

Two months after Dr. Fauci made these public assurances, an explosion of grave side effects, including miscarriages, narcolepsy, and febrile convulsions, was causing carnage in multiple countries.[83] According to the European Medicines Agency (EMA),[84] Pandemrix caused more than 980 cases of severe neurological injuries, paralysis from Guillain-Barré syndrome, debilitating narcolepsy, and cataplexy, including in more than 500 children. The Glaxo vaccine killed and injured so many children and health workers with various forms of brain damage that it forced Glaxo to withdraw the vaccine.[85, 86]

The 2009 H1N1 swine flu pandemic was another hyped global contagion fraud that never materialized.

Epidemiologist Dr. Wolfgang Wodarg, chairman of the Health Committee, of the Parliamentary Assembly of the Council of Europe (PACE), declared that the 2009 "false pandemic" was "one of the greatest medicine scandals of the century."[87] The director of the WHO Collaborating Center for Epidemiology in Munster, Germany, Dr. Ulrich Kiel, labeled the pandemic a meticulously planned hoax. "We are witnessing a gigantic misallocation of resources ($18 billion so far) in terms of public health," Kiel said.[88] Writing in *Forbes* magazine, medical journalist Michael Fumento concluded that "This wasn't merely overcautiousness or simple misjudgment. The pandemic declaration, and all the Klaxon-ringing since, reflect sheer dishonesty motivated not by medical concerns but political ones."[89]

Wolf-Dieter Ludwig, medical professor and chairman of the Drug Commission of the German Medical Profession, declared that "The boards of Health have been taken in by a campaign of the pharmaceutical companies that simply wanted to earn money with the supposed threat."[90]

As usual, there was no investigation of Dr. Fauci or the other medical officials who choreographed this multibillion-dollar fraud. The pharmaceutical companies walked away with billions, sticking governments and taxpayers with the ruinous cost of compensating flu shot injuries.

In his 2011 article about the scandal in the journal of *Dr. Med. Mabuse*, "The Power of Money: A Fundamental Reform of the WHO is Overdue," psychologist Thomas Gebauer wrote that "Increasingly, private money or earmarked donations from individual states are deciding on the goals and strategies of the WHO." The extent of their influence was recently demonstrated by the way the WHO dealt with the "swine flu." The article opens with a photo of Bill Gates.

In his book *Virus Mania*, journalist Torsten Engelbrecht quotes epidemiologist Angela Spelsberg, an expert on pandemic manipulation and drug industry corruption, that the "swine flu pandemic was deliberately used by the pharmaceutical industry for marketing purposes."[91]

2016 Zika

In March of 2016, Dr. Fauci again misled the public—this time into believing that the Zika virus was causing an epidemic of microcephaly among newborn babies in Brazil. One thing we know for sure: Zika doesn't cause microcephaly. Dr. Fauci had to have learned this. Zika was endemic to Central America and much of South Asia for many generations with no reported association with microcephaly. Dr. Fauci's critics claimed that an experimental DPT vaccine administered to pregnant women in 2015–2016 in the slums of northeast Brazil was the likely culprit for the wave of microcephaly. Extensive use of highly toxic pesticides in that corner of the nation may have also contributed. They accused Dr. Fauci of pointing the finger at Zika to distract attention from the more likely culprits, and to extract billions of dollars from Congress to develop yet another chimeric vaccine. The servile media, fattening on pharma advertising, delighting in the frightening epidemic that yielded children with tiny heads and great big ratings for the networks, obligingly heaped fuel onto Dr. Fauci's Zika terror crusade. Fear drives viewership. As CNN Technical Director Charlie Chester explained to industry analysts during the COVID-19 crisis, "COVID? Gangbusters with ratings, right? Which is why we have the death toll on the side."[92]

Dr. Fauci announced that he was pulling funds from malaria, influenza, and tuberculosis research programs in order to fund "a series of four or five vaccines" to rescue America from Zika. By fanning the flames of pandemic panic, Dr. Fauci, buttressed by his partner Bill Gates,[93] requested an additional nearly $2 billion congressional appropriation to NIAID to develop a Zika vaccine.[94, 95] That money swelled his agency's Zika budget to about $2 billion and enriched his Pharmaceutical partners.[96] Dr. Fauci funneled $125 million to a new Cambridge, Massachusetts, startup then called Moderna Therapeutics, to develop an mRNA vaccine for Zika. Gates appeared on CNBC to tout Moderna and promote its efforts to deliver a Zika jab.[97] He put $18 million into a project with the Wellcome Trust to fund a US-owned company, Oxitec, headquartered near Oxford University in the UK,[98] to release millions of genetically modified mosquitoes in Brazil and the communities[99] to exterminate the mosquito species blamed for spreading Zika.[100, 101] This was a follow-up to an even slightly more sinister 2008 Gates-funded study by Professor Hiroyuki Matsuoka at Jichi Medical University in Japan to engineer mosquitoes that can act as "flying syringes" to inject malaria vaccine into people—both the willing and the unwilling.[102] In 2021, Gates would expand on this macabre project by investing $25 million in an effort to genetically modify mosquitoes to stealthily deliver coronavirus vaccine to the vaccine-hesitant.[103, 104] I'm not joking.

The feverish predictions of a microcephaly scourge in Brazil soon fizzled. World Health Organization spokesman Christopher Dye told NPR that while "we apparently saw a lot of cases of Zika virus in 2016, there was no microcephaly."[105] Peaking at a high of about 5,200 cases in 2016,[106] the United States has recorded a total of about 550 Zika cases since then, with roughly 80 percent of those occurring in 2017,[107] with no reported microcephaly. The disease never spread beyond Florida and Texas, and no cases of Zika-associated microcephaly ever materialized.

Undaunted, Dr. Fauci warned that the disease "will come again" to the United States and that the country "absolutely [has] to be prepared" for it.[108]

In 2019, health officials reported only 15 cases of Zika in the United States, all of them microcephaly-free. The Mayo Clinic, meanwhile, reported in December[109] that, despite Dr. Fauci's $2 billion expenditure, there is no functional vaccine for the disease. By 2020, Dr. Fauci could no longer credibly blame the microcephaly epidemic on Zika, and he stopped talking about his vaccine. In June 2020, Dr. Fauci, under questioning before Congress, sheepishly explained, "It was never brought to full fruition because Zika disappeared."[110]

2016 Dengue

The Gates/Fauci Zika scam squandered billions of taxpayer money. But the Gates/Fauci dengue vaccine collaboration had a far graver outcome: this time, their "lifesaving vaccine" was a deathtrap in a syringe. Over a span of two decades, NIAID worked with the Gates Foundation to develop a vaccine against the mosquito-borne dengue virus, the most widespread tropical disease after malaria. Only a month after Fauci's agency filed its first of 305 patent applications in November 2003, toward "development of mutations useful for attenuating dengue viruses and chimeric dengue viruses, the Gates Foundation announced a $55 million grant to support the Pediatric Dengue Vaccine Initiative.[111] In September 2006, Sanofi Pasteur entered a partnership with the Initiative.[112]

By July 2007, NIAID's prototype dengue vaccine candidate emerged out of preclinical trials with what Dr. Fauci called "a promising future." NIAID awarded "several industry sponsors in Europe and Brazil" nonexclusive licenses for its formulations. Early the following year, Dr. Fauci issued another of his hysterical pandemic warnings in a commentary for the American Medical Association's journal, "[A] disease most Americans have never heard of could soon become more prevalent if dengue, a flu like illness that can turn deadly, continues to expand into temperate climates and increase in severity." Efforts to control the transmitting mosquitoes had fallen short, Fauci said, and "widespread appearance of dengue in the continental United States is a real possibility." To fight the disease, "the formidable challenges of understanding dengue pathogenesis and of developing effective therapies and vaccines must be met."[113]

NIAID announced its dengue virus vaccine clinical trial in August 2010, at the Gates-funded Johns Hopkins Bloomberg School of Public Health in Baltimore and at the University of Vermont. Fauci said: "With increasing infection rates and disease severity around the world and the discovery of dengue in parts of Florida, finding a way to prevent dengue infection is an important priority."[114]

Gates's WHO fueled Dr. Fauci's feverish dengue furor, warning: "In 2012, dengue ranks as the most important mosquito-borne viral disease with an epidemic potential in the world. There has been a 30-fold increase in the global incidence of dengue during the past 50 years, and its human and economic costs are staggering." However, referring to the Gates/Fauci projects, WHO predicted progress on vaccines that induce "long-lasting protective immunity."[115]

Dr. Ralph Baric, the gain-of-function guru, was the American darling of both NIAID and the Defense Advanced Research Project Agency (DARPA). His lab at UNC–Chapel Hill received $726,498 from the Gates Foundation for using recombinant dengue viruses to advance dengue vaccine development. Originating in February 2015, the three-year grant was scheduled to conclude early in 2018.[116]

In July 2014, Lance Gordon, the BMGF's director for Neglected Infectious Diseases in its Global Health Program, released news that the Sanofi Pasteur experimental dengue vaccine that Gates and Dr. Fauci funded was showing positive clinical results. Amidst his sunny forecast, Gordon made an ominous allusion that would have sounded DEFCON 1 to anyone decoding its implication. NIAID's clinical trials in Brazil, he acknowledged, showed signals of "pathogenic priming." That foreboding phrase describes an enhanced immune response that can trigger system-wide inflammation and death when the vaccinated individual is reexposed to the wild virus.

Infectious disease experts and health regulators had recognized the deadly potential of pathogenic priming since the 1980s, when one study showed that "more severe responses were found to be 15–80 times more likely in secondary dengue infections than in primary infections."[117] In 2004, an experimental MERS vaccine had produced robust antibody response in children during an NIH trial and then catastrophic illness and death when researchers exposed the children to wild virus.[118] Similarly in 2012 and 2014, a collaborative of Chinese and US researchers had developed coronavirus vaccines that produced antibodies in ferrets and cats, and then killed them when they encountered the actual wild coronavirus.

But Gordon's admission didn't set off an alarm. The WHO, under Gates's firm control, was bent on accelerating development of the Gates/Fauci dengue project. Dr. Fauci was also undeterred. Omitting any mention of the danger signals, Dr. Fauci proclaimed in January 2016 that the project would proceed: "Researchers in NIAID's Laboratory of Infectious Diseases spent many years developing and testing dengue vaccine candidates designed to elicit antibodies against all four dengue virus serotypes."[119]

An article published in the *American Ethnologist* bore a curious title: "Chimeric globalism: Global health in the shadow of the dengue vaccine" (April 2015).[120] The piece described the NIAID effort: "A laboratory-engineered, 'chimeric' dengue fever vaccine entered late-stage clinical trials in the late 2000s." The article asked readers to consider the implications when vaccine development is not entirely driven by a public health aspiration, but by "the divergent logics of pharmaceutical capital, humanitarianism, and biosecurity."

The dengue venture didn't proceed smoothly for Sanofi Pasteur. With Gates Foundation support,[121] the French pharma company spent twenty years and some $2 billion to develop Dengvaxia, testing the vaccine in several large trials on over 30,000 children globally.[122] When Dr. Scott Halstead, who studied dengue for more than fifty years with the US military, read the clinical safety data trial in the *New England Journal of Medicine*,[123] he immediately knew something was very wrong. Some children who caught dengue after receiving the vaccination experienced dramatically worsened symptoms. For kids never before exposed to dengue, Dengvaxia also appeared to increase the lifelong risk of a deadly complication known as plasma leakage syndrome, which catapults a person into profound shock before killing them. Dr. Halstead was so worried that he raised alarm bells in six separate editorials for scientific journals.[124] He even made a video warning the Philippine government, which was about to start a mass vaccination campaign.[125] Gates, Dr. Fauci, and Sanofi ignored Halstead's frantic warnings.

Sanofi responded by publishing a rebuttal to Dr. Halstead and promising more studies. Without waiting for the research, in April, 2016, Bill Gates's minions at WHO moved to recommend Dengvaxia for all children ages 9 to 16.[126] Already the previous December, the Dengue Vaccine Initiative—supported by Gates Foundation funding—had announced that the Philippine government would soon become the second country (after Mexico) to approve Dengvaxia shots.

A year and a half later, Sanofi announced that it had new information about the vaccine's safety. Confirming Dr. Halstead's fears, the company made the alarming admission that Dengvaxia did indeed increase the risk of hospitalization and cytoplasmic leakage syndrome.[127] By this time, health officials had already inoculated some 800,000 Filipino children. At least 600 had died.[128]

The WHO eventually changed its recommendation, saying that Dengvaxia was safe only for kids who'd had a prior dengue infection and admitting that 100,000 should not have received the shot. Following autopsies on 600 deceased children, the Philippine Public Attorney indicted fourteen Philippines government officials and six Sanofi executives for criminal homicide.[129]

Accustomed as he was to this sort of collateral damage in his war against the bugs, Dr. Fauci put a sunny face on the dead children, telling the *Wall Street Journal* in January 2018, "We do not think this is going to be a showstopper in any way or form." Although, he added, "clearly there's going to be not as smooth a trip." Operating on his consistent strategy that the best defense is a good offense, Dr. Fauci announced full speed ahead in Dengvaxia trials in Brazil—pathogenic priming be damned! He boasted that "NIAID's dengue vaccine candidate is in a late-stage clinical trial involving 17,000 participants in Brazil" and it had "induced an immune response in tests against all four dengue types." NIAID's vaccine "has been licensed to several companies, including Merck, which said it plans to start its own trial this year."[130]

In December 2018, Merck and the Instituto Butantan—the main producer of vaccines in Brazil—announced a collaboration agreement after licensing "certain rights from National Institute of Allergy and Infectious Diseases (NIAID)" to develop live attenuated tetravalent vaccines for dengue. The nonprofit Instituto Butantan "will receive a $26 million upfront payment from Merck and is eligible to receive up to $75 million for the achievement of certain milestones related to the development and commercialization of Merck's investigational vaccine as well as potential royalties on sales. . . . It acts in partnership with various universities and entities such as the Bill & Melinda Gates Foundation for the achievement of its institutional objectives."[131]

In May 2019, the FDA approved Sanofi's Dengvaxia vaccine for use in the United States, Puerto Rico, Guam, and the British Virgin Islands—with the caveat that doctors first have proof of a prior dengue infection to make sure the jab wouldn't pose any risks to the child.[132]

The 600 Philippine children died as the result of "pathogenic priming," or "antibody dependent enhancement." Padron-Regalado et al. report on dozens of papers where SARS and MERS vaccines under development led to antibody dependent enhancement (ADE) in animal trials upon viral challenge.[133] An inactivated SARS virus vaccine platform led to immunopathologies consistent with ADE in mice challenged with the virus.[134] A vaccine candidate based on a SARS N-protein resulted in immunopathology with eosinophilic lung infiltrates in mice upon SARS-CoV challenge.[135] Vaccinia virus expressing the SARS S-protein showed strong inflammatory responses leading to hepatitis in the livers of vaccinated ferrets upon challenge with SARS-CoV.[136] Vaccines based on soluble S-protein alone elicited antibody dependent enhancement within in vitro studies involving human B-cells leading the authors to warrant concern regarding human vaccine development.[137] A chemically inactivated virus MERS vaccine led to lung pathology (eosinophilic infiltrates) with a virus challenge in mouse studies by Agrawal et al.[138] A vaccine based on the transgenic spike protein of MERS when administered to mice led to pulmonary hemorrhage after a challenge with MERS-CoV virus.[139] Conclusion: "The development of highly effective and safe vaccines for COVID-19 should consider aspects such as the possibility of ADE and other adverse effects previously observed with SARS and MERS. Even though these features have only been seen in some animal models and vaccination regimens, the possibility is still there to be considered for COVID-19."

In April 2020, soon after the COVID-19 pandemic began, vaccine tycoon and Merck spokesperson, Dr. Paul Offit, Director of the Vaccine Education Center at Philadelphia's Children's Hospital, warned about similar effects from a SARS-CoV-2 vaccine. "We saw that with the dengue vaccine," Offit told an interviewer. "In children who've never been exposed to dengue before, [it] actually made them worse when they were then exposed to the natural virus. Much worse, causing something called dengue hemorrhagic shock syndrome. Children died, vaccinated children who were less than 9 years of age."[140]

A warning about the tendency of coronavirus vaccines to induce pathogenic priming appeared in a 2009 article in *Expert Review of Vaccines* republished on NIH's website in January 2014: "The greatest fear among vaccinologists is the creation of a vaccine that is not only ineffective, but which exacerbates disease. Unfortunately, CoV vaccines have a history of enhancing disease, notably with feline CoVs."[141]

Pandemic Championships

There is an old saw about a cuckold with a murderous grudge against a lion tamer but no gumption for homicide. For years, he follows the circus hoping to be in attendance on that inevitable day when an aggrieved feline turns on the trainer. When decades of frustrated waiting for divine justice finally exhaust his patience, he sneaks into the lion tamer's dressing room to sprinkle pepper in his wig powder. That evening, the lion sneezes and decapitates its philandering handler.

The compelling evidence suggesting that COVID-19 emanated from a Fauci-funded Little Shop of Horrors in Wuhan, China, raises the ironic possibility that the man whom two US presidents have charged with leading the global response to the COVID-19 pandemic may be the same man who spawned it. That strange paradox might cause some to cynically ponder the logic behind Dr. Fauci's peculiar decisions to defy President Obama's 2014 gain-of-function moratorium, to dodge NIH's internal safety review committee, to launder money to Chinese scientists with military affiliations through a sketchy bioweapons grifter, to finance criminally reckless experiments minting souped-up pathogens in a shabby Chinese lab with lax safety

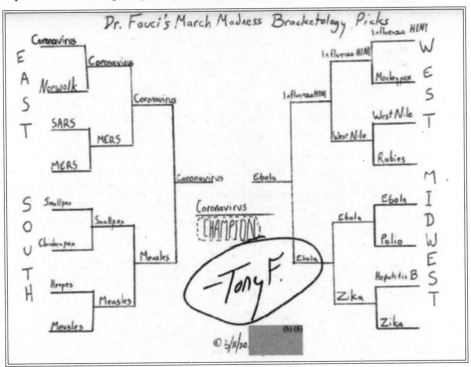

protocols. Are we justified in asking ourselves whether Tony Fauci, after decades of concocting toothless pandemics, was finally peppering the wig?

But putting aside Dr. Fauci's involvement with Wuhan and his decades of fashioning flop contagions, we must acknowledge that in 2020, he finally hit the jackpot with COVID-19. Among the more revealing documents in Dr. Fauci's June 2021 email dump is a rough schematic (oppostite) that Dr. Fauci signed "Tony F."[142] depicting a March Madness–style tournament bracket scoring the pestilential contestants during two decades of mostly phony contagions. COVID-19 finally emerges as champion. The doodle is titled "Dr. Fauci's March Madness Bracketology Picks" and dated March 11, 2020. In his macabre pool, Coronavirus—top-seeded out of the East region—defeats the entire field that includes his long litany of contrived diseases including smallpox, chickenpox, bird flu, swine flu, Zika, hepatitis B, smallpox, MERS, and measles. The drawing suggests Dr. Fauci's pride in a final, satisfying victory after a long, often-triumphant career engineering global pandemics:

Endnotes

1   Richard Krause, "The Swine Flu Episode and the Fog of Epidemics," *Emerging Infectious Diseases* (Jan., 2006), antimicrobe.org/h04c.files/history/cdc-05-1132.pdf

2   Gerald Posner, *Pharma: Greed, Lies, and the Poisoning of America* (Avid Reader Press, 2020)

3   Matt Taibbi, "Big Pharma's Covid-19 Profiteers: How the race to develop treatments and a vaccine will create a historic windfall for the industry—and everyone else will pay the price," *Rolling Stone* (Aug. 13, 2020), rollingstone.com/politics/politics-features/big-pharma-covid-19-profits-1041185/

4   Richard Neustadt and Harvey Fineberg, *The Swine Flu Affair: Decision-Making on a Slippery Disease: Swine Flu Chronology January 1976—March 1977* (National Academies Press, 1978), ncbi.nlm.nih.gov/books/NBK219595/

5   Ibid.

6   Bart Barnes, "Vaccine specialist J. Anthony Morris dies at 95," *Washington Post*, (Jul. 3, 2014), washingtonpost.com/local/obituaries/vaccine-specialist-j-anthony-morris-dies-at-95/2014/07/03/e786b9b8-0228-11e4-8572-4b1b969b6322_story.html

7   Torsten Engelbrecht, Claus Köhnlein, et al., *Virus Mania: How the Medical Industry Continually Invents Epidemics, Making Billions at Our Expense* (Books on Demand 3rd ed, 2021), 118

8   "Swine Flu," *60 Minutes*, CBS (1976), https://childrenshealthdefense.org/video/60-minutes-swine-flu-1976-vaccine-warning/

9   Harvey Fineberg, "Swine flu of 1976: lessons from the past," *World Health Organization* (Jun., 2009), who.int/bulletin/volumes/87/6/09-040609.pdf

10   Alliance for Human Research Protection, John Anthony Morris, MD (1919–2014) (Sep. 27, 2014), https://ahrp.org/john-anthony-morris-md/

11   Ibid.

12   Ibid.

13   Ibid.

14   Abbie Boudreau and Scott Zamost, CNN, "Ex-CDC head recalls '76 swine flu outbreak," https://edition.cnn.com/2009/HEALTH/04/30/swine.flu.1976/index.html

15   Victor Cohn, "U.S. Agrees to Pay Those Paralyzed By Swine Flu Shots," *Washington Post* (Jun. 21, 1978), washingtonpost.com/archive/politics/1978/06/21/us-agrees-to-pay-those-paralyzed-by-swine-flu-shots/26c65a54-e3c9-4e4c-a23f-b8a411b563b3/

16   Arnold W. Reitze, Jr., "Federal Compensation for Vaccination Induced Injuries," *Boston College Environmental Affairs Law Review*, vol. 12, issue 2, article 2 (Dec. 1, 1986), lawdigitalcommons.bc.edu/cgi/viewcontent.cgi?article=1614&context=ealr

17   U.S. House of Representatives, Hearing of the Subcommittee on Select Revenue Measures of the Committee on Ways and Means, "Funding of the Childhood Vaccine Program," Testimony of J. Anthony Morris, 114-115, (Mar. 5, 1987), https://play.google.com/books/reader?id=s_8sAAAAMAAJ&pg=GBS.PA114&hl=en

18   Alliance for Human Research Protection, https://ahrp.org/trzboard/john-anthony/

19   Stephen Lendman, "Early and Current Fears About Vaccine Dangers," *Dissident Voice* (Aug. 29, 2009), dissidentvoice.org/2009/08/early-and-current-fears-about-vaccine-dangers/

20   Alliance for Human Research Protection, John Anthony Morris, https://ahrp.org/trzboard/john-anthony/

21   Ibid.

22   Ibid.

23   Ibid.

24   Robert F. Kennedy Jr., "Foreword," Kent Heckenlively and Judy Mikovits, *Plague of Corruption* (Skyhorse, 2020)

25   Regis Vilchez and Janet Butel, "Emergent Human Pathogen Simian Virus 40 and Its Role in Cancer," Clinical Microbiology Reviews (Jul. 17, 2004), ncbi.nlm.nih.gov/pmc/articles/PMC452549/

26   Anthony Fauci, "Infectious Diseases: Considerations for the 21st Century," *Clinical Infectious Diseases* (Mar 1, 2001), sci-hub.do/10.1086/319235

27   Annie Maccoby Berglof, "Wellcome Trust Director Jeremy Farrar," *Financial Times* (Jun. 20, 2014), ft.com/content/e6e8b13a-f2df-11e3-a3f8-00144feabdc0

28   Associated Press, "Glaxo Wins Takeover Bid for Wellcome," Los Angeles Times (March 8, 1995), latimes.com/archives/la-xpm-1995-03-08-fi-40255-story.html

29   Daniel Thomas, "Wellcome Trust Hails 'Spectacular Year' For Science," *Financial Times* (December 15, 2020), ft.com/content/fd002907-c7df-4b24-b277-83a5e5d93b2b

30   John Stone, "Promotional Healthscares," *BMJ Rapid Response* (Jun. 28, 2007), bmj.com/rapid-response/2011/11/01/promotional-healthscares

31   Kate Kelland, "Sobering Coronavirus Study Prompted Britain to Toughen Its Approach," Reuters (March 17, 2020), reuters.com/article/us-health-coronavirus-britain-research/sobering-coronavirus-study-prompted-britain-to-toughen-its-approach-idUSKBN2141EP

32   Amy Maxmen, "Jeremy Farrar: When Disaster Strikes," *Journal of Experimental Medicine* (Jan. 19, 2009), ncbi.nlm.nih.gov/pmc/articles/PMC2626674/

33   James Sturcke, "Bird flu pandemic 'could kill 150m,'" *The Guardian* (Sep 30, 2005), theguardian.com/world/2005/sep/30/birdflu.jamessturcke

34   Robert Roos, "Indonesia reports 20 H5N1 cases—19 fatal—since January," *CIDRAP* (Dec. 30, 2009), cidrap.umn.edu/news-perspective/2009/12/indonesia-reports-20-h5n1-cases-19-fatal-january

35   John Fund, "'Professor Lockdown' Modeler Resigns in Disgrace," *National Review* (May 6, 2020), nationalreview.com/corner/professor-lockdown-modeler-resigns-in-disgrace/

36   Ibid.

37   Ibid.

38   Alan Reynolds, "How One Model Simulated 2.2 Million U.S. Deaths from COVID-19," Cato Institute (Apr. 21, 2020, 3:05 p.m.), https://www.cato.org/blog/how-one-model-simulated-22-million-us-deaths-covid-19

39   Ibid.

40   Jerry Dunleavy, "Fauci Worked Behind the Scenes to Cast Doubt on Wuhan Lab Leak Hypothesis, Emails Show," *Washington Examiner* (June 2, 2021 12:49 p.m.), washingtonexaminer.com/news/fauci-worked-behind-scenes-cast-doubt-wuhan-lab-leak-hypothesis

41   Tucker Carlson, "Tucker: Fauci deserves to be under 'criminal investigation,'" *Tucker Carlson Tonight* (June 2, 2021), youtube.com/watch?v=yp6btJhS66c&t=154s

42   Bill Moyers, "H5N1—Killer Flu—Interview: Dr. Anthony Fauci," *PBS—Wide Angle* (Sept. 20, 2005), pbs.org/wnet/wideangle/interactives-extras/interviews/h5n1-killer-flu-dr-anthony-fauci/2519/

43   Engelbrecht, Köhnlein, et al., 257

44   Ibid., 447

45   Ibid., 253

46   Michael Specter, "Bird Flu: Nature's Bioterrorist," *New Yorker* (Feb. 20, 2005), newyorker.com/magazine/2005/02/28/natures-bioterrorist

47   Ibid.

48   Timothy Williams, "Bush announces $7.1 billion bird flu plan," *New York Times* (Nov. 1, 2005), nytimes.com/2005/11/01/world/americas/bush-announces-71-billion-bird-flu-plan.html

49   Robert Roos, "Bush asks $7.1 billion to prepare for flu pandemic," University of Minnesota CIDRAP (Nov. 1, 2005), cidrap.umn.edu/news-perspective/2005/11/bush-asks-71-billion-prepare-flu-pandemic

50   Engelbrecht, Köhnlein, et al., 291

51   D. M. Morens and A. S. Fauci, "The 1918 influenza pandemic: insights for the 21st century." *Journal of Infectious Diseases* 2007 Apr 1;195(7):1018–1028. doi: 10.1086/511989. Epub 2007 Feb 23. PMID: 17330793, pubmed.ncbi.nlm.nih.gov/17330793/

52   David M. Morens, Jeffery K. Taubenberger, and Anthony S. Fauci, "Predominant Role of Bacterial Pneumonia as a Cause of Death in Pandemic Influenza: Implications for Pandemic Influenza Preparedness," *Journal of Infectious Diseases*, vol. 198, Iss. 7 (Oct. 1, 2008): 962–970, doi.org/10.1086/591708, academic.oup.com/jid/article/198/7/962/2192118

53   Engelbrecht, Köhnlein, et al., 290

54   Ibid.

55   Ibid., 291

56   U.S. House of Representatives, Committee on Government Reform, "The Next Flu Pandemic: Evaluating U.S. Readiness," Testimony of Anthony Fauci (Jun., 30, 2005), govinfo.gov/content/pkg/CHRG-109hhrg22808/html/CHRG-109hhrg22808.htm

57   WHO, "Cumulative number of confirmed human cases for avian influenza A(H5N1) reported to

WHO, 2003–2020" (Jan. 20, 2020), who.int/influenza/human_animal_interface/2020_01_20_tableH5N1.pdf?ua=1

58   Michael Fumento, "Dr. Fauci's Recurring Disease 'Nightmares' Often Don't Materialize" (Jun. 12, 2020), fumento.com/articles/dr_faucis_recurring_disease_nightmares_often_dont_materialize/

59   Michael Fumento, "Why the WHO Faked a Pandemic," *Forbes* (Feb. 5, 2010), pvvb7zurnn5i2d3zip pt7ddtfizkuffxgwmhba6eedxcbbhoj4bq.arweave.net/fWof5pFreo0PeUPfP4xzKjKqFLc1mHCDxCD uITuTwM

60   Matthew Weaver, "Swine flu could affect third of world's population, says study," *The Guardian* (May 12, 2009), theguardian.com/world/2009/may/12/swine-flu-report-pandemic-predicted

61   David Adam, "Special report: The simulations driving the world's response to COVID-19," *Nature* (Apr. 2, 2020), nature.com/articles/d41586-020-01003-6

62   Vaccine Impact Modelling Consortium, Imperial College London, Key Partners, vaccineimpact.org/partners/

63   Michael Fumento, "The Pandemic Is Political," *Forbes* (Oct. 16, 2009), forbes.com/2009/10/16/swine-flu-world-health-organization-pandemic-opinions-contributors-michael-fumento/?sh=6dad9c7f8ce9

64   Vaccines: Sleeping Contracts, UK Parliament (Oct. 9, 2007), archive.is/0dWrs

65   Extracts of statements made by the leading participants at the public hearing on "The handling of the H1N1 pandemic: more transparency needed?," organised by the Committee on Social, Health and Family Affairs of the Parliamentary Assembly of the Council of Europe (PACE) in Strasbourg on Tuesday 26 January 2010, Council of Europe (Jan. 26, 2010), archive.is/0dWrs

66   Fumento, "Why the WHO Faked a Pandemic"

67   Organic Consumers Union, "WHO's Handling of Swine Flu Deeply Marred by Conflict of Interest with Drug Companies, Top Medical Journal Accuses" (Jun. 7, 2010), organicconsumers.org/news/whos-handling-swine-flu-deeply-marred-conflict-interest-drug-companies-top-medical-journal

68   David Derbyshire, "Government virus expert paid £116k by swine flu vaccine manufacturers," *Daily Mail* (Jul. 27, 2009), dailymail.co.uk/news/article-1202389/Government-virus-expert-paid-116k-swine-flu-vaccine-manufacturers.html

69   Alistair Dawber, "Nations scrap orders for GSK swine flu jab," *Independent* (Jan. 16, 2010), independent.co.uk/news/business/news/nations-scrap-orders-gsk-swine-flu-jab-1869653.html

70   Richard Mihigo et al., "2009 Pandemic Influenza A Virus Subtype H1N1 Vaccination in Africa—Successes and Challenges," *Journal of Infectious Diseases* (Dec. 15, 2012), academic.oup.com/jid/article/206/suppl_1/S22/984162

71   Andreas Kruck et al., "What Went Wrong? The World Health Organization from Swine Flu to Ebola," *Nature Public Health Emergency Collection* (Oct. 29, 2017), ncbi.nlm.nih.gov/pmc/articles/PMC7122988/#CR56

72   Deborah Cohen and Philip Carter, "WHO and the pandemic flu 'conspiracies,'" *BMJ* (June 12, 2010), bmj.com/content/bmj/340/7759/Feature.full.pdf

73   Richard Knox, "What You Need to Know About Swine Flu Vaccine," NPR (Oct. 5, 2009), npr.org/templates/story/story.php?storyId=113446539

74   Brien Williams, PhD, "The American Association of Immunologists Oral History Project—Transcript," The American Association of Immunologists (Dec. 9, 2015), aai.org/AAISite/media/About/History/OHP/Transcripts/Trans-Inv-034_Fauci_Anthony_S-2015_Final.pdf

75   "Swine flu could sicken over 2 billion in 2 years," NBC (Jul. 24, 2009), nbcnews.com/id/wbna32122776

76   Engelbrecht, Köhnlein, et al., 289

77   Marc Van Ranst, "How to Join Forces in Influenza Pandemic Preparedness: Communication and Public Engagement," Stakeholder Conference, Chatham House (Jan. 22, 2019), vimeo.com/320913130.

78   April Fulton, "Taking a Pregnant Pause on Swine Flu Vaccine," NPR (Sept. 16, 2009), npr.org/sections/health-shots/2009/09/taking_a_pregnant_pause_on_flu.html

79   G. S. Goldman, "Comparison of VAERS fetal-loss reports during three consecutive influenza seasons: was there a synergistic fetal toxicity associated with the two-vaccine 2009/2010 season?" (2013), *Human & Experimental Toxicology*, 32(5):464–475, doi.org/10.1177/0960327112455067.

80   James G. Donahue et al., "Association of spontaneous abortion with receipt of inactivated influenza vaccine containing H1N1pdm09 in 2010–11 and 2011–12," *Vaccine* vol. 35, iss. 40 (2017): 5314–5322, sciencedirect.com/science/article/pii/S0264410X17308666

81   "2009 H1N1 Influenza update," National Institutes of Health, youtube.com/watch?v=hsXEgJqR_vY

82   Peter Doshi, "Pandemrix vaccine: why was the public not told of early warning signs?," *BMJ* (Sep. 20, 2019), archive.hshsl.umaryland.edu/bitstream/handle/10713/8270/Doshi_Pandermrix2018.pdf?sequence=1&isAllowed=y

83   Ibid.

84   Committee for Medicinal Products for Human Use (CHMP), CHMP Type II variation assessment report (Jun. 26, 2014), ema.europa.eu/en/documents/variation-report/pandemrix-h-c-832-ii-0069-epar-assessment-report-variation_en.pdf

85   Shaun Lintern, "Doctors Orders," Buzzfeed (Feb. 23, 2018), buzzfeed.com/shaunlintern/these-nhs-staff-were-told-the-swine-flu-vaccine-was-safe

86   Kate Kelland, "GSK says science does not link pandemic H1N1 flu vaccine to sleep disorder," Reuters (May 28, 2020), reuters.com/article/health-coronavirus-gsk-vaccine/update-1-exclusive-gsk-says-science-does-not-link-pandemic-h1n1-flu-vaccine-to-sleep-disorder-idUSL1N2DA1FS

87   F. William Engdahl, "European Parliament to Investigate WHO and 'Pandemic' Scandal," *Healthcare in Europe* (Jan. 26, 2010), healthcare-in-europe.com/en/news/european-parliament-to-investigate-who-pandemic-scandal.html

88   Marta Michels, "Why did the WHO simulate a swine flu pandemic in 2009?" (Mar. 9, 2021), int.artloft.co/why-did-the-who-simulate-a-swine-flu-pandemic-in-2009/

89   Fumento, "Why the WHO Faked a Pandemic"

90   Engelbrecht, Köhnlein, et al., 343

91   Ibid., 357

92   Emily Jacobs, "CNN staffer tells Project Veritas network played up COVID-19 death toll for ratings," *New York Post* (Apr. 14, 2021), nypost.com/2021/04/14/cnn-staffer-tells-project-veritas-network-played-up-covid-19-death-toll-for-ratings/

93   FNIH, "Foundation for the National Institutes of Health Celebrates 20 Years of Outstanding Partnerships to Advance Biomedical Research with Inaugural Charles A. Sanders, M.D.," Partnership Award, FNIH Press Release, (October 19, 2016), ("The Foundation for the National Institutes of Health (FNIH) has selected two longstanding, private-sector partners—Bill & Melinda Gates Foundation and Pfizer—as recipients of the inaugural Charles A. Sanders, M.D., Partnership Award. Honoring 20 years of advancing biomedical research under the leadership of former Chairman of the Board Charles A. Sanders, M.D., the award recognizes persons and/or organizations that have made significant contributions to the FNIH's work to build, implement and nurture private-public partnerships in support of the mission of the National Institutes of Health (NIH)."); fnih.org/news/press-releases/foundation-national-institutes-health-celebrates-20-years-outstanding; fnih.org/sites/default/files/press-release10-19-16.pdf

94   Anthony S Fauci, "Research Conducted and Supported by the National Institutes of Health (NIH) in Addressing Zika Virus Disease," Testimony before the House Committee on Foreign Affairs, Subcommittee on Africa, Global Health, Global Human Rights, and International Organizations and Subcommittee on the Western Hemisphere, February 10, 2016; docs.house.gov/meetings/FA/FA16/20160210/104450/HHRG-114-FA16-Wstate-FauciA-20160210.pdf

95   Puneet Kollipara, "Funding shift for Zika helps NIH, but more research money requested," *Science* (Apr. 8, 2016), https://www.sciencemag.org/news/2016/04/funding-shift-zika-helps-nih-more-research-money-requested

96   Kaitlyn M. Morabito and Barney S. Graham, "Zika Virus Vaccine Development," *Journal of Infectious Diseases*, vol. 216, iss. suppl. 10 (Dec. 15, 2017): S957–S963, doi.org/10.1093/infdis/jix464

97   Andrew Zaleski, "Bill and Melinda Gates are placing bets on this biotech in the race to develop a Zika vaccine," CNBC (May 19, 2017), cnbc.com/2017/05/18/bill-and-melinda-gates-bet-on-this-biotech-to-develop-zika-vaccine.html

98   Oxitec, oxitec.com/en/our-company-and-culture ("We are a US-owned company with headquarters and R&D facilities in the UK, just a few miles from where we started at Oxford University.")

99   Martin Enserink, "GM Mosquito Trial Strains Ties in Gates-Funded Project," *Science* (Nov. 16, 2010), sciencemag.org/news/2010/11/gm-mosquito-trial-strains-ties-gates-funded-project

100   Sarah Boseley, "Mosquitos to be infected with bacteria in fight against Zika virus," *The Guardian* (Oct 26, 2016), theguardian.com/world/2016/oct/26/mosquitos-infected-wolbachia-bacteria-zika-virus

101   Emily Waltz, "GM mosquitoes fire first salvo against Zika virus," *Nature Biotechnology* 34 (2016): 221–222 (2016), https://www.nature.com/articles/nbt0316-221.

102   "Production of a Transgenic Mosquito, as a Flying Syringe, to Deliver Protective Vaccine via Saliva," Global Grand Challenges, Bill and Melinda Gates Foundation. Funding date: October 1, 2008, gcgh.grandchallenges.org/grant/production-transgenic-mosquito-flying-syringe-deliver-protective-vaccine-saliva

103   Martin Enserink, "GM Mosquito Trial Strains Ties in Gates-Funded Project," *Science* (Nov. 16, 2010), sciencemag.org/news/2010/11/gm-mosquito-trial-strains-ties-gates-funded-project

104   Humans Are Free, "Bill Gates Wants to Release Genetically Modified Mosquitoes to Inject You with Vaccines," Humans Are Free (Feb. 26 2021), humansarefree.com/2021/02/bill-gates-wants-to-release-genetically-modified-mosquitoes-to-inject-you-with-vaccines.html

105   Michaeleen Doucleff, "Why Didn't Zika Cause A Surge In Microcephaly In 2016?" NPR (Mar. 30, 2017), npr.org/sections/goatsandsoda/2017/03/30/521925733/why-didnt-zika-cause-a-surge-in-microcephaly-in-2016

106   2016 Case Counts in the US, Centers for Disease Control (Apr. 24, 2019), cdc.gov/zika/reporting/2016-case-counts.html.

107   2017 Case Counts in the US, Centers for Disease Control (Apr. 24, 2019), cdc.gov/zika/reporting/2017-case-counts.html.

108   "NIAID's Fauci urges continued global vigilance on Zika, other diseases," NIH Fogarty International Center (Mar. 22, 2016), fic.nih.gov/News/Pages/2016-niaid-fauci-zika-virus-talk.aspx.

109   Gregory A Poland et al., "Zika Vaccine Development: Current Status," *Mayo Clinic Proceedings*, vol.

9, iss. 12 (Dec. 1, 2019): P2572–2586, mayoclinicproceedings.org/article/S0025-6196(19)30483-5/fulltext

110   "White House Coronavirus Task Force Members Testify on Federal Response to Pandemic," C-SPAN 3:36:32 (June 23, 2020), c-span.org/video/?473229-1/white-house-coronavirus-task-force-members-testify-federal-response-pandemic

111   Scott Halstead, "Gates Foundation Commits $55 Million to Accelerate Dengue Vaccine Research," Gates Foundation (Sept. 2003), gatesfoundation.org/ideas/media-center/press-releases/2003/09/dengue-vaccine-research

112   Len Lavenda, "Sanofi Pasteur and the Pediatric Dengue Vaccine Initiative Partner against Dengue Fever," SANOFI (Sept. 29, 2006), news.sanofi.us/press-releases?item=137021

113   "NIAID Experts See Dengue as Potential Threat to U.S. Public Health," National Institutes of Health, U.S. Department of Health and Human Services (Jan. 8, 2008), nih.gov/news-events/news-releases/niaid-experts-see-dengue-potential-threat-us-public-health

114   Robert Roos, "NIAID launches clinical trial of Dengue vaccine," *CIDRAP* (Aug. 9, 2010), cidrap.umn.edu/news-perspective/2010/08/niaid-launches-clinical-trial-dengue-vaccine

115   Monica A. McArthur et al., "Dengue vaccines: recent developments, ongoing challenges and current candidates." *Expert Review of Vaccines*, vol. 12, iss. 8 (2013): 933–953, doi:10.1586/14760584.2013.815412 https://www.ncbi.nlm.nih.gov/pmc/articles/PMC3773977/

116   UNC, Baric grant from BMGF: Ralph S. Baric Curriculum Vitae, sph.unc.edu/wp-content/uploads/sites/112/2016/09/CV_Ralph_Baric.pdf

117   Lynne Peeples, "Avoiding pitfalls in the pursuit of a COVID-19 vaccine," *Proceedings of the National Academy of Sciences* vol. 117, iss. 15 (Apr. 2020): 8218–8221, doi: 10.1073/pnas.2005456117, pnas.org/content/117/15/8218.

118   Chien-Te Tseng et al., "Immunization with SARS Coronavirus Vaccines Leads to Pulmonary Immunopathology on Challenge with the SARS Virus" (Apr. 20, 2012), ncbi.nlm.nih.gov/pmc/articles/PMC3335060/

119   NIH, "Dengue vaccine enters phase 3 trial in Brazil" (Jan. 14, 2016), nih.gov/news-events/news-releases/dengue-vaccine-enters-phase-3-trial-brazil

120   Alex M. Nading, "Chimeric globalism: Global health in the shadow of the dengue vaccine," *American Ethnologist* (Apr. 9, 2015), anthrosource.onlinelibrary.wiley.com/doi/abs/10.1111/amet.12135

121   Sabine Institute Press Release, "Statement by the Dengue Vaccine Initiative on the Philippines Regulatory Approval of Sanofi Pasteur's Dengue Vaccine, Dengvaxia®,"(Dec. 23, 2015), sabin.org/updates/pressreleases/statement-dengue-vaccine-initiative-philippines-regulatory-approval-sanofi

122   Michaeleen Doucleff, "Rush To Produce, Sell Vaccine Put Kids in Philippines at Risk" (May 3, 2019), npr.org/sections/goatsandsoda/2019/05/03/719037789/botched-vaccine-launch-has-deadly-repercussions

123   Sri Rezeki Hadinegoro et al., "Efficacy and Long-Term Safety of a Dengue Vaccine in Regions of Endemic Disease," *New England Journal of Medicine* (Sep. 24, 2015), nejm.org/doi/full/10.1056/NEJMoa1506223

124   Scott B. Halstead, "Dengvaxia concern flagged nearly 2 years before Philippines pulls dengue vaccine," *Infectious Disease News* (Dec. 5, 2017), healio.com/news/infectious-disease/20171205/dengvaxia-concern-flagged-nearly-2-years-before-philippines-pulls-dengue-vaccine

125   Scott B. Halstead, "Scott Halstead on Dengvaxia," YouTube Video (Dec. 16, 2017), youtube.com/watch?v=835rZQ7v8yw

126   Scott B. Halstead, "Critique of World Health Organization Recommendation of a Dengue Vaccine," *Journal of Infectious Diseases* (Dec. 15, 2016), watermark.silverchair.com/jiw340.pdf

127   Betsy McKay et al., "Safety Fears Threaten Global Dengue Vaccine Effort," *Wall Street Journal* (Jan. 8, 2019), wsj.com/articles/safety-fears-threaten-global-dengue-vaccine-effort-1515330002

128   Doucleff

129   Ibid.

130   "Safety Fears Threaten Global Dengue Vaccine Effort, *Wall Street Journal* (Jan. 8, 2018), wsj.com/articles/safety-fears-threaten-global-dengue-vaccine-effort-1515330002

131   "Merck and Instituto Butantan Announce Collaboration Agreement to Develop Vaccines to Protect Against Dengue Infections," businesswire.com/news/home/20181212005533/en/Merck-and-Instituto-Butantan-Announce-Collaboration-Agreement-to-Develop-Vaccines-to-Protect-Against-Dengue-Infections

132   Doucleff

133   Padron-Regalado et al., "Vaccines for SARS-CoV-2: Lessons from Other Coronavirus Strains" *INFECT DIS THER* (April 23, 2020), doi: 10.1007/s40121-020-00300-x

134   Tseng et al., "Immunization with SARS coronavirus vaccines leads to pulmonary immunopathology on challenge with the SARS virus," *PLOS ONE* (April 20, 2012), doi: 10.1371/journal.pone.0035421

135   Deming et al., "Vaccine efficacy in senescent mice challenged with recombinant SARS-CoV bearing epidemic and zoonotic spike variants," *PLOS MED* (Dec. 26, 2006), doi: 10.1371/journal.pmed.0030525

136   Weingartl et al., "Immunization with modified vaccinia virus Ankara-based recombinant vaccine

against severe acute respiratory syndrome is associated with enhanced hepatitis in ferrets," *J VIROL* (Nov. 2004), doi: 10.1128/JVI.78.22.12672-12676.2004

137   Kam et al., "Antibodies against trimeric S glycoprotein protect hamsters against SARS-CoV challenge despite their capacity to mediate FcgammaRII-dependent entry into B cells in vitro," *VACCINE* (August 22, 2006) doi: 10.1016/j.vaccine.2006.08.011

138   Agrawal et al., "Immunization with inactivated Middle East Respiratory Syndrome coronavirus vaccine leads to lung immunopathology on challenge with live virus," *HUM VACCIN IMMUNOTHER* (June 7, 2016), doi: 10.1080/21645515.2016.1177688

139   Hashem et al. "A highly immunogenic, protective, and safe adenovirus-based vaccine expressing middle east respiratory syndrome coronavirus S1-CD40L fusion protein in a transgenic human dipeptidyl peptidase 4 mouse model". J INFECT DIS. (July 4, 2019) doi: 10.1093/infdis/jiz137

140   Children's Health Defense, "The Dengue Vaccine: A Cautionary Tale" (Aug. 25, 2020), childrenshealthdefense.org/news/the-dengue-vaccine-a-cautionary-tale/

141   Rachel L. Roper and Kristina E Rehm, "SARS vaccines: where are we?," *Expert Review of Vaccines*, vol. 8, no. 7 (2009): 887–898, doi:10.1586/erv.09.43 (Jan. 9, 2014), ncbi.nlm.nih.gov/pmc/articles/PMC7105754/

142   Conor Skelding, "Fauci email dump includes 'sick' March Madness-style virus bracket," *New York Post,* (Jun 5, 2021). https://nypost.com/2021/06/05/fauci-files-include-sick-march-madness-style-virus-bracket/

ChildrensHealthDefense.org/fauci-book

childrenshd.org/fauci-book

For updates, new citations and references, and new information about topics in this chapter:

CHAPTER 12
GERM GAMES

War Games: Genesis of the Biosecurity State

"Those who would give up essential liberty, to purchase a little temporary safety, deserve neither liberty nor safety."

—Benjamin Franklin

"Many of us are pondering when things will return to normal. The short response is: never. Nothing will ever return to the "broken" sense of normalcy that prevailed prior to the crisis because the coronavirus pandemic marks a fundamental inflection point in our global trajectory."

—Klaus Schwab, *The Great Reset* (July 2020)

"I want to be straight with you: There will be no return to the old normal for the foreseeable future."

—Tedros Adhanom Ghebreyesus, World Health Organization Director-General

History of Bioweapons

The United States began its first large-scale offensive bioweapons research during World War II in the spring of 1943 on orders from President Franklin Roosevelt, as a collaboration between the US military and its pharmaceutical industry partners. Pharma titan George W. Merck ran the Pentagon's offensive bioweapons program while simultaneously directing his drug manufacturing behemoth. Merck boasted that his team could deliver biowarfare agents without vast expenditures or constructing huge facilities. Another advantage of bioweapons, he remarked, was that their development could proceed under the guise of legitimate medical research.

The intelligence agencies were involved in the top secret program from the outset. George Merck's hands-on employee, Frank Olson, was an American bacteriologist, biological warfare scientist, and CIA officer.[1] He worked for the United States Army Biological Warfare Laboratories (USBWL) at Fort Detrick with Merck and the US military developing the US bioweapons and psywarfare arsenal. Project Artichoke was an experimental CIA interrogation program that used psychoactive drugs like LSD in pursuit of "enhanced" interrogation methods. The project was part of a larger CIA program exploring approaches for controlling both individuals and populations. Olson was involved with Project Artichoke with moral misgivings, beginning in May 1952: after watching a documentary on Protestant reformation leader Martin Luther, a conscience-stricken Olson informed his bosses he intended to quit the biowarfare program.[2]

Around the time of that announcement, Olson's CIA colleague, Sidney Gottlieb, head of the CIA's MKUltra program, covertly dosed him with LSD. A week later, on November 28, 1953, Olson plunged to his death from a window of New York's Hotel Statler. The US government first described his death as a suicide, and then as misadventure. In 1975, the government admitted its guilt in the murder and offered Olson's family an out-of-court settlement of $1,250,000, later reduced to $750,000, which they accepted with an official apology from President Gerald Ford and then-CIA Director William Colby.[3]

By 1969, the US bioweapons program had developed weapons of a "nuclear equivalence," according to David Franz, who, for twenty-three years, served as commander of the US Army Medical Research Institute of Infectious Diseases (USAM-RIID).[4] The principal limitation, Franz acknowledged, was the difficulty of managing bioweapons so as to prevent accidental escape. Ironically, Franz would later play a key role in the Pentagon/Fauci gain-of-function programs leading up to the COVID-19 pandemic.

It all ended—seemingly—in late 1969, when President Nixon traveled to Fort Detrick to announce the closure of America's bioweapons program for moral and strategic reasons. America signed the Biological Weapons Convention in 1972—forbidding development, use, and stockpiling of biological weapons—and mothballed most of its labs.[5] But the agreement—a supplement to the Geneva Convention—left thousands of scientists, military contractors, and Pentagon caliphs as stranded assets yearning for the program's revival.

The treaty also included a yawning loophole: it allowed production of anthrax and other biological warfare agents for vaccine production. The Pentagon and CIA spooks continued to cultivate bioweapon seed stock. Between 1983 and 1988, Searle Pharmaceuticals CEO Donald Rumsfeld, acting as Ronald Reagan's envoy in Iraq, arranged for the top-secret shipment of tons of chemical and biological armaments, including anthrax and bubonic plague, to Iraqi President Saddam Hussein, hoping to reverse his looming defeat by Iran's million-man army. Ayatollah Khomeini's victorious Iranian forces were then routing Saddam in their war over the Persian Gulf. The Bush administration feared the impact on global oil supplies if Iran prevailed in that conflict.[6]

The Birth of the Biosecurity Agenda

Following the collapse of the Soviet Union in 1988–1991, the military-industrial complex began rummaging about for a more reliable enemy to permanently justify its hefty share of the GDP. While most Americans eagerly awaited the ballyhooed "peace dividend," Pentagon mandarins and their emporium of contractors may have considered with dismay that someone else would be spending money that was rightfully theirs. The peace dividend never materialized. Beginning with the first World Trade Center bombing in 1993 and culminating in 9/11, Islamic terrorism replaced the Soviets as the essential adversary in US foreign policy. It may have provided solace to the military and its contractors that "terrorism" was a more reliable long-term foe than the Soviets. Since terrorism is a tactic, not a nation, an imprecisely defined "terrorism" had the allure of an enemy that could never be vanquished. We can imagine the defense contractors' relief when Vice President Dick Cheney declared the "Long War"[7]—one, he promised, would last for generations—with battlegrounds "scattered in more than 50 nations."[8]

Military contractors held tight to their gravy train with the mission of building an expensive new arsenal of anti-terror technologies. But terrorism had its own shortfall, namely, the challenge of sustaining public fear sufficient to justify spending substantial portions of GDP to meet a threat that killed fewer Americans annually than lightning strikes. By 1999, some farsighted Pentagon planners were already looking ahead to the more exuberant and sustainable prosperity that would come with a war on germs.

Most historians date the nativity of the modern "Biosecurity Agenda" to the October 2001 anthrax attacks. But years earlier, military and medical industrial complex planners were already conceptualizing biosecurity as a potent strategy for leveraging potential pandemics or bioterrorism into vast funding increases, and as a device for metamorphosing America, the world's exemplary democracy, into a national security state with global dominance.

Robert Kadlec: "Let The Games Begin"

Bioweapons expert Robert P. Kadlec[9] is an American physician and retired Colonel in the United States Air Force who served as Assistant Secretary of Health and Human Services for Preparedness and Response from August 2017 until January 2021, and who managed the COVID-19 crisis during the Trump administration. Second only to his longtime crony and comrade in arms Anthony Fauci, Robert Kadlec played an historic leadership role in fomenting the contagious logic that infectious disease posed a national security threat requiring a militarized response. Since the 1993 World Trade Center terror attack, Kadlec had been evangelizing about an imminent anthrax attack that would doom the American way of life. In the mid-1990s, Kadlec served as part of an elite Air Force operations unit of UN weapons inspectors fruitlessly hunting the Iraqi desert for Saddam Hussein's suspected stores of anthrax and botulism following the first Persian Gulf War.

* * *

At 2:47 in the early morning of February 1, 2020, four hours after his loyal grantee, virologist Kristian Andersen, informed Dr. Fauci that he and other leading biologists believed that the genetic sequence responsible for the "furin cleave" on the virus's "spike protein"—the peculiar structure that allows the organism to bind tightly to, and infect cells with the ACE-2 receptor—was highly unlikely to be the product of natural selection, Dr. Anthony Fauci fired a carefully worded email to Kadlec. Dr. Fauci's other emails from that evening suggest that he was intensely worried that the Chinese experiments that may have created this striation in the novel coronavirus would bear his fingerprints. If Dr. Fauci's gain-of-function research had indeed minted COVID-19, then Kadlec would also be implicated. Kadlec served on the small so-called P3CO Committee charged with approving NIH's gain-of-function experiments, and it is clear from Dr. Fauci's email that the subject was also on Kadlec's mind. Dr. Fauci attached an article[10] to his email to Kadlec. It was "Bat Lady" Shi Zhengli's deceitful effort to downplay the laboratory leak hypothesis. "Bob: This just came out today," Dr. Fauci told his gain-of-function confederate. "Gives a balanced view."[11] Subsequent events proved that the author of that article was deliberately lying to conceal the Wuhan lab's manipulation of coronavirus pathogens that were nearly identical to the microbe that caused COVID-19. Both Kadlec and Fauci had been involved, for over a decade, in promoting and funding these dangerous experiments through NIAID and the Biomedical Advanced Research and Development Authority (BARDA), the biosecurity funding agency that Kadlec had helped create, including funneling millions of dollars in US funding to Zhi, the hapless writer of the exculpatory article. Dr. Fauci's email shows these two technocrats, and others, patching together evidence for the dubious official story that they would tell the world. Over the next few weeks, Dr. Fauci would pull the reliable old levers that he had manipulated for decades to transform convenient canards into official orthodoxies. The contrived cosmologies he thereby constructed would hold for a full year before they finally began to unravel.

* * *

Kadlec is a Dr. Strangelove knockoff with deep ties to spy agencies, Big Pharma, the Pentagon, and military contractors who profiteer from the spread of bioweapons alarmism. Intelligence agency historian and journalist Whitney Webb describes Kadlec as a man "enmeshed in the world of intelligence, military intelligence and corporate corruption, dutifully fulfilling the vision of his friends in high places and behind closed doors."[12] In 1998, Kadlec created an internal strategy paper for the Pentagon, promoting the development of pandemic pathogens as a stealth weapon that the Pentagon could deploy against its enemies without leaving fingerprints:

> Biological weapons under the cover of an endemic or natural disease occurrence provides an attacker the potential for plausible denial. Biological warfare's potential to create significant economic losses and consequent political instability, coupled with plausible deniability, exceeds the possibilities of any other human weapon.[13]

* * *

Kadlec, in 1999, organized his paranoia into several "illustrative scenarios" to demonstrate the United States' vulnerability to biological attack. In one of Kadlec's doomsday fantasies dubbed "Corn Terrorism," China clandestinely sprays corn seed blight over the Midwest from commercial airliners. Kadlec warns, "China gains significant corn market share and tens of billions [of] dollars of additional profits from their crop," while leaving the US Corn Belt in ruin. Another Kadlec scenario, titled "'Lousy Wine,'" envisions "disgruntled European winemakers" covertly releasing grape lice concealed in cans of pâté to target California wine producers.[14]

In an April 2001 study for the National Defense University National War College, Kadlec urgently recommended the creation of a Strategic National Stockpile to warehouse countermeasures including vaccines and antibiotics, and recommended regulatory changes to provide for mandatory vaccinations and coercive quarantines in the event of a pandemic. Those ideas helped win him an appointment as Special Assistant for Biodefense Planning to President George W. Bush after the post–September 11 anthrax attacks later that same year.[15] From this sinecure, Kadlec's fervent lobbying persuaded Congress to establish a Strategic National Stockpile, whose contents are currently worth $7 billion. Kadlec would come to control purchases for that stockpile, and—following the lead of his comrades, Bill Gates and Tony Fauci—he would use that power to enrich his vaccine industry friends and sideline public health.

As journalist Alexis Baden-Mayer observed, "Kadlec created the biodefense industrial complex as we know it. And he rules it like a czar."[16]

The Bill Gates/Anthony Fauci-Funded Biosecurity: "Let the War Games Begin"

In 1999, Dr. Kadlec organized a simulation of a smallpox terrorist attack on US soil for a joint exercise by the newly formed Johns Hopkins Center for Civilian Biodefense Strategies and the Department of Health & Human Services (HHS). The founder of the Center was D.A. Henderson, famed for leading the program that eradicated smallpox in 1977. The Senior Fellow and cofounder of the Johns Hopkins Center was a CIA spook and pharmaceutical industry lobbyist named Tara O'Toole. She took over as chief when Henderson left. The third Center Director was Tom Inglesby, who remains in that role. In 1999, the Bill & Melinda Gates Foundation committed $20 million to Johns Hopkins to establish the Bill & Melinda Gates Institute for Population and Reproductive Health.[17] For the next two decades, Gates would direct a vast stream of funding to the enterprise of elevating biosecurity as the national priority. Some of his most visible investments funded a series of simulations presided over by Inglesby at his Johns Hopkins Center. Those simulations would make Inglesby the congenial face of biosecurity paranoia, feed the burgeoning biodefense industry, and help lay the foundation for the modern security state.

The deal pipeline from NIH and NIAID to Johns Hopkins—an astonishing $13 billion since 2001—dwarfs Gates's contributions to the school.[18] But shoddy or perhaps deliberately obscure reporting makes it nearly impossible to determine how many of these dollars flowed to Inglesby and his center.

Kadlec's simulations, and over a dozen that would succeed it over the next twenty plus years—many under Bill Gates's direction—shared common features. None of them emphasized protecting public health by showing Americans how to bolster their immune systems, to eat well, to lose weight, to exercise, to maintain vitamin D levels, and to avoid chemical exposure. None of these focused on devising the vital communications infrastructures to link frontline doctors during a pandemic or to facilitate the development and refinement of optimal treatment protocols. None of these dealt seriously with the need to identify off-the-shelf (now known as "repurposed") therapeutic drugs to mitigate fatalities and to shorten a pandemic's duration. None of them considered ways to isolate the sick and protect the vulnerable—or how to shield people in nursing homes and other institutions from infection. None of them questioned the efficacy of masks, lockdowns, and social distancing in reducing casualties. None of them engaged in soul-searching about how to preserve constitutional rights during a global pandemic.

Instead, the simulations war-gamed how to use police powers to detain and quarantine citizens, how to impose martial law, how to control messaging by deploying propaganda, how to employ censorship to silence dissent, and how to mandate masks, lockdowns, and coercive vaccinations and conduct track-and-trace surveillance among potentially reluctant populations.

"Coercion should be the last strategy to consider in a pandemic," remarked physician and biological warfare expert Meryl Nass, MD. "If you have a remedy that works, people will flock to get it. It's troubling that the first and only option was creating a police state."

The Still-Unsolved Mystery of the Post-9/11 Anthrax Attacks

Contemporaneously with Johns Hopkins' smallpox simulation, the Pentagon launched a top-secret project at a former nuclear weapons site in the Nevada desert to test the feasibility of building a small anthrax production facility using off-the-shelf equipment easily available in hardware stores and biological supply catalogs.[19] Code-named Project Bacchus, a small cohort of faux terrorists—military weapons experts—succeeded in producing a few pounds of anthrax. Two years after the Pentagon's Nevada anthrax project, someone associated with the United States Army mounted a far-reaching campaign of sending anthrax to members of Congress and key media figures, officially launching the "Biosecurity Era."

In the light of subsequent events, we cannot exclude the possibility that someone in our government carried out a false flag attack against Americans as a provocation for some larger agenda. This is not an outlandish conspiracy theory. During my uncle's presidential administration, the Joint Chiefs of Staff submitted a plan—termed Operation Northwoods—proposing false flag attacks, including mass murders of random American citizens, to justify an invasion of Cuba. My uncle reacted with horror to Joint Chief Chairman Lyman Lemnitzer's Northwoods briefing pitch and abruptly walked out of the presentation. "And they call us the human race," he remarked to his secretary of state, Dean Rusk.[20]

US intelligence agencies and military industrial complex insiders initially (and ultimately wrongly) blamed the 2001 anthrax letter attacks on Saddam Hussein or al-Qaeda and later used similarly incorrect pretexts to launch a war against Iraq. The mailing of anthrax introduced Americans to a new enemy more frightening than garden-variety terrorism. While terrorists could destroy key buildings and airliners, the biosecurity narrative warns that pathogens could enter any American home and invisibly slay its occupants. Germs, therefore, easily outgunned al-Qaeda as a reliable wellspring of terror. This was the lesson Kadlec had been broadcasting for five years. The delivery of anthrax through the mail brought home his jeremiads. By 2020, biosecurity would altogether eclipse Islamic terrorism as the spear tip of US military and foreign policy. The topic of "infectious diseases" suddenly became the most effective way to open government pockets.[21]

Meet the El-Hibri Family

In 1998, Lebanese-born financier Ibrahim El-Hibri and his son, Fuad, with former chairman of the Joint Chiefs of Staff Admiral William Crowe, Jr., established a corporation called BioPort and paid the state of Michigan $25 million for its aging vaccine manufacturing campus. The purpose the El-Hibris intended to use the factory for was to manufacture anthrax vaccine for sale to the US military. El-Hibri Sr. was a longtime associate of both Robert Kadlec and Admiral Crowe—who chaired the Joint Chiefs under Presidents Reagan and George H. W. Bush. The El-Hibris had previous success in the anthrax vaccine business, having made a small fortune by purchasing anthrax vaccines made by the UK government and reselling them at 100 times the purchase price to the Saudi Arabian government.[22] Less than a month after taking over the Michigan-based business, BioPort signed an exclusive $29 million contract with the Pentagon to "manufacture, test, bottle, and store the anthrax vaccine" for American troops stationed abroad.[23] The secretary of the Army indemnified the factory the day before signing the contract on September 3, 1998. The El-Hibris never safety-tested their concoction. They didn't have to—they had no liability for injuries.

Ten months before the El-Hibris bought the plant, an FDA audit uncovered contamination problems, suspect record keeping, and assorted security breaches at their laboratory, as well as nine million stored doses that were adulterated. Almost as soon as BioPort was formed, it began receiving large sums from the US Army to rehabilitate the anthrax plant. But it was still unable to pass an FDA audit. In 1999, they bulldozed the factory and rebuilt it at taxpayer expense. The state of Michigan sweetened the deal. But the FDA would not give its stamp of approval to the new manufacturing facility. BioPort, with a hefty lobbying team and designer furniture in its executive offices, kept crying poor and coming back to the US government for additional handouts[24] before finally falling into a death spiral around the bankruptcy drain in mid-2001.[25] The October 2001 anthrax incidents proved the El-Hibris' salvation. The Pentagon leveraged the strange attacks, turning them into the long-awaited provocation, justifying the crusade to expand the battlefront in bioweapons research.

The 1972 Biological Weapons Convention meant neither the brass nor the spooks could legally research or produce bioweapons. But the convention left open the loophole that signatories could develop "dual use" vaccine and weapon technologies so long as the projects had a defensive rationale. After the anthrax attacks, "vaccines" suddenly became a euphemism for bioweapons and a ticket back to deep water for a beached biowarfare industry. Military planners at the Pentagon, BARDA, DARPA, and the CIA (through USAID) began pouring money into "gain-of-function" experiments. "Dual use" research was suddenly in vogue.

Dark Winter 2001

During June 22 and 23, 2001, less than three months before the 9/11 attacks, the Pentagon launched a war game code-named Operation Dark Winter at Andrews Air Force Base that emphasized the military's earnest commitment to bioweapon vaccines. Robert Kadlec, the lead organizer of this pandemic simulation, also coined its code name.[26]

The "tabletop" scenario simulated a smallpox attack on US locations, beginning in Oklahoma City (the site of a real domestic terror attack in 1995). Dark Winter participants explored strategies for imposing coercive quarantines; censorship; mandatory masking, lockdowns, and vaccination; and expanded police powers as the only rational responses to the pandemic. The failure, in the Dark Winter case, to quickly implement such countermeasures allowed the galloping spread of the Pentagon's imaginary smallpox epidemic to overwhelm America's response capabilities, precipitating massive civilian casualties, widespread panic, societal breakdown, and mob violence. The Pentagon summary of the exercise concluded that scarcity of vaccines to curtail the contagion's spread proved the most severe limitation on management options.

The Dark Winter exercise eerily predicted many aspects of what would follow just months later with the anthrax letter attacks. Such uncanny miracles of foreshadowing became a recurring feature of each subsequent Germ Game.

The Spooks and the Simulations

By playing the role of US president, the Senate Defense Committee's longtime chairman, Senator Sam Nunn, a dyed-in-the-wool war hawk, brought prestige, urgency, and a militaristic gestalt to Kadlec's Dark Winter exercise.

Most of the other key participants shared Kadlec's intelligence agency pedigrees. CIA involvement was a consistent feature of this and all the subsequent simulations. Other participants included: Robert Kadlec's fellow intelligence officer and War College professor, Colonel Randall Larsen (USAF), another career bioweapons expert, who helped choreograph the exercise and appeared in its fictional, scripted news clips; CIA's former director, James Woolsey, was a participant and organizer, as was a pharmaceutical industry lobbyist and biological weapons expert; Tara O'Toole, a Director of the CIA hedge fund In-Q-Tel;[27] the CIA's former deputy director for Science and Technology, Ruth David; Hopkins bioterrorism expert Tom Inglesby; and *New York Times* journalist Judith Miller also participated.[28]

James Woolsey's presence and that of Col. Larsen, Ruth David, and Tara O'Toole signaled the intelligence community's ubiquitous but shadowy presence in biosecurity and all things vaccine. (I sat on a board with Woolsey for several years and am familiar with his deep anxieties about germ warfare.) Woolsey's germophobia rivals Kadlec's; Woolsey calls a biological weapons attack "the single most dangerous threat to US national security in the foreseeable future."[29]

O'Toole is a biodefense enthusiast, cofounder of the Johns Hopkins Center for Civilian Biodefense Studies, and executive vice president at the CIA's investment arm, In-Q-Tel. That shady firm is the vector by which US intelligence services infiltrate start-up firms on the cutting edge of technological innovation. O'Toole, like her longtime confederate Kadlec, juggles deep and disturbing relationships with the same retinue of rapacious pharmaceutical industry and military contractors that Kadlec also cultivated.

In 2009, when President Obama nominated O'Toole for undersecretary for Science and Technology at the Department of Homeland Security, Sen. John McCain criticized her for concealing her role as strategic director of a pharmaceutical industry lobbying outfit, Alliance Biosciences.[30] Alliance is an unincorporated corporate front group created by Ibrahim El-Hibri and his partner, former Joint Chiefs Chair Admiral William Crowe, and funded by other bioweapons firms. Alliance has no tax filing and operates out of a K Street influence shop. The Congressional Record shows that the Alliance is a so-called "stealth lobbying" firm that spent $500,000 over 2005 to 2009 pitching Congress and the Homeland Security department for greater biodefense expenditures, and particularly for anthrax vaccines. Alliance's other funders include Pfizer; the International Pharmaceutical Aerosol Consortium; and Sig Technologies, a biodefense military contractor.[31]

O'Toole's nomination to undersecretary at the Department of Homeland Security also prompted objections from more mainstream bioweapons experts, including the preeminent Rutgers microbiologist Richard Ebright: "She was the single most extreme person, either in or out of government, advocating for a massive biodefense expansion and relaxation of provisions for safety and security." Ebright added, "She makes Dr. Strangelove look sane. O'Toole supported every flawed decision and counterproductive policy on biodefense, biosafety, and biosecurity during the Bush Administration. O'Toole is as out of touch with reality, and . . . paranoiac. . . . It would be hard to think of a person less well-suited for the position."[32]

During those same 2009 confirmation hearings, Democratic Senator Carl Levin of Michigan added to the voices of skepticism: "Dr. O'Toole fell short of the strict adherence to scientific principles when she was the director of the Johns Hopkins Center for Civilian Biodefense Strategies." Noting that "Dr. O'Toole was one of the principal designers and authors of the June 2001 Dark Winter exercise that simulated a covert attack on the United States by bioterrorists," Levin faulted O'Toole for using the exercise to promote her biosecurity agenda with hyperbolic pandemic fantasies: "But many top scientists have said that the Dark Winter exercise was based on faulty and exaggerated assumptions about the transmission rate of smallpox."[33]

Dr. James Koopman of the Department of Epidemiology at the University of Michigan made the ungenerous assessment that O'Toole's enthusiasm for germ warfare had clouded her scientific judgment. Koopman, an expert at modeling the transmission rates of infectious diseases who participated in the smallpox eradication program, complained that Dr. O'Toole "has not sought balanced scientific input in her thinking, that she shows a lack of analytic orientation to scientific issues, and that she has generated hype about bioterrorism that she will feel obligated to defend rather than pursue a balanced approach."[34]

Dr. Michael Lane, the former director of the Centers for Disease Control Smallpox Eradication Program, likewise condemned O'Toole for padding her assumptions about smallpox transmission rates in Dark Winter, which he characterized as "improbable" and even "absurd."[35]

Ironically, even Dr. Fauci, who by then was already the king of embellishing and fabricating pandemics, voiced his disapproval of O'Toole and Kadlec's extreme Dark Winter exaggerations, which Dr. Fauci declared "much, much worse than would have been the case" in real life.[36]

The transmission rate of smallpox was not the only area where Dr. O'Toole and Kadlec ignored facts. On February 19, 2002, O'Toole wrote that "Many experts believe that the smallpox virus is not confined to these two official repositories [one in the United States and one in Russia] and may be in the possession of states or subnational groups pursuing active biological weapons programs." O'Toole cited a June 13, 1999, *New York Times* article as the source for her alarming assertion that "subnational groups" controlled smallpox stocks. But that article included no reference to any non-state group actors possessing any biological weapons.[37]

Another key Dark Winter planner and participant was Ruth David, a former deputy director at the CIA. In 1998, David became president of ANSER, a nonprofit corporation with deep ties to the CIA. ANSER played a key role in pushing the government toward "homeland security" post-9/11 and became a primary promoter of biometric and facial recognition software for US law enforcement agencies. Among other functions, ANSER funds a mysterious defense contractor from South Carolina called Advanced Technology International.[38] ATI somehow became the vector through which the government arranged at least $6 billion of secretive Operation Warp Speed vaccine contracts with Pfizer, Bill Gates's Novavax vaccine, Johnson & Johnson, and Sanofi.[39] Those contracts, comprising the majority of Operation Warp Speed's $10 billion budget, suggest a deep CIA involvement with the COVID-19 vaccine enterprise's cozy deals with Big Pharma. As assistant secretary for Preparedness and Response with HHS, Robert Kadlec personally signed off on those sweetheart deals. The terms allow Operation Warp Speed to completely "bypass the regulatory oversight and transparency of traditional federal contracting mechanisms," as NPR put it.[40]

In a January 2021 exposé, the *New York Times* dug into Kadlec's secretive vaccine contracts, observing that "available documents . . . suggest that drug companies demanded, and received, flexible delivery schedules, as well as patent protection and immunity from liability if anything goes wrong. In some instances, countries are prohibited from donating or reselling doses, a ban that could hamper efforts to get vaccines to poor countries."[41]

Dark Winter Aftermath

Despite all its hiccups, Dark Winter was an extraordinary success. It foreshadowed the real bioweapons incidents occurring less than three months later, inflamed public germophobia, and fortified the official narrative after the first September 18 anthrax attack letters, which pointed fingers at Saddam Hussein and/or al-Qaeda as the probable culprits. Several Dark Winter participants displayed extraordinary prescience in the weeks leading up to the anthrax attacks, along with a relentless determination to pin the caper on Saddam. The anthrax attack's first casualty, Robert Stevens, was hospitalized and diagnosed with anthrax on October 2. Highly publicized and laudatory Senate hearings on the Dark Winter simulation that began on October 1, 2001—three days before the anthrax attacks became public knowledge—functioned to imbue US government officials, the national press, and the public with Dark Winter's paranoid assumptions and to assign the blame to Saddam.

Another Dark Winter planner, Jerome Hauer, along with spymaster James Woolsey and *New York Times* reporter Judith Miller, spent the three weeks between 9/11 and 10/4 banging the gong about imminent anthrax attacks, carpet-bombing the television talk shows, kibitzing on the nightly news, and gabbing up the Sunday morning TV gasbags. Judith Miller received special assistance in this task from her employer, the *New York Times*, which published her numerous alarmist reports and warnings about coming biological attacks on American soil. Incredibly, the attack arrived exactly as Miller, Hauer, and Woolsey predicted and with exquisite timing—smack in the middle of the US Senate hearings over America's vulnerability to an anthrax attack. Hauer, a bioterrorism expert and pharmaceutical industry operative, is currently an executive with Teneo, a consulting firm that counsels corporations on security matters and is one of the leading advocates of mandatory vaccines for employees as a condition for employment.[42]

Members of the think tank the Project for a New American Century (PNAC) also played a key role in sounding the alarm that a biological weapons attack was certain to follow on the heels of 9/11 and then simultaneously amplified the panic and blamed Iraq following the anthrax letter attacks. PNAC's core doctrine was that, as the Cold War victor, America and US-based multinationals—particularly petroleum and pharmaceutical companies—had earned the right to rule the world for a century or so. PNAC members populated virtually all of the key foreign policy posts in the Bush White House. The warmongering cabal called themselves "The Vulcans" in honor of their belligerent brand of US imperialism. Their members included Dick Cheney, Scooter Libby, Donald Rumsfeld, Douglas Feith, Elliott Abrams, John Bolton, and Rumsfeld's advisers Richard Perle and Paul Wolfowitz. Critics called them the "Chicken Hawks" because ironically, each one of them had draft-dodged the Vietnam War.[43]

Osama bin Laden, the author of the World Trade Center attacks, supposedly directed that operation from an Afghan cave. But Donald Rumsfeld complained, "There aren't any good targets in Afghanistan."[44] The PNAC chicken hawks were determined to use 9/11 as a pretext for a war against Iraq, beneath which God had mischievously stockpiled so much of America's oil. Anthrax provided that provocation. Control of global oil resources was, for PNAC, a key stepping-stone for the coming century of American imperialism, and a bioweapon attack against America became the ideal provocation for preemptive invasion.

It's noteworthy that Judith Miller not only covered the Dark Winter exercise for the *New York Times*, she was also an active planner and participant in the simulation, playing the part of a reporter.[45] Miller was an O.G. germaphobe and veteran biosecurity booster.

On September 4, 2001, exactly one week before the 9/11 attacks, Miller, excerpting from a paranoid book, *Germs*, she had written with *Times* reporters William Broad and Steve Engelberg, reported approvingly in the *New York Times* that the Pentagon had green-lighted "a project to make a potentially more potent form of anthrax bacteria."[46] Miller did not explain why this response seemed rational or even sane.

Miller's articles repeating Pentagon and CIA claims about Saddam's bioweapons cache and his probable involvement with the anthrax attacks helped fuel the US invasion of Iraq. According to *New York Magazine*:

> During the winter of 2001 and throughout 2002, Miller produced a series of stunning stories about Saddam Hussein's ambition and capacity to produce weapons of mass destruction . . . almost all of which have turned out to be stunningly inaccurate.[47]

Miller's jingoistic reporting—*New York Magazine* dubbed her "Chicken Little"— played such a decisive role in validating the White House warmongers' Iraq invasion agenda that the *New York Times* afterward made an unprecedented apology for its role in what then was, arguably, the worst foreign policy decision in United States history.

Miller was so keen to facilitate an Iraq invasion that she illegally leaked the identity of CIA agent Valerie Plame, to punish Plame's husband, State Department diplomat Joseph Wilson, who had publicly challenged White House and CIA narratives about Iraq obtaining yellowcake uranium from Niger.

The CIA, at that time, was aggressively pushing for war. George W. Bush later said that his worst mistake during his White House years was swallowing the CIA's guarantees: "The biggest regret of all the presidency has to have been the intelligence failure in Iraq. A lot of people put their reputations on the line and said the weapons of mass destruction is a reason to remove Saddam Hussein."[48] In 2003, during the run-up to the war, CIA Director George Tenet assured President Bush that Saddam had a secret arsenal of Weapons of Mass Destruction (WMDs): "Don't worry, it's a slam dunk."[49]

Miller served three months in jail for contempt before she agreed to disclose the identity of her confederate, Lewis "Scooter" Libby, V.P. Cheney's chief of staff. Libby, who told Miller that Plame was a clandestine CIA agent and directed her to publish the revelation, subsequently went to prison for the crime. It will be many years before the CIA releases documents explaining the agency's true relationships, if any, with Miller and Libby. Libby, a PNAC founder and key visionary and promoter of America's 100-Year Reich, was an early champion of the modern biosecurity agenda, with multiple personal connections with the intelligence community at Yale, Rand, Northrop Grumman, and the Pentagon. The State Department's Bureau of East Asian and Pacific Affairs—his employer in the early 1980s—had, and still has, deep CIA ties. His obsession with bioterrorism led Libby to write a novel about a smallpox pandemic and earned him the White House nickname "Germ Boy." Following his pardon and subsequent prison release by President Donald Trump, Libby joined Robert Kadlec's Blue Ribbon Panel for Biodefense (BRPB), which promotes: biosecurity as the fulcrum of US foreign policy, the twenty-first century as the age of US empire, and mass vaccination as a foreign policy tool. Libby's fellow BRPB director, William Karesh, is the executive vice president of Peter Daszak's EcoHealth Alliance, the organization through which Dr. Fauci, Kadlec, and the Pentagon— through DARPA—were laundering gain-of-function payments to Chinese scientists in Wuhan. Libby also serves as senior vice president of the Hudson Institute, a think tank with deep connections to the pharmaceutical industry, Monsanto, and the CIA. He guides the institute's program on national security and defense issues. In 2021, former CIA Director Mike Pompeo joined the Hudson Institute.

The pervasive CIA involvement in the global vaccine putsch should give us pause. There is nothing in the CIA's history, in its charter, in its composition, or in its institutional culture that betrays an interest in promoting either public health or democracy. The CIA's historical preoccupations have been power and control. The CIA has been involved in at least seventy-two attempted and successful coups d'état between 1947 and 1989,[50] involving about a third of the world's governments. Many of these were functioning democracies. The CIA does not do public health. It does not do democracy. The CIA does coups d'état.

Smallpox: Biosecurity Blossoms

Dark Winter was part of a persistent campaign by the intelligence agencies and the bioweapons lobby to keep smallpox fears alive in the public consciousness. Even before the disease was eradicated in 1977, public health regulators had discontinued smallpox vaccinations in the United States. Public health advocates urged the federal bureaucracies and the military to destroy their smallpox stockpile,[51] to prevent the disease from escaping and, possibly, decimating humanity. Ignoring these warnings, the George W. Bush administration purchased even more. During the run-up to the Iraq war, President Bush aimed to inoculate the US population with smallpox vaccines. Skeptics charged that the reckless scheme was PNAC's transparent gimmick for hyping fear of Saddam Hussein's mythological bioweapons program. Dr. Meryl Nass, writing on the history of smallpox vaccine, later reported:

> The smallpox vaccine was known to be highly reactogenic. . . . When the vaccine was given to healthcare workers and first responders in 2003, episodes of heart failure, heart attacks, myocarditis, and death quickly mounted. Doctors and nurses learned that they could not sue for damages if injured, and at first there was no federal compensation either. They began refusing to be vaccinated.[52]

The Clinton administration continued to stockpile millions of smallpox vaccines and Congress allotted money for a compensation program, but the maximum award was only $250,000 for a permanent disability or death. After distributing 40,000,000 inoculations, the wave of alarming injuries caused the government to abandon the project's civilian arm. The military continued vaccinating soldiers with the untested, unapproved, deadly vaccine, with catastrophic results.[53] The vaccine caused symptomatic myocarditis in one in every 216 soldiers, and subclinical myocarditis in one in thirty-five soldiers, according to a 2015 US Army study. Government officials have since recognized vaccines as a probable culprit in the era's epidemic of Gulf War Syndrome, which affected vaccinated soldiers, both deployed and those vaccinated in preparation for deployment, but never deployed. (The court observed that "Absent an informed consent or presidential waiver, the United States cannot demand that members of the armed forces also serve as guinea pigs for experimental drugs."[54,55])

10/4 Anthrax Attack

Less than four months after the Dark Winter simulation and three weeks after 9/11, a mysterious spate of letters containing fine white anthrax spores arrived by mail at several news media outlets and the Capitol Hill offices of two senators, Tom Daschle and Patrick Leahy. Those two senators had been the most vocal in condemning the post-9/11 infringements on civil liberties pushed by the PNAC crowd. Administration and press accusations pegging Saddam Hussein as the probable culprit in the anthrax attacks, which killed five Americans, fueled Congress' hasty passage of the Patriot Act—as Michael Moore proved, not a single elected member had read the bill—and its jingoistic declaration of war on Iraq.

By abolishing traditional privacy protection, the Patriot Act created "an entire terror industry," according to a 2021 report by Action Center on Race and Economy.

The biggest beneficiaries have been Silicon Valley tech companies, particularly Amazon, Microsoft, and Google, who have partnered with federal intelligence agencies to mine data and "profit from the war on terror by at least $44 billion since 2001." The Patriot Act passage, the report says, "opened the door for Big Tech to become, first and foremost, the brokers of our personal data, selling it to secret agencies and private companies at home and abroad unleashing the era of the digital economy."[56]

Second only to Vice President Dick Cheney, the staunchest war hawk among George W. Bush's beltway coterie was his secretary of defense, former Searle Pharmaceutical CEO and PNAC chieftain Donald Rumsfeld—the very man who, fourteen years earlier, had given Saddam his anthrax arsenal. While no one has ever proven the origin of the anthrax in those letters, the FBI concluded that the powder had come from a US military lab.[57]

Robert Kadlec was first among the large coterie of pharmaceutical companies and military contractors to benefit from the anthrax scare. Immediately after the anthrax letters arrived, Kadlec became a special adviser on biological warfare to then-secretary of defense Donald Rumsfeld and his PNAC deputy, Paul Wolfowitz.

Three Suspects—All Linked to the US Military

The PNAC cabal was determined to blame the anthrax attack on Saddam Hussein, and Rumsfeld's deputy, Paul Wolfowitz, tasked Kadlec with confirming the presence of bentonite in the anthrax used in the attacks. Experts had advised Rumsfeld and Wolfowitz that bentonite was a "fingerprint" unique to Iraqi anthrax stocks; its presence would therefore put the blame on Saddam. Kadlec did not succeed in finding bentonite in any of the anthrax samples that the FBI tested. But repeated media reports claiming otherwise allowed warmongers to drum up jingoistic hysteria against Saddam. By late October 2001, one nationwide poll found that 74 percent of respondents wanted the United States to take military action against Baghdad, despite a complete lack of evidence connecting Iraq to either 9/11 or the anthrax attacks.[58]

Instead of pointing the finger at Saddam, the FBI lab found that the anthrax spores originated from one of three US Army labs; Fort Detrick; a lab at the University of Scranton; or Battelle's West Jefferson facility, owned by an El-Hibri business partner.[59]

The FBI closed its investigation after its leading suspect, a vaccinologist, Dr. Bruce Ivins, who ran the US Army lab at Fort Detrick, allegedly took his own life. A multitude of critics of the shoddy and haphazard FBI investigation complained that Ivins was the victim of a ham-handed FBI frame. According to the FBI's former lead investigator, Richard Lambert, the FBI team hid a "mountain" of evidence that would have exonerated Ivins.[60]

In 2008, following Ivins's untimely "suicide," Department of Justice civil attorneys in Florida, defending a claim by the widow of anthrax victim Robert Stevens, publicly challenged the FBI's assertions that Ivins had been the culprit and instead pointedly "suggested that a private laboratory in Ohio" managed by Battelle and linked to the El-Hibris "could have been involved in the attacks."[61] DOJ headquarters quickly had its Florida attorneys rewrite their brief, omitting this claim.

An Italian publication, *Il Manifesto*, reported in its October 2001 issue that the FBI had placed the El-Hibris on its suspects list for sending the anthrax spores through the US mail.[62]

Cui Bono

Since 1995, Kadlec had been frothing about bioterrorism to war college students and urging the creation of a Strategic National Stockpile (SNS) to warehouse vaccines and other countermeasures. In 2004, with Kadlec now working for Secretary Rumsfeld at the Bush White House, Congress passed the Public Health Security and Bioterrorism Preparedness Act—which Kadlec drafted—directing the secretary of HHS to maintain a "Strategic National Stockpile (SNS)" managed jointly by DHS and HHS.[63]

The same week, Congress passed the Project BioShield Act—which Kadlec also helped draft—launching the Biomedical Advanced Research and Development Authority (BARDA), a government-operated investment bank that would germinate new technologies for Kadlec's stockpile. With Kadlec's guidance, BARDA would become a federal ATM machine for Big Pharma, biodefense contractors, and gain-of-function researchers. Along with Dr. Fauci's NIAID and the Pentagon's DARPA, BARDA would be the other big-league funder for experiments to create pandemic superbugs in Wuhan and elsewhere. Kadlec's statute authorized the purchase of $5 billion of matériel—including vaccines—for the stockpile, creating a gold mine, as we shall see, for Kadlec's friends the El-Hibris.

Another conspicuous beneficiary of the Stockpile was then-Secretary of State Donald Rumsfeld, and Kadlec's boss, who made a killing during the 2004 fake bird flu pandemic, which Tony Fauci ginned up—with his confederate, an ambitious young British physician and Wellcome Trust researcher, Jeremy Farrar. Sixteen years later, as Director of Wellcome Trust, Farrar would play a key role in the 2020 Wuhan cover-up. The Pentagon, in 2004[64] and 2005, in response to Farrar's concocted contagion, stockpiled 80 million doses of Gilead's flu remedy Tamiflu. Secretary Rumsfeld had served on the board of Gilead from 1988 to 2001 and was its chairman from 1997 until he joined the Bush administration as defense secretary. He retained stock in the pharmaceutical company, which netted him a $5 million profit from the Tamiflu run-up. George Shultz, another PNAC war hawk, also hit the jackpot, cashing in $7 million of Gilead stock during the Tamiflu run-up.[65]

The biggest winners, however, were the El-Hibris: the anthrax attacks brought them exoneration, salvation, and extravagant windfalls.

BioPort's Rebirth and Reinvention as Emergent BioSolutions

Anthrax arrived just in time for the El-Hibris. BioPort was by then on the ropes. The El-Hibris' anthrax vaccine facility was facing bankruptcy and the loss of its operating license. BioPort's Pentagon contract expired in August 2001, with a host of outstanding accounting mysteries impeding its renewal. The Pentagon had given BioPort millions to renovate its factory, but much of that money instead financed senior management bonuses and an opulent makeover for the El-Hibris' executive offices. Millions more simply "disappeared," according to journalist Whitney Webb. In 2000, not long after receiving its first Pentagon bailout, BioPort contracted none other than Battelle Memorial Institute to cultivate its anthrax seed stock.

Kadlec's boss, Donald Rumsfeld, told aides that his biosecurity priority after the incidents of anthrax sent through the mail was rescuing BioPort: "We're going to try to save it, and try to fashion some sort of an arrangement whereby we give one more crack at getting the job done with that outfit. It's the only outfit in this country that has anything under way, and it's not very well under way, as you point out."[66]

Gold Rush

In the summer of 2001, two months before the 9/11 World Trade Center attacks, the Department of Defense officially launched its drive to revive bioweapons research by sending a report to Congress, authored by Kadlec, pleading that the military's system for developing vaccines to protect troops from anthrax, smallpox, and other exotic bioweapons "is insufficient and will fail."[67]

Beginning with the 9/11 attack, the War on Terror triggered a tectonic shift in global security priorities and elephantine ripples in defense spending patterns across the globe as open democracies began shifting to a security state footing. The revival of US government interest in germ warfare opened new opportunities. The US biodefense budget went from $137 million in 1997 to $14.5 billion for 2001–2004.[68] Every agency with a colorable claim to a National Security function paddled out frantically to barrel the money tsunami. Between 2001 and 2014, the United States spent around $80 billion on biodefense. Since germ weaponry was still illegal, vaccines became a critical euphemism for the revival of the multibillion-dollar bioweapons industry. Pentagon sources told *Science Magazine* that the military was applying for "a sweeping overhaul of how the federal government develops vaccines to protect both the military and civilians."[69] The Pentagon's assault on the vaccine space was both an opportunity and threat to Dr. Fauci and NIAID.

US Vice President Cheney and his PNAC confederates found some convenient loopholes in the Geneva Convention through which they drove a fortyfold expansion in spending in biological weapons research.

The Department of Defense had strict systems in place to ensure compliance with the Biological Weapons Convention. Those restrictions limited the Pentagon's freedom to undertake new research programs, particularly those referred to as "the leading edge of biodefense." Cheney's response, recalls Professor Richard Ebright, "was to transfer this research from the Department of Defense to the National Institutes of Health, specifically to the National Institute of Allergy and Infectious Diseases (NIAID). By about 2004, this transfer was complete, and NIAID had been transformed into an arm of the defense sector."[70] This made the NIAID Director Anthony Fauci a major player in biodefense and germ warfare.

Dr. Fauci sharpened his elbows and began maneuvering for a leading role for NIAID in milking the BARDA/Homeland Security's cash cows. NIAID's biosecurity budget went from zero dollars in 2000 to $1.7 billion after the 2001 anthrax letters, much of that for bioweapons vaccines.[71]

Within five months following the anthrax postal incidents, Dr. Fauci had created two new sub-agencies to capture his share of the cheese: the NIAID Strategic Plan for Biodefense Research and the NIAID Biodefense Research Agenda for CDC Category A agents, which were those microorganisms designated by CDC to be potential pandemic pathogens. To populate the sub-agencies, he assembled a cadre of his loyal deputies and infectious disease principal investigators from the HIV bonanza. Their mission was to brand contagions as pressing terror threats, drum up pandemic panic, and lobby for government support for NIAID's new battery of biodefense vaccinations.

Dr. Fauci and the El-Hibris found common cause. Dr. Fauci could run interference for the El-Hibris at FDA, overriding regulatory anxieties about BioPort's laboratory and product safety. The El-Hibris, in turn, provided Dr. Fauci with a ready-made biodefense vaccine and a beachhead into the arcane maze of military contracting. Taking to the airwaves, Dr. Fauci made himself the face of biodefense. In a style now familiar to Americans, Fauci warned the public that postal workers who had handled the letters containing anthrax spores "might still be harboring these in their lungs even after taking two months of antibiotics," spreading plague with the morning mail. Taking the El-Hibris' vaccine prophylactically, Dr. Fauci advised, might help.[72, 73] Dr. Fauci's signature fearmongering was, of course, his trademark science-free speculation.

Nestling the El-Hibris under his protective wing, Dr. Fauci swept aside FDA's safety concerns and publicly praised BioPort's experimental anthrax vaccine, BioThrax. He brushed aside the reservations of critics that the El-Hibris never established BioThrax's safety with some of his prototypical dissembling. Dr. Fauci said, "The vaccine is designed to get the immune system to recognize the proteins—and therefore the bacteria—and destroy both."[74]

In a December 2001 PBS interview, Fauci promised to deliver BioThrax—which had failed to pass a single FDA audit during the prior four years—at record pace. Fauci explained, "In usual times, that is a process that takes years and years," but he committed that his project for delivering BioThrax "is going to be markedly truncated because of the urgency of the situation."[75]

PBS observed that because of BioPort's production problems, the Pentagon had dramatically scaled back its plan to vaccinate US forces, and there were insufficient anthrax vaccines in the Pentagon's stockpile to conduct the mass civilian inoculation program that had been Dr. Fauci's ultimate aim.[76] But BioPort still possessed the only military contract, and Fuad El-Hibri announced that he was primed to ramp up production.

Practically every veteran federal bureaucrat was jockeying to ride the War on Terror into the high stakes winner's circle. The military's medical corps, maneuvering for its share of the overflowing stream of bioterrorism funding, had proposed that each American soldier should receive seventy-five new vaccines upon enlistment, to cover every potential bioweapon. The brass asked President Bush to finance the development of this inoculation fusillade. Not to be outgunned by the military doctors, Dr. Fauci announced in an October 2002 speech that within ten years, "his institute would produce a vaccine, a therapeutic drug and an adjuvant drug for each of some two dozen bioweapons diseases, such as plague and hemorrhagic fever." According to an article in *Scientific American*, "one scientist who requested anonymity said that

Dr. Fauci told him that the Bush administration had demanded this goal and that he accepted it to prevent the Department of Defense or the Department of Homeland Security from getting the job." Dr. Fauci was openly competing with the military in an escalating campaign to soak the taxpayers using the risk posed by anthrax as a pretext. NIAID's biodefense budget alone increased sixfold between 2002 and 2003—from $270 million to $1.75 billion.[77]

When no further bioterror attacks occurred over the next ten years, Dr. Fauci skillfully maintained his annual $1.7 billion biosecurity funding by deftly recalibrating his rhetoric away from bioterrorism hype. Instead, he invoked the new panic of natural but emerging infectious diseases. Dr. Fauci's pivot to conflate infectious disease with terrorism proved a milestone inflection point in the militarization of pandemic response and in overcoming the traditional revulsion among Western democracies—codified in the Nuremberg Charter—against coercive medical interventions.

Despite the fact that they collectively killed only 800 people globally,[78] the SARS coronavirus outbreaks between 2002 and 2004 were therefore a godsend to Dr. Fauci. The NIAID Director ignored the most compelling caveat from those incidents: the fact that coronavirus lab escapes in China, Taiwan, and Singapore had precipitated several of the outbreaks.[79] Fauci boasted in 2011, "Through the anthrax response, we built both a physical and an intellectual infrastructure that can be used to respond to a broad range of emerging health threats."[80] By that time, the escalating intramural arms race to capture Pentagon, CIA, BARDA, DARPA, and HHS biosecurity funding was pulling the military, CIA, and NIAID deeper and deeper into the dicey alchemy of "gain-of-function research" that would ultimately culminate inside the BSL-4 Pandora's box in Wuhan.[81]

The CIA Dips In Its Toe

The CIA had a long, sordid history of secretly promoting the US bioweapons program. One of the agency's first projects was establishing a network of so-called "ratlines" that Army intelligence officers used to smuggle some 1,600 chemicals and bioweapons and WMD experts—many of them Nazi Party kingpins and notorious war criminals—out of the reach of the Allies' Nuremberg prosecutors following World War II. The directors of a notorious operation, code-named Paperclip, provided these researchers with new identities and put them to work developing US germ warfare capacity at Ft. Detrick and elsewhere even after 1972. As late as 1997, the CIA defied the Bioweapons Treaty to launch a top-secret—and highly illegal—effort to create a doomsday "bacteria bomblet."[82]

The CIA officially made its open debut in the biosecurity racket in 2004, with its launch of Argus, a project that monitors biological, terrorist, and pandemic threats in 178 nations.[83] CIA operative and pediatrician Jim Wilson set up the program at Georgetown University with funding from DHS and the Intelligence Innovation Center to create and implement global foreign biological event detection and tracking capability, capable of assessing millions of pieces of information about social behavior daily and to train government officials in pandemic preparedness.[84] One of the key figures in this global surveillance effort was CIA officer Dr. Michael Callahan.

Dr. Michael Callahan is one of the biggest names in bioweapons research. Dr. Callahan ran a biosecurity program for the former CIA surrogate USAID before serving as Director of DARPA's bioweapons research program. At DARPA, he competed to outdo NIH in laundering money through Peter Daszak's EcoHealth Alliance to perform bioweapons research, including at the Wuhan lab.[85]

And as DARPA director, Callahan launched the PREDICT project in 2009 following Jeremy Farrar's fake bird flu pandemic. PREDICT appeared to be a reincarnation of the CIA's Argus project under the cover of USAID. PREDICT is the largest single source of funding to Daszak, with a $3.4 million subgrant routed through the University of California (2015–2020). PREDICT became the largest funder of gain-of-function studies and served as the principal funding vehicle through which the gain-of-function cartel evaded Barack Obama's 2014 presidential moratorium.[86]

When, during the height of the presidential gain-of-function moratorium, Ralph Baric and the UTMB lab's Vineet Menachery brazenly published their alarming 2015 study—describing their reckless experiments to breed pandemic bat coronaviruses that could spread via respiratory droplets in humanized mice—they omitted mentioning, in their initial online version of the article, that one of the funding sources was USAID-EPT-PREDICT. Apparently hoping to cover its tracks, PREDICT had laundered its grant through Peter Daszak's EcoHealth Alliance.

USAID's PREDICT program boasts that it has identified almost a thousand new viruses, including a new strain of Ebola, and trained some 5,000 people. In October 2019, not long before COVID-19 emerged, USAID abruptly ceased funding PREDICT, a decision bemoaned by Daszak in the *New York Times* as "definitely a loss."[87]

Callahan had a chummy relationship with Daszak, with whom he coauthored several articles—including throughout the gain-of-function moratorium. In April 2015, for example, the names of Michael V. Callahan and Peter Daszak appeared as coauthors on a paper published in the *Virology Journal* and titled "Diversity of Coronavirus in Bats from Eastern Thailand."[88]

Callahan was well aware that he and his confederates were toying with fire. In 2005, Callahan testified before Congress as he was moving into his new office at DARPA. He concluded the hearing with a chilling warning about the nation's new commitment to Janus-faced gain-of-function science that Drs. Fauci, Robert Kadlec, Callahan himself, and many others would proceed to blithely ignore:

> the dark science of biological weapon design and manufacture parallels that of the health sciences and the cross mixed disciplines of modern technology. Potential advances in biological weapon lethality will in part be the byproduct of peaceful scientific progress. So, until the time when there are no more terrorists, the US Government and the American people will depend on the scientific leaders of their field to identify any potential dark side aspect to every achievement.[89]

Even after leaving DARPA and USAID, Callahan periodically boasted of his continuing influence over US pandemic response policies at the highest levels of government. He alluded to his confidence in these mysterious connections in 2012: "I still have federal responsibilities to The White House for pandemic preparedness and exotic disease outbreak which will continue for the near future."[90]

On January 4, 2020, Callahan called Dr. Robert Malone from China just as the coronavirus began taking its first wave of casualties. Malone, a former contractor to the US Army Medical Research Institute of Infectious Diseases and the chief medical officer at Alchem Laboratories, is the inventor of the mRNA vaccine technology platform. Malone first met Callahan in 2009 through Malone's sometime business partner, Daryl Galloway, a CIA officer who formerly served in the US Navy and at one point held the post of director of JSTO in the Defense Threat Reduction Agency. To Malone, Galloway introduced Callahan as a fellow CIA officer. During his January 4 phone call, Callahan told Malone that he was just outside Wuhan. Malone assumed that Callahan was visiting China under cover of his Harvard and Massachusetts General Hospital appointments. Callahan told Malone that he had been treating "hundreds" of COVID-19 patients. Callahan subsequently described to *National Geographic* how he had pored through thousands of case studies at the outbreak's epicenter. He giddily reported his amazement at the virus's "magnificent infectivity," and its capacity to explode "like a silent smart bomb in your community."[91] Callahan later confessed to Malone that he lacked authority to be in Wuhan and had escaped by boat when the government imposed its quarantine. Callahan repeated parts of this story to Brendan Borrell, a writer for *Science*. Later, DTRA scientist Davis Hone, a GS15 officer, warned Malone to stop talking about Callahan, saying that "We had no military personnel in Wuhan at the time of the outbreak and Michael was lying about his presence." Malone told me, "That would mean that Michael also lied to Brendan Borrell." On leaving China, Callahan returned to Washington to brief federal officials and then went directly to work as a "special adviser" to Robert Kadlec, managing the government's response to the coronavirus.

Robert Kadlec as "Bad Santa": The El-Hibris Cash In

By 2011, BioPort was already profiting handsomely in the bioweapons/vaccine space. After 9/11, President Bush—presumably at the urging of Secretary Rumsfeld, Robert Kadlec, and Dr. Fauci, whose advice he valued—had placed BioPort's Michigan lab under protection "in the national interest."[92] El-Hibri and his son, gnawing on gristle prior to 10/4, began fattening themselves on NIAID and BARDA contracts. With friends like Fauci and Kadlec in high places, BioPort, which changed its name to Emergent BioSolutions in 2004 to escape its checkered past, was enjoying the first bright days of the charmed journey that would place the El-Hibris among the elite army of COVID-19 nouveaux-billionaires in 2021.[93]

After 2001, Rumsfeld's Pentagon agreed to hike BioPort's compensation by 30 percent—from $3.35 in its 1998 contract to $4.70 per dose—and to purchase anthrax shots for 2.4 million members of the armed forces, each of whom the military would require to receive six doses over an eighteen-month period.[94] That was $60 million worth of poorly performing and unapproved vaccines for a threat that never again surfaced. The anthrax threat was always phantasmagoric; since anthrax does not spread through human-to-human transmission, terrorists plotting an anthrax epidemic would need to somehow simultaneously release spores over dozens of US cities.

The anthrax deal was exceptionally ridiculous, since antibiotics are a far safer, more elegant, and more useful defense against anthrax. The prescribed remedy, ciprofloxacin, is a cheap, commonly used antibiotic that Tony Fauci himself recommended after the 2001 postal incidents. "The best approach toward anthrax is antimicrobial therapy," Dr. Fauci admitted to Congress in 2007.[95] Indeed, the night of the 9/11 attacks, the White House Medical Office thoughtfully and presciently dispensed ciprofloxacin to select White House staff who were accompanying Dick Cheney to the safety of Camp David.[96]

Furthermore, the El-Hibris' anthrax jab was by far the worst of a bad lot. According to the *Times*, "Emergent's anthrax vaccine was not the government's first choice. It was more than 30 years old and plagued by manufacturing challenges and complaints about side effects. Officials instead backed a company named VaxGen, which was developing a vaccine using newer technology licensed from the military."[97]

In 2004, the El-Hibris cofounded, with their partner and former Joint Chiefs Chair Admiral William Crowe, a lobbying group called the Alliance for Biosecurity, as part of their strategy to secure lucrative BARDA-funded BioShield contracts and beat back upstart competitors like VaxGen. That lobbying group recruited two of the Johns Hopkins Center for Biosecurity spooks with whom Kadlec had written the Dark Winter simulation, Tara O'Toole and Col. Randall Larsen, and enlisted more than fifty lobbyists to successfully block VaxGen from muscling in on its locked-up anthrax government monopoly. With these sorts of friends in high places, Emergent made the National Strategic Stockpile an exclusive captive market. By 2006, VaxGen had lost its $800 million contract and was bankrupt, and Emergent remained the government's sole source monopoly. Emergent then purchased VaxGen's anthrax vaccine for $2 million, at pennies on the dollar.

A 2021 *New York Times* exposé titled "How One Firm Put an 'Extraordinary Burden' on the US's Troubled Stockpile" documented Emergent's airtight domination of stockpile purchases: "As Emergent prospered, other companies working on pandemic remedies for the stockpile were squeezed out of government spending decisions." Several federal health officials anonymously told the *Times* that "preparations for an outbreak like Covid-19 almost always took a backseat to Emergent's anthrax vaccines."[98]

By 2011, the El-Hibris' connections had put Emergent in the driver's seat. Despite its vaccine's glaring and dangerous deficiencies, Emergent received $107 million in 2010 from Kadlec's baby, BARDA,[99] and up to $29 million from Fauci's NIAID to develop NuThrax (its old anthrax vaccine with a new adjuvant) for large-scale manufacture in 2014.[100] By 2010, Emergent's anthrax vaccine price had risen to about $28 (now closer to $30 per dose), with 75 percent gross profit margin for the El-Hibris.[101] As with BioThrax, the El-Hibris never performed functional safety testing for NuThrax, and the FDA has never approved the vaccine, but BARDA recently contracted for $261 million of this experimental and notoriously dangerous unlicensed anthrax vaccine. By then, the company had grown from a single corporate office in Rockville, Maryland, to headquarters in Seattle, Munich, and Singapore. Its projects include developing vaccines for pandemic flu and tuberculosis, in partnership with Oxford University and with funding from the Gates Foundation.

Despite NuThrax's failure to win FDA approval, almost half of the Strategic National Stockpile's half-billion-dollar annual budget prior to 2020 went to Emergent's two anthrax vaccines—a cost that, according to the *New York Times*, "left the government with less money to buy supplies needed in a pandemic."[102]

Some guardian angels with invisible hands seemed to catch the El-Hibris every time they stumbled. In March 2021, two federal officials anonymously told the *New York Times* that "One year, the government increased its order of Emergent's main anthrax vaccine by $100 million after the company insisted it needed the additional sales to stay in business. . . . At the time that order was announced in 2016, the [federal vaccine stockpile] reserve already had enough to vaccinate more than 10 million people. The stockpile has long been the company's biggest and most reliable customer for its anthrax vaccines, which expire and need to be replaced every few years."[103] After that, the cards really started breaking for the El-Hibris.

When Kadlec left the federal government, the El-Hibris did not forget the man who rescued them from bankruptcy and possibly from arrest. In the summer of 2012, Fuad El-Hibri made Robert Kadlec managing director and part owner of his own biodefense company, East West Protection.[104] The company received Pentagon backing that year to build a US biodefense site in Utah, in partnership with the HHS. CEO Bob Kramer told *Forbes*, "It was designed" to prevent a future pandemic.[105] With El-Hibri financing, Kadlec founded a company, RPK Consulting, which provided consulting services to Emergent until 2015. The firm paid Kadlec $451,000 in 2014 alone.

In 2015, the El-Hibris bought out Kadlec's shares of East West, allowing him to take the post of deputy staff director for the United States Senate Select Committee on Intelligence. Two years later, President Donald Trump nominated Kadlec to become assistant secretary for Preparedness and Response (ASPR), an office within Health and Human Services. During his confirmation process, Kadlec neglected to disclose his financial entanglements with the El-Hibris on the Senate nomination forms.

The El-Hibris apparently anticipated a windfall for Emergent from Kadlec's new posting. In July 2017, four days after Kadlec's nomination, Emergent announced that it was acquiring the rights to the smallpox vaccine from Sanofi Pasteur, the government's previous supplier.[106]

On August 3, the Senate confirmed Kadlec, and, sure enough, although the US taxpayers were now paying his salary, Kadlec never really stopped working for the El-Hibris. And that year, Christmas arrived early for the Lebanese arms dealers. Immediately after his appointment, Kadlec maneuvered deftly to move management of the Strategic National Stockpile, which he had conceived and created, from the Centers for Disease Control and Prevention to his own office, giving him authority over all acquisitions for the $7 billion contents.[107]

As soon as Emergent completed its acquisition of the Sanofi smallpox jab, Kadlec moved to increase the government's stockpile of these worthless and dangerous vaccines. Sanofi Pasteur had been charging the stockpile $4.27 per dose and had five years remaining on a ten-year government contract worth about $425 million. The

El-Hibris initially sought only a modest price increase, but Kadlec generously finalized a sweetheart deal with his friends and former business partners, doubling the five-year term that the El-Hibris had requested to ten years. Kadlec also doubled the number of doses per year—from 9 to 18 million—and gave the El-Hibris twice the price per dose that Sanofi received. Kadlec's new contract for the El-Hibris promised Emergent $9.44 per dose in the first year, with that figure rising annually throughout the contract term. In the end, Kadlec awarded the El-Hibris a 10-year, $2.8 billion no-bid contract to purchase their smallpox vaccines.[108]

The stockpile was already overflowing with smallpox vaccines in 2018. The CDC reported on its website in June 2019—and continues to say—that the stockpile already had sufficient smallpox vaccine for every American. Kadlec explained that his large purchase was necessary to "keep the production base warm"[109]—another way of saying, to keep the El-Hibris fat. Kadlec wrapped his gift with a red ribbon, Kadlec's brazenly corrupt announcment that the stockpile would no longer fund Emergent's competitors.

Emergent BioSolutions received more than $1.2 *billion* in contracts from Kadlec during the Trump years, with millions more coming from NIAID and DARPA.[110]

Kadlec's brassy approach inspired awestruck admiration within the pharmaceutical industry; in March 2020, President Donald Trump's HHS secretary, Alex Azar, former Eli Lilly President and Pharma lobbyist, designated Kadlec to lead the department's response to the COVID-19 pandemic. Kadlec's appointment was a signal to Big Pharma of the impending orgy of ransack, pillage, and plunder. Naturally the El-Hibris would enjoy the king's share of booty. That same year, Kadlec invoked the Emergency Use Authorization to purchase $370 million worth of the El Hibris' licensed and unlicensed anthrax vaccines. 2020 was the year with the largest sales of Emergent's anthrax vaccines to date.[111]

After the FDA authorized Johnson & Johnson's COVID-19 vaccine for emergency use in February 2021, Kadlec pressured the pharmaceutical giant to sign a $480 million contract with the El-Hibris to perform the manufacturing of J&J's COVID-19 jabs. *Forbes* headlined: "Little-Known Publicly Traded Company Given Massive Deal to Manufacture One-Shot Covid-19 Vaccine."[112]

By June, Kadlec's BARDA upped the ante with another $628 million gift to Emergent BioSolutions, for scaling up production of targeted vaccine candidates. Emergent signed separate deals worth hundreds of millions with AstraZeneca and Bill Gates's Novavax to manufacture vaccine doses at its Gaithersburg, Maryland, factory.[113, 114]

A March 7, 2021, *New York Times* exposé about Emergent's crooked relationship with the government reported that a billion dollars in payments to the company for anthrax and smallpox vaccines took up almost half the Strategic National Stockpile's budget.[115] Emergent had become the #1 vendor to the stockpile.

To finance these windfalls for the El-Hibris, Kadlec needed to short other stockpile supplies. By the time the novel coronavirus emerged, the stockpile had only 12 million N95 respirators. Kadlec also scuttled an Obama-era initiative to spend a relatively trivial $35 million to build a machine that could produce 1.5 million N95 masks per day. To rationalize their inventory gaps, Kadlec pled poverty.[116] The *New York Times* reported shocking shortfalls in protective gear for health care workers, ventilators, and masks just as the COVID-19 crisis called for them. Well aware of the situation, Kadlec "was unwilling to free up money by reducing the supply of anthrax vaccines."[117]

The El-Hibris' second sugar daddy, Dr. Anthony Fauci, was also raining down manna on Emergent.

At the beginning of the pandemic, Emergent signed a development deal with NIAID for a plasma-derived therapy. Dr. Fauci aimed to incorporate the company's COVID-HIG product into one of NIAID's clinical studies, with initial funding of $14.5 million coming from Kadlec through BARDA. In turn, Kadlec supported Dr. Fauci's pet project, Moderna, the mRNA jab caper that Dr. Fauci and Bill Gates considered their Holy Grail. In mid-April 2020, Kadlec arranged for BARDA to provide Moderna up to $483 million to accelerate the Fauci/Gates vaccine's development and manufacturing. That amounted to about half of what BARDA doled out to all of Moderna's competitors combined, including Johnson & Johnson, Pfizer, and Astra-Zeneca.[118]

Kadlec was also generous to Bill Gates, arranging a $1.6 billion grant—the largest to date—from Operation Warp Speed to Gates's biotech selection, Novavax. Although the company, based in Gaithersburg, Maryland, had never brought a vaccine to market in its thirty-three-year history, and was then on the verge of collapse, Gates and his obedient minions at the Coalition for Epidemic Preparedness (CEPI) had placed a bet on Novavax's technology, which uses moth cells to pump out crucial molecules at a faster rate than typical vaccines.[119] Kadlec's generosity with his Warp Speed wampum caused Novavax's stock to surge 30 percent. John J. Trizzino, Novavax's chief business and financial officer, said the company did nothing inappropriate but acknowledged that it used its connections to Gates to help win the deals.

In September 2019, less than a month before COVID began circulating, the Gates Foundation made a $55 million pre-IPO equity investment in BioNTech. The company also had never brought a single product to market.[120] Soon afterward, the German government followed Gates with a $445 million infusion into BioNTech.[121] On July 21, 2020, when Robert Kadlec committed Operation Warp Speed to a $2 billion purchase of 100 million doses of BioNTech/Pfizer's COVID-19 vaccine,[122] the company's stock value soared, with Bill Gates's equity shares increasing to an evaluation of $1.1 billion.

In October 2020, Emergent became one of four companies collaborating on a clinical trial for a combination treatment regimen that included Dr. Fauci's drug remdesivir as a "background therapy." The company said in a statement: "Emergent is proud to continue our partnership with NIAID/NIH and . . . BARDA to advance potential therapeutic solutions for COVID-19 in hospitalized patients."[123]

Bill Gates owned a large stake in remdesivir's manufacturer, Gilead.[124] WHO's own studies showed clearly—as even WHO acknowledged—that remdesivir was useless against COVID.[125] Worse, the drug's extreme toxicity—remdesivir's side effects mimic the late-stage symptoms of COVID[126, 127]—may actually aggravate the severity of the illness.[128] To overcome these obstacles, Dr. Fauci financed and rigged a suite of flawed studies to suggest—deceptively—that remdesivir might slightly reduce the number of days a patient would stay in the hospital.[129] The WHO's much larger studies proved that there was no reduction in length of hospital stay. Nevertheless, using his blatantly orchestrated "research," Dr. Fauci then forced remdesivir's approval through FDA as "Standard of Care" for COVID. At the same time, Dr. Fauci and Bill Gates were financing and promoting studies to discredit chloroquine and hydroxychloroquine and sabotage ivermectin—two effective COVID remedies that posed an existential threat to remdesivir and the entire Fauci/Gates COVID vaccine enterprise.

Emergent's CEO, Robert Kramer, boasted to Wall Street analysts in February, 2021, that the year had been "the strongest year in our 22-year history."[130] The *New York Times* reported that Emergent's stock had reached such a zenith that Fuad El-Hibri "cashed in shares and options worth over $42 million, more than he had redeemed in the previous five years combined."[131]

When in April 2021, Emergent BioSolutions ruined 15 million Johnson & Johnson COVID-19 vaccines due to quality-control mishaps at its poorly managed Baltimore production facility, Congress launched an investigation into whether Emergent used high-level connections to get billions of dollars in federal contracts despite a history of failing to deliver satisfactorily on its contracts.[132] Congressional investigators also raised concerns about Emergent's inadequate staff training, persistent quality-control issues, and the company stiffing the government with an "unjustified" 800 percent price increase for its anthrax vaccine. The Democratic chairs of the House Committee on Oversight and Reform and Select Oversight Subcommittee on the Coronavirus Crisis focused their inquiry on Kadlec's role. In a letter, the committee chairs complained that Kadlec "appears to have pushed for" the $628 million award to Emergent to develop a Covid vaccine factory "despite indications that Emergent did not have the ability to reliably fulfill the contract."[133]

As the top dog among the COVID-19 pandemic's government managers, Kadlec had promoted Emergent as the United States' primary vaccine manufacturing facility. In April 2021, the *Times* published another extensive exposé reporting that Emergent had not yet been able to produce a single acceptable dose of any COVID-19 vaccine.[134] Following exposés in the *New York Times* and the *Washington Post*, J&J took over the production at that plant. The FDA stepped in after inspecting the facility and ordered Emergent to halt all production of materials for COVID-19 vaccines pending a review and remediation, and to quarantine all existing materials.[135]

HHS ordered Emergent to discard millions of contaminated doses. Instead, in March 2021, the company shipped millions of doses of its defective vaccines to Canada, Europe, South Africa, and Mexico. The House Select Subcommittee on the Coronavirus Crisis held a hearing on May 19, 2021, and ordered Emergent to turn over all its federal contracts since 2015 and all communications with Robert Kadlec.[136] Emergent's political invincibility left the company unbowed by all those scandals. In July 2020, Emergent announced a five-year, $450 million deal to manufacture COVID drugs for Johnson & Johnson.[137] In February 2021, HHS awarded Emergent another contract, this one worth up to $22 million to develop a COVID-19 therapy.[138]

Atlantic Storm 2003, 2005

In January of 2003 and again in 2005, a cabal of US and European military, intelligence, and medical officials germ-gamed another exercise they called Atlantic Storm.

Thomas V. Inglesby and the spooks, Tara O'Toole and Col. Randall J. Larsen, were the simulation's principal authors.[139]

Both the 1999 HHS smallpox simulation and the June 2001 Dark Winter smallpox simulation focused, ominously, not on public health, but on the quandary of how to impose control over US and global populations during public health emergencies, how to sweep away civil rights and impose mass obedience to military and medical technocrats. Atlantic Storm further probed these sinister disquisitions. High-level government figures, including Madeleine Albright playing the president of the United States and WHO Director-General Gro Harlem Brundtland playing herself, hosted a summit of transatlantic military and intelligence agency planners coordinating responses after a radical terrorist band unleashes smallpox.

According to the After-Action Report, the key issues for summit principals were "coping with scarcity of critical medical resources such as vaccines" and assuring a uniform coordinated response among all governments in the world. The simulation stressed the inadequacy of current multilateral frameworks like NATO and the EU to cope with social, economic, and political disruption from an international epidemic, "be it natural or the result of a bioterrorist attack," and emphasized the importance of developing systems to coordinate global lockstep security protocols that went beyond "just stockpiling vaccines or training more doctors."[140]

Characteristically, the assembled eminences bypassed any discussion of bolstering people's immune system response or testing and distributing off-label therapeutics and went directly to recommending militarized strategies including police state controls, mass propaganda and censorship, and the suspension of civil rights and due process rulemaking in favor of diktats by health authorities, all aimed at coercive vaccination of the population. These scenarios, which health officials and spooks conceived of and gamed back in 2005, became our collective reality in 2020 and 2021.

Global Mercury 2003

Between September 8 and 10 of that same year, the spooks at the US State Department Office of the Coordinator for Counterterrorism organized another scenario exercise dubbed Global Mercury with the CDC, the NIH, the FDA, the WHO, and the Department of State. Over a fifty-six-hour period, public health technocrats coordinated communications and lockstep response between "trusted agents" from the GHSAG nations (the United States, the UK, Canada, France, Germany, Italy, Japan, and Mexico), during a simulated outbreak after self-inoculated terrorists spread smallpox to countries around the world.[141]

The SCL Simulation 2005

Atlantic Storm and Global Mercury were additional loud notes amplifying persistent Pentagon signals that biosecurity was the emerging growth sector for national defense. In response to such tocsins, private military contractors began thronging to the pandemic "surveillance and psyops" sector like hogs to a corncrib.

Long before Robert Mercer (with his daughter Rebekah) became Donald Trump's biggest private donors, and before they launched the right-wing social media platform

Parler, he created the first private-sector provider of psychological warfare services in 1993. The Mercers' Strategic Communication Laboratories (SCL) Group was the parent company to the notorious data-manipulating firm Cambridge Analytica. This brand new psyops firm, headquartered in the UK, drew some of the largest crowds in 2005 when it set up a high-tech propaganda "ops center" at the UK's annual military technology showcase.[142]

As a contemporary article in *Slate* described the SCL simulation, "classic signs of smallpox" are "threatening a pandemic of epic proportions" when "a shadowy media firm steps in to help orchestrate a sophisticated campaign of mass deception." SCL takes on the task of convincing the entire country's population to comply with lockdown rules by inventing a lie about an unleashed cloud of toxic chemicals. The mission's objective is to prevent mass panic and casualties from the classified threat of smallpox. SCL feeds disinformation to the press and manufactures medical data. "Londoners stay indoors . . . convinced that even a short walk into the streets could be fatal."[143]

The article continues: "If SCL weren't so earnest, it might actually seem to be mocking itself, or perhaps George Orwell. At the end of the smallpox scenario, dramatic music fades out to a taped message urging buyers to 'embrace' strategic communications, which it describes as 'the most powerful weapon in the world.' . . . What makes SCL's strategy so unusual is that it proposes to propagate its campaign domestically, at least some of the time, and rather than influence just opinion, it wants people to take a particular course of action."[144]

The company based its psyops strategies on propaganda techniques developed by a virtual lab called the Behavioral Dynamics Institute, run out of Leeds University by Professor Phil Taylor, a consultant to UK and American defense agencies until his death at 56 in 2010. The article identified SCL only as "funded by private investors."[145] Company chief Nigel Oakes described its nefarious skullduggery as "mind-bending" for political purposes.[146] In a March 20, 2018 interview with Yahoo Finance, Oakes described himself as a man "without much of an ethical radar."[147]

According to SCL's public affairs director Mark Broughton, "Basically, we're launching ourselves . . . on the defense market and homeland security market at the same time." Aware that the company might face criticism over its promotion of totalitarian security states, Broughton emphasized to *Slate* the company's role in saving lives. "There is some altruism in it," he said grudgingly, "but we also want to earn money."[148]

How War Games Became Instruments for Imposing Obedience

Dark Winter, Atlantic Storm, and Global Mercury were only three of over a dozen Germ Games staged by military, medical, and intelligence planners leading up to COVID-19. Each of these Kafkaesque exercises became uncanny predictors of a dystopian age that pandemic planners dubbed the "New Normal." The consistent feature is an affinity among their simulation designers for militarizing medicine and introducing centralized autocratic governance.

Each rehearsal ends with the same grim punchline: the global pandemic is an excuse to justify the imposition of tyranny and coerced vaccination. The repetition of these exercises suggests that they serve as a kind of rehearsal or training drill for an underlying agenda to coordinate the global dismantlement of democratic governance.

Military intelligence analysts first introduced scenario planning, as a strategic device during World War II. RAND's iconic military planner, Herman Kahn, used sophisticated war game simulations to model nuclear engagement strategies in the Cold War era.[149] Working for Royal Dutch/Shell, futurologists Pierre Wack and Peter Schwartz of the Global Business Network (GBN) pioneered scenario-planning simulations as a strategic device for their corporate clients in the 1970s and 1980s.[150] By the millennium, simulations had evolved into an indispensable vehicle for military policy makers, intelligence agency planners, public health technocrats, and the petroleum and pharmaceutical multinationals for reinforcing prescribed responses that allow predictable and rigid control of the outcomes of future crises.

After 9/11, the rising biosecurity cartel adopted simulations as signaling mechanisms for choreographing lockstep response among corporate, political, and military technocrats charged with managing global exigencies. Scenario planning became an indispensable device for multiple power centers to coordinate complex strategies for simultaneously imposing coercive controls upon democratic societies across the globe.

Virtually all of the scenario planning for pandemics employ technical assumptions and strategies familiar to anyone who has read the CIA's notorious psychological warfare manuals for shattering indigenous societies, obliterating traditional economics and social bonds, for using imposed isolation and the demolition of traditional economies to crush resistance, to foster chaos, demoralization, dependence and fear, and for imposing centralized and autocratic governance.[151]

In particular, the exercises incorporate psyop techniques gleaned from the notorious "Milgram Obedience Experiments." In those 1960s exercises, Yale social psychology professor Dr. Stanley Milgram was able to show that researchers could formulaically manipulate "ordinary citizens" from all walks of life to violate their own conscience and commit atrocities, so long as an authority figure (a doctor in a white lab coat) ordered them to do so. The subjects believed they were torturing fellow volunteers, by electrocution, out of sight in an adjacent room. As a doctor instructed them to rev up the juice, the recruits could hear the nightmarish screaming of actors pretending to be suffering electrocution and their pleadings for mercy. Of Milgram's forty subjects, some 65 percent administered the full-bore 450-volt shocks they had been told were potentially fatal. Milgram describes his experiments as proof that "obedience to authority" trumps morality and conscience:

> Stark authority was pitted against the subjects' strongest moral imperatives against hurting others, and, with the subjects' ears ringing with the screams of the victims, authority won more often than not. The extreme willingness of adults to go to almost any lengths on the command of an authority constitutes the chief finding of the study.[152]

In his book *A Question of Torture: CIA Interrogation, from the Cold War to the War on Terror*, University of Wisconsin historian Alfred W. McCoy suggests that the Yale obedience experiments were funded by the CIA as part of MKUltra's studies on the control of human behavior.[153] During that time, the CIA funneled money through various federal agencies to fund 185 independent researchers to perform sinister behavioral manipulation studies at universities across North America.[154] Milgram first proposed his obedience research in a 1960 solicitation to the Group Psychology Branch of the Office of Naval Research (ONR), a key conduit for the CIA's MKULTRA mind control experiments. The dean who hired Milgram later as a professor at City University of New York was a former deputy director of ONR. Milgram's Yale mentor was Irving L. Janis, who wrote the seminal Air Force study of Soviet mind-control and hypnosis for the Rand Corporation. Milgram's other connections to the CIA's Psychological Warfare program are too numerous to mention here.

In an equally important revelation, the CIA mind-control experiments identified social isolation as the *primary* protocol for controlling societal and individual behavior: "In 1960, one of the agency's most active contractors, Lawrence Hinkle of Cornell, confirmed the significance of [social isolation] . . . for the CIA mind-control effort . . . in light of the neurological literature, the most promising of all known techniques."[155]

The CIA's research found that "the effect of isolation on the brain function [on an individual] is much like that which occurs if he is beaten, starved, or deprived of sleep."[156]

Social isolation affects organic brain development, and the human body, length of life, cardiovascular health, and so on. Social isolation doubles the risk of death in Blacks while increasing the risk of early death in Caucasians by 60–84 percent, while other studies show that it is safer to smoke fifteen cigarettes a day—or be an alcoholic—than to be socially isolated:

> Meta-analysis co-authored by Julianne Holt-Lunstad, PhD, a professor of psychology and neuroscience at Brigham Young University, [found that] lack of social connection heightens health risks as much as smoking 15 cigarettes a day or having alcohol use disorder. [Holt-Lunstad] also found that social isolation is twice as harmful to physical and mental health as obesity. . . . "There is robust evidence that social isolation significantly increases risk for premature mortality, and the magnitude of the risk exceeds that of many leading health indicators."[157]

NIH's collaboration with the CIA in these odious torture, obedience, and brainwashing experiments heaps additional ignominy on the agency. During the 1950s, NIH scientist Dr. Maitland Baldwin conducted social isolation experiments on monkeys and humans at NIH headquarters and CIA safehouses. MKUltra's experiments used "expendables"—people whose deaths or disappearances would go unnoticed—including "a rather gruesome experiment" in which Baldwin had subjected a soldier to forty hours of isolation, causing him to go insane and to kick apart the box in which Maitland imprisoned him. Maitland, who told his "Operation Artichoke" case officer that isolating subjects for over forty hours could cause "irreparable damage" and perhaps be "terminal," nevertheless agreed to go forward if the agency could provide cover and subjects.[158]

The various scenario-planning simulations provided a unique forum to convene key decision makers, and to introduce, and then to sanction, with authoritative voices, previously unspeakable conduct that violated democratic and ethical norms. That conduct included the forced isolation and quarantine of entire populations, including the healthy; censoring free speech; violating privacy with track and trace surveillance systems; trampling property rights and religious freedoms; and obliterating traditional economies via nationwide business lockdowns, enforced masking, coercive medical interventions, and other assaults on human rights, civil rights, constitutions, and democracies. With each new simulation, the staccato repetition of the message by "trusted experts"—doctors in lab coats and authoritative collectives like Secretary of State Madeleine Albright, Sen. Sam Nunn, WHO Director-General Gro Harlem Brundtland, and Sen. Tom Daschle—reinforced the lesson that censorship, isolation, the militarization of medicine, totalitarian controls, and coercive vaccine mandates are the only appropriate response to pandemics. Scenario planning, in other words, is a potent brainwashing technique for creating and fortifying anti-democratic orthodoxies among key political leaders, the press, and the technocracy, and preparing the nation to tolerate a coup d'état against its Constitution without resistance.

Lockstep Simulation 2010

In 2009, President Obama declared biosecurity as the spear tip of US foreign policy, dispersing memos to all government agencies instructing them to integrate biosecurity into their mission. By 2010, US spy agencies were demonstrating a growing interest in vaccines as a foreign policy instrument. Just as the Cold War, and later on, the "War on Terror," had rationalized US military presence across the world as a bulwark against brushfire nationalist rebellions purportedly orchestrated by a communist monolith, vaccination programs could justify interventions in developing countries with high disease burdens as a tool for social and political control. In 2010, the WHO pronounced biosecurity as the centerpiece of its approach for managing global risks.[159]

That same month, as Bill Gates delivered his Decade of Vaccines speech at the UN, biosecurity—the war on microbes—was already eclipsing the "War on Islamic Terrorism" as the preferred driver of the security state cartel. A few days later, Peter Schwartz authored a scenario report funded by the Rockefeller Foundation titled "Scenarios for the Future of Technology and International Development."[160] A section called "Lockstep" reinforced the burgeoning orthodoxy that rigid global tyranny was the antidote to infectious disease:

> In 2012, the pandemic that the world had been anticipating for years finally hit. Unlike 2009's H1N1, this new influenza strain—originating from wild geese—was extremely virulent and deadly. Even the most pandemic-prepared nations were quickly overwhelmed when the virus streaked around the world, infecting nearly 20 percent of the global population and killing 8 million in just seven months. . . .
>
> The pandemic also had a deadly effect on economies: international mobility of both people and goods screeched to a halt, debilitating industries like tourism and breaking global supply chains. Even locally, normally bustling shops and office buildings sat empty for months, devoid of both employees and customers.
>
> *During the pandemic, national leaders around the world flexed their authority and imposed airtight rules and restrictions, from the mandatory wearing of face masks to body-temperature*

checks at the entries to communal spaces like train stations and supermarkets. Even after the pandemic faded, this more authoritarian control and oversight of citizens and their activities stuck and even intensified. In order to protect themselves from the spread of increasingly global problems—from pandemics and transnational terrorism to environmental crises and rising poverty—*leaders around the world took a firmer grip on power.*[161] (Emphasis added)

Schwartz's chilling document goes on to predict that citizens terrified by germs and orchestrated propaganda willingly relinquish their civil and constitutional rights. The population, Schwartz predicts, will not start rebelling against the new tyranny and authoritarian clampdowns for more than ten years.

Intelligence agencies left their fingerprints all over these scenario-planning exercises. Schwartz—like O'Toole, Larsen, Kadlec, Woolsey, and David—is one of the many leading promoters of weaponized vaccines as a foreign policy tool with deep connections to the Intelligence Apparatus. Schwartz's résumé chronicles multiple touchpoints with spy agencies before and after he authored the "Lockstep" scenario. In 1972, Schwartz joined the Stanford Research Institute (later SRI International), an early pioneer in computer technology and artificial intelligence. Schwartz rose to run SRI's Strategic Environment Center, at a time when SRI was hosting the CIA's notorious MKUltra program and actively researching psychological warfare including the sophisticated use of propaganda, torture, and psychiatric chemicals to shatter societies and impose centralized control. Schwartz left to become head of Scenario Planning for Royal Dutch/Shell. He then cofounded the Global Business Network (GBN) in 1987 as a corporate consultant specializing in analyzing intelligence and in "future-think" strategies. Shell Oil was GBN's highest-revenue client.

In the early 1990s, Ken McCarthy, who would become an early pioneer of practical efforts to commercialize the Internet, met Schwartz at a large Thanksgiving gathering in a remote location in rural Harris, California. Schwartz introduced himself to McCarthy, an anthropology graduate from Princeton, and Schwartz began probing McCarthy's interest in being recruited for a contract with an unnamed West African country that involved "weakening tribal and family structures on behalf of a federal government." Recalling the encounter, McCarthy told me, "I found Schwartz's proposal intensely disturbing." Schwartz dismissed McCarthy's qualms as "naive." McCarthy says, "It made a lasting impression on me—so much so that I've recounted the story many times over the years."[162]

Schwartz's client, Shell Oil, had extensive oil holdings in the Ogoni region of Nigeria. In 1995, the Nigerian government executed Ogoni environmental leader, writer, and television producer Ken Saro-Wiwa and eight other environmental organizers based on charges that they had "incited violence." Saro-Wiwa's arrest, trial by a military tribunal, and subsequent execution followed a harassment campaign against him and other Ogoni environmental leaders, which started in 1993 after they repeatedly mobilized peaceful demonstrations against Shell, attracting over 300,000 of the region's total population of 600,000.[163] The United Nations General Assembly and the European Union condemned Saro-Wiwa's execution, and the United States recalled its ambassador to Nigeria.[164]

In 1993, Schwartz, along with Stewart Brand and Nicolas Negroponte, was one of the driving forces behind the founding of *Wired* Magazine, which became the central clearinghouse for mainstream news coverage of the burgeoning online ecosystem. *Wired* quickly earned notoriety as a clearinghouse for intelligence agency chatter. Prior to *Wired*, *Mondo 2000*, the Bay Area's original tech and culture magazine, reflected the progressive, idealistic viewpoints of many of the pioneer tech innovators. In contrast, *Wired*, which appropriated *Mondo 2000*'s look and feel and no small number of its employees, glorified military and intelligence agency celebrities and corporate CEOs who happened to be clients of Nicholas Negroponte's MIT Lab. *Wired* gained snowballing prominence in the early 2000s at the same time that the CIA launched its notorious investment firm, In-Q-Tel, to infiltrate the tech industry and put Silicon Valley on steroids with easy terms and government contracts.[165] (Scenario planner Tara O'Toole served as In-Q-Tel's executive vice president.)

It's worth recalling here that the defense and intelligence agencies had a beachhead in the tech industry from its birth: the Defense Advanced Research Project Agency, DARPA, created the Internet by building the ARPANET grid in 1969.[166] DARPA is the Pentagon's angel investor and venture fund. In addition to creating the Internet, DARPA developed GPS, stealth bombers, weather satellites, pilotless drones, and the M16 rifle. DARPA was, perhaps, the largest funder of gain-of-function research, outstripping even Dr. Fauci's NIH in some years. In 2017 alone, DARPA laundered at least $6.5 million through Peter Daszak's EcoHealth Alliance to fund experiments[167] at the Wuhan lab. DARPA funded additional gain-of-function experiments at Fort Detrick and other biosecurity research at Battelle's laboratory at St. Joseph, Missouri.[168] Beginning in 2013, DARPA also financed the key technologies for the Moderna vaccine.[169]

In 2002, DARPA set off a firestorm among human rights advocates from the Left and Right by creating a comprehensive data mining system under President Reagan's National Security Advisor, Admiral John Poindexter. Public protests forced DARPA to scuttle that project, but critics have accused the agency of using the technology to help launch Facebook.[170] By remarkable coincidence, DARPA shut down its Facebook-like project LifeLog, a venture that involved MIT contractors, the very same month—February 2004—that Mark Zuckerberg started Facebook just a thirty-minute walk up the Charles River in Cambridge, Massachusetts, on the campus of Harvard University.

In 2010, DARPA's visionary director, Dr. Regina Dugan, moved to Google as an executive, and in 2016, she transferred to Google's competitor, Facebook, running a mysterious project called Building 8.[171] In 2018, she moved again, to run Wellcome Leap, a health technology breakthrough innovation project of Wellcome Trust. Her peregrinations offer another example of the incestuous links between Big Tech, Big Pharma, and the military and intelligence agencies.

According to veteran CIA officer Kevin Shipp, Silicon Valley CEOs who accepted In-Q-Tel contracts would become some of the 4.8 million Americans subsequently pressured into signing CIA "State Secret Contracts," which subject signatories to twenty-year prison sentences, property forfeitures, and other draconian reprisals imposed by secret courts for even minor violations of arbitrary provisions—including admitting to signing the contract: "Once he signs that secrecy agreement, that Silicon

Valley entrepreneur is now functionally the indentured servant of the agency. It binds him and his company for life, and the agreement itself is classified."[172]

Wired's seed funding came from MIT Media Lab founder Nicholas Negroponte, whose brother, John Negroponte, was the first Director of National Intelligence, notorious for his support of Central American death squads. *Wired*'s central function was to "scrub every last particle of progressive thinking from reporting on the then-developing online world and to promote a pro-military/pro-corporate/pro-intelligence agency view within the digital media and technology community,"[173] according to McCarthy, who lived and worked in San Francisco in the 1990s and organized the first conference on monetizing the web. When he saw his first copy of *Wired*, Dr. Timothy Leary reportedly called it "the CIA's answer to *Mondo 2000*."[174]

In 2015, *Wired* emerged as a promoter of a particular brand of autism epidemic denial known as "Neurodiversity." By normalizing autism as "neurodiversity," this movement seeks to dilute autism numbers, deny the vaccine association, and promote the larger view that all vaccines are safe and vaccine injuries are the delusions of crackpots. This "movement" has spawned an army of "activist" trolls weaponized to attack autism researchers, advocacy groups, and even families of vaccine-injured children. Steve Silberman, a writer for *Wired* since 2010, published the book *Neurotribes* in 2015 to massive acclaim and highly orchestrated publicity. It became the manifesto for the new "autism rights" movements, which also demonize medical freedom and food safety advocates. Its tactics include online attacks and aggressive disruption of public events, including conferences and film screenings.

Wired is also the fountainhead of the equally sinister movement transhumanism, which advocates for the integration of human beings and machines. The movement's ancillary aims include extending the lifespans of key Silicon Valley billionaires indefinitely and "liberating humanity from biological restraints"—using AI, novel therapies like stem cells and nanobots, vaccination, and subdermal chips. Jacques Ellul, an early pioneer, described transhumanism's elegant capacity for top-down control of humanity:

> For the psychocivilized society, the complete joining of man and machine will be calculated according to a strict system, the so-called "biocracy." It will be impossible to escape this system of adaption because it will be articulated with so much scientific understanding of the human being. The individual will have no more need of conscience and virtues. His moral and mental furnishing will be a matter of the biocrats' decisions.[175]

Transhumanism, in its various doctrinal approaches, has fervent acolytes among the Silicon Valley elites, including C-suite titans at Microsoft, Facebook, Tesla's Elon Musk, Google Engineering Director Raymond Kurtzweil, PayPal founder Peter Thiel, satellite and biotechnology titan Martine Rothblatt, and Bill Gates. In-Q-Tel has made transhumanism one of the persistent themes of its long-term investment strategies.

A celebration of transhumanism, from In-Q-Tel's website.[176]

Not everyone is a fan: Francis Fukuyama has called the transhumanism movement "the greatest threat to humanity."[177]

Schwartz served as a consultant on the 1998 sci-fi disaster film *Deep Impact* and the 1992 futuristic film *Minority Report*, which follows a special PRE-CRIME police unit able to arrest murderers before they commit their crimes. Emerging reality seldom disappoints Schwartz's past predictions; in 2020, a Chinese whistleblower revealed that the Chinese government has widely deployed facial recognition technologies that can detect guilty thoughts against dissident minority groups. A March 3, 2021, article in the *Guardian* predicts that demands by government enforcement agencies will make remote emotion detection technologies a $36 billion industry by 2023.[178]

Schwartz's auguring skills are legendary. One of his early plots for GBN scenarios tested strategies by a major airline for surviving a coronavirus pandemic. *TIME* Magazine's 2004 profile focused on Schwartz's unerring prognosticating: "Very rarely have we really missed," he told *TIME* of his forecasting. "More often our failure is in getting people to take it seriously." The *TIME* article mentioned one of his most impressive fortune-telling stunts: In 2000, as part of a study for a Senate commission, Schwartz predicted "the horrifying possibility of terrorists flying planes into the World Trade Center."[179]

In 2016, as senior vice president of strategic planning at Salesforce.com, Schwartz chaired a session at the World Government Summit titled "How governments get ready for the unthinkable."[180] Three thousand participants from 125 countries attended that year. Barack Obama delivered the keynote speech; Klaus Schwab, president of the World Economic Forum and the head of the World Bank, put a happy face on global crisis as a potential path to the cashless society so coveted by international banksters: "The Digital Currency: Is It the Way of the Future?"

In 2014, Schwartz conducted an offstage interview with Schwab at a Salesforce conference on "The Future of Global Governance,"[181] following a speech by Hillary Clinton, in which the two men forecast the merger of new devices with the human brain allowing machinery to control "our brains, with our souls and our hearts." They applauded the concept of biology as part of the new technological and scientific revolution and praised the capacity of the Internet to integrate in a continuous interaction with the machinery of the human mind, to control aberrant and criminal behavior [i.e., dissent], and to challenge people's sacred identities. Schwartz describes a machine-driven evolution that will supplant emotional intelligence with knowledge and data. According to Schwab, a new intelligence will be distributed and of course will accelerate even more technological progress: "If you combine, let's say, brain research with big data, you have fantastic new areas with tremendous application [for controlling behavior]." Schwartz lauds Salesforce as a participant in this process.[182]

As chief futures officer for Salesforce, Schwartz currently markets a "vaccine management" software platform that allows governments to track, trace, monetize, and enforce vaccine compliance among global populations. An autumn 2021 video describes "the latest factors impacting our ability to move out of multiple, pandemic-driven global crises" and promotes Saleforce's software as the solution. Schwartz predicts a dystopian future in which ever-evolving mutant strains of SARS-CoV-2 drive skyrocketing death rate curves—and, presumably, ballooning pharma profits—making "the race between the vaccines and a virus" the conflict that will define the world economy and civilization's future.[183]

The Salesforce system is elaborate and provides local governments the ability to establish a credential ID system. In-Q-Tel markets a competitive technology, B.Next, for tracking and tracing, facilitating pandemic management. "Given the reality of the capacity of most government information technology (IT) departments, national, state and local it's fair to say that without Salesforce.com, In-Q-Tel, and other companies like IBM, the planning and execution of population-wide vaccination programs of the kind Dr. Fauci and others called for would have been logistically impossible," says McCarthy.

Training Day for Tyranny

By 2010, the Fauci/Gates partnership was spearheading the globalist biosecurity agenda. Bill Gates began partnering with military and intelligence planners to stage regular follow-up simulations. Each successive drill repeated the narrative of Schwartz's "Lockstep" scenario for different audiences of key power brokers. These exercises served as devices for planners to rehearse their schemes with critical functionaries and to coordinate communications and choreograph the actions of diverse government, industry, military, intelligence, energy, and financial power centers in their lockstep march to replace constitutional democracy with authoritarian plutocracy. The "global war" against infectious diseases provided the rationale for oppressive government and corporate interventions. The arsenal for this war is the endless batteries of mandated vaccines to combat the diseases weaponized by gain-of-function experiments and marketed by sophisticated government/corporate propaganda.

In February 2017, Gates told the Munich Security Conference—the leading global convention on international security policy—that "we ignore the link between health security and international security at our peril." He warned that "a highly lethal global pandemic will occur in our lifetimes" by "a quirk of nature or at the hand of a terrorist." The world needs to "prepare for epidemics the way the military prepares for war."[184]

MARS 2017

By mid-2017, the Rockefeller Foundation and intelligence agency planners had passed to Bill Gates their baton as the primary funder and front man for the military/intelligence community's increasingly regular pandemic simulations. In May, the health ministries for the world's wealthiest twenty (G20) nations assembled for the first time, gathering in Berlin to participate in a Joint Exercise Scenario with an imagined China responding to a contagion dubbed MARS, for "Mountain Associated Respiratory Virus."[185] (Mars is also the Roman god of war.) German governmental institutions collaborated to produce the simulation with the Gates Foundation, the Rockefeller Foundation, the World Bank, the WHO, and the Robert Koch Institution (RKI). The ministers hailed from the United States, Russia, India, China, Britain, France, Germany, Canada, Argentina, Brazil, Korea, Mexico, Saudi Arabia, Indonesia, South Africa, Turkey, and the European Union.

The exercises' two moderators also worked closely with the Gates Foundation; David Heymann served simultaneously as chair of the UK's Centre on Global Health Security and as an epidemiologist with the Gates-funded London School of Hygiene and Tropical Medicine. Heymann also sits with Moderna CEO Stéphane Bancel on the Mérieux Foundation USA Board. BioMérieux is the French company that built the Wuhan lab.[186] Throughout the COVID-19 pandemic, Heymann has chaired the WHO's Scientific Technical Advisory Group for Infectious Hazards. The other moderator of the 2017 simulation was Professor Ilona Kickbusch, a member of Gates's Global Preparedness Monitoring Board.

Over two days, the global health ministry officials and other "guest countries and international representatives" bore witness to a "timeline of the unfolding pandemic," known as MARS, a novel respiratory virus, spread from busy markets in a mountainous border region of an unnamed but China-like country—to nations around the globe. Only draconian clampdowns by neighboring governments and heroic WHO technocrats orchestrating a tightly choreographed centralized global response save humanity from a chaotic dystopian apocalypse.

In an hour-long documentary about that event, German journalist Paul Shreyer shows the health ministers intently studying the simulation exercises: "When we look at that picture," Shreyer says, "we might comprehend a bit better why in today's crisis, all or at least most of the countries are proceeding very coordinatedly, and why in every country, more or less the same is acted out. . . . They were given the same general recipes and procedural instructions that are now being realized in a synchronized way."[187]

SPARS 2017

Five months later, in October 2017, Gates convened yet another tabletop pandemic at the Johns Hopkins Center for Health Security, the global biosecurity command center. Gates's foundation, along with NIAID and NIH, are major funders of the Johns Hopkins Bloomberg School of Public Health.[188] "SPARS 2017" chronicled an imaginary coronavirus pandemic that would, supposedly, run from 2025 to 2028. The exercise turned out to be an eerily precise predictor of the COVID-19 pandemic exactly three years later.

Gates's working group, which staged the exercise, was a collection of characters with deep connections to intelligence agencies and NIH. They included Luciana Borio, vice president of the CIA's In-Q-Tel; and Joseph Buccina, director of Intelligence Community Support and B.Next Operations at In-Q-Tel. Prior to joining B.Next, Buccina was a Program Manager for In-Q-Tel's biotech portfolio, which works with tech start-ups specializing in enhanced products for the intelligence and defense communities. Matthew Shearer, a Senior Analyst at the Johns Hopkins Center for Health Security and Associate Editor of the peer-reviewed journal *Health Security*, would discover the first US cases of coronavirus in Seattle in February 2020.[189] Walter Orenstein, MD, is a former surgeon general who managed CDC's fraudulent efforts to suppress the science linking autism to vaccines, from 1999 to 2004. He left HHS to serve as Deputy Director for Immunization Programs at the Bill & Melinda Gates Foundation, and adviser to WHO. Another Working Group Member was vaccine developer Dr. Gregory Poland, whom the National Institutes of Health has continuously funded since 1991.[190]

Building on the Pentagon's anthrax simulation (1999) and the intelligence agency's "Dark Winter" (2001), Atlantic Storm (2003, 2005), Global Mercury (2003), Schwartz's "Lockstep" Scenario Document (2010), and MARS (2017), the Gates-funded SPARS scenario war-gamed a bioterrorist attack that precipitated a global coronavirus epidemic lasting from 2025 to 2028, culminating in coercive mass vaccination of the global population. And, as Gates had promised, the preparations were analogous to "preparing for war."[191]

Under the code name "SPARS Pandemic," Gates presided over a sinister summer school for globalists, spooks, and technocrats in Baltimore. The panelists role-played strategies for co-opting the world's most influential political institutions, subverting democratic governance, and positioning themselves as unelected rulers of the emerging authoritarian regime. They practiced techniques for ruthlessly controlling dissent, expression, and movement, and degrading civil rights, autonomy, and sovereignty. The Gates simulation focused on deploying the usual psyops retinue of propaganda, surveillance, censorship, isolation, and political and social control to manage the pandemic. The official eighty-nine-page summary is a miracle of fortune-telling—an uncannily precise month-by-month prediction of the 2020 COVID-19 pandemic as it actually unfolded.[192] Looked at another way, when it erupted five years later, the 2020 COVID-19 contagion faithfully followed the SPARS blueprint. Practically the only thing Gates and his planners got wrong was the year.

Gates's simulation instructs public health officials and other collaborators in the global vaccine cartel exactly what to expect and how to behave during the upcoming plague. Reading through the eighty-nine pages, it's difficult not to interpret this stunningly prescient document as a planning, signaling, and training exercise for replacing democracy with a new regimen of militarized global medical tyranny. The scenario directs participants to deploy fear-driven propaganda narratives to induce mass psychosis and to direct the public toward unquestioning obedience to the emerging social and economic order.

According to the scenario narrative, a so-called "SPARS" coronavirus ignites in the United States in January 2025 (the COVID-19 pandemic began in January 2020). As the WHO declares a global emergency, the federal government contracts a fictional firm that resembles Moderna. Consistent with Gates's seeming preference for diabolical cognomens, the firm is dubbed "CynBio" (Sin-Bio) to develop an innovative vaccine using new "plug-and-play" technology. In the scenario, and now in real life, Federal health officials invoke the PREP Act to provide vaccine makers liability protection.[193]

Another company in this scenario receives an Emergency Use Authorization for a remdesivir-like antiviral named Kalocivir that federal officials previously evaluated as a therapeutic for SARS and MERS.[194]

This item seems to predict Dr. Fauci and Bill Gates's aggressive promotion of a failed Ebola drug, remdesivir, during the pandemic as "Standard of Care" for COVID-19. Dr. Fauci helped develop the drug, and Gates has a substantial equity stake in its manufacturer, Gilead. The two men promoted remdesivir during the earlier Ebola and Zika pandemics, despite its stunning inadequacy as a remedy for these ailments. Promotion of remdesivir, and the simultaneous Gates/Fauci orchestrated suppression of ivermectin and hydroxychloroquine, collectively—as we shall see—caused hundreds of thousands of deaths in the United States alone.

According to the scenario, by late January, SPARS has spread to every state and forty-two countries. In record speed, a coalition of ingenious corporate and heroic government officials miraculously produce a new vaccine, "Corovax," just in time for a July 2026 Emergency Use Authorization rollout.

This medical marvel meets resistance from several nuisance groups who complain that the companies have not adequately tested the jab. Among these ingrates are African Americans, alternative medicine enthusiasts, and a rapidly growing members of an anti-vaccination movement who bellyache on social media. But government and industry leaders depicted in those eighty-nine pages have plans to silence and censor these dangerous elements and to crush all resistance.[195]

The SPARS team responds with a flood of propaganda to drown doubt with vaccine plugola, public shaming of the vaccine-hesitant, and patriotic appeals.

While allies in government and the media boost public acceptance with propaganda, impose censorship, and muzzle dissent, Gates's minions recruit trusted "interlocutors," familiar community and medical leaders, to mollify the public that the experimental, unapproved, hastily tested, zero-liability vaccine is "safe and effective." The most effective "interlocutor" is Dr. Paul Farmer, Harvard's esteemed medical anthropologist and cofounder of Partners in Health, which provides medical care to impoverished regions around the globe. The simulation report states: "Paul Farmer, the renowned global health expert . . . lauded the safety and efficacy of Corovax and underscored the dangers of SPARS. His only regret, he said, was that the vaccine could not yet be made available to everyone on the planet."[196] (The real-life Farmer lists Gates as his organization's top funding partner.)

By springtime 2026, with the EUA vaccine rollout in full swing, public reservations about the vaccine are multiplying. The scenario blueprint predicts waves of severe neurological vaccine injuries soon appearing among children and adults. The CDC is meeting escalating skepticism toward its exaggerated predictions of coronavirus lethality; official fatality number indicates that coronavirus mortalities are comparable to the seasonal flu:

> By May 2026, public interest in SPARS had begun to wane. In late April the CDC had publicized an updated case fatality rate estimate, suggesting that SPARS was only fatal in 0.6 percent of cases in the United States.[197] (Note: the 2020 COVID-19 case fatality rate was a mere 0.26 percent according to CDC).

The SPARS organizers warn that dropping death rates will spark "public sentiment, widely expressed on social media, that SPARS was not as dangerous as initially thought." This perilous drop in popular fear jeopardizes the vaccine enterprise.

The SPARS team turns to pandemic porn—constantly repeated death counts and case counts—to amplify the panic decibel so as to assure the success of their mass inoculation program. To overcome the public's dangerous complacency, the CDC and FDA, in concert with other government agencies and their social media experts, begin developing a new public health propaganda campaign:

> create a core set of messages that could be shared by all public health and government agencies over the next several months during which time the SPARS vaccine could be introduced.[198]

In a section headed "Food for Thought," the scenario challenges participants to devise their own strategies for disabling common sense so as to achieve broad vaccine coverage:

> How might federal health authorities avoid people possibly seeing an expedited SPARS vaccine in development and testing process as somehow "rushed" and inherently flawed. . . . How might federal health authorities respond to critics who propose that liability protection for SPARS vaccine manufacturers jeopardizes individual freedom and well-being? . . . What are the potential consequences of health officials over-reassuring the public about the potential risks of a novel SPARS vaccine when long-term effects are not yet known?[199]

Even a casual read of the Foundation's planning document makes clear that Gates's preparation has little to do with public health and everything to do with limiting freedom and aggressively marketing vaccines.

The planners tell their intended audience—"public health providers and pandemic communicators"—that public concerns over worrisome reactions and vaccine side effects can be drowned out by flooding the airwaves with good news about vaccine successes: The dismaying role of mainstream media in these exercises is to broadcast propaganda, impose censorship, and manufacture consent for oppressive policies. In their projections, the social planners project absolute confidence that news media and social media companies will fully cooperate with this coup d'état. The simulation planners presciently assume their capacity to undermine the Fourth Estate in its role as the gladiatorial champion of free speech and democracy, and their ability to subvert the social media, which once promised to democratize the flow of information. Both mainstream and social media titans, it turns out, are predisposed to serve globalist elites. Gates and his cronies somehow intuited that these institutions would obligingly shape news coverage so as to manufacture obedience with compulsory vaccination and the dismemberment of the Constitution:

> In the following months . . . the WHO began developing an enhanced international vaccine program based on the expanded financial support of the United States and other countries. As time passed and more people across the United States were vaccinated, claims of adverse side effects began to emerge. . . . Given the positive reaction to the federal government's response and the fact that the majority of US citizens willing to be vaccinated had already been immunized, the negative publicity surrounding adverse reactions had little effect on nationwide vaccination rates.[200]

Gates and his team assure pandemic planners that they will easily avoid culpability for the wave of long-term neurological injuries that they cause by their experimental vaccines:

> While the federal government appeared to have appropriately addressed concerns around the acute side effects of Corovax, the long-term, chronic effects of the vaccine were still largely unknown. Nearing the end of 2027, reports of new neurological symptoms began to emerge. After showing no adverse side effects for nearly a year, several vaccine recipients slowly began to experience symptoms such as blurry vision, headaches, and

numbness in their extremities. Due to the small number of these cases, the significance of their association with Corovax was never determined.[201]

According to organizers, the purpose of Gates's simulation was to prepare "public health communicators" with a step-by-step strategic playbook for the upcoming pandemic. Eighteen months into the COVID-19 pandemic, it is difficult to peruse Gates's detailed 2018 planning document without feeling that we are all being played.

Laying Pipe for Totalitarianism

Following the success of the SPARS simulation, Gates projected a progressively darker and more martial tone and stepped up his declarations about the need for authoritarian coercion to cinch compliance with vaccination against the impending pandemic.

On April 18, 2018, Gates delivered a speech at the Malaria Summit in London, warning that a deadly new disease could arise within a decade, taking the world "by surprise," spreading globally and killing tens of millions. Hinting at the need for increased coordination between health officials and militaries, Gates reiterated: "The world needs to prepare for pandemics in the same serious way it prepares for war."[202] Gates's simulations invoke the concept of "total war," meaning the mobilization of entire populations, the sacrifice of global economies, and the obliteration of democratic institutions and civil rights.

Appreciating the challenges of imposing tyrannical controls in a democracy, Gates increasingly focused his efforts on enrolling critical allies in Big Tech and the military.

On April 27, Gates told the *Washington Post* that he had warned President Trump about "the increasing risk of a bioterrorism attack."[203] Emphasizing his frequent contacts with the president and military advisers, he publicly disclosed having regular meetings with H. R. McMaster, Trump's former national security advisor.

Gates was simultaneously building bridges with social media tycoons, including Amazon's Jeff Bezos, whose support he would need for his master plan. Like all totalitarian capers, Gates's gambit would require some book burning, and Bezos would be there to oblige. Beginning in March 2020, Amazon would outright ban or throttle the delivery of entire categories of books and videos that questioned official orthodoxies—including the scientific basis for the lockdown that would multiply Bezos's wealth by tens of billions. In the finest Operation Mockingbird tradition, Bezos's *Washington Post* also pitched in, including a shrill yet adoring propaganda tract under the headline "Bill Gates calls on US to lead fight against a pandemic that could kill 33 million." That month Gates announced a $12 million Grand Challenge, in partnership with the family of Google's cofounder Larry Page, to accelerate developing a universal flu vaccine.[204] Google's parent company, Alphabet, was already heavily investing in vaccine manufacturing start-ups and had signed a $76 million partnership with GlaxoSmithKline. Apparently anticipating rich returns to Big Tech from the lockdown he would orchestrate, Gates was, by then, among the largest shareholders of Amazon, Google, Facebook, and, of course, Microsoft.

The day after the *Post* story ran, a board member of the EcoHealth Alliance emailed zoologist and bioweapons expert Peter Daszak: "Any connections with Bill Gates we could [re]-activate given this perfect alignment in mission?"

Daszak responded: "re: gates and google—we have good connections at both orgs . . ." We'll definitely be reaching out to them again. . . . Ever since the Ebola outbreak [G]ates [foundation] are now getting more into pandemic preparedness."[205]

Daszak, at that juncture, was acting as a conduit through which Tony Fauci, Robert Kadlec, the Pentagon (DARPA), and USAID—formerly a CIA cover and nowadays reporting to the National Security Council—were laundering grants to fund gain-of-function experiments, including at the Wuhan Institute of Virology Biosafety Lab. In 2018, the French government had warned US government officials that the Wuhan lab, which the French helped build, was shoddily maintained and inadequately staffed and secured. For example, the French construction company, bioMérieux, which built the lab, had neglected to properly complete the negative airflow system—a critical piece of infrastructure to prevent the escape of viruses deliberately enhanced to create pandemics. Dr. Fauci ignored the warning.

When in May 2021 I emailed bioMérieux's ex-CEO (2007–2011), Stéphane Bancel, to ask him if he knew that his company had violated its contract to provide a functional system, he did not reply. Bancel by that time was CEO of Moderna and a partner of Bill Gates and Tony Fauci, operating a company that would be the primary beneficiary of the lab leak, quickly making Bancel's 9 percent stake worth over $1 billion and counting. In March 2019, eight months before COVID-19 began circulating, Bancel had reapplied for a patent for an mRNA technology for Moderna's new vaccine. The US patent office had previously rejected his application. But this time he approached the patent office with special urgency, expressing "a concern for reemergence or a deliberate release of the SARS coronavirus."[206, 207, 208, 209, 210]

Between Germ Game simulations, Gates continued his barnstorming tour laying pipe for mass panic and authoritarian rule. At the annual Shattuck Lecture on April 27, 2018, in Boston, he warned: "We can't predict when, but given the continual emergence of new pathogens, the increasing risk of a bioterror attack, and how connected our world is through air travel, there is a significant probability of a large and lethal, modern-day pandemic occurring in our lifetimes." Biological weapons of mass destruction, he warned, had "become easier to create in the lab." Gates went on to add that "we are supporting efforts by others, including the National Institute of Allergy and Infectious Diseases, whose vaccine candidate [presumably Moderna] is expected to advance to human safety trials in about a year."[211]

Clade X 2018

Then, on May 15, 2018, inside the darkened ballroom of Washington's Mandarin Oriental Hotel, foreboding military music introduced another "pandemic/biowarfare preparation exercise" hosted by the Johns Hopkins Center for Health Security (formerly the Hopkins Population Center, which Gates and NIH fund). The daylong event, dubbed Clade X, "simulate[d] the response to a fictitious bioengineered pathogen for which there is no vaccine."[212] Hoping to reduce world population, an elite cult released their genetically engineered bug from a Zurich lab. The disease spreads first to Germany and

Venezuela and then to the United States, killing 100 million people globally as "health-care systems collapsed, panic spread, the US stock market crashed."[213]

The simulation included "a series of National Security Council-convened meetings of ten US government leaders, played by individuals prominent in the fields of national security or epidemic response."[214] The exercise emphasized the need for militarized pandemic responses and explored strategies for controlling media and social media. It was a training drill to prepare political, bureaucratic, military, and intelligence officials to support the coup d'état against American democracy and the US Constitution. Participating were a kitchen cabinet of former top leaders of the FDA and CDC, as well as a former CIA general counsel. Playing themselves were ex-Senate Majority Leader Tom Daschle and Indianapolis Congresswoman Susan Brooks. Daschle, a former Army Intelligence officer who was among the targets of the 2001 anthrax-laced letters, became a pharmaceutical industry lobbyist by 2018. Susan Brooks, the so-called "Member from Eli Lilly," founded the Congressional Biodefense Caucus. She also introduced a successful bill in 2015—the Social Media Working Group Act of 2014—to establish a Social Media Bureau within the Department of Homeland Security to facilitate censorship of social media during national emergencies. Another of her bills in 2015 sought to streamline implementation of coercive vaccination programs by the federal government during pandemics.

Clade X livestreamed on Facebook before about 150 invited guests, including carefully selected representatives of major media. The simulation left his adulatory press quaking with fear. "This mock pandemic killed 150 million people. Next time it might not be a drill," Jeff Bezos's *Washington Post* headlined.[215] The *New York Post* assured readers that "the world is completely unprepared for the next pandemic."[216, 217]

As the *Post*'s reporter summarized:

> The simulation mixed details of past disasters with fictional elements to force government officials and experts to make the kinds of key decisions they could face in a real pandemic. It was a tense day. The exercise was inspired in part by the troubled response to the Ebola epidemic of 2014. Unlike Ebola, "which spreads through direct contact and bodily fluids," this latest "was a flulike respiratory virus, which would spread far more easily from person to person through coughing and sneezing . . ."
>
> In the exercise, schools closed, the demand for surgical masks and respirators far exceeded supply, and hospitals in the United States were quickly overwhelmed." Among the "difficult questions": "An entry ban on flights from other countries?" "Who should get the vaccines first?"[218]

It's noteworthy that none of the Hopkins simulations contemplate the efficacy of repurposed medications to mitigate or end the pandemic. And none of them allow for soul-searching about the abolition of constitutional rights and the wholesale destruction of America's political and judicial systems in favor of a tyrannical medical and military junta. None of them recognize that there is no pandemic exception in the United States Constitution. Instead, they were too busy war-gaming a high-level mutiny against American democracy.

All of the Hopkins simulation stories end with the same affirmations: the advisability of militarized police state response and the dire need for broadly deployable mRNA vaccines upon which Gates and Fauci had already invested billions of dollars: "Players underscored the need for the United States to 'go from bug to drug' faster."[219]

And each simulation highlighted the so-called "need" to quarantine and isolate the healthy, censor criticism of the Gates/Fauci vaccines and coerce the population into receiving vaccines rushed into distribution, all in opposition to logic, common sense, and previous public health practices.

Hopkins Center Director Tom Inglesby explained that the event's immediate purpose was to "provide experiential learning" for new decision makers in the Trump Administration.[220] Of course, the event's embedded press corps lauded Gates as the hero of the day—the beneficent billionaire whose genius, alone, would save us from the murderous contagion.

An adulatory *New Yorker* article, "The Terrifying Lessons of a Pandemic Simulation," giddily embraced the images of a nation at war with Gates as the general atop his gleaming white steed: "Philanthropist-in-chief Bill Gates drew on models developed by the [Gates-funded] Institute for Disease Modeling [IMHE], a venture founded by his former Microsoft colleague Nathan Myhrvold, to warn that, at our current state of readiness, roughly thirty-three million people would die within the first six months of a global pandemic similar to the 1918 flu."[221] (Gates would deploy his IMHE minions in January 2020 to grotesquely exaggerate the COVID-19 predicted mortalities—22 million dead in 12 months—to justify Tony Fauci's draconian lockdown.)

Where did the mock virus originate? In this scenario, "someone has genetically modified a mostly harmless parainfluenza virus to kill," recounted *MIT Technology Review*. "The fictional culprit is A Brighter Dawn, a shadowy group promoting the philosophy that fewer people—a lot fewer—would be a good thing for planet Earth." Johns Hopkins pandemic specialist Eric Toner created the scenario after carrying out "meticulous research to come up with a plausible threat using real virology and epidemiological models. The result was so realistic that the organizers chose not to present too many details."[222]

A clear strategic objective for Gates and Fauci was the repetition of the message that a global pandemic was inevitable, that only mandatory vaccines could avert catastrophe, and that obliteration of civil rights will be required. Most astonishing was their capacity to mobilize the obliging global media to uncritically swallow and promote these propositions in complete contradiction of all previously accepted science and history.

That same month, PBS's *NewsHour*—once revered as the most incorruptible of all US television media—ran an adoring feature on Dr. Fauci prominently, touting the need for a universal flu vaccine in a two-part report on "Why another flu pandemic is likely just a matter of when."

PBS cut to a tour of Fauci's Vaccine Research Center with Dr. Barney Graham, coinventor of Moderna's mRNA vaccine. In the next segment, the PBS reporter asked Dr. Fauci about "a shot to protect against all known and unknown strains of the [flu] virus." Dr. Fauci replied: "Several years ago, I wouldn't have been able to give you even an approximation of when that would be, because the science wasn't giving us the clues that we could actually do that. Now with these exquisite techniques of structure-based vaccine design, I think we are in shooting distance."[223] Dr. Fauci continued, "We have got to be able to have something that, when a new pandemic virus emerges, we already have something on the shelf to do something about it, something that you could make and it would be useable so that, when you stockpile it, it really is a stockpile."[224]

The show was functionally an infomercial for Moderna and mRNA vaccines. PBS didn't mention that Dr. Fauci's NIAID had pumped massive funding into Moderna's vaccine or that NIAID claimed patent rights and stood to profit handsomely from its approval. Nor did PBS acknowledge that the Bill & Melinda Gates Foundation had previously given PBS *NewsHour* millions of dollars,[225] or that, by 2019, Gates had also bet millions on Moderna's mRNA vaccine. Gates owns a substantial equity stake in the company.

In September 2019, the Gates-funded John Hopkins Center for Health Security followed up on its Clade X event by issuing an eighty-four-page report, "Preparedness for A High-Impact Respiratory Pathogen Epidemic." The report focused on the only end point that seemed to really concern Gates—the Gates/Fauci mRNA vaccine project. If there was any doubt that pushing mRNA vaccine was the entire purpose of the exercise, the white paper cleared that up. The Clade X summary called for making the top priority of all government, media, and biosecurity players the coordinated drive for:

> R&D aimed at rapid vaccine development for novel threats and distributed surge manufacturing. . . . Nucleic acid (RNA and DNA)–based vaccines are widely seen as highly promising and potentially rapid vaccine development pathways, though they have not yet broken through with licensed products.[226]

Both Gates and Fauci had already invested such enormous financial resources in that technology. In this light, the simulations can be interpreted as marketing and public relations exercises designed to recruit and train political, military, media, and public health officials to advance their enterprise using censorship, propaganda, and state-sponsored violence, if necessary.

The report concluded with a revealing warning about biosafety, "particularly for countries that are funding research with the potential to result in accidents with pathogens that could initiate high-impact respiratory pandemics."[227] The report warned that the possibility of deliberate release "could substantially add to the extraordinary consequences that would follow a naturally occurring pandemic event with the same agent.[228] Mass vaccination strategies should be developed and put in place to increase immediate access."[229]

Put simply, through the medium of this sponsored report Gates, is saying that we need a rapid mass vaccination strategy in place to anticipate the accidental or deliberate release of the kind of enhanced pathogens that his working partner, Dr. Fauci, was funding the development of in Wuhan, under the pretext of vaccine research.

Though Gates's simulation highlighted the need for masks and respirators, Gates,

Dr. Fauci, and Kadlec ignored stockpiling these items, and the same for any antiviral drugs that might successfully treat sick people.[230] Instead, they were laser-focused on next-gen vaccines, on compulsory administration to healthy uninfected populations, on censorship and other coercive devices, on constructing and controlling global health agencies, and on surveillance technologies.

Global Preparedness Monitoring Board

Later, in May 2018—with imprimatur from the WHO and the World Bank Group—Gates created a kind of permanent standing committee called the Global Preparedness Monitoring Board (GPMB), including some of the most powerful global public health kingpins, to institutionalize the lessons derived from all these scenario planning drills.[231] The global committee would serve as the real-life authoritative collective for imposing rules during the upcoming pandemic. This so-called "independent" monitoring and accountability body's purpose was to validate the imposition of police state controls by global and local political leaders and technocrats, endorsing their efforts to take the kind of harsh actions that Gates's simulation modeled: subduing resistance, ruthlessly censoring dissent, isolating the healthy, collapsing economies, and compelling vaccination during a projected worldwide health crises. GPMB's board includes a pantheon of technocrats whose cumulative global power to dictate global health policy is virtually irresistible: Anthony Fauci; Sir Jeremy Farrar of Wellcome Trust; Christ Elias of BMGF; China's CDC director, George Gao; Russian health minister, Veronika Skvortsova; WHO's health director, Michael Ryan; its former director, Gro Harlem Brundtland; its former programming director, Ilona Kickbusch; and UNICEF's Henrietta Holsman Fore, who is former director of USAID, that used to be a reliable CIA front.

In June 2019, about twenty weeks before the start of the COVID pandemic, Dr. Michael Ryan, executive director of the WHO's health emergencies program, summarized the conclusions of GPMB's pandemic report, warning that "we are entering a new phase of high impact epidemics" that would constitute "a new normal" where governments worldwide would strengthen control and restrict the mobility of citizens.[232]

Crimson Contagion 2019

That August—not even ten weeks before the first COVID-19 infections were reported in Wuhan—a 2019 war game code-named Crimson Contagion capped eight months of planning overseen by Robert Kadlec, who was, by then, President Trump's Disaster Response Leader. Also involved in this virus war game scenario was Anthony Fauci representing the NIH, Dr. Robert R. Redfield of the CDC, and HHS Secretary Alex Azar.[233] The HHS Office of Preparedness and Response teamed with the top spooks at the National Security Council to lead the four-day nationwide "Functional Exercise."[234]

So now Kadlec—who had, for twenty years, been writing scripts for using a pandemic to overthrow democracy and curtail constitutional rights—was in a perfect position to do just that. With this virus simulation, he included all the key players who would manage what was to become a de facto coup d'état sixty days hence.

While earlier simulations functioned as training drills for high-level political, military, press, intelligence agency, and regulatory commissars, the 2019 Crimson Contagion simulation functioned as a nationwide crusade to evangelize state-level health bureaucracies, municipal officials, hospital and law enforcement agencies across America with the messages developed in the preceding simulations.

Under a veil of enforced secrecy, organizers staged the Crimson Contagion exercise nationwide at over 100 centers. "Participation included 19 federal departments and agencies, 12 key states, 15 tribal nations and pueblos, 74 local health department and coalition regions, 87 hospitals, and over 100 healthcare and public health private sector partners."[235] The simulation scenario envisioned a "novel influenza" pandemic originating in China labeled H7N9. As with COVID-19, air travelers rapidly spread the deadly respiratory illness across the globe.

In this scenario, by the time US health officials first identify the virus in Chicago, it is already galloping like the Grim Reaper across other metropolitan areas, forcing the HHS Secretary to declare a national public health emergency. The WHO delays a month before declaring a pandemic. The multistate, multiregional exercise that took place just months before the real-world COVID-19 pandemic focused on "critical infrastructure protection; economic impact; social distancing; scarce resource allocation; prioritization of vaccines and other countermeasures."[236] (Again not including therapeutic medicines.) The Crimson Contagion exercise achieved eerily accurate forecasting with numbers that precisely predicted the official casualty data for COVID-19: 110 million forecasted illnesses, 7.7 million predicted hospitalizations, and 568,000 deaths in the United States alone.

The draft report, dated October 19, 2019, and marked "not to be disclosed," didn't become public until the *New York Times* obtained a copy under the Freedom of Information Act and published a front-page article on March 19, 2020, eight days after the WHO declared COVID-19 a pandemic.[237] Only under pressure from another FOIA request did Kadlec's HHS Office of the Assistant Secretary of Preparedness and Response release the January 2020 After-Action Crimson Contagion Report the following September. It is available online here: governmentattic.org/38docs/HHSaarCrimsonContAAR_2020.pdf.

The *Times* story contained this paragraph: "The October 2019 report documents that officials at the Department of Homeland Security and Health and Human Services, and even at the White House's National Security Council, were aware of the potential for a respiratory virus outbreak originating in China to spread quickly to the United States and overwhelm the nation."[238] The *New York Times* takeaway missed altogether the larger and more significant stories: that the Crimson Contagion's planners precisely predicted every element of the COVID-19 pandemic—from the shortage of masks to specific death numbers—months before COVID-19 was ever identified as a threat and that their overarching countermeasure was the preplanned demolition of the American Constitution by a scrupulously choreographed palace coup.

The Crimson Contagion draft report complains that existing federal funding sources were insufficient to combat a pandemic and concluded, predictably, that government officials needed far more money and far more power: "A significant topic of concern centered around the inadequacies of existing executive branch and statutory authorities to provide HHS with the requisite mechanisms to serve successfully as the lead federal agency in response to an influenza pandemic."[239]

The team noted that "The group . . . concluded they would soon need to move toward aggressive social distancing, even at the risk of severe disruption to the nation's economy and the daily lives of millions of Americans."[240]

TOPOFF 2000–2007

In the course of researching this book, I discovered that, beginning in 2000, the security, military, police, and intelligence agencies have been secretly staging other mass simulations, under the codename TOPOFF, of which the public is almost entirely unaware. Each of these functioned as training exercises for the lockstep imposition of global totalitarianism. Many of these drills have involved tens of thousands of local police, health officials, and emergency responders across the United States, Canada, Mexico, and Europe, as well as representatives from the FBI, the State Department, the intelligence agencies, and private corporations from chemical, petroleum, financial, telecom industries, and health sectors.

Four TOPOFF (Top Official) exercises between May 2000 and 2007 mobilized DOJ, FBI, and FEMA officials staging scenario planning around chemical and bioweapons attacks. The first of them, in May 2000, modeled chemical biological attacks in Denver, Colorado, and Portsmouth, New Hampshire, exploring logistics for quarantining an entire state (Colorado). The executive summary complains that "stronger measures to protect the local Colorado citizens were not implemented"[241] and warns that to survive such a disaster, the state must immediately take quick and decisive action to quarantine the population, including the enforcement of an unprecedented "no contact out of your home"[242] policy that became the hallmark of the response to COVID-19 twenty years later.

The Department of Homeland Security sponsored TOPOFF 2, in May 2003, including more than 8,000 participants in Seattle and Chicago, as well as significant participation by the Canadian government.[243]

TOPOFF 3, in April 2005, simulated biological and chemical attacks in New Jersey and Connecticut, involving more than 20,000 participants from over 250 federal, state, and local agencies, private businesses, volunteer groups, and international organizations. Canada and the UK coordinated simultaneous exercises.[244]

TOPOFF 4, running from October 15 to October 24, 2007, involved more than 23,000 participants from government and the private sector, simulating attacks in Guam, Portland, and Phoenix. In Washington, DC, the State Department activated an Exercise Task Force and participated in high-level meetings with other Department and agency decision makers, including American embassies in Canberra, Ottawa, and London.[245]

"These are brainwashing exercises," says former CIA officer and whistleblower Kevin Shipp. "Getting all of these thousands of public health and law enforcement officials to participate in blowing up the US Bill of Rights in these exercises, you basically have obtained their prior sign-off on torpedoing the Constitution to overthrow its democracy. They know that none of these participants are going to suddenly start soul-searching when the real thing happens. The CIA has spent decades studying exactly how to control large populations using these sorts of techniques." Shipp adds: "We are all subjects now being manipulated in a vast population-wide Milgram experiment, with Dr. Fauci playing the doctor in the white lab coat instructing us to ignore our virtues and our conscience and obliterate the Constitution."[246]

Event 201: October 2019

Under Gates's direction in mid-October 2019, only two months after Crimson Contagion and three weeks after US intelligence agencies believe that COVID-19 had begun circulating in Wuhan, the cabal of potentates and institutions that compose the Biosecurity Cartel began preparing decision makers for the mass eviction of informed critics of the vaccine industry from social media. That month, Gates personally organized yet another training and signaling exercise for government biosecurity functionaries. This war game consisted of four "tabletop" simulations of a worldwide *coronavirus* pandemic. Participants included a group of high-ranking kahunas from the World Bank, the World Economic Forum, Bloomberg/Johns Hopkins University Populations Center, the CDC, various media powerhouses, the Chinese government, a former CIA/NSA director, vaccine maker Johnson & Johnson, the globe's largest pharmaceutical company; finance and biosecurity industry chieftains, and the president of Edelman, the world's leading corporate PR firm. Conspiracy-minded critics dub this cabal the "Deep State." The World Economic Forum Director Klaus Schwab has christened their agenda the "Great Reset."[247]

Event 201 was a signaling exercise, but it was also, as we shall see, a training run for a "government in waiting." Its principals would quickly move into key positions to run pandemic response a few months later.

At Gates's direction, the participants role-played members of a Pandemic Control Council, war-gaming a contagion that serves as pretext for this insurgency against American democracy. They drilled a retinue of psychological warfare techniques for controlling official narratives, silencing dissent, forcibly masking large populations, and leveraging the pandemic to promote mandatory mass vaccinations. Needless to say, there was little talk of building or fortifying immune systems, existing off-the-shelf remedies, or off-patent therapeutic drugs and vitamins. Instead, there was abundant palaver about expanding government's authoritarian powers, imposing draconian restrictions, curtailing traditional civil rights, which might include of rights of assembly, free speech, private property, jury trials, due process, and religious worship, as well as promoting and coercing the uptake of new, patentable, antiviral drugs and vaccines. The participants walked through imaginary global coronavirus contagion scenarios that focused on fear-mongering, blanket censorship, mass propaganda, and police state strategies culminating in compulsory mass vaccination.

As with the Clade X simulation, the most trusted Pharma-friendly media attended. *Forbes* and *Bloomberg* participated in the exercise, which focused on war-gaming the medical cartel's censorship initiative. The Bloomberg Foundation is a major funder of the Johns Hopkins Center. Oddly, Gates later claimed that this simulation didn't occur. On April 12, 2020, Gates told BBC, "Now here we are. We didn't simulate this, we didn't practice, so both the health policies and economic policies, we find ourselves in uncharted territory."[248] Unfortunately for that whopper, the videos of the event are still available across the Internet. They show that Gates and team did indeed simulate health and economic policies. It's hard to swallow that Gates had forgotten.

Organizers billed Event 201 as a vehicle for delineating "areas where public/private partnerships will be necessary during the response to a severe pandemic in order to diminish large-scale economic and societal consequences." They reminded attendees that "experts agree" that it is only a matter of time before one of these epidemics becomes "global."[249]

Event 201 was as close as one could get to a "real-time" simulation. It was a meeting of a hypothetical Pandemic Emergency Board, in the same week that COVID-19 was already claiming its first victims in Wuhan. "We're not sure how big this could get, but there's no end in sight,"[250] warns one hypothetical physician in an opening briefing. Gates's simulated coronavirus epidemic was far worse than the authentic COVID-19 outbreak that would hit America just weeks later. The simulated version caused 65 million deaths at the eighteen-month end point and global economic collapse lasting up to a decade.[251] Compared to the Gates simulation, therefore, the actual COVID-19 crisis is a bit of a dud. Public health officials claim 2.5 million deaths "attributed to COVID" globally over 13 months. The death counts from COVID in our real-life COVID-19 predicament are highly inflated and questionable. Further, the death of 2.5 million must be put in the context of a global population of 7.8 billion, with around 59 million deaths expected annually in any event. Event 201's predictions of decade-long economic collapse will probably prove more accurate—but only because of the draconian lockdown promoted by both Gates and Dr. Fauci.

The theme of Event 201 was that such a crisis would prove an opportunity to promote new vaccines and tighten information and behavioral controls through propaganda, censorship, and surveillance. Gates's script anticipates vast anti-vaccine resistance triggered by mandates and fanned by Internet posts.

Muzzling Talk of Lab Generation

Five months before WHO declared a global pandemic, at a time when 99.999 percent of Americans had never heard the phrase "gain-of-function," key government officials were already planning strategies for suppressing public discussion of the potential that a coronavirus might have been deliberately manipulated to enhance its pathogenicity and transmissibility in humans.

One of their central fixations was how to silence "rumors" that the coronavirus was laboratory-generated. Event 201's fourth simulation anticipated the manipulation and control of public opinion and muzzling any colloquy about artificially enhanced pathogens. Everyone voiced their urgent concerns that authorities must instantly squelch and discredit any speculation that someone deliberately or accidentally released a lab-made bug. This segment is most revealing for its uncannily accurate prediction of democracy's current crisis. The fundamental assumption of all participants was that censorship and propaganda are legitimate exercises of Federal power. The participants discussed mechanisms for stamping out "disinformation" and "misinformation," by "flooding" the media with propaganda ("good information"), imposing penalties for spreading falsehoods, and discrediting dissent ("the anti-vaccination movement").

What follows are thumbnail portraits of some of the participants in this aspect of the operation, along with accounts of their specific comments and actions:

- **George Gao**, the director of the Chinese Center for Disease Control (CCDC), worried about how to suppress the inevitable "rumors" that the virus is laboratory generated: "People believe, 'This is a manmade' . . . [and that] some pharmaceutical company made the virus." Two months after speaking those words, Gao himself would lead the Chinese effort to tamp down rumors of lab creation. Gao also orchestrated the Chinese government drive to vaccinate a billion Chinese citizens.[252]

- **Dr. Tara Kirk Sell**, a senior scholar at Bloomberg School of Health's Johns Hopkins Center for Health Security, worried that pharmaceutical companies are being accused of introducing the virus so they can make money on drugs and vaccines: "[We] have seen public faith in their products plummet." She notes with alarm that "Unrest, due to false rumors and divisive messaging, is rising and is exacerbating spread of the disease as levels of trust fall and people stop cooperating with response efforts. This is a massive problem, one that threatens governments and trusted institutions."[253]

  Sell reminds her confederates that "We know that social media is now the primary way that many people get their news, so interruptions to these platforms could curb the spread of misinformation." There are many ways, Dr. Sell advises, for government and industry allies to accomplish this objective: "Some governments have taken control of national access to the Internet. Others are censoring websites and social media content and a small number have shut down Internet access completely to prevent the spread of misinformation. Penalties have been put in place for spreading harmful falsehoods, including arrests."[254]

  Like many other Event 201 collaborators, Sell moved into government service soon after declaration of the pandemic.

  Since the beginning of the COVID-19 pandemic, Dr. Sell has worked as a kind of United States "Minister of Truth" coordinating US government and WHO efforts to quash, to dissent and discredit, vilify, and gaslight dissenters. She calls her occupation by the Orwellian term "Infodemiology" which she describes as tracking spread of misinformation (dissenting opinions) and curtailing its spread through risk communication and censorship.[255]

- **Jane Halton** served Australia as both Health and Finance ministers and is a board member of Australia's ANZ Bank. ANZ funds Australia's large and influential vaccine sector.[256] Halton is one of the authors of Australia's oppressive "no jab, no pay" policy. She was the former president of WHO's World Health Assembly. Today, she is chair of Gates's global Coalition for Epidemic Preparedness (CEPI), which serves the role of diverting philanthropic and government financing toward the development of pandemic vaccines by profit-making pharmaceutical companies. She assured her fellow Event 201 participants that, behind the scenes, the Gates Foundation was already creating algorithms "to sift through information on these social media platforms"[257] to protect the public from dangerous thoughts and information. In March of 2020, Halton joined the Executive Board of the Australian National COVID-19 Coordination Commission, which imposed the world's most draconian lockdown and the most dramatic abridgements of civil rights in that nation's history.

- **Chen Huang**, an Apple research scientist, Google scholar, and the world's leading expert on tracking and tracing and facial recognition technology, role-plays the newscaster reporting on government countermeasures. He blames riots on anti-vaccine activists and, approvingly, predicts that Twitter and Facebook will cooperate in "identify[ing] and delete[ing] a disturbing number of accounts dedicated to spreading this information about the outbreak" and to implement "Internet shutdowns . . . to quell panic."[258]

- **Matthew Harrington**, director of Global Operations and Digital Communications, Edelman—the world's largest public relations firm, which represents Pfizer, AstraZeneca, Johnson & Johnson, and Microsoft—agrees that social media must fall in line to promote government policy: "I also think we're at a moment where the social media platforms have to step forward and recognize the moment to assert that they're a technology platform and not a broadcaster is over. They in fact have to be a participant in broadcasting accurate information and partnering with the scientific and health communities to counterweight, if not flood the zone, of accurate information. Because to try to put the genie back in the bottle of the misinformation and disinformation is not possible."[259]

- **Stephen Redd**, the admiral of the United States Public Health Service and assistant surgeon general, has the sinister notion that government should mine social media data to identify and collect data on Americans with negative beliefs: "I think with the social media platforms, there's an opportunity to understand who it is that's susceptible . . . to misinformation, so I think there's an opportunity to collect data from that communication mechanism."[260] A couple of months after expressing these ideas, Redd assumed his new post: deputy director of CDC managing the COVID countermeasures.

- **Adrian Thomas**, VP Global Strategy, Programs & Public Health, for Johnson & Johnson, the world's largest pharmaceutical company, announced "some important news to share from our member companies [Pharma]. . . . We are doing clinical trials in new antiretrovirals, and in fact, in vaccines!" He recommends a strategy to address the problems that will inevitably badger these companies when "rumors were actually spreading" that their shoddily tested products "are causing deaths and so patients are not taking them anymore." He suggests, "maybe we're in the mistake of reporting and counting all the fatalities and infections."[261] This worry may explain why federal regulators chose to deliberately maintain a dysfunctional surveillance system designed to hide more than 99 percent of vaccine injuries. Thomas has manned Johnson & Johnson's Pandemic Response and Vaccine Development program since March 2021.

- Former CIA Deputy Director **Avril Haines** unveiled a strategy to "flood the zone" with propaganda from "trusted sources," including "influential community leaders, as well as health workers." She warns about "false information that is starting to actually hamper our ability to address the pandemic, then we need to be able to respond quickly to it."[262] On April 11, 2021, President Biden appointed Haines as director of National Intelligence, now the highest official in charge of pandemic response.

- **Matthew Harrington** (Edelman CEO) observes that the Internet—which once promised to decentralize and democratize information—now needs to be centralized: "I think just to build a little bit on what Avril said is, I think as in previous conversations where we've talked about centralization around management of information or public health needs, there needs to be a centralized response around the communications approach that then is cascaded to informed advocates, represented in the NGO communities, the medical professionals, et cetera."[263] Edelman boasts that tech is its biggest client, followed closely by Pharma. Microsoft is Edelman's most important account.

- **Dr. Tom Inglesby** is director of the Johns Hopkins Center for Health Security. He is an adviser to NIH, the Pentagon, and Homeland Security. Like many other Event 201 participants, Inglesby migrated immediately into real-life management of the crisis. Three months later, he would move over to HHS as senior adviser of the COVID-19 response. Inglesby agrees that greater centralized control is needed: "You mean centralized international?"[264]

- **Matthew Harrington** (Edelman) replies that information access should be: "Centralized on an international basis, because I think there needs to be a central repository of data facts and key messages."[265]

- **Hasti Taghi** (media adviser) sums up: "The anti-vaccine movement was very strong and this is something specifically through social media that has spread. So as we do the research to come up with the right vaccines to help prevent the continuation of this, how do we get the right information out there? How do we communicate the right information to ensure that the public has trust in these vaccines that we're creating?"[266]

- **Kevin McAleese**, a communications officer for Gates-funded agricultural projects, observes that "To me, it is clear countries need to make strong efforts to manage both mis- and disinformation. We know social media companies are working around the clock to combat these disinflation campaigns. The task of identifying every bad actor is immense. This is a huge problem that's going to keep us from ending the pandemic and might even lead to the fall of governments, as we saw in the Arab Spring. If the solution means controlling and reducing access to information, I think it's the right choice."[267]

- **Dr. Tom Inglesby** (Johns Hopkins) concurs, asking if "In this case, do you think governments are at the point where they need to require social media companies to operate in a certain way?"[268]

- **Lavan Thiru**, Singapore's finance minister, suggests that the government might make examples by arresting dissidents with "governments on enforcement actions against fake news. Some of us, this new regulations are come in place about how we deal with fake news. Maybe this is a time for us to showcase some cases where we are able to bring forward some bad actors and leave it before the courts to decide whether they have actually spread some fake news."[269]

- **Sofia Borges**, head of the New York office of the UN, spoke of putting out positive stories about people who'd beaten the disease and "having a centralized source of information and a world body that could garner the respect of everyone. I think the WHO, in this instance, might be that source of information."[270]

- **Adrian Thomas** added, "It's important to think about what atypical players in the private sector can we bring to bear in this? Bringing multinational pharmaceutical companies to talk about… why their products are safe could be seen as non-credible."[271]

* * *

Gates's Event 201 global pandemic war game quickly demonstrated that it was reaching and indoctrinating its intended audiences—the globe's top-level decision makers. A week after Event 201, presidential aspirant Joe Biden read a *Washington Post* article about the follow-up to the Event 201 report coauthored by the Hopkins Center for Health Security. According to a newly invented Global Health Security Index assessing 195 countries, "No country—the United States included—is fully prepared to respond to a deliberate or accidental threat with the potential to wipe out humanity."[272] Biden tweeted a response on October 25, 2019: "We are not prepared for a pandemic. . . . We need leadership that . . . focuses on real threats, and mobilizes the world to stop outbreaks before they reach our shores."[273]

At the end of November 2019, Robb Butler—the head of WHO/Europe's Vaccine Preventable Diseases and Immunization Program from 2014 to 2018—told the European Scientific Conference on Applied Infectious Disease Epidemiology that "vaccine hesitancy" must be tackled and "immunization is a best buy."[274]

The Triumph of the Military/Intelligence Complex: Intelligence Agencies and COVID-19

In November 2020, the British spy agency MI6 announced that its spooks would be surveilling foreigners all over the world (presumably including Americans) who questioned official orthodoxies about COVID-19 vaccines. Declaring the launch of an "offensive cyber-operation to disrupt anti-vaccine propaganda,"[275] the Foreign Branch hinted that it would henceforth target individuals who asked awkward or impudent questions about vaccines or questioned official COVID proclamations or countermeasures. The agency promised to deploy the same arsenal of monitoring and harassment weaponry and dirty tricks that it formerly reserved for terrorists. According to *The Times*, "The spy agency is using a toolkit developed to tackle disinformation and recruitment peddled by Islamic state."[276] A government source assured the paper they weren't kidding around: "GCHQ has been told to take out anti-vacciners online and on social media. There are ways they have used to monitor and disrupt terrorist propaganda."[277]

Federal law forbids US spy agencies from spying on or surveilling US citizens, but the Western intelligence bureaucracies work in collaboration with one another, and the CIA often deploys European, Israeli, and Canadian agencies as surrogates to skirt US laws.

In August 2020, after I appeared as a keynote speaker before an estimated crowd of 1.2 million democracy and civil rights advocates from every European nation protesting COVID restrictions at a Peace and Justice Rally in Berlin, Germany's domestic intelligence agency announced that it would begin monitoring the top leaders of the group that invited me. The spy agency accused COVID protesters of trying to "permanently undermine trust in state institutions and their representatives,"[278] according to the news agency AFP. "Now, the definition of terror is so broad," says former CIA official Kevin Shipp, "that any mention of COVID vaccines comes under their purview."[279]

These were the first explicit acknowledgments of the pervasive involvement by Western intelligence agencies in the vaccine enterprise that the global press has long overlooked. As two decades of Germ Game simulations foreshadowed, US and foreign clandestine agencies have a secretive but dominating presence in the COVID-19 pandemic response. Intelligence community alumni and active officers occupy key positions in the international agencies that promote global vaccinations. For example, President Biden's director of USAID is former WHO Ambassador Samantha Power. Power is an imperialist war hawk who as President Obama's National Security Advisor persuaded him to intervene militarily in Libya. She has declared that her primary goal at USAID is "to restore US prestige by getting American-made vaccines 'into arms' around the world."[280] UNICEF's Director, Anthony Lake, was President Bill Clinton's national security advisor and his nominee to be CIA Director until corruption charges derailed his appointment. In January 2020, UNICEF telegraphed its brave new embrace of authoritarianism by cheerleading the Maldives legislature's passage of a bill making it a criminal offense for parents to decline any government-recommended vaccine for their children. UNICEF's unsheathed enthusiasm makes clear that the organization regards the Maldives innovation as a pilot program for humanity.[281]

GlaxoSmithKline's spinoff, the Wellcome Trust, has played a central role in the marriage of Big Pharma to the Western spy agencies. From 2015 until October 2020, the Chair of Wellcome Trust—the UK version of the Gates Foundation—was the former director general of MI5, Dame Eliza Manningham-Buller, a thirty-five-year counterespionage veteran who also functioned as official liaison between British and US intelligence agencies. Anthony Fauci's emails reveal that Wellcome Trust Director Sir Jeremy Farrar worked directly with Dr. Fauci to orchestrate the cover-up of the Wuhan lab leak evidence, assigning a staff of five Wellcome Trust operatives to manage the fraud.[282]

Dame Manningham-Buller has served as chair of the Imperial College London since 2011. Anthony Fauci and Western health officials widely cited the Imperial College's inaccurate COVID-19 fatality projections—ginned up by the Wellcome Trust's notorious epidemiologist, Neil Ferguson—to justify the draconian global lockdowns.[283] Ferguson's guileful projections overestimated fatality rates by more than an order of magnitude. He did the same with mad cow disease and other diseases du jour. MI6 spy Christopher Steele is a leader of the British organization "Independent SAGE," a sketchy, yet highly influential collective of social scientists, psychologists, and professional propagandists who use the news media to relentlessly pressure the UK government anytime it hesitates to deploy the flinty authoritarianism needed to achieve "zero COVID."[284]

Steele is only one of many former intelligence officers who cheerlead draconian responses to COVID and applaud the onset of totalitarianism. One of the early promoters of the marginalization, demonization, and officially sanctioned abuse of vaccine-hesitant parents is Juliette Kayyem, the former assistant secretary for Homeland Security under President Obama, and former member of the Council on Foreign Relations and the National Committee on Terrorism. Kayyem was forced out of her high-level job at the *Washington Post* when critics leaked her involvement with the Israeli spyware company that makes the software system used to track and murder Saudi journalist Jamal Khashoggi.[285] As early as April 2019, she was editorializing for the *Washington Post* that parents who declined measles vaccines for their children should face "isolation, fines, arrests" and be treated to the same sanctions that government uses against terrorists and sex offenders.[286]

As early as 1977, Watergate journalist Carl Bernstein documented the CIA's control over 400 leading American journalists and institutions, including the *New York Times* and *TIME* Magazine. The CIA's long and pervasive domination of the *Washington Post* via Project Mockingbird, beginning with its owners Katharine and Phil Graham and leading editors and reporters, is well documented. There is little evidence that its new owner, Jeff Bezos, has pruned away these corrupting influences. The *Post* and the *Times* have been the leading media cheerleaders for draconian pandemic response. On September 5, Max Blumenthal—son of frequent *Washington Post* contributor Sidney Blumenthal—exposed the *Post* for publishing a phony "Doctor on the Street Interview" in which a supposedly typical DC physician called for extrajudicial murder of the vaccine-hesitant parents through medical neglect. Blumenthal pointed out that the physician was actually the vice president of Technical Staff at In-Q-Tel.[287]

The CIA and other intelligence agencies aggressively recruit scientists like Jeremy Farrar, whose research involves postings in foreign countries.[288] Additionally, it uses vaccination drives as a cover for broader strategic actions. Between 2011 and 2014, for example, the CIA used the WHO's Global Eradication Program to conduct fake polio and Hepatitis B vaccine programs in Pakistan as a way to surreptitiously collect DNA from individuals in its efforts to track down Osama bin Laden.

These are only a few of the myriad examples of the closely kept involvements by spy agencies in treating vaccination as a foreign policy tool and as an instrument of fear, suppression, and control independent of any genuine health concerns.

* * *

In July 2021, one year and four months into the misery of the global lockdown, the FAA had to divert air traffic over a section of the country stretching from the West Coast to Michigan to make room for the fleets of private jets converging on Sun Valley, Idaho, for the thirty-eighth annual meeting of the world's most exclusive conclave, sometimes called the Summer Camp for Billionaires, or "Mogul Fest."[289] The 2021 meeting included Bill Gates, Apple CEO Tim Cook, Mark Zuckerberg, Amazon founder Jeff Bezos, Mike Bloomberg, Google founders Larry Price and Sergey Brin, Warren Buffett, Netflix CEO Reed Hastings, Disney Chair Robert Iger, Viacom/CBS Chair Shari Redstone, and one of the lockdown's most influential propagandists, Anderson Cooper, who has acknowledged that he responded to a CIA recruitment poster while attending Yale and worked an indeterminate number of summers thereafter in Langley.

All the discussions at the event were, as usual, closely guarded, but participants acknowledged conversing about cryptocurrencies and artificial intelligence. This year, the robber barons hosted, as their guest of honor, CIA Director William Joseph Burns, and by all reports, the mood among the titans was bullish.[290] By that time, US billionaires were well on their way to increasing their collective wealth by $3.8 trillion in a single year, while obliterating the American middle class, which permanently lost about the same amount. These tech and media magnates, who had magnified their billions from the lockdown, were the same men who had used their media and social media platforms to censor complaints about the lockdown, even as it filled their coffers past the bursting point.

Each of these fat cats had helped grease the skids for the calamitous collapse of America's exemplary constitutional democracy. The Bill of Rights was, by then, indefinitely suspended. The participants of that event had privatized the public square and then obstructed the free flow of information and open debate—the oxygen and sunlight of democracy. Their censorship allowed their allies in the technocracy to effect the most extraordinary curtailment of American constitutional rights in history: closing churches across the country, shuttering a million businesses without due process or just compensation, suspending jury trials for corporate malefactors, passing regulations without constitutionally guaranteed transparency public hearings or comment, violating privacy through warrantless searches, and track-and-trace surveillance and abolishing the rights of assembly and association.

After twenty years of modeling exercises, the CIA—working with medical technocrats like Anthony Fauci and billionaire Internet tycoons—had pulled off the ultimate coup d'état: some 250 years after America's historic revolt against entrenched oligarchy and authoritarian rule, the American experiment with self-government was over. The oligarchy was restored, and these gentlemen and their spymasters had equipped the rising technocracy with new tools of control unimaginable to King George or to any other tyrant in history.

* * *

COVID-19: A Military Project

In the councils of government, we must guard against the acquisition of unwarranted influence, whether sought or unsought, by the military-industrial complex. The potential for the disastrous rise of misplaced power exists and will persist.

We must never let the weight of this combination endanger our liberties or democratic processes. We should take nothing for granted. Only an alert and knowledgeable citizenry can compel the proper meshing of the huge industrial and military machinery of defense with our peaceful methods and goals, so that security and liberty may prosper together.

—Dwight Eisenhower, 1961

With all the preparations for a coordinated military response, with deep involvement from intelligence agencies, it should come as no surprise that the government's COVID-19 response quickly emerged as a military project.

On Sept. 28, 2020, science journalist Nicholas Florko published in *STAT* a leaked organizational schematic[291] exposing the $10 billion Operation Warp Speed project as a highly structured Defense Department campaign with "vast military involvement." The byzantine flowchart[292] shows four generals and sixty other military officials commanding Operation Warp Speed, badly outnumbering civilian health technocrats from HHS, who represented a mere twenty-nine of the roughly ninety leaders on the chart.

HHS's Deputy Chief of Staff for Policy, Paul Mango, told *STAT* that the Department of Defense was deeply enmeshed in every aspect of the project, including creating more than two dozen vaccine pop-up manufacturing plants, airlifting in equipment and raw materials from across the globe, and erecting cybersecurity and physical security operations "to ensure an eventual vaccine is guarded very closely from 'state actors who don't want us to be successful in this.'" This paranoid addendum seems like a pretextual effort to link vaccine-hesitant Americans to sinister foreign governments, thereby justifying a military and intelligence agency response. It is, in short, a "conspiracy theory," albeit an official one. Mango told *STAT* that Warp Speed planning and debriefing occurs "in protected rooms used to discuss classified information." A senior federal health official told *STAT* he was struck by the sight of soldiers in military uniforms ambling about HHS's headquarters in downtown Washington, including over 100 soldiers in the HHS corridors wearing "Desert Storm fatigues."

Health officials complained to *STAT* that they found themselves marginalized as Warp Speed devolved into a partnership between the military and the pharmaceutical industry, presided over by Robert Kadlec—who, according to Mango, personally signed off on every business agreement made by HHS for Operation Warp Speed.

Warp Speed has secret deals with six major drug companies developing COVID-19 vaccines. The operation's chief adviser is Moncef Slaoui, a former GlaxoSmithKline official who prior to the pandemic served as chairman of Moderna, the Fauci/Kadlec/Gates collaboration that would be Warp Speed's primary beneficiary. By characterizing his post as an "outside contractor," Slaoui, who holds roughly $10 million in GSK stock, dodged the application of federal ethics rules. Slaoui has since promised to donate any increase in the value of his stock.[293]

"The first person to be fired should be Dr. Slaoui," Sen. Elizabeth Warren (D-Mass.) responded at a hearing. "The American people deserve to know that COVID-19 vaccine decisions are based on science, and not on personal greed."[294]

Dr. Fauci had direct hands-on involvement with Warp Speed through his employee Larry Corey, who described himself as an "ex-officio" member of the Warp Speed governance. Corey runs Dr. Fauci's COVID-19 prevention network, which transforms HIV clinical trial networks into Phase 3 COVID-19 clinical trials.[295]

Dr. Fauci was undaunted by the military takeover of US health policy, applauding the Operation as a "talent show." Dr. Fauci told *STAT* he was untroubled by the dearth of public health experience among Warp Speed's Pentagon leadership: "If you go through the organizational boxes of Operation Warp Speed, they're very,

very impressive." Tom Inglesby also lauded the military involvement. "There is deep knowledge of science and on how to manage complex government operations," said Inglesby. "It's clearly operating in a challenging pandemic and political environment, and we won't know if we have a safe and effective vaccine until the trials are finished. But it's a highly competent group of people working to make it happen."[296]

HHS secretary Alex Azar—a former Pharma CEO and lobbyist—and defense secretary Mark Esper share top billing as the organizational chairs. Slaoui, the project's formal civilian leader, and Gen. Gustave Perna serve as Operation Warp Speed's CEO.

Immediately beneath Perna and Slaoui are Lieutenant General (Retired) Paul Ostrowski,[297] a former Special Forces soldier who manages distribution of an eventual vaccine, and Matt Hepburn, who specializes in futuristic warfare projects for the Pentagon, including a program to implant high-tech sensors into soldiers to detect illnesses and for other purposes.

"This should be a medical and not be a military operation," Holocaust survivor and medical ethics advocate Vera Sharav told me. "It's a public health problem. Why are the military and the CIA so heavily involved? Why is everything a secret? Why can't we know the ingredients of these products, which the taxpayers financed? Why are all their emails redacted? Why can't we see the contracts with vaccine manufacturers? Why are we mandating a treatment with an experimental technology with minimal testing? Since COVID-19 harms fewer than 1 percent, what is the justification for putting 100 percent of the population at risk? We need to recognize that this is a vast human experiment on all of mankind, with an unproven technology, conducted by spies and generals primarily trained to kill and not to save lives." What could possibly go wrong?

Endnotes

1   Stephen Kinzer, "From mind control to murder? How a deadly fall revealed the CIA's darkest secrets," *The Guardian* (Sep. 6, 2019), theguardian.com/us-news/2019/sep/06/from-mind-control-to-murder-how-a-deadly-fall-revealed-the-cias-darkest-secrets

2   Michael Ignatieff, "Who Killed Frank Olson?," *The Guardian* (Apr. 6, 2001), theguardian.com/books/2001/apr/07/books.guardianreview4

3   H. P. Albarelli, Jr., "Part One: The Mysterious Death of CIA Scientist Frank Olson," *Crime Magazine* (Dec. 14, 2002), crimemagazine.com/part-one-mysterious-death-cia-scientist-frank-olson

4   David Franz, "The Dual Use Dilemma: Crying out for Leadership," *Saint Louis University Journal of Health Law and Policy* 6 (Vol. 7:5 2013), slu.edu/law/academics/journals/health-law-policy/pdfs/issues/v7-i1/david_franz_article.pdf

5   Franz

6   William Lowther, "Rumsfeld 'helped Iraq get chemical weapons,'" *Daily Mail* (Dec. 31, 2002), dailymail.co.uk/news/article-153210/Rumsfeld-helped-Iraq-chemical-weapons.html

7   Christopher G. Pernin et al., "Unfolding the Future of the Long War: Motivations, Prospects, and Implications for the U.S. Army," Rand Corporation (2008), rand.org/content/dam/rand/pubs/monographs/2008/RAND_MG738.pdf

8   James Sterngold, "Cheney's grim vision: decades of war / Vice president says Bush policy aimed at long-term world threat," *San Francisco Chronicle* (Jan. 15, 2004), sfgate.com/politics/article/Cheney-s-grim-vision-decades-of-war-Vice-2812372.php

9   Wikispooks: Robert Kadlec. https://wikispooks.com/wiki/Robert_Kadlec

10   Jon Cohen, "Mining coronavirus genomes for clues to the outbreak's origins," *Science* (Jan. 31, 2021), sciencemag.org/news/2020/01/mining-coronavirus-genomes-clues-outbreak-s-origins

11   Hans Mahncke and Jeff Carlson, "Fauci Team Scrambled in January 2020 to Respond to Lab Leak Allegations, Emails Show," *Epoch Times* (Jun. 2, 2021), theepochtimes.com/fauci-team-scrambled-in-january-2020-to-respond-to-lab-leak-allegations-emails-show_3842427.html

12   Whitney Webb and Raul Diego, "Head of the Hydra—The Rise of Robert Kadlec," The Last American Vagabond (May 14, 2020), thelastamericanvagabond.com/head-hydra-rise-robert-kadlec/

13   Neville Hodgkinson, "Covid's Dark Winter: How Biological War Games Stole Our Freedom," *Conservative Woman* (June 30, 2021), conservativewoman.co.uk/covids-dark-winter-how-bio-war-gaming-robbed-us-of-our-liberty

14   Barry R. Schneider and Lawrence E. Grinter, eds., *Battlefield of the Future: 21st Century Warfare* Issue (Air University Press Rev. Ed. 2008), 261–262 airuniversity.af.edu/Portals/10/CSDS/Books/battlefield_future2.pdf

15   Jon Swaine, Robert O'Harrow Jr., and Aaron Davis, "Before pandemic, Trump's stockpile chief put focus on biodefense. An old client benefited," *Washington Post* (May 4, 2020), washingtonpost.com/investigations/before-pandemic-trumps-stockpile-chief-put-focus-on-biodefense-an-old-client-benefited/2020/05/04/d3c2b010-84dd-11ea-878a-86477a724bdb_story.html

16   Alexis Baden-Mayer, "Dr. Robert Kadlec: How the Czar of Biowarfare Funnels Billions to Friends in the Vaccine Industry," *Organic Consumers* (Aug. 13, 2020), organicconsumers.org/blog/dr-robert-kadlec-how-czar-biowarfare-funnels-billions-friends-vaccine-industry

17   "Gates Foundations Give Johns Hopkins $20 Million Gift to School of Public Health for Population, Reproductive Health Institute," BMGF (May 1999), gatesfoundation.org/ideas/media-center/press-releases/1999/05/johns-hopkins-university-school-of-public-health

18   NIH Reporter, Johns Hopkins Funding 2001–2021, https://reporter.nih.gov/search/W2pb_quLtkOEn58czHh1wA/projects/charts?fy=2021;2020;2019;2018;2017;2016;2015;2013;2014;2012;2011;2010;2009;2008;2007;2006;2005;2004;2003;2002;2001&org=JOHNS%20HOPKINS%20UNIVERSITY

19   "Secret project manufactured mock anthrax," *Washington Times* (Oct. 26, 2001), washingtontimes.com/news/2001/oct/26/20011026-030448-2429r/

20   Robert F. Kennedy, Jr. *American Values: Lessons I Learned from My Family*, (Harper Collins, 2018), 215

21   Engelbrecht, Köhnlein, et al., 368

22   Marjorie Censer, "CEO took roundabout path to Emergent," *Washington Post* (Jan. 3, 2011). washingtonpost.com/wp-dyn/content/article/2010/12/30/AR2010123003293.html

23   Tim Reid, "The needle and the damage done," *London Times* (Nov. 26, 2002), vaccinetruth.org/gulf-war-syndrome.html

24   Subcommittee on National Security, Veterans Affairs, and International Relations of the Committee on Government Reform (Jun. 30, 1999), hsdl.org/?view&did=2088

25   Martin Meyer Weiss, "Anthrax Vaccine and Public Health Policy," *American Journal of Public Health* (Nov., 2007), ncbi.nlm.nih.gov/pmc/articles/PMC2040369/

26   Whitney Webb and Raul Diego, "Head of the Hydra—The Rise of Robert Kadlec," The Last American Vagabond (May 14, 2020), thelastamericanvagabond.com/head-hydra-rise-robert-kadlec/

27   Tara O'Toole, MD, MPH, Professional Profile, Center for Health Security, https://www.centerforhealthsecurity.org/our-people/otoole/

28   Webb and Diego

29   Susan Peterson, "Epidemic disease and national security," *Security Studies* vol. 12, no. 2 (2002): 74, DOI: 10.1080/09636410212120009, researchgate.net/publication/232909887_Epidemic_Disease_and_National_Security

30   Congressional Record Senate 155, pt. 20 (Nov. 4, 2009), govinfo.gov/content/pkg/CRECB-2009-pt20/html/CRECB-2009-pt20-Pg26672.htm

31   Jim McElhatten, "Exclusive: Obama Nominee Omitted Ties to Biotech," *Washington Times* (Sept. 8, 2009), washingtontimes.com/news/2009/sep/8/obama-nominee-omitted-ties-to-biotech/

32   Noah Shachtman, "DHS's New Chief Geek Is a Bioterror 'Disaster,' Critics Charge," *Wired* (May 6, 2009), wired.com/2009/05/dhs-new-geek-in-chief-is-a-biodefense-disaster-critics-say/

33   Congressional Record Senate 155, pt. 20

34   Ibid.

35   Ibid.

36   Ibid.

37   Ibid.

38   Richard Abott, "ANSER Acquires Advanced Technology International," *Defense Daily* (Jan. 31, 2017), defensedaily.com/anser-acquires-advanced-technology-international/business-financial/

39   Sydney Lupkin, "How Operation Warp Speed's Big Vaccine Contracts Could Stay Secret," NPR (Sept. 29, 2020), npr.org/sections/health-shots/2020/09/29/917899357/how-operation-warp-speeds-big-vaccine-contracts-could-stay-secret

40   Ibid.

41   Matt Apuzzo and Selam Gebrekidan, "Governments Sign Secret Vaccine Deals. Here's What They Hide," *New York Times* (Jan. 28, 2021), nytimes.com/2021/01/28/world/europe/vaccine-secret-contracts-prices.html?referringSource=articleShare

42   Kevin Kajiwara and Jerome Hauer, "Teneo Insights Webinar: COVID-19 Pandemic and Vaccines," TENEO, (Jan 8, 2021), teneo.com/teneo-insights-webinar-covid-19-pandemic-and-vaccines/

43   Jim Lobe, "Chicken Hawks as Cheer Leaders," Foreign Policy in Focus Advisory Committee (2002), globalization.icaap.org/content/v2.2/lobe.html

44   Matt Duss, "Iraq: Because Rumsfeld Needed Better Targets," ThinkProgress (Jul. 28, 2009), archive. thinkprogress.org/iraq-because-rumsfeld-needed-better-targets-a4dcb1335c29/

45   Washington's Blog, "The Pentagon's 'Operation Dark Winter': June 2001 Bioterror Exercise Foreshadowed 9/11 and Anthrax Attacks," Global Research (Oct 12, 2014), globalresearch.ca/ the-pentagons-operation-dark-winter-june-2001-bioterror-exercise-foreshadowed-911-and-anthrax-attacks/5407575

46   Judith Miller, "A National Challenged: Spores; U.S. Agrees To Clean Up Anthrax Site In Uzbekistan," *New York Times* (Oct. 23, 2001), nytimes.com/2001/10/23/world/a-nation-challenged-spores-us-agrees-to-clean-up-anthrax-site-in-uzbekistan.html

47   Franklin Foer, "The Source of the Trouble," *New York Magazine* (May 28, 2004), nymag.com/ nymetro/news/media/features/9226/#print

48   Reuters Staff, "Bush calls flawed Iraq intelligence biggest regret," Reuters (Dec. 1, 2008), reuters.com/ article/vcCandidateFeed2/idUSN01511412

49   Simon Jeffrey, "The slam-dunk intelligence chief," *The Guardian* (Jun. 3, 2004), theguardian.com/ world/2004/jun/03/usa.simonjeffery

50   Lindsey A. O'Rourke, "The U.S. tried to change other countries' governments 72 times during the Cold War," *Washington Post* (Dec 23, 2016), washingtonpost.com/news/monkey-cage/ wp/2016/12/23/the-cia-says-russia-hacked-the-u-s-election-here-are-6-things-to-learn-from-cold-war-attempts-to-change-regimes/

51   CDC, Research—Smallpox (Jan. 22, 2019), cdc.gov/smallpox/research/index.html

52   Dr. Meryl Nass, "When mass vaccination programs are mounted in a hurry, bad outcomes and liability are invariably big issues," (Apr.17, 2021), anthraxvaccine.blogspot.com/2021/04/when-mass-vaccination-programs-are.html

53   Ibid.

54   John Doe #1 et al., v. Donald H. Rumsfeld, et al., 297 F. SUPP., U.S. Dist., (2003), biotech.law.lsu. edu/cases/vaccines/Doe_v_Rumsfeld_I.htm

55   Yet to this day, the CDC website puts forth a favorable view of the smallpox vaccine, starting with the well-worn assurances of safety: The smallpox vaccine is safe, and it is effective at preventing smallpox disease.

56   Andrea Germanos, "Big Tech War Profiteers Raked in $44 Billion During 'Global War on Terror," THE DEFENDER, (Sep 13, 2021). https://childrenshealthdefense.org/defender/big-tech-sells-war-amazon-google-microsoft-44-billion/

57   Rick Weiss and Susan Schmidt, "Capitol Hill Anthrax Matches Army's Stocks," *Washington Post* (Dec. 16, 2001), washingtonpost.com/archive/politics/2001/12/16/capitol-hill-anthrax-matches-armys-stocks/ccc7d65b-9235-4ccb-84a6-c9d5064ada91/

58   Webb and Diego

59   Ibid.

60   Ibid.

61   Jerry Markon, "Justice Dept. Takes on Itself in Anthrax Attacks," *Washington Post* (Jan. 27, 2012), washingtonpost.com/politics/justice-dept-takes-on-itself-in-probe-of-2001-anthrax-attacks/2012/01/05/gIQAhGLlVQ_story.html

62   Ian Gurney, "Bin Laden Profits from U.S. Anthrax Vaccine Manufacture?," What Really Happened (2002), whatreallyhappened.com/WRHARTICLES/binladenprofits.html

63   Webb and Diego

64   Jeffrey Lean and Jonathan Owen, "Donald Rumsfeld makes $5m killing on bird flu drug," *Independent* (Mar. 12, 2016), independent.co.uk/news/world/americas/donald-rumsfeld-makes-5m-killing-bird-flu-drug-6106843.html

65   Nelson D. Schwartz, "Rumsfield's Growing Stake in Tamiflu," CNN Money (October 31, 2005), https://money.cnn.com/2005/10/31/news/newsmakers/fortune_rumsfeld/

66   Ibid.

67   Jon Cohen and Eliot Marshall, "Vaccines for Biodefense: A System in Distress," *Science* (Oct. 19, 2001): 498–501, science.sciencemag.org/content/294/5542/498

68   Franz

69   Cohen and Marshall

70   Paul D. Thacker, "A Continued Candid Conversation with Richard Ebright on the History of U.S. Research Funding for Biological Agents in America and Abroad That Lack Critical Safety Overview," Disinformation Chronicle (Aug. 7, 2021), disinformationchronicle.substack.com/p/a-continued-candid-conversation-with?token=eyJ1c2VyX2lkIjozMzgzNzMwOSwic G9zdF9pZCI6NDAwMzQ0 MTMsIl8iOiIzaC81byIsImlhdCI6MTYyOTMyODIyMiwiZXhwIjoxNjI5MzMxODIyLCJpc3MiO iJwdWItMjY0Mjk5Iiwic3ViIjoic G9zdC1yZWFjdGlvbiJ9.gP5u8VH5vzokw_6Sjmre4Fm70GEegdm rFp2hbqfwTmg

71   Anthony S. Fauci, "The global challenge of infectious diseases: the evolving role of the National Institutes of Health in basic and clinical research," V.6 No. 8 *NATURE* 745 (August 2005), https:// www.nature.com/articles/ni0805-743.pdf?origin=ppub

72   Shankar Vedantam, "Lingering Worries Over Vaccine," *Washington Post* (Dec. 20, 2001), washingtonpost.com/archive/politics/2001/12/20/lingering-worries-over-vaccine/d05e4188-5672-4821-b68d-b979964a3ba4/

73　*ABC This Week* (October 28, 2001).

74　Ibid.

75　"The Anthrax Vaccine: Officials are taking another look at the controversial anthrax vaccine, Betty Ann Bowser reports," PBS (Dec. 4, 2001), pbs.org/newshour/show/the-anthrax-vaccine

76　Ibid.

77　John Dudley Miller, "Postal Anthrax Aftermath: Has Bio-Defense Spending Made Us Safer?," *Scientific American* (Nov. 1, 2008), scientificamerican.com/article/postal-anthrax-aftermath/

78　Frequently Asked Questions About SARS, CDC (Mar. 3, 2005), cdc.gov/sars/about/faq.html

79　Fiona Fleck, "SARS outbreak over, but concerns for lab safety remain," *Bulletin of the WHO* 82 (6): 470 (June 2004), who.int/bulletin/volumes/82/6/news.pdf?ua=1

80　"Anthony S. Fauci Reflects on the 2001 Anthrax Attacks," RWJF (Oct. 3, 2011), rwjf.org/en/blog/2011/10/anthony-s-fauci-reflects-on-the-2001-anthrax-attacks.html

81　Franz

82　Judith Miller, Stephen Engelberg, and William J. Broad, "U.S. Germ Warfare Research Pushes Treaty Limits" (September 4, 2001), nytimes.com/2001/09/04/world/us-germ-warfare-research-pushes-treaty-limits.html

83　Hsinchun Chen, et al. Argus, Vol. 21 *Infectious Disease Informatics: Syndromic Surveillance for Public Health and BioDefense*, 177–181 (July 14, 2009), doi:10.1007/978-1-4419-1278-7_13

84　"Local Challenges of Global Proportions: Evaluating Roles, Preparedness for, and Surveillance of Pandemic Influenza," Committee on Homeland Security and Governmental Affairs, US Senate 110th Congress (September 28, 2007), https://www.govinfo.gov/content/pkg/CHRG-110shrg38846/pdf/CHRG-110shrg38846.pdf

85　Raul Diego, "DARPA's Man in Wuhan," Unlimited Hangout (July 31, 2020), unlimitedhangout.com/2020/07/investigative-reports/darpas-man-in-wuhan/

86　Ibid.

87　Donald G. McNeil Jr.,"Scientists Were Hunting for the Next Ebola. Now the U.S. Has Cut Off Their Funding," *New York Times* (October 25, 2019), https://www.nytimes.com/2019/10/25/health/predict-usaid-viruses.html

88　Daszak/Callahan paper: "Diversity of coronavirus in bats from Eastern Thailand," *Virol J*. 2015; 12: 57. (April 11, 2012), doi: 10.1186/s12985-015-0289-1, ncbi.nlm.nih.gov/pmc/articles/PMC4416284/

89　Engineering Bio-Terror Agents: Lessons from the Offensive U.S. and Russian Biological Weapons Program, House of Representatives, Subcommittee on Prevention of Nuclear and Biological Attack, Committee on Homeland Security (July 13, 2005), fas.org/irp/congress/2005_hr/bioterror.html

90　Matt Windsor, "The War on Bugs, Michael Callahan: Alumni Profile," *University of Alabama Medicine Magazine*, Winter, 2013, https://www.uab.edu/medicine/magazine/winter-2013/alumni-profile-michael-callahan

91　Richard Conniff, "How devastating pandemics change us," *National Geographic* (Jul. 14, 2020), https://www.nationalgeographic.com/magazine/article/how-devastating-pandemics-change-us-feature

92　Steve Watson, "Anthrax Coverup: A Government Insider Speaks Out," 911 Blogger (2002), 911blogger.com/node/9774

93　Scott Lilly, "Getting Rich on Uncle Sucker: Should the Federal Government Strengthen Efforts to Fight Profiteering?," Center for American Progress (Oct., 2010), cdn.americanprogress.org/wp-content/uploads/issues/2010/10/pdf/unclesucker.pdf

94　Laura Rozen, "The anthrax vaccine scandal: Why did the Pentagon allow BioPort Corp. to remain the sole U.S. supplier of a crucial weapon against bioterror, despite years of failure to deliver the vaccine?," Salon (Oct. 15. 2001), salon.com/2001/10/15/anthrax_vaccine/

95　"Can Bioshield Effectively Procure Medical Countermeasures That Safeguard the Nation?," House of Representatives, Subcommittee on Emerging Threats, Cybersecurity and Science and Technology of the Committee on Homeland Security 49 (April 18, 2007), govinfo.gov/content/pkg/CHRG-110hhrg43559/pdf/CHRG-110hhrg43559.pdf

96　Sandra Sobieraj, "White House mail sorters anthrax-free," Associated Press (October 24, 2001), web.archive.org/web/20020111104128/http://www.phillyburbs.com/terror/news/1024beth.htm

97　Chris Hamby and Sheryl Gay Stolberg, "How One Firm Put an 'Extraordinary Burden' on the U.S.'s Troubled Stockpile," *New York Times* (Mar. 6, 2021), nytimes.com/2021/03/06/us/emergent-biosolutions-anthrax-coronavirus.html

98　Hamby and Stolberg

99　"HHS Contract Valued at Up to $107 Million to Develop Large-Scale Manufacturing for BioThrax," Emergent Biosolutions (Jul. 14, 2010), investors.emergentbiosolutions.com/news-releases/news-release-details/emergent-biosolutions-awarded-hhs-contract-valued-107-million?ID=1447267&c=202582&p=irol-newsArticle

100　"Emergent BioSolutions Awarded Contract to Develop a Dry Formulation of NuThrax, a Next Generation Anthrax Vaccine," Business Wire (Sept. 8, 2014), businesswire.com/news/home/20140908006391/en/Emergent-BioSolutions-Awarded-Contract-to-Develop-a-Dry-Formulation-of-NuThrax-a-Next-Generation-Anthrax-Vaccine

101　Hamby and Stolberg

102 Ibid.

103 Ibid.

104 Webb and Diego

105 Alex Knapp, "Little-Known Publicly Traded Company Given Massive Deal to Manufacture One-Shot Covid-19 Vaccine," *Forbes* (Mar. 1, 2021), forbes.com/sites/alexknapp/2021/03/01/little-known-publicly-traded-company-given-massive-deal-to-manufacture-one-shot-covid-19-vaccine/?sh=53c4c3b53217

106 "Emergent Buys Sanofi Pasteur's Smallpox Vaccine for Up-to-$125M," *Gen Eng News* (Jul. 14, 2017), genengnews.com/topics/drug-discovery/emergent-buys-sanofi-pasteurs-smallpox-vaccine-for-up-to-125m/

107 Webb and Diego

108 "Emergent BioSolutions Awarded 10-Year HHS Contract to Deliver ACAM2000®, (Smallpox (Vaccinia) Vaccine, Live) Into the Strategic National Stockpile," Emergent Biosolutions (Sep. 3, 2019), investors.emergentbiosolutions.com/news-releases/news-release-details/emergent-biosolutions-awarded-10-year-hhs-contract-deliver

109 Swaine, O'Harrow, and Davis

110 Zhiyuan Sun, "Is There Any Hope Left for Emergent BioSolutions?," Motley Fool (Apr. 27, 2021), fool.com/investing/2021/04/27/is-there-any-hope-left-for-emergent-biosolutions/

111 Hamby and Stolberg

112 Knapp

113 Reuters Staff, "U.S. awards new $628 million contract to boost output of potential COVID-19 vaccine," Reuters (Jun. 1, 2020), reuters.com/article/us-health-coronavirus-usa-vaccine/u-s-awards-new-628-million-contract-to-boost-output-of-potential-covid-19-vaccine-idUSKBN2382DO

114 "Emergent BioSolutions Signs Agreement with AstraZeneca to Expand Manufacturing for COVID-19 Vaccine Candidate," Emergent BioSolutions (Jul. 27, 2020), investors.emergentbiosolutions.com/news-releases/news-release-details/emergent-biosolutions-signs-agreement-astrazeneca-expand

115 Hamby and Stolberg

116 Swaine, O'Harrow, and Davis

117 Hamby and Stolberg

118 "Moderna Announces Award from U.S. Government Agency BARDA for up to $483 Million to Accelerate Development of mRNA Vaccine (mRNA-1273) Against Novel Coronavirus," Modern (Apr. 16, 2020), investors.modernatx.com/news-releases/news-release-details/moderna-announces-award-us-government-agency-barda-483-million

119 Katie Thomas and Megan Twohey, "How a Struggling Company Won $1.6 Billion to Make a Coronavirus Vaccine," *New York Times* (Jul. 16, 2020), nytimes.com/2020/07/16/health/coronavirus-vaccine-novavax.html

120 "BioNTech Announces New Collaboration to Develop HIV and Tuberculosis Programs," BioNTech (Sep. 4, 2019), investors.biontech.de/news-releases/news-release-details/biontech-announces-new-collaboration-develop-hiv-and

121 Reuters Staff, "BioNTech wins $445 million German grant for COVID-19 vaccine," Reuters (Sep. 15, 2020), reuters.com/article/health-coronavirus-germany-vaccine/biontech-wins-445-million-german-grant-for-covid-19-vaccine-idUSKBN2661KP

122 Sydney Lupkin, "U.S. To Get 100 Million Doses of Pfizer Coronavirus Vaccine in $1.95 Billion Deal," NPR (Jul. 22, 2020), npr.org/sections/coronavirus-live-updates/2020/07/22/894184607/u-s-to-get-100-million-doses-of-pfizer-coronavirus-vaccine-in-1-95-billion-deal

123 "Emergent BioSolutions' COVID-19 Human Immune Globulin Product Candidate to be Included in NIH-Sponsored Phase 3 Clinical Trial of Hyperimmune Intravenous Immunoglobulin to Treat COVID-19," Emergent BioSolutions (Oct. 8, 2020), investors.emergentbiosolutions.com/news-releases/news-release-details/emergent-biosolutions-covid-19-human-immune-globulin-product

124 Gilead Sciences, Inc., U.S. Securities and Exchange Commission, Form 10-K (Dec. 31, 2005), sec.gov/Archives/edgar/data/882095/000119312506045128/d10k.htm

125 World Health Organization, "WHO recommends against the use of remdesivir in COVID-19 patients" (Nov 20, 2020), who.int/news-room/feature-stories/detail/who-recommends-against-the-use-of-remdesivir-in-covid-19-patients

126 Drugs.com, Remdesivir Side Effects (Feb 13, 2021), drugs.com/sfx/remdesivir-side-effects.html

127 Centers for Disease Control and Prevention, Symptoms of COVID-19 (Feb. 22, 2021), cdc.gov/coronavirus/2019-ncov/symptoms-testing/symptoms.html

128 Jason D. Goldman et al., "Remdesivir for 5 or 10 Days in Patients with Severe Covid-19," *New England Journal of Medicine* (May 27, 2020), nejm.org/doi/full/10.1056/NEJMoa2015301

129 John H. Beigel et al., "Remdesivir for the Treatment of Covid-19—Final Report," *New England Journal of Medicine* (Oct. 8, 2020), nejm.org/doi/10.1056/NEJMoa2007764

130 Hamby and Stolberg

131 Sheryl Gay Stolberg and Chris Hamby, "Shake-Up at Covid Vaccine Manufacturer That Tossed Millions of Doses," *New York Times* (Apr. 29, 2021), nytimes.com/2021/04/29/us/emergent-biosolutions-covid-vaccine-manufacturing.html

132   U.S. House of Representatives House Committee on Oversight and Reform and Select Oversight Subcommittee on the Coronavirus Crisis, "Examining Emergent BioSolutions' Failure to Protect Public Health and Public Funds," (May 19, 2021), docs.house.gov/meetings/VC/VC00/20210519/112641/HHRG-117-VC00-Transcript-20210519.pdf

133   Rich Mendez, "Congressional investigation launched into Emergent BioSolutions' federal vaccine contracts," CNBC (Apr. 20, 2021), cnbc.com/2021/04/20/congressional-investigation-launched-into-emergent-biosolutions-federal-vaccine-contracts-.html

134   Matt Richtel, "Covid News: New Cases in Nursing Homes Fall Dramatically After Vaccinations," *New York Times* (May 19, 2021), nytimes.com/live/2021/05/19/world/covid-vaccine-coronavirus-mask

135   Berkeley Lovelace Jr., "FDA asks Emergent plant to pause manufacturing while it investigates botched Covid vaccines," CNBC (Apr 12, 2021), cnbc.com/2021/04/19/fda-asks-emergent-plant-to-pause-manufacturing-while-it-investigates-botched-covid-vaccines.html

136   U.S. House of Representatives House Committee on Oversight and Reform and Select Oversight Subcommittee on the Coronavirus Crisis, "Examining Emergent BioSolutions' Failure to Protect Public Health and Public Funds," youtu.be/7fFAYqHQ5Xg

137   "Emergent BioSolutions Signs Five-Year Agreement for Large-Scale Drug Substance Manufacturing for Johnson & Johnson's Lead COVID-19 Vaccine Candidate," Emergent BioSolutions (Jul. 6, 2020), investors.emergentbiosolutions.com/news-releases/news-release-details/emergent-biosolutions-signs-five-year-agreement-large-scale-drug

138   Swaine, O'Harrow, and Davis

139   Atlantic Storm, Johns Hopkins Center for Health Security (2005), centerforhealthsecurity.org/our-work/events-archive/2005_atlantic_storm/

140   Bradley T. Smith et al., "Navigating the storm: report and recommendations from the Atlantic Storm exercise," *Biosecurity and Bioterrorism* (2005), pubmed.ncbi.nlm.nih.gov/16181048/

141   Global Mercury Post Exercise Report, Homeland Security Digital Library (2003), hsdl.org/?abstract&did=234582

142   Sharon Weinberger, "You Can't Handle the Truth: Psy-ops propaganda goes mainstream," *Slate* (Sept. 19, 2005), slate.com/news-and-politics/2005/09/psy-ops-propaganda-goes-mainstream.html

143   Ibid.

144   Ibid.

145   Ibid.

146   Josh Meyer, "Cambridge Analytica boss went from 'aromatics' to psyops to Trump's campaign," Politico (Mar. 22, 2018), politico.com/story/2018/03/22/cambridge-analytica-trump-campaign-479351

147   Isobel Asher Hamilton, "The founder of Cambridge Analytica's parent company admits he lacked an 'ethical radar,'" Yahoo Finance (Aug. 20, 2018), finance.yahoo.com/news/founder-cambridge-analytica-apos-parent-112639746.html

148   Weinberger

149   Louis Menand, "Fat Man: Herman Kahn and the Nuclear Age," *New Yorker* (Jun. 27, 2005), newyorker.com/magazine/2005/06/27/fat-man

150   Joel Garreau, "Conspiracy of Heretics: The Global Business Network was founded in 1988 as a think tank to shape the future of the world. It's working," *Wired* (Nov. 1, 1994), wired.com/1994/11/gbn/

151   See, e.g., John Perkins, *Confessions of an Economic Hit Man* (2004) and Marks, Manchurian Candidate chapter 8, CIA Mind Control Experiments, ahrp.org/category/scientific-racism/cia-mind-control-experiments/

152   Saul McLeod, "The Milgram Shock Experiment," Simply Psychology (2017), simplypsychology.org/milgram.html

153   Alfred W. McCoy, *A Question of Torture: CIA Interrogation, from the Cold War to the War on Terror* (Metropolitan Books/Henry Holt, 2006), 33

154   Ibid., 29

155   Ibid., 41

156   Ibid., 33

157   Amy Novotney, "The Risks of Social Isolation," *Monitor on Psychology* 50:5 (May 2013): 32, apa.org/monitor/2019/05/ce-corner-isolation

158   Alliance for Human Research Protection, "Dr. Maitland Baldwin, a student of Harold Hebb explored sensory deprivation at NIH" (Jan 18, 2015), ahrp.org/dr-maitland-baldwin-a-student-of-harold-hebb-explored-sensory-deprivation-at-nih/

159   World Health Organization, "Biosecurity: An integrated approach to manage risk to human, animal and plant life and health" (Mar. 3, 2010), who.int/foodsafety/fs_management/No_01_Biosecurity_Mar10_en.pdf

160   The Rockefeller Foundation and Global Business Network, "Scenarios for the Future of Technology and International Development" (May, 2010), nommeraadio.ee/meedia/pdf/RRS/Rockefeller%20Foundation.pdf

161   Ibid., 18.

162   RFK Jr. email from Ken McCarthy, June 2021

163   Michael Birnbaum QC, *Nigeria—Fundamental Rights Denied Report Of The Trial Of Ken Saro-wiwa And Others*, Article 19 & Bar Human Rights Committee of England and Wales and the Law Society of England and Wales (June 1995), article19.org/data/files/pdfs/publications/nigeria-fundamental-rights-denied.pdf

164   Carl Hartman, "U.S. Recalls Ambassador After Nigeria Executes Nine Activists," AP (Nov. 10, 1995), apnews.com/article/7df4b1a7b045dbb9d73e8e75a4c18e6a

165   Yasha Levine, *Surveillance Valley: The Secret Military History of the Internet* (Public Affairs, 2018), 174

166   Sean Gallager, "50 Years Ago Today, the Internet was born. Sort of," Ars Technica (Oct. 29, 2019), arstechnica.com/information-technology/2019/10/50-years-ago-today-the-internet-was-born-sort-of/

167   USA Spending. Gov, 12.351—Scientific Research—Combating Weapons of Mass Destruction (Oct 2, 2017), usaspending.gov/award/ASST_NON_HDTRA11710064_9761

168   Shoshana Bryen and Stephen Bryen, "Was the U.S. Complicit in China's Covid Research?," Jewish Policy Center (May 27, 2021), jewishpolicycenter.org/2021/05/27/was-the-u-s-complicit-in-chinas-covid-research/

169   "DARPA Awards Moderna Therapeutics a Grant for up to $25 Million to Develop Messenger RNA Therapeutics™," Moderna (Oct. 2, 2013), investors.modernatx.com/news-releases/news-release-details/darpa-awards-moderna-therapeutics-grant-25-million-develop

170   Renée DiResta, "How the Tech Giants Created What Darpa Couldn't," *Wired* (May 29, 2018), wired.com/story/darpa-total-informatio-awareness/

171   Miguel Helft, "Google's ATAP Head Regina Dugan Joins Facebook To Start DARPA-Inspired Team," *Forbes* (Apr. 13, 2016), forbes.com/sites/miguelhelft/2016/04/13/googles-atap-head-regina-dugan-joins-facebook-to-start-darpa-inspired-team/?sh=2f8293fa1996

172   Interview with Robert F. Kennedy Jr.

173   Interview with Robert F. Kennedy Jr.

174   SF Weekly Staff, "Mondo 1995 (Part II)" (Oct 11, 1995), sfweekly.com/news/mondo-1995-part-ii/

175   Jacques Ellul, *The Technological Society* (French 1954; Vintage Books, 1964)

176   Fred Hapgood, "Transhumanism: Securing the Post-Human Future," CSO (Jan. 1, 2005), csoonline.com/article/2118249/transhumanism--securing-the-post-human-future.html

177   Fred Hapgood, "Transhumanism: Securing the Post-Human Future," CSO (Jan. 1, 2005), csoonline.com/article/2118249/transhumanism--securing-the-post-human-future.html

178   Michael Standaert, "Smile for the camera: the dark side of China's emotion-recognition tech," *The Guardian* (March 3, 2021), theguardian.com/global-development/2021/mar/03/china-positive-energy-emotion-surveillance-recognition-tech

179   Chris Taylor, "Forecasting: The Futurologist: Looking Ahead in a Dangerous World," *Time Magazine* (Oct. 11, 2004), content.time.com/time/subscriber/article/0,33009,995373,00.html

180   Peter Schwartz, "How Governments Get Ready for the Unthinkable," World Government Summit (Feb. 8, 2016), worldgovernmentsummit.org/events/annual-gathering/2016/session-detail/how-governments-get-ready-for-the-unthinkable

181   2014 Sales Force Conference on "The Future of Global Governance," sfdc.hubs.vidyard.com/watch/lemzpanyZA5yQfed0poDTQ

182   The Salesforce Conversations with Peter Schwartz and Klaus Schwab, Salsesforce (2014), sfdc.hubs.vidyard.com/watch/lemzpqnyZA5yQfedOpoDTQ

183   "Winter 2021 COVID-19 Scenarios to Inform Your Business Decisions," Salesforce, salesforce.com/resources/videos/covid-scenarios/

184   Bill Gates, Munich Security Conference, Prepared Statement (Feb. 17, 2017), gatesfoundation.org/ideas/speeches/2017/02/bill-gates-munich-security-conference

185   Federal Ministry of Health, Germany, et al., "The 5C Health Emergency Simulation Exercise Package" (May, 2017), bundesgesundheitsministerium.de/fileadmin/Dateien/3_Downloads/G/G20-Gesundheitsministertreffen/5C_Health_Emergency_Simulation_Exercise_Manual.pdf

186   France24, "The Wuhan lab at the core of a virus controversy" (Apr. 17, 2020), france24.com/en/20200417-the-wuhan-lab-at-the-core-of-a-virus-controversy

187   Paul Schreyer, "Pandemic simulation games—Preparation for a new era?," YouTube 45:57 (Mar. 10, 2021), youtube.com/watch?v=d3WUv5SV5Hg

188   Bill & Melinda Gates Foundation, "Gates Foundations Give Johns Hopkins $20 Million Gift to School of Public Health for Population, Reproductive Health Institute," Gates Notes (May, 1999), gatesfoundation.org/ideas/media-center/press-releases/1999/05/johns-hopkins-university-school-of-public-health

189   Matthew Shearer, MPH, Professional Profile, Johns Hopkins Center for Health Security, centerforhealthsecurity.org/our-people/Shearer/

190   Gregory A. Poland, MD, Clinical Profile, Mayo Clinic, mayo.edu/research/faculty/poland-gregory-a-m-d/bio-00078220

191   Sandi Doughton, "Bill Gates: We must prepare for the next pandemic like we prepare for war," *Seattle Times* (Jan. 27, 2001), seattletimes.com/seattle-news/health/bill-gates-we-must-prepare-for-the-next-pandemic-like-we-prepare-for-war/

192   Monica Schoch-Spana et al., "The SPARS Pandemic, 2025–2028: A Futuristic Scenario for

Public Health Risk Communicators," Johns Hopkins Center for Health Security (October 2017), jhsphcenterforhealthsecurity.s3.amazonaws.com/spars-pandemic-scenario.pdf

193   Ibid., 12

194   Ibid., 9

195   Ibid., 23

196   Ibid., 47

197   Ibid., 25

198   Ibid., 25

199   Ibid., 13

200   Ibid., 53

201   Ibid., 60

202   Jade Scipioni, "Bill Gates in 2018: The World Needs to Prepare for Pandemics Just Like War," CNBC (Jan. 27, 2020), cnbc.com/2020/01/27/bill-gates-in-2018-world-needs-to-prepare-for-pandemics-just-like-war.html#:~:text=Bill%2520Gates%2520in%25202018%253A%2520The,for%2520pandemics%2520just%2520like%2520war&text=Speaking%2520at%2520an%2520event%2520hosted,way%2520it%2520prepares%2520for%2520wa

203   Lena H. Sun, "Bill Gates calls on U.S. to lead fight against a pandemic that could kill 33 million," *Washington Post* (Apr 27, 2018), washingtonpost.com/news/to-your-health/wp/2018/04/27/bill-gates-calls-on-u-s-to-lead-fight-against-a-pandemic-that-could-kill-millions/

204   Ibid.

205   Biohazard FOIA Maryland Emails, 29, 36 (Nov. 6, 2020), usrtk.org/wp-content/uploads/2020/11/Biohazard_FOIA_Maryland_Emails_11.6.20.pdf

206   United States Patent Application, Ciaramella, et al., Betacoronavirus mRNA vaccine (Mar 28, 2019), patft.uspto.gov/netacgi/nph-Parser?Sect1=PTO1&Sect2=HITOFF&d=PALL&p=1&u=%2Fnetahtml%2FPTO%2Fsrchnum.htm&r=1&f=G&l=50&s1=10702600.PN.&OS=PN/10702600&RS=PN/10702600

207   United States Patent Application, Ciaramella; et al., Betacoronavirus MRNA Vaccine (Feb 28, 2020), appft.uspto.gov/netacgi/nph-Parser?Sect1=PTO2&Sect2=HITOFF&p=1&u=%2Fnetahtml%2FPTO%2Fsearch-bool.html&r=2&f=G&l=50&co1=AND&d=PG01&s1=%22deliberate+release%22&s2=SARS&OS=%22deliberate+release%22+AND+SARS&RS=%22deliberate+release%22+AND+SARS

208   United States Patent Application, Ciaramella; et al., Respiratory Virus Vaccines (Mar 28, 2019), appft.uspto.gov/netacgi/nph-Parser?Sect1=PTO2&Sect2=HITOFF&p=1&u=%2Fnetahtml%2FPTO%2Fsearch-bool.html&r=3&f=G&l=50&co1=AND&d=PG01&s1=%22deliberate+release%22&s2=SARS&OS=%22deliberate+release%22+AND+SARS&RS=%22deliberate+release%22+AND+SARS

209   United States Patent Application, HPIV3 RNA VACCINES (Jul 20, 2018), appft.uspto.gov/netacgi/nph-Parser?Sect1=PTO2&Sect2=HITOFF&p=1&u=%2Fnetahtml%2FPTO%2Fsearch-bool.html&r=4&f=G&l=50&co1=AND&d=PG01&s1=%22deliberate+release%22&s2=SARS&OS=%22deliberate+release%22+AND+SARS&RS=%22deliberate+release%22+AND+SARS

210   United States Patent Application, Ciaramella; et al., HMPV RNA VACCINES (Jul 20, 2018), appft.uspto.gov/netacgi/nph-Parser?Sect1=PTO2&Sect2=HITOFF&p=1&u=%2Fnetahtml%2FPTO%2Fsearch-bool.html&r=5&f=G&l=50&co1=AND&d=PG01&s1=%22deliberate+release%22&s2=SARS&OS=%22deliberate+release%22+AND+SARS&RS=%22deliberate+release%22+AND+SARS

211   Bill Gates, Shattuck Lecture Innovation for Pandemics, BMGF (Apr. 27, 2018), gatesfoundation.org/Ideas/Speeches/2018/04/Shattuck-Lecture-Innovation-for-Pandemics

212   Johns Hopkins: Pandemic Exercise Highlights Policies to Prevent or Reduce Worst Possible Outcomes in Future Pandemics, ASPPH (May 23, 2018), aspph.org/johns-hopkins-pandemic-exercise-highlights-policies-to-prevent-or-reduce-worst-possible-outcomes-in-future-pandemics/

213   Lena H. Sun, "This Mock Pandemic Killed 150 Million People. Next Time It Might Not Be a Drill," *Washington Post* (May 30, 2018), washingtonpost.com/news/to-your-health/wp/2018/05/30/this-mock-pandemic-killed-150-million-people-next-time-it-might-not-be-a-drill/

214   Nick Alexopulos, "Clade x pandemic exercise highlights policies needed to prevent or reduce the worst possible outcomes in future pandemics," Johns Hopkins Center for Health Safety (May 15, 2018), centerforhealthsecurity.org/news/center-news/2018/2018-05-15_clade-x-policy-recommendations.html

215   Lena H. Sun, op. cit.

216   Jamie Seidel, "The world is completely unprepared for the next pandemic," *New York Post* (Jul. 30, 2018), nypost.com/2018/07/30/the-world-is-completely-unprepared-for-the-next-pandemic/

217   Nicola Twilly, "The Terrifying Lessons of a Pandemic Simulation," *New Yorker* (Jun. 1, 2018), newyorker.com/science/elements/the-terrifying-lessons-of-a-pandemic-simulation

218   Lena H. Sun, op. cit.

219   Ibid.

220   Ibid.

221   Twilly

222   Antonio Regalado, "It's fiction, but America just got wiped out by a man-made terror germ," *MIT*

Technology Review (May 30, 2018), technologyreview.com/2018/05/30/2746/its-fiction-but-america-just-got-wiped-out-by-a-man-made-terror-germ/

223 William Brangam, "A Universal Flu Vaccine Could Finally Be Within Sight," Part 3, *PBS NewsHour* 1:22 (Jun. 20, 2019), pbs.org/newshour/show/how-close-are-scientists-to-a-universal-flu-vaccine

224 Ibid., 3:42

225 FirangiAffairs, "Reputation Laundering By Bill Gates? Continuous Donations to Mainstream Media Outlets," Kreately (May 6, 2021), kreately.in/reputation-laundering-by-bill-gates-continuous-donations-to-mainstream-media-outlets/

226 "Preparedness for a High-Impact Respiratory Pathogen Pandemic," Johns Hopkins Center for Health Security (Sept. 2019), page 11, apps.who.int/gpmb/assets/thematic_papers/tr-6.pdf

227 Ibid., 13

228 Ibid., 14

229 Ibid., 11

230 Alexis Baden-Mayer, "Dr. Robert Kadlec: How the Czar of Biowarfare Funnels Billions to Friends in the Vaccine Industry," Organic Consumers Union (Aug 13, 2020), organicconsumers.org/blog/dr-robert-kadlec-how-czar-biowarfare-funnels-billions-friends-vaccine-industry

231 World Health Organization, "A World in Disorder" (2020), apps.who.int/gpmb/assets/annual_report/2020/GPMB_2020_AR_EN_WEB.pdf

232 James Gallagher, "Large Ebola Outbreaks New Normal, says WHO," BBC (Jun. 7, 2019), bbc.com/news/health-48547983

233 Eric Lipton et al., "He Could Have Seen What Was Coming: Behind Trump's Failure on the Virus," *New York Times* (Apr. 11, 2020), nytimes.com/2020/04/11/us/politics/coronavirus-trump-response.html

234 "Crimson Contagion 2019 Functional Exercise After-action Report," Department of HHS (Jan. 2020), int.nyt.com/data/documenthelper/6824-2019-10-key-findings-and-after/05bd797500ea55be0724/optimized/full.pdf#page=1

235 Ibid., 5

236 Ibid.

237 David E. Sanger et al., "Before Virus Outbreak, a Cascade of Warnings Went Unheeded," *New York Times* (Mar. 19, 2020), nytimes.com/2020/03/19/us/politics/trump-coronavirus-outbreak.html

238 Ibid.

239 *Crimson Contagion 2019 Functional Exercise After-action Report*

240 Lipton et al.

241 Exercise TOPOFF 2000 And National Capital Region (Ncr) After-action Report, National Response Team 12 (Aug. 2011), nrt.org/sites/2/files/TOPOFF.pdf

242 Ibid.

243 Top Officials (TOPOFF), U.S. Dept. of State archive, 2001-2009.state.gov/s/ct/about/c16661.htm

244 Ibid.

245 Ibid.

246 Interview with Robert F. Kennedy Jr.

247 Klaus Schwab, "Now Is the Time for a 'Great Reset,'" World Economic Forum (Jun. 3, 2020), weforum.org/agenda/2020/06/now-is-the-time-for-a-great-reset/

248 "Bill Gates: Few countries will get 'A grade' for coronavirus response," BBC 1:11 (Apr. 12, 2020), bbc.com/news/av/world-52233966

249 Event 201, Johns Hopkins Center for Health Security, centerforhealthsecurity.org/event201/

250 Event 201 Pandemic Exercise Segment 4, Transcript, 1 (Aug 7, 2020), childrenshealthdefense.org/wp-content/uploads/Event-201-Pandemic-Exercise-Segment-4-Communications-Discussion-and-Epilogue-Video-bill-gates.pdf

251 Ibid., 13

252 Ibid., 7

253 Ibid., 2

254 Ibid., 2

255 Tara Kirk Sell, PhD, MA, Professional Profile, Johns Hopkins Center for Health Security, centerforhealthsecurity.org/our-people/sell/

256 Biography of Jane Halton, who.int/docs/default-source/documents/about-us/jane-halton.pdf?sfvrsn=b3200427_2

257 Event 201 Pandemic Exercise Segment 4, Transcript, 4 (Aug 7, 2020), childrenshealthdefense.org/wp-content/uploads/Event-201-Pandemic-Exercise-Segment-4-Communications-Discussion-and-Epilogue-Video-bill-gates.pdf

258 Ibid., 1

259 Ibid., 3

260 Ibid., 6

261 Ibid., 7

262   Ibid., 8
263   Ibid.
264   Ibid.
265   Ibid.
266   Ibid., 9
267   Ibid., 1–2
268   Ibid., 5
269   Ibid., 11
270   Ibid., 10
271   Ibid., 13
272   Lena H. Sun, "None of these 195 countries—the U.S. included—is fully prepared for a pandemic, report says," *Washington Post* (Oct. 24, 2019), washingtonpost.com/health/2019/10/24/none-these-countries-us-included-is-fully-prepared-pandemic-report-says/
273   Brooke Seipel, "Obama highlights Biden's tweet from a year ago warning Trump wasn't ready for pandemic," The Hill (Oct. 25, 2020), thehill.com/homenews/campaign/522695-obama-highlights-bidens-tweet-from-a-year-ago-warning-trump-wasnt-ready-for
274   Robb Butler, "Vaccination refusal, hesitancy, acceptance and demand," ECDC YouTube 16:15 (Feb. 28, 2020), youtube.com/watch?v=1LCcaYoaZWg&t=981s
275   Lucy Fisher, "GCHQ in cyberwar on anti-vaccine propaganda," *Times* (London) (Nov. 9, 2020), thetimes.co.uk/article/gchq-in-cyberwar-on-anti-vaccine-propaganda-mcjgjhmb2
276   Ibid.
277   Ibid.
278   WION Web Team, "Germany ropes in spy agency to monitor anti-vaxxers, Covid deniers" (Apr. 28, 2021), wionews.com/world/germany-ropes-in-spy-agency-to-monitor-anti-vaxxers-covid-deniers-381096
279   Interview with Robert F. Kennedy Jr.
280   Karen DeYoung, "Samantha Powers Wants to Restore U.S. Prestige by Getting American Made Vaccines 'into Arms' Around the World," *Washington Post* (May 10, 2021), washingtonpost.com/national-security/samantha-power-usaid-vaccine-diplomacy/2021/05/10/69fd20d2-af7c-11eb-b476-c3b287e52a01_story.html
281   Jeremy R. Hammond, "UN Praises Maldives Bill Outlawing Informed Consent for Pharmaceuticals," CHD (Nov. 19, 2019), childrenshealthdefense.org/news/un-praises-maldives-bill-outlawing-informed-consent-for-pharmaceuticals/
282   Samuel Chamberlain, Mark Moore, and Bruce Golding, "Fauci was warned that COVID-19 may have been 'engineered,' emails show," *New York Post* (Jun. 2, 2021), nypost.com/2021/06/02/fauci-was-warned-that-covid-may-have-been-engineered-emails/
283   Vanessa Beeley, "Who controls the British Government response to Covid–19?," OffGuardian (May 9, 2020), off-guardian.org/2020/05/09/who-controls-the-british-government-response-to-covid-19/
284   "A Better Way To Go—Towards A Zero Covid UK," Independent Sage (Jul. 7, 2020), independentsage.org/independent-sage-on-achieving-a-zero-covid-uk-i-e-the-elimination-of-the-virus-from-the-uk/
285   Stephanie Kirchgaessner, "Ex-Obama official exits Israeli spyware firm amid press freedom row," *The Guardian* (Feb. 4, 2020), theguardian.com/world/2020/feb/04/ex-obama-official-juliette-kayyem-quits-israeli-spyware-firm-amid-press-freedom-row
286   Juliette Kayyem, "Anti-vaxxers are dangerous. Make them face isolation, fines, arrests," *Washington Post* (Apr. 30, 2019), washingtonpost.com/opinions/2019/04/30/time-get-much-tougher-anti-vaccine-crowd/
287   Max Blumenthal, Twitter, (Sep 5, 2021). https://twitter.com/maxblumenthal/status/1434623485612023813?lang=en
288   Central Intelligence Agency, "How does my Biology background fit at the CIA?" (Mar 23, 2018), youtube.com/watch?v=puajFEb7TTQ
289   David Gura, "Moguls, Deals and Patagonia Vests: A Look Inside 'Summer Camp for Billionaires,'" NPR (Jul. 5, 2021), npr.org/2021/07/05/1012587989/moguls-deals-and-patagonia-vests-a-look-inside-summer-camp-for-billionaires
290   Cynthia Littleton, "Sun Valley Conference Has Subdued Start as Moguls Convene," *Variety* (Jul. 6, 2021), variety.com/2021/tv/news/sun-valley-allen-co-bobby-kotick-bill-gates-bill-burns-1235013104/
291   Nicholas Florko, "New document reveals scope and structure of Operation Warp Speed and underscores vast military involvement," STAT (Sept. 28, 2020), statnews.com/2020/09/28/operation-warp-speed-vast-military-involvement/
292   Operation Warp Speed flowchart 7-30-2020 (from STAT article 9/28-2020 by Nicholas Florko) https://www.statnews.com/wp-content/uploads/2020/09/Slide1_WarpSpeed-1024x761.jpg
293   Isaac Arnsdorf, "Trump's Vaccine Czar Refuses to Give up Stock in Drug Company Involved in His

Government Role," Propublica (Sept. 23, 2020), propublica.org/article/trumps-vaccine-czar-refuses-to-give-up-stock-in-drug-company-involved-in-his-government-role

294   "At HELP Hearing, Warren Draws Attention to Vaccine Czar Conflicts of Interest; FDA Director Agrees Financial Conflicts of Interest Could 'Affect Public Perception' of the Vaccine Development Process," Elizabeth Warren Press Release (Sept. 23, 2020), warren.senate.gov/newsroom/press-releases/at-help-hearing-warren-draws-attention-to-vaccine-czar-conflicts-of-interest-fda-director-agrees-financial-conflicts-of-interest-could-affect-public-perception-of-the-vaccine-development-process

295   Nicholas Florko, "New document reveals scope and structure of Operation Warp Speed and underscores vast military involvement," STAT (Sept. 28, 2020), statnews.com/2020/09/28/operation-warp-speed-vast-military-involvement/

296   Ibid.

297   Washington Post Live, "Coronavirus: Vaccine Distribution with Lt. Gen. (Ret.) Paul Ostrowski," *Washington Post* (December 7, 2020 at 3:30 p.m. EST), washingtonpost.com/washington-post-live/2020/12/07/coronavirus-vaccine-distribution-with-lt-gen-paul-ostrowski/

ChildrensHealthDefense.org/fauci-book
childrenshd.org/fauci-book

For updates, new citations and references, and
new information about topics in this chapter:

AFTERWORD

What I have described in the preceding chapters can seem overwhelming and dispiriting. The forced-vaccine campaign and other cruel actions by Dr. Fauci and his acolytes might seem "too big to fail." But that is up to the citizens of our country.

We can bow down and comply—take the jabs, wear the face coverings, show our digital passports on demand, submit to the tests, and salute our minders in the Bio-surveillance State.

Or we can say No. We have a choice, and it is not too late.

COVID-19 is not *the* problem; it is a problem, one largely solvable with early treatments that are safe, effective, and inexpensive.

The problem is endemic corruption in the medical-industrial complex, currently supported at every turn by mass-media companies. This cartel's coup d'etat has already siphoned billions from taxpayers, already vacuumed up trillions from the global middle class, and created the excuse for massive propaganda, censorship, and control worldwide. Along with its captured regulators, this cartel has ushered in the global war on freedom and democracy. Playwright and essayist C. J. Hopkins describes the moment all too well:

> There is nothing subtle about this process. Decommissioning one 'reality' and replacing it with another is a brutal business. Societies grow accustomed to their 'realities.' We do not surrender them willingly or easily. Normally, what's required to get us to do so is a crisis, a war, a state of emergency, or . . . you know, a deadly global pandemic.
>
> During the changeover from the old 'reality' to the new 'reality,' the society is torn apart. The old 'reality' is being disassembled and the new one has not yet taken its place. It feels like madness, and, in a way, it is. For a time, the society is split in two, as the two 'realities' battle it out for dominance. 'Reality' being what it is (i.e., monolithic), this is a fight to the death. In the end, only one 'reality' can prevail.
>
> This is the crucial period for the totalitarian movement. It needs to negate the old 'reality' in order to implement the new one, and it cannot do that with reason and facts, so it has to do it with fear and brute force. It needs to terrorize the majority of society into a state of mindless mass hysteria that can be turned against those resisting the new 'reality.' It is not a matter of persuading or convincing people to accept the new 'reality.' It's more like how you drive a herd of cattle. You scare them enough to get them moving, then you steer them wherever you want them to go. The cattle do not know or understand where they are going. They are simply reacting to a physical stimulus. Facts and reason have nothing to do with it.

As we consider the unprecedented bludgeoning of our Constitution over the past two years, it's worth pausing to remember the smallpox epidemic that stalled Washington's army during the Revolution and the malaria contagion that culled the Army of Virginia. Though both alerted the Framers to the deadly and disruptive potential of infectious disease epidemics, the Framers nevertheless opted to include no pandemic exception to the United States Constitution.

Yet today, the pandemic is being used to create a string of new exceptions to our Constitution. We are given just one rationale to explain everything that is happening:

COVID. For just a brief moment, let's look away from the ostensible reason things are happening, and focus instead on what is happening.

Those controlling the levers of power vilify dissenters and punish every attempt at questioning, skepticism, and debate. Like all tyrants in history, they ban books, silence artists, condemn writers, poets, and intellectuals who question the new orthodoxies. They have outlawed gatherings and forced citizens to wear masks that instill fear and divide communities, and atomized any sense of solidarity by preventing the most subtle and eloquent nonverbal communication for which God and evolution gave humans forty-two facial muscles.

Predictably, the pandemic became a pretense for expanded tyranny across the globe—making changes that have nothing to do with a virus. Hungary clamped down on free speech and banned public depictions of homosexuality. China shuttered Hong Kong's last pro-democracy newspaper and jailed its executives, editors, and journalists. In Belarus, President Lukashenko subdued protests with mass arrests and even hijacked a passenger plane to arrest a dissident journalist. Cambodia abolished due process and arrested political opponents. Poland's government abolished rights for women and gays and effectively banned abortion. India's Prime Minister arrested journalists and ordered Twitter to remove critical posts. Russia's President Vladimir Putin used the pandemic as (another) pretext for jailing powerful opponents and banning mass gatherings. And democracies were not much different: France required its citizens to show a signed declaration to travel more than 1 kilometer from home. Australia was more liberal, allowing citizens to venture up to 5 kilometers from home— but then again, Australia also built new detention centers. Britain banned its citizens from traveling abroad.

Many similar things happened in the United States, including New York's Senate passing a law to allow for the forcible and indefinite detention of residents deemed to be a threat to "public health." But for America, freedom of speech has been the biggest casualty of the emerging tyranny. The now-popular term "misinformation" has come to mean any expression that departs from official orthodoxies. Social media and news media companies serve as stenographer and defender of any position pronounced by government. The intentional failure of journalistic inquiry, curiosity, and investigation, the failure to probe, to ask tough questions (or any questions) of those in power—has enabled the madness and the sadness of 2020 and 2021. There is a web of motives at work, but I'll cite a simple one:

Big pharmaceutical companies are the biggest advertisers on news and television outlets. Their $9.6 billion annual advertising budget buys more than commercials—it buys obeisance. (In 2014, network president Roger Ailes told me he would fire any of his news show hosts who allowed me to talk about vaccine safety on air. "Our news division," he explained, "gets up to 70 percent of ad revenues from pharma in non-election years.")

I know the role of the news media is not news to you, so I'll cite just one example: Vaccine mandates are ostensibly based upon the idea that vaccines will prevent transmission of COVID-19. If they don't prevent transmission, if both the vaccinated and unvaccinated can spread the virus, then there is no relevant difference between the two groups—other than that one group is not complying with government commands.

Forcing an entire population to accept an arbitrary and risky medical intervention is the most intrusive and demeaning action ever imposed by the United States Government, and perhaps any government.

And it is based upon a lie.

The Director of the CDC, Dr. Fauci, and the WHO have all had to reluctantly acknowledge that the vaccines cannot stop transmission.

When Israel's Director of Public Health addressed the FDA Advisory Panel, she left no doubt about the vaccines' inability to stop transmission of the virus, or stop sickness, or stop death. Describing Israel's situation as of September 17th, 2021, she said:

> Sixty percent of the people in severe and critical condition were, um, were immunized, doubly immunized, fully vaccinated. Forty-five percent of the people who died in this fourth wave were doubly vaccinated.

Even so, three weeks later, on October 7th—just days before this book went to press—the President of the United States announced that he was ensuring healthcare workers are vaccinated, "because if you seek care at a healthcare facility, you should have the certainty that the people providing that care are protected from COVID and cannot spread it to you."

The President just told Americans that being vaccinated provides "*certainty*" that vaccinated people are "*protected from COVID and cannot pass it to you.*"

Not one question was posed to the President about this stunning disconnect, about the obvious untruth—and that speech gives us a stark example of what's going on.

A televised image of an unchallenged leader mouthing untrue pronouncements to mislead and control the population—that is the world of George Orwell's sadly prophetic novel, *1984*.

It is a hopeful sign that halfway into 2021, Orwell's seventy-year-old book suddenly became a top-20 bestseller in the United States. Apparently, more people are aware of what's going on than the powerful give them credit for.

That awareness, that basic common sense, reminds us that democracies can reassert legislative control over rogue dictators—whether mayors, governors, presidents or prime ministers. Rational legislatures can choke off funding that supports few and harms many. They can initiate investigations, spur criminal prosecutions, and restore freedom.

Even without government engagement, it is ordinary people who can rescue us from tyranny. We can say *No* to compliance with jabs for work, *No* to sending children to school with forced testing and masking, *No* to censored social media platforms, *No* to buying products from the companies bankrupting and seeking to control us. These actions are not easy, but living with the consequences of inaction would be far harder. By calling on our moral courage, we can stop this march towards a global police state.

* * *

I founded Children's Health Defense (CHD) long before COVID-19. Our goal was to put an end to the epidemic of childhood diseases arising from toxic exposures of all types, including some vaccines. CHD seeks to educate the public and hold bad actors accountable in order to help ensure a healthy future for our children. As this book goes to press, the campaign to force unsafe COVID vaccines into children's bodies is reaching its peak. If our children are to enjoy the blessings of liberty and health, we must end this COVID-19 nightmare. We can no longer "trust the experts" or follow their warped version of science. That's what got us here.

With the information in this book, I hope you'll educate others, engage more effectively with your local government, school board, health department, legislators, police (and often more promising, your elected sheriff). CHD has chapters around the country and the world; join any of many health freedom groups. Sign up for the free Children's Health Defense online news site, *The Defender*. Stay informed. Stay active. We can jettison this insanity if enough people refuse to participate in a new apartheid based upon forced medical procedures.

The United States still suffers from the brutal and ugly history of slavery, segregation, racism, and alas, forced medical procedures. Let us not start this all over again, condemning African-Americans more than any other racial group, to second-class citizen status.

As I was writing this book, I reread Martin Luther King, Jr.'s majestic "I Have a Dream" speech at the Lincoln Memorial in 1963. Reverend King reaches out to us through all these years:

> But we refuse to believe that the bank of justice is bankrupt. We refuse to believe that there are insufficient funds in the great vaults of opportunity of this nation. And so, we've come to cash this check, a check that will give us upon demand the riches of freedom, and the security of justice. We have also come to this hallowed spot to remind America of the fierce urgency of Now. This is no time to engage in the luxury of cooling off or to take the tranquilizing drug of gradualism. Now is the time to make real the promises of democracy.

Join with us to take back our democracy and our freedom. I'll see you on the barricades.

Robert F. Kennedy Jr.

AUTHOR'S NOTE

Though this book appears to end here, it cannot end here—since the story is far from over. Every day brings new information, new data, new revelations, and new whistleblowers. Accordingly, I will continue writing chapters and making them available via the URL and QR Code below. On behalf of everyone with Children's Health Defense, thank you for reading this book and for continuing to follow the crucial topics addressed within its pages.

ChildrensHealthDefense.org/fauci-book
childrenshd.org/fauci-book

For updates, new citations and references, and
new information about topics in this chapter:

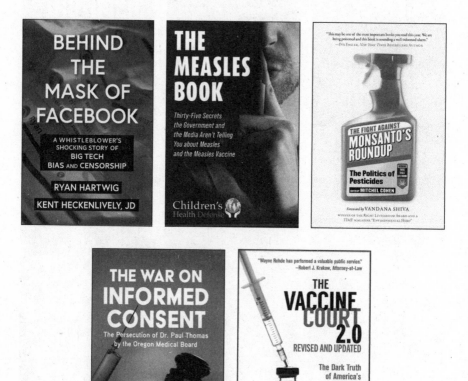